JACARANDA

SCIENCE QUEST 9

AUSTRALIAN CURRICULUM | THIRD EDITION

GRAEME LOFTS | MERRIN J. EVERGREEN

jacaranda
A Wiley Brand

Third edition published 2018 by
John Wiley & Sons Australia, Ltd
42 McDougall Street, Milton, Qld 4064

First edition published 2011
Second edition published 2015

Typeset in 11/14 pt Times LT Std

ISBN: 978-0-7303-4692-0

Front cover image: Asmus Koefoed / Shutterstock

Cartography by Spatial Vision, Melbourne and MAPgraphics Pty Ltd, Brisbane

Illustrated by various artists, diacriTech and the Wiley Art Studio.

Typeset in India by diacriTech

All activities have been written with the safety of both teacher and student in mind. Some, however, involve physical activity or the use of equipment or tools. **All due care should be taken when performing such activities**. Neither the publisher nor the authors can accept responsibility for any injury that may be sustained when completing activities described in this textbook.

A catalogue record for this book is available from the National Library of Australia

Printed in Singapore
M WEP237380 021123

CONTENTS

OVERVIEW

Jacaranda Science Quest 9 Australian Curriculum Third Edition has been completely revised to help teachers and students navigate the Australian Curriculum Science syllabus. The suite of resources in the *Science Quest* series is designed to enrich the learning experience and improve learning outcomes for all students.

Science Quest is designed to cater for students of all abilities: no student is left behind and none is held back. *Science Quest* is written with the specific purpose of helping students deeply understand science concepts. The content is organised around a number of features, in both print and online through Jacaranda's *learnON* platform, to allow for seamless sequencing through material to scaffold every student's learning.

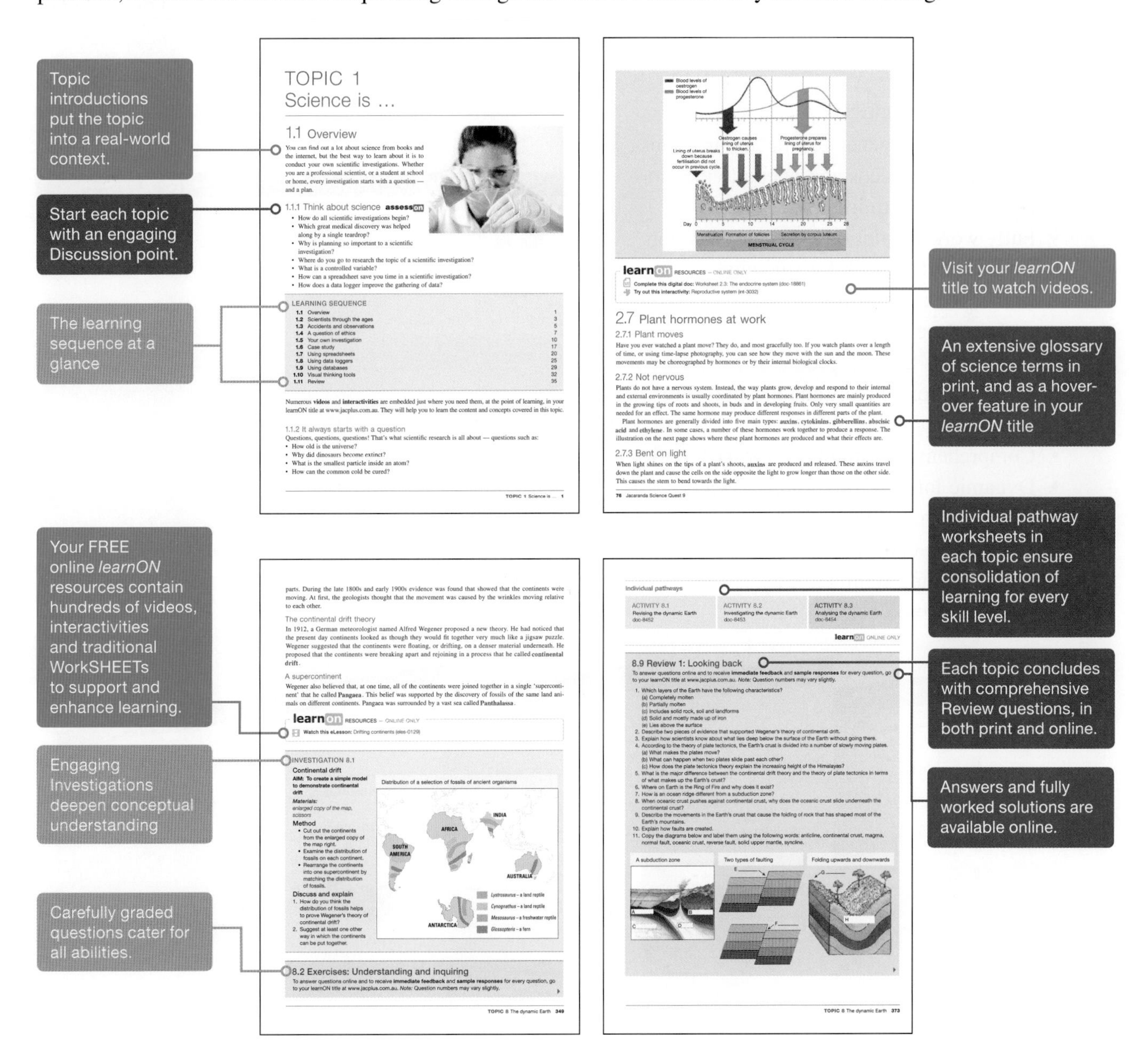

LearnON is Jacaranda's immersive and flexible digital learning platform that transforms trusted Jacaranda content to make learning more visible, personalised and social. Hundreds of engaging videos and interactivities are embedded just where you need them — at the point of learning. At Jacaranda, our 'learning made visible' framework ensures immediate feedback for students and teachers, with customisation and collaboration to drive engagement with learning.

Science Quest contains a free activation code for *learnON* (please see instructions on the inside front cover), so students and teachers can take advantage of the benefits of both print and digital, and see how *learnON* enhances their digital learning and teaching journey.

learnon includes:

- Students and teachers connected in a class group
- Hundreds of videos and interactivities to bring concepts to life
- Fully worked solutions to every question
- Immediate feedback for students
- Immediate insight into student progress and performance for teachers
- Dashboards to track progress
- Collaboration in real time through class discussions
- Comprehensive summaries for each topic
- Dynamic interactivities help students engage with and work through challenging concepts.
- Formative and summative assessments
- And much more …

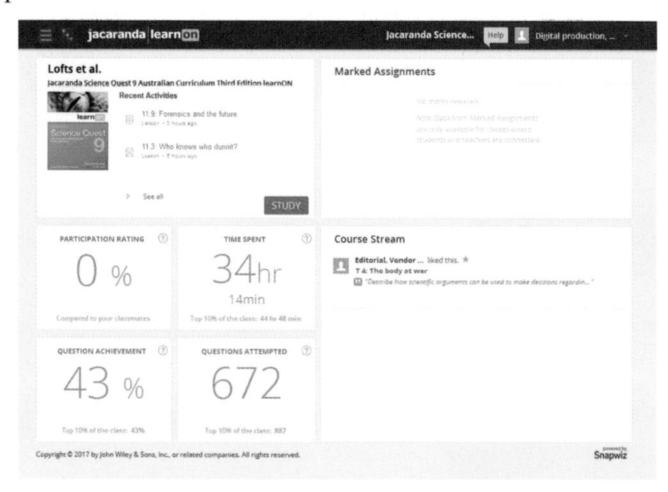

PREFACE

To the science student

Science is both a body of knowledge and a way of learning. It helps you to understand the world around you: why the sun rises and sets every day, why it rains, how you see and hear, why you need a skeleton and how to treat water to make it safe to drink. You can't escape the benefits of science. Whenever you turn on a light, eat food, watch television or flush the toilet, you are using the products of scientific knowledge and scientific inquiry.

Global warming, overpopulation, food and resource shortages, pollution and the consequences of the use of nuclear weapons are examples of issues that currently challenge our world. Possible solutions to some of these challenges may be found by applying our scientific knowledge to develop new technologies and creative ways of rethinking the problems. It's not just scientists who solve these problems; people with an understanding of science, like you, can influence the future. It can be as simple as using a recycling bin or saving energy or water in your home.

Scientific inquiry is a method of learning. It can involve, for example, investigating whether life is possible on other planets, discovering how to make plants grow faster, finding out how to swim faster and even finding a cure for cancer. You are living in a period in which knowledge is growing faster than ever before and technology is changing at an incredible rate.

Learning how to learn is becoming just as important as learning itself. *Science Quest* has been designed to help you learn how to learn, enable you to 'put on the shoes of a scientist' and take you on a quest for scientific knowledge and understanding.

To the science teacher

This edition of the *Science Quest* series has been developed in response to the Australian curriculum for Science. The Australian curriculum focuses on seven **General capabilities** (literacy, numeracy, ICT competence, critical and creative thinking, ethical behaviour, personal and social competence, and intercultural understanding). The history and culture of Aboriginal and Torres Strait Islanders, Australia's engagement with Asia, and sustainability have been embedded with the general capabilities where relevant and appropriate.

Science Quest interweaves **Science understanding** with **Science as a human endeavour** and **Science inquiry skills** under the umbrella of six **Overarching ideas** that 'represent key aspects of a scientific view of the world and bridge knowledge and understanding across the disciplines of science'.

The Australian Science curriculum provides the basis for the development of a Science curriculum in schools throughout Australia. However, it does not specify what you do in your classroom and how to engage individual classes and students.

We have attempted to make the *Science Quest* series a valuable asset for teachers, and interesting and relevant to the students who are using it. *Science Quest* comes complete with online support for students, including answers to questions, interactivities to help students investigate concepts, and video eLessons featuring real scientists and real-world science.

Exclusively for teachers, the online *Science Quest* teacher resources provides teaching advice and suggested additional resources, testmaker questions with assessment rubrics, and worksheets and answers.

Graeme Lofts and Merrin J. Evergreen

ACKNOWLEDGEMENTS

The authors and publisher would like to thank the following copyright holders, organisations and individuals for their assistance and for permission to reproduce copyright material in this book.

Images

• AAP Newswire: **165**/AP Photo • Alamy Australia Pty Ltd: **4** (right), **312**/Pictorial Press Ltd; **114**/World History Archive; **187** (middle)/© Zoonar GmbH; **190** (bottom right)/Chronicle; **308** (top)/BSIP SA; **326**/fStop Images GmbH; **342**/Sueddeutsche Zeitung Photo; **383**/age fotostock • Alamy Stock Photo: **38**/© PHOTOTAKE Inc.; **150** (top)/Helen Sessions; **211** (B)/Roy Wylam; **219**/World History Archive; **390** (bottom)/© sciencephotos; **390** (top)/Sebastian Loram; **402** (top)/© doc-stock; **454** (bottom)/Sueddeutsche Zeitung Photo • Australian Bureau Meteorol.: **231**/Bureau of Meteorology • australscope Pty Ltd: **335**, **336** (a)/Austral International • Bruno Annetta: **170** (A), **170** (B) • Carol Grabham: **388** (top)/David Grabham • Chemical Heritage Foundation: **194** (top)/Courtesy Othmer Library of Chemical History, CHF. Photo by James R. Voelkel • Corbis Images: **153**/KTSDESIGN/Science Photo Library • Creative Commons: **212** (C)/Hewa / Creative Commons Attribution 3.0 • CSIRO Marine Research: **275** (bottom left)/© CSIRO Oceans and Atmosphere • David Ferguson: **200** • Doris Taylor: **160**/Provided courtesy of Texas Heart Institute • Getty Images: **61**/Anatomical Travelogue; **80**/VLADGRIN; **96** (C)/momentimages; **126**/AMR Image; **186** (bottom)/Eraxion; **202** (B)/Cosmin Manci/Shutterstock; **417** (bottom)/© guy harrop / Alamy Stock Photo • Getty Images Australia: **4** (left), **178**, **207**, **207**, **311** (B)/Science Photo Library; **56** (top)/Science Photo Library/STEVE GSCHMEISSNER; **57**/CLOUDS HILL IMAGING LTD; **60**/STEVE GSCHMEISSNER; **85** (top)/WELLCOME DEPT. OF IMAGING NEUROSCIENCE/SCIENCE PHOTO LIBRARY; **103**/FABRICE COFFRINI; **106** (bottom left)/Medical Body Scans; **130** (left)/BSIP/UIG; **130** (right)/Bodenham LTH NSH Trust / SPL; **155** (A)/Martin M. Rotker; **155** (B)/Biophoto Associates; **180** (B)/MA ANSARY DR; **187** (A)/CATH ELLIS/SPL; **195** (C)/SSPL; **204** (B)/Science Picture Co; **211** (A)/CDC/SPL; **211** (C)/Edward Gooch; **212** (A)/Bettmann; **247** (bottom)/Andrew Syred/SPL; **247** (top)/STEVE GSCHMEISSNER/SPL; **262**/P Motta,G Macchiarelli; **275** (bottom right)/Norbert Wu/Minden Pictures; **380** (middle)/MATT MEADOWS/SPL; **388** (bottom)/Hank Morgan; **394**/Purestock; **401**/BSIP/UIG Via Getty Images; **406** (right), **413**/Fox Photos; **410** (bottom)/dpmike / iStock; **462**/Bloomberg; **471** (top)/Bloomberg / Contributor • iStockphoto: **126** (D)/© Sarah Lee; **402** (bottom left)/© Eraxion • James A. Sullivan: **220**/© cellsalive.com • John Wiley & Sons Australia: **324** (bottom)/Photo by Neale Taylor; **450** (bottom a), **450** (bottom b)/Werner Langer; **456** (bottom left)/Photo by Werner Langer; **473** (top)/Photo taken by Renee Bryon • John Wiley & Sons, Inc.: **126** (E), **126** (F), **126** (G) • Kino Studio: **472** (top)/Fred Adler • MAPgraphics: **364** (top) • Michael E Phelps, Dr: **106** (top A), **106** (top B), **106** (top C)/© Courtesy of Drs. Michael Phelps & John Mazziotta • Microsoft Corporation: **21**, **30** (B), **31** (A), **31** (C), **31** (E), **32** (B)/Screenshot reprinted by permission from Microsoft Corporation; **30** (A), **31** (B), **31** (D), **31** (F), **32** (A)/Screen shot reprinted by permission from Microsoft Corporation • Neale Taylor: **2** (bottom), **473** (bottom) • Newspix: **144**; **164**/Bill McAuley; **212** (B)/News Ltd; **212** (D)/Andy Baker; **221**/Sally Harding • Nobel Foundation: **327** (top)/Image provided courtesy of the Nobel Foundation • Out of Copyright: **219** (bottom left), **219** (bottom middle), **219** (bottom right)/The John Curtin School of Medical Research, ANU; **286**/© National Library of Australia • Pascale Warnant: **12** (bottom)/Photograph in banner c Julie Stanton • Perth Zoo: **279** (C) • Photodisc: **126** (A), **126** (B), **126** (C), **406** (left) • Public Domain: **191** (bottom), **194** (bottom), **195** (D), **195** (E), **311** (C); **202** (C)/Wikipedia; **204** (A)/https://commons.wikimedia.org/wiki/File:Landing_of_Columbus_2.jpg • Science Photo Library: **190** (bottom left)/SHEILA TERRY • Shutterstock: **1**/Drazen; **2** (top)/Peter Bernik; **3**/Elena Korn; **6**/mffoto; **8**/robin2; **9**, **186** (middle)/royaltystockphoto.com; **11**, **204** (C)/dotshock; **12** (top)/wawritto; **19**/Toa55; **39** (A)/ESB Professional; **39** (B)/Aaron Amat; **50** (middle)/Evgeny Litvinov; **50** (top)/Peter Waters; **51**/ChameleonsEye; **56** (bottom)/PhotoHouse; **62**/Daniel Wiedemann; **85** (bottom)/ktsdesign; **89**/Juan Gaertner; **94**/Creativa Images; **96** (A)/Sielemann; **96** (B)/luna4; **96** (D)/Hung Chung Chih; **98**/Sebastian Kaulitzki; **101**/Featureflash Photo Agency; **106** (bottom right)/withGod; **107**/James Steidl; **126**/BlueSkyImage; **129**/Steve Cukrov; **132**/Science Photo Library; **140**/Dmitry Lobanov; **142**/Kotomiti Okuma; **150** (bottom)/Kerdkanno; **152** (A)/TADDEUS; **152** (B)/Maceofoto; **152** (C)/Amero; **152** (D)/stable; **152** (E)/NinaM; **154** (A)/Svitlana-ua; **154** (B)/David Orcea; **154** (C)/Mega Pixel; **156** (right)/Rick Becker-Leckrone; **163** (right)/Ericsmandes; **163** (left)/LeonP; **180**/Susan Law Cain; **180**, **180** (I)/Istvan Csak; **180**/Fenton one; **180**/IVY PHOTOS; **180**/aceshot1; **180**/schankz; **180** (C)/Henrik Larsson; **180** (h)/Gines Romero; **190** (top)/Vasileios Karafillidis; **191** (top)/AridOcean; **195** (B), **195** (bottom)/Georgios Kollidas; **198**/mathagraphics; **202** (A)/javarman; **218**/Kokhanchikov; **230**/Olga Sapegina; **233**/Janelle Lugge; **234**/Cloudia Spinner; **236** (A)/kingfisher; **236** (B)/goodcat; **249**/D. Kucharski K. Kucharska; **250**/Serg64; **275** (A)/Johan Larson; **278**/Kristian Bell; **279** (B)/fivespots; **280** (A)/SINITAR; **280** (B)/most popular; **281** (A)/© Jan Hopgood; **281** (B)/© Kenneth William Caleno; **283**/Neale Cousland; **291**/© MaxFX; **296**/magnetix; **307**/Marcin Balcerzak;

311 (A)/Sergey Kamshylin; **315**/NEstudio; **316** (left)/MaluStudio; **319**/yurok; **322**/rosesmith; **324** (top)/Danny74; **327** (bottom)/Unknown; **336** (B)/Anticiclo; **336** (C)/Jose Arcos Aguilar; **338**/Chepko Danil Vitalevich; **339**/3Dsculptor; **346**/ Anders Peter Photography; **348** (top)/Teerapun; **356** (bottom left)/Olga Danylenko; **356** (top)/Tim Pryce; **361** (top)/Adwo; **364** (bottom)/yankane; **371**/NigelSpiers; **375**; **376**/YanLev; **383** (top)/Natursports; **387**/Tomislav Pinter; **387**/Peter Hermes Furian; **400**/Tomatito; **409** (top left)/Olga Popova; **409** (top right)/DeshaCAM; **410** (top)/Adrio Communications Ltd; **414** (bottom), **414** (bottom)/Denys Prykhodov; **414** (top left)/TebNad; **423** (middle)/Monkey Business Images; **426**/WilleeCole Photography; **433**/Krasowit; **437** (bottom)/s-ts; **442**/kanvag; **443** (top)/fztommy; **450** (bottom c)/Georgi Roshkov; **450** (bottom d)/dcwcreations; **450** (bottom e)/koka55; **452**/Darren Brode; **460**/Jinning Li; **461**/Etaphop photo; **472** (bottom)/ Alexandr Makarov; **477** (bottom)/Alexey Boldin; **478**/Andrew Jalbert • Sterling K. Clarren: **157**/Centre for Community Child Health, Research / www.bcricwh.bc.ca • Vernier Software & Technology: **26** (A), **26** (B), **27** (A), **27** (B), **28** (A), **28** (B) • Walter & Eliza Hall Institute: **220** • Wikipedia: **105**/© Creative Commons; **156** (left), **192** (middle), **196**, **275** (B), **308** (middle)/Public Domain; **275** (C)/JJ Harrison; **279** (A)/Public Domain

Text

TOPIC 1
Science is ...

1.1 Overview

You can find out a lot about science from books and the internet, but the best way to learn about it is to conduct your own scientific investigations. Whether you are a professional scientist, or a student at school or home, every investigation starts with a question — and a plan.

1.1.1 Think about science **assessON**

- How do all scientific investigations begin?
- Which great medical discovery was helped along by a single teardrop?
- Why is planning so important to a scientific investigation?
- Where do you go to research the topic of a scientific investigation?
- What is a controlled variable?
- How can a spreadsheet save you time in a scientific investigation?
- How does a data logger improve the gathering of data?

LEARNING SEQUENCE

Numerous **videos** and **interactivities** are embedded just where you need them, at the point of learning, in your learnON title at www.jacplus.com.au. They will help you to learn the content and concepts covered in this topic.

1.1.2 It always starts with a question

Questions, questions, questions! That's what scientific research is all about — questions such as:

- How old is the universe?
- Why did dinosaurs become extinct?
- What is the smallest particle inside an atom?
- How can the common cold be cured?

Every science investigation, whether it is conducted in a government research laboratory, a hospital, a museum or a space shuttle, begins with at least one question.

Although you are unlikely to even attempt to try to answer the preceding questions in your school science laboratory, there are many scientific questions that you can answer. Here are some examples.

- Does an audience affect the performance of an athlete?
- What is the best shape for a boomerang?
- Which type of soil do earthworms prefer?
- How do heating and cooling affect the way that rubber stretches?

When do you perform at your best?

INVESTIGATION 1.1

What can I investigate?

- In groups, brainstorm a list of questions that could be answered by doing an investigation in a school science laboratory. Record all the questions that are suggested even if they seem silly or difficult. The examples above might help you to think of some other ideas.
- From your list, remove any questions that the group feels are not likely to be answered because of a lack of the right equipment. Keep a record of the questions that are removed for this reason to submit to your teacher. You may find that equipment you thought was unavailable can be obtained, or that the question can be answered with different equipment.
- From your list, remove any questions that the group feels would be unsafe to try to answer, or that would be cruel to animals.
- Submit the remaining questions to your teacher for discussion by the whole class.

Scientific understanding, including theories and models (like this one of DNA made by Watson and Crick), is contestable and refined over time.

1.2 Scientists through the ages

Science as a human endeavour

1.2.1 Stepping stones

When you think of scientists, what image do you have in your mind? Albert Einstein? Marie Curie? Or do you have a picture of a small, absent-minded man wearing a lab coat — without hobbies, without friends and with no personality. Unfortunately, that's often the way scientists are portrayed in the media. The fact is, scientists are normal people who live similar lives to the rest of us.

Before putting on 'the shoes of a scientist' to conduct your own investigation, it's worth asking the question 'What, or who, is a scientist?' The answer to that question has been changing constantly for more than 2000 years.

The ancient Greek 'scientists' were very different from the scientists of today. They were called philosophers. The ancient Greek philosophers were curious and made accurate observations but they didn't perform experiments to test their ideas. They were thinkers, who tried to explain the structure of matter, the sun and the night sky. They walked the streets, discussing their ideas about nature, politics and religion with each other and their followers.

Although the ideas of the ancient Greek philosophers were limited by a lack of technology, they provided a stepping stone for the more recent growth in scientific knowledge.

One of the early Greek philosophers was Democritus who, in about 500 BC, suggested that all matter was made of tiny particles.

Aristotle, born in Greece fourteen years after the death of Democritus, reasoned that all matter was composed of four elements — earth, air, fire and water. About 2000 years later, Scottish scientist Joseph Black (1728–1798) discovered a fifth 'element'. He had discovered a new gas that he called 'fixed air'. We now call the gas carbon dioxide and know that it is not an element.

There are many other examples, including Hippocrates, born in the same year as Democritus, who taught his medical students to use observation rather than theory to diagnose illness. Hippocrates is regarded by many as the father of modern medicine.

Almost without exception, present-day scientific discoveries depend on work done previously by other scientists.

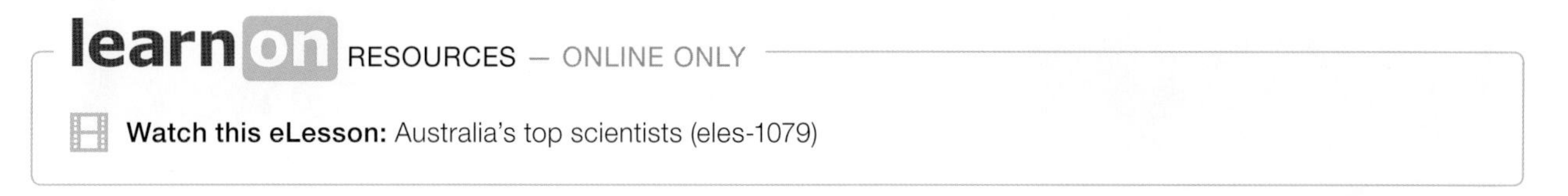

1.2.2 The scientific revolution

The way in which scientists worked changed greatly during the lifetime of Galileo Galilei (1564–1642), who is probably best known for being the first person to use a telescope to study the moon, planets and stars. Galileo also performed many experiments to investigate the motion of objects on the Earth's surface.

Galileo wrote about the need for controlled experiments and the importance of accurate observations and mathematical analysis. In fact, Galileo is described by many scientists and historians as the founder of the scientific method.

Galileo Galilei

Galileo's legacy

Some of the great scientists of the seventeenth century who followed Galileo and used the scientific methods he wrote about were:

- Johannes Kepler (1571–1630), who developed a number of laws about the motion of planets around the sun
- William Harvey (1578–1657), who used scientific methods to discover how blood circulates through the human body
- Robert Boyle (1627–1691), who applied the scientific method in chemistry to investigate the structure of matter more than 200 years before the current model of the atom was developed
- Robert Hooke (1635–1703), who used the newly invented microscope to observe and investigate the cells that make up living organisms.

These scientists were followed by Sir Isaac Newton (1642–1727), who was born in the same year that Galileo died. Newton was able to use mathematics to describe and explain the role of gravity in the motion of the Earth and other planets around the sun. He also explained much of the behaviour of light.

The work of the scientific pioneers of this era has influenced the thinking of those that followed and continues to influence scientists in the twenty-first century.

Modelling DNA

James Watson and Francis Crick won the Nobel Prize in 1962 for a piece of work that was a key discovery in biochemistry. In 1953 they established the structure of deoxyribonucleic acid, or DNA, the substance that makes up genes.

The model of DNA developed by Watson and Crick was based on the results of other scientists; for example:

- the work of Erwin Chargaff, who determined the basis of parts of DNA in 1951
- the X-ray diffraction photographsw (taken using X-rays rather than light) developed in 1949 by Rosalind Franklin and Maurice Wilkins.

Watson and Crick's breakthrough with DNA was possible thanks to the earlier discoveries of other scientists. Scientists today continue to build on the work of Watson and Crick. Their breakthrough has allowed other scientists to understand inherited diseases, and enabled the new field of genetic engineering to emerge.

Other branches of science work in a similar way. There are many examples of scientists furthering the work done by their colleagues, such as the recent achievements of genetic researchers.

Watson and Crick and their model of DNA

Rosalind Franklin provided an important stepping stone in the discovery of DNA.

1.2.3 Working in teams

Until the twentieth century, most scientists worked alone, with little or no financial support. Communication between individual scientists was difficult. Many of them wrote to each other and read the work of their fellow scientists. However, the telephone was not invented until 1876 and, of course, there was no email, no computers and no overseas travel except by ship.

Since the early twentieth century, most scientists have worked in teams. Their work is almost always supported and funded by organisations, industry or governments. Communication and teamwork between scientists all over the world are easier to achieve because of phones, the internet, email and jet aircraft.

1.2 Exercises: Understanding and inquiring

To answer questions online and to receive **immediate feedback** and **sample responses** for every question, go to your learnON title at www.jacplus.com.au. *Note:* Question numbers may vary slightly.

Remember

1. Identify two aspects of what is now called 'the scientific method' that the ancient Greek philosophers practised.
2. According to Aristotle, all matter was composed of four elements. What were those elements?
3. Why was Galileo described by many as the founder of the scientific method?

Think

4. Why was the period of the seventeenth century labelled 'the scientific revolution'?
5. Name some major technologies that were not available to the early Greeks and that have helped modern scientists to test their hypotheses.
6. Which technologies did seventeenth-century scientists have available to them that the early Greek scientists did not have?
7. List the qualities that you would expect a present-day scientist to have.

Investigate

8. Research and report on the Hippocratic Oath and its importance to medical practitioners.
9. Joseph Black made other discoveries as well as the one described here. Research some of these and find out how they have affected everyday life.

Imagine

10. Imagine that Galileo Galilei could return to a university in Italy today and observe the way in which scientists at the university worked. Write a one-page account of the observations that he might enter into his diary at the end of the day.

1.3 Accidents and observations

Science as a human endeavour

1.3.1 A matter of luck?

Some of the greatest scientific discoveries have been made by accident. The development of batteries, penicillin and X-rays began with 'accidents' in laboratories. However, was it all just a matter of luck?

1.3.2 Jumping frog's legs

The very first electric cell was created by accident over 200 years ago. Luigi Galvani, an Italian physician, was dissecting the leg of a recently killed frog. The leg was held by a brass hook. When he cut the leg with an iron knife, the leg twitched. Galvani investigated further by hanging the frog's legs on an iron railing with brass hooks. Whenever the frog's legs came into contact with the iron railing, they twitched. Galvani incorrectly proposed a theory of 'animal electricity' as the reason behind the muscle spasms.

Reports of Galvani's observations reached his friend Alessandro Volta, another Italian scientist. Volta suggested that the twitch was caused by a sudden movement of electric charge between the two different metals. The frog's flesh, he suggested, conducted the charge. Galvani had, without realising it, produced the world's first electric cell. The **galvanometer**, an instrument used to measure small electric currents, was named after Luigi Galvani.

1.3.3 Outside looking in

X-ray images allow doctors, dentists and veterinarians to 'see' through living flesh. The pictures taken with X-rays, called radiographs, are obtained by passing X-rays through objects onto a photographic plate.

Unlike light, X-rays pass through the human body. Some parts of the body absorb more of the X-rays than others, leaving a shadow on the plate. Bones leave the sharpest shadows, making it possible to detect fractures and abnormalities.

X-rays have many other uses. They are used in metal detectors at airports and to detect weaknesses and cracks in metal objects. X-rays can be used by archaeologists to examine ancient objects (including Egyptian mummies) found under the ground or in ruins without touching and damaging them.

X-rays were discovered by accident in 1895 while German physicist Wilhelm Röntgen (pronounced 'Rentjen') was experimenting with a glass tube that glowed as electrons moved through it at high voltage. He had, by chance, left a photographic plate on a nearby bench. Röntgen noticed that whenever electrons were passing through the tube, the photographic plate glowed. This was puzzling because the glass tube was wrapped in heavy black paper and, since the room was in total darkness, there was no light to expose the photographic plates.

Röntgen investigated his puzzling observations further. He found that these mysterious rays that seemed to be coming from the tube could pass through human flesh as well as black paper. He obtained a clear image of the bones in his wife's hand as she rested it on the photographic film.

Röntgen's accidental discovery changed the face of medical practice in many ways.

X-ray pictures can reveal broken bones and disease in internal organs.

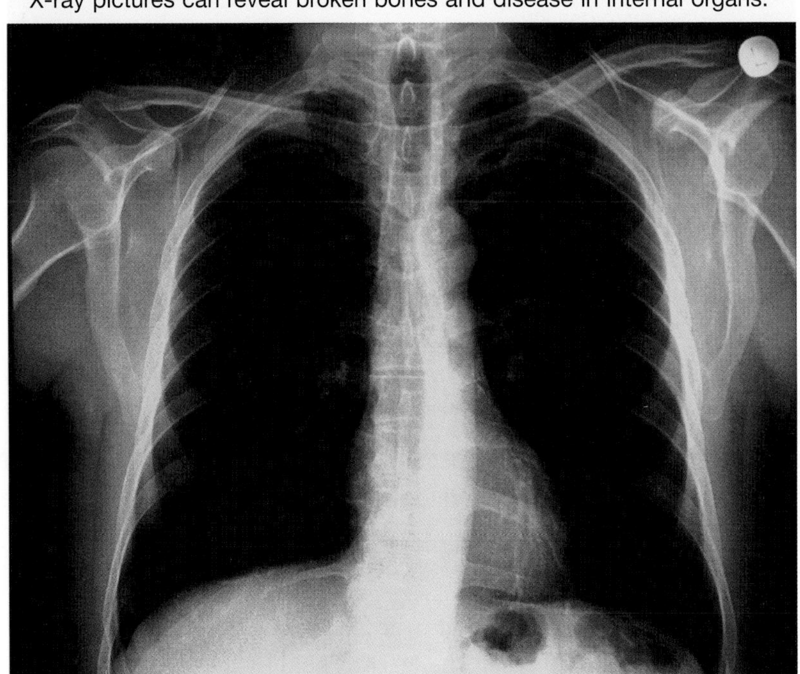

1.3.4 One accident after the other

Penicillin is one of the most commonly used drugs in the treatment of diseases caused by bacteria. The discovery and production of penicillin followed a series of accidental observations. The first observation of penicillin was made in 1928 by Scottish bacteriologist Sir Alexander Fleming.

Fleming's interest in bacterial diseases intensified during World War I, when he was treating wounded soldiers. He noticed that the antiseptics used to treat wounds killed white blood cells more quickly than the harmful bacteria they were designed to kill. The white blood cells form part of the body's natural resistance to bacteria.

After the war, Fleming began searching for substances that would kill bacteria without harming the body's natural defences.

One day during his search, a teardrop fell into a dish containing a layer of bacteria. When he checked the dish the following day, he noticed a clear layer where the teardrop had fallen. Fleming then found that a chemical in human teardrops, which he named **lysozyme**, was able to kill some types of bacteria without harming the body's natural defences. Unfortunately, lysozyme was not effective against most disease-causing bacteria.

Fleming's greatest discovery occurred in 1928 when he was trying to find a cure for influenza. A tiny piece of mould had fallen into a Petri dish in which he was growing bacteria before the lid was put on. Fleming noticed that there was no further growth of bacteria around the mould. He later admitted that if it had not been for his earlier experience with the teardrop, he may have thrown the dish away because it had been spoiled.

The mould, *Penicillium notatum*, contained a substance called **penicillin**, which kills many disease-causing bacteria without harming the body's natural defences. A new problem arose — how to separate and purify the substance. It was an Australian scientist, Howard Florey (1898–1968), who succeeded in separating and purifying the penicillin antibiotic. Together with Boris Chain, a Jewish refugee from Germany, Florey found a way of producing enough penicillin to treat a number of diseases. Their success came just in time for use in treating the many wounded in World War II. Fleming, Florey and Chain shared the Nobel Prize in Medicine in 1945.

1.3 Exercises: Understanding and inquiring

To answer questions online and to receive **immediate feedback** and **sample responses** for every question, go to your learnON title at www.jacplus.com.au. *Note:* Question numbers may vary slightly.

Remember

1. Which modern-day device was accidentally created by Luigi Galvani?
2. What form of radiation was discovered by Wilhelm Röntgen?
3. Which drug was later produced as a result of Alexander Fleming's accidental observation?

Think

Your answers to questions 4 and 5 could be presented in a table.

4. Consider the discoveries made by Galvani, Röntgen and Fleming. In each case, describe the skills and scientific knowledge used in making and developing their discovery.
5. Make a list of the personal qualities that enabled Galvani, Röntgen and Fleming to take advantage of their chance observations.
6. Were the discoveries of the electric cell, X-rays and penicillin really just accidents? Explain your answer.

Imagine

7. Do you think that the electric cell, X-rays and penicillin would have been discovered if it had not been for the chance observations of Galvani, Röntgen and Fleming? Explain your answer.

Investigate

8. Find out about some other scientific discoveries that were the result of accidents or chance observations.

1.4 A question of ethics

Science as a human endeavour

1.4.1 Science and ethics

Ethics is the system of moral principles on the basis of which people, communities and nations make decisions about what is right or wrong. Scientific inquiry takes place in communities that have political, social and religious views and is undertaken by people who have personal views about all sorts of issues. It is a human endeavour and therefore cannot be separated from ethics and questions about right and wrong.

Ethical values vary between countries, religions, communities and individuals — even between members of the same family. For example, capital punishment, the execution of a person for committing a crime, is considered by some to be right and by others to be wrong.

Science interacts with ethics in several ways, including:

- affecting the way in which science is conducted
- affecting the types of scientific research carried out
- in the conflict or match between scientific ideas and religious beliefs
- providing scientific community practices that act as a model for ethical behaviour.

1.4.2 Animal testing

Animals are used in scientific research in many ways, including: to test the effects of potential drugs; to test cosmetics for allergies; to understand the functioning of parts of the body; and to test new surgical techniques. In some research and testing, the animals die. Animals used may include monkeys, bees, mice, worms and dogs, among others.

There are ethical issues about whether animals should ever be used in scientific research, or if some types of animals shouldn't be used, or if some types of research shouldn't be carried out at all.

The use of animal testing is increasingly becoming an ethical issue around the world. The European Union has banned the sale of cosmetics tested on animals.

1.4.3 Medical research

Medical research is carried out partly by public institutions such as universities and specialist research departments, and partly by private companies. The main purpose of research in public institutions has been to increase understanding and to provide solutions to existing problems; while private companies aim to provide new products or services that can be sold for profit. Some research in public institutions is done with the aim of making money, and some private research aims at increasing scientific knowledge, which raises ethical questions about whether new drugs are produced more for their profitability than for the benefit of the community.

Life expectancy varies greatly around the world, as do patterns of disease. Cancer, heart disease and diabetes kill many Australians and billions of dollars are spent on researching their causes and treatment. Diarrhoeal diseases and malaria are readily treated in Australia, but kill millions of people each year in Africa, Asia and South America — sometimes because of lack of information and sometimes because of lack of low-cost products. This raises ethical and social questions, such as:

- Is it right that effective drugs are unavailable to millions because of their cost?
- What is the fundamental purpose of developing pharmaceuticals?
- Should the type of treatment be determined by the profit it generates?

Another source of ethical concern in medical research relates to the testing of new drugs. When pharmaceutical companies design new drugs, they need to test these thoroughly before being able to sell them. Some people argue that the testing regime is too lengthy and that new drugs that have the potential to treat deadly diseases should be supplied to the people dying from these diseases even if the drug has not been fully tested.

HOW ABOUT THAT!

When Barry Marshall and Robin Warren came to the conclusion that stomach ulcers were probably caused by a bacteria, they were faced with some tricky ethical and safety considerations. A stomach ulcer occurs when the lining of the walls of the stomach becomes damaged and the acid inside the stomach eats away at the stomach wall. It is a very painful condition. Previously it was thought that ulcers were caused by lifestyle factors, including stress, so it was difficult to treat ulcers. People were usually told to avoid stress, for example by changing job or cutting their work hours, and to cut out particular foods, sometimes with no improvement to their health.

Helicobacter pylori bacteria in the human stomach cause stomach ulcers. They move their hair-like structures to travel around the stomach lining.

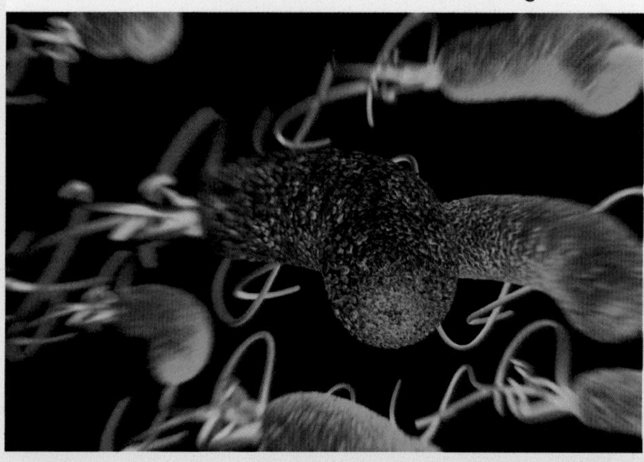

Barry Marshall and Robin Warren suspected that ulcers were actually caused by bacteria called *helicobacter pylori*. They had found these bacteria in the stomachs of people suffering from stomach ulcers but not in the stomachs of healthy individuals. They had also studied the bacterium. The only way to know for sure would be to deliberately infect someone with the bacteria and find out whether they developed a painful ulcer. There were risks involved; for instance, the bacteria could cause other health problems. It could even kill the patient. There were also ethical issues associated with deliberately trying to make a healthy person sick. In the end, Barry Marshall carefully weighed up the risks involved and decided to test his hypothesis on himself. He swallowed a solution of the bacteria and soon became ill and developed the early symptoms associated with the development of stomach ulcers. He then treated himself with antibiotics. Now when a patient is diagnosed with a stomach ulcer, treatment is simple — a course of antibiotics usually fixes the problem.

1.4.4 Agriculture

Traditional plant breeding methods — manually putting pollen from one plant into the flower of another to produce a 'cross'— were once the only means of modifying plant types; a slow and laborious process. Now, using techniques for moving genes from one plant to another, it is possible to design plants that have certain characteristics. This technique of **genetic modification (GM)** is controversial. GM crops are greatly restricted in Australia. GM techniques have been used to produce crops that:

- are resistant to herbicide so that weed control is more effective (canola)
- produce their own pesticides to reduce insect attack (cotton)
- contain added nutrients (rice).

Discussion about the ethics of GM crops often focuses on the role of companies in developing GM crops for the profit they are expected to bring. Ethical issues are also raised about whether GM techniques should be used by public research laboratories and international agencies to improve food supply in regions where many people are undernourished.

1.4 Exercises: Understanding and inquiring

To answer questions online and to receive **immediate feedback** and **sample responses** for every question, go to your learnON title at www.jacplus.com.au. *Note:* Question numbers may vary slightly.

Remember

1. Explain why scientific inquiry should not take place without considering whether it is right or wrong.
2. Identify an illness that affects people worldwide and kills millions in poor countries but almost no-one in Australia.
3. Explain how GM crops are different from other crops.

Think

4. Describe the ethical issues associated with the experiment carried out by Barry Marshall.

Discuss

5. In small groups, discuss the following statements.
   (a) Cosmetics should never be tested on animals.
   (b) All medical research, including research into new drugs, should be done by non-profit organisations rather than by companies aiming to make a profit.
   (c) Food made from genetically modified crops should have a special label to show that it contains GM ingredients.

Investigate

6. What does a bioethicist do? What training does a bioethicist require?
7. Research and report on alternatives to using animals in research.
8. Outline some of the arguments against using genetically modified crops.

1.5 Your own investigation

1.5.1 Begin with a plan

Whenever you take a trip away from home, you need to plan ahead and have some idea of where you are going. You need to know how you are going to get there, what you need to pack and have some idea of what you are going to do when you get there.

It's the same with an experimental investigation. Planning ahead increases your chances of success. It's easier if you can break an investigation down into steps as shown below.

1.5.2 Finding a topic

Your investigation is much more likely to be of high quality if you choose a topic that you will enjoy working on. These steps might help you choose a good topic.

1. Think about your interests and hobbies. They might give you some ideas about investigation topics.
2. Make a list of your ideas.
3. Brainstorm ideas with a partner or in a small group. You might find that exchanging ideas with others is very helpful.
4. Find out what other students have investigated in the past. Although you will not want to cover exactly the same topics, investigations performed by others might help you to think of other ideas.

Think about your interests.

Think about your hobbies.

Brainstorm your ideas with others.

5. Do a quick search in the library or at home for books or newspaper articles about topics that interest you. Search the internet. You might also find articles of interest in magazines or journals. You could use a table like the one below to organise your ideas.

Topic area	Name of book, magazine, website etc.	Chapter or article	Topic ideas

From observations to ideas

Many ideas for scientific investigations start with a simple observation. Some well-known investigations and inventions from the past started that way. Even though the discoveries by Galvani, Röntgen and Fleming described in sections 1.3.2, 1.3.3 and 1.3.4 were made by accident, they would not have been made without observation skills. There is also another important 'ingredient' in these discoveries — curiosity and the ability to ask questions and form ideas that can be tested by experiment and further observation.

Danish scientist Hans Oested discovered the connection between electric current and magnetism when, in 1819, he noticed that a compass needle pointed in the wrong direction every time it was placed near a wire carrying an electric current. He went on to design experiments to find out exactly how different electric currents affected compass needles. The results of his experiments started a flood of inventions, including electric generators and motors.

An investigation by 15-year-old student Catherine Pippos began with an observation that her friends seemed to perform better in track and field events when there was an audience cheering them on. You have probably seen this yourself. Her investigation 'Does an audience affect the performance of an athlete?' involved three different sporting activities and compared the performance of a large group of students under three different conditions:

- no audience
- a quiet audience
- a cheering audience.

The sporting activities were:

- goal shooting in basketball
- sit-ups
- shot-put.

What do you think she found out? Perhaps you could try a similar investigation.

Could an audience really affect this athlete's performance? To answer the question scientifically, an investigation is needed.

Defining the question

Once you have decided on your topic, you need to determine exactly what you want to investigate. It is better to start with a simple, very specific question than a complicated or broad question. For example, the topic 'earthworms' is very broad. There are many simple questions that could be asked about earthworms.

For example:

- Which type of soil do earthworms prefer?
- How much do earthworms eat?
- Do earthworms prefer meat or vegetables?
- How fast does a population of earthworms grow?

Your question needs to be realistic. In defining the question, you need to consider whether:

- you can obtain the background information that you need
- the equipment that you need is available
- the investigation can be completed in the time you have available
- the question is safe to investigate.

There are many problems relating to earthworms that could be investigated.

1.5.3 Keeping records

A **logbook** is an essential part of a long scientific investigation. It provides you with a complete record of your investigation, from the time you begin to search for a topic. Your logbook will make the task of writing your report very much easier.

A logbook is just like a diary. Make an entry whenever you spend time on your investigation. Each entry should be clearly dated. It's likely that the first entry will be a mind map or list of possible topics. Other entries might include:

- notes on background research conducted in the library. Include all the details you will need for the **bibliography** of your report (see section 1.5.9).
- a record of the people that you asked for advice (including your teacher), and their suggestions
- diagrams of equipment, and other evidence that you have planned your experiments carefully
- all of your 'raw' results, in table form where appropriate
- an outline of any problems encountered and how you solved them
- first drafts of your reports, including your thoughts about your conclusions.

An online logbook

An exercise book can be used as a logbook, but there are several advantages in maintaining your logbook online in the form of a **blog**. There are many sites that will allow you to set up a free blog. Your teacher might be able to provide some suggestions. Once you set up a blog, every entry you make will be dated automatically. You can upload documents, diagrams, photos and short videos. You can also add links to other sites and invite friends, family and teachers to post comments about your progress.

There are some precautions that you should take if you decide to use a blog as a logbook.

A blog used as a logbook for a student research investigation.

- Limit your posts to those related to your science investigation. Don't use your logbook blog for social networking.
- Do not include your address or phone number.
- If your blog is on the internet (rather than a school intranet):
  - do not post any photos of yourself in school uniform or any other clothing that will identify where you go to school
  - do not include your full name, address, phone number or the name of your school in the blog. Use only your first name or a nickname.
  - use privacy settings or use a password to ensure that only trusted school friends, family and your teacher have access to the blog.

WHAT DOES IT MEAN?

The word *blog* is a recent addition to the English language. It is an abbreviation of the words *web log*.

1.5.4 Designing the experiments

In order to complete a successful investigation, you need to make sure that your experiments are well designed. Once you've decided exactly what you are going to investigate, you need to be aware of:

- which variables need to be controlled and which variables can be changed
- whether a control is necessary
- what observations and measurements you will make and what equipment you will need to make them
- the importance of repeating experiments (replication) to make your results more reliable
- how you will record and **analyse** your data.

A poorly designed investigation is likely to produce a conclusion that is not **valid**.

Controlling variables

A **variable** is an observation or measurement that can change during an experiment. You should change only one variable at a time in an experiment. The variable that you deliberately change during an experiment is called the **independent variable**. The variable that is being affected by the independent variable — that is, the variable you are measuring — is called the **dependent variable**. For example, if you were performing an experiment to find out which brand of fertiliser was best for growing a particular plant, the independent variable would be the brand of fertiliser. The dependent variable would be the heights of the plants after a chosen number of days.

When you are testing the effect of an independent variable on a dependent variable, all other variables should be kept constant. Such variables are called **controlled variables**. For example, in the fertiliser experiment, the type of plant, amount of water provided to each plant, soil type, amount of light, temperature and pot size are all controlled variables. The process of controlling variables is also known as **fair testing**.

The need for a control

Some experiments require a **control**. A control is needed in the fertiliser experiment to ensure that the result is due to the fertilisers and not something else. The control in this experiment would be a pot of plants to which no fertiliser was added. All other variables would be the same as for the other three pots.

Replication

Replication is the repeating of an experiment to make results more reliable. In the case of the fertiliser experiment, a more reliable result could be obtained by setting up two, three or four pots for each brand of fertiliser. An average result for each brand or the control could then be calculated.

1.5.5 Using information and communications technology

Computer hardware and software are important tools used by scientists during their investigations. For example:

- spreadsheets can be used to organise and analyse data
- data loggers can be used to collect large numbers of measurements of variables that are difficult to collect in other ways
- databases can be used to arrange data or information so that it is easier to locate.

These tools are described in subtopics 1.7, 1.8 and 1.9 in this chapter.

1.5.6 Getting approval

You should now be ready to write a plan for your investigation. You should not commence any experiments until your plan has been approved by your science teacher. Your plan should include the following information.

1. Title

The likely title — you may decide to change it before your work is completed.

2. The problem

A statement of the question that you intend to answer. Include a hypothesis. A hypothesis is an educated guess about the outcome of your experiments. It is usually based on observations and able to be tested by further observations or measurements.

3. Outline of your experiments

Outline how you intend to go about answering the question. This should briefly outline the experiments that you intend to conduct.

4. Equipment

List here any equipment that you think will be needed for your experiments.

5. Resources

List here the sources of information that you have already used and those that you intend to use. This list should include library resources, organisations and people.

Write out a plan for your investigation.

1.5.7 Gathering data

Once your plan has been approved by your teacher, you may begin your experiments.

Details of how you conducted your experiments should be recorded in your logbook. All observations and measurements should be recorded. Use tables where possible to record your data.

Where appropriate, measurements should be repeated and an average value determined. All measurements — not just the averages — should be recorded in your logbook.

Photographs should be taken if appropriate.

You might need to change your experiments if you get results you don't expect. Any major changes should be checked with your teacher.

All observations and measurements should be recorded.

WHAT DOES IT MEAN?

The word *data* is the plural of *datum*, which means 'a piece of information'. It comes from the Latin word meaning 'something given'.

1.5.8 Graphing variables

If you use a graph to show your results, you would normally graph the independent variable (the one you changed) on the x-axis, and the dependent variable (the one you measured) on the y-axis. When the dependent variable changes with time, you can graph time on the x-axis and the dependent variable on the y-axis. For example, in the fertiliser experiment, two types of graphs could be used, a line graph or a column graph (bar chart).

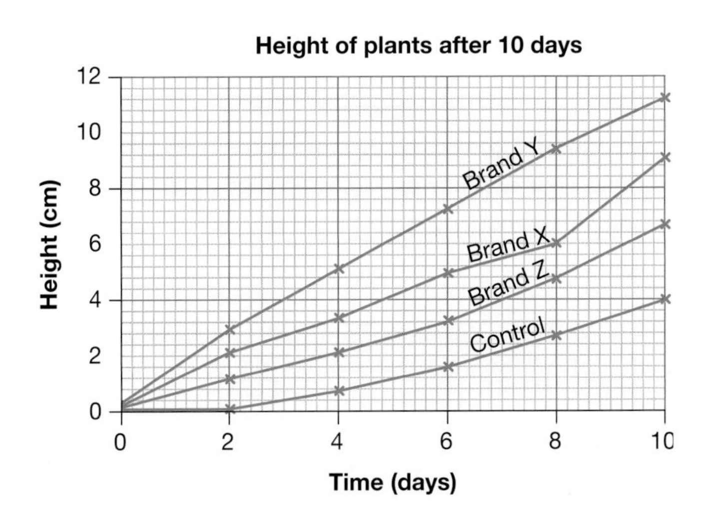

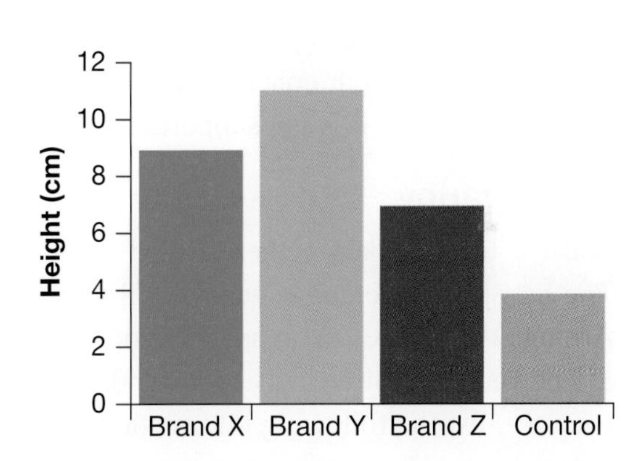

1.5.9 Writing your report

You can begin writing your report as soon as you have planned your investigation, but it cannot be completed until your observations are complete. Your report should be typed or neatly written on A4 paper and presented in a folder. It should begin with a table of contents, and the pages should be numbered. Your report should include the following headings (unless they are inappropriate for your investigation).

Abstract

The abstract provides the reader with a brief summary of your whole investigation. Even though this appears at the beginning of your report, it is best not to write it until after you have completed the rest of your report.

Introduction

Present all relevant background information. Include a statement of the problem that you are investigating, saying why it is relevant or important. You could also explain why you became interested in the topic.

Aim

State the purpose of your investigation: that is, what you are trying to find out. Include the hypothesis.

Materials and methods

Describe in detail how you did your experiments. Begin with a list or description of equipment that you used. You could also include photographs of your equipment if appropriate. The method description must be detailed enough to allow somebody else to repeat your experiments. It should also convince the reader that your investigation is well controlled. Labelled diagrams can be used to make your description clear. Using a step-by-step outline makes your method easier to follow.

Results

Observations and measurements (often referred to as data) are presented here. Data should, wherever possible, be presented in table form so that they are easy to read. Graphs can be used to help you and the reader interpret data. Each table and graph should have a title. Make sure you use the most appropriate type of graph for your data. Some examples of graphs are shown bottom of previous page.

Discussion

Discuss your results here. Begin with a statement of what your results indicate about the answer to your question. Explain how your results might be useful. Any weaknesses in your design or difficulties in measuring could be outlined here. Explain how you could have improved your experiments. What further experiments are suggested by your results?

Conclusion

This is a brief statement of what you found out. It is a good idea to read your aim again before you write your conclusion. Your conclusion should also state whether your hypothesis was supported. You should not be disappointed if it is not supported. In fact, some scientists deliberately set out to reject hypotheses!

Bibliography

Make a list of books, other printed or audio-visual material and websites to which you have referred. The list should include enough detail to allow the source of information to be easily found by the reader. Arrange the sources in alphabetical order.

The way a resource is listed depends on whether it is a book, magazine (or journal) or website. For each resource, list the following information in the order shown:

- author(s), if known (book, magazine or website)
- title of book or article, or name of website
- volume number or issue (magazine)
- URL (website)
- publisher (book or magazine), if not in title
- place of publication, if given (book)
- year of publication (book, magazine or website)
- chapter or pages used (book).

Some examples of different sources are listed below:

- John Thomas Moore, *Chemistry Essentials for Dummies*, Wiley Publishing, Inc.; Indianapolis, 2010
- Justin Gregg, 'How Smart are Dolphins?' *Focus Science and Technology*, Issue 264, February 2014, BBC, pages 52–57
- Australian Marine Wildlife Research & Rescue Organisation, http://www.amwrro.org.au, 2014.

Acknowledgements

List the people and organisations who gave you help or advice. You should state how each person or organisation assisted you.

1.5.10 Everyone has talent

In most states and territories, there are competitions or events that provide opportunities for you to present reports of your own scientific research. Each year, tens of thousands of dollars in prizes are awarded to hundreds of entrants. Information about these competitions and events can be obtained from your science teacher.

1.5 Exercises: Understanding and inquiring

To answer questions online and to receive **immediate feedback** and **sample responses** for every question, go to your learnON title at www.jacplus.com.au. *Note:* Question numbers may vary slightly.

Remember

1. Construct a flowchart to show the steps that you need to take before beginning your experiments.
2. What is the advantage of repeating an experiment several times?
3. Describe how each of the following computer tools can be used in a scientific investigation.
   (a) Spreadsheets
   (b) Data loggers
   (c) Databases
4. Describe the difference between an independent and a dependent variable.
5. Outline an example of the use of a control in an experiment.

Think

6. Outline why it is important to plan a research investigation carefully.
7. Discuss the advantages and disadvantages of using a blog as a logbook for your investigation.
8. For each problem described below, identify the independent and dependent variable and three other variables that would need to be controlled.
   (a) Josie wanted to find out whether the water in her drink bottle would stay cold for longer if she wrapped the bottle in foil or a towel.
   (b) Charlotte would like to know whether ice blocks made from green coloured water melt at the same temperature as uncoloured ice blocks.
   (c) Jayden is testing the hypothesis that tall people are faster long-distance runners than short people.
   (d) Shinji is testing the myth that plants grow faster if you play them music for at least 2 hours a day.
   (e) Nikita has heard that most people shrink slightly (in height) throughout the day and stretch out at night. She would like to know whether this is true.
9. Why is it better to write the abstract of a scientific report last, even though it appears at the beginning?
10. In which section of your report do you describe possible improvements to your experiments?
11. The television show *Mythbusters* involves a team led by Adam and Jamie carrying out investigations to test various myths.
    (a) Define the term *myth*. (Use a dictionary if necessary.)
    (b) Look at the list of myths Adam and Jamie have investigated and pick at least three that you could test using equipment available at home or at school.
    (c) If your school has any episodes of *Mythbusters* available, watch an episode. Make a list of the myths tested in the episode and discuss the validity of the experiments carried out by Adam, Jamie and their team.

learnon RESOURCES — ONLINE ONLY

Complete this digital doc: Worksheet 1.1: Setting up a logbook (doc-18850)

Complete this digital doc: Worksheet 1.2: Variables and controls (doc-18851)

1.6 Case study

1.6.1 Investigating muddy water

Sean, a Year 9 student, conducted an experimental investigation to compare the turbidity (cloudiness) of water in the following three locations:

- a creek near his school
- a creek near his home
- a river near his home.

His search for information in the library revealed that the cloudiness was caused by particles of soil (and sometimes pollution) suspended in the water. Sean chose his topic because he was interested in the environment. He felt that clean water was the right of all living things. His research and background knowledge led him to form the hypothesis that 'the clearest water will be in the river'.

Sean took water samples from each of the three locations on four days. He found a method of measuring turbidity from a library book. It involved adding a chemical called potash alum to a sample of water in a jar. The potash alum makes the particles of suspended soil clump together and fall to the bottom of the jar. A layer of mud is formed. The height of the mud at the bottom is then measured.

SEAN'S INVESTIGATION

Materials:

- 4 large jars or bottles with lids for collecting water samples (capacity of about 1 L each)
- 4 identical jam jars with lids, labelled 1, 2, 3 and 4
- metal teaspoon (not plastic, in case it breaks)
- potash alum (potassium aluminium sulfate)
- 4 water samples from different locations
- ruler with 1-millimetre graduations
- 100 mL measuring cylinder
- permanent marker

Method

1. Water samples (about 1 litre each) were collected from a specific part of the creeks and river on the same day.
2. Each of three clean jars was filled to the same level with the water samples — a labelled jar for each location. A fourth labelled jar was filled to the same level with distilled water.
3. One level teaspoon of potash alum was added to each jar. Lids were put on the jars and the jars were shaken.
4. The jars were left for 30 minutes to allow the particles to settle.
5. The height of the layer of mud on the bottom of each jar was measured and recorded.
6. The jars were emptied and washed and the experiment was repeated three more times.
7. Water samples were collected from the same locations on three other days over a ten-day period and the entire experiment was repeated three more times.

A summary of Sean's method, including a list of materials and equipment required, is given on the above. You will notice that Sean used a fourth sample. It was needed as a control and contained distilled water. This was to ensure that there was nothing in the pure water to cause a layer at the bottom of the jar when the potash alum was added. His results are in the table below.

Results table measuring the levels of mud in water samples from three areas

	Height of mud (mm)															
	Day 1				Day 2				Day 3				Day 4			
	Test				Test				Test				Test			
Water sample	1	2	3	Average	1	2	3	Average	1	2	3	Average	1	2	3	Average
1. Home creek	3.5	4.0	5.0	4.2	5.0	4.5	5.0	4.8	4.5	5.0	4.5	4.3	5.0	4.5	4.0	4.5
2. School creek	2.5	2.0	2.0	2.2	3.0	2.5	2.5	2.7	2.0	2.5	2.5	2.3	2.0	2.0	2.5	2.2
3. Barnes River	1.0	0.5	0	0.5	2.0	1.0	1.5	1.5	0.5	1.0	0.5	0.7	0.5	0.5	0.5	0.5
4. Distilled water	0	0	0	0	0	0	0	0	0	0	0	0	0	0	0	0

1.6.2 Analysing the data

Sometimes it is necessary to refine the raw data (the data initially collected), presenting them in a different way. Sean was planning to use his average measurements to make a column graph. He decided to simplify his table so that it was easier to construct the column graph. The simplified table and column graph make it easier for others to read the results, and easier for Sean to see patterns and draw conclusions.

Table showing average heights of mud in water from three different areas

Sample number and source	Height of mud (mm)			
	Day 1	Day 2	Day 3	Day 4
1. Home creek	4.2	4.8	4.3	4.5
2. School creek	2.2	2.7	2.3	2.2
3. River	0.5	1.5	0.7	0.5

Sean's graph makes it easier to see patterns and draw conclusions.

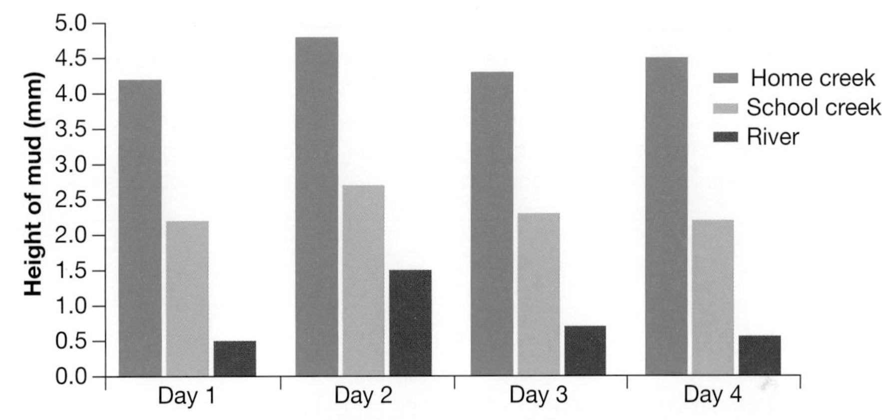

1.6.3 Being critical

Sean was pleased with his results and was able to draw conclusions. In the discussion section of his report, he suggested that further studies be done. The turbidity was affected by weather conditions and the sampling needed to be done over a longer period, and in different weather conditions. Sean had recorded weather details on each day that he sampled water and was able to explain the very high mud level in the river on day 2. It is almost always possible to suggest improvements to your experiments.

1.6.4 Drawing conclusions

Sean's hypothesis, that the clearest water would be in the river, was supported. His conclusion was written in point form.

1. The home creek has the muddiest water, with sample values ranging from heights of 4.2 to 4.8 mm of mud per 200 mL of water. The school creek has moderate amounts of mud compared to the other two samples. Sample values ranged from 2.2 to 2.7 mm of mud per 200 mL of water. The river water is the clearest, with sample values of 0.5 to 1.5 mm of mud per 200 mL of water.
2. Weather conditions can alter the amount of mud in water bodies by either adding run-off from drains or stirring up the water. This was particularly noticeable in the samples taken from the river site on day 2, which followed a period of rain.

Sean's teacher was pleased, and suggested that Sean carry out further research and rewrite his material. And what did he think about entering his project for competitions?

Chemical waste running into a river. How might you test for such materials in a water sample from this site?

The last word comes from Sean. After successfully completing his student research project, he said: 'It all depends on the experimental design — get that right and the rest is likely to run smoothly.'

1.6 Exercises: Understanding and inquiring

To answer questions online and to receive **immediate feedback** and **sample responses** for every question, go to your learnON title at www.jacplus.com.au. *Note:* Question numbers may vary slightly.

Think

1. For Sean's experiment, identify:
   (a) the independent variable
   (b) the dependent variable
   (c) the variables he controlled.
2. Explain why a sample of distilled water was included in Sean's experiment.
3. Explain why Sean repeated the experiment 3 times on 4 separate days.
4. Suggest how Sean could improve the reliability and accuracy of his experiment.
5. Why did Sean use a column graph rather than another type of graph to present his results?
6. In your opinion, is Sean's conclusion valid? Give reasons for your answer.

1.7 Using spreadsheets

1.7.1 The advantages of spreadsheets

A spreadsheet is a computer program that can be used to organise data into columns and rows.

Once the data are entered, mathematical calculations, such as adding, multiplying and averaging, can be carried out easily using the spreadsheet functions.

Spreadsheets have many advantages over handwritten or word-processed results. For example, with spreadsheets you can:

- make calculations quickly and accurately
- change data or fix mistakes without redoing the whole spreadsheet
- use the spreadsheet's charting function to present your results in graphic form.

1.7.2 Elements of a spreadsheet

Although there are a number of spreadsheet programs available, they all have the same basic features and layout, as shown in example 1 below. The data shown are from a student research project about the different factors on the growth of bean plants.

Example 1

- At the top of the spreadsheet are the toolbar and formula bar.
- A *row* is identified by a number; for example, 'row 1' or 'row 2'.
- A *column* is identified by a letter; for example, 'column A' or 'column B'.
- A *cell* is identified by its column and row address. For example, 'cell G3' refers to the cell formed by the intersection of column G with row 3. In this example, cell G3 is the active cell (shown by its heavy border). The active cell address and its contents (once data are entered) are shown to the left of the formula bar.
- A *range* is a block of cells. For example, 'range C3:F4' includes all the cells in columns C through to F and rows 3 through to 4.

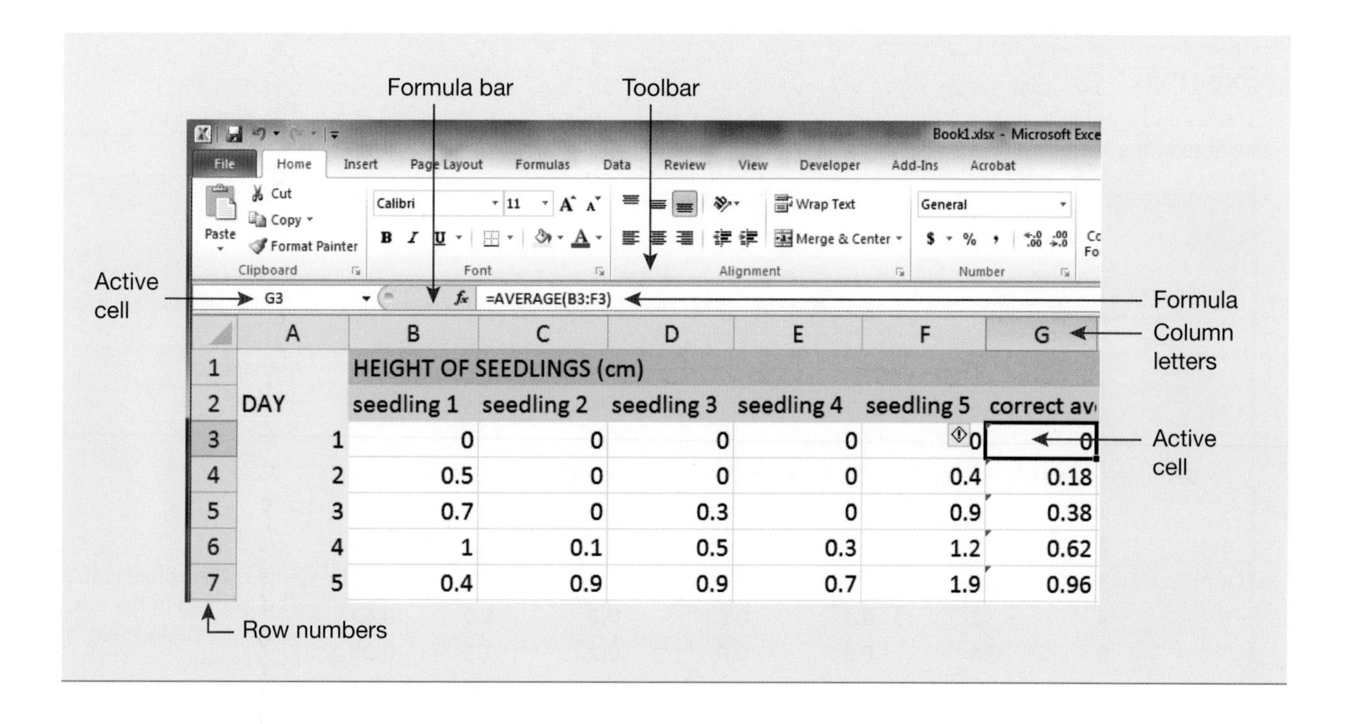

1.7.3 Entering data into cells

You can enter different types of data into a cell:

- a number or value
- a label, that is, text (for titles and headings)
- a formula (an instruction to make a calculation).

Decide in which cell you want to insert the data (the active cell). Type the data in the cell and press 'Enter'. To edit or change the data, simply highlight the cell and type in the new data — it will replace the old data when you press 'Enter'. Example 2 on the next page shows a spreadsheet in which data have been entered.

1.7.4 Creating formulae

To create a formula, you need to start with a special character or symbol to indicate that you are keying in a formula rather than a label or value. This is usually one of the symbols =, @ or +, depending on the spreadsheet program. For example, a formula to add the contents of cell B1 to cell C1 would take one of the following forms: =B1+C1 or @B1+C1 or +B1+C1.

Once you have entered the formula in a cell, the result of the calculation, rather than the formula, will be shown. The formula can be seen in the status bar when the cell is active. (See example 2 on the next page.) If you subsequently needed to change the values in B1 or C1, the spreadsheet will automatically use the formula to recalculate and show the new result.

The symbols used for mathematical operations in spreadsheets are:

+ for addition
– for subtraction
* for multiplication
/ for division.

Example 2

The spreadsheet in example 1 has been further developed. Formulae have now been entered to average the heights of the seedlings.

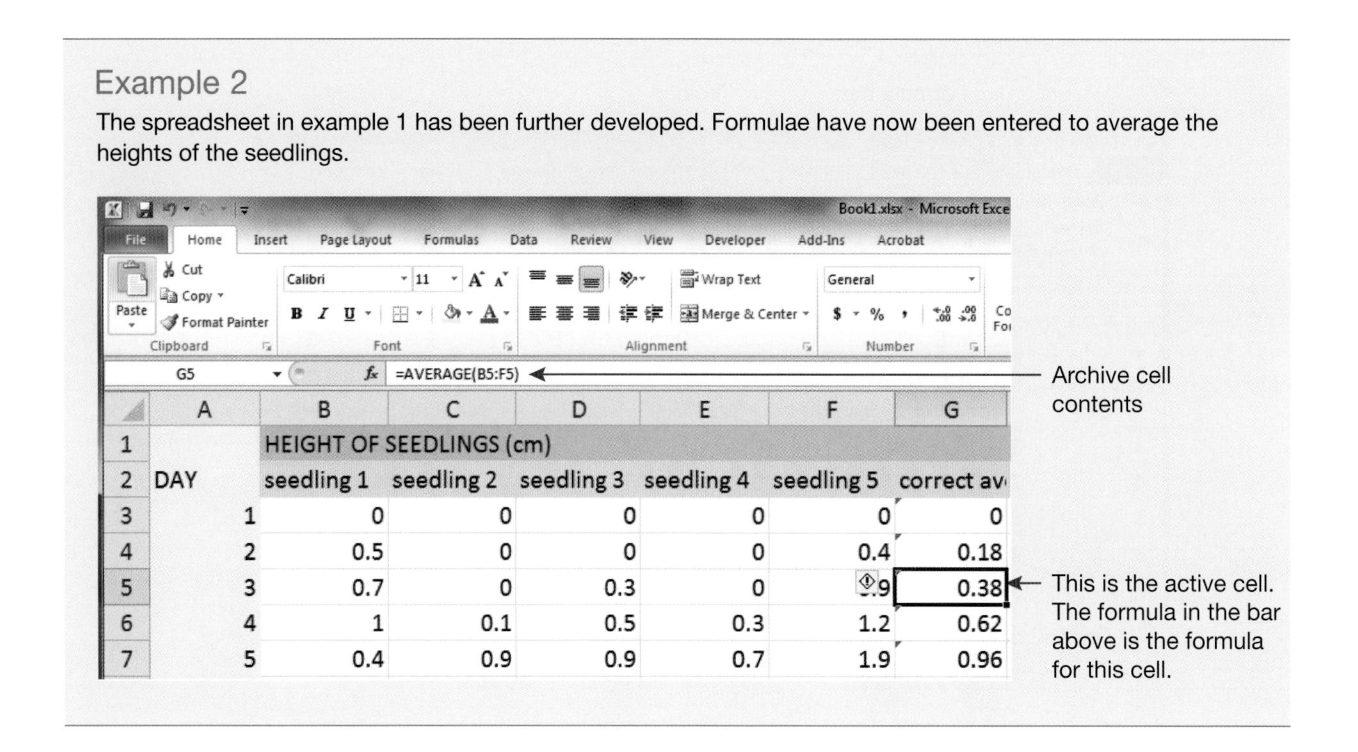

1.7.5 Using functions

Some common types of calculations are built into the spreadsheet, so that you don't always need to type out the full formulae. These are called **functions**. All functions have two parts: the name and a value (called the **argument**) that the function will operate on. The value is normally placed in parentheses, ( ), and can be written as a set of numbers or as a range (a block of cells). For example, a function to calculate the average of the amounts entered in cells B1, B2, B3 and B4 would be written: =AVERAGE(B1:B4).

Some of the common functions found in spreadsheets are shown in the table below.

Common spreadsheet functions

Name	Application	Example	Result
AVERAGE	calculates the average of the argument values	=AVERAGE(1,2,3,4)	2.5
COUNT	counts the number of values in the argument	=COUNT(A3:A6)	4
MAX	returns the largest value in the argument	=MAX(1,9,5)	9
MIN	returns the smallest value in the argument	=MIN(1,9,5)	1
MODE	returns the most common value in the argument	=MODE(1,1,5,5,1)	1
ROUND	rounds the argument to the number of decimal places specified	=ROUND(12.25,1)	12.3
SUM	calculates the sum of the values in the argument	=SUM(1,9,5)	15

1.7.6 Copying cells

Spreadsheets have a command that allows you to copy a formula or value from one cell to another cell (or into a range of cells). This is usually found in the *Edit* menu (*Fill Down* or *Fill Right*). The way a formula is copied depends on whether the cell references use:

- **relative referencing**, which you use when you want the cell address in the formula to change according to the relative location of the cell that you have copied it to. Example 2 in section 1.7.4 uses relative referencing. The formula AVERAGE(B5:F5) in the active cell G5 was copied downwards, so that there was no need to type the formulae in the rest of the column. The formula in the next cell (G6) is therefore AVERAGE(B6:F6) and so on.

- **absolute referencing**, which you use when you want a cell address in the formula to be constant, no matter where it is copied to. Absolute referencing is denoted by the symbol $ placed in the cell address. For example, $B$3 (see example 3 below).

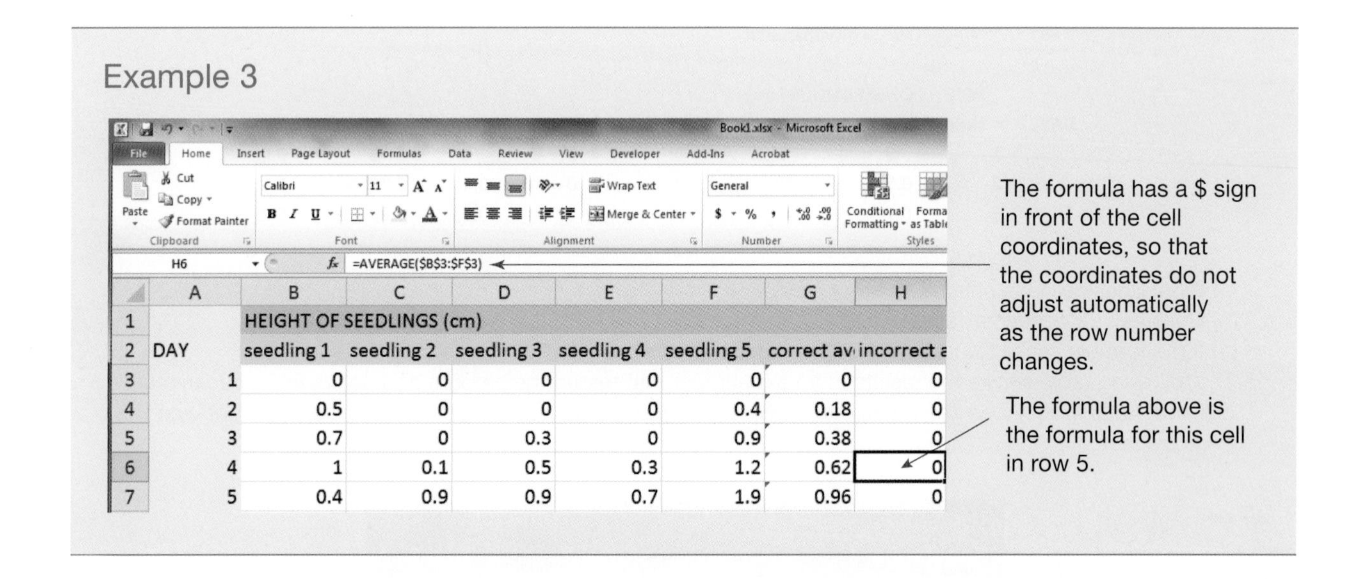

	A	B	C	D	E	F	G	H
1		HEIGHT OF SEEDLINGS (cm)						
2	DAY	seedling 1	seedling 2	seedling 3	seedling 4	seedling 5	correct av	incorrect a
3	1	0	0	0	0	0	0	0
4	2	0.5	0	0	0	0.4	0.18	0
5	3	0.7	0	0.3	0	0.9	0.38	0
6	4	1	0.1	0.5	0.3	1.2	0.62	0
7	5	0.4	0.9	0.9	0.7	1.9	0.96	0

1.7.7 Formatting cells

Investigate your spreadsheet program (most come with a tutorial) to learn how to use other useful features such as:

- adding and deleting rows or columns (useful if you have forgotten to include some calculations in your planning or decide you don't need some items)
- changing column widths (to show the full cell contents when the data are longer than the default column width) and changing row heights so that you can use larger font sizes for titles and headings
- inserting horizontal or vertical lines to improve the presentation of your spreadsheet
- changing cell formats to control how the data are to be displayed, such as using different fonts and character styles (underlining, bold, italic).

You can also format numeric values in a variety of ways. For example, the *Fixed* or *Number* format will display values to the number of specified decimal places. The *Percent* format will display values as a percentage, to the number of specified decimal places.

Once you have keyed in your data and included any necessary calculations, print out your spreadsheet and save it to a disk so that you can store the document and use it later.

1.7.8 Spreadsheet graphics

The three main types of graphs — pie, bar and line graphs — can usually be produced by a spreadsheet. It means that you can easily display your results graphically, but you still need to decide which is the most appropriate type of graph for your data.

The first step in producing a spreadsheet graph is to select the block of the cells that contains the data to be graphed. Use the spreadsheet's charting function, which usually brings up a window where you can indicate the type of graph, and add title and label details. When you are satisfied with the result, you can display and print out your graph.

1.7 Exercises: Understanding and inquiring

To answer questions online and to receive **immediate feedback** and **sample responses** for every question, go to your learnON title at www.jacplus.com.au. *Note:* Question numbers may vary slightly.

Using data

1. Look at the section of a spreadsheet presented below and answer the following questions:

H6 fx =AVERAGE($B$3:$F$3)

A		B	C	D	E	F	G	H
		HEIGHT OF SEEDLINGS (cm)						
DAY		seedling 1	seedling 2	seedling 3	seedling 4	seedling 5	correct av	incorrect a
	1	0	0	0	0	0	0	0
	2	0.5	0	0	0	0.4	0.18	0
	3	0.7	0	0.3	0	0.9	0.38	0

(a) What does cell G3 contain?
(b) Does cell E2 contain a value or a label?
(c) If the formula in cell G4 is AVERAGE(B4:F4), what would the formula be in cells G5 and G6?

2. The following table shows the results of an experiment that tested the amount of time taken for eucalyptus oils and other substances (0.1 mL of each) to evaporate at a constant temperature. The experiment was done twice.

	Time (s)	
Substance	**Trial 1**	**Trial 2**
Methylated spirits	4.17	1.85
Turpentine	63.48	43.02
Water	54.42	57.05
Oil from *E. rossi*	195.92	191.23
Oil from *E. nortonii*	103.99	105.39

(a) Enter the data into a spreadsheet.
(b) Use the spreadsheet function to calculate the average time that each substance took to evaporate.

3. The table below shows the distance travelled by Jesse at 3-second intervals during a 100-metre sprint. The data were recorded during the sprint by attaching a paper tape to Jesse's waist. As he ran, the tape was pulled through a timer that printed a dot every 3 seconds.

Time (s)	Distance travelled in time interval (m)	Average speed for time interval (m/s)
0	0	
3	35	
6	25	
9	15	
12	15	
15	10	

(a) Enter the data into a spreadsheet. Calculate the average speed travelled in each 3-second interval by applying a formula to the first cell in the column, and then copying it down. Remember that average speed can be calculated by dividing the distance travelled by the time taken:

$$\text{Speed} = \frac{\text{distance}}{\text{time}}$$

(b) What was Jesse's average speed over the total time?

4. The following data were collected by two car servicing centres in Canberra in 1998, at the request of a student. The table shows the level of carbon monoxide and carbon dioxide emissions (as a percentage of total emissions) from cars of various ages.

Year car manufactured	Carbon monoxide (%)	Carbon dioxide (%)
1977	3.17	11.8
1983	2.48	13.6
1985	3.7	11.4
1987	1.6	13.1
1989	1.08	10.2
1996	0.19	15.2

   (a) Enter the data into a spreadsheet and create a graph to display these results.
   (b) Create formulae to work out the average carbon monoxide and carbon dioxide emissions for:
      (i) cars manufactured up to 1985
      (ii) cars manufactured from 1987 onwards.
   (c) Car manufacturers were required to install catalytic converters in cars made after 1986. Catalytic converters cut down carbon monoxide emissions by converting some of the carbon monoxide to carbon dioxide. What can you conclude from these data about the success of catalytic converters?
5. Find the results of a science experiment this year that required some calculations. Enter the data and use spreadsheet formulas or functions to complete the calculations. If the data are suitable, create a graph to display the results.

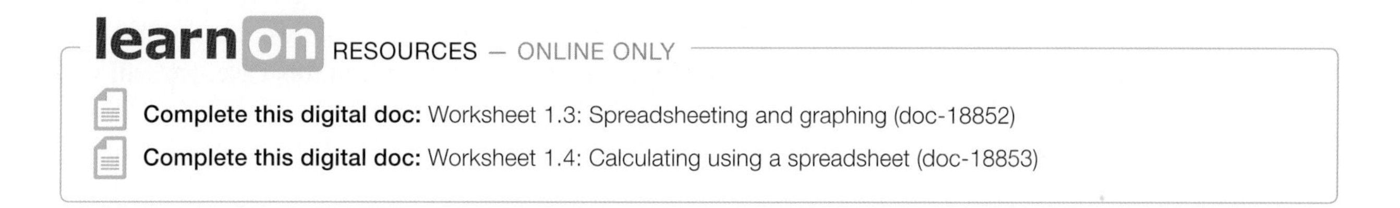

1.8 Using data loggers

1.8.1 The data logger

A data logger is a device that stores a large number of pieces of information (data) sent to it by sensors attached to it.

The data logger can transfer this data to another device, such as a graphing calculator or, more commonly, a computer, which can use data logger software or a spreadsheet program to manipulate the data (see section 1.7.1). Usually the computer or calculator graphs the collected data, and we can use these graphs to see patterns and trends easily.

1.8.2 When can a data logger be used?

Data loggers are particularly useful whenever an experiment requires several successive measurements. Sometimes, these measurements will take place over several hours or days — such as when measuring the way air pressure varies with the weather. Sometimes, many measurements must be taken over a short time interval — such as when measuring changes in air pressure as sound waves pass by. Data loggers are very flexible and can help scientists gather and analyse data for these types of experiments, as well as many others. As an example of how a data logger might help you in your scientific investigations, let's consider the following common exothermic and endothermic experiments.

1.8.3 Exothermic and endothermic processes

In an experiment, we investigate temperature changes in chemical processes. In part 1, we observe the reaction of magnesium metal with dilute hydrochloric acid and, in part 2, citric acid and baking soda. In addition to the laboratory equipment required for this experiment, including safety glasses, we will need a data logger with a temperature sensor attached to it. The data logger will need to be attached to a computer on which the data logger software has been installed.

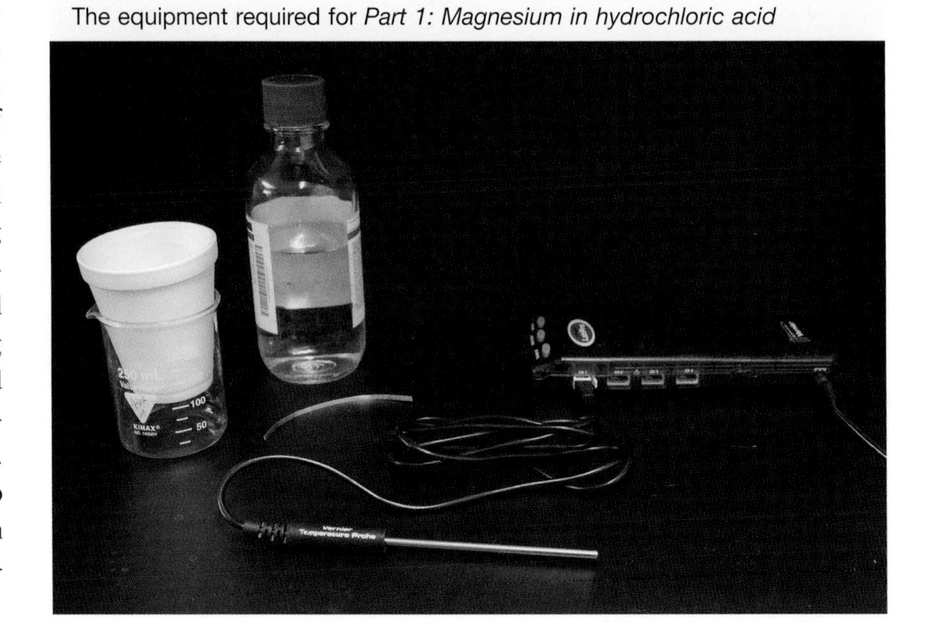

The equipment required for *Part 1: Magnesium in hydrochloric acid*

Part 1: Magnesium in hydrochloric acid

Active metals react with dilute acids to give off hydrogen gas and leave behind a salt that usually stays dissolved in the water in which the acid was dissolved. To investigate whether heat is given off or taken in during the reaction, we will need the equipment shown in the photo above. We could use a test tube or a beaker as shown in the photo. If we use a beaker, we will have to use more acid; in this case, we will use 100 mL of 0.5 mol/L hydrochloric acid.

We now set up the data logger to collect data for the length of time that we need and at the rate we require. The data logger itself or its software allows us to do this. The first screen below shows the data logger being

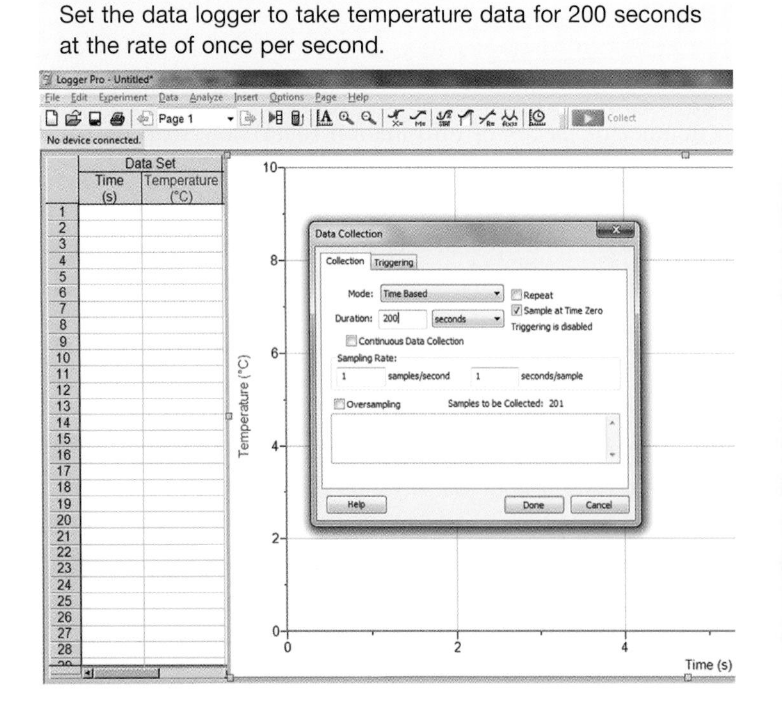

Set the data logger to take temperature data for 200 seconds at the rate of once per second.

Place the temperature sensor into the beaker of dilute acid and add the magnesium.

set to collect temperature data for 200 seconds at the rate of once per second. Now it's a simple matter of putting the temperature sensor in the dilute acid, pressing the button on the data logger to start data collection and adding the magnesium.

The reaction proceeds for 200 seconds and the sensor sends a temperature measurement every second to the data logger. When the selected time has passed (that is, after 200 seconds), the data logger sends all the data to the computer, which (via the software) displays it as a graph, as shown below.

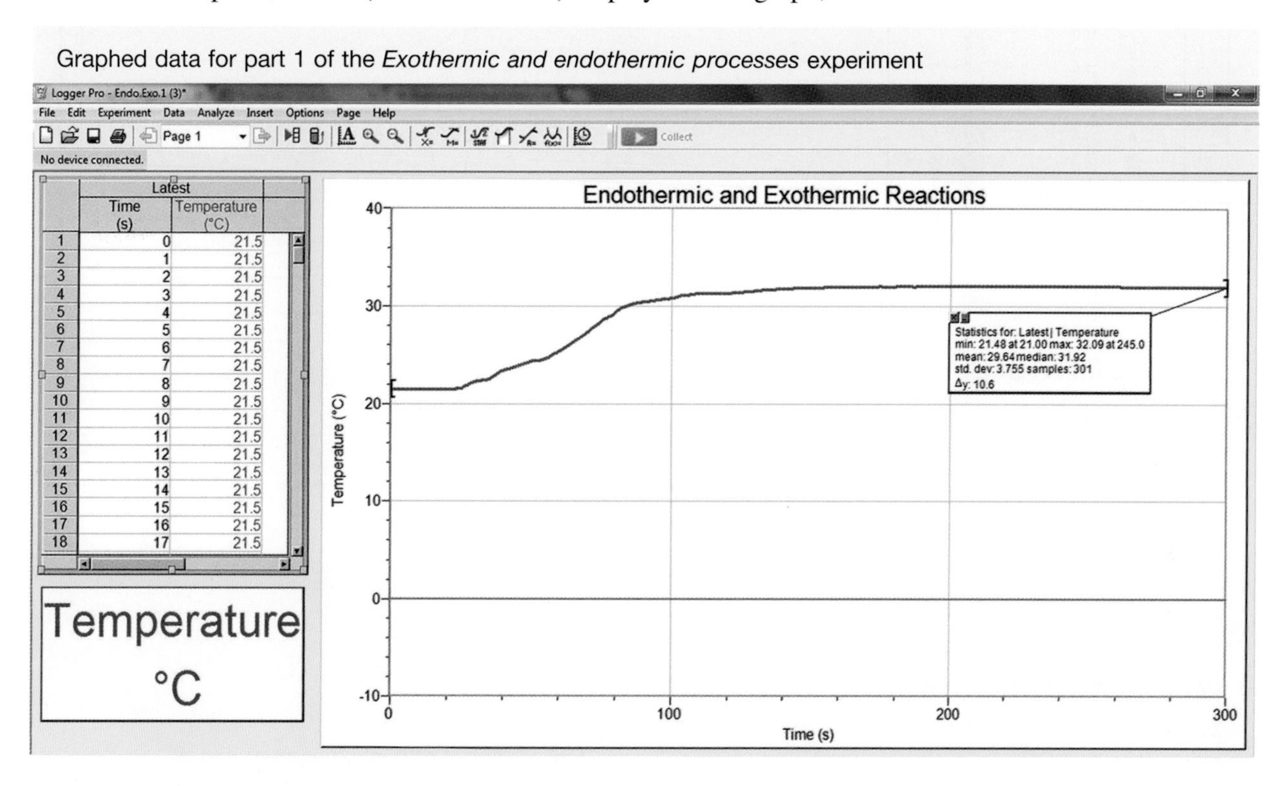

Graphed data for part 1 of the *Exothermic and endothermic processes* experiment

Part 2: Citric acid and baking soda

For this part of the experiment, we will need baking soda, citric acid, a beaker, a foam cup, other necessary laboratory equipment such as safety glasses, as well as a data logger and temperature sensor. We will use 30 mL of citric acid and 10 g of baking soda. These items are shown in the photograph at the right.

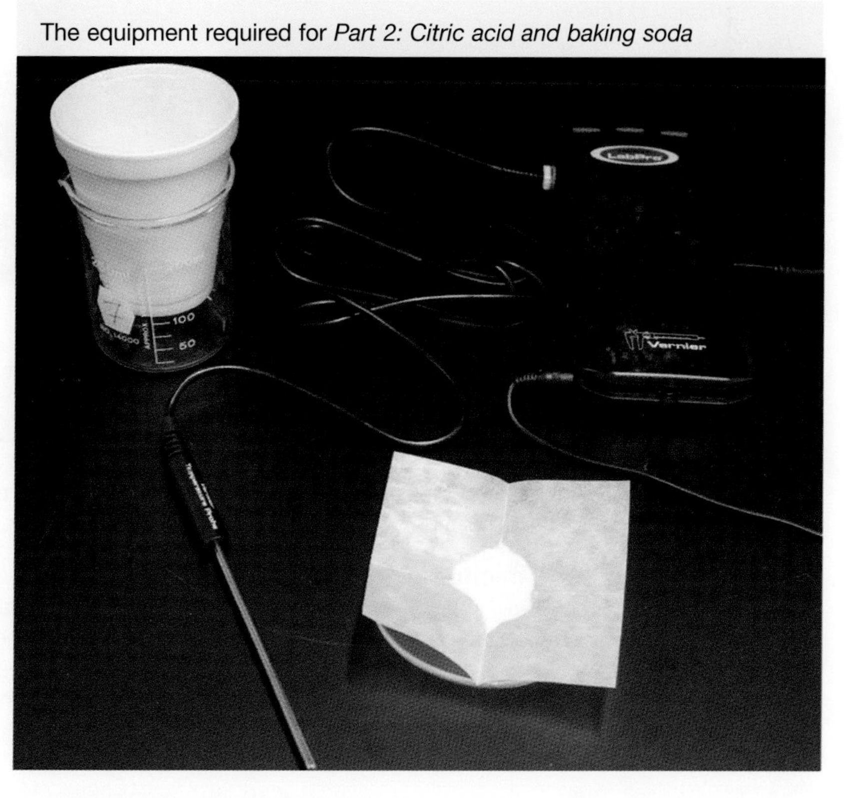

The equipment required for *Part 2: Citric acid and baking soda*

Once again, we set the run time to 200 seconds and the data collection rate to once per second. We insert the temperature sensor into the acid, press a button on the data logger to start data collection and then add the baking soda to the acid. The data logger collects the data, which the computer software automatically graphs after completion of the run, as shown on next page.

Graphed data for part 2 of the *Exothermic and endothermic processes* experiment

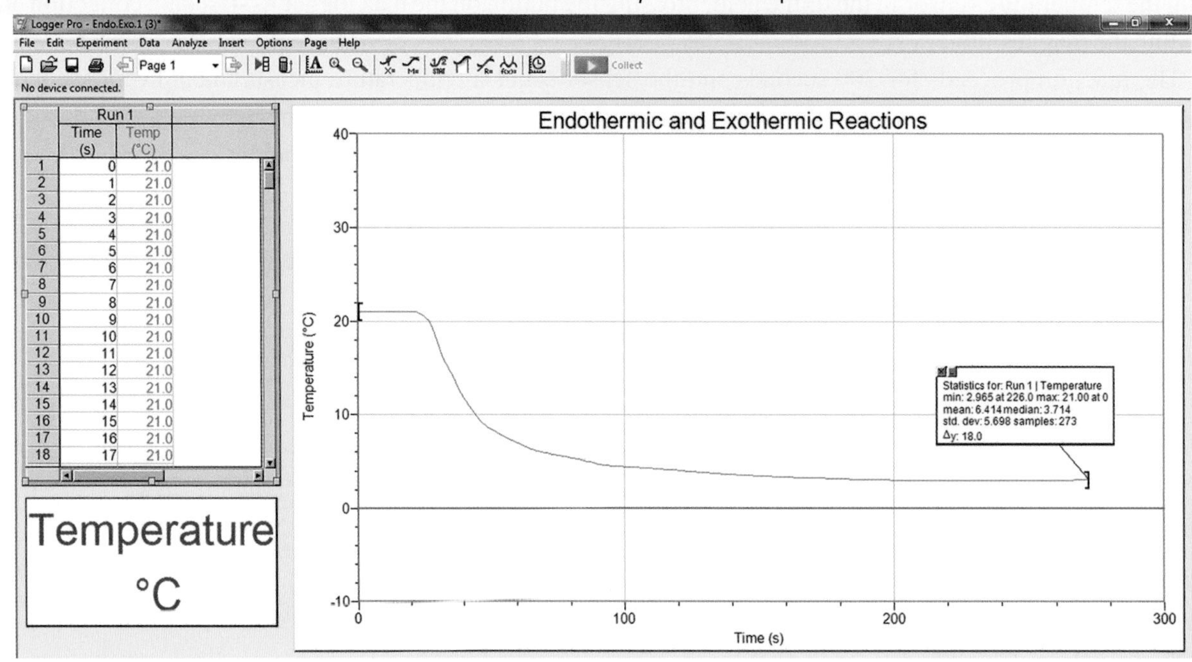

	Time (s)	Temp (°C)
1	0	21.0
2	1	21.0
3	2	21.0
4	3	21.0
5	4	21.0
6	5	21.0
7	6	21.0
8	7	21.0
9	8	21.0
10	9	21.0
11	10	21.0
12	11	21.0
13	12	21.0
14	13	21.0
15	14	21.0
16	15	21.0
17	16	21.0
18	17	21.0

INVESTIGATION 1.2

Exothermic and endothermic processes

PART 1 IS FOR TEACHER DEMONSTRATION ONLY

Materials:

safety glasses
bench mat
4 large test tubes and test-tube rack
10 mL measuring cylinder
balance
thermometer (–10 °C to 110 °C)
stirring rod
magnesium ribbon
sandpaper
0.5 mol/L hydrochloric acid
30 mL citric acid solution (10 g dissolved in 100 mL water)
10 g baking soda (dissolved in 100 mL water)

Parts 1 and 2 of this experiment can be demonstrated with the aid of a data logger.

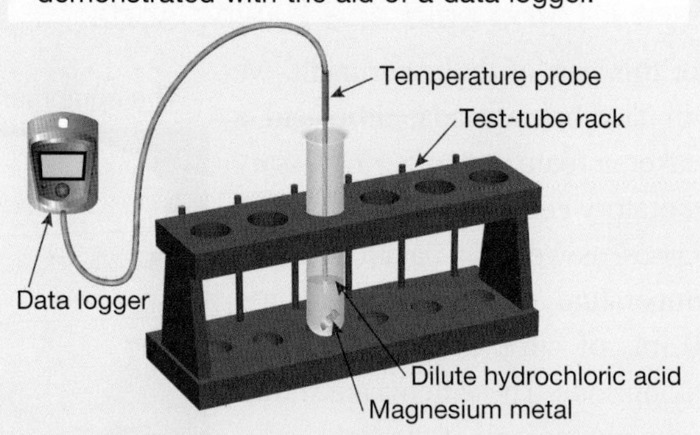

Part 1: Magnesium in hydrochloric acid

- Pour 10 mL of 0.5 mol/L hydrochloric acid into a test tube in a test-tube rack. Place a thermometer or probe in the test tube and allow it to come to a constant temperature. Record the temperature of the solution.
- Clean a 10 cm piece of magnesium ribbon using the sandpaper until it is shiny on both sides. Coil the magnesium ribbon and place it into the test tube of hydrochloric acid.
- Observe the temperature of the solution as the magnesium reacts with the hydrochloric acid. Record the final temperature of this solution.

Part 2: Baking soda in citric acid

- Pour 10 mL of citric acid solution into a test tube in a test-tube rack. Place a thermometer in the water in the test tube and allow it to come to a constant temperature. Record the temperature of the water.
- Use a balance to weigh 3 g of baking soda; add it to the water in the test tube and stir gently.
- Observe the temperature of the solution as the baking soda dissolves in the water. Record the final temperature of this solution.

1.8 Exercises: Understanding and inquiring

To answer questions online and to receive **immediate feedback** and **sample responses** for every question, go to your learnON title at www.jacplus.com.au. *Note:* Question numbers may vary slightly.

Think

1. Look back at *Part 1: Magnesium in hydrochloric acid* in section 1.8.3.
   (a) Write a word equation for the reaction that occurs.
   (b) Look at the graph of temperature vs time for this reaction. Was the reaction exothermic or endothermic? How do you know?
   (c) How long after data collection began was the magnesium ribbon added to the acid? How do you know?
   (d) How did the person who conducted this investigation know when the reaction was complete?
   (e) What was the initial temperature of the dilute acid used in this experiment?
   (f) What change in temperature did this reaction cause in the liquid in the beaker?
2. Look at the graph of the collected data produced by the computer for *Part 2: Citric acid and baking soda* in section 1.8.3.
   (a) What was the temperature of the acid at the start of the experiment?
   (b) What was the lowest temperature that the solution of citric acid and baking soda reached? How long after first adding the baking soda did this occur?
   (c) Is dissolving baking soda in citric acid an exothermic or endothermic process? How do you know?
3. Sensors are the devices that take the measurements that the data logger collects. Think of scientific investigations that could use data collected by sensors that measure:
   (a) electric current
   (b) acidity of solutions
   (c) concentration of carbon dioxide in the air
   (d) total dissolved solids (salt content)
   (e) light intensity.

Investigate

4. Acids are corrosive substances; they react with most metals, such as the magnesium in part 1 of the experiment. The temperature probe is made of metal but it doesn't react with acids. What sort of metal is it and what protects it from the acid?
5. Describe a data logger and what it does in a way that a Year 7 student would understand.
6. List the advantages of using a data logger over taking the measurements manually. Describe an experiment in which using a data logger provides an advantage over manual data collection.

1.9 Using databases

1.9.1 Databases

Databases are simply information or data arranged in one or more tables. We use databases every day; for example, when we look up information in a phone book, a timetable or the index of a book.

An electronic database is one of the most powerful computer applications and is an important tool for a business, an organisation or a scientist.

A database's design is crucial to its usefulness, so a database must be designed with ease of searching uppermost in mind. In the *Understanding and inquiring* section below, you will create a database using some of the features of Microsoft Access.

1.9 Exercises: Understanding and inquiring

To answer questions online and to receive **immediate feedback** and **sample responses** for every question, go to your learnON title at www.jacplus.com.au. *Note:* Question numbers may vary slightly.

Analyse and evaluate

Creating a database of Nobel prize winners

Before creating your database, you will need to find some information to put in it. This is best done as a class activity with each student in the class researching one or two Nobel prize winners.

- Each student in the class should research one or two different Nobel prize winners. Choose people who have won a Nobel prize for work in the categories of Chemistry, Physics or Medicine.
- For each prize winner, collect the data listed below. Ideally the data should be written on cards that can be passed around the class, or they could be displayed in large writing on large sheets of paper around the room.
  - First name
  - Last name
  - Country of birth
  - Year of birth
  - Category of award (such as Chemistry, Physics and Medicine)
  - Organisation (where the person worked)
  - Nobel prize awarded for (one sentence or phrase that outlines the work for which the scientist received the award)
  - Share received (if the award was shared by a group of people)
- Microsoft Access software is commonly used to create databases. The following instructions are for the 2003 edition of this software. Other editions are similar to use but the screens are not exactly the same. You can start Access by clicking on *Start*, then *Programs* and then the Access icon shown at right.

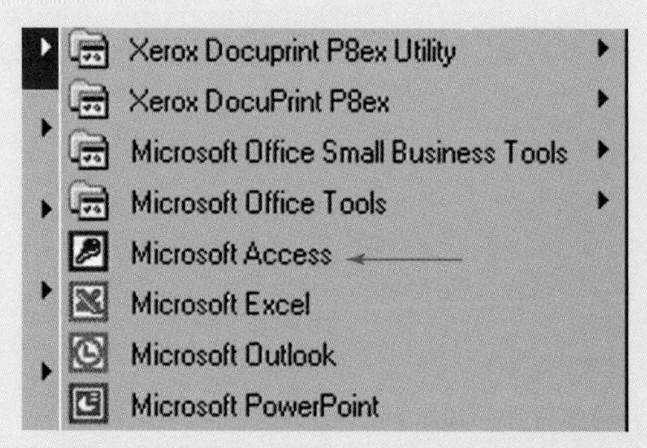

- When you open the software, click on *File* and then *New*. A list of options will appear in the task pane on the right-hand side of the screen. Choose the option *Blank database*. A dialog box will appear for you to enter a name for your database and navigate to the folder where you want to save the database. Choose a sensible name (such as 'Nobel prize winners') and save it where you normally save your science work. This is shown in the screenshot below.

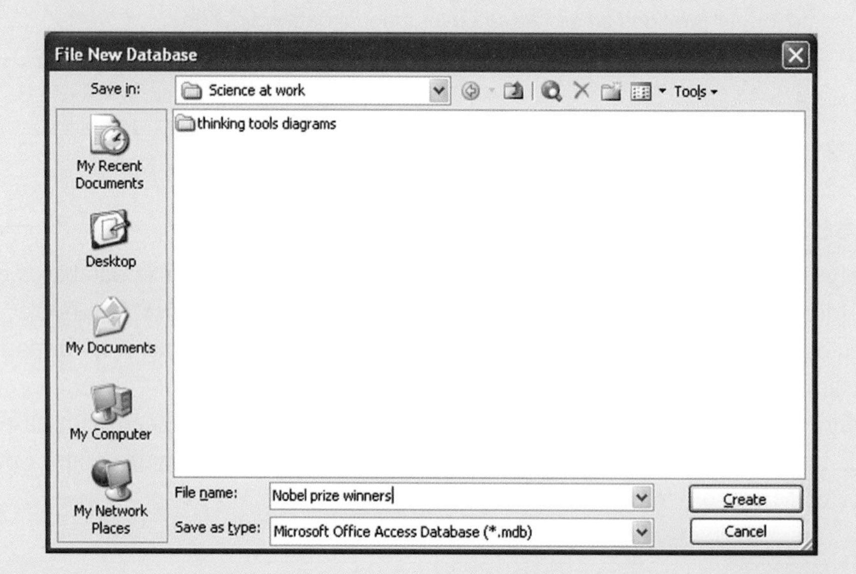

- A new dialog box will be displayed. Choose the option *Create table in Design view* and press [Enter]. A new screen will appear where you can enter field names, which are the column headings for the database. Enter the field names as shown on next page. You will note that, by default, the data type is Text even though some of the fields are numbers. This is not important for this database.

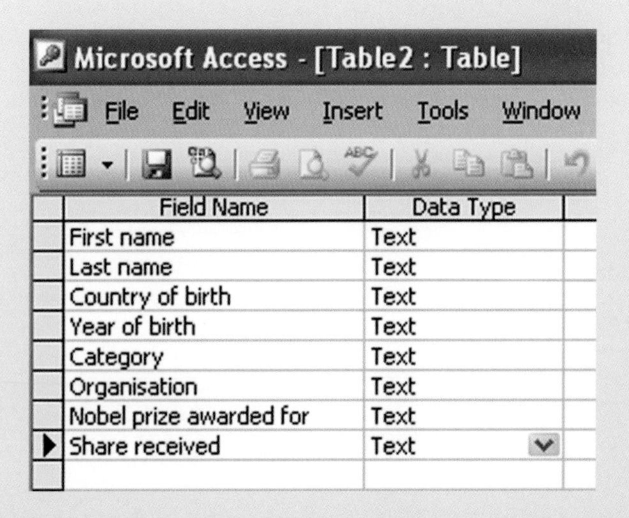

- Now that you have designed the database, it is time to change to datasheet view. Click on the Datasheet view button in the top left-hand corner of the screen. You will be prompted to save the table. Give the table a suitable name (such as 'Table 1') and click *Save*. When you are asked if you want to create a primary key, click *No*. The table shown below should appear. You are now in datasheet view. Note that the Design view button now appears in the left-hand corner of the screen.

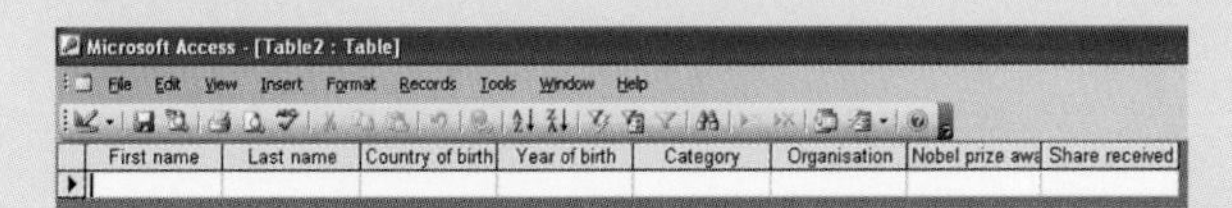

- Enter the data that you and your classmates found into the table. When you have done this, save your database.

The great thing about databases is that they allow you to search for data that match particular criteria. This is called running a *query*. We are going to create a query to find all the Nobel prize winners in our database who were awarded a prize for Medicine and were born in the United States.

- Make sure you are in datasheet view. Click on the arrow next to the New object button. Select *Query* and then *Query wizard* and click *OK*. The fields in your table will be displayed; click on the ones you want to appear in the query then click on the single arrow to move them into the *Selected Fields* box. Select the following fields: first name, last name, country of birth and category. When you have done this, click *Next*. In the next dialog box, enter a name for your query, select *Modify the query design* and click on *Finish*.

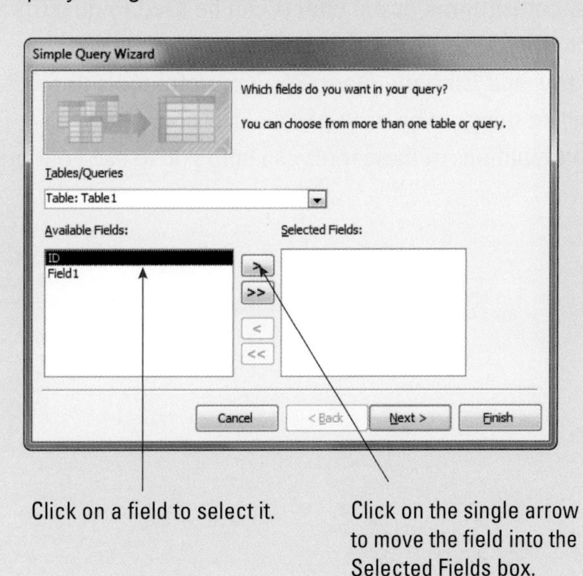

- The screen below will appear. Now enter the criteria you want the query to look for in the appropriate boxes. In the Category column, type 'Chemistry' (without the quotation marks) in the *Criteria* row. In the Country of birth column, type 'United States' in the *Criteria* row. Quotation marks will automatically appear when you press [Enter]. This is shown below.

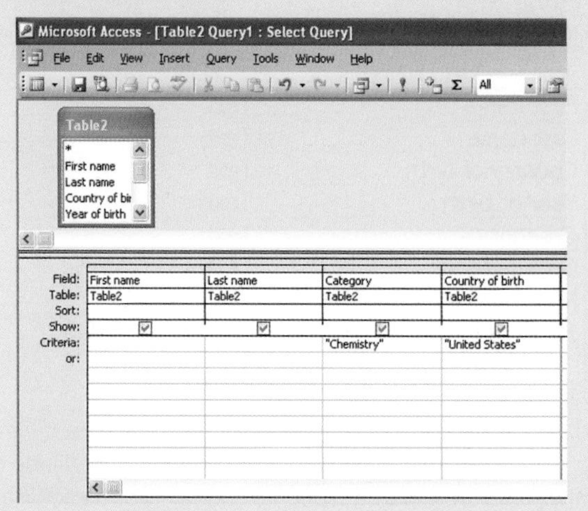

- Now click on the Run button in the toolbar near the top of the screen. The query will run and a table displaying the Nobel prize winners that match your criteria will appear.
- Create a new query to display the Nobel prize winners who won the Nobel prize for Physics and were born in England.

Run

1.10 Visual thinking tools

1.10.1 Structuring your thinking

There are so many different ways to see and share what is happening inside your brain. Here are some tools that can be used to make your thinking visible so that you can share and discuss it with others.

Like a builder, it is important for you to use the right tool to get the job done.

- **Storyboards**, **flowcharts**, **timelines** and **cycles** are useful tools to sequence your thoughts.
- **Matrixes** and **bubble maps** are useful when you want to classify or organise your thoughts.
- **Priority grids**, **target maps**, **continuums** or **pie charts** can be used to quantify or rank ideas.
- **PMI charts** and **Y charts** help you to visualise or reflect on an idea.
- **Concept maps**, **Venn diagrams** and **fishbone diagrams** are useful tools to focus your thoughts, such as when you need to analyse and compare things in order to make a decision.

There are also times when combinations of these tools can help you to use your brain and time more effectively.

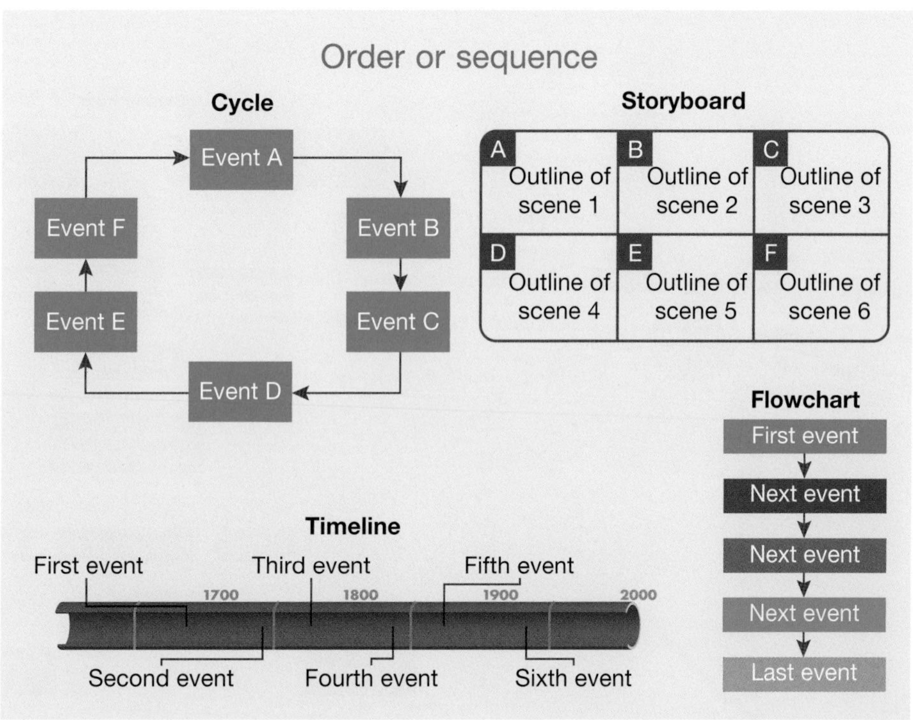
Order or sequence
Cycle
Event A
Event B
Event C
Event D
Event E
Event F
Storyboard
A Outline of scene 1
B Outline of scene 2
C Outline of scene 3
D Outline of scene 4
E Outline of scene 5
F Outline of scene 6
Flowchart
First event
Next event
Next event
Next event
Last event
Timeline
First event
Third event
Fifth event
1700
1800
1900
2000
Second event
Fourth event
Sixth event

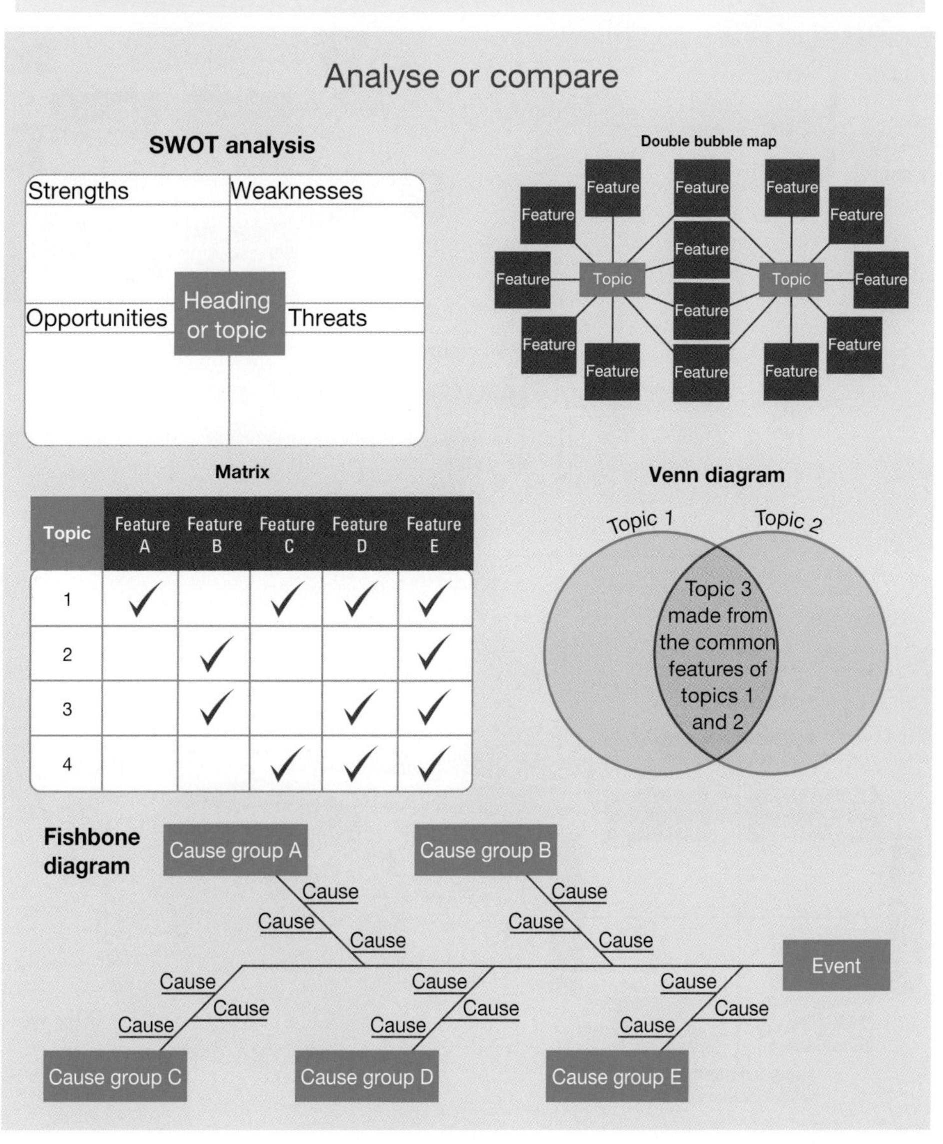
Analyse or compare
SWOT analysis
Strengths
Weaknesses
Opportunities
Heading or topic
Threats
Double bubble map
Feature
Topic
Topic
Matrix
Topic
Feature A
Feature B
Feature C
Feature D
Feature E
1
2
3
4
Venn diagram
Topic 1
Topic 2
Topic 3 made from the common features of topics 1 and 2
Fishbone diagram
Cause group A
Cause group B
Cause
Event
Cause group C
Cause group D
Cause group E

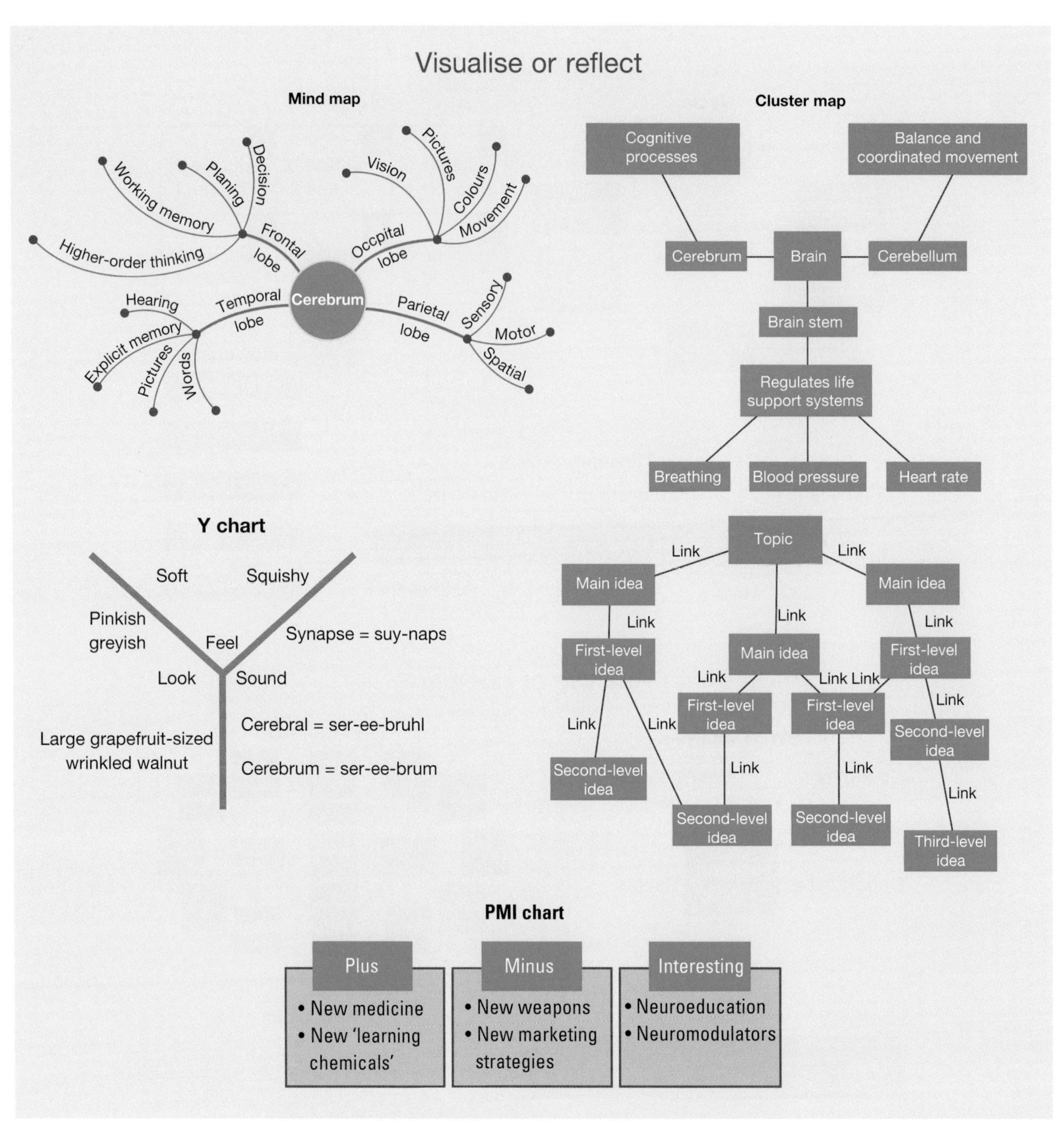

Visualise or reflect
Mind map
Cerebrum
Frontal lobe
Occipital lobe
Temporal lobe
Parietal lobe
Decision
Planing
Working memory
Higher-order thinking
Vision
Pictures
Colours
Movement
Hearing
Explicit memory
Pictures
Words
Sensory
Motor
Spatial
Cluster map
Cognitive processes
Balance and coordinated movement
Cerebrum
Brain
Cerebellum
Brain stem
Regulates life support systems
Breathing
Blood pressure
Heart rate
Y chart
Soft
Squishy
Pinkish greyish
Feel
Synapse = suy-naps
Look
Sound
Cerebral = ser-ee-bruhl
Large grapefruit-sized wrinkled walnut
Cerebrum = ser-ee-brum
Topic
Link
Main idea
First-level idea
Second-level idea
Third-level idea
PMI chart
Plus
• New medicine
• New 'learning chemicals'
Minus
• New weapons
• New marketing strategies
Interesting
• Neuroeducation
• Neuromodulators

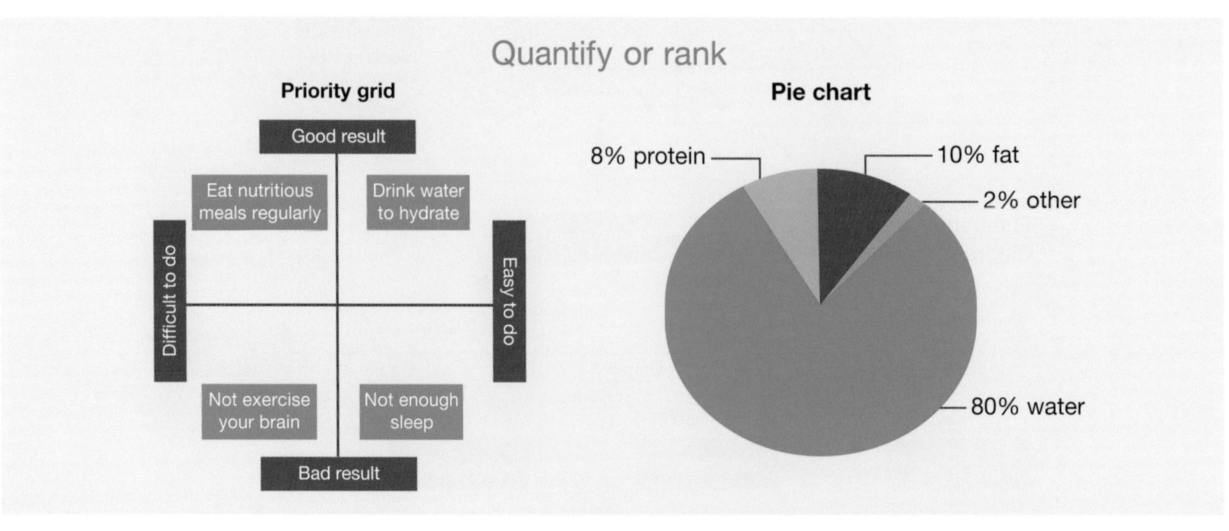

Quantify or rank
Priority grid
Good result
Eat nutritious meals regularly
Drink water to hydrate
Difficult to do
Easy to do
Not exercise your brain
Not enough sleep
Bad result
Pie chart
8% protein
10% fat
2% other
80% water

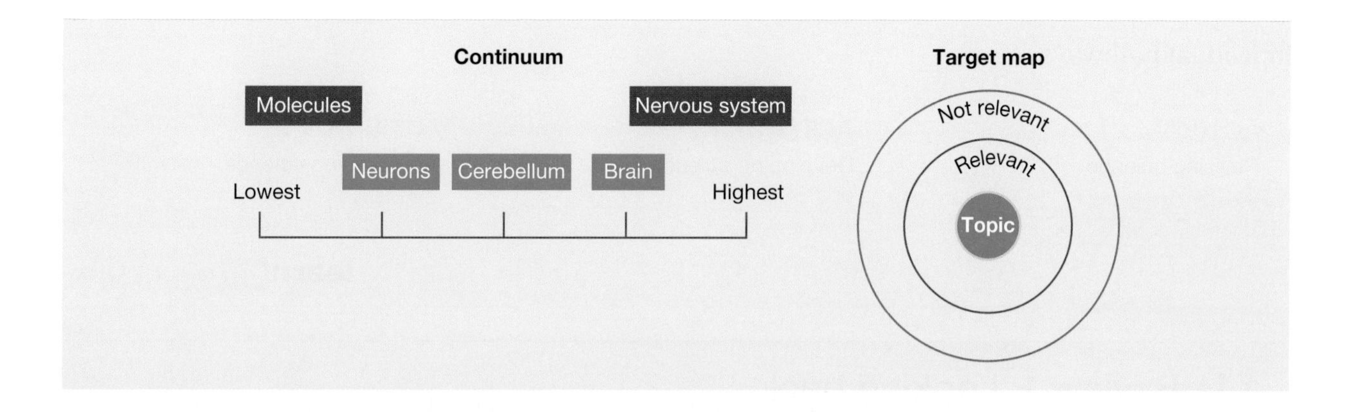

1.10 Exercises: Understanding and inquiring

To answer questions online and to receive **immediate feedback** and **sample responses** for every question, go to your learnON title at www.jacplus.com.au. *Note:* Question numbers may vary slightly.

Remember

1. State the types of visual thinking tools that are best suited to:
   (a) quantifying or ranking ideas
   (b) visualising or reflecting
   (c) analysing or comparing
   (d) ordering or sequencing.

Think and discuss

2. Select and copy five of the visual tools from the previous pages and then add more information to each.
3. Use a visual thinking tool to summarise key or interesting points from each subtopic within this chapter.

1.11 Review

1.11.1 Study checklist

Science as a human endeavour

- describe the changes in the way in which scientists have worked since ancient times
- acknowledge that present-day scientists usually work in teams and that their investigations are influenced by the work done by other scientists
- describe some scientific discoveries that have been made through a combination of accidents and observation
- explain how ethical values affect scientific endeavour and discuss some examples of ethical dilemmas in scientific endeavours

Conducting your own investigation

- identify questions to be investigated through group discussion, the internet or other resources
- recognise the importance of observations in formulating questions to investigate
- identify dependent, independent and controlled variables in an investigation
- recognise the need for a control in some investigations
- select and use appropriate equipment for the collection of data, including data loggers where appropriate
- record the progress of an investigation in a logbook
- use tables, spreadsheets and databases for the recording of data
- design and construct appropriate graphs to represent data and assist in the identification of trends and patterns and the formation of conclusions
- compare conclusions with earlier predictions or hypotheses
- evaluate the success of an investigation and suggest improvements that would make the findings more accurate or reliable
- write a scientific report using the appropriate headings, terminology, tables and graphs

Individual pathways

ACTIVITY 1.1	ACTIVITY 1.2	ACTIVITY 1.3
Revising science doc-8431	Developing science skills doc-8432	Investigating science doc-8433

learnON ONLINE ONLY

1.11 Review 1: Looking back

To answer questions online and to receive **immediate feedback** and **sample responses** for every question, go to your learnON title at www.jacplus.com.au. *Note:* Question numbers may vary slightly.

1. Match the words in the list below with their meanings.

Words	Meanings
(a) Conclusion	(i) Concerns that deal with what is morally right or wrong
(b) Abstract	(ii) The variable that is deliberately changed in an experiment
(c) Discussion	(iii) The part of a journal article where a brief overview of the article is given
(d) Results	(iv) A list of steps to follow in an experiment
(e) Hypothesis	(v) The answer to the aim or the problem
(f) Ethical considerations	(vi) A list of equipment needed for the experiment
(g) Independent variable	(vii) The variable that is measured in an experiment
(h) Dependent variable	(viii) States what was seen or measured during an experiment. May be presented in the form of a table or graph.
(i) Method	(ix) A sensible guess to answer a problem
(j) Apparatus	(x) The part of a report where problems with the experiment and suggestions for improvements are discussed

2. Miranda wanted to test the following hypothesis: Hot soapy water washes out tomato sauce stains better than cold soapy water.
   (a) List the equipment she will need.
   (b) Identify the independent and dependent variables in this investigation.
   (c) List 5 variables that will need to be controlled.
   (d) Outline a method that could be used to test the hypothesis.
   (e) Miranda's results are shown in the table above right.
   (f) Write a conclusion based on Miranda's results.

Water temperature (°C)	Observations
20	Dark stain left after washing
40	Faint stain left after washing
60	No stain left after washing
80	No stain left after washing

3. Gemina and Habib wanted to investigate whether the type of surface affects how high a ball bounces. Habib thought the ball would probably bounce the highest off a concrete floor. They dropped tennis balls from different heights onto a concrete floor, a wooden floor and carpet. Their results are shown in the table at right.
   (a) Write a hypothesis for this experiment.
   (b) Construct a line graph of Gemina and Habib's results.
   (c) Use your graph to estimate the values X, Y and Z.
   (d) Identify two variables that had to be kept constant in this experiment.

Distance ball dropped (cm)	Average height of bounce (cm)		
	Concrete	Wood	Carpet
25	22	14	8
50	46	34	18
75	70	50	26
100	94	66	34
125	X	85	Z
150	128	94	48
175	129	Y	50
200	130	100	51

(e) Identify two trends in the results.
(f) Do the results support the hypothesis you wrote?
(g) Predict how high the tennis ball would bounce off each floor if it was dropped from a height of 225 cm.
4. List some of the factors affecting the decision about whether money is spent on finding a cure for a particular disease.
5. Should farmers be allowed to plant the type of crop they believe produces the best yield, irrespective of whether others object to the manner in which the crop was bred?
6. In the film *Super Size Me*, the film-maker Morgan Spurlock gains weight and suffers health problems after thirty days of eating from only one fast-food chain. The film suggests that this fast food is unhealthy.
(a) What factors should be taken into account when considering the effects of a fast-food diet compared with a broader eating pattern?
(b) Was this a controlled experiment?
(c) Is Spurlock's argument valid? Explain your answer.
(d) What type of arguments could the fast-food chain put forward in response to the film *Super Size Me*?

learnON RESOURCES — ONLINE ONLY

Complete this digital doc: Worksheet 1.5: Investigating (doc-18854)
Complete this digital doc: Worksheet 1.6: Organising and evaluating results (doc-18855)
Complete this digital doc: Worksheet 1.7: Drawing conclusions (doc-18856)
Complete this digital doc: Worksheet 1.8: Summarising (doc-18857)
Complete this digital doc: Worksheet 1.9: Evaluating media reports (doc-18858)

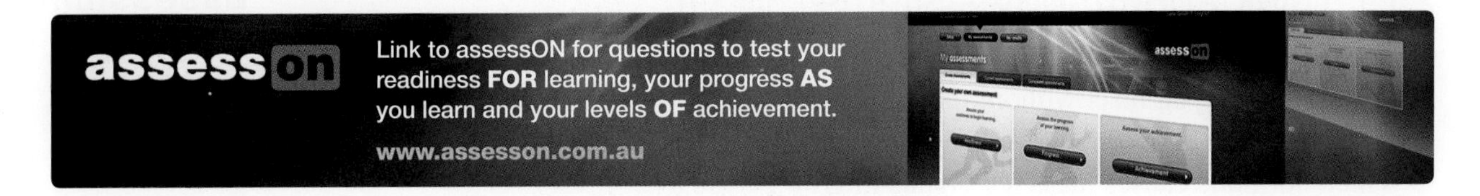

TOPIC 2
Control and coordination

2.1 Overview

You are a multicellular organism made up of a number of body systems that work together to keep you alive. Your body systems are made up of organs, which are made up of tissues, which are made up of particular types of cells. Your cells communicate with each other using electrical impulses and chemicals such as neurotransmitters and hormones. The coordination of this communication is essential so that the requirements of your cells are met and a stable internal environment is maintained.

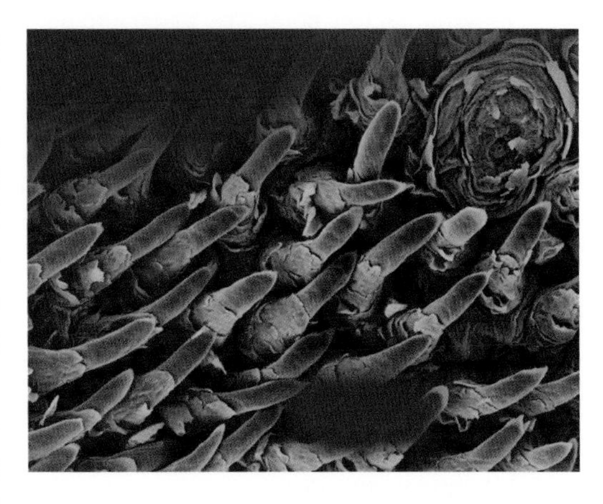

This scanning electron micrograph of the tongue surface shows the papillae that give the tongue its texture. The papillae also contain the tastebuds, part of the sensory system that sends information to the brain.

2.1.1 Think about body systems

assessON

- What's so good about negative feedback?
- Ouch! How do you react without thinking about it?
- What's the link between glucose, glycogen and glucagon?
- Which hormone causes male sex organs to grow?
- What's the link between hormones and the menstrual cycle?
- Which neurotransmitter acts like the brakes on your emotions?

LEARNING SEQUENCE

Numerous **videos** and **interactivities** are embedded just where you need them, at the point of learning, in your learnON title at www.jacplus.com.au. They will help you to learn the content and concepts covered in this topic.

2.1.2 Speedy reactions?

How fast can you react to a potentially threatening situation? Imagine you are in the situation at right. How would you feel? How would you react? Would everything start to happen in slow motion and then quickly speed up? Can an incident be avoided?

When you first see the danger, you detect it using receptors in your eyes. This message is then sent to your nervous system, which will tell your body what to do. As there is potential danger in this situation, your endocrine system may also react by producing hormones such as adrenaline to trigger your body to 'get up and go'. Hopefully this all happens fast enough to avoid anyone getting hurt!

INVESTIGATION 2.1

Fast or slow

AIM: To increase awareness of different types of responses to stimuli

Method and results

- Carefully observe each situation below and then answer questions 1 to 4.
  - (a) A mobile has lost a piece and is hanging crooked. When a fly lands on the mobile, it becomes balanced again. Given the masses in the diagram, what is the mass of the fly? *Response:* Solving the puzzle
  - (b) Ouch! You step on a sharp object. *Response:* You lift your foot quickly.
  - (c) You have been in three classes before lunch. You had very little breakfast and you feel that you have no energy. Your friend Janine, who knows everything, tells you that you have low blood sugar and must eat your lunch so that your blood sugar level can get back to normal. The bell rings, and you rush to the canteen to get lunch. *Response*: Getting your blood sugar back to normal

1.5 g

1.2 g

2.0 g

1. Order the responses from fastest to shortest response time.

Discuss and explain

2. Using your current understanding of how you respond to your environment, suggest reasons for the different types of responses and how your body processes the information to bring about the response.
3. Suggest a question or hypothesis for each scenario that you could investigate.
4. Propose another scenario and predict what your body's response would be. Suggest why and how it would respond in this way.
5. (a) Find out how seeing danger quickly approaching can result in a change of behaviour (such as running faster, stopping or screaming). Outline the involvement of both nerves and hormones.
   (b) Construct a cartoon or comic strip to summarise your findings.
6. Find some activities that can be used to determine your reaction time. Test your classmates and record times to see who is the fastest.

2.2 Coordination and control

2.2.1 Working together

You are a **multicellular organism** made up of many cells that need to be able to communicate with each other. They need to be able to let other cells know when they need help and support, when they need more of something or when they need to get rid of something.

Your cells can be organised to form tissues. These tissues make up organs, and the organs make up systems, which perform particular jobs to keep you alive.

The cells of multicellular organisms cannot survive independently of each other. They depend on each other and work together. Working together requires organisation, coordination and control.

Multicellular organisms show a pattern of organisation in which there is an order of complexity.

Multicellular organisms → contain Systems → contain Organs → contain Tissues → contain Cells

While one of your cells may be a part of one of the systems in your body, it may also need to communicate and interact with other systems to stay alive. It may depend on your digestive system to break down nutrients so that its chemical requirements are in a form that can be used, and your circulatory system to deliver these. Your respiratory system will also be involved in ensuring a supply of oxygen and removal of carbon dioxide. Other organs in the excretory system will also be involved in removing wastes that may otherwise be toxic. Your systems need to work together so that a comfortable stable environment for the cells is maintained.

All body systems work together.

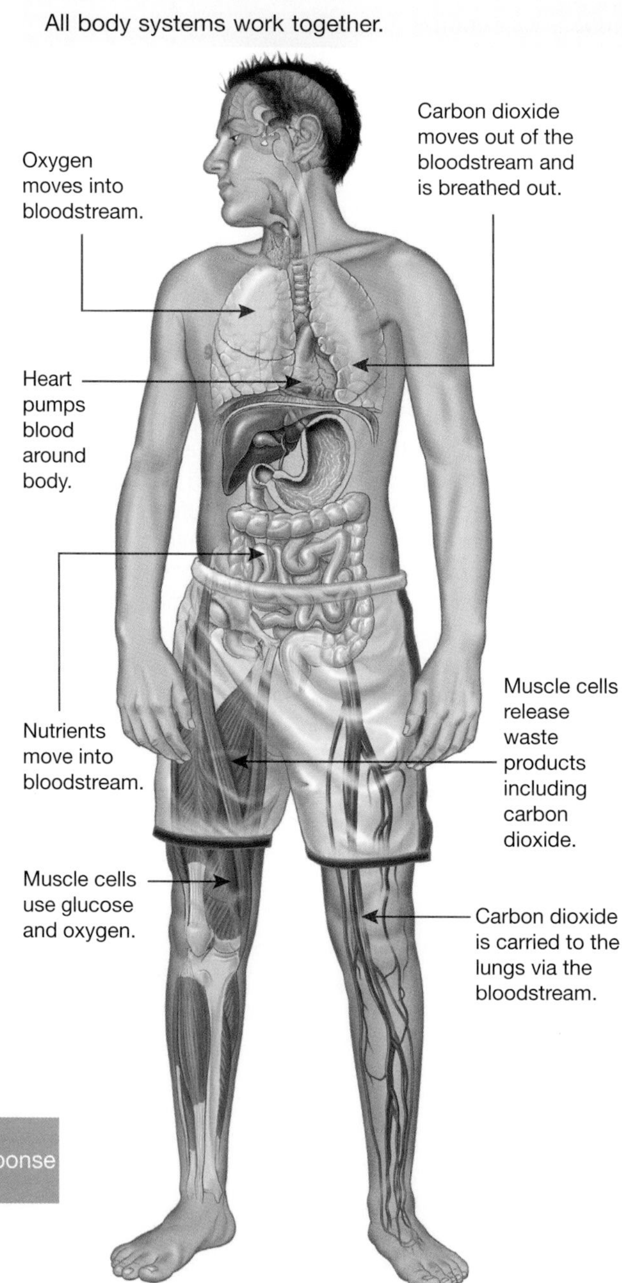

2.2.2 Homeostasis

The internal environment in which your cell lives needs to be kept constant. Temperature, pH and concentrations of ions, glucose, water and carbon dioxide need to be within a particular range. Maintenance of this constant internal environment is called **homeostasis**.

2.2.3 Stimulus–response model

To be able to achieve homeostasis, any changes or variations (stimuli) in the internal environment need to be detected (by receptors). If a response is required, this needs to be communicated to effectors to bring about some type of change or correction so the conditions can be brought back to normal. This is described as a **stimulus–response model**.

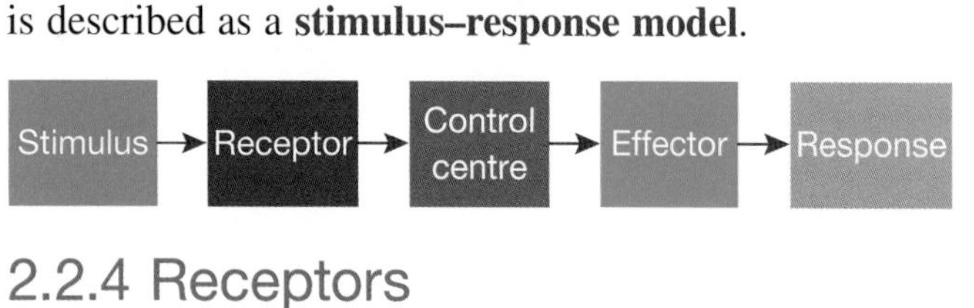

2.2.4 Receptors

Receptors identify changes inside and outside your body. These special types of nerve cells may be located in sense organs such as your eyes, ears, nose, tongue and skin. Different types of receptors respond to particular stimuli.

Type of receptor	Which stimuli does it respond to?	Where in your body?
Photoreceptor	Light	Eye
Mechanoreceptor	Sound	Ear
Chemoreceptor	Chemicals	Tongue, nose
Thermoreceptor	Temperature	Skin

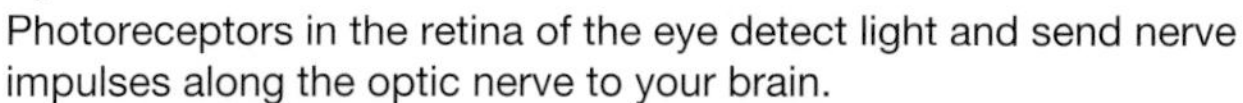
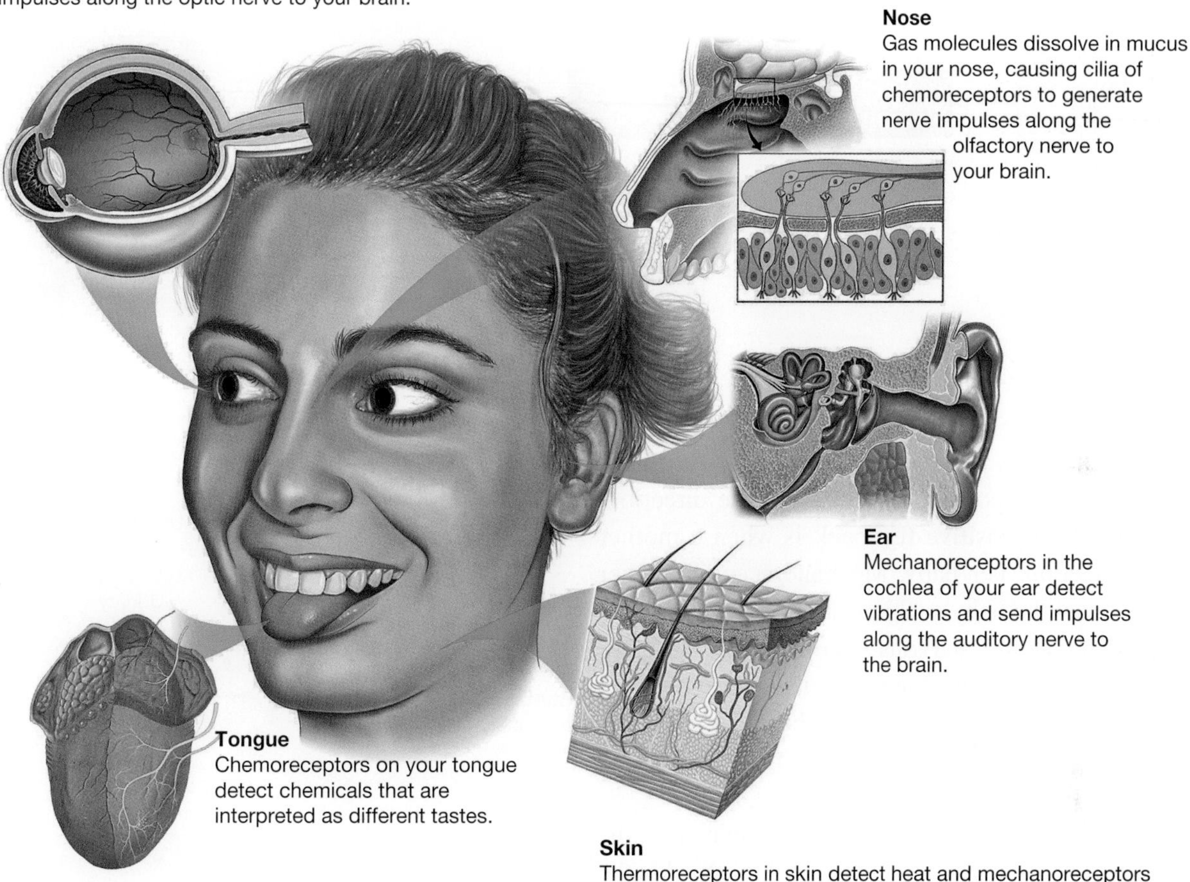

2.2.5 Control centre

Once a stimulus has been detected by a receptor, a message in the form of a nerve impulse travels to the central nervous system (brain and spinal cord). It is here that the message is processed to determine which response will be appropriate. A message is then sent to the appropriate effector.

2.2.6 Effectors

Effectors such as muscles or glands receive the message from the central nervous system to respond in a particular way. Their response depends on the original stimulus. For example, if your hand is too close to a candle flame, then muscles in your arm may respond to move your hand away from it. Some other examples are shown in the thermoregulation figure on page 44.

2.2.7 Giving feedback?

Stimulus–response models can also involve negative or positive feedback. Most biological feedback systems involve negative feedback.

2.2.8 Negative feedback

Negative feedback occurs when the response is in an opposite direction to the stimulus. For example, if levels of a particular chemical in the blood were too high, then the response would be to lower them. Likewise, if the levels were too low, then they would be increased.

The regulation of **glucose** levels in your blood involves negative feedback. If an increase in blood glucose levels has been detected by receptors, the **pancreas** responds by secreting **insulin**, which may trigger an increased uptake of glucose by liver and muscle cells and the conversion of glucose into **glycogen** for storage. This lowers the blood glucose levels.

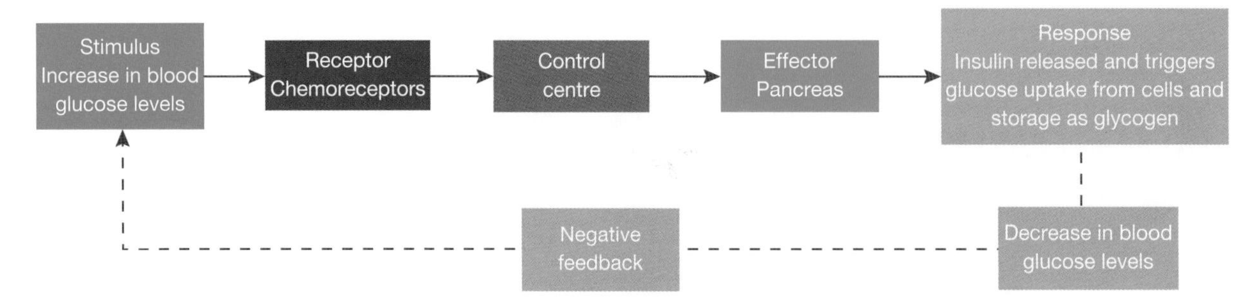

2.2.9 Positive feedback

Whereas negative feedback involves a response in an opposite direction to the stimulus, **positive feedback** results in the response going in the same direction. An example of positive feedback is when a mother is breastfeeding her baby. Mechanoreceptors in her nipple detect the baby sucking. The message is transferred to her central nervous system (in this case, her spinal cord) which then sends a message to muscles lining the milk glands to respond by releasing milk. The response continues until the baby stops sucking and the stimulus is removed.

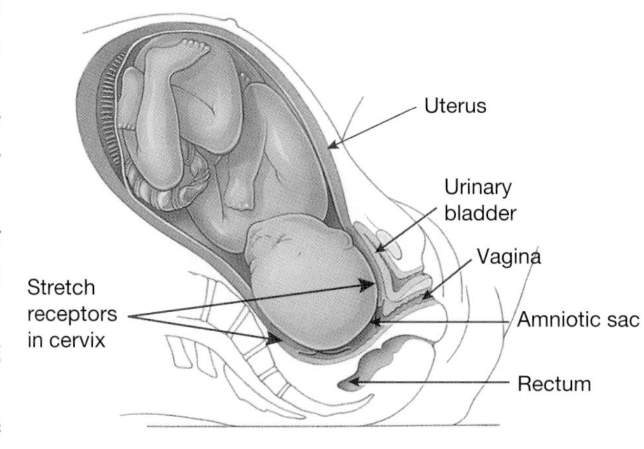

Positive feedback is also responsible for the increase in contractions during childbirth. As the baby moves into the cervix (the area that connects the uterus to the vagina), it causes stretching of receptors that result in the release of a hormone called **oxytocin**. Oxytocin causes the uterus to contract. As the baby is pushed further into the cervix, the stretch receptors are further stretched, resulting in more oxytocin release and stronger contractions. Once the baby is born, the stimulus disappears, so oxytocin levels drop, as do the contractions.

2.2.10 All under control

To work together effectively, these systems require coordination. The two systems with this responsibility are the **nervous system** and the **endocrine system**. While both of these systems require **signalling molecules** to communicate messages throughout the body, they have different ways of going about it.

Feature	Endocrine system	Nervous system
Speed of message	Slow	Fast
Speed of response	Usually slow	Immediate
Duration of response	Long lasting	Short
Spread of response	Usually slow	Very localised
How message travels through body	In circulatory system — in bloodstream	In nervous system — along nerves and across synapses
Types of message	Hormones (chemicals)	Electrical impulse and neurotransmitters (chemicals)

2.2.11 Nervous system

Your nervous system is composed of the **central nervous system** (brain and spinal cord) and the **peripheral nervous system** (the nerves that connect the central nervous system to the rest of the body). Messages are taken to the central nervous system by **sensory neurons** and taken away from it by **motor neurons**. The nervous system sends the message as an electrical impulse along a neuron and then as a chemical message (**neurotransmitters**) across the gaps (synapses) between them.

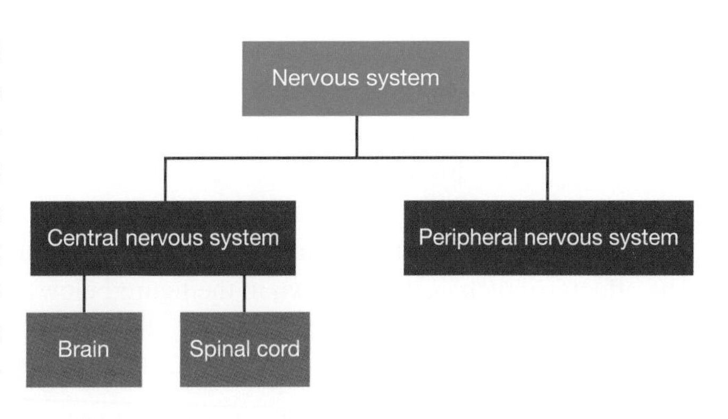

2.2.12 Endocrine system

Your endocrine system is composed of **endocrine glands** that secrete chemical substances called **hormones** into the bloodstream. These chemical messages are transported throughout the circulatory system to specific target cells in which they bring about a particular response.

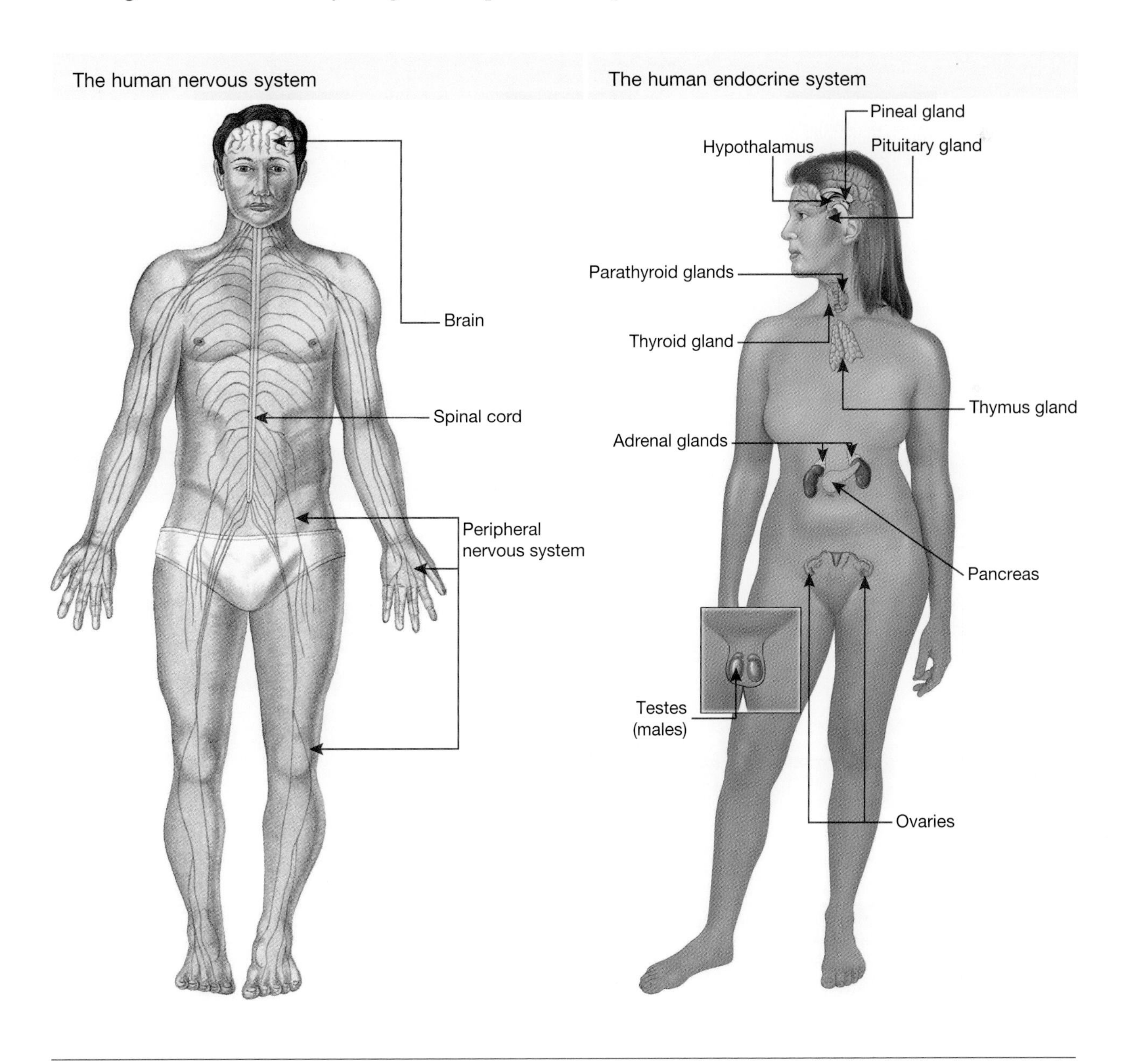

2.2.13 Working together

The control of body temperature, referred to as **thermoregulation**, provides an example of the nervous and endocrine systems working together. Evidence suggests that a part of your brain called the hypothalamus contains a region that acts as your body's **thermostat**. It contains thermoreceptors that detect the temperature of blood that flows through it.

If your body temperature increases or decreases from within a particular range, messages from thermoreceptors in your skin or hypothalamus trigger your hypothalamus to send messages to appropriate effectors. The effectors (such as those shown in the figure below) then bring about a response that may either increase or decrease body temperature.

Temperature regulation is an example in which the nervous system and the endocrine system work together to maintain your body temperature within a range that is healthy for your cells. Can you suggest terms to describe the links in the figure below?

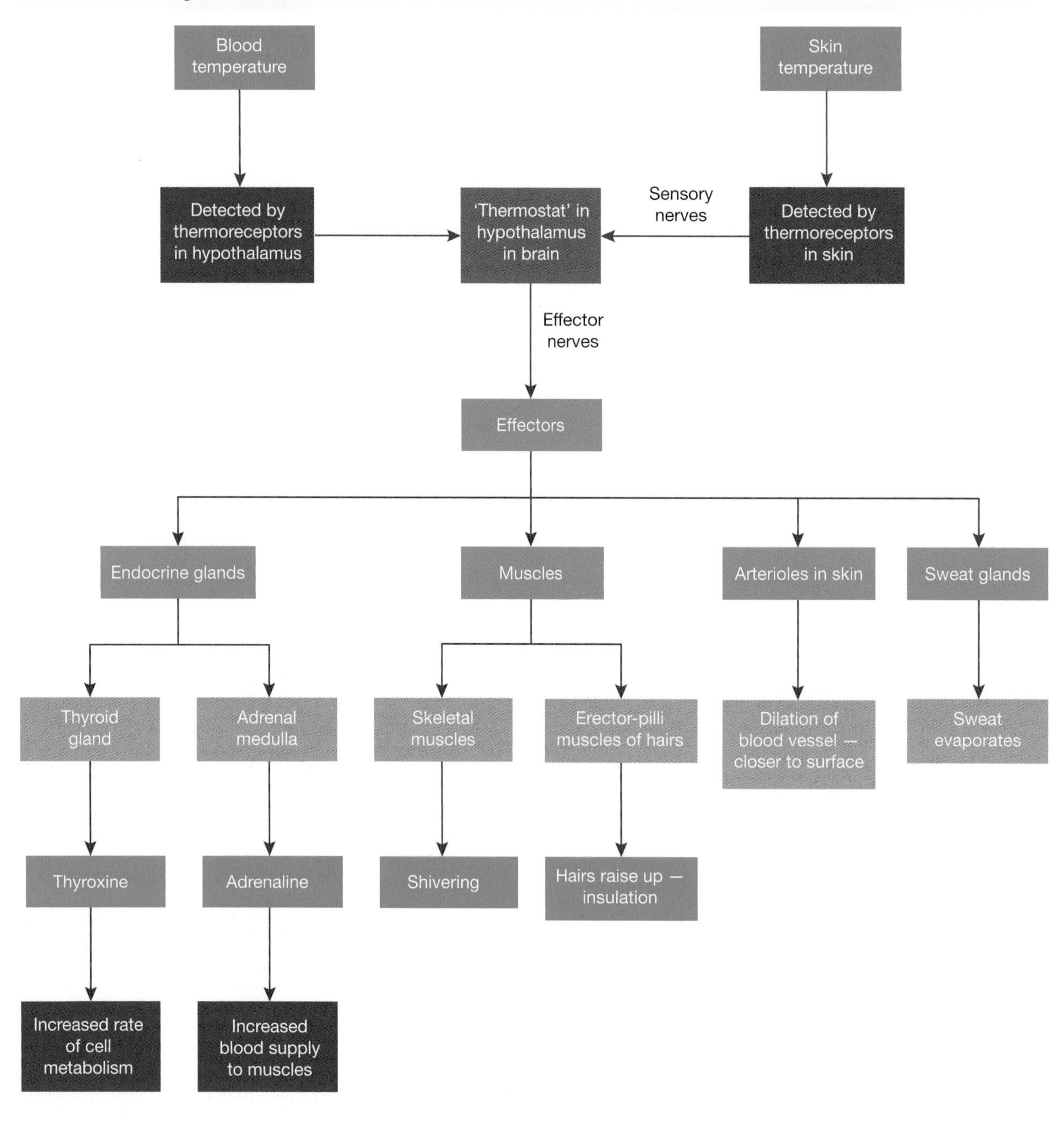

2.2 Exercises: Understanding and inquiring

To answer questions online and to receive **immediate feedback** and **sample responses** for every question, go to your learnON title at www.jacplus.com.au. *Note:* Question numbers may vary slightly.

Remember and think

1. Construct a flowchart to show the relationship between the following:
   (a) cells, organs, multicellular organisms, tissues
   (b) effector, response, control centre, stimulus, receptor.
2. Identify the type of receptor that would respond to the following stimuli:
   (a) light
   (b) sound
   (c) chemicals
   (d) temperature.
3. Organise the terms below into a Venn diagram so that they are grouped into their families.

Central nervous system	Electrical impulse
Endocrine gland	Glucagon
Insulin	Homeostasis
Hormone	Motor neuron
Neurotransmitter	Pancreas
Peripheral nervous system	Sensory neuron
Stimulus–response model	

4. What is a stimulus–response model?
5. Give an example of a negative feedback mechanism in the human body.
6. Distinguish between:
   (a) receptors and effectors
   (b) negative and positive feedback
   (c) sensory neurons and motor neurons
   (d) central nervous system and the peripheral nervous system
   (e) the endocrine system and the nervous system
7. Label the endocrine system and nervous system in the figures at right.

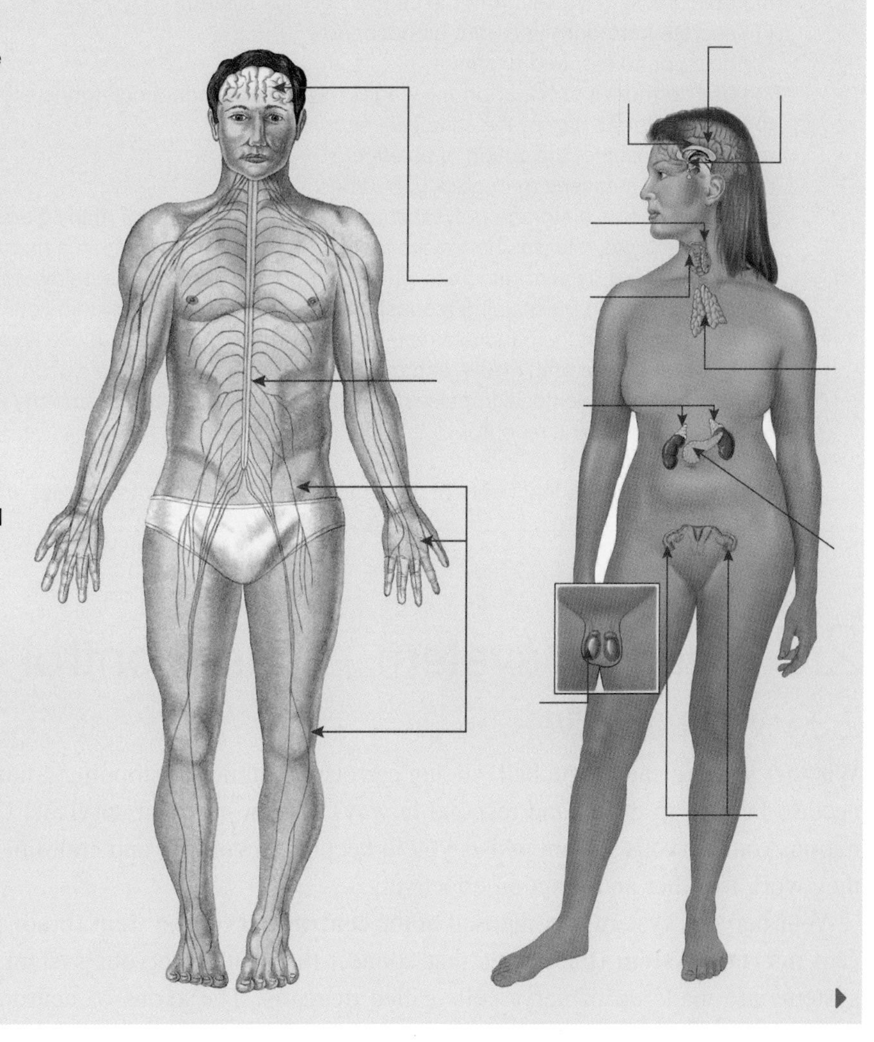

8. Match the terms listed in the box with the correct description below.

Central nervous system	Chemoreceptors
Control centre	Effector
Electrical impulse	Endocrine gland
Endocrine system	Glucagon
Glucose	Glycogen
Homeostasis	Hormone
Insulin	Motor neuron
Negative feedback	Nervous system
Neurotransmitter	Pancreas
Peripheral nervous system	Receptor
Response	Sensory neuron
Stimulus	Stimulus–response model

(a) I describe a model that helps you to achieve homeostasis.
(b) I am a response that is in the opposite direction to the stimulus.
(c) I consist of the brain and the spinal cord.
(d) I take messages to the central nervous system.
(e) Insulin is an example of what I am.
(f) I detect chemicals and can detect a decrease in blood glucose levels.
(g) I carry chemical messages within the nervous system.
(h) I take messages away from the central nervous system.
(i) I secrete hormones into your bloodstream.
(j) I help you to respond to stimuli.
(k) I use hormones to transport messages throughout your circulatory system to a target cell.
(l) I am a 'central' part of the stimulus–response model.
(m) I am the result of the action of an effector.
(n) I carry the message through your neurons.
(o) I am known as a storage polysaccharide and am made up of many glucose subunits.
(p) Your pancreas releases hormones to keep my blood levels within a narrow range.
(q) I am released by your pancreas when blood glucose levels are below normal.
(r) I am involved in maintaining a constant internal environment within your body.
(s) My release from the pancreas results in a lowering of blood glucose levels.
(t) My cells release both insulin and glucagon.
(u) I use neurons, electrical impulses and neurotransmitters to pass on messages.
(v) I am detected by a receptor.
(w) I detect a stimulus.
(x) I contain neurons that connect the central nervous system to the rest of the body.

2.3 Nervous system — fast control

2.3.1 What a nerve!

Whether you are catching a ball, slicing carrots, breathing or stopping a fall, you need to be in control. You need to be able to detect and respond in ways that ensure your survival. This requires control and coordination. Your nervous system assists you in keeping in control, and coordinating other body systems, so that they work together and function effectively.

Your nervous system is composed of the **central nervous system** (brain and spinal cord) and the **peripheral nervous system** (the nerves that connect the central nervous system to the rest of the body). These systems are made up of nerve cells called **neurons**. The axons of neurons are grouped together to form **nerves**.

2.3.2 Neurons

There are three different types of neurons: **sensory neurons**, which carry the impulse generated by the stimulus to the central nervous system; **interneurons**, which carry the impulse through the central nervous system; and **motor neurons**, which take the impulse to effectors such as muscles or glands.

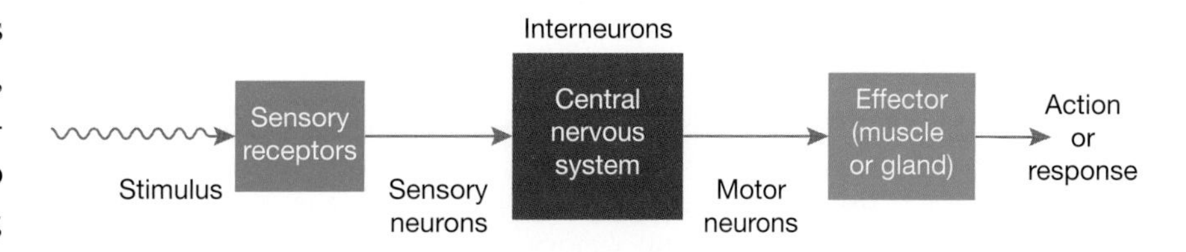

2.3.3 Structure of a neuron

Neurons, like most other eukaryotic cells, contain a **nucleus** and other cell **organelles**, **cytosol** and a **cell membrane**. However, the various types of neurons are all quite different. These differences mean that each particular neuron type is suited to its specific communication role in the nervous system.

The structure of different types of neurons

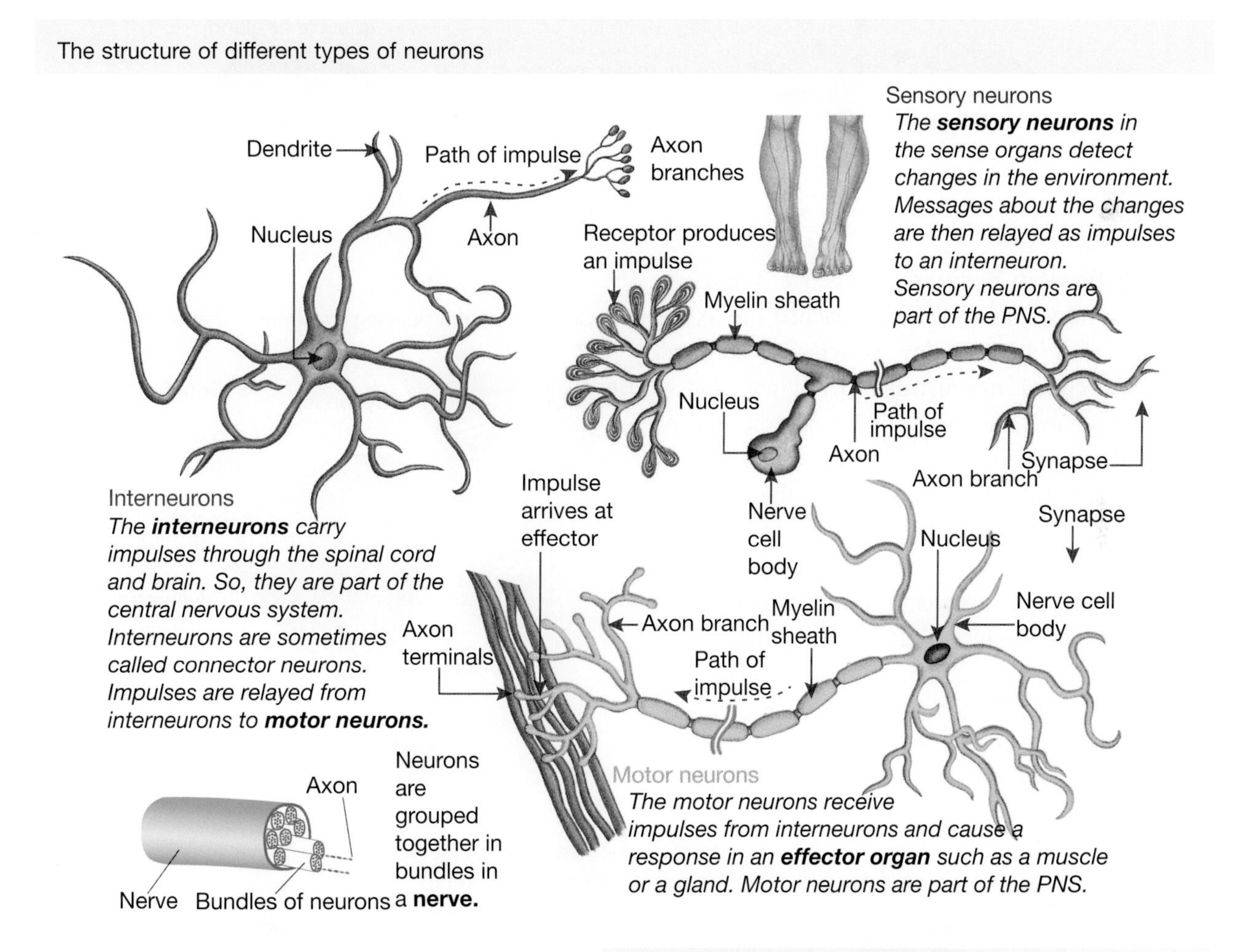

Neurons are made up of three main parts: a cell body, dendrites and an axon. On the cell membrane of the **cell body** of a neuron are highly sensitive branching extensions called **dendrites**. These dendrites possess numerous receptors that can receive messages from the other cells. Once this message has been received it moves as an electrical impulse in one direction from the

An electrical impulse moves in only one direction through a neuron.

dendrite, through the cell body and then through a long structure called an **axon**. This structure is often covered with a white insulating substance called **myelin**, which helps speed up the conduction of the message through the neuron.

Relationship between the different kinds of neurons

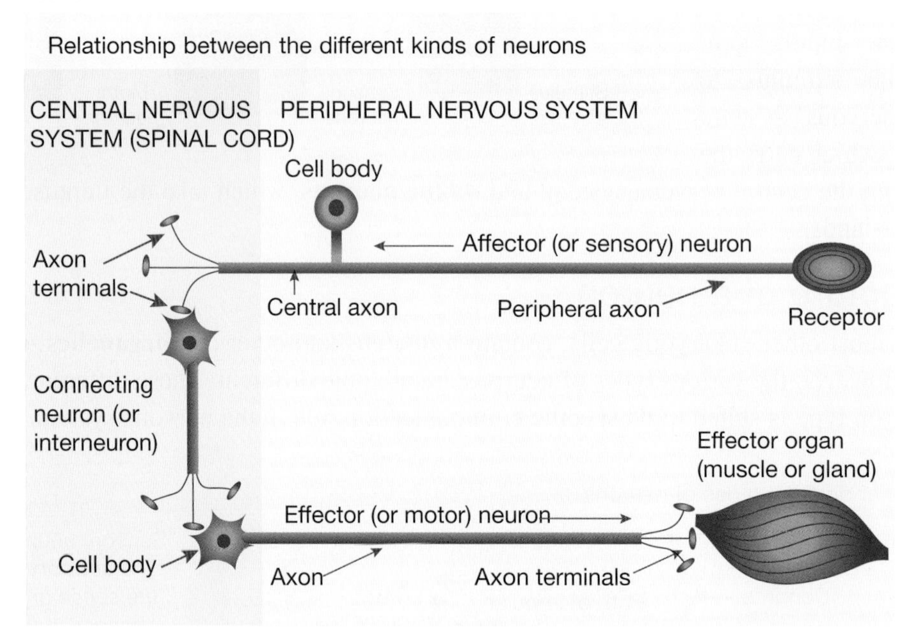

2.3.4 Synapses

The gap between neurons is called a **synapse**. When the nervous impulse has reached the axon terminal of a neuron, tiny **vesicles** containing chemicals called **neurotransmitters** are transported to the cell membrane of the neuron. These chemicals are then released into the synapse.

The neurotransmitters move across the synapse and bind to receptors on the membrane of the dendrites of the next neuron. This may result in triggering the receiving neuron to convert the message into a nervous impulse and conduct it along its length. When it reaches the axon terminal, neurotransmitters are released into the synapse to be received by the dendrites of the next neuron. This continues until the message reaches a motor neuron

Neurotransmitters passing along the synapse to the next neuron

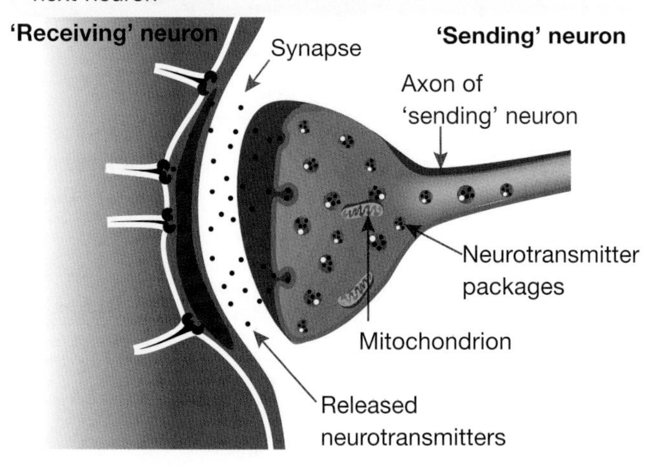

Your nervous system involves the use of both electrical signals (nerve impulses) and chemical signals (neurotransmitters).

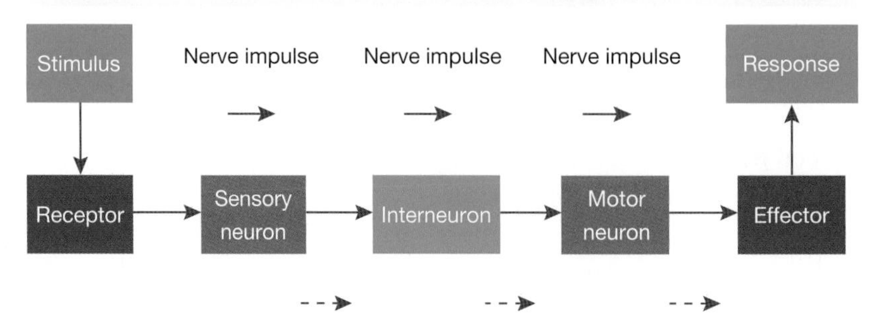

which then communicates the message to an **effector**, such as a muscle or gland. The effector may then respond to the message; for example, a muscle cell may contract or a gland may secrete a chemical.

2.3.5 Need to think about it?

Sometimes, you need to consciously think about what your body does. At other times, however, actions happen without you having to think about them.

2.3.6 Reflex actions — Act! No need to think!

Have you ever had sand thrown in your eyes or touched something too hot? Sometimes you don't have time to think about how you will react to a situation. Some actions need to be carried out very quickly — it may be a matter of survival! These actions are examples of reflex actions.

Watch out! Sometimes you need to react very quickly and there is no time to think.

Stimulus → Receptor → Sensory neuron → Interneuron in spinal cord → Motor neuron → Effector → Response

Reflex actions may involve only a few neurons and require no conscious thought. Once the stimulus is detected by a receptor, the message is sent via the sensory neuron to the interneuron in the spinal cord and then from the motor neuron to the effector to bring about a response. The message does not have to go the brain. This type of pathway, which involves only a few neurons and travels only to and from the spinal cord, is called a **reflex arc**.

You also react to many internal stimuli using reflex actions. Breathing, for example, is a response regulated by chemoreceptors detecting changes in carbon dioxide levels in your blood. It's very helpful that you don't have to remember to breathe — imagine what would happen if you forgot to!

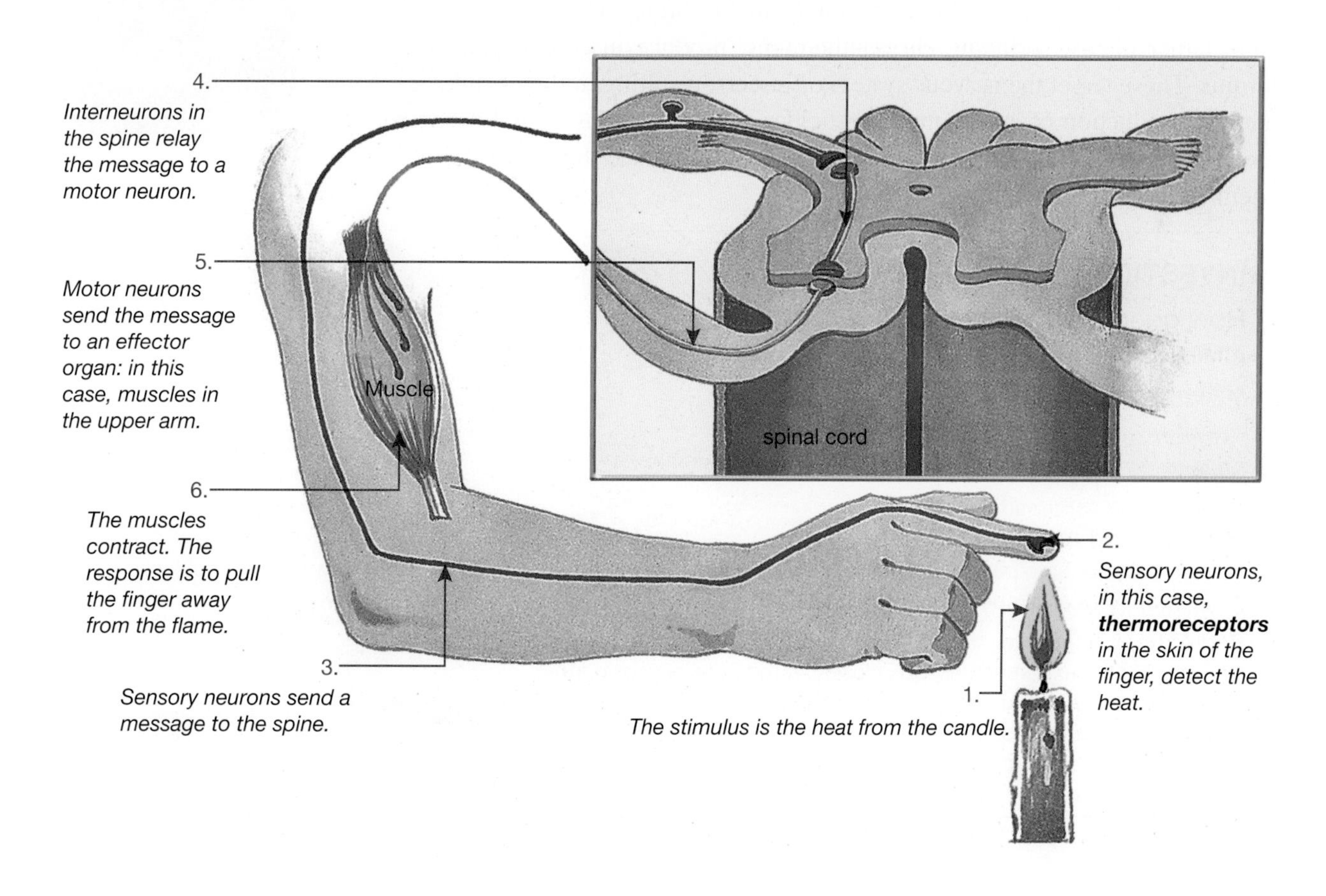

2.3.7 Think about it

More complex actions involve many interconnecting neurons and specialised parts of the brain. The messages pass into and along the spinal cord to be interpreted. When *thinking* takes place, we can make decisions about which responses are needed. Impulses are then sent along appropriate motor neurons to the effectors. This is called a *conscious response*. Many learned actions can become automatic if the same pathways are used often enough. Skill development and control in playing musical instruments and sport, for example, depend on practice during which the same pathways are often used.

2.3.8 Chemical warfare

Beware of toxic ticks, stinging trees, nasty nettles or jellyfish! Many plants and animals have ways of repelling boarders or paralysing their prey.

2.3.9 Pests and poisons

How do they do it? Blue-ringed octopuses, paralysis ticks, tiger snakes and other animals and plants produce cocktails of poisons that block the production and action of neurotransmitters at synapses. The poison from a red-back spider, for example, empties the impulses out of the neurotransmitters. Interfering with the neurotransmitters' job of carrying the message to the next neuron interferes with the transference of the message and can cause spasms and paralysis.

Many plants produce chemicals that sting by strongly stimulating the pain receptors in the skin. Messages are sent rapidly to the brain, which interprets them as pain. Other plants, including chrysanthemums, produce insecticides such as pyrethrums. These target the nervous system of insects, resulting in their death. The commercial production of such natural pesticides is a large industry and is regarded as environmentally friendly because natural pesticides replace the use of more harmful chemicals.

INVESTIGATION 2.2

How good are your reflexes?

AIM: To investigate some automatic responses

Materials:

well-lit room
chair
stopwatch or clock with a second hand

Method and results

Work in pairs for both parts of this activity. Decide who will be the experimenter and who will be the subject. Then swap roles and repeat both parts.

Part A: Kept in the dark

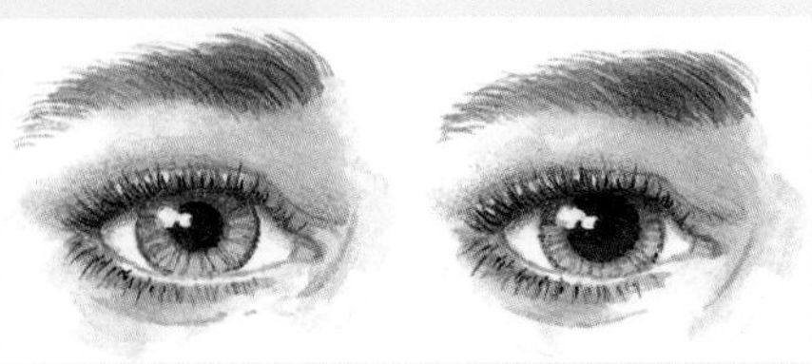

1. If you are the experimenter, look closely at the eyes of your partner, noting the size of the pupils.
   - Ask your partner to close his or her eyes for 60 seconds.
2. At the end of this time, monitor your partner's eyes for any changes.

Part B: Knee jerk

- Have your partner sit on a chair with one leg crossing over the other knee.

3. Use the edge of your hand to gently strike the crossed leg of your partner just below the knee in the joint.
   - You may need to repeat this a few times to get a response from your partner.
   - Record your observations.

Discuss and explain

Part A: Kept in the dark

4. What changes did you notice?
5. Identify the (a) stimulus and (b) response.
6. Why do you think this reflex action is important to our survival?
7. Can you control the size of your pupil?
8. Describe your observations.
9. Identify the (a) stimulus, (b) response and (c) effector.
10. Did you get the response the first time? Why or why not?
11. Can you control a knee-jerk response?

Conclusion and evaluation

12. Summarise your overall findings.
13. Suggest possible improvements to this experiment and suggest further relevant investigations that could be carried out.

2.3.10 Nerve nasties

Similar chemicals have been used as agents of human warfare. These chemicals specifically target the nervous system. Nerve gas, for example, contains a substance which prevents neurotransmitters functioning properly at the synapses. The neurotransmitters accumulate, causing the nervous system to go haywire. Such chaos can result in death.

Some chemicals can specifically target your nervous system and disrupt its functioning. Scientists and experts working with dangerous materials use suits such as those shown here for protection.

The first nerve gas, tabun, was initially developed when German scientists were developing a better insecticide. This has led to more deadly agents such as sarin and VX. All nerve gases block the body's production of an enzyme called acetylcholinesterase. This enzyme regulates the nerves controlling the action of particular muscles. A deficiency of acetylcholinesterase leads to tightening of your diaphragm, convulsions and death.

2.3 Exercises: Understanding and inquiring

To answer questions online and to receive **immediate feedback** and **sample responses** for every question, go to your learnON title at www.jacplus.com.au. *Note:* Question numbers may vary slightly.

Remember

1. Match the term with its description in the table below.

Term	Description
Central nervous system	Gap between neurons
Motor neuron	Made up of neurons
Nerves	Nerves that connect the central nervous system to the rest of the body
Neuron	Takes messages to the central nervous system
Neurotransmitter	Made up of a cell body, dendrites and axon
Peripheral nervous system	Brain and spinal cord
Sensory neuron	Chemical messenger that carries messages from one neuron to another across a synapse
Synapse	Takes messages away from central nervous system

2. Suggest how you could link the nervous system terms in the flowchart at right.

Electrical impulse	Receptor
Motor neuron	Neurotransmitter
Sensory neuron	Stimulus
Response	Effectors

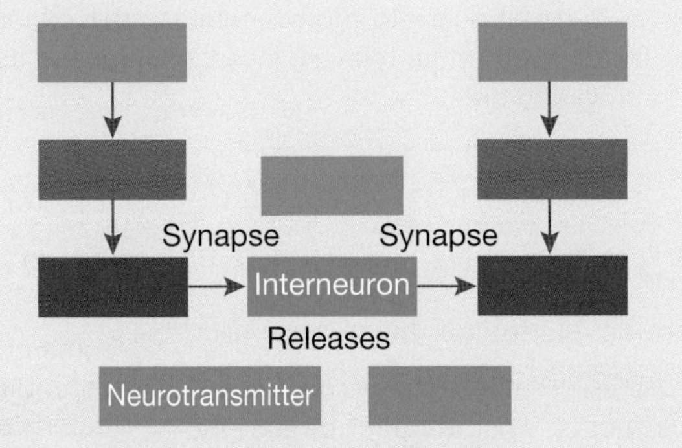

3. Label the cell body, dendrites and axon on the motor neuron bottom right and show the direction in which the impulse travels.
4. Distinguish between:
   (a) a receptor and an effector
   (b) a sensory neuron, an interneuron and a motor neuron
   (c) a neuron and a nerve.
5. Construct a diagram to describe how impulses are transmitted between sensory and motor neurons.
6. Distinguish between a reflex action and a conscious response. Provide an example of each.
7. Describe one way in which animals can cause paralysis.
8. Describe how some plants defend themselves against:
   (a) humans
   (b) insects.

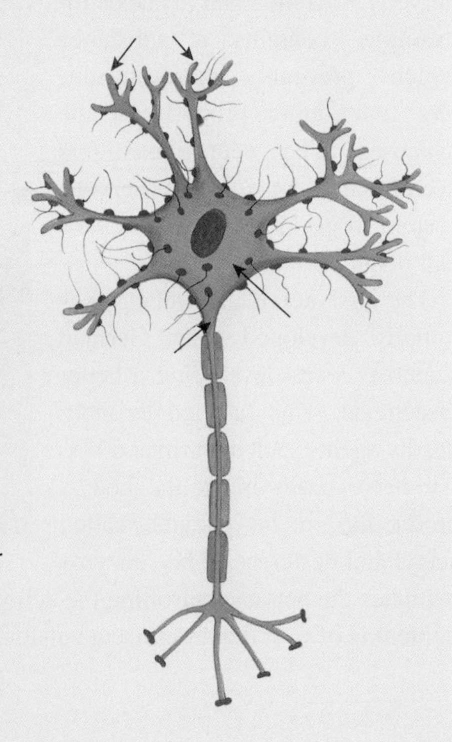

Think and discuss

9. Suggest how the structure of the different types of neurons suits them to their function.
10. Describe the advantage of the presence of myelin on the axon of a neuron.
11. Identify which of the following responses are reflex actions and which are conscious responses: sneezing, blinking, scratching your head, clapping, breathing.
12. Suggest a reason why the pupil of your eye increases in size in dim light.
13. How does blocking the production and action of neurotransmitters cause paralysis?
14. What could be the effect of toxins on aquatic food chains?

15. Suggest ways in which chemicals that affect the nervous system may be mopped up.
16. What do insecticides do?
17. Suggest why nerve gas is used in warfare.

Investigate and discuss

18. Conduct a survey of insecticides at your local nursery, garden supplies shop or supermarket. Construct a table in which to record:
    (a) the names of commercial brands of insecticides
    (b) the target organisms
    (c) the active chemical ingredients
    (d) information given about safety precautions.
    Find out how the main ingredients act in each of the insecticides and include them in a report in your survey.
19. Researchers studying Gulf War syndrome carried out experiments on chickens to discover the cause of the illness.
    (a) Find out about their experiments and conclusions.
    (b) Do you think the researchers were justified in carrying out these experiments? List arguments for and against.
20. There is a danger that chemical and biological weapons may one day be used in acts of terrorism.
    (a) Search the media for relevant examples of chemicals and their effects.
    (b) Report on your findings and discuss them with your team.
    (c) Is the use of chemical warfare ever justifiable? Discuss this with your team, recording all the various opinions and views.
    (d) What sorts of strategies do we have in Australia to cope with threats of chemical warfare? How effective do you consider these to be? In your teams, brainstorm other strategies that could be used.
    (e) On your own or in a team, write a story, newspaper article or diary entry that describes the effects of a chemical warfare attack in Australia.
21. Working in groups of four, make a list of about 10 different poisonous or venomous Australian plants and animals. Each person is to research and report on at least one. As a group, decide what aspects to include in the report. Present your findings as a PowerPoint presentation or poster with a taped commentary.

Create

22. Make models of the different neuron types using balloons, string or cotton, straws and tape.
23. In a group, act out a simple reflex arc and a conscious response.

Investigate, think and discuss

24. Imagine that you are a scientist involved in researching the nervous system. Propose a relevant question or suggest a hypothesis for a scientific investigation and outline how you would design your investigation. Search the internet for relevant research or information.
25. Find examples of how developments in imaging technologies have improved our understanding of the functions, interactions and diseases of the nervous system. Share your findings with your class.
26. Investigate how technologies using electromagnetic radiation are used in medicine, for instance in the detection and treatment of cancer of the brain or other parts of the nervous system.
27. There are often claims made about particular products in the media relating to the nervous system. Examples include drugs that regulate moods or enhance memory. Select one of these examples (or one of your own), find out more about it and then evaluate the claims being made in advertisements or in the media.

RESOURCES — ONLINE ONLY

Try out this interactivity: A nervous response (int-0670)

Try out this interactivity: A bundle of nerves (int-0015)

Complete this digital doc: Worksheet 2.1: The nervous system (doc-18859)

2.4 Getting the message

2.4.1 Five senses

Watch out! Your survival can depend on detecting changes in your environment.

Imagine not being about to see, hear, feel or sense the world outside your body. No sound, no colour, no taste or smell — just darkness and silence. Without senses, you might not even be able to sense that!

Sense organs are used to detect **stimuli** (such as light, sound, touch, taste and smell) in your environment. Examples of human sense organs are your eyes, ears, skin, tongue and nose. These sense organs contain special cells called receptors. These receptors are named according to the type of stimuli that they respond to (as shown in the table below). You have photoreceptors for vision, chemoreceptors for taste and smell and mechanoreceptors for pressure, touch, balance and hearing.

2.4.2 Five receptors

Thermoreceptors enable you to detect variations in temperature and are located in your skin, body core and part of your brain, called the hypothalamus. **Mechanoreceptors** are sensitive to touch, pressure, sound, motion and muscle movement and are located in your skin, skeletal muscles and inner ear. **Chemoreceptors** are sensitive to particular chemicals and are located in your nose and tastebuds and **photoreceptors** are sensitive to light and are located only in your eyes.

Pain receptors enable you respond to chemicals released by damaged cells. Detection of pain is important because it generally indicates danger, injury or disease. Although these receptors are located throughout your body, they are not found in your brain.

2.4.3 Touch — sharp or hot?

Your skin contains a variety of types of receptors. Pain receptors and mechanoreceptors enable you to detect whether objects are sharp and potentially dangerous. There are also **hot thermoreceptors** that detect an increase in skin temperature above the normal body temperature (37.5 °C) and **cold thermoreceptors** that detect a decrease below 35.8 °C. These thermoreceptors can also protect you from burning or damaging your skin. The sensitivity of these receptors can depend on how close together they are and their location in your skin.

Your skin contains a variety of receptors that provide you with a sense of touch.

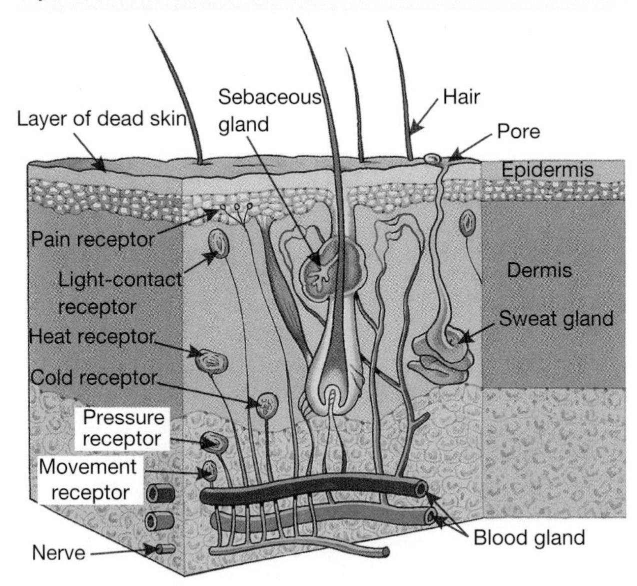

Examples of different types of receptors

Sense	Sense organ	Stimulus	Receptor	Type of receptor
Sight	Eye	Light	Rods and cones in the retina	Photoreceptor
Hearing	Ear	Sound	Hairs in the cochlea	Mechanoreceptor
Touch	Skin	Heat, cold Pressure, movement	Separate receptors for each type of stimulus	Thermoreceptor Mechanoreceptor
Taste	Tongue	Chemical substances: sweet, salty, bitter and sour	Tastebuds	Chemoreceptors
Smell	Nose	Chemicals: odours	Olfactory nerves inside nose	Chemoreceptors

INVESTIGATION 2.3

Touch receptors in your skin

AIM: To detect where the skin is most sensitive to light contact

Materials:

2 toothpicks
ruler
2 rubber bands
blindfold

Method and results

	Distance (cm) between two points when only one point is felt	
Part of the skin	Your partner	You
Inside forearm		
Palm of hand		
Calf		
Finger		
Back of neck		

1. Construct a table with the headings shown above.
   - Use rubber bands to attach two toothpicks to a ruler so that they are 2 cm apart.
2. Predict in which areas of the body the skin will be most sensitive and least sensitive.
   - Blindfold your partner. Gently touch your partner's inside forearm with the points of the two toothpicks.
   - Ask your partner whether two points were felt. Move one toothpick towards the other in small steps until your partner is unable to feel both points. To make sure that there is no guesswork, use just one point from time to time.
3. Record the distance between the toothpicks when your partner can feel only one point when there are really two points in contact.
4. Repeat this procedure on the palm of one hand, a calf (back of lower leg), a finger and the back of the neck.
5. Swap roles with your partner and repeat the experiment.

Discuss and explain

6. Which touch receptors were being used in this experiment?
7. Construct a graph to represent your data and comment on observed patterns.
8. Which area of the skin was (a) most sensitive and (b) least sensitive?
9. Suggest why the skin is not equally sensitive all over the body.
10. Which parts of the skin are likely to have the most contact receptors?
11. Discuss how your predictions compared to your experimental results.
12. Suggest improvements to this investigation and further experiments to investigate contact receptors.

Can you feel one point or two?

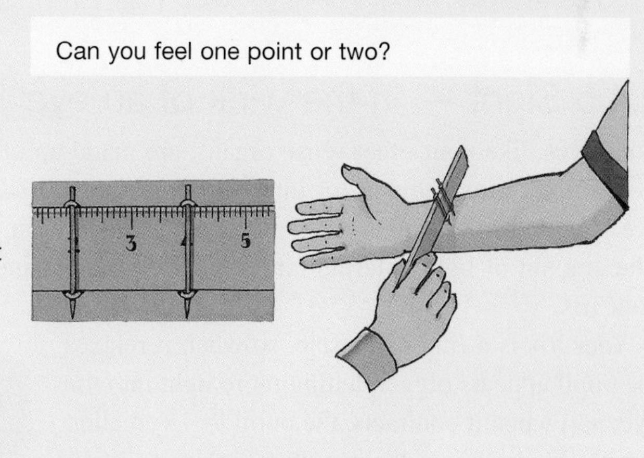

2.4.4 Smell — sweet or stinky?

The sweet scent of a rose or the stink of garbage? Gaseous molecules from the air are breathed in through your nose. When dissolved in the mucus of your nasal cavity, the hair-like cilia of your nasal chemoreceptors are stimulated to send a message via your **olfactory nerve** to your brain to interpret it, giving you the sensation of smell.

Chemoreceptors in your nose enable you to have a sense of smell.

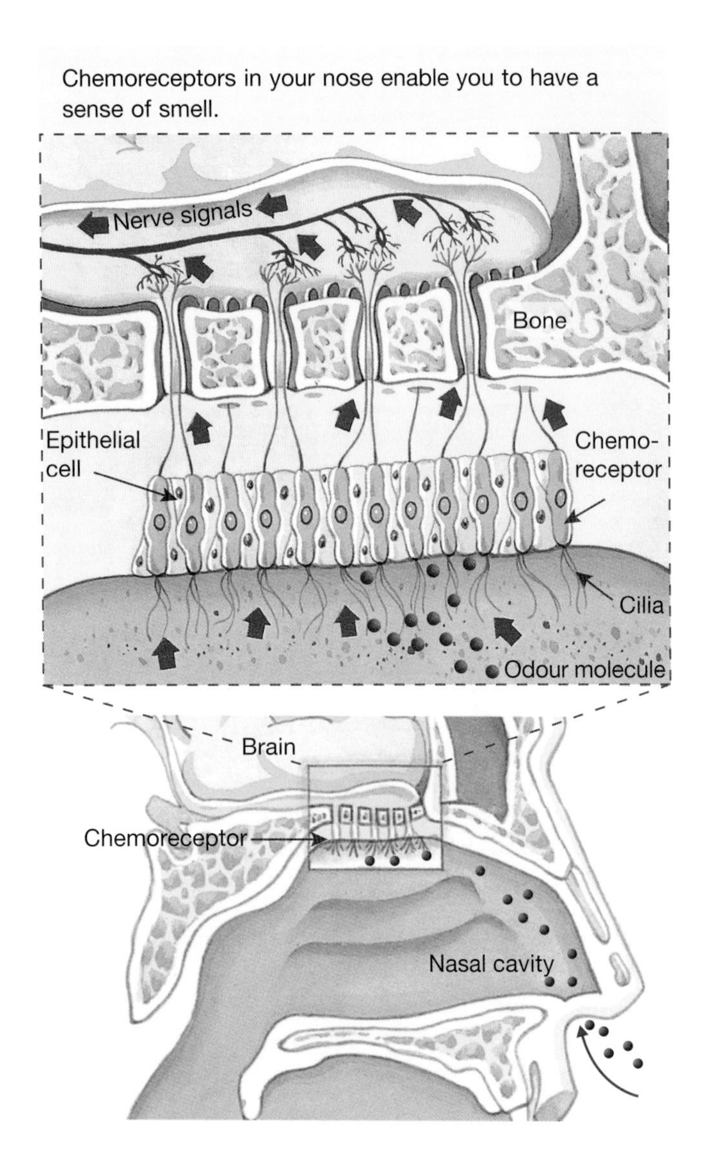

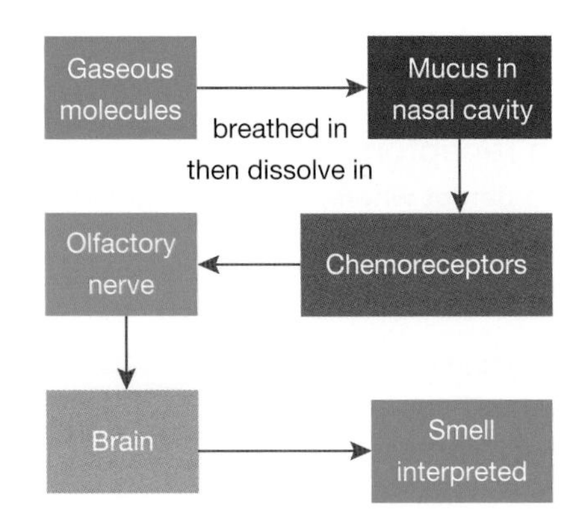

Transmission electron microscope (TEM) image of a chemoreceptor

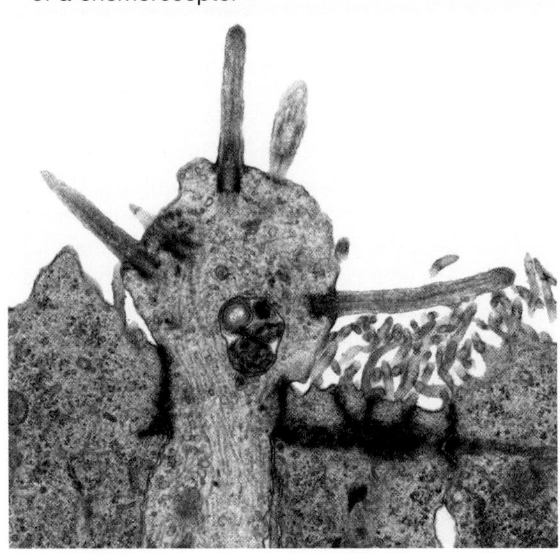

2.4.5 Sight — in the wink of an eye

Your eyes, like your other sense organs, are made up of many different parts, each with its own special job to do. Look into a mirror (or into the eyes of the person next to you). The dark spot in the centre of your eye is called the **pupil**. Your pupil is simply a hole in the **iris**. Your iris is the coloured part of your eye. The amount of light entering into your eye is determined by the size of your pupil, which is controlled by your iris.

Your iris is a ring of muscle, so when it relaxes the pupil appears bigger, letting more light into the eye; and when it contracts, the pupil looks smaller, letting less light into the eye. In a dark room, your pupil is large so that as much light as possible can enter your eye. If you were to move outside into bright light, your pupil would become smaller. This **reflex action** helps to protect your eyes from being damaged from too much light.

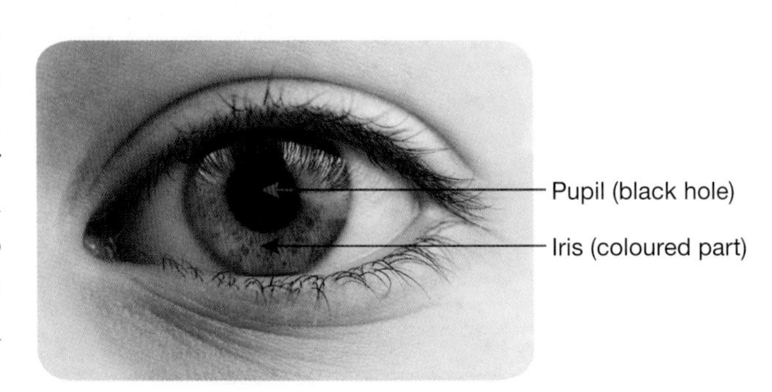

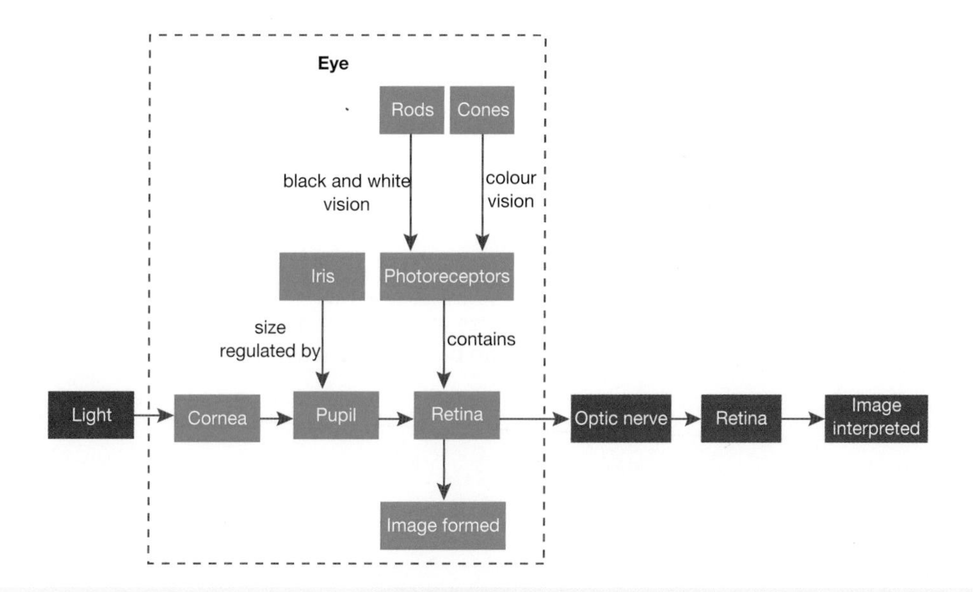

There are a number of structures within your eyes that function together so that you can detect and respond to light.

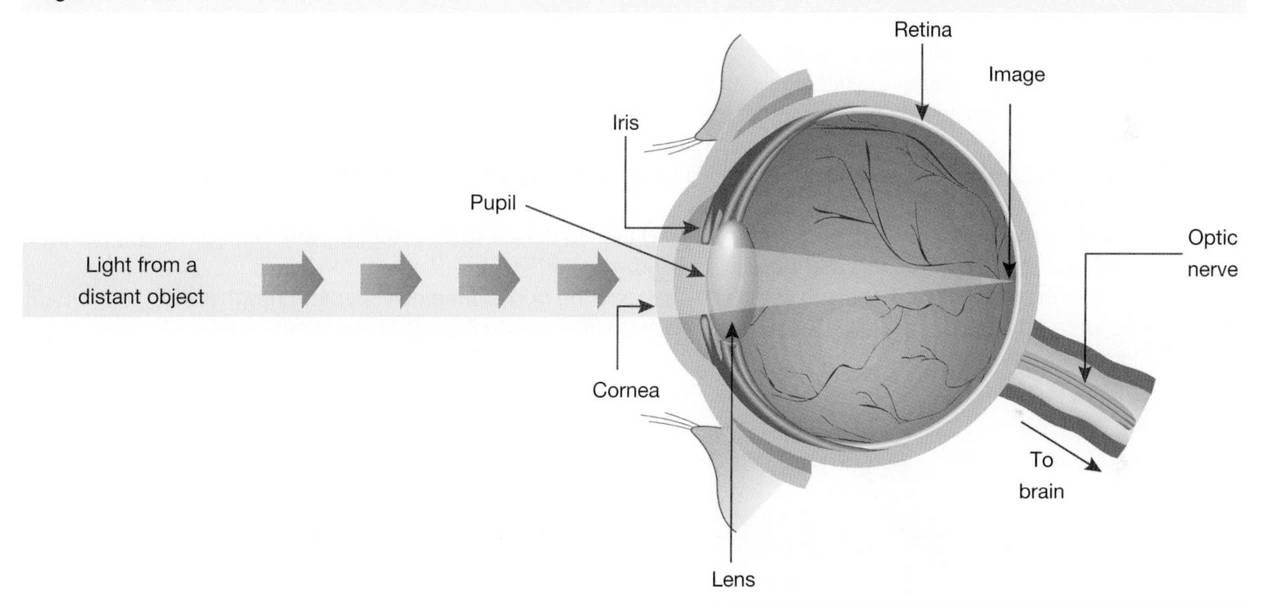

2.4.6 Getting the picture

The **cornea** is the clear outer 'skin' of your eye. It is curved so that the light approaching your eye is bent towards the pupil. The clear, jelly-like **lens** bends or focuses light onto a thin sheet of tissue that lines the inside of the back of your eye called the **retina**. The lens is connected to muscles which can make it thick or thin. This allows your retina to receive a sharp image of distant or nearby objects. **Short-sightedness** and **long-sightedness** are conditions in which a sharp image is not received on the retina. In these cases, the image can be sharpened by using artificial lenses such as those in glasses.

Although your eye receives light and produces an image of what you see, it is your brain that interprets

A scanning electron micrograph of the rod (purple) and cone (blue) cells in the eye. The rod cells detect light intensity and the cone cells respond specifically to colour.

and makes sense of the image. The photoreceptors in the retina respond to the light stimuli by sending signals to your **optic nerve** which then forwards them to your brain for interpretation.

INVESTIGATION 2.4

Getting an eyeball full!

AIM: To investigate the structure of an eye

Caution

In this activity you will be using sharp instruments. Discuss with your teacher and other members of the class a list of safety rules that should be followed carefully before beginning this activity.

Materials:

bull's eye or similar
dissection board
newspaper
paper towelling
scalpel or razor blade
safety glasses
forceps
stereo microscope
water

Method and results

- Carefully place the bull's eye on a dissection board covered with newspaper and paper towelling.

1. Draw and label the structures of the bull's eye before and after your dissection. (Use the diagrams below to help you to label your drawing.)
2. Add descriptive comments to your labels as you make your observations throughout this activity.
   - Put on safety glasses just in case any of the aqueous or vitreous humour squirts out at you. Aqueous and vitreous humour are jelly-like liquids which give eyes their shape.
3. Carefully cut a small window just behind the iris using a razor blade or scalpel. Record your observations regarding the toughness of the sclerotic coating.
   - From this window, cut towards and then all the way around the iris so that you have cut the eye into two parts.
   - Lift off the top part of the eye and examine the iris.
   - Remove the lens with forceps and see if you can read the print on the newspaper through it.
   - Use water to rinse out the jelly-like material (humour) from inside the eye and examine the retina.
4. Record your observations.
   - Follow your teacher's instructions regarding the cleaning of equipment and disposal of the dissected eye.

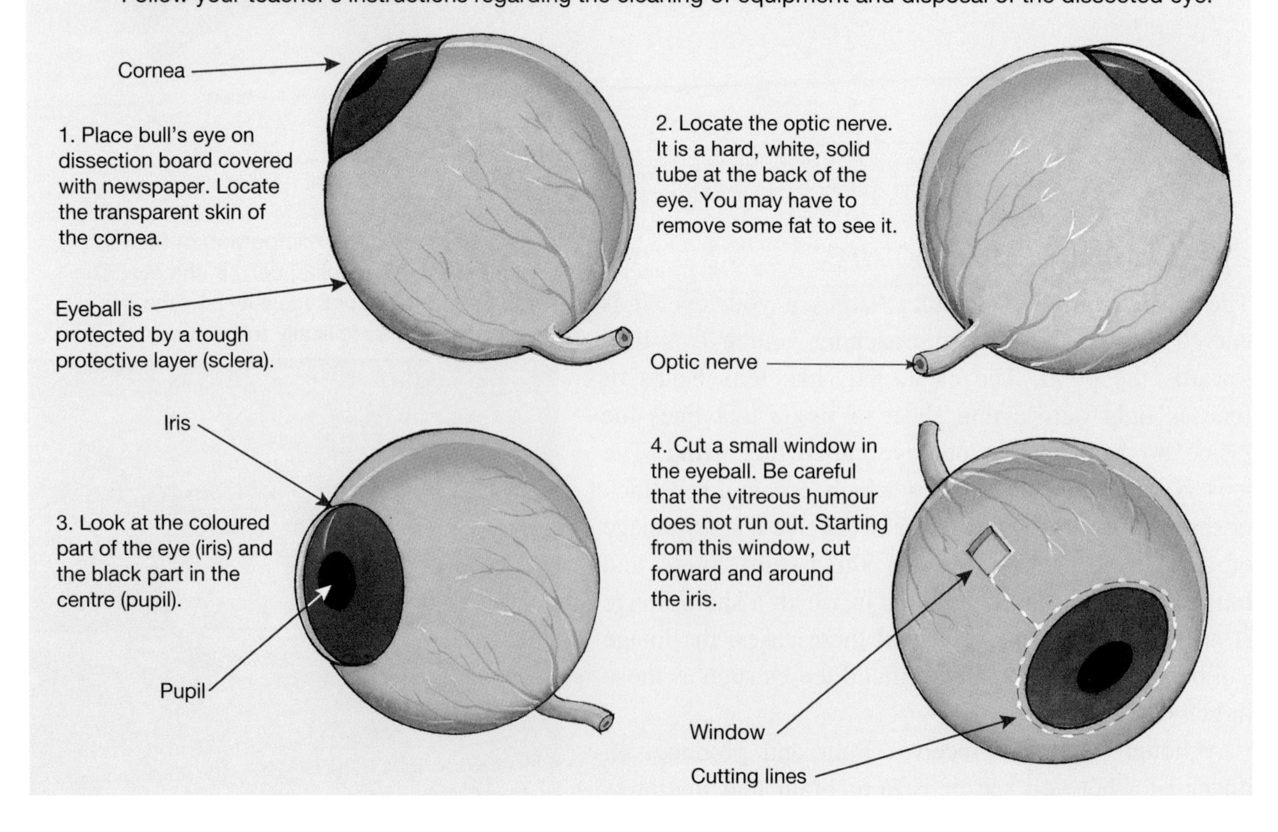

Discuss and explain

5. What is the black part in the middle of the iris?
6. What did you observe when you looked at the newspaper through the lens?
7. What did the retina look like? Could you find the optic nerve?
8. Summarise your findings in a table underneath your labelled bull's eye drawings.
9. What does the diaphragm in a microscope do? Which part of the eye does the diaphragm in a monocular microscope most resemble?
10. Find out more about one of the parts of the eye that you have observed, such as its function, related diseases or surgery.

INVESTIGATION 2.5

In the dark

AIM: To investigate the effect of light intensity on the iris of a human eye

Method and results

- Cup your hands loosely over both eyes so that you cannot see anything but your hands. Keep your eyes open. Look at the insides of your hands.
- After about one minute, have your partner look carefully at your pupils.

Discuss and explain

1. What happens to the iris as your hands are removed?
2. Explain your observations.

2.4.7 Black and white or colour?

Why do you see in black and white at night and in colour during the day? It is because of **rods** and **cones**. These are two different types of photoreceptors located in your retina. Rods are more sensitive to light and allow you to see in black and white in dim light. Cones are responsible for colour vision, are less sensitive to light and operate best in daylight. At night, there is not enough light for your cones to sense colour.

Are you colour blind? **Colour blindness** is an inherited condition that is generally more common in males; however, females can also be colour blind, due to the way in which the condition is inherited. There are also different types of cones. If you have a deficiency of one or more of these it may mean that you find it difficult to see a particular colour or combinations of colours.

The receptor cells in the retina respond to brightness and colour.

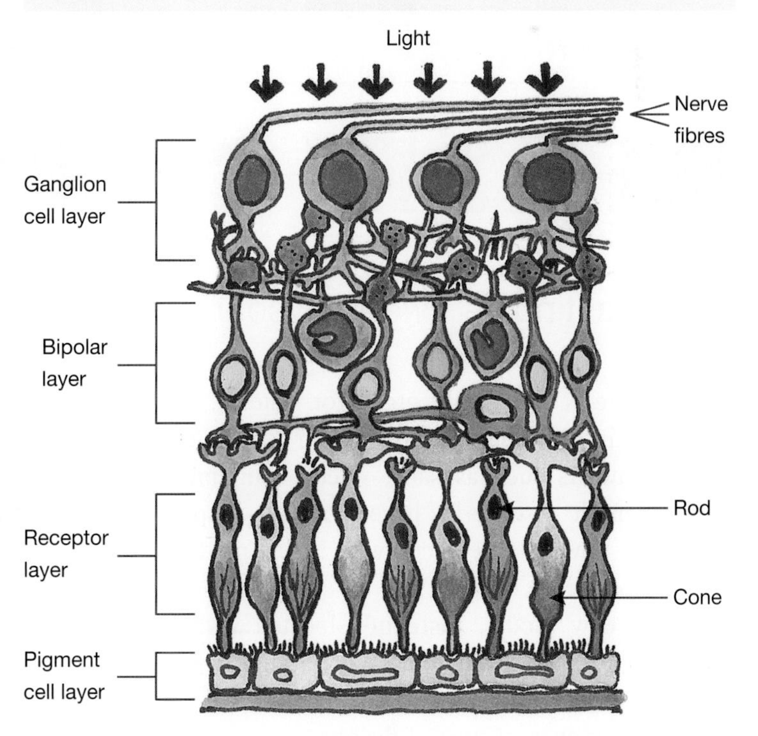

2.4.8 Hearing — catching vibrations

The ear is your sense organ that detects sound. When the air inside your **ear canal** vibrates, it causes your **eardrum** to vibrate at the same rate. Three tiny bones known as **ossicles** in your **middle ear** receive this vibration from your eardrum and then pass it to your inner ear. Inside your inner ear, thousands of tiny hairs attached to nerve cells of the snail-shaped **cochlea** detect the vibration and send a message to your brain via your **auditory nerve**. Your brain interprets the message as hearing sounds.

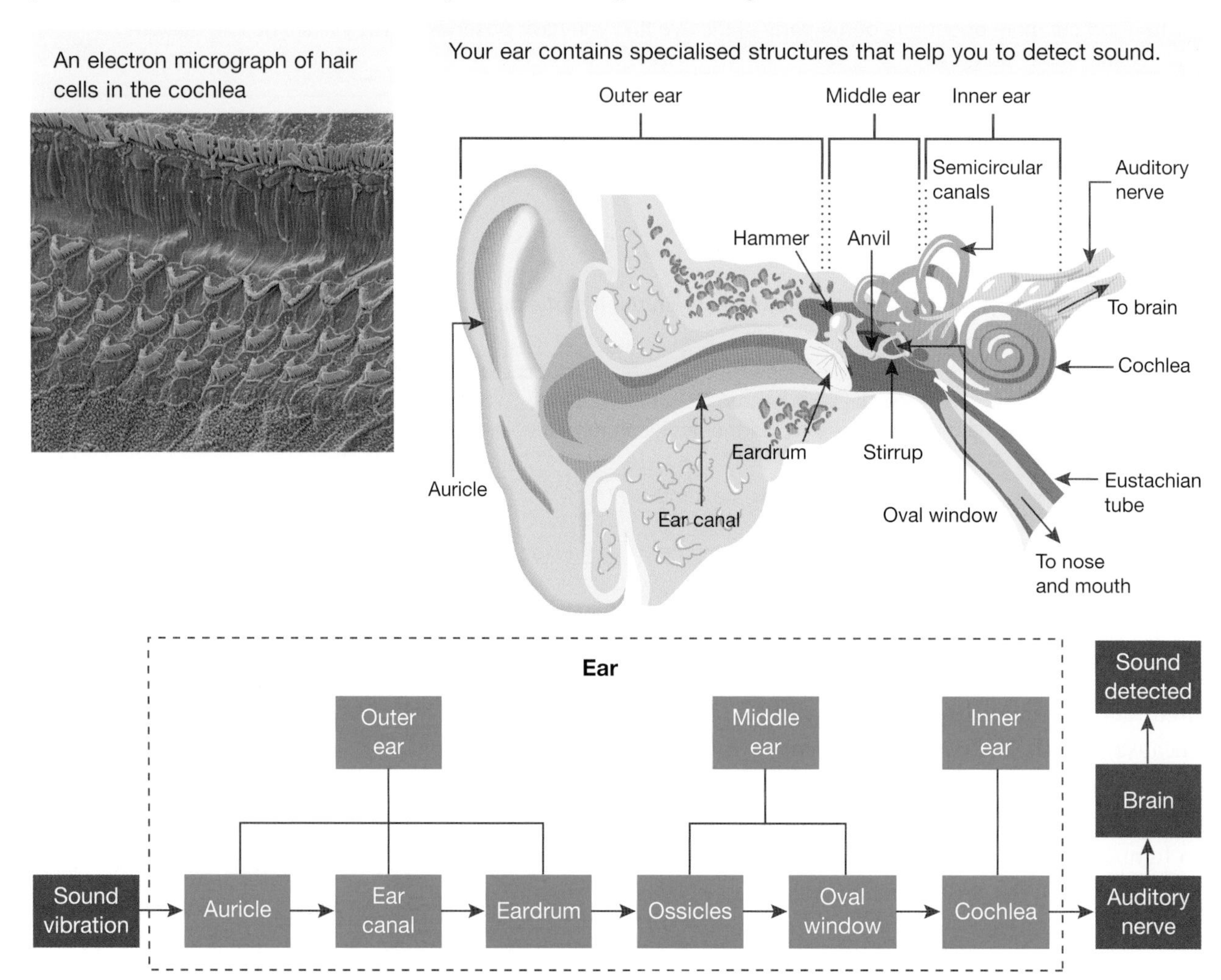

An electron micrograph of hair cells in the cochlea

Your ear contains specialised structures that help you to detect sound.

2.4.9 Tasting — sweet or sour?

Change of taste

The **tongue** is your sense organ for taste. It was once thought tastebuds in different regions of your tongue could detect particular flavours such as salty, sweet, sour, bitter and savoury. New scientific discoveries have, however, disproved this model and it has now been replaced with a new model to explain how we gain our sense of taste.

In the new model, **tastebuds** located within bumps called **papilla** across your tongue have the ability to sense all flavours. This is because each of these tastebuds contains taste cells with receptors for each of type of flavour.

This model of taste is now obsolete. Current research suggests that we do not have different areas on our tongue for different taste sensations.

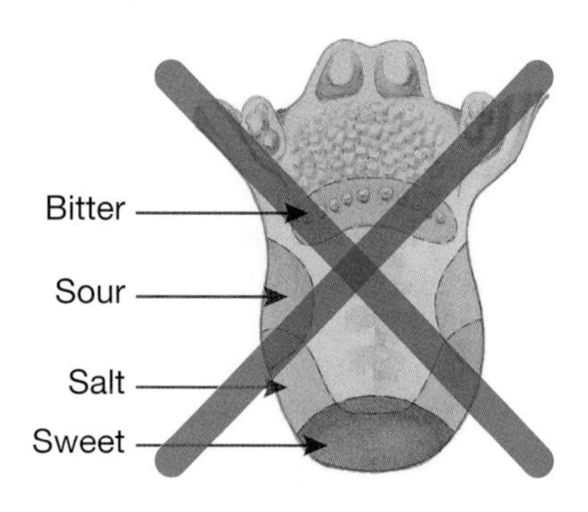

2.4.10 Hardwired for flavour

Our brains are wired so that we enjoy sweet, savoury and salty foods so that we can obtain the energy, protein and nutrients that we need to survive. Mass-produced foods, however, are often packed with high amounts of sugar and salt. This has resulted in our sense of taste increasing our chance of suffering from conditions such as diabetes, heart disease and obesity.

Researchers have discovered tiny compounds that can magnify the taste of foods, so that they can taste saltier and sweeter than they really are. The use of these taste enhancers could lead to reduced sugar, salt and monosodium glutamate (MSG) being added to foods and fewer taste-related diseases.

Your tongue contains tastebuds containing chemoreceptors (as shown here) sensitive to particular chemicals.

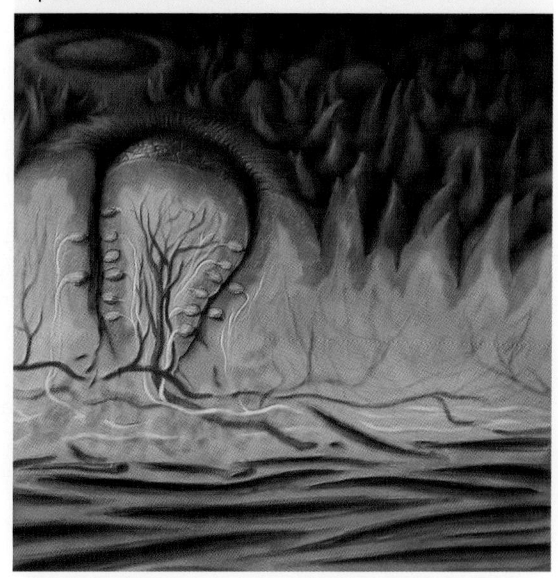

2.4 Exercises: Understanding and inquiring

To answer questions online and to receive **immediate feedback** and **sample responses** for every question, go to your learnON title at www.jacplus.com.au. *Note:* Question numbers may vary slightly.

Remember

1. State the purpose of the sense organs.
2. List examples of:
   (a) five stimuli detected by human sense organs
   (b) five human sense organs
   (c) five types of receptors.
3. Provide an example of a type of receptor. State where you would find it and the stimulus that it detects.
4. Identify the location of the:
   (a) optic nerve
   (b) olfactory nerve.
5. Describe the difference and relationship between the pupil and iris in the eye.
6. Construct a mind map to show the structures that are important to vision.
7. In which part of the human body is an observed image:
   (a) formed
   (b) interpreted?
8. Outline the differences between the functions of rods and cones in the eye.
9. Construct a flowchart that shows structures involved in:
   (a) smell
   (b) vision
   (c) sound.
10. Describe the new model that is used to explain the involvement of our tongues in the sensation of taste. How is this different to the previous model?
11. Suggest a reason why we are 'hardwired for flavour'.
12. Suggest how the discovery of taste enhancers may reduce the chances of getting 'lifestyle' diseases such as some types of diabetes, heart disease and obesity.

Think and discuss

13. If cats have rods, but no cones, what does that mean in terms of how they see the world?
14. Why is the thickest part of your skin on the soles of your feet?
15. Why are some parts of your skin, such as the back of your hand, more sensitive to heat than others?
16. How do movement receptors receive a sensation of movement when they are well below the surface of the skin?
17. When you walk into a dark room at night, you cannot see anything. A minute later, without any additional light, you can see. What behaviour of the eye allows this to happen?
18. Olfactory receptor cells are important to enable us to smell things. A human has about 40 million, whereas a rabbit has 100 million and a dog has 1 billion! What effect might this difference have on the chances of survival for these animals?
19. Make some spectacles out of cardboard and use red-coloured cellophane for the lenses. If you wear the spectacles when a light or torch is turned on, your eyes will be using only their cones. When you return to the dark and remove the spectacles, your rods will be unaffected by the red light. Therefore you won't need to adapt to the dark, and can use your eyes immediately.

Investigate

20. Investigate and report on the work of Australian scientists such as Fiona Wood and Marie Stoner on artificial skin.

RESOURCES — ONLINE ONLY

Complete this digital doc: Worksheet 2.2: Skin (doc-18860)

2.5 Getting in touch with your brain

2.5.1 What is in your brain?

Take a guess. What looks like a grey wrinkled walnut and is about the size of a large grapefruit? *Hint:* you are using it to figure out what the mystery object is!

The average brain weighs around 1.5 kilograms and is made up of about 80 per cent water, 10 per cent fat and 8 per cent protein. Although our brains contain about a billion brain cells, only about 10 per cent are active neurons (nerve cells); the remaining brain cells are there to nourish and insulate the neurons. These neurons can grow extensions called **dendrites**, which reach out like branches on a tree, allowing communication between other neurons. This communication is very important in relaying information about your environment and deciding what to do with it.

2.5.2 More than just a bag of chemicals!

Your brain is more than just a mix of chemicals and cells. It is the control centre of all of your body's functions and is responsible for intelligence, creativity, perceptions, reaction, emotions and memories. It can be said that your brain is at the wheel, steering your body's systems so that it continues to function correctly, whether it's remembering the taste of chocolate, working out a crossword puzzle, controlling your heartbeat or monitoring the glucose level in your blood.

When you think, you are using your brain. Another name for thinking is **cognition**. You also 'feel' with your brain. Happiness, sadness and anger are examples of feelings or **emotions** that are interpreted by your brain. Your brain also interprets messages about your internal and external environments, and plays a key role in **regulating** processes that keep you alive.

2.5.3 Patterns of organisation

Your brain cells are organised into different areas within your brain. Although they may have different functions, they communicate and work together to keep you alive. There are a number of different models that are used to describe the structure of the human brain.

2.5.4 From back to front

Your **hindbrain** is really a continuation of your spinal cord. It develops into the **pons** and cerebellum, and the **medulla oblongata** (medulla). Extending through your hindbrain and midbrain is a network of fibres called the **reticular formation** — a network of neurons that opens and closes to increase or decrease the amount of information that flows into and out of the brain. The reticular formation helps regulate alertness (from being fully awake or deeply asleep), motivation, movement and some of the body's reflexes (such as sneezing and coughing). The **forebrain** develops into the cerebrum, **cerebral cortex** (outer, deeply folded surface of the cerebrum) and other structures such as the **thalamus**, **hypothalamus** and **hippocampus**.

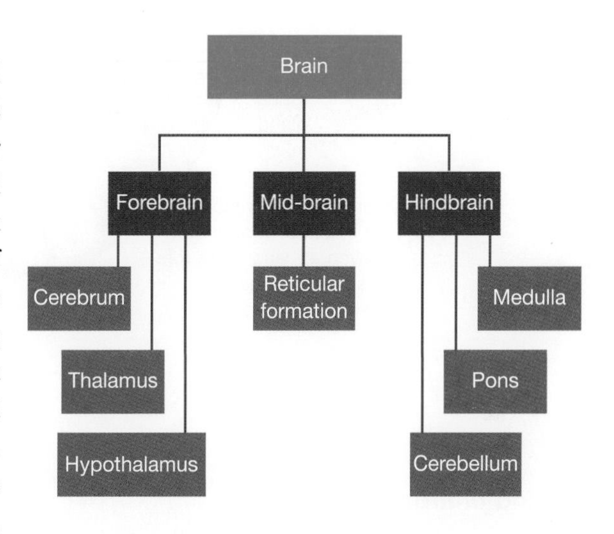

2.5.5 Brain stem or medulla

Not all actions in your body require conscious thought. These are called involuntary actions and you don't need to think about them for them to occur. Breathing, heartbeat, blood pressure, coughing, vomiting, sneezing and salivating are all examples of involuntary actions controlled by your **brain stem**.

Your brain stem (or medulla) is located between your spinal cord and your cerebrum. If this vital structure is damaged, death may result. One of the reasons drugs such as heroin and cocaine are so dangerous is that they can impair the functioning of this structure, causing interruptions to heartbeats or breathing.

2.5.6 Cerebellum

What's grey on the outside, looks like two clams side by side, is about the size of your fist and without it you'd fall over? The answer is your **cerebellum**.

Your 'little' brain

Your cerebellum is located near the brain stem, underneath the cerebrum. Although it takes up only about 10 per cent of your brain's volume, the cerebellum contains over half of all of your brain's neurons. Your cerebellum has key roles in posture, coordination, balance and movement. Current research also suggests that it may also be involved in memory, attention, spatial perception and language.

The word *cerebellum* means 'little brain' in Latin and that's just what it looks like. There are two halves (or hemispheres), one for each side of the brain. Each of these hemispheres consists of three lobes. There is a lobe that receives sensory input from your ears to help you to maintain your balance. Another lobe gets messages from your spinal cord to let your brain know what some other moving parts of your body are up to. There is even a lobe that communicates with your cerebrum, the thinking part of your brain.

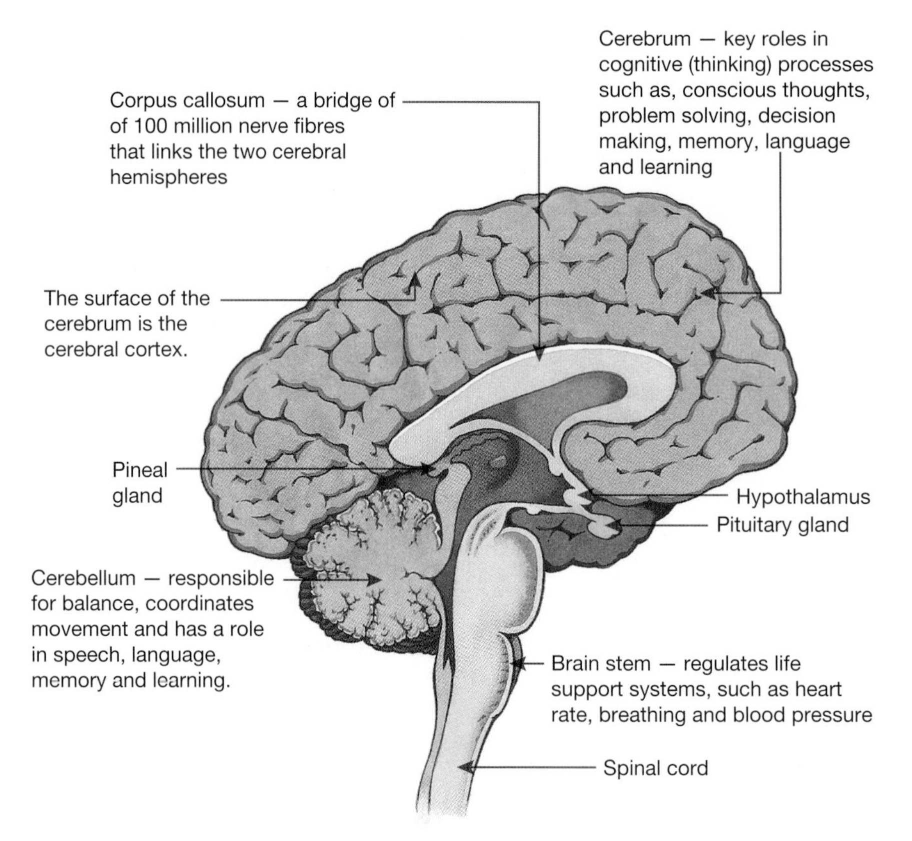

Taking charge

When you start learning a new skill, you have to think carefully about what you are doing. Once you have got the hang of it, your cerebellum takes over from your thinking context to tell your body what to do. Research has shown that when the cerebellum is in charge, you can move faster and are less clumsy. Other research suggests that long-term memory traces for motor learning are located in your cerebellum and that movement may help your thinking because of increased signals travelling between your cerebrum and cerebellum.

2.5.7 Cerebrum

The **cerebrum** is the largest part of the brain and makes up about 90 per cent of your brain's total volume. The cerebrum is responsible for **higher-order thinking** (such as problem solving and making decisions) and controls speech, conscious thought and voluntary actions (actions that you control by thinking about them). The cerebrum is also involved in learning, remembering and personality.

The cerebrum is made up of four primary areas called lobes. Each of these lobes is associated with particular functions.

You can use a piece of paper to model how the cerebrum can fit into such a small area. If you screw up the piece of paper so that it is roughly the size of your fist you can see how the cerebrum, with its large surface area, can fit into a small area within your skull. Its many wrinkles and folds are the reason that only about one-third of this structure is visible when you look at the outside of a brain.

2.5.8 Left and right — two brains in one?

Your cerebrum is divided into two grey wrinkly **cerebral hemispheres** — the right cerebral hemisphere (mainly responsible for the left side of your body) and the left cerebral hemisphere (mainly responsible for the right side of your body). While each hemisphere is specialised to handle different tasks they work together as an integrated whole, communicating with each other through a linking bridge of nerve fibres called the **corpus callosum**.

Although each cerebral hemisphere processes information differently they are both involved in putting together the total picture of what you sense around you. During your learning, it is important to employ learning activities that utilise the strengths of both hemispheres (even if it can feel a little uncomfortable sometimes). This will allow you to focus on 'whole-brained' learning.

Although the left and right cerebral hemispheres have specialised functions, they communicate with each other. Depending on the task, you can switch from one side of your brain dominance to the other a number of times each day.

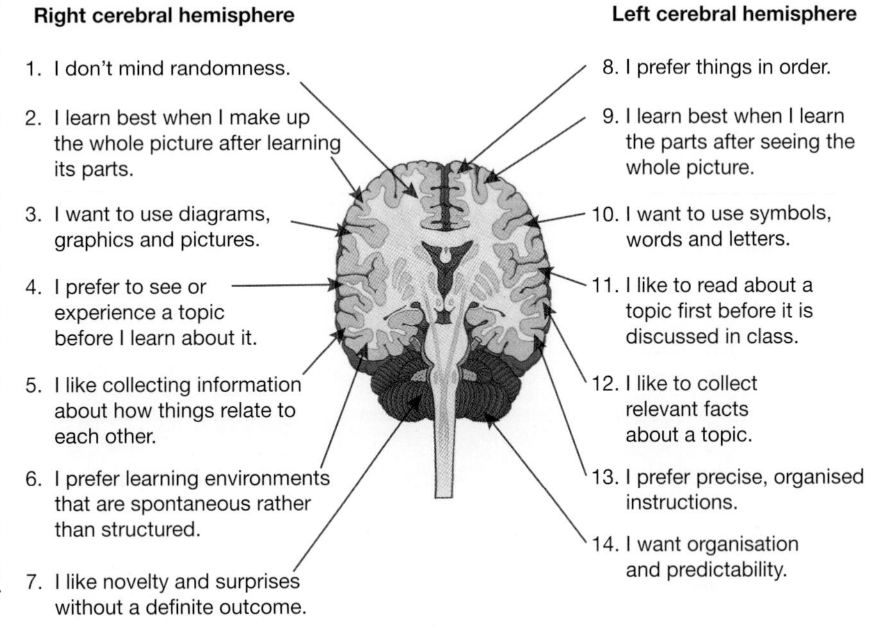

HOW ABOUT THAT!

Like your thumbprint, your brain is unique. Not only may it be a different size and weight from your friends, but the learning connections between cells in your brain are different. These connections are made as a result of your experiences and this forms your own personal 'cognitive map', which can change over time as you build up more experiences. This difference in our brain's 'internal wiring' can explain why people at the scene of the same accident can have such different eyewitness reports.

2.5.9 Tasty words and colourful letters?

A small percentage of the population have their senses crossed and associate letters with a flavour, numbers with a gender and sounds with colour. This condition is known as **synaesthesia**. It has been likened to receiving information in one sense and it triggering an experience in another. So while you might hear music, the sounds trigger seeing particular colours! There are thought to be at least 54 documented types of this condition. Currently there is exciting research being conducted in this area, investigating how people with this condition form and remember memories. Some of these investigations involve the use of **functional magnetic resonance imaging (fMRI)** to get a 3-dimensional image of the brain so that the areas of the brain that are activated during different mental tasks can be recorded.

Someone with synaesthesia might perceive certain letters and numbers as they are shown here.

SYNAESTHESIA
0123456789

INVESTIGATION 2.6

Brain dissection

AIM: To investigate the structure of the brain

CAUTION

Handle dissecting instruments with care and ensure they are placed in a sterilising solution after use. Wear gloves throughout the dissection and wash your hands thoroughly at the end.

Materials:

a semi-frozen sheep's brain
dissecting board
dissecting instruments (scalpel, forceps, scissors)
plastic ruler
paper towel
gloves

Method and results

1. Construct a table with the headings shown below.

	Appearance			
Brain structure	**Colour**	**Texture**	**Other features**	**Size**
Cerebrum				
Cerebellum				
Brain stem				

- Place the brain so that the cerebral hemispheres are at the top of the board and the brain stem is at the bottom.
- Identify the external features of the brain: the cerebral hemispheres, cerebellum and brain stem.
- Use your forceps and try to lift the meninges (membranes protecting the brain). You may be able to observe the cerebral fluid between these membranes and the hemispheres.

2. Carefully observe the overall appearance of each structure and, using a plastic ruler, measure its size (length, width and height). Include this information in a table like the one above.

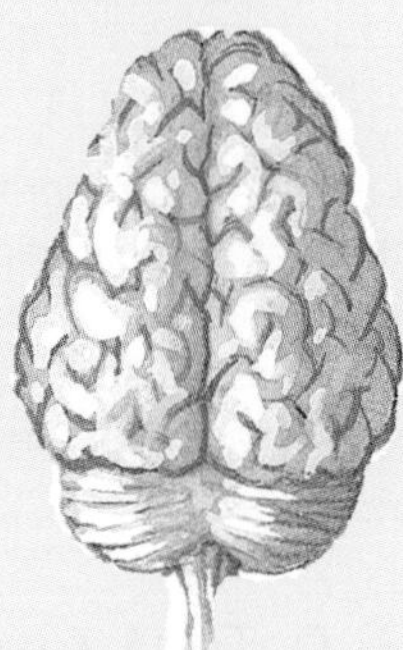

3. Draw a diagram of the sheep's brain, labelling the external features.
   - On your diagram, identify and label the part of the brain that controls the sheep's:
     (a) heart rate
     (b) balance required for walking
     (c) ability to locate its lamb.
4. Using your scalpel, cut the brain in half between the right and left hemispheres, and separate the two cerebral hemispheres.
5. Draw a cross-section of the brain. Be sure to label it!
   - Now make a second cut down through the back of one of the hemispheres to see inside the cerebellum and brain stem.
6. Record your observations.

Discuss and explain

7. (a) Which structures contained the grey and white matter?
   (b) Find out why these structures are different colours.
8. Which part of the sheep's brain is the biggest? Is this the same pattern in human brains?
9. The brain is usually protected by a bony skull. It is also covered with three layers of connective tissue called meninges and surrounded by cerebral fluid. Suggest how the meninges and cerebral fluid help protect the brain.
10. Summarise your findings.
11. Identify strengths and limitations of your investigation of the brain and suggest improvements.

2.5.10 It's a bit like a ...

Using **analogies** and **metaphors** can be very useful in helping you to connect information that you already know to new information. An example that has been used in the past is 'the brain is like a computer'. This provides a framework of known ideas to relate to new ideas.

While analogies and metaphors and models can be very useful in your learning, they also have limitations. The more we find out about the brain, the less suitable a previously used metaphor may be. Examples of other analogies that have been used for the brain include comparing it to a hydraulic system, a telephone switchboard and, more recently, an ecosystem in a jungle! These analogies often reflect the most current technological innovation of the time.

Did you notice examples of analogies and metaphors mentioned throughout these pages? How effective have they been in helping you 'get a handle' on new information about the brain?

2.5 Exercises: Understanding and inquiring

To answer questions online and to receive **immediate feedback** and **sample responses** for every question, go to your learnON title at www.jacplus.com.au. *Note:* Question numbers may vary slightly.

Remember

1. Name the organ that has been described as the control centre of your body.
2. Identify the part of your brain that:
   (a) takes up the greatest volume
   (b) regulates heartbeat, breathing and blood pressure
   (c) generates the most complex thoughts
   (d) coordinates movement
   (e) manages communication between left and right hemispheres.
3. Use analogies to describe the appearance of the:
   (a) brain
   (b) cerebrum
   (c) cerebellum.
4. Distinguish between:
   (a) cerebrum and cerebellum
   (b) left and right cerebral hemispheres
   (c) cerebrum and cerebral cortex
5. Copy the cluster map shown and insert 'cerebrum', 'cerebellum' and 'brain stem' into their appropriate location.

Investigate, think and discuss

6. Christopher Reeve, the actor who played Superman in the early Superman movies, damaged his brain stem when he fell off a horse. Find out and report on:
   (a) the effect this had on his brain function
   (b) medical research that may help others with such damage.
7. Find out more about the cerebellum and how it may be involved in learning.
8. In teams, research the structure and functions of different parts of the brain.
9. *Brains react to music like a drug.* This was a claim made in the media in 2011. It was based on a scientific study that used PET (positron emission tomography) and fMRI brain scans

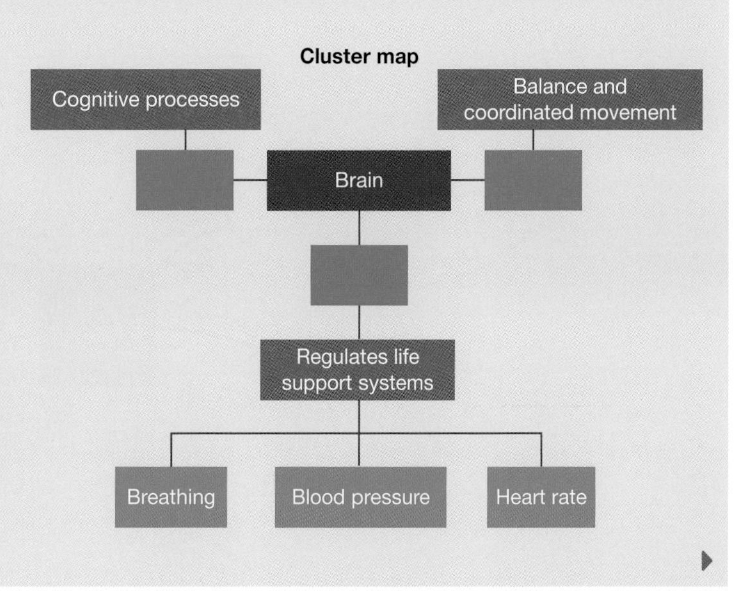

to record brain activity of volunteers while they listened to their favourite piece of music. The PET scan detected a release of dopamine (a neurotransmitter responsible for feeling a sense of reward and pleasure) in their brains and the fMRI scan showed increased blood flow to the emotional response areas.

(a) For this investigation suggest:

(i) a hypothesis

(ii) the dependent variable(s) and independent variable

(iii) an appropriate control group

(iv) controlled variables.

(b) Find out more about similar investigations. Is the media claim supported by your findings? Explain.

10. Find out more about the Nobel Prize winning physicist Richard Feynman, who described seeing equations in colour, and the expressionist artist Wassily Kandinsky, who associated musical tones with specific colours.
11. Recently a technique called diffusion tensor imaging (DTI) has been used to compare the connectivity between the brains of grapheme–colour synaesthetes and non-synaesthetes. Find out more about this research and the findings.
12. Work out whether you are left- or right-brain dominant:
    (a) Give a mark out of 5 for each of the statements shown in the diagrams for each hemisphere of the brain.
    (b) Add up the total score for each side. In which hemisphere of the brain did you score higher?
    (c) What does this mean in terms of your learning?
13. Find out about different hypotheses regarding synaesthesia and the types of research that scientists are currently involved in. On the basis of your findings, what hypothesis would you suggest?
14. Investigate why damage to the right side of the brain often affects the left side of the body.
15. What is meningitis and how does it affect the brain?

Investigate, think, create and design

16. Design an instruction manual to help someone learn a new physical skill. Be creative and make it fun and exciting. Evaluate the effectiveness of your manual by trying it out on other students in the class.
17. Design an activity that uses the cerebellum to learn a more about the brain.
18. Nobel Prize winner Roger Sperry described the hemispheres of the brain as 'each with its own memory' and 'competing for control'.
    (a) Find out why Roger Sperry was awarded the Nobel Prize in medicine in 1981.
    (b) Do you agree with his comments about the brain's hemispheres? Explain.
    (c) Construct a model of the right and left hemispheres that creatively shows the types of tasks that they are involved in.
19. Construct a labelled model of the brain using food, coloured plasticine or other materials.
20. The cerebrum is made up of four primary areas called lobes. Each of these lobes is associated with particular functions. Find out which of the following terms is associated with each lobe and then construct a completed mind map. Terms to use include words, pictures, explicit memory, hearing, sensory, motor, spatial, decision, planning, working memory, vision, colours, movement and higher-order thinking.

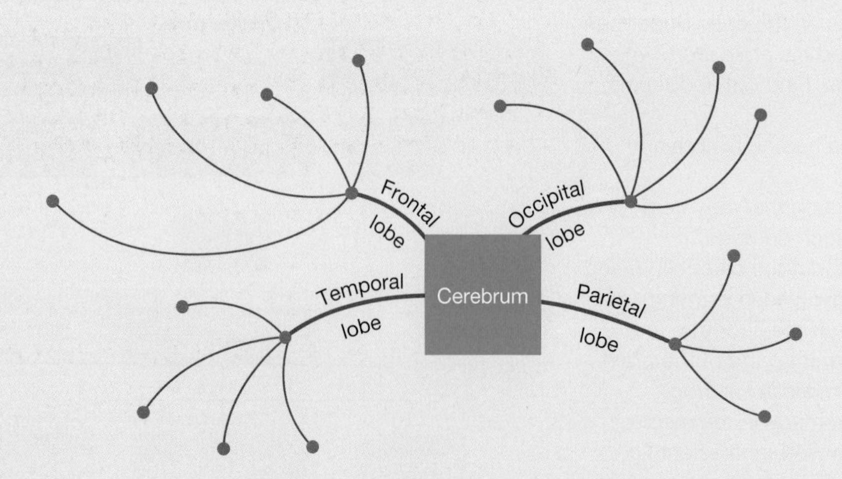

Analyse and discuss

21. The table at right shows the brain size as a percentage of body mass for a number of animals. Use this data, your own knowledge (and other resources if required) to answer the following questions.
    (a) Construct a graph using the information in the table.
    (b) Comment on any trends or patterns in the graph.
    (c) Suggest and discuss two possible explanations for the observed patterns.
    (d) Suggest a relevant hypothesis that could be investigated.
    (e) Formulate three questions about how the data was collected (method used).
    (f) Suggest three relevant questions that could be further investigated.

Animal	Brain size as % of body mass
Mouse	10
Chimp	0.8
Elephant	0.1
Dolphin	1
Cat	1
Human	2

2.6 Endocrine system — slow control

2.6.1 Helpful hormones

Thirsty? Too hot or too cold? Feeling different or noticing changes in how you look, feel or act? Chemicals in your blood not only help to keep you balanced, but are also very important in controlling and coordinating your growth and development.

Your nervous system is not the only means of controlling and coordinating activities in your body. Your endocrine system uses chemical messengers called **hormones**. They are produced in your **endocrine glands** and are released directly into your bloodstream. Although hormones are carried to all parts of your body, only particular cells have receptors for particular hormones. It is a little like radio signals, which are sent out in all directions but picked up only by radios attuned to a particular signal. These **target cells** are attuned to the hormones carried through your body and respond in a particular way.

Some of the main glands of the endocrine system

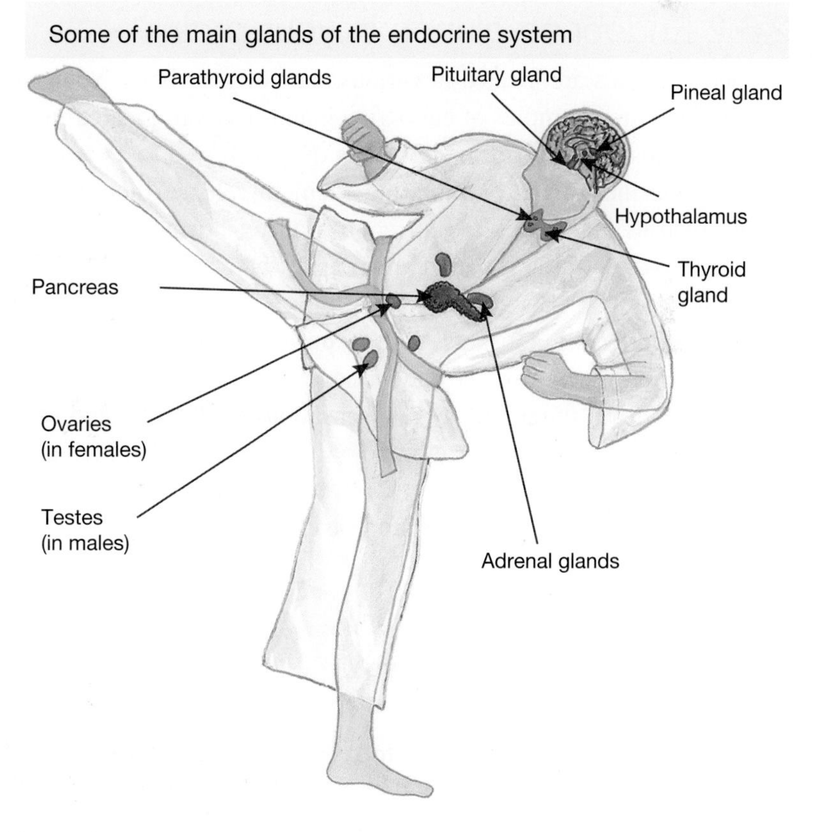

Hormones control and regulate functions such as metabolism, growth, development and sexual reproduction. Like the nervous system, the endocrine system detects a change in a variable, and often acts using a negative feedback mechanism to counteract the initial change. The endocrine system also works with the nervous system to regulate your body's responses to stress. The effects of the endocrine system are usually slower and generally longer lasting than those of the nervous system.

Examples of endocrine glands and their hormones

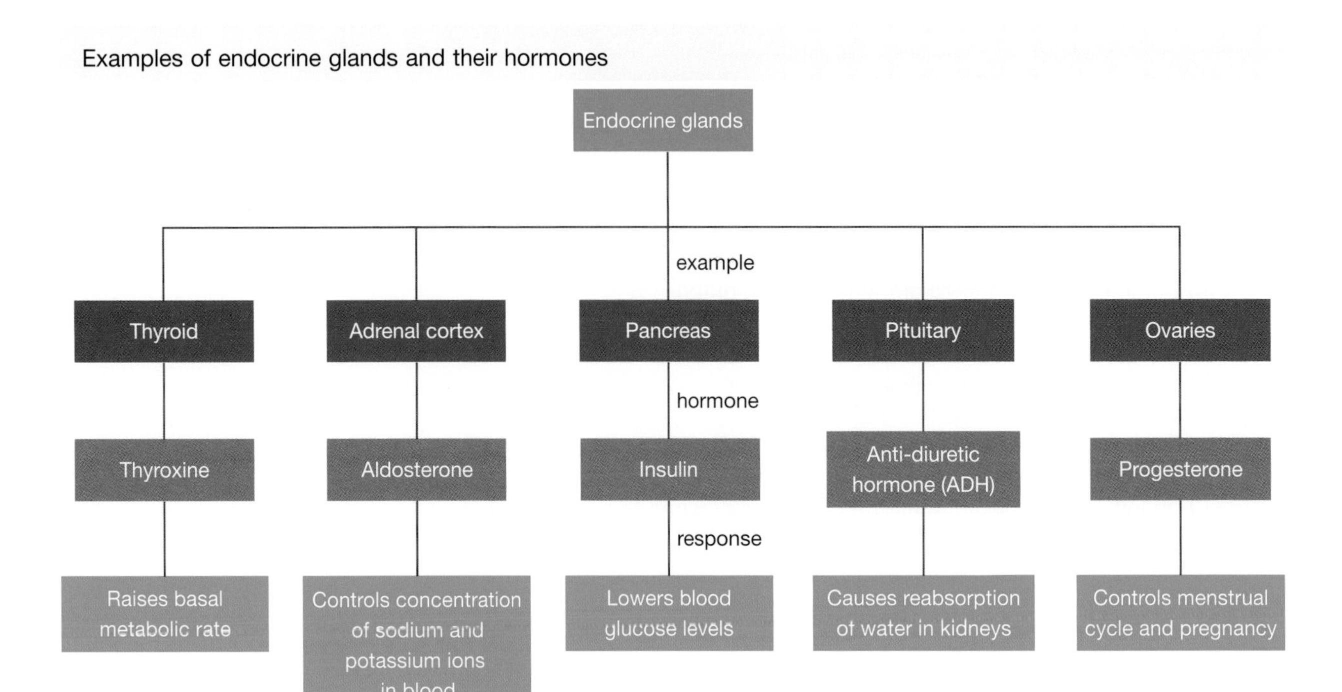

2.6.2 Endocrine glands in your brain

Endocrine glands are located in various parts of your body. Your pituitary gland, the hypothalamus and the pineal gland are examples of endocrine glands that are located in your brain.

I'm the boss! The **pituitary gland** is often referred to as your 'master gland' because it controls many other endocrine glands, stimulating them to release their own hormones. For example, your thyroid gland, ovaries and testes are all controlled by hormones released by this endocrine gland. Hormones released by the pituitary gland can control water balance, growth, development and reproduction-related processes.

Feeling hungry or too hot? Your **hypothalamus** sends hormones to the pituitary gland to control its release of hormones to other endocrine glands. It also releases hormones that control body temperature, growth, sex drive, thirst, hunger and sensations of pleasure and pain. The hypothalamus links your nervous system to your endocrine system and is involved in reflex actions such as those involved in the beating of your heart and breathing.

Feeling sleepy? Your **pineal gland** produces the hormone melatonin which controls body rhythms such as waking and sleeping.

2.6.3 Keeping balance

Keeping warm

Negative feedback helps our body to keep its internal conditions stable so that you can function effectively. An example of this is if your body temperature is too low. The decrease in body temperature acts as the stimulus, which is detected by thermoreceptors in your body. This message is taken to the hypothalamus, which activates warming mechanisms. One of these mechanisms involves the thyroid gland. It responds by secreting the hormone thyroxine, which increases the metabolic rate of cells, releasing heat to warm you. Raising body temperature reduces the need for the hypothalamus to direct the thyroid gland to secrete thyroxine. Regulation of body temperature is referred to as **thermoregulation**. This process shown on the next page (and in more detail on page 44) provides an example of how the endocrine and nervous systems both work together to keep your body functioning effectively.

Sweet control

The regulation of blood glucose levels involves your endocrine system and negative feedback. After you have eaten a lot of sugary food, your blood glucose levels increase. This rise is detected by cells in your **pancreas**, which then secretes the hormone **insulin.** Insulin travels in the bloodstream and specific target cells in your liver and muscles respond by increasing the uptake of glucose into the cells and the conversion of **glucose** into **glycogen**, which is then stored. The result is that blood glucose levels return to their 'normal' levels (see page 42).

If a decrease in blood glucose levels is detected, the pancreas secretes the hormone **glucagon**. This hormone also travels in the blood to the liver and muscle cells, but in this case the response is that glycogen is broken down into glucose. Glucose is released into the blood, increasing blood glucose levels back to their 'normal' level (see figure below).

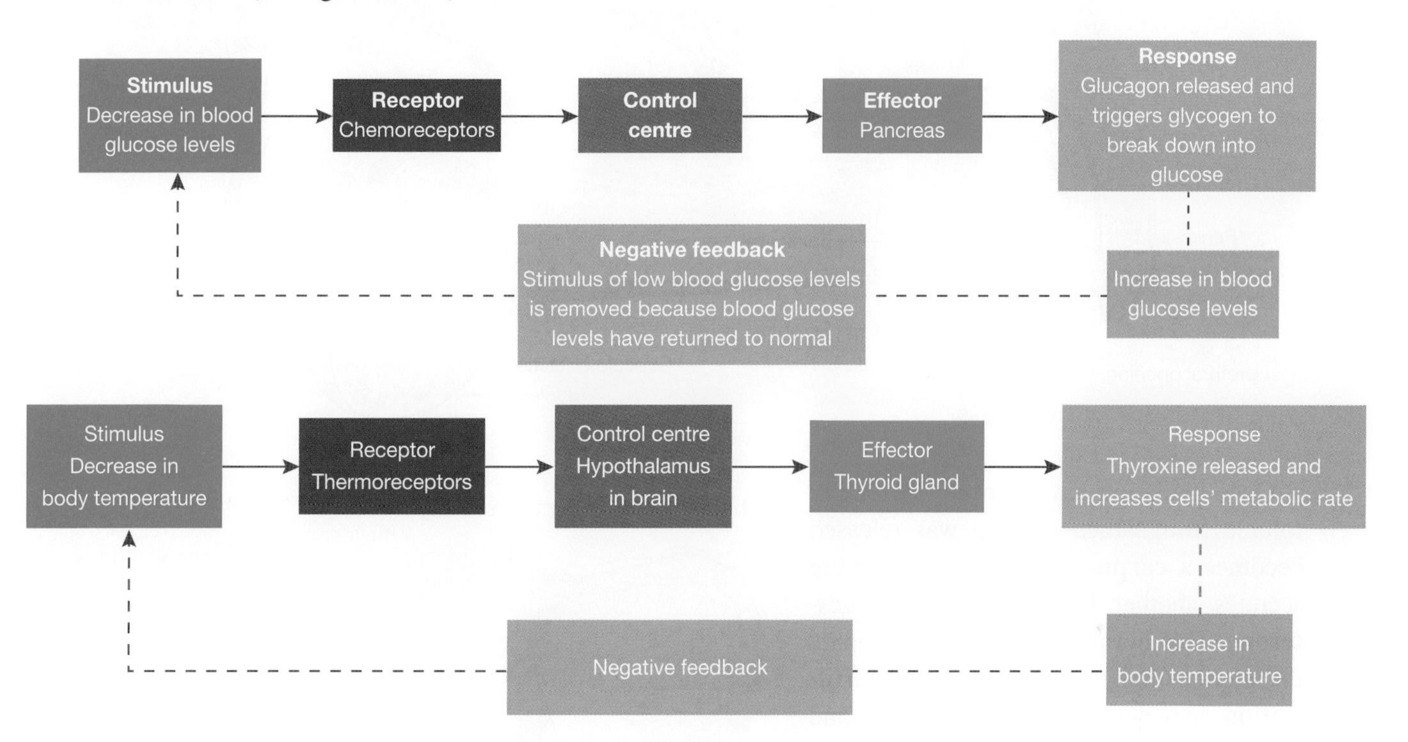

Reproductive control?

The endocrine system also plays a key role in controlling and coordinating human reproduction and development.

His hormones ...

When a male has reached puberty, his **pituitary gland** secretes **luteinising hormone** (or **LH**). LH acts on his **testes** to produce another hormone called **testosterone**. An increase in testosterone levels causes his sex organs to grow and testes to begin to produce **sperm**. Other secondary sex characteristics are increased muscle development, changes in his voice, muscle and hair growth and hormones.

Her hormones ...

When a female has reached puberty, her **pituitary gland** secretes **follicle-stimulating hormone** (or **FSH**). FSH then acts on her **ovaries**, and **follicles** (structure in which the egg develops) begin to grow. A hormone called **oestrogen** is secreted by the growing follicles, which causes the thickening of the lining of the **uterus** to prepare it for a potential fertilised egg. Increased levels of oestrogen also stimulate the **hypothalamus** to produce more FSH and LH. Increasing levels of LH cause the follicle to swell. The mature follicle bulges on the surface of the ovary, ruptures, and the **ovum** (unfertilised egg cell) is released from the **ovary** into the **fallopian tube**. This process is called **ovulation**.

(a) The male reproductive system
(b) The internal structure of the testes
(c) An increase in the level of testosterone during puberty triggers the testes to produce sperm cells.

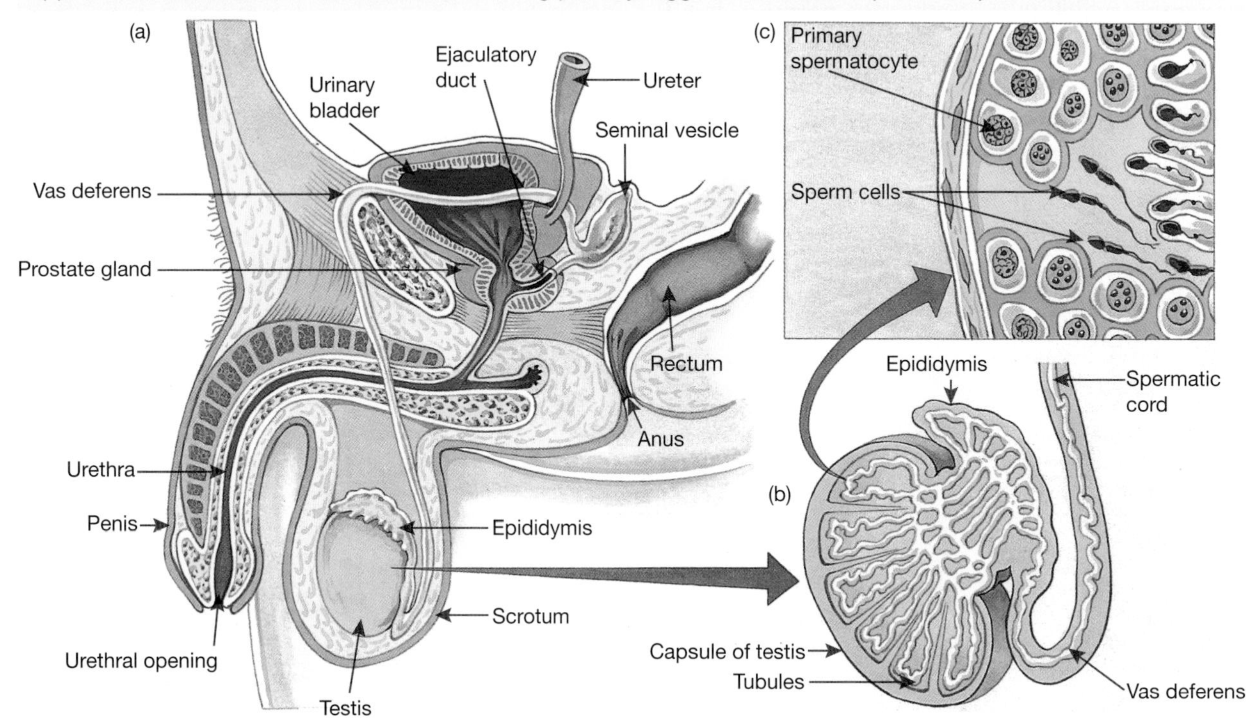

Once ovulation has occurred, the empty follicle from which the egg was released becomes a **corpus luteum**. This structure secretes another hormone called **progesterone**. This hormone continues to prepare the uterine lining for pregnancy. If **fertilisation** does not occur, both the ovum and corpus luteum break down. This causes the progesterone levels to drop and hence the lining of the uterus (endometrium) to break down. Blood and uterine lining are discharged through the vagina in a process called **menstruation**. When progesterone levels drop, the pituitary gland produces FSH and the cycle begins again. These cyclic changes in the ovaries and lining of the uterus as a result of changing hormone levels in the blood are called the **menstrual cycle**.

The female reproductive system

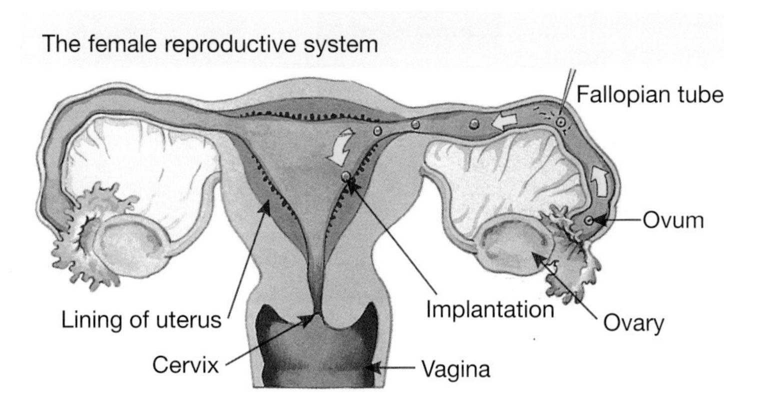

2.6.4 Harnessing hormones

Hormones can be harnessed to either increase or decrease fertility. In some situations, hormones can even be involved in aborting embryos. There are a number of issues that have been expressed about the production, availability, uses and consequences of these hormones.

2.6.5 Harnessing hormones for her

A commercially produced hormone RU486 (Mifeprex), also known as the abortion pill, is one such example. RU486 not only offers possibilities of contraception, but it also can terminate a pregnancy by blocking the action of progesterone. This causes the lining of the uterus to break down so that the embryo is unable to implant into it. This pill is less invasive and has fewer side effects than a surgical abortion and it enables termination at a much earlier stage. The possibility of using this pill as an abortion option, however, has resulted in a division of opinions as to whether it should be made widely available in Australia. There have been some reports suggesting that it is being over-used in other countries.

While other hormone-based contraceptives are increasingly available, they are no longer seen only in a pill form. They are now appearing in patches, gels, implants and insertable vaginal rings. There is also research on the development of a 'morning-before' pill. This pill works by altering the ion content of the woman's reproductive tract for about 36 hours. The changes that it produces make it more difficult for the sperm to swim and hence less likely for them to reach the ovum to fertilise it.

There are also plans to develop contraceptive drugs that target hormone receptors rather than altering hormone levels. These new contraceptives may work by tricking the egg into thinking that it is already fertilised so that it blocks sperm from penetrating it. Other new contraceptives may involve the development of hormones that prevent the fertilised egg from implanting in the uterus.

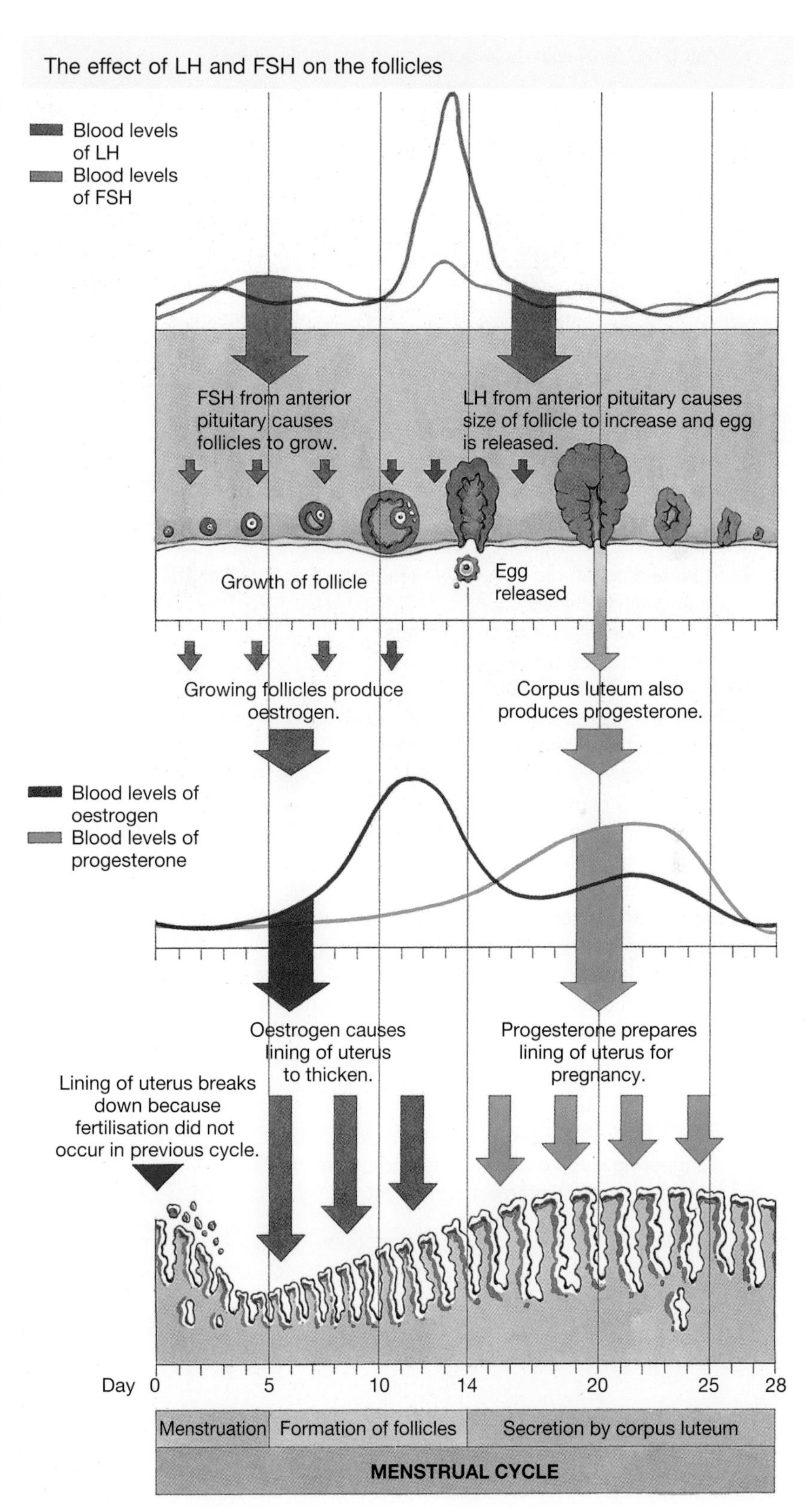

2.6.6 Harnessing hormones for him

Scientists are working on developing male contraceptive pills. These are based on combinations of androgen and progesterone. Androgen blocks sperm development and progesterone blocks testosterone production. While combinations of these hormones may be used to prevent fertility, there are possible side effects that need to be considered.

2.6 Exercises: Understanding and inquiring

To answer questions online and to receive **immediate feedback** and **sample responses** for every question, go to your learnON title at www.jacplus.com.au. *Note:* Question numbers may vary slightly.

Remember

1. Match the term with its description in the table below.

Term	Function
Anti-diuretic hormone (ADH)	Increases blood glucose levels
Glucagon	Lowers blood glucose levels
Insulin	Increases metabolic rate of cells
Oestrogen	Causes testes to produce sperm
Progesterone	Controls menstrual cycle and pregnancy
Testosterone	Causes thickening of the uterine lining
Thyroxine	Causes reabsorption of water in kidneys

2. Identify three endocrine glands located in your brain.
3. What are hormones and where are they produced?
4. Are all parts of the body affected by a particular hormone? Explain.
5. Describe how hormones are transported throughout the body.
6. Outline why the pituitary gland is often referred to as your 'master gland'.
7. Provide an example of negative feedback that includes the involvement of a hormone.
8. Distinguish between:
   (a) hormones and endocrine glands
   (b) menstruation and ovulation
   (c) endometrium and uterus
   (d) testes and sperm.
9. Describe the relationship between the:
   (a) pancreas, liver, glucose, glucagon, glycogen and insulin
   (b) pituitary gland, LH, testes, testosterone and sperm
   (c) pituitary gland, FSH, ovary, oestrogen, follicles, uterine lining, hypothalamus, LH, ovum, fallopian tube and ovulation
   (d) corpus luteum, uterine lining, progesterone and menstruation.
10. Explain why adrenaline is referred to as the 'fight or flight' hormone.
11. (a) What are other names for RU486?
    (b) Why do people use RU486 and how does it work?
12. Other than pills, in which forms can hormone-based contraceptives be used?
13. (a) Name the two hormones that may be used in a male contraceptive pill.
    (b) Outline how these hormones can be used to prevent fertility.

Think and discuss

14. Suggest how you could link the endocrine system terms in the flowchart at right.

Glucagon	Insulin	Pancreas
Glycogen	Glucose	

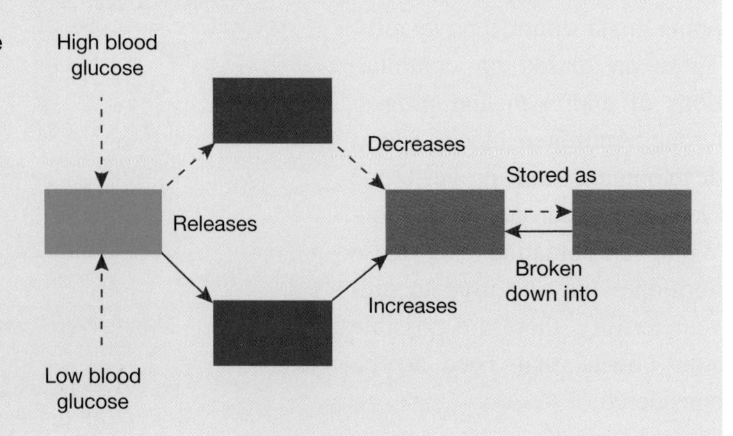

15. Use the information provided in this subtopic to make up your own summary mind map on the endocrine system.
16. Suggest some advantages and disadvantages of the effects of adrenaline in modern-day living.
17. How might hormone replacement therapy help reduce the effects of menopause in women?
18. What three things do the endocrine system and the nervous system have in common?

Create

19. Use the question 14 diagram and the diagrams to help you write and act out a play about how blood glucose is controlled in your body.

Investigate, discuss and present

20. Discover more about the hormones used to increase milk or food production (for example, lactation in cows and goats or growth in cows, sheep or chickens). Gather information on the advantages and disadvantages of these hormones. Use the information in a class debate entitled: *Hormones should be used to increase food production for humans.*
21. Find out and report on hormones that could be used to the advantage of humans. Present your information in an advertising brochure.
22. A synthetic chemical called pyrethrin is increasingly being used in sheep dip. It breaks down within a few days, but during that time it can kill many types of invertebrates in the waterways.
    (a) Why are sheep dipped?
    (b) How could sheep dip reach waterways?
    (c) Suggest implications for the deaths of invertebrates on other organisms.
23. (a) Find out about the history, development and side effects of RU486.
    (b) Search media resources and the internet for arguments for and against the availability and use of RU486.
    (c) Share and discuss your information with others in your team.
    (d) Reflect on your findings and discussions and then state your opinion on the availability and use of RU486. Give reasons for your opinion.
    (e) As a class, be involved in a debate on RU486.
24. Find out more about research on male contraceptives. Prepare a newspaper article or brochure outlining your findings.
25. Male and female fertility patterns are different. Find out the key differences and comment on how they may affect the development and use of effective hormone-based contraceptives.
26. Find out about other ways in which your body temperature is regulated.
27. How are hormones involved in the balance of water in your body?
28. Find out about the effects of having deficiencies in any of the hormones listed in the table at the beginning of this subtopic.
29. Investigate the statement: *Too much adrenaline can cause stress-related diseases*.

Analysing data

30. Refer to the diagram on page 73 and the figure on the next page to answer the following questions.
    (a) Which hormone in the graph is at the highest level just prior to ovulation?
    (b) When is ovulation likely to occur?
    (c) When is progesterone at its highest levels?
    (d) At what stage in the cycle is the endometrium the thickest?
    (e) Describe the changes in the concentrations of each of the hormones throughout the menstrual cycle.
    (f) Research the changes in the levels of FSH (follicle stimulating hormone) and LH (luteinising hormone) throughout the menstrual cycle.

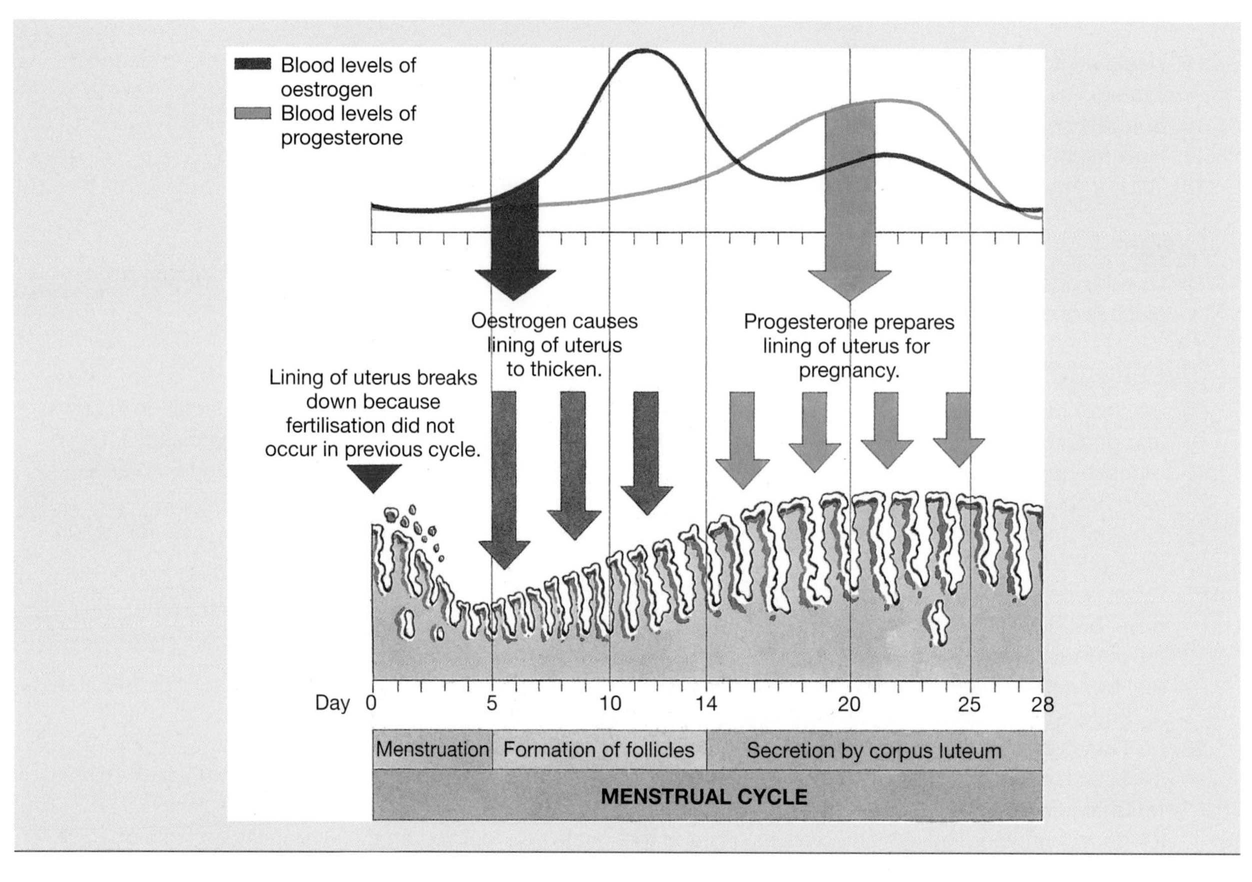

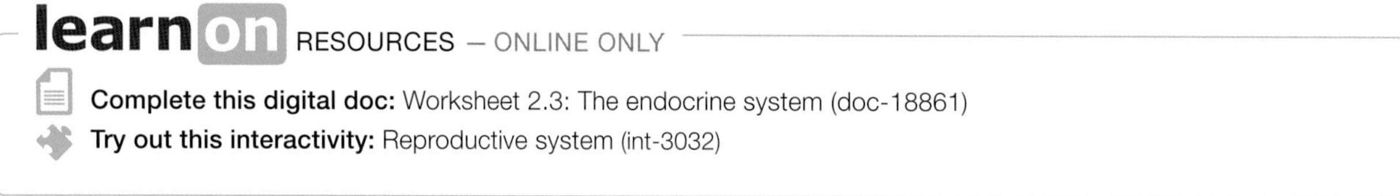

2.7 Plant hormones at work

2.7.1 Plant moves

Have you ever watched a plant move? They do, and most gracefully too. If you watch plants over a length of time, or using time-lapse photography, you can see how they move with the sun and the moon. These movements may be choreographed by hormones or by their internal biological clocks.

2.7.2 Not nervous

Plants do not have a nervous system. Instead, the way plants grow, develop and respond to their internal and external environments is usually coordinated by plant hormones. Plant hormones are mainly produced in the growing tips of roots and shoots, in buds and in developing fruits. Only very small quantities are needed for an effect. The same hormone may produce different responses in different parts of the plant.

Plant hormones are generally divided into five main types: **auxins**, **cytokinins**, **gibberellins**, **abscisic acid** and **ethylene**. In some cases, a number of these hormones work together to produce a response. The illustration on the next page shows where these plant hormones are produced and what their effects are.

2.7.3 Bent on light

When light shines on the tips of a plant's shoots, **auxins** are produced and released. These auxins travel down the plant and cause the cells on the side opposite the light to grow longer than those on the other side. This causes the stem to bend towards the light.

Hormone journeys

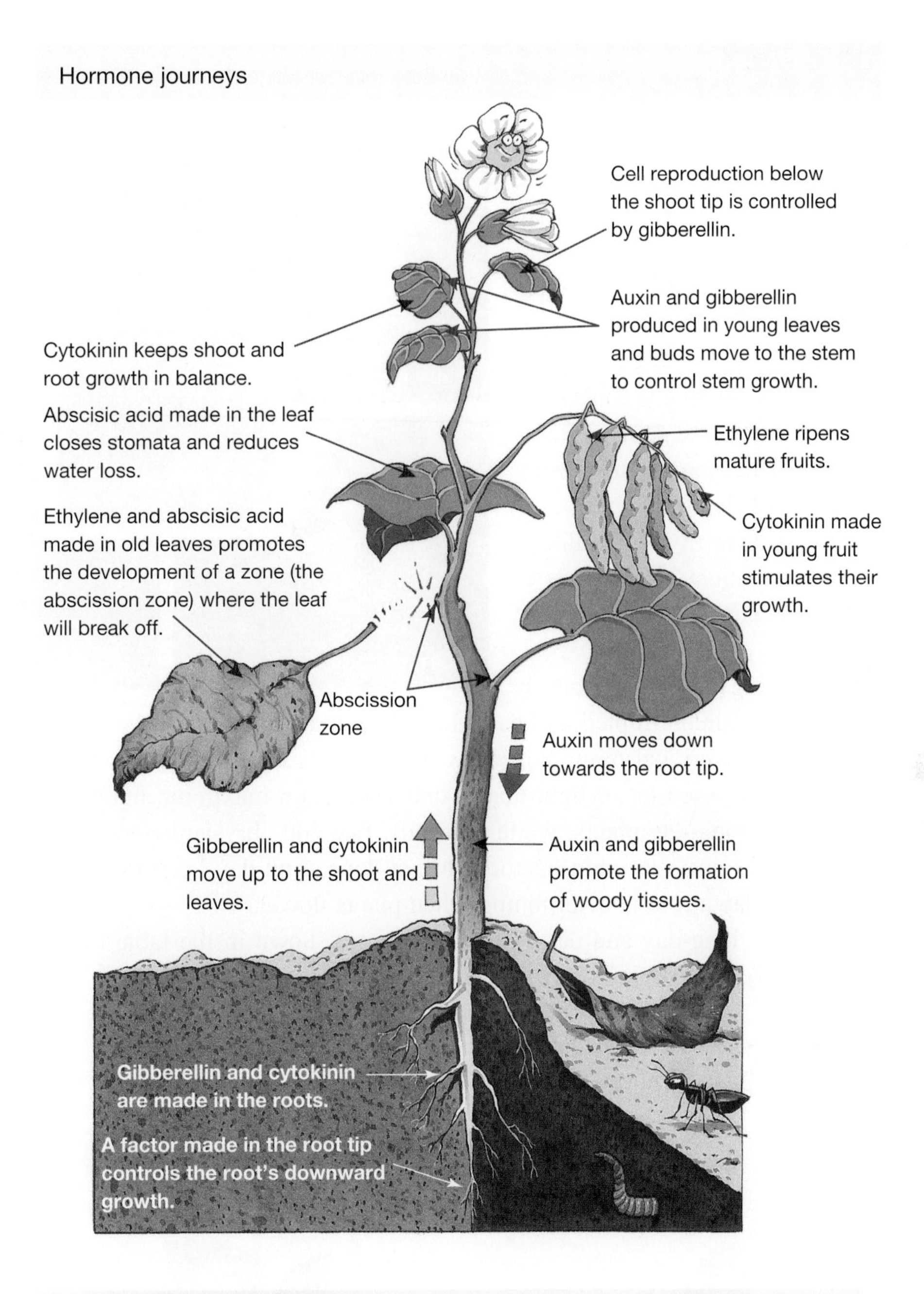

Auxins cause plants to grow towards the light.

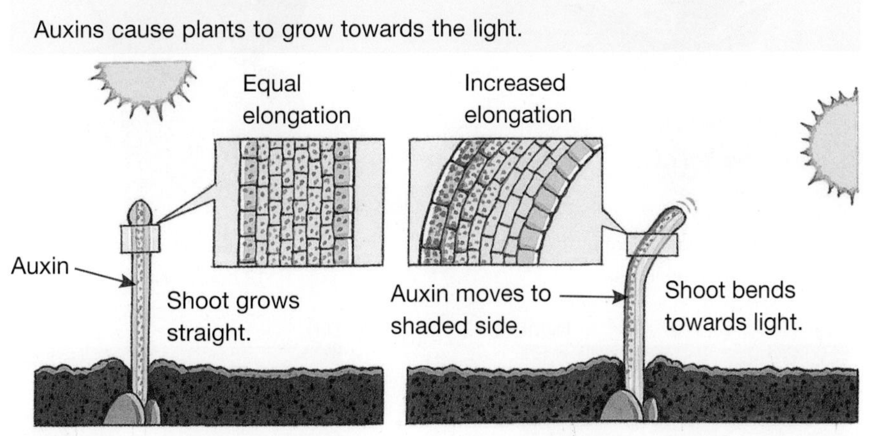

2.7.4 Is it time yet?

Like us, plants have internal biological clocks and may have different patterns of movement in a 24-hour cycle. The leaves of some plants, for example, may be horizontal during the day and then drop into a 'sleeping' position at night. If you were to place these plants in 24 hours of daylight or darkness, they would continue their 'sleeping' movements because the sleep pattern of these plants is internally controlled. Such a 24-hour cycle is referred to as a **circadian rhythm** or cycle. The opening and closing of flowers is another example of a plant's activities that involves a circadian rhythm. Unlike a plant's movement towards light, these kinds of movements are independent of the direction of the stimulus.

Getting into circadian rhythm: a flower performs its daily dance. ***Source:*** © Biozone International Ltd/Richard Allan.

The timing of flowering of many plants is controlled by the length of uninterrupted darkness. Long-day plants flower only when the number of daylight hours is over a certain critical minimum (or when darkness is less than a critical value). Short-day plants flower only when exposed to daylight that is under a certain maximum number of hours. Gladioli, cabbage and hibiscus are long-day plants, while daffodils, rice and chrysanthemums are short-day plants. Day-neutral plants, such as potatoes and tomatoes, do not depend on day length to flower. Hormones also are currently thought to play a role in determining when plants flower.

Examples of short-day, long-day and day-neutral plants are shown in the table on the next page. How would you tend to classify plants that flower in spring?

Different plants flower in response to different day and night lengths.
(a) The poinsettia (*Euphorbia pulcherrima*) is a short-day flower and flowers only when the day length becomes less than 12.5 hours.
(b) The hibiscus (*Hibiscus spp.*) is a long-day plant and flowers only when the day length becomes greater than 12 hours.

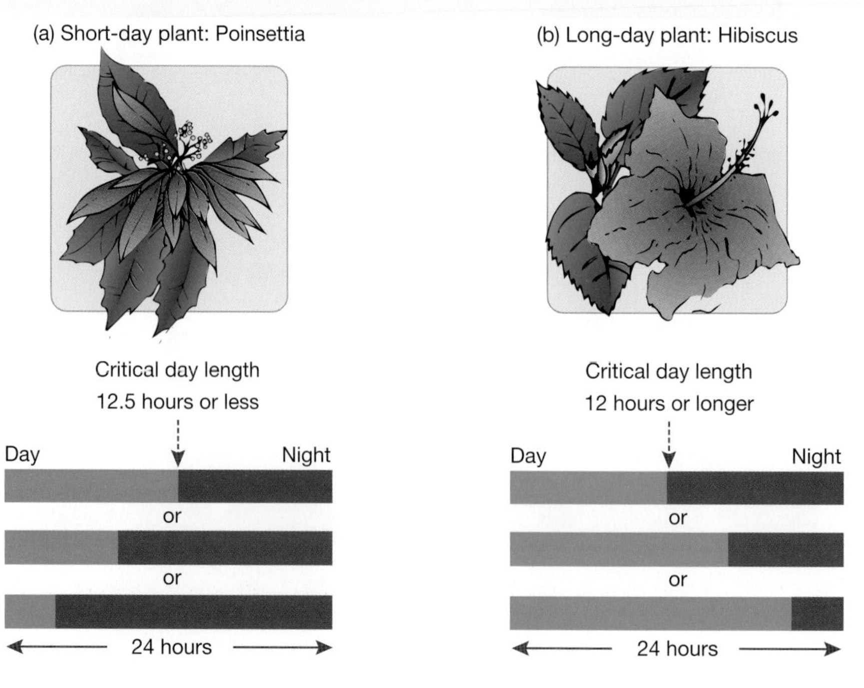

Some examples of short-day, long-day and day-neutral plants

Short-day	Long-day	Day-neutral
Monocotyledons *Rice* (*Oryza sativa*)	Wheat (*Triticum aestivum*)	Corn (*Zea mays*)
Dicotyledons *Chrysanthemum* (*Chrysanthemum spp.*)	Cabbage (*Brassica oleracea*)	Potato (*Solanum tuberosum*)
Poinsettia (*Euphorbia pulcherrima*)	Hibiscus (*Hibiscus syriacus*)	Rhododendron (*Rhododendron* spp.)
Violet (*Viola papilionacea*)	Spinach (*Spinacea oleracea*)	Tomato (*Lycopersicon esculentum*)

2.7.5 Harvesting hormones

Observations made in the days of gas street lamps caused scientists to think about the trees downwind of the lamps that shed their leaves. Experiments led to the discovery that the ethylene gas used in the lamps was responsible. Further research showed that ethylene was a plant hormone that promotes a plant's ability to shed leaves.

Such investigations have increased our level of understanding and allowed us to put some hormones to work. The mind map below shows some of the many uses of hormones.

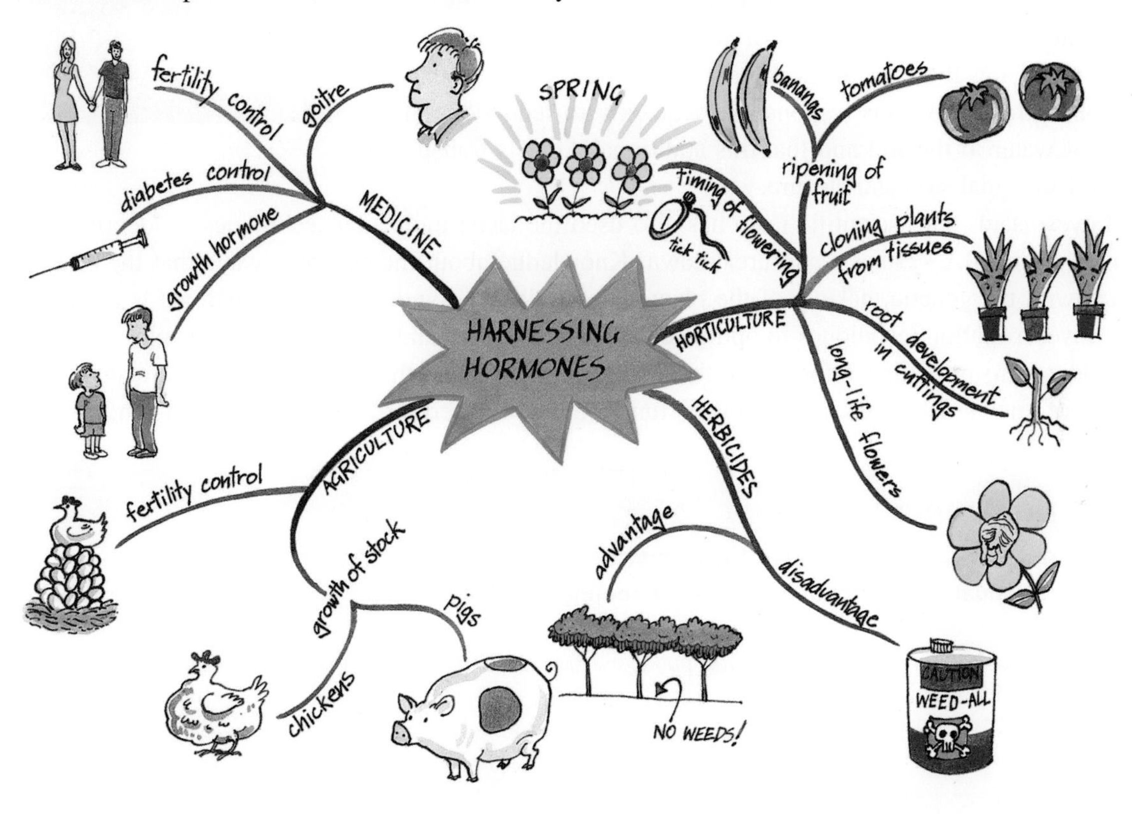

DID YOU KNOW?

Weed killers

Some selective weed killers which use plant hormones work on particular plants but not others. A type of weed called a dandelion is a common problem in lawns. It produces a familiar yellow flower which turns into a large puffball containing small seeds attached to a feathery umbrella-shaped structure, which helps the plant to disperse its seeds on the wind. Lawns containing the dandelions can be sprayed with selective weed killer without killing the grass. The selective weed killer contains a growth hormone that causes the weeds to grow too quickly. The weed killer is absorbed by the weeds in larger quantities than grass.

2.7.6 Linking ABA to saving our precious water resources

While virtually all of Australia's horticultural crops are grown using some form of irrigation to supplement rainfall, crops grown under cover are entirely dependent on irrigation. Water for irrigation is becoming an increasingly scarce and expensive commodity. It is therefore vital that we understand the ways that plants use this water so that we can optimise its use and improve economic returns to farmers.

Nearly all of the water used by plants passes through pores on the leaf surface called stomata. This causes plants a dilemma. How can they keep their stomata open to obtain carbon dioxide for photosynthesis, while at the same time restrict excessive water loss through them? The answer lies in a surprisingly complex set of control mechanisms.

Dr Brian Loveys investigated the role of the plant hormone abscisic acid in the regulation of water use in grapevines.

Right at the heart of the control mechanisms is the plant hormone abscisic acid (ABA). ABA induces stomatal closure and if it is not present plants very quickly die from excessive water loss. Understanding how plants control the amounts of ABA in roots and leaves has therefore been a research priority for scientists over the last few decades.

It has been necessary to develop sensitive methods for measuring ABA because, like most hormones, it is present only at very low concentrations. Using these methods we have been able to show that the ABA concentration in roots responds very quickly to reductions in the amount of water in the soil and that this additional ABA is transported to the leaf to signal stomatal closure.

Dr Loveys said, 'Our scientific team has also used the latest molecular techniques to identify the genes responsible for ABA synthesis and breakdown. Knowledge about the complex ways that the environment interacts with the genetic makeup of the plant to control ABA synthesis and breakdown is allowing us to devise novel irrigation techniques to optimise these mechanisms and improve the efficiency of water use.

'In addition to providing information that is useful to farmers, the research has significantly increased our body of knowledge about how all plants function, and furthermore, has been a lot of fun.'

INVESTIGATION 2.7

The effect of a commercial rooting powder on cutting development

AIM: To investigate the effect of a commercial rooting powder on the development of plant cuttings

Materials:

10 cuttings (daisies and geraniums work well, but so will lavender and rosemary — they'll just take a little longer)
2 × 12-centimetre flowerpots or tubs
a micro test tube of rooting powder
potting mix
2 labels

Method and results

1. Construct a table to summarise your data.
   - Trim the lower leaves off all ten cuttings.
   - Fill the flowerpots with soil and label one *Control* and the other *Test*.
   - Plant five cuttings in the control pot and water them.
   - Dip the other five cuttings in the rooting powder, plant them in the test pot and water them.
   - Place them in a warm position and keep the pots watered equally.
2. After one week, dig one cutting up from each pot and compare their root growth.
3. Continue doing this for the next four weeks, adding observations to your table as you go.

Discuss and explain

4. Sketch the features of plants from each pot.
5. Identify the pot in which the plants grew the best.
6. Identify which pot contained the plants with the most developed roots.
7. Suggest the plant hormone that you think may be present in the rooting powder.
8. Discuss reasons horticulturalists use rooting powders.
9. Suggest improvements to the design of this experiment.
10. Suggest a hypothesis that could be investigated using similar equipment. (You may use internet research to identify relevant problems to investigate.)
11. Design an investigation to test your hypothesis. Include an explanation for your choice and treatment of variables.

INVESTIGATION 2.8

Plant responses to hormones

AIM: To investigate the effects of plant hormones

Materials:
2 pieces of holly
a small apple
plasticine
labels
2 small bottles of water
2 × 1 L beakers or large jars
tray or board

Method and results

- Select two pieces of holly that are similar (size, age, number of leaves).
- Set up the experiment as shown in the diagram at right and leave for 24 hours. The plasticine is used to cover the pouring lip of the beaker. Label each beaker with your name and date.

1. Observe and record the holly in the two beakers every 24 hours until there are some obvious differences between them.

Discuss and explain

2. What differences did you observe in the two beakers over the period of the experiment?
3. What do you think might have caused these differences?
4. Why was it important to select twigs of holly that were similar?
5. What do you think is responsible for the different responses?
6. Construct a hypothesis that relates to this investigation.
7. Identify strengths and limitations of the investigation and suggest ways in which it could be improved.
8. Suggest a research question about plant hormones that could investigated.
9. Design an investigation to further explore your research question.

2.7 Exercises: Understanding and inquiring

To answer questions online and to receive **immediate feedback** and **sample responses** for every question, go to your learnON title at www.jacplus.com.au. *Note:* Question numbers may vary slightly.

Remember

1. If plants do not have a nervous system, what coordinates their life cycle?
2. List the five types of plant hormones.
3. List the main locations in plants where hormones are produced.
4. Describe how hormones allow a plant to grow towards light.
5. What is a circadian rhythm? Give an example.
6. Which two factors are thought to control the timing of flowering?
7. Give examples of:
   (a) long-day plants
   (b) short-day plants
   (c) day-neutral plants.
8. Identify the relationship between street gas lamps and the discovery of ethylene.
9. State examples of ways in which hormones could be used for:
   (a) horticulture
   (b) agriculture
   (c) medicine.

Investigate, think and discuss

10. Which types of plants do you think would flower in winter and summer?
11. By knowing the effects of plant hormones, horticulturists are able to control the timing of the flowering of plants and the ripening of fruits. Why do you think they do this?
12. Why would gardeners put bulbs of some kinds of plants, such as tulips, in a refrigerator for some weeks before planting?

Investigate, discuss and present

13. Find out more about the Dutch biologist Friedrich Went (1863–1935) and his research that led to the isolation of the auxin that causes plants to bend towards light.
14. Find out more about one of the following types of hormone: auxins, ethylene, abscisic acid or gibberellins. Summarise your information in a poem, poster or newspaper article.
15. Find out which group of plant hormones is responsible for the carpets of colour created by trees losing their leaves during autumn.
16. Find out more about the effects of ethylene on plants. Present your findings in a creative format.
17. The presence of chemical wastes in water supplies and our environment has caused some concern. Some of these chemical wastes contain hormones or chemicals that interfere with hormones. Find out and report on two examples of these.
18. It takes just one bad apple to spoil the whole bagful. Investigate the validity of this statement.
19. The herbicide Agent Orange was used in the Vietnam War. Find out and report on (a) why it was used, (b) how it worked, and (c) issues surrounding its use.
20. Find out more about Australian research that involves plant hormones. In your report identify the hormone and its function and the reason for the research.

2.8 Keeping emotions under control?

Science as a human endeavour

2.8.1 Just survive!

Feeling happy, sad, scared, disgusted or angry? Did you know that these five emotions are caused by the effects of chemicals binding to receptor sites on your cells?

Imagine a situation in which you have felt threatened. How did you feel? How did you react? Did you want to run, or did you want to stay and fight?

Your emotions enable you to react to situations. They influence your behaviour. Our ancestors relied on their emotions to survive. Sometimes there is no time to think about how to react to a situation. This is when your emotional brain can get into the driver's seat and take control.

Your 'emotional brain' or **limbic system** is made up of a collection of structures within your brain. The limbic system is involved in memory, controlling emotions, decision making, motivation and learning. These include parts of your thalamus, hypothalamus, hippocampus and **amygdala**.

2.8.2 Feeling angry?

Feeling angry? Is your heart racing; are your hands cold; do you have a sick feeling in your stomach? Anger can be one of our most primitive emotions. It is certainly a powerful one. Uncontrolled anger can lead to physical fights, arguments and self-harm. Controlled anger, however, can be a very useful emotion that can help motivate you to make positive changes.

When you feel angry, your hypothalamus responds by sending messages to your **pituitary** to instruct your **adrenal glands** to release **adrenaline**. This hormone acts to increase your heart rate, dilate your pupils, constrict skin blood vessels and shut down digestion. This helps you to see any threats better and provides your muscles with more glucose and oxygen just in case you need to face the danger and *fight*, or take *flight* and escape it by running away.

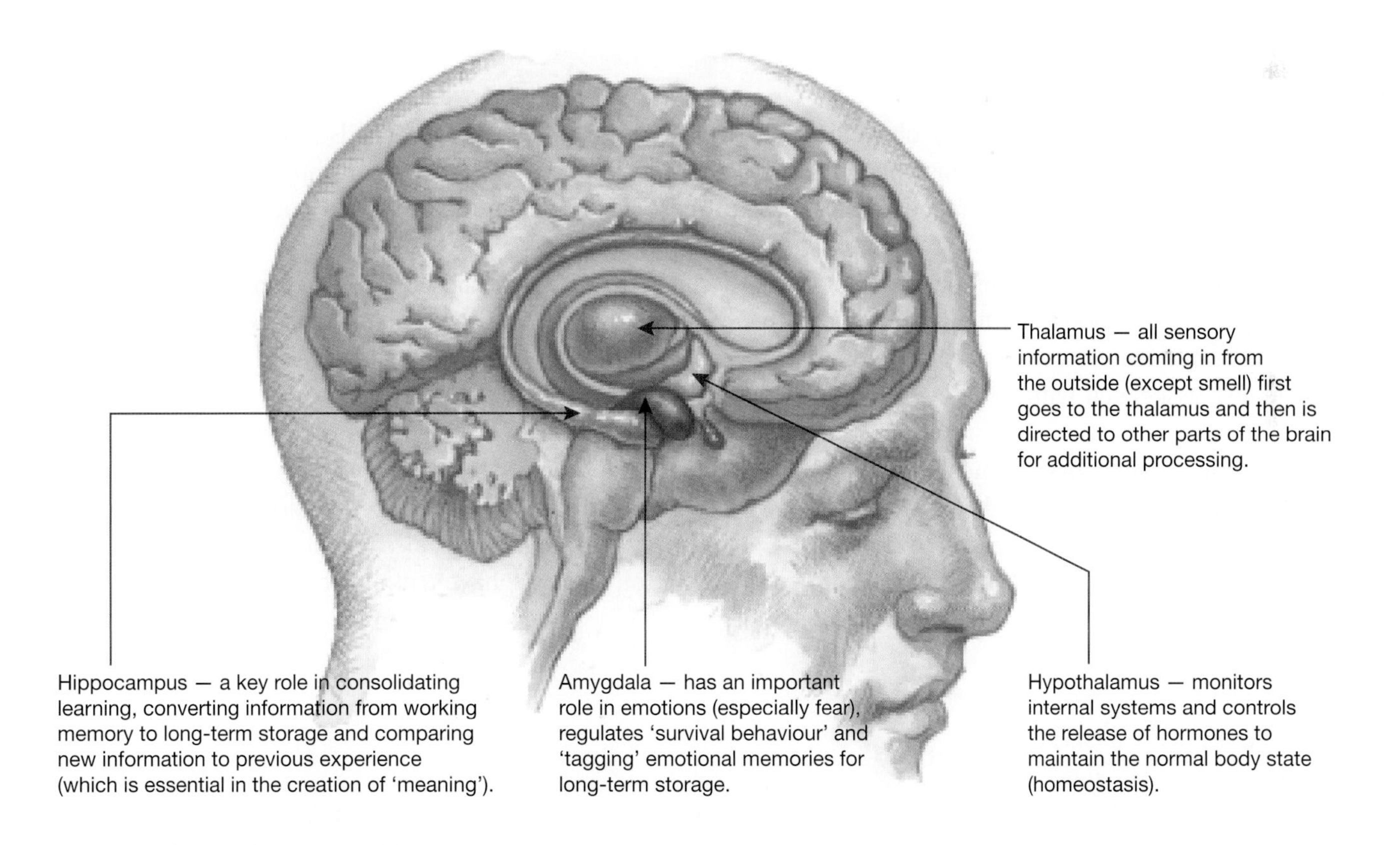

2.8.3 Tagged as a threat

Your 'flight or fight' response actually originates in your amygdala. It is this tiny part of your limbic system (about the size of your thumbnail) that decides the emotional value of what is happening. It asks: 'Does this mean something significant to me?' It may sense a particular facial expression or tone as being threatening, or it may detect an event that was previously 'tagged' as being a negative experience.

2.8.4 Keeping the anger

Staying angry, or long periods of stress, can damage another part of your limbic system called your hippocampus. If the stress or anger lasts more than a few minutes, your adrenal glands also release cortisol. Sustained high levels of this hormone can lead to the death of hippocampus neurons, which may result in diminished learning, spatial recall and memory.

2.8.5 False alarms

Your prefrontal cortex or thinking brain is also involved in assessing a threat and placing it in context. If your thinking brain considers it to be a false alarm, it sends a message to your hypothalamus to trigger actions to calm things down; it does this by sending out messages to decrease your stress hormone levels and their effects.

2.8.6 Mirrored feelings

Feel upset, or feel upset for someone else? **Mirror neurons** are a group of neurons that activate when you perform an action and when you see or hear others performing the same action. Research is suggesting that these neurons are important in being able to feel **empathy** towards other people. If this theory is further supported, how could this connection increase the chances of the survival of our species?

2.8.7 Mood chemistry

Neurotransmitters are chemicals involved in passing messages between your nerve cells (neurons). Within your brain there are many neurotransmitters that influence how you feel and react; **serotonin**, **norepinephrine** and **dopamine** are three examples. Imbalances of these neurotransmitters can contribute to a variety of mental illnesses.

Serotonin acts like the brakes on your emotions. It can produce a calming effect and is important for maintaining a good mood and feelings of contentment. It also plays a role in regulating memory, appetite and body temperature. Low levels of serotonin can produce insomnia, depression and aggressive behaviour and are also associated with obsessive–compulsive and eating disorders.

Norepinephrine can act like the accelerator. It can promote alertness, better focus and concentration. Your brain also needs this chemical to form new memories and to transfer them to your long-term storage.

Dopamine is important for healthy assertiveness and autonomic nervous system function. Dopamine levels can be depleted by stress or poor sleep. Too much alcohol, caffeine and sugar may also lead to reduced dopamine activity in your brain. People with Parkinson's disease have a diminished ability to synthesise dopamine.

HOW ABOUT THAT!

MRI

Early brain research used dead or diseased brains. Advancements in scientific applications of technology have enabled researchers to examine living brains. One such technology is magnetic resonance imaging (MRI) which allows scientists to actually see which parts of the brain are active when various tasks are performed; these parts 'light up' with different colours to show brain activity.

Brain studies using MRI have produced some very interesting findings! These images show the responses of a person in pain (left) and a person watching someone in pain (right).

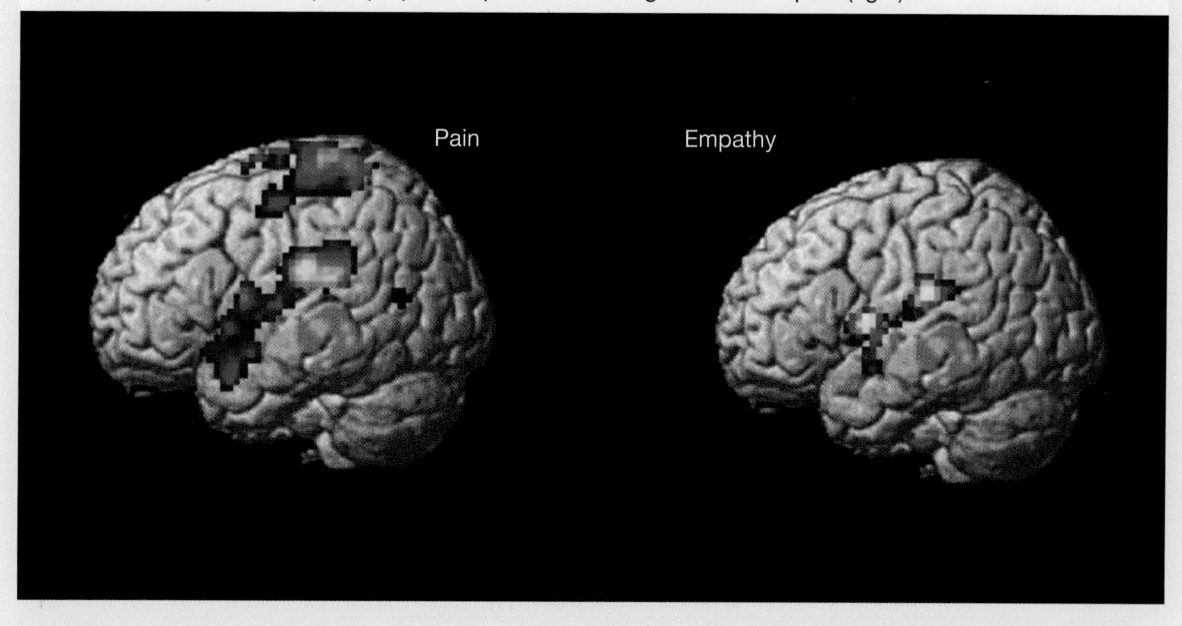

2.8.8 Emotions and learning

Are emotions gatekeepers to your intellect? Are emotions important to your learning too? If emotions are important to your learning, are some emotions better than others? Can some emotions actually interfere with your learning?

If this is the case your learning can be enriched if you are in a safe, caring and inviting climate for learning. If you were to describe your ideal learning environment, what would it look like, feel like and sound like?

Feeling safe and taking risks

In a safe and caring environment, learners can learn by trial and error, ask questions and feel safe enough to risk making mistakes or getting something wrong. When the learner experiences stress or feels threatened, survival instincts can take over. Chemicals are released that place their body in a heightened alert phase, to help prepare them for a possible dangerous situation. If a learner is in this stressed state it is difficult to use higher-order thinking, and it can be difficult to learn effectively.

While your hippocampus has an important role in forming long-lasting memories, your amygdala can act as a memory filter, labelling information to be remembered by tying it to events or emotions that are experienced at the time.

When you are experiencing a time of stress, your survival instincts can take over. You produce chemicals that place your body in a heightened alert phase, to help prepare you for a possible dangerous situation. When you are in a stressed state it is difficult to use your higher-order thinking and you may find it difficult to learn effectively.

Neurotransmitters are chemicals that carry messages between neurons.

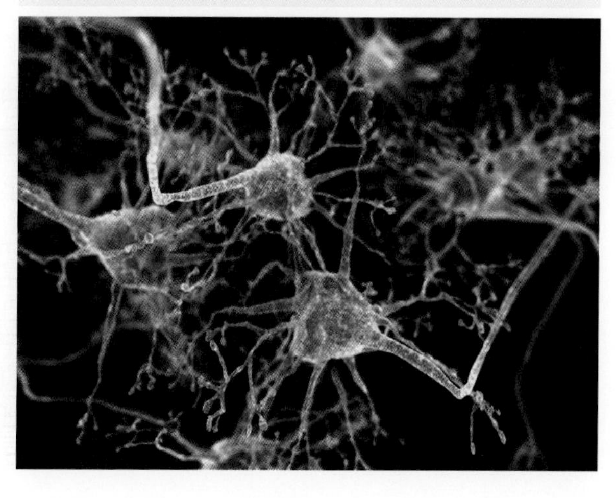

Not all challenges and stresses are bad for learning. When the brain is faced with a challenging, intricate and complex problem, all of its parts can be involved and attention, meaning and relevance for learning can result.

2.8 Exercises: Understanding and inquiring

To answer questions online and to receive **immediate feedback** and **sample responses** for every question, go to your learnON title at www.jacplus.com.au. *Note:* Question numbers may vary slightly.

Remember

1. Name six basic human emotions.
2. Match the term with the most appropriate description in the table below.

Term	Description
Amygdala	Collection of structures within your brain involved in memory, controlling emotions, decision-making, motivation and learning
Hippocampus	Organ involved in sensory memory and 'attention'
Limbic system	Involved in converting information from working memory to long-term storage and helping to make 'meaning' of new information by comparing it to previous experience
Thalamus	Plays a key role in emotions and regulates 'survival behaviour' and 'tagging' emotional memories for long-term storage

3. Describe the function of your amygdala.
4. Describe the relationship between your pituitary, your adrenal glands and adrenaline.
5. Outline the relationship between prolonged stress or anger, cortisol and learning.
6. Describe how your hypothalamus can be involved in controlling stress.
7. Explain how the release of adrenaline can increase your chances of survival.
8. (a) Identify three neurotransmitters in your brain that can influence how you feel and react.
   (b) Describe the effects that each of these neurotransmitters can have on your behaviour.

Think and discuss

9. Suggest how mirror neurons may increase the survival of our species.
10. Respond to each of the questions asked in 'Emotions and learning' section:
    (a) Are emotions gatekeepers to your intellect?
    (b) Are emotions important to your learning?
    (c) If emotions are important to your learning, are some emotions better than others?
    (d) Can some emotions actually interfere with your learning?
11. What if no-one ever got angry? Would this be a good thing? Imagine what the world would be like. Construct a PMI chart about your imagined world.
12. (a) List some examples of angry behaviour that you have seen.
    (a) Suggest ways in which this angry behaviour could have been managed.
13. Discuss appropriate ways of managing behaviour. Which of these appeal to you? Why?
14. (a) If you were angry with one of your team members or classmates, suggest appropriate ways of managing your anger.
    (b) With your team, agree on a set of rules or strategies that could be used to manage anger or conflicts if they occur.
15. (a) Suggest questions to find out viewpoints, perspectives and opinions of others.
    (b) With your team or class, discuss strategies that could be used to deal with situations when viewpoints differ.
16. If anger is one of our most primitive emotions, it must have some survival advantages. Discuss with your team what these advantages might be. Present your findings in a visual tool.

Investigate, think and discuss

17. (a) Research the effects of at least three different human hormones, such as testosterone, adrenaline, cortisol and oestrogen, and then report your findings back to your team.

(b) Use this information and your own opinions to discuss the following question: *Do our hormones determine who we are and what we do, or can we have some conscious control over this?*
(c) In your team, decide on a brief statement that summarises the opinion.
(d) How strongly do you agree with your group's opinion? Rate your response on a scale of 0 to 5, with 0 meaning 'Strongly disagree' and 5 'Strongly agree.' Give reasons for your response.
(e) Survey your class or do a class spectrogram to determine how many of, or the degree to which, your class members agree with this statement.
(f) Find out and record differing opinions of as many of your class as you can.
(g) Have you changed your initial opinion or has it stayed the same? Explain.

18. Have you seen a young child throw a tantrum? This is a case of not being able to control emotions. Although the child's amygdala is fully mature, the necessary links with the cortex are not yet fully developed. Find out more about these links between different parts of the brain and their effects on behaviour. How could you explain this to the parent of a toddler?
19. Find examples of music that helps relax you and calm you down when you are feeling stressed. Share your music with others to see if it has the same effect on them.
20. Some convicted murderers may have killed in a 'fit of rage'. Find out if there are any documented links between committing murder and frontal lobe activity in the brain.
21. Find out about the connections between brain neurotransmitters, behaviour and the following medications: Prozac, Zoloft, Topamax, Provigil and Abilify. Report your findings to the class.
22. In 1947, the Swedish biologist Ulf von Euler discovered norepinephrine and later won a Nobel prize for his research. Find out more about research into this neurotransmitter and how it may be involved in helping you to learn.
23. *Our emotions are our personalities*. Do you agree with this statement? Discuss your opinion with others in your team. Present a summary of your discussion to the class.
24. Select one of the following statements, then find out what information you need to know in order to make a decision as to whether the statement is correct or incorrect.
    - Males need competition so that they feel stimulated and know their place in the hierarchy, whereas females first do things to be liked and, if that doesn't work, then use a 'victim strategy'.
    - Boys are more interested in objects, and girls in human relations.
    - In order for boys to achieve at school, they need to compete and struggle through the class hierarchy.
    - Male thinking is more competition-driven whereas female thinking is more security-driven.
    - Males collect facts whereas females are more interested in the relationship between the facts.
25. Can fears or phobias be unlearned? Find out more about research involving chemicals such as glutamate to achieve this.
26. Investigate and report on problems associated with extreme emotions.
27. Select one of the statements below and use both your own experience and that of others expressed in the media to discuss it from different perspectives.
    - To an extent, emotions can justify our actions.
    - Emotion has its own language.
    - Emotion is more powerful than reason.

Create and construct

28. (a) On your own, in a pair or in a team, write a story about anger management.
    (b) Present your story to the class as a puppet play, picture storybook or song.

learnon RESOURCES — ONLINE ONLY

Complete this digital doc: Worksheet 2.4: Body continuum (doc-18862)

Complete this digital doc: Worksheet 2.5: The brain (doc-18863)

2.9 Total recall?

Science as a human endeavour

2.9.1 What is memory?

While learning is about gaining new knowledge, memory is about retaining and then retrieving that knowledge. To achieve this you need to coordinate your thinking.

If a friend gave you her phone number, how long could you remember it without writing it down? While learning is about gaining new knowledge, memory is about retaining and then retrieving that learned information. That is, for us to remember something, we have to be able to *record* the experience and *store* it in an appropriate part of the brain. If we are unable to *retrieve* or pull out that information, we have forgotten it.

2.9.2 Building memories

Modelling memory?

Scientists construct models to communicate ideas. Models can be concrete (for example, a plastic model of the brain) or symbolic (for example, a map or diagram). Models provide the opportunity for learners to bring previous knowledge into their working memory. This helps learners to attach meaning to, and make sense of, their new learning.

A model can be used to represent various stages in how information is processed by your brain. Shown bottom right is a simplified version of an information-processing model. While this model has many limitations, it provides a framework that can be used to help you attach previous knowledge to new learning about the stages of memory.

2.9.3 Filtering

What was that? You use your senses (for example, sight and hearing) to detect various stimuli in your environment. Incoming information is filtered through a system called the **sensory register** (shown in the model as venetian blinds). This system filters incoming information on the basis of its importance to you. Your sensory register involves your **thalamus** (a part of the limbic system of your brain) and a portion of your brain stem called the reticular activating system (RAS). The more important the information is to your survival, the higher the chance that it will get through for further processing in your brain.

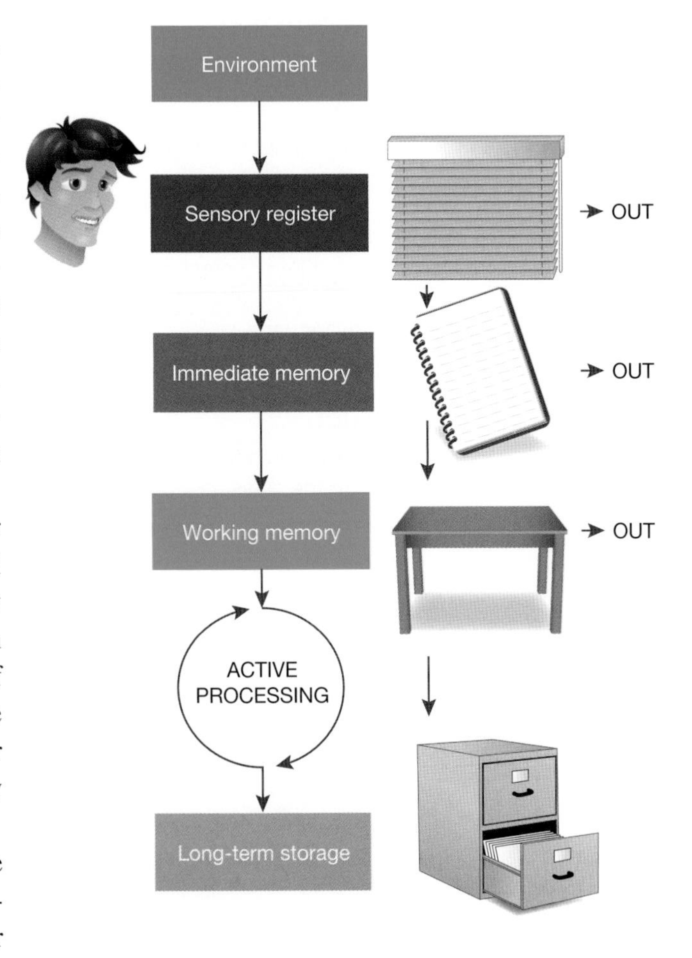

Even if information has made it through your sensory register, it doesn't mean that you will remember it. You will remember information only if you have stored it in long-term storage (shown in the model as a filing cabinet). It is the job of your **hippocampus** to encode it and send it to one or more of the long-term storage areas in your brain. This encoding takes time and is usually done during deep sleep.

Memories are not stored as a whole or in one place. When you retrieve and reconstruct memories, storage areas distributed throughout your brain are activated. While long-term storage can be thought of as where your memories are stored in your brain, your long-term memory relates to the dynamic process of sorting and retrieving the information.

2.9.4 Synaptic patterns

Memory is not stored in just one place in your brain. It is currently thought that memories exist as patterns of connections at the synapses between the brain's neurons. To store a particular memory, nerve signals travel along a specific pathway through certain synapses. Each time this memory is remembered, nerve signals are reactivated to again travel along this pathway.

2.9.5 Just visiting

Before any information is stored in long-term storage, it needs to pass through your temporary short-term memory. Examples of short-term memory include immediate memory (shown as a notepad in the model) and working memory (shown as a table in the model).

2.9.6 'Notepad' memory

Information that has made it past your thalamus moves to your immediate memory where a decision is made about what to do with it. Your past experience helps to determine its importance. An example of the length of time information will stay in this type of memory is when you temporarily remember a phone number and ring it. After this time the information may be lost or, if considered important enough, moved to your working memory.

2.9.7 'Working table' memory

It is within your working memory that information generally captures your focus and demands attention. There is a limited capacity (amount of information dealt with) and time limit for this type of memory. Research suggests that this capacity changes with age. Between the ages of 5 and 14 years there is a range of about 3–5 'chunks' that can be dealt with at one time; after this age it increases to about 5–7 chunks. This limited capacity is one of the reasons why you need to memorise songs, poems or other information in stages. By memorising a few lines at a time and repeating them (or rehearsing them) you are able to increase the number of items in your working memory. This is an example of chunking.

Studies have suggested that the time limit in working memory is about 10–20 minutes. This is often the amount of time you can spend on one activity. This time, however, can be influenced by interest and motivation. Both of these can have emotional elements and also involve a special part of your brain called the **amygdala**.

2.9.8 Remember to learn

Your past experiences influence new learning. What you already know acts as a filter to help you focus on things that have meaning and ignore those that don't. Your self-concept (how you see yourself in the world) is also shaped by your past experiences. It is your self-concept that often determines how much attention you will give to new information.

You can transfer things from your short-term memory into your long-term memory by rehearsing information (practising) and applying meaning to it. The two key questions asked in the decision of whether to move information into long-term memory are:

- does it make sense?
- does it have meaning?

I don't understand! This is the type of comment made when a learner is having trouble making sense of new learning. Determining whether new information 'makes sense' is related to whether the new information fits in with what you already know.

Why do I have to know this? Whether the new information 'has meaning' relates to whether it is relevant to you and whether you consider that the purpose of remembering it is worthwhile. You can improve the chance that you remember something by making connections between the new learning and your previous knowledge.

If both sense and meaning are present, there is a very high chance that the information will be sent to long-term storage.

Is meaning present?		Is sense present? No	Is sense present? Yes
	Yes	Moderate to high	Very high
	No	Very low	Moderate to high

2.9.9 Unlocking memory doors

There are keys that you can use to unlock your memory doors. Seven of these are primacy, recency, repetition, standing out, association, chunking and visuals.

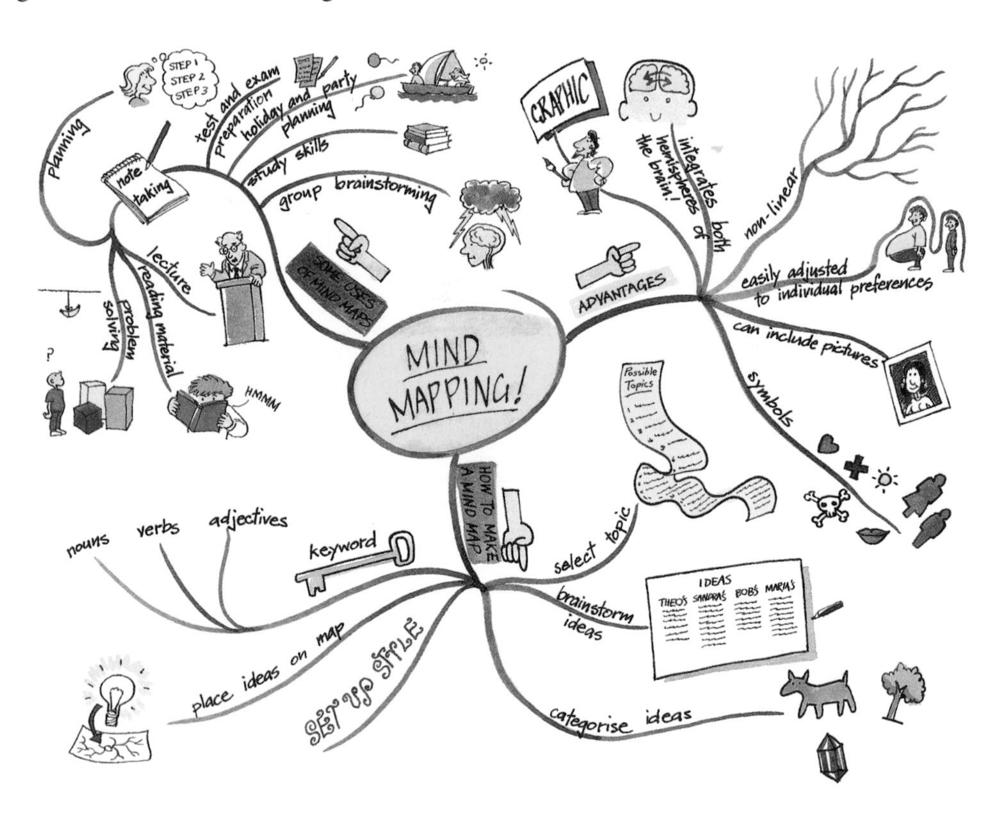

2.9.10 Primacy and recency

When you read a book or see a movie you will usually remember the beginning and the ending. **Primacy** is about recalling and remembering the first time that you do something. **Recency** is the opposite. It is remembering the last time or the ending.

2.9.11 Repetition

Repetition, or regularly reviewing information, is needed to reactivate your stored memory and prevent it from being buried under layers of other information. Research suggests that you can achieve about 90 per cent recall if you review content within 24 hours. This drops to 30 per cent if you review after 72 hours (3 days). Repetition can be achieved visually, by reading, playing games with the new information, highlighting or using visual thinking tools.

2.9.12 Standing out

Think about a lesson that you remember well. What made it more memorable than other lessons? Was it fun? Was there something different or new about the experience? Did you use mnemonics or analogies? A **mnemonic** is a technique that helps you remember something. This may involve telling a tale (using key terms within a story), linking (linking terms and images) or using acronyms (using the first letters of words; for example, SPEWS). Some of these ideas are very effective because they overlap with other memory keys. All of these things can help content stand out and make it easier for you to remember it.

2.9.13 Association

If new knowledge is linked to previous knowledge your recall is greatly enhanced. This is called learning by **association**. It helps you to anchor the information in time and space. Using real-life examples or metaphors can assist in this, as can the use of smell, music and colour.

2.9.14 Chunking

How do you eat a whole elephant? The answer of course is 'a bit at a time'. Learning is similar. You don't have to learn it all at once. The short-term memory of teenagers can usually contain only five (plus or minus two) bits of information at once. By organising information into small **chunks**, it is easier to remember it.

2.9.15 Visuals

Reading text in colour can help you to use both sides of your brain. The same can be said for a dramatic acting out; for example, performing the story of how blood flows through your body.

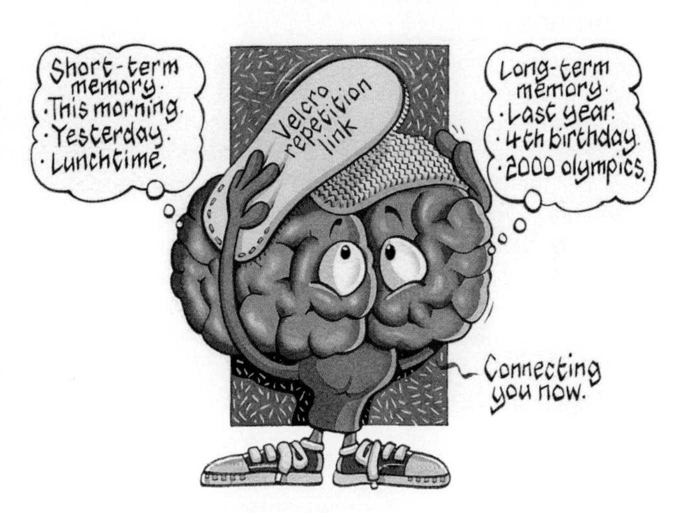

2.9.16 Memory neurotransmitter

A key neurotransmitter involved in learning and memory formation is **acetylcholine**. This neurotransmitter is released in the brain during learning. Acetylcholine is involved in the strengthening of connections between neurons in the brain and hence in the formation of new memories. Consequently, drugs that boost the amount of acetylcholine release are used as an effective treatment for diseases such as Alzheimer's, that impair cognition.

2.9.17 Memory blockers

Scientists are working on drugs to improve or even erase memory. Drugs that can enhance learning are being sought as an easy way to do well in tests and exams. However, there are disadvantages and advantages to drugs designed to block memories.

Current research includes studies on drugs that specifically block or erase problem memories at the molecular level. While this can be a great advantage to those who suffer post-traumatic stress disorder (PTSD), there are concerns that other memories could also be erased.

Researchers are exploring the possibility of using chemicals called beta-blockers, cortisol and hydrocortisone to alter our memory processes. Beta-blockers can bind to the receptors on the cell surface that would usually bind to adrenaline and noradrenaline. By blocking these hormones, beta-blockers may stop the hormones' stressful effects and prevent deep memory formation.

While all this research is exciting and innovative, what are the ethical considerations? Who controls which memories are to be erased and when? What do bad memories have to do with our consciences and our perceptions of right and wrong? Will there be global rules and regulations? If so, who will write them and make sure that they are maintained?

2.9.18 Stressful memories down deep

Your hippocampus and amygdala are also involved in emotional responses to an experience or memory. When your sense organs pick up a stimulus it goes to your thalamus and is then dispatched to your amygdala to assess its emotional quality. If it is recognised as potentially threatening, it triggers your body to release adrenaline and noradrenaline to set you up for fight or flight. The hippocampus then processes the memory and imprints it deeper than it would other memories. This will allow you to be primed quickly for action if it occurs again.

In this way, memories of traumatic or highly emotional events are burned into your brain more deeply and are remembered for longer. While in evolutionary terms this may have increased our chances of survival, traumatic events can result in PTSD.

2.9 Exercises: Understanding and inquiring

To answer questions online and to receive **immediate feedback** and **sample responses** for every question, go to your learnON title at www.jacplus.com.au. *Note:* Question numbers may vary slightly.

Remember

1. How can you transfer short-term memory into long-term memory?
2. Construct a mind map to summarise the five memory systems.
3. Sketch seven keys. On each key, describe a memory key strategy.
4. Name the part of the brain that transfers information from short-term memory to long-term memory.
5. List the five different memory systems described in this subtopic and write a brief description of each in your own words.
6. The colour red is directly stored in your long-term memory. List examples of vehicles, signs and symbols that have applied this knowledge. Suggest why your brain processes the colour red in this way.
7. Create and present a rhyme, song or poem about:
   (a) your memory systems
   (b) memory keys.
8. Suggest the advantage of traumatic or emotionally charged events being remembered more deeply.

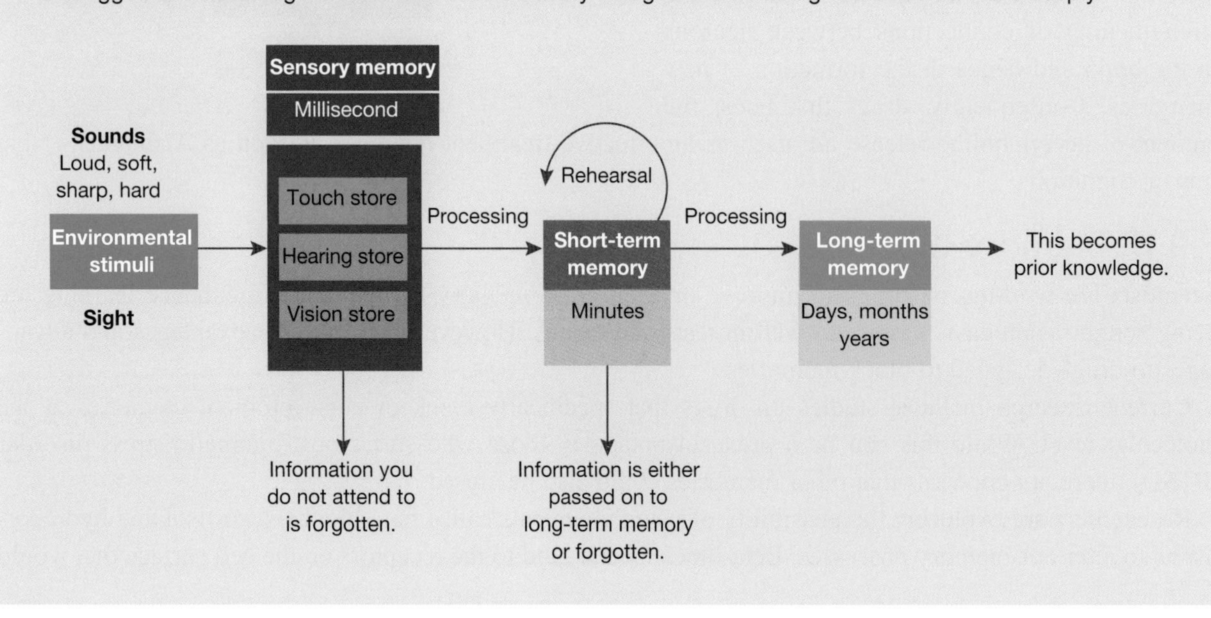

Investigate, think and discuss

9. Carefully observe the information processing memory model on the previous page.
   (a) Identify strengths and limitations of this model in its communication of how memories are formed.
   (b) Suggest improvements to this model.
   (c) Identify the similarities and differences between this model and the one.
   (d) Which model best helps you understand the concepts related to memory? Why?
   (e) Explain why scientists used models to communicate their understanding of our brain's function.
10. On the basis of what you know about how the brain functions, suggest why one of the 'memory door' strategies helps you to remember.
11. Get a pencil and paper and then concentrate on the number below for 7 seconds. After 7 seconds, look away and write the number down. Did you get it right? Compare with others in your team.
    5167340
    Now repeat the procedure again with the number below. Did you get it right? Compare with others in your team. Did you get the same results for each number? Discuss your results with your team. Suggest a reason for any differences between results.
    3847918362
12. Although we all use the same senses to collect information from our environment, they do not contribute equally to our learning. Learners develop preferences for certain senses over others. This is where terms such as 'visual', 'auditory' and 'kinaesthetic' learners originate.
    (a) Research each of these and develop a set of questions that can be used to determine which preferences you and others in your class have.
    (b) Discuss the impact that these differences can have on your learning.
    (c) Suggest how you can use this knowledge to be a more effective learner.
13. Find out the possible effects of the following chemicals on learning: adrenaline, phenylalanine, norepinephrine, calpain and choline.
14. Find out more about memory-enhancing drugs. Construct a PMI chart to summarise and share your findings.
15. Write a newspaper article, cartoon or web page on ways to improve your memory.
16. Find out more about research on memory and chemicals that may be used to enhance or erase it. Organise a class debate on one of the following statements.
    (a) Drugs that have an effect on memory should be illegal.
    (b) Everyone should have access to drugs that erase memories.
    (c) Research on drugs that alter memories should be stopped.
17. The colour red is directly stored in your long-term memory. List examples of vehicles, signs and symbols that have applied this knowledge. Suggest why your brain processes the colour red in this way.

Investigate, create and present

18. Find out more about the information-processing model; then, in teams of 8, discuss how you could act it out. Include the following roles: sensory register, immediate memory, working memory, long-term storage (two people), incoming information (three people).
19. Research the structure and function of the thalamus, amygdala or hippocampus and construct a model to communicate your findings to others.

2.10 Sleep on it

Science as a human endeavour

2.10.1 Sleep on it

Are you a night owl or an early bird? Do you get sleepy during the day or find it hard to wake up in the mornings? Did you know that sleeping is as essential to your health as food and water?

2.10.2 A very old network ...

One of the oldest portions of your brain is your **reticular formation**. This network of fibres and cell bodies is located in the central core of your brain stem (medulla oblongata) and extends through other areas of your brain. It can be considered a network of neurons that opens and closes to increase or decrease the amount of information that flows into and out of your brain. It helps regulate your alertness (from being fully awake or deeply asleep), motivation, movement and some of your reflexes.

2.10.3 What's your rhythm?

Your **circadian rhythm** is the regular pattern of mental and physical changes that happen to you throughout a 24-hour time period. This rhythm may be controlled by your body's biological clock. This clock is really a pair of pin-sized structures made up of about 20 000 neurons called your **suprachiasmatic nucleus (SCN)**, which is located in your hypothalamus, near where your optic nerves cross.

2.10.4 Catch that yawn

Why do you often get drowsy when it is dark and wake up when it is light? The answer lies in your nervous system and levels of chemicals in your brain. **Photoreceptors** in the retina of your eye detect light and create signals that travel along your optic nerve to your SCN. Your SCN then sends signals to a number of different parts of your brain.

In the evening, the signal that light is decreasing travels from your SCN to your **pineal gland**, which then produces a hormone called **melatonin**. Increased levels of melatonin in the evening tell your body that it's time to sleep and you begin to feel drowsy. During adolescence, these levels peak later in the day, which may explain why you get tired later at night and want to sleep in the next morning.

There is also evidence that the accumulation of a chemical called **adenosine** in your blood while you are awake may cause drowsiness. While you sleep, this chemical gradually breaks down.

Your brain develops from back to front!

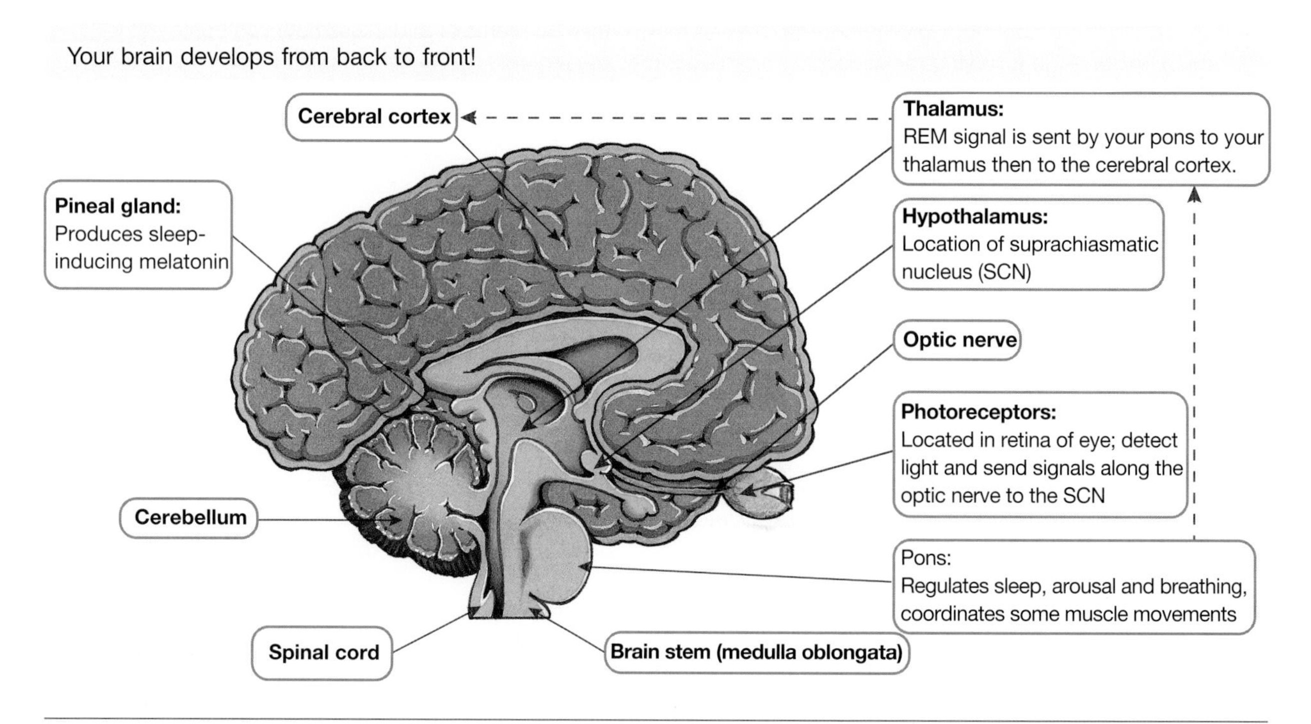

2.10.5 Sleeping switches

Some examples of ways in which you can improve the quality of your sleep

Neurotransmitters can also control whether you are asleep or awake by acting on particular groups of neurons in your brain. The neurotransmitters serotonin and norepinephrine keep some parts of your brain active while you are awake. During sleep, the production of these neurotransmitters is switched off. As these chemicals are involved in logical and consequential thinking, your judgement of time and location can become distorted.

Some foods and medicines can change the balance of your neurotransmitters and affect how alert or drowsy you are and also how well you sleep. Drinks or foods that contain caffeine stimulate some parts of your brain and can cause insomnia (inability to sleep).

Neurons involved in controlling sleep also interact closely with your immune system. Infectious diseases like the flu can make you feel sleepy. This may be because of the powerful sleep-inducing chemicals of our immune system called **cytokines**. Sleep may also help you to conserve energy and other resources that the immune system may need.

2.10.6 Catching sleep waves

During the night, your body experiences sleep cycles lasting 90–110 minutes, with periods of REM (rapid eye movement) and non-REM sleep. You might have three to five sleep cycles each night.

2.10.7 Dropping off

There are four stages of non-REM sleep, and about 75 per cent of your night's sleep is spent in non-REM sleep. Stage one lasts for about 5 per cent of your sleep and is a transition period from wakefulness to sleep. During this stage, your muscles may contract and you may feel 'jumps' or 'twinges' in your legs. In the second stage (45 per cent of an average night's sleep) your brainwaves become larger and eye movements cease. In your third (12 per cent) and fourth (13 per cent) stages of non-REM sleep, your brain will show delta wave activity. You will be in a deep sleep and be difficult to arouse.

2.10.8 Dream time

Your REM sleep is your dream time, and usually makes up about 20–25 per cent of the night's sleep. In REM sleep your breathing becomes more rapid, irregular and shallow and your eyes flick in different directions. Your first REM sleep each night lasts about 70–90 minutes. If you are woken during REM sleep, you can often describe your dreams.

REM sleep is triggered by the pons in your brain. Your pons also shuts off neurons in your spinal cord to temporarily paralyse your limbs so that you don't act out your dreams. The REM sleep signal is sent by your pons to your thalamus, then to the cerebral cortex. As REM sleep stimulates the regions of your brain used in learning, some believe that dreams are the cortex's attempt to interpret and put meaning to new information and experiences.

Heavy smokers may have reduced amounts of REM sleep and sleep lightly. Although alcohol can help you to fall into a light sleep, it also reduces REM and deep restorative stages of sleep.

Your brain emits electrical impulses at different frequencies when it is engaged in different activities.

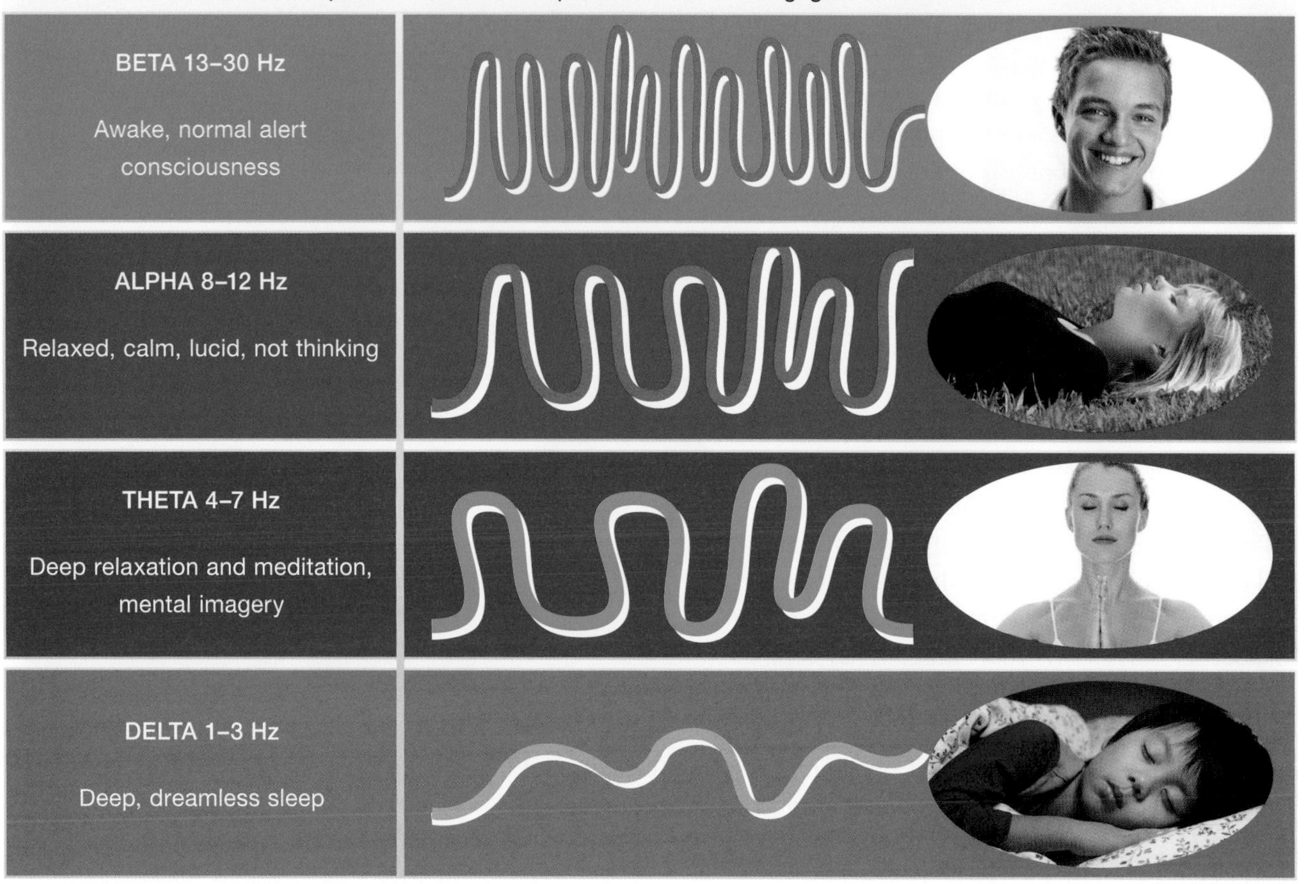

2.10.9 Sleep learning

Recent research has shown that, while you are asleep, your brain consolidates and practises what has been learned during the day. This suggests that learning continues to take place while you sleep. If this is true, it is another reason for getting a good night's sleep before a test or exam, rather than staying up all night studying!

2.10.10 Catching brain waves

Your brain emits waves of electrical impulses at different frequencies when it is engaged in different activities. These frequencies are measured in cycles per second (cps) or Hertz (Hz). Technologies such as an **electroencephalogram (EEG)** can be used to measure the patterns of this electrical activity.

REM sleep is triggered by a structure in the brain called the pons.

Beta (β) waves (13–30 Hz) are the fastest waves with the shortest wavelength. When your brain is emitting beta waves you are using many of your senses and are strongly engaged. An example of this may be if you were involved in an active conversation at a party or playing sport. This type of brainwave is associated with short-term memory, alertness and concentration and is in very high levels if you are anxious about something.

When your brain is emitting **alpha (α) waves** (8–12Hz) it is likely that you are calm and relaxed, but still aware of your environment. If you are involved in solving a problem, reflecting on an experience or creatively visualising something, you may be emitting this type of wave. When your brain is in this state you may be processing information and activating your long-term memory.

When you are in a deep dreamless sleep, your brain will be emitting **delta (δ) waves.**

HOW ABOUT THAT!

An electroencephalograph (EEG) can be used to measure the overall patterns of electrical activity of your brain. When you are asleep, theta and delta wave activity is present. When you are awake, your brain tends to show alpha waves if you are relaxed and beta waves if you are alert.

2.10 Exercises: Understanding and inquiring

To answer questions online and to receive **immediate feedback** and **sample responses** for every question, go to your learnON title at www.jacplus.com.au. *Note:* Question numbers may vary slightly.

Remember

1. What is a circadian rhythm?
2. Where is your suprachiasmatic nucleus (SCN) located and what does it do?
3. How is light involved in whether or not you are sleepy?
4. What effect do increased levels of melatonin have on your body?
5. What effect can the switching off of serotonin and norepinephrine have on you?
6. Suggest why infectious diseases like the flu might make you feel sleepy.
7. Do you spend more time in REM or non-REM sleep? In which one are you likely to dream?
8. What stops you from acting out your dreams?
9. Which types of brainwaves are seen in deep, dreamless sleep?

Think and discuss

10. Discuss the effect of light pollution in your bedroom.
11. Why might you be more vulnerable to asthma at night-time?

Investigate

12. Travelling from one time zone to another can disrupt your circadian rhythm and you can experience a condition known as jet lag. Find out more about how light therapy has been used to help reduce the effects of jet lag by helping to reset biological clocks.
13. While most adults need about 7 or 8 hours sleep, teenagers usually require about 9 hours. Find out more about research into adolescence and sleep.
14. Investigate and report on one of the following sleep conditions: sleep apnoea, narcolepsy, restless leg syndrome, talking in your sleep, sleepwalking, night terrors.
15. If you don't get enough sleep, you may be drowsy and unable to concentrate. Severe sleep deprivation may result in hallucinations and mood swings. What are some other consequences of sleep deprivation?
16. If someone is in a coma or under anaesthesia, are they really asleep?
17. There is an early morning dip in blood pressure at about 2 or 3 am. Investigate and discuss why there are more records of heart attacks within the first six hours of waking than at any other time.
18. Select one of the following, research it and:
    (a) summarise your findings into a poster or multimedia presentation to share with others
    (b) describe how scientific evidence or knowledge can be used to validate your findings
    (c) use internet research to identify two problems related to this topic that could be investigated.
    - The effects of decongestants and antidepressants on sleep
    - Theories for why we yawn. Do you agree with any of these? Why?
    - Ways to sleep more effectively
    - Theories of how sleep may affect learning
    - Patterns of age and sleep
    - The effects of 'sleep debt'
    - Microsleeps
    - Driver fatigue
    - The effects of shift work on sleep
    - The effects of total blindness on sleep

19. (a) Describe the pattern observed in the melatonin levels in the graph at right.
    (b) Suggest an interpretation of the observed pattern.
    (c) Use other resources to find out more about melatonin and its effects on your body.
    (d) Suggest a link between light, melatonin and the body's resulting responses.
    (e) Suggest how melatonin levels may affect your learning.
    (f) Research seasonal affective disorder (SAD) and determine a possible link to melatonin levels.
    (g) Find out about and report on at least one example of research related to melatonin.

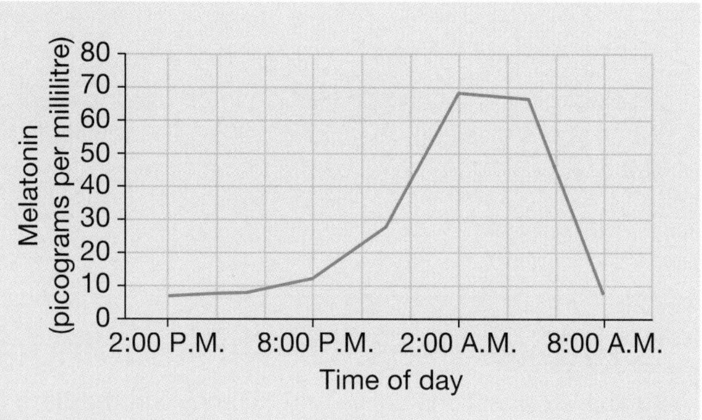

2.11 The teen brain

Science as a human endeavour

2.11.1 Growth spurts and pruning

Did you know that you had more neurons in your brain before you were born? Most of your brain development occurs in two stages: growth spurts and pruning. Throughout the first months of your life, your brain grew rapidly, producing millions of brain cells. A few months before you were born, there was dramatic pruning of your brain cells to remove unnecessary cells.

2.11.2 Like pruning a tree

Between the ages of about 6 and 11, neurons grow bushier and make dozens of connections (synapses) to other neurons, creating new pathways for nerve signals. This process peaks at around ages 11–12.

Use it or lose it! **Synapses** are the connections between the neurons where the message is passed from one neuron to the next. The synapses that carry the most messages get stronger and those that are not used much grow weaker. **Synaptic pruning** is the elimination of the weakest connections between neurons in the brain's cortex (grey matter). During this adolescent pruning up to 30 000 synapses may be eliminated each second. Only the connections that experience has shown to be useful are preserved. It is a bit like pruning a tree. The weaker branches are cut back to allow the other branches to flourish.

There are thousands of different neural pathways that can be travelled in the brain.

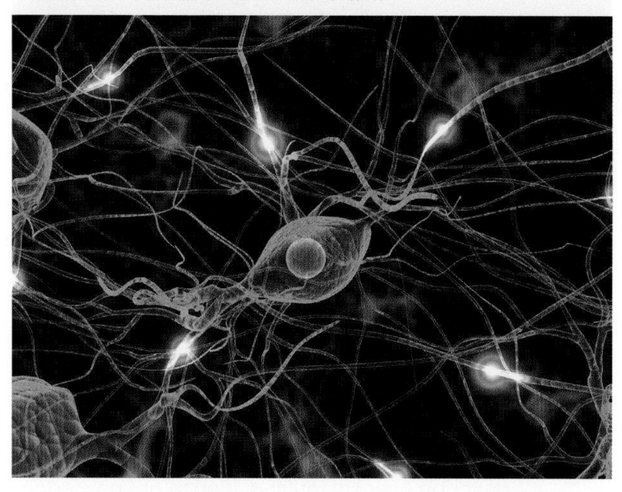

Nerve proliferation . . .

. . . and pruning

2.11.3 Wrapped in a white coat?

Your brain uses synaptic pruning to consolidate your learning by pruning away less travelled connections and myelinating neurons involved in the busy connections so that they become fixed as synaptic pathways.

In the process of **myelination**, neurons are coated with in a white material called **myelin**. The myelin coat acts like the plastic material wrapped around electrical wires for insulation. While myelination of neurons insulates, it also increases the speed at which the nerve impulse can move through it and hence the speed at which the message is communicated.

Images of the brain using MRI technology show that the amount of grey matter in the brain is reduced throughout childhood and adolescence and the amount of white matter increases. Does this suggest a link between increased cognitive (thinking) abilities and myelination?

These composite MRI brain images show that as you mature, your brain contains less grey matter. The increased amount of white matter is due to myelination, a process that also increases the speed at which your neurons can communicate messages.

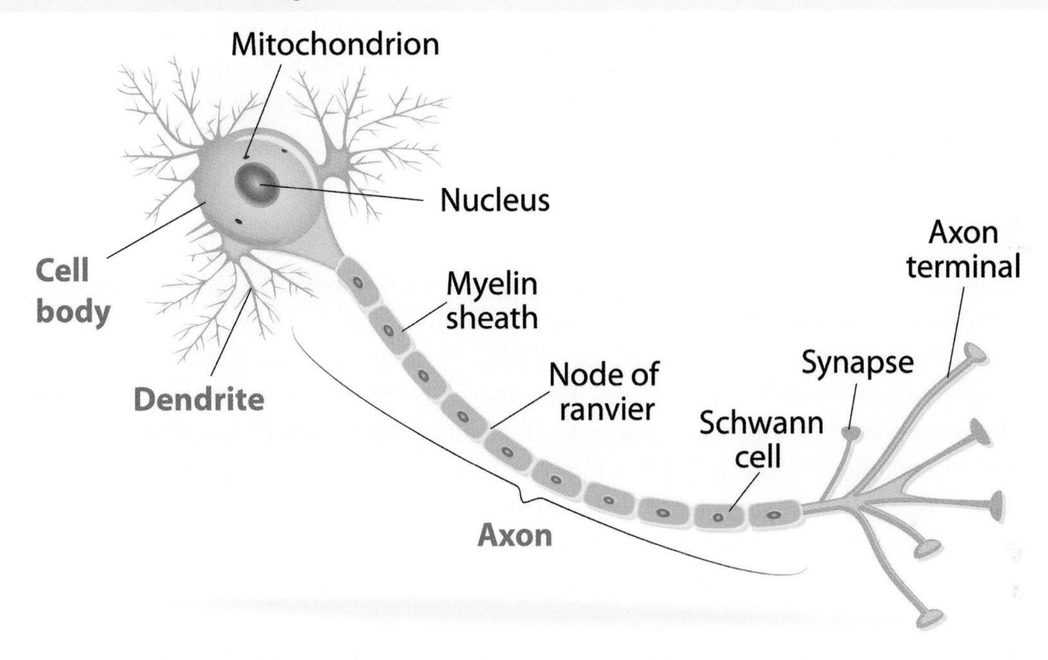

2.11.4 'Teeny' neuroscience discoveries

Knowledge about the teen brain is definitely a work in progress. Neuroscience research is providing us with many new discoveries due to the development of technologies that can provide images of living, growing brains. Using technologies such as PET (positron emission tomography) scans and fMRI (functional magnetic resonance imaging), scientists can observe growth spurts and losses, and map our brain's activity while we are involved in a variety of experiences.

2.11.5 Prefrontal cortex

It was once though that brains had finished their development by the end of childhood, but we now know that adolescence is a very busy time for brain growth and change. The prefrontal cortex in the brain undergoes a growth spurt at about 11–12 years of age, followed by a period of pruning and organisation of new neural connections. It is often referred to as the 'area of sober thought', and is now thought not to reach full maturity until the age of around 25. The prefrontal cortex is responsible for impulse control, planning, decision making, strategising and judgement. Is this why some teenagers act before they think about the possible consequences of their actions?

2.11.6 Basal ganglia

The basal ganglia act like a personal assistant to the prefrontal cortex, helping it to prioritise information. They grow neural connections at about the same time as the prefrontal cortex, and then prune them.

2.11.7 Corpus callosum

This bundle of nerves that connects the left and right hemispheres of the brain is thought to also be involved in problem solving and creativity. During your teens, the nerve fibres thicken and increase the effectiveness of information processing.

2.11.8 Amygdala

The amygdala is the emotional centre of your brain. This is the brain's area for primal feelings such as fear and rage. Since a teenager's prefrontal cortex may not yet have matured, they may use their amygdala and associated gut instincts when making decisions. Teenagers also tend to rely more on this part of the brain when processing emotional information, which may lead to impulsive behaviour. Adults are more likely to rely on their more developed and rational prefrontal cortex, which can balance out inappropriate emotions and impulses from their amygdala.

2.11.9 Dopamine spikes

Are adolescents neurally wired to take risks? In 2010, scientists observed fMRI images of participants involved in particular learning activities. Their research results led them to hypothesise that risk-taking in adolescents may be due to over-activity in the mesolimbic dopamine system of their brains.

What are the implications of this possible new knowledge? Is the risk-taking observed in many teenagers due to a spike in their levels of the neurotransmitter dopamine? While further research may support or disprove this hypothesis, the possibility that this may be the case opens many new possibilities for research and consequent issues. If it is supported, how accountable are teenagers for their behaviour? It can be seen that from research exciting new knowledge can be developed, but the implication of this knowledge also needs be considered or explored. Who determines the future uses of new knowledge in scientific discoveries?

The neurotransmitter dopamine is known to be important for motivation to seek rewards.

2.11.10 Back-to-front brain development

Did you know that your brain develops from bottom to top, from back to front, and from right to left? The development of your brain has been 'programmed' for the two tasks that confront survival of the human race (staying alive and getting into the gene pool). In the first 10 years of life, you learn the skills to stay alive. In the next 10 years, you learn how to be a productive and reproductive human. This wiring of your brain is essential to the survival of our species.

Your brain develops from back to front!

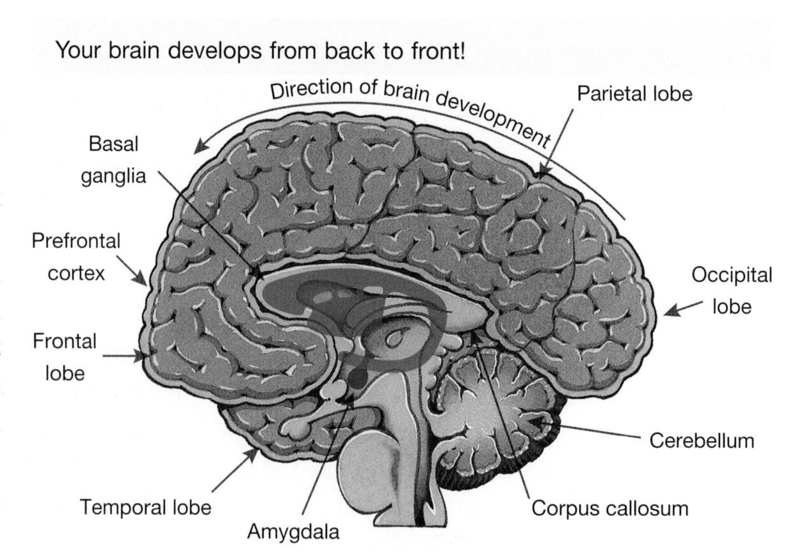

2.11 Exercises: Understanding and inquiring

To answer questions online and to receive **immediate feedback** and **sample responses** for every question, go to your learnON title at www.jacplus.com.au. *Note:* Question numbers may vary slightly.

Investigate

Select one of the following statements and claims and list five questions that it raises. Investigate these questions and present your findings in a creative and interesting way.

(i) You were born with a very immature brain (about 1/3 adult size) because of your mother's upright stance (walking on two legs) and her relatively narrow birth canal.
(ii) That we are a cooperative, social species with a rich language-driven culture is due to our limited and 'helpless' early brain development and long dependence as children.
(iii) Some research suggests that the corpus callosum is bigger and more developed in women than in men. Other studies contradict this.
(iv) Female brains may be smaller, but they mature a lot faster and have more synapses.
(v) Girls are better at ... than boys because their brains are better.
(vi) Boys are better at ... than girls because of the way their brains develop.
(vii) Teenagers get into so much trouble because they think and act through their amygdala.
(viii) A drug should be developed so that the brains of teenagers are more like those of adults.
(ix) It is important to survival of the species for adolescents to be wired to take risks so that they can learn new ways of doing things.
(x) Schools should start later in the day because teenagers need more sleep than those of other ages.

2.12 Getting back in control

Science as a human endeavour

2.12.1 Is your body ignoring you?

Imagine not being able to move. Your brain tells your legs or arms to do something but they ignore you. How would it feel? What can you and science do to help?

Damage to the spinal cord of the nervous system may be the result of a disease or an accident or be congenital (already present at birth). Whatever the cause, this type of damage can be devastating and debilitating.

Although there is currently no cure for spinal injury, teams of scientists around the world are involved in research that is aimed at improving the quality of life for those with this injury. Perhaps a technology not yet developed may one day lead to a cure.

2.12.2 Paralysis and spinal injury

All of the nerves in your peripheral nervous system throughout your body connect to your spinal cord. Damage to this cord can prevent communication of messages between your brain and your body. This loss of communication can lead to **paralysis** (loss of movement).

Damage to different parts of the spinal cord results in different types of paralysis. For example, if you were in an accident in which the lower back section of your spine was completely crushed, messages would not be able to travel between your legs and feet and your brain. This loss of communication would mean that you would not be able to sense pain, heat, cold or touch in these parts of your body. You would also be unable to stand or walk as you would not be able to control the muscles in your legs and feet.

Actor Christopher Reeve raised awareness of the consequences and need for research related to spinal injuries.

Christopher Reeve, an actor who played Superman in a series of early movies, damaged his spinal cord in the neck region in a sporting accident. The consequence was that he was paralysed below the neck and required the use of a machine to breathe air into and out of his lungs as he was unable to breathe for himself. In the years following his accident he raised awareness of spinal injuries and increased public and political interest in related research.

2.12.3 Paralysis and disease

A number of diseases can also result in paralysis. One such condition is motor neuron disease. Although the cause of this disease is still unknown, its effects are devastating. While the brain and the senses are usually unaffected, the person with the disease becomes increasingly paralysed.

Motor neuron disease, as the name suggests, targets motor neurons and progressively destroys them. Sensory neurons, however, remain unaffected. This means that a person paralysed with motor neuron disease could hear and see a mosquito, feel it biting their arm, feel the itchiness, but be unable to move to scratch it or talk to tell someone to scratch it for them.

People with motor neuron disease sense their environment, but increasingly cannot respond to it. This paralysis eventually involves all muscles within the body. Sadly, motor neuron disease is fatal.

Stem cells — a possible treatment?

Embryonic stem cells (a topic introduced in *Science Quest 8*) have many properties that scientists find exciting. They can produce new cells for longer than other cells and under the right conditions they can be made to differentiate into particular cell types. Some current research is investigating the injection of nerve cells produced from embryonic stem cells into the site of spinal injury. Although it is early days for this research, it is hoped that it may lead to the recovery of muscle function in some cases.

The use of stem cells to treat (and possibly even cure) a variety of diseases is being investigated. The research is, however, accompanied by much debate.

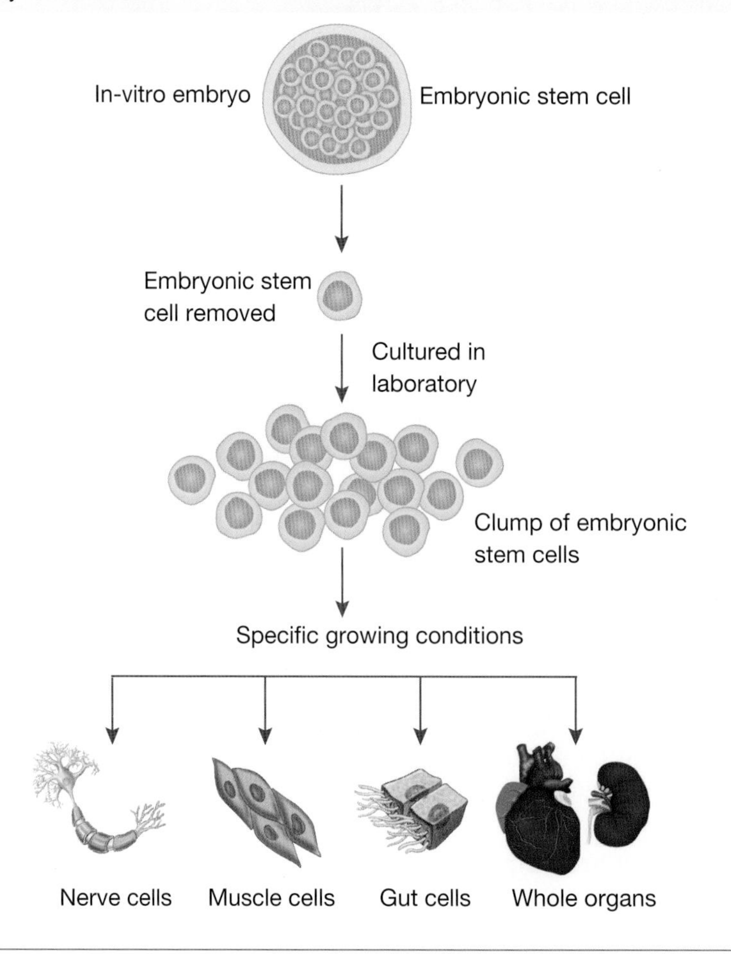

Although the possible applications of this research are exciting, technologies involving the use of human embryonic stem cells are fraught with issues and controversy. Most of this debate centres on the source of the stem cells — human embryos that have been obtained from the surplus embryos of couples undergoing IVF treatment.

2.12.4 Brain-control interface technology

Currently making an entrance into the mass market are games and toys which utilise **brain-control interface technology**. In these applications, computer software in 'mindsets' are used to decode brain wave patterns and facial movements to bring about particular responses in the external environment (for example, moving an object by just thinking about it).

Brain-control interface technology has many possible applications.

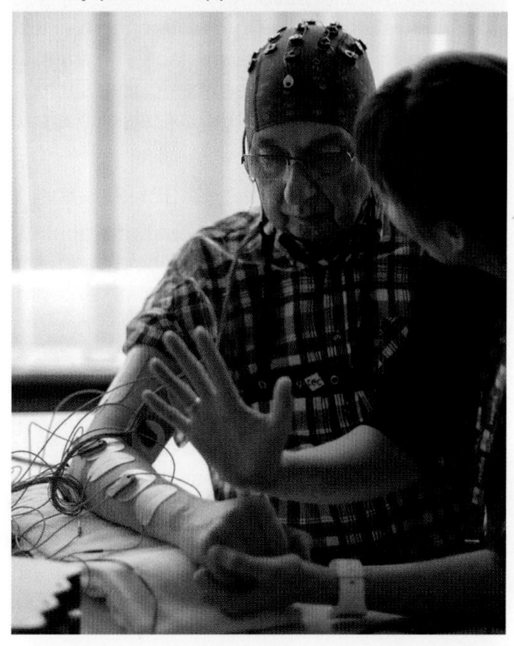

Broader applications of this technology, for example **implanted electrodes** and **neural prostheses**, are being researched and developed in order to provide assistance to people with a variety of disabilities. There have already been cases in which paralysed people have been able to move their wheelchairs by just thinking about the movement, or those who are unable to talk have been able to use their brain to result in their thoughts being spoken aloud.

Could such technology be used in other ways — could it be used to help blind people to see, and deaf people to hear? What other senses could be assisted using this technology? Could it be used to enable us to experience senses that humans do not currently possess?

2.12 Exercises: Understanding and inquiring

To answer questions online and to receive **immediate feedback** and **sample responses** for every question, go to your learnON title at www.jacplus.com.au. *Note:* Question numbers may vary slightly.

Remember

1. Define the terms *spinal cord, paraplegia, quadriplegia* and *paralysis*.
2. Outline the properties that make stem cells interesting to researchers.
3. Describe evidence that suggests that stem cells may one day be used to restore some mobility after a spinal injury.
4. Outline how brain-control interface technology can bring about body responses.
5. Describe an application of implanted electrodes or neural prostheses.

Think

6. Use the graph on the next page showing the causes of spinal injury to answer the following questions.
   (a) What are the two leading causes of spinal injury?
   (b) What percentage of spinal injuries are sports related? Suggest which sports might have the highest risk of spinal injury.
7. Explain why an injury in the neck region of the spinal cord may result in quadriplegia, whereas an injury in the lower back region of the spinal cord may result in paraplegia.
8. Imagine that you are on an ethics committee for a university. An ethics committee is a group of people that decides whether experiments should be carried out on ethical grounds. Discuss whether you would allow scientists to do experiments that involve deliberately crushing the spinal cord of rats.

Investigate

9. Draw a map to show the location and type of wheelchair access available at your school. Do the same for your local shopping centre. Are there any places it would be impossible for a disabled person to get to? Find out from your local council what regulations there are relating to wheelchair access to parks and public and commercial buildings.
10. Investigate some improvements to wheelchair design that have been made in the last 20 years.
11. Find out about wheelchair designs for particular sports such as wheelchair basketball and racing.

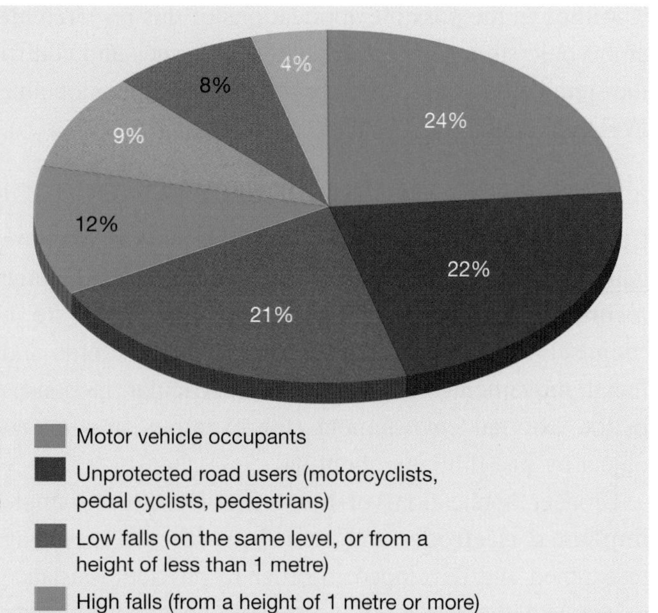

Think, investigate, discuss and create

12. Formulate questions about brain-control interface technology that could be investigated scientifically and then design an experiment that could be used to investigate one of your questions.
13. Intercranial electrodes may be very useful in brain-control interface applications and benefit those who are paralysed or have conditions such as motor neuron disease.
    (a) As a group, formulate questions about intercranial electrodes, brain-control interface applications, paralysis and motor neuron disease.
    (b) Research questions proposed by your group and then discuss your findings.
    (c) Suggest further questions that could be explored and research these.
    (d) Use your imagination and creativity to share your findings as a model, a science fiction story, an animation or a multimedia documentary.

Standard disability parking and access signs

14. (a) Find out more about modern brain implant technologies.
    (b) Suggest ethical issues that may be raised about the research and application of these technologies.
    (c) Organise a class debate that considers various perspectives on these technologies.

2.13 Opening up your brain

Science as a human endeavour

2.13.1 Your amazing brain

Your brain is amazing, mysterious and powerful. While you use it to formulate, ask and investigate questions, sometimes these questions are about the brain itself! What do you know about your brain? Why not open your brain up to new ideas, new discoveries and new questions about brains?

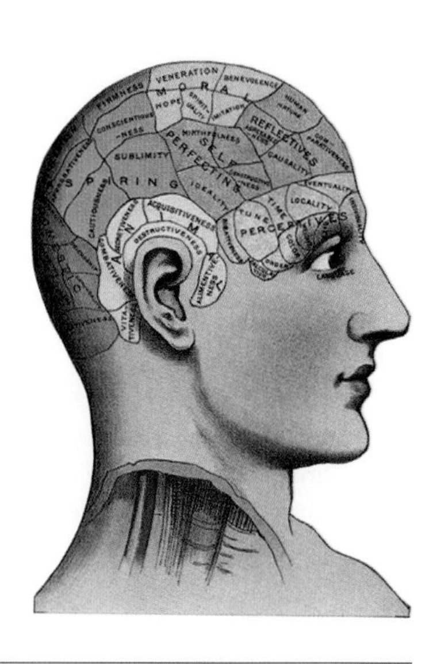

2.13.2 Brain in mind

Throughout history, humans have asked many questions about the human brain and there have been varied theories about its structure and how it works. Some questions have been about how brain cells interact with each other and what happens when the brain grows, ages or is damaged. Other questions relate to how it is involved in our learning, experiences and emotions, or how it contributes to make us who we are. There have even been investigations to design and construct artificial brains!

2.13.3 Phrenology

Frantz Joseph Gall, a German physician, developed the theory of phrenology in 1796. He believed that the brain was made up of a number of individual 'organs' which could be detected by visible inspection of the skull. This led to the belief that the size, shape and bumps of a person's skull determined their character and mental capacity. This theory was particularly popular between 1810 and 1840. While phrenology is now dismissed as a **pseudoscience**, some of its assumptions are still valid. The idea that mental processes can be localised in the brain is one such claim and is supported by our modern neuroimaging techniques.

WHAT DOES IT MEAN?

The word phrenology comes from the Greek terms *phren*, meaning 'mind', and *logos*, meaning 'knowledge'.

2.13.4 Neurology

Guillaume-Benjamin-Amand Duchenne de Boulogne (1806–1875) was a French neurologist who greatly advanced the science of muscular electrophysiology and electromyography. In 1835, he began experimenting on therapeutic electropuncture — which involved applying an electric shock under the skin with sharp electrodes to stimulate the muscles. This increased his understanding of the conductivity of neural pathways. Some refer to Guillaume as the father of modern neurology and in recognition of his research (and discovery), Duchenne muscular dystrophy is named after him.

2.13.5 An integrated approach

Our interest in brains has given rise to a variety of new branches of science. Examples of these include neurobiology, neuroscience, neurophysiology, neuropsychology and neuroanatomy. The frontiers of brain science also require an integrated approach that combines approaches and technologies from various scientific fields. Scientists in medical, biological, molecular biological, theoretical science, psychology, biophysics and various computer technologies can all be involved in trying to find out more about particular aspects of our brains.

Duchenne and an assistant give a demonstration of the mechanics of facial expression using electropuncture.

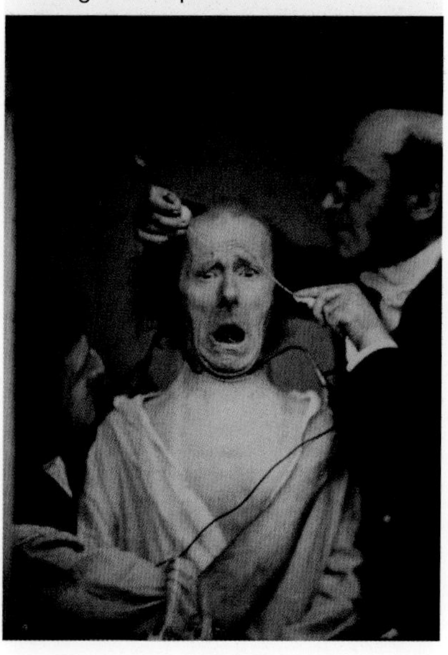

2.13.6 Brain on display

It is no wonder that some scientific terms are often referred to in an abbreviated form! This is especially the case with some of the names of imaging technologies used to look at the structure and function of the brain. **Computerised axial tomography (CAT)** and **magnetic resonance imaging (MRI)** produce computer images of the brain's internal structure. Scanning technologies that provide information about brain function include: **electroencephalography (EEG)**; **magnetoencephalography (MEG)**; **positron emission tomography (PET)**; **functional magnetic resonance imaging (fMRI)**; and **functional magnetic resonance spectroscopy (fMRS)**. A key advantage of these scanning technologies is that they can analyse the brain while its owner is alive — and using it!

2.13.7 PET

PET was the first technology used to observe brain functions. It involves injection of a radioactive solution into the brain. The amount of radiation measured in particular regions indicates levels of activity in those parts at that time.

PET scans of people with normal brain activity participating in different tasks. Red indicates the greatest level of brain activity, whereas blue indicates the brain areas that are least active.

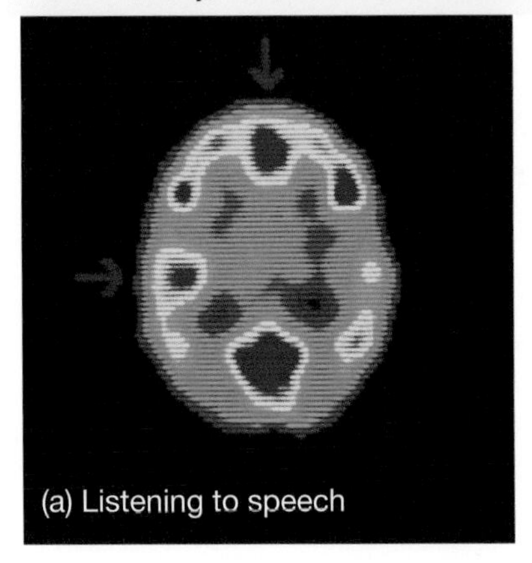

(a) Listening to speech

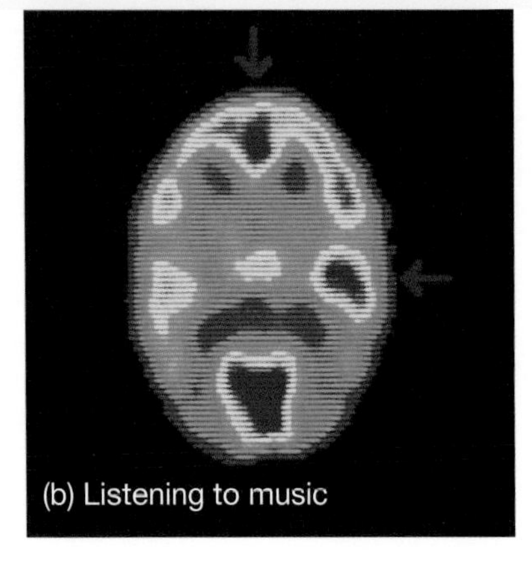

(b) Listening to music

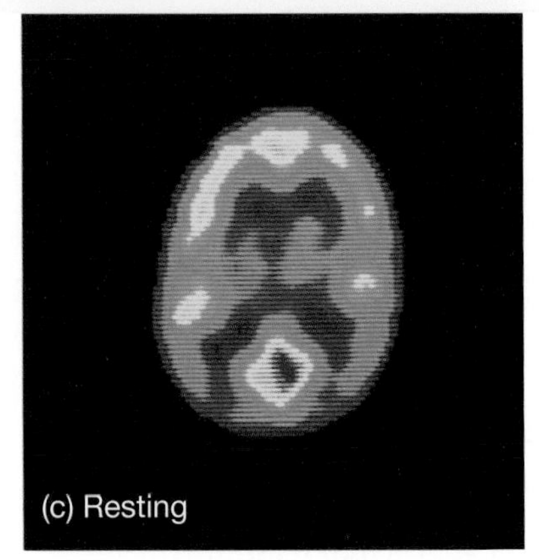

(c) Resting

2.13.8 EEGs and MEGs

EEGs and MEGs involve the attachment of multiple electrodes to be the scalp and the measurement of either electrical or magnetic activity occurring in the brain during mental processing. These technologies record activation of groups of neurons responding to a specific event and help to determine how quickly this occurs in the brain.

2.13.9 fMRI and fMRS

Areas in your brain involved in thinking require more oxygen than the parts not involved. This oxygen is transported by haemoglobin, a molecule that contains iron, which is magnetic.

fMRI uses a large magnet to compare the amount of oxygenated haemoglobin entering brain cells with the amount of deoxygenated haemoglobin that is leaving them. The computer generated images colour the regions with greater oxygenated blood. This allows the pinpointing of the activated brain regions to be located within a centimetre.

While fMRS uses the same equipment as fMRI it uses different computer software. This technology can record and identify levels of specific chemicals during brain activity and has been used to study language function in the brain.

MRI scan of a benign brain tumour

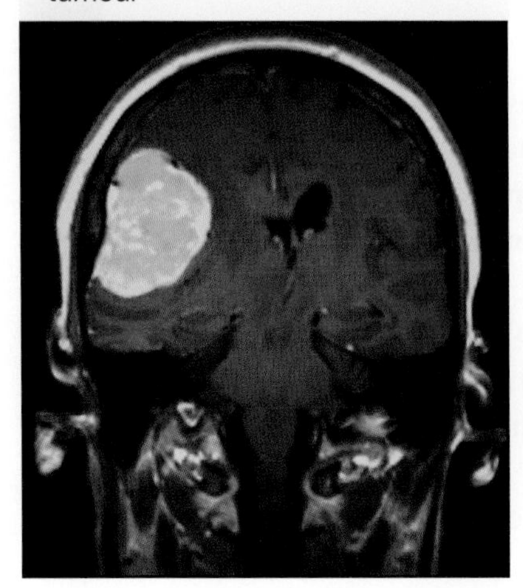

Scientists use various technologies to study the structure and function of brains. These technologies provide us with information that enables us to develop models so that the knowledge can be communicated to others.

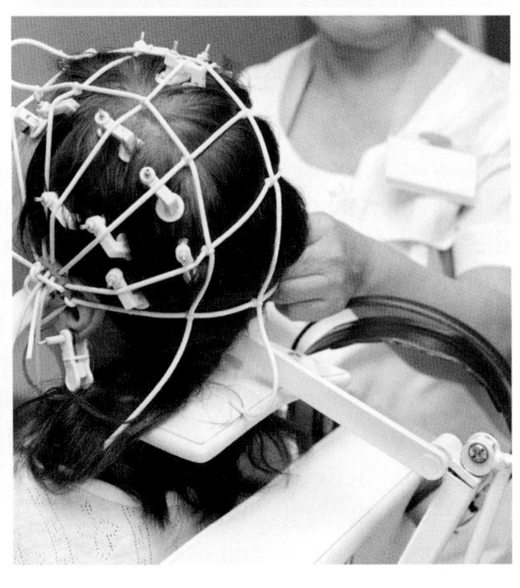

2.13.10 Neurotechnology

We have learned more about the human brain in the past 10–15 years than we have in the rest of recorded history. This new information is leading revolutionary changes in how we use our brains and think about them. New technologies are providing us with new knowledge about the brain and how it works. With new knowledge, previously held ideas often need to be modified. In some cases, the previous understanding or theories have needed to be discarded completely so that new theories can be developed to replace them.

2.13.11 Neuroplasticity and neurogenesis

Contrary to what was believed in the past, our brains and brain connections, or neural pathways, are not static and unchanging. They are constantly wiring and rewiring. Stimulation and challenging your brain encourages the growth of dendrites and the production of new neurons. Lack of stimuli can result in weakening of existing connections and possible pruning of them. You may also lose new neurons in the process.

Currently there are some exciting research projects on **neurogenesis** (meaning 'the birth of new neurons'). This research is investigating whether factors such as exercise and different moods can influence how many neurons are being 'born' each day and how many survive.

A technician monitoring a patient undergoing an MRI scan of the brain

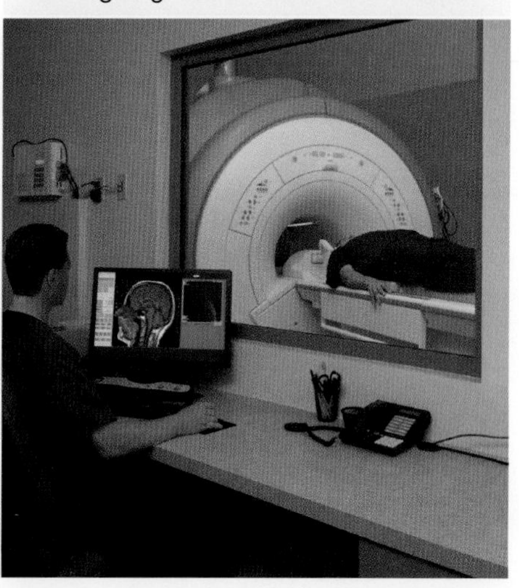

PMI chart on neurotechnology

Plus	Minus	Interesting
• **New medicines** For example, these may be individually styled and used to cure or treat mental illnesses with high efficacy and negligible side effects. This may lead to a major decrease in mental illnesses such as depression, schizophrenia, bipolar disorders, substance abuse and obsessive–compulsive disorders.	• **New weapons** For example, neuroweapons that influence how the brain (and person) responds to particular stimuli or situations	• **Neuroeducation** For example, a tablet to ingest or a connection to an implant in your brain to pass on the knowledge of a new language, topic or skill
• **New chemicals** For example, chemicals that may be ingested to: • — influence learning, memory processes, decision making and attention • — influence moods, motivation, feelings and awareness • — restore and extend the capacity of our senses (enhancing the sense of sight, hearing, smell or taste)	• **New marketing strategies** For example, neuromarketing that uses knowledge of the brain to prioritise their products over others	• **Neuromodulators** For example, a tablet to ingest to help people feel happy all of the time

2.13.12 Shaped by neurotechnology

Our society shapes the development of new technologies. It is also shaped by these technologies. Discoveries in neurotechnology have been enhanced by developments in information technologies. Development of nanobiochips and brain-imaging technologies increase the accuracy of biological and neurological analysis. Nano-imaging techniques will enable analysis of events at the neuromolecular level in the brain. Knowledge of these events will enhance our understanding about how our brains work and give us power to modify their function.

In the future, neurotechnology may provide us with knowledge that may lead to the development of new treatments for diseases, new industries — and new problems to consider and solve. How will new neurotechnologies change human societies? How will they change us?

2.13 Exercises: Understanding and inquiring

To answer questions online and to receive **immediate feedback** and **sample responses** for every question, go to your learnON title at www.jacplus.com.au. *Note:* Question numbers may vary slightly.

Remember

1. Phrenology has a colourful history and varied interpretations. Give an example of how it has been used.
2. Phrenology is considered by many to be a pseudoscience. What is meant by this term? Do you agree? Explain your response.

Investigate, think and discuss

3. (a) Use the internet to identify claims regarding the effects of exposure to types of electromagnetic radiation (such as X-rays, microwaves and gamma rays) on humans.
   (b) Find evidence that supports or negates the claims.
   (c) Identify issues that are relevant to human exposure to these types of radiation and construct PMI charts to summarise the key points and concerns.
   (d) Analyse the language used by media reports in terms of bias and perspective.
   (e) Select one of the issues and organise a class debate.
4. Research ways in which the development of imaging technologies has improved our understanding of the structure and function of the human brain.
5. Investigate how electromagnetic radiation technologies are used in the detection and treatment of cancer. Report your findings as a television documentary, podcast or newspaper article.
6. There are claims that brain scans can reveal personality types and the type of career that you are best suited to. Find out more about these claims.
   (a) On the basis of your findings, do you agree that brain scans are capable of this? Justify your response.
   (b) Find out and discuss issues related to the use of brain scans in this way. Do you agree with using brain scans in this way? Explain why.
7. Frantz Gall believed that the brain was made up of a number of individual 'organs' that created one's personality. Find out examples of these 'brain organs' and how this information was used. Not everyone agreed with the ideas of phrenology. Find examples of arguments for and against phrenology, summarising your findings in a PMI chart.
8. If you were to hear about a new model or theory about the brain in the media, describe how you would use scientific knowledge to determine its possible validity.
9. Use your knowledge of science to test claims made in advertising or expressed in the media (or in this text) with regard to any of the following.
   - The Mozart effect increases the depth of learning.
   - Mobile phones can cause brain cancer.
   - Neuro-linguistic programming (NLP) helps people lead better, fuller and richer lives.
   - Some people are real left-brainers!
   - Faulty mirror neurons can lead to autism.
   - Sleep enhances memory.
   - Zapping the brain using transcranial direct current stimulation can spark new ideas.

Investigate, think and create

10. Phrenology gave rise to the invention of the psychograph. Research what it is and about its history.
11. Carefully observe the neurotechnology PMI chart in this subtopic. Select one of the boxes and research a particular aspect of it that is of interest to you. Develop an advertising or political campaign (complete with multimedia aspects) to either promote or criticise your neurotechnology application.

2.14 Review

2.14.1 Study checklist

Nervous system

- state another name for a nerve cell
- draw a neuron, labelling the dendrites, cell body and axon
- state the function of receptors and provide examples of at least three types
- use a flowchart to show how receptors are involved in your ability to sense your environment
- use a flowchart to describe the stimulus–response model
- describe how negative feedback can assist you in maintaining homeostasis
- outline the overall function of the nervous system
- outline the key components of the nervous system
- draw a labelled diagram of the structure of a neuron
- use a flowchart to show how a message is conducted and transmitted in the nervous system
- compare and contrast nervous impulses and neurotransmitters
- use a flowchart to describe the process involved in a reflex action
- explain the need for some reactions to be reflex actions
- compare reflex actions with those under conscious control
- describe how damage to the nervous system can result in paralysis
- outline the effects of motor neuron disease on the ability to sense and respond to the environment
- recall the types of brain waves and the chemicals the body produces that are associated with sleep drugs
- identify the cerebrum, cerebellum and brain stem and outline their key functions
- list three neurotransmitters in the brain that can influence feelings and actions

Endocrine system

- outline the overall function and key components of the endocrine system
- recall the main glands of the endocrine system and some of the hormones they produce
- use a diagram to show how the stimulus–response model can be used to describe the involvement of the endocrine system in homeostasis

Science as a human endeavour

- consider how the development of imaging technologies has improved our understanding of the functions and interactions of body systems
- use knowledge of science to test claims made in advertising or expressed in the media
- recognise aspects of science, engineering and technology within careers such as medicine, medical technology, biomechanical engineering, pharmacy and physiology
- comment on how scientific understanding, including models and theories, are contestable and are refined over time through a process of review by the scientific community

Individual pathways

ACTIVITY 2.1	ACTIVITY 2.2	ACTIVITY 2.3
Investigating control systems doc-14549	Analysing control systems doc-14550	Investigating control systems further doc-14551

learnon ONLINE ONLY

2.14 Review 1: Looking back

To answer questions online and to receive **immediate feedback** and **sample responses** for every question, go to your learnON title at www.jacplus.com.au. *Note:* Question numbers may vary slightly.

1. Construct a flowchart to show:
   (a) the following terms in order from smallest to largest:
      organs
      organelles
      cells
      molecules
      tissues
      systems
      atoms
   (b) the stimulus–response model.
2. Complete the following table.

Stimulus	Receptor	Sense organ
		Eye
	Chemoreceptor	
Vibrations, pressure		
	Thermoreceptor	

3. Place the following labels in the correct places on the diagrams below.
   dendrite
   sensory neurons
   nerve cell body
   effector
   axon
   motor neurons

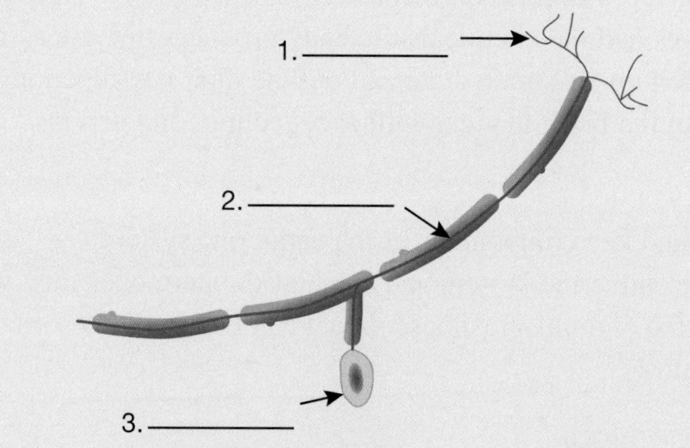

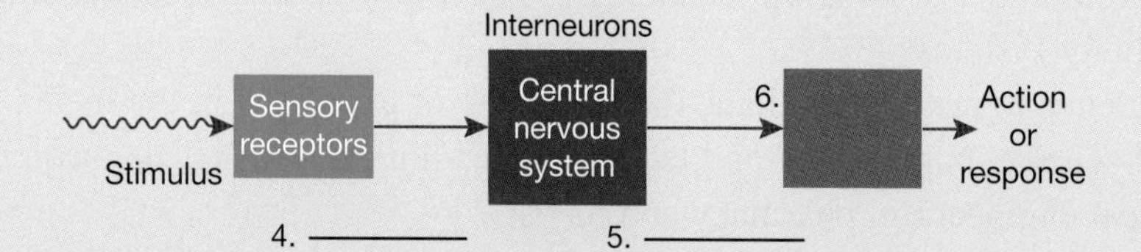

4. Label each of the parts of the brain at right and state one of the functions of each.
5. Underline the incorrect term in each sentence and replace it with the correct term. Write definitions of the incorrect words you replaced.
   (a) The neuron carries hormones to target cells.
   (b) The master gland of the endocrine system is the adrenal gland.
   (c) The brain and spinal cord make up the peripheral nervous system.
   (d) Each molecule has tissues which carry out particular functions.
6. Construct a table to summarise the differences between the nervous and endocrine systems. Make sure you include the

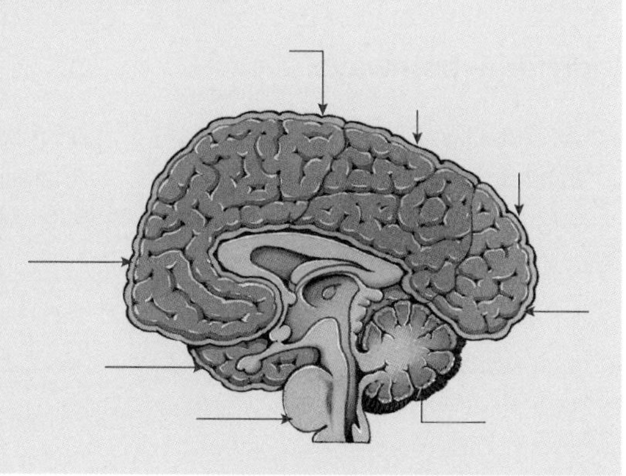

name of the information each system produces, how that information is carried throughout the body, and the speed and length of each system's response.
7. Draw a flowchart that outlines what happens when you sit down on a chair that has a sharp object on it. Include both nervous and endocrine responses.
8. Construct a continuum to show the following from smallest to largest:
   - nervous system
   - cerebellum
   - molecules
   - brain
   - neurons.
9. Describe functions of the following parts of your brain.
   (a) Cerebrum
   (b) Cerebellum
   (c) Brain stem or medulla
10. The flowchart at right shows a series of events that may occur when you encounter a stressful event. Suggest descriptions or labels for each of the links.

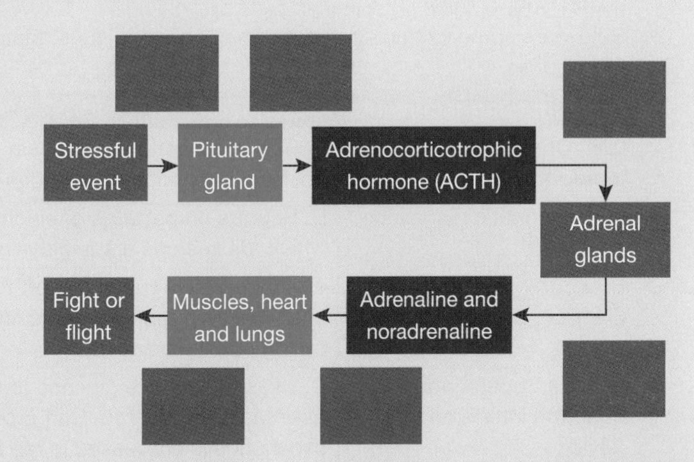

11. Match the terms with their appropriate description in the table below.

Term	Description
Central nervous system	Made up of a cell body, dendrites and axon
Motor neuron	Takes messages away from the central nervous system
Nerves	Takes messages to the central nervous system
Neuron	Brain and spinal cord
Neurotransmitter	Chemical messenger that carries messages from one neuron to another across a synapse
Peripheral nervous system	Nerves that connect the central nervous system to the rest of the body
Sensory neuron	Gap between neurons
Synapse	Made up of neurons

12. Suggest how you could link the nervous system terms in the flowchart at right.
   - Electrical impulse
   - Motor neuron
   - Sensory neuron
   - Response
   - Receptor
   - Neurotransmitter
   - Stimulus
   - Effectors

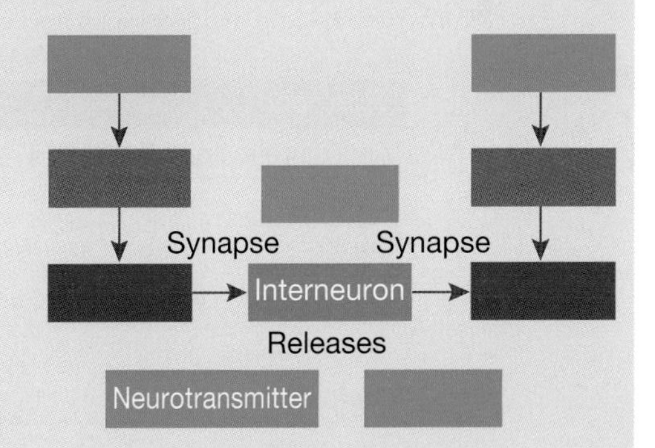

13. Place the terms in their appropriate position in the flowchart at bottom right: cell body, axon, dendrite, stimulus
14. Describe the relationship between adrenaline, pituitary, adrenal cortex, heart rate, stress.
15. Use a diagram to show how blood glucose levels are controlled.
16. Recall three endocrine glands and hormones they produce. Describe a function of each of the hormones.
17. Provide an example of a negative feedback mechanism.

An electrical impulse moves in only one direction through a neuron.

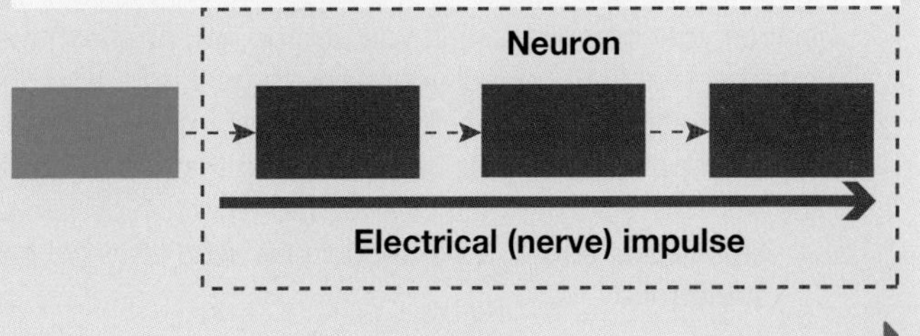

18. Suggest how analogies and metaphors can be useful in helping you connect information that you know to new information. Provide an example.
19. Neurolaw? How do you feel about the idea of the determination of guilt or innocence on the basis of a brain scan? There have been suggestions that brain scans (e.g. fMRI) should be used within our legal system. Do you think that these should be allowed as evidence in courts? Discuss and share your opinion with others. Justify your opinion.
20. Below are some examples of brain imaging techniques. Match the name of the technique to what it looks like or does.

Brain imaging technique	Description
CAT: Computerised axial tomography	A. Records electrical activity in defined areas, using colour to represent positive and negative locations in the cerebral cortex.
MRI: Magnetic resonance imaging	B. Reports on patterns of electrical transmission within an active brain which are seen as a squiggly line graph.
EEG: Electroencephalogram	C. Image that focuses on soft tissue and can show differences in chemical composition; some MRI techniques can monitor brain activity during cognitive activity.
SQUID: Superconductivity quantum interference device	D. Uses radioactive glucose to monitor blood flow through the brain as areas are activated. Can provide information of how and where an experience is processed in the brain.
PET: Positron emission tomography	E. Responds to small magnetic fields caused by electrical current of firing neurons and can identify source of electrical activity in the brain.
BEAM: Brain electrical activity mapping	F. Shows 3D graphical images of the density of tissue such as bone and tumours.

21. When is 'dead enough', good enough? There have been claims in the media that some organ donations have occurred when people were in the process of dying rather than being completely dead. Is near enough, good enough? On the basis of your scientific knowledge, what do you think about this issue? What is your personal opinion on this issue? Does your opinion on this issue match that of your scientific understanding? Discuss this issue with others in the class.
22. Tasty words and colourful letters? It is thought that about 4 per cent of the population have their senses crossed and associate letters with a flavour, numbers with a gender or sounds with a colour. This is called synaesthesia. It would be hard for people with synaesthesia to imagine a world without this extra perception. Find out more about this process and suggest how it might lead to the perception of a different world. Imagine having synaesthesia and describe what the world might be like.

Hormone	Function
Anti-diuretic hormone (ADH)	Causes reabsorption of water in kidneys
Glucagon	Causes testes to produce sperm
Insulin	Causes thickening of the uterine lining
Oestrogen	Controls menstruation cycle and pregnancy
Progesterone	Increases blood glucose levels
Testosterone	Increases metabolic rate of cells
Thyroxine	Lowers blood glucose levels

23. Match the hormone with the appropriate function.
24. Overheating can lead to heat exhaustion. This is your body's response to an excessive loss of water and salt (in your perspiration). If you get too hot, heat exhaustion may lead to heatstroke. When this occurs you may be unable to control your body temperature and death may result.
    (a) The relations diagram on the next page shows some causes of overheating. In a team, suggest some other possible causes of overheating that could be added to the diagram.
    (b) Construct a mind map of a team brainstorm on symptoms of overheating, heat exhaustion and heatstroke, strategies that could be used to avoid heatstroke, and treatments for heat exhaustion and heatstroke.

25. Having extremely low body temperature is also potentially life threatening. This condition is called hypothermia. Find out more about this condition and present your findings in a relations diagram.
26. Thermoregulation is the process whereby your body tries to keep your internal body temperature stable. Find out how your voluntary behaviour and your nervous and endocrine systems can help to cool or heat you. Present your findings in a mind map with diagrams or hyperlinks.
27. (a) Use the information entitled *Hairy stuff!* below to construct a relations diagram on some causes of hairiness.
    (b) Find out more about one of the disorders or diseases that has increased hairiness as a symptom. Present your findings in a mind map.

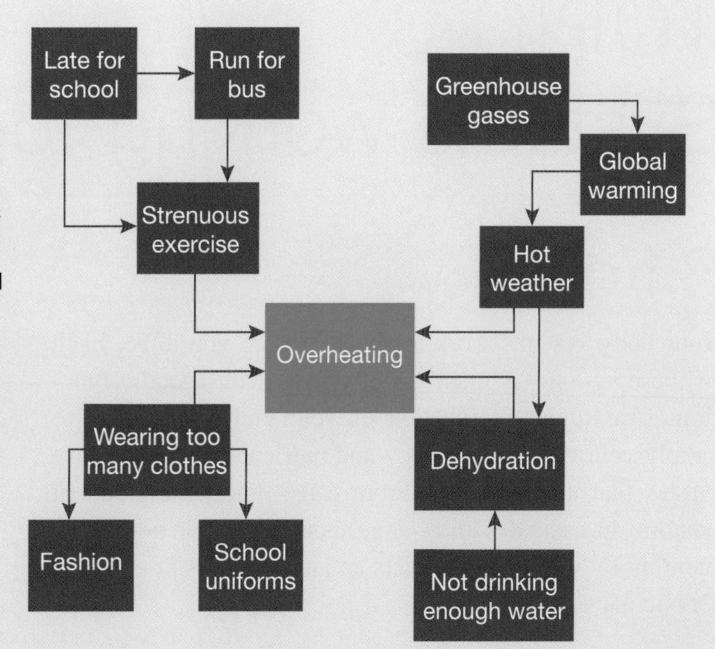

Hairy stuff!

Hormonal changes throughout life can cause changes in hair type and how it's produced. For example, prior to puberty, facial hair is a fine, thin type (vellus hair). Hormones released once puberty occurs can transform facial hair into a coarse, pigmented variety (terminal hair). While the growth of vellus hair is not affected by hormones, the growth of terminal hair is.

When females are experiencing menopause, there may be changes in the ratio of male and female hormones (androgens to oestrogen). This hormonal ratio change can produce an increase in facial hair. Heredity can also play a part in facial hair as it determines how thickly hair follicles are distributed throughout your skin.

Some medications and substances can cause hairiness (hirsutism). These include testosterone, steroids, Minoxidil, Rogaine and some blood pressure medication. Hairiness can also be a symptom of a number of disorders or diseases such as adrenal disorders (including Cushing's syndrome), anorexia nervosa, polycystic ovary syndrome and some pituitary disorders.

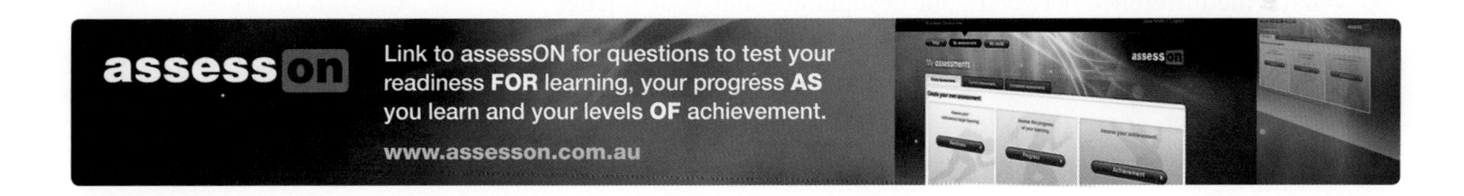

TOPIC 3 Systems working together

3.1 Overview

Your body systems work together to keep you alive. Each of these systems is made up of organs with specific functions. An important function of your body systems is to supply your cells with energy and nutrients, and to remove wastes that are produced. From early times, the study of anatomy has added to knowledge about the human body. This painting by Rembrandt depicts an anatomy lesson given by Dr Nicolaes Tulp in 1632.

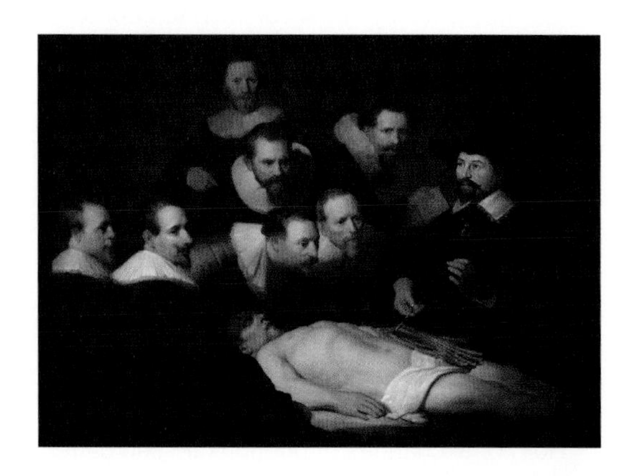

3.1.1 Think about body systems

assesson

- Is it the amount of oxygen or carbon dioxide in your blood that influences your breathing rate?
- In what form are old red blood cells excreted in faeces?
- Which vitamin deficiency may result in poor blood clotting?
- Why is being slimy a good thing for a bolus?
- What's the link between your urethra and your ureter?
- What's wrong with glucose in your urine?
- Does eating food stop you from getting drunk?
- What's the link between cocaine and neurotransmitters in the brain?
- What's the difference between your oesophagus and your trachea?
- Which is better, high or low GI?
- Which type of blood vessels take blood to the heart?
- Is chicken soup good for fevers?
- Should the government be able to control what and how much you eat and drink?

LEARNING SEQUENCE

Numerous **videos** and **interactivities** are embedded just where you need them, at the point of learning, in your learnON title at www.jacplus.com.au. They will help you to learn the content and concepts covered in this topic.

3.1.2 Ingredients for life?

Have you ever wondered what the recipe for life is? Which ingredients would you blend together to make up a living thing? How could this mixture result in life?

Many of the mysteries of life are being explored. Scientists have developed a whole range of different instruments and technologies to discover more about life processes. This has helped develop our knowledge and understanding of the structure of living things and how they work. Investigations provide us with more information about chemical processes that occur in cells and keep living things alive.

Scientists from different fields study the ingredients for life in different ways. **Biologists** may be interested in the cells that make up the organism and how these help the organism to grow, maintain itself and respond to internal and external environments. Chemists may see the organism as an amazingly complicated combination of atoms and molecules that are involved in millions of chemical reactions every second. Physicists and engineers, however, may see organisms as incredibly sophisticated self-controlling and self-repairing machines.

Your quest — Recipe for life

1. (a) Identify an environment in which your organism will live.
   (b) Describe the temperature, light intensity, water availability, food sources and other factors that you consider to be important to the survival of your organism.
2. Design your organism.
   (a) Identify how your organism obtains
   (i) nutrients and (ii) oxygen, and (iii) how its wastes will be removed.
   (b) Identify how nutrients, oxygen and its wastes are transported within its body.
   (c) Identify how the organism senses and responds to its environment.
3. Draw labelled diagrams of your organism's cells, tissues, organs and systems. Remember to take the function of each of these into account when you are designing its structure.
4. Describe how each of your organism's systems work together to keep it alive.
5. Construct a model of your organism.
6. Construct an electronic or hard-copy brochure that advertises what a magnificent life form your organism is.

3.2 Respiratory and circulatory systems

3.2.1 In and out . . .

Although you don't have to think about breathing, it is essential for your survival. It is a necessary function so that you can meet oxygen requirements for cellular respiration and for the removal of waste products such as carbon dioxide.

Human respiratory system

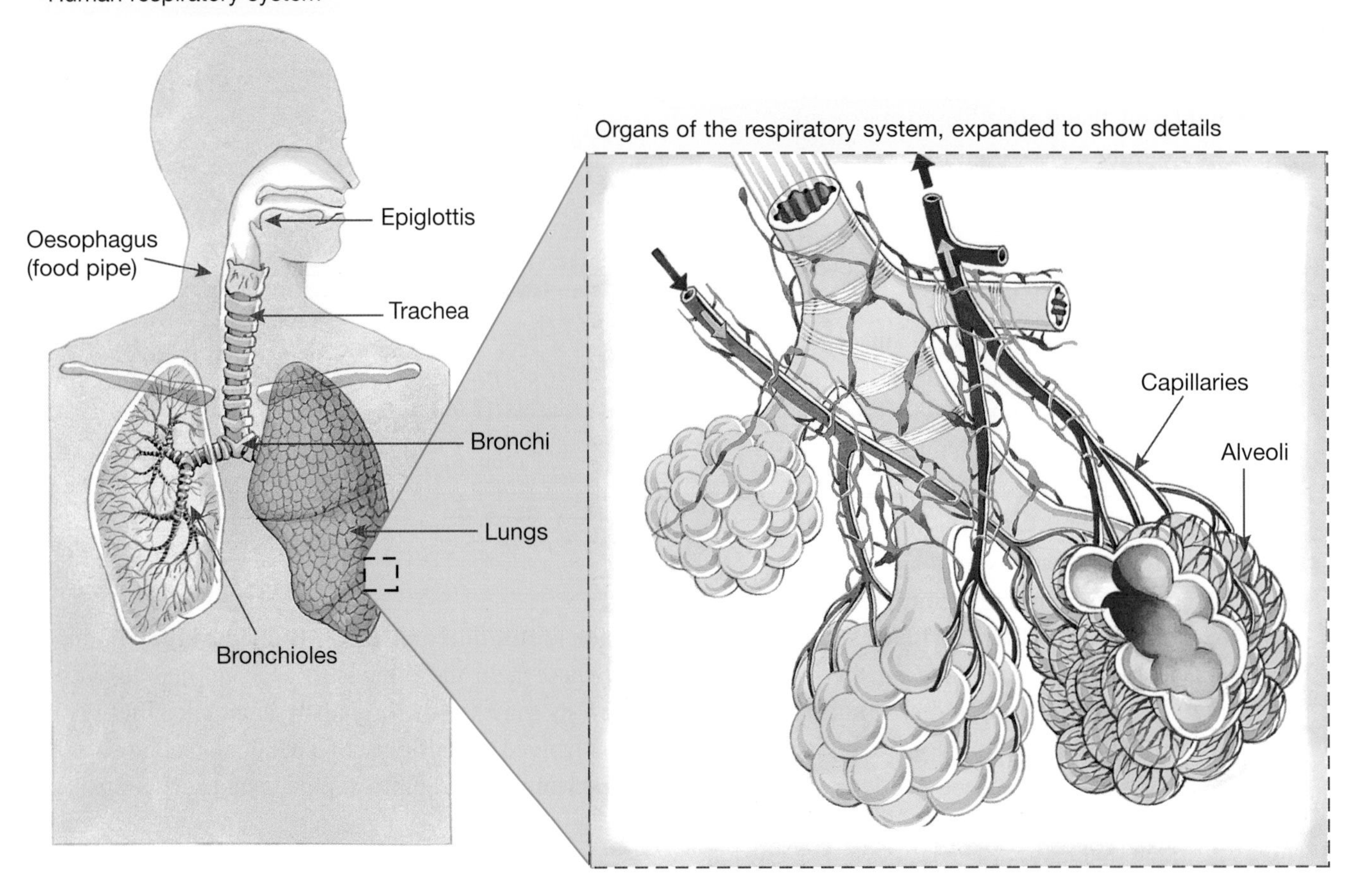

3.2.2 Cells need energy!

Your circulatory and respiratory systems work together to provide your cells with **oxygen** which is essential for **cellular respiration**. This process involves the breaking down of **glucose** so that energy is released in a form that your cells can then use. As can be seen in the cellular respiration equation below, **carbon dioxide** is produced as a waste product. The carbon dioxide then needs to be removed from your cells or it would cause damage or death to them.

$$\text{Glucose} + \text{oxygen} \rightarrow \text{carbon dioxide} + \text{water} + \text{energy}$$

3.2.3 Getting oxygen into your respiratory system

Your **respiratory system** is responsible for getting oxygen into your body and carbon dioxide out. This occurs when you inhale (breathe in) and exhale (breathe out).

When you breathe in, you actually take in a mixture of gases (of which about 21 per cent is oxygen) from the air around you. The air moves down your **trachea** (or windpipe), then down into one of two narrower tubes called **bronchi** (bronchus), then into smaller branching tubes called **bronchioles** which end in tiny air sacs called **alveoli** (alveolus).

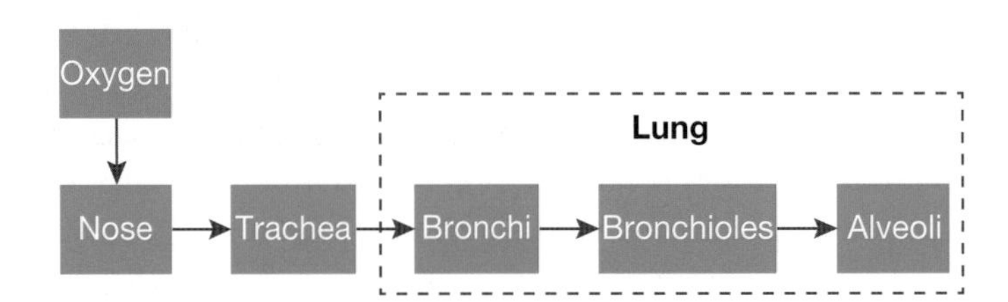

3.2.4 Transporting around

Your **circulatory system** is responsible for transporting oxygen and nutrients to your body's cells, and wastes such as carbon dioxide away from them. This involves blood cells that are transported in your blood vessels and heart. The three major types of blood vessels are **arteries**, which transport blood from the heart, **capillaries**, in which materials are exchanged with cells, and **veins**, which transport blood back to the heart (as shown below).

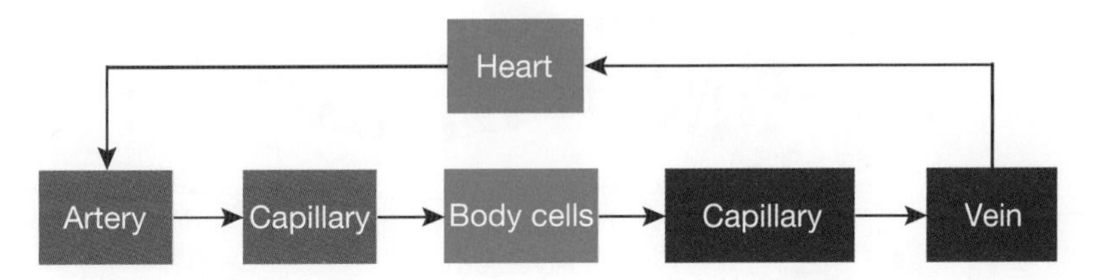

3.2.5 Getting oxygen into your circulatory system

Your alveoli are surrounded by a network of capillaries. These capillaries contain **red blood cells** (or **erythrocytes**) that contain **haemoglobin**, an iron-based pigment that gives your blood its red colour. Oxygen moves from the alveoli into the red blood cells in the surrounding capillaries and binds to the haemoglobin to form oxyhaemoglobin. It is in this form that the oxygen is transported to your body cells.

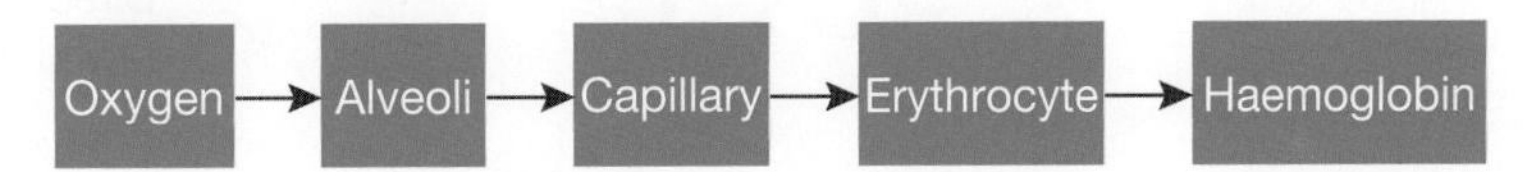

3.2.6 Getting oxygen to your cells

Oxygenated blood travels from your lungs via the **pulmonary vein** to the **left atrium** of your **heart**. From here, it travels to the **left ventricle** where it is pumped under high pressure to your body through a large artery called the **aorta**.

The arteries transport the oxygenated blood to smaller vessels called **arterioles** and finally to capillaries through which oxygen finally diffuses into body cells for use in cellular respiration.

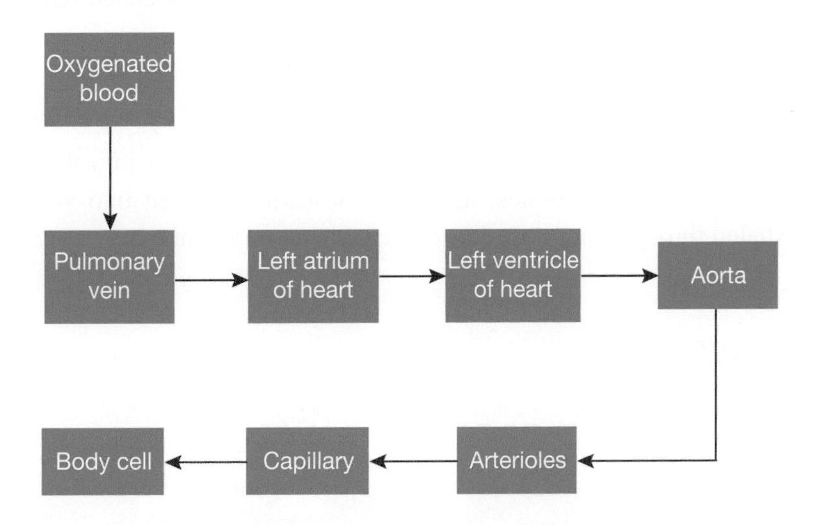

In an alveolus, oxygen diffuses into the blood and carbon dioxide diffuses out of the blood.

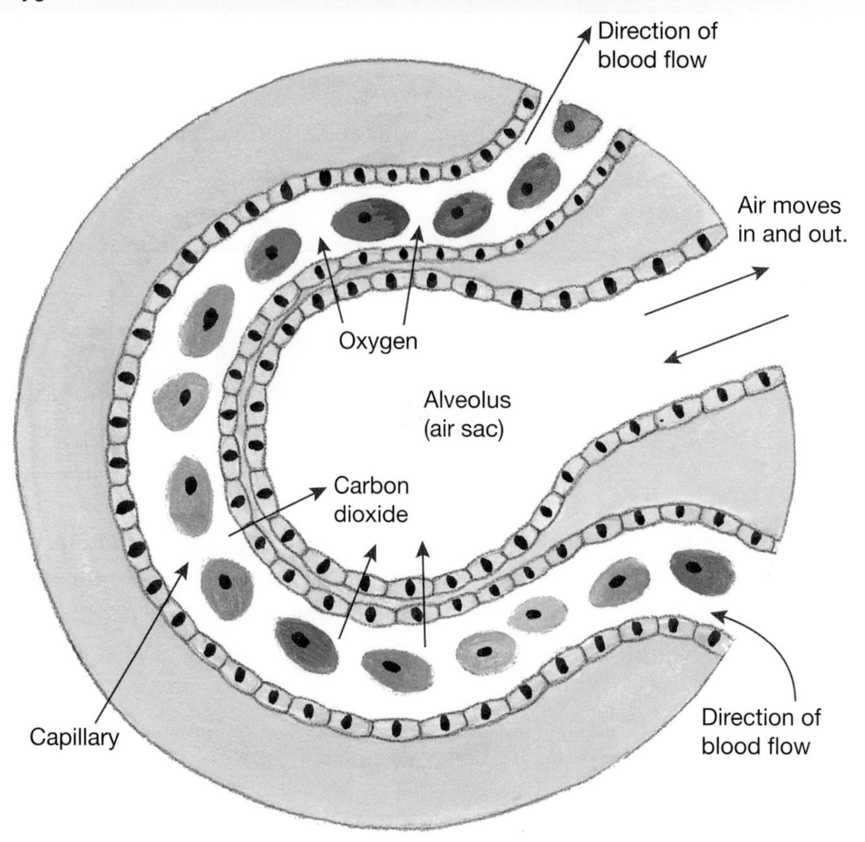

3.2.7 Getting carbon dioxide away from your cells

When oxygen has diffused into the cell and the waste product of cellular respiration, carbon dioxide, has diffused out of the cell into the capillary, the blood in the capillary is referred to as **deoxygenated blood**. This waste-carrying blood is transported via capillaries to **venules** (small veins) to large veins called **vena cava**, then to the **right atrium** of your heart. From here it travels to the **right ventricle** where it is pumped to your lungs through the **pulmonary artery**, so called because it is associated with your lungs. The pulmonary artery is the only artery that does not contain oxygenated blood.

3.2.8 Getting carbon dioxide into your respiratory system

Carbon dioxide from the deoxygenated blood in your capillaries diffuses into the alveoli in your lungs. It is then transported into your bronchioles, then your bronchi, and then into your trachea. From here, carbon dioxide is exhaled through your nose (or mouth) when you breathe out.

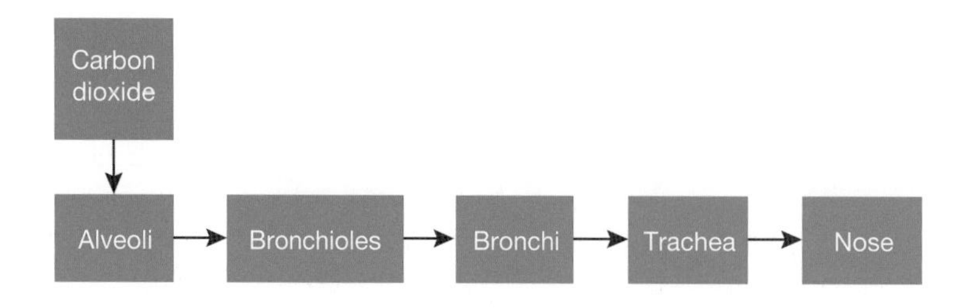

The movement of blood through the heart

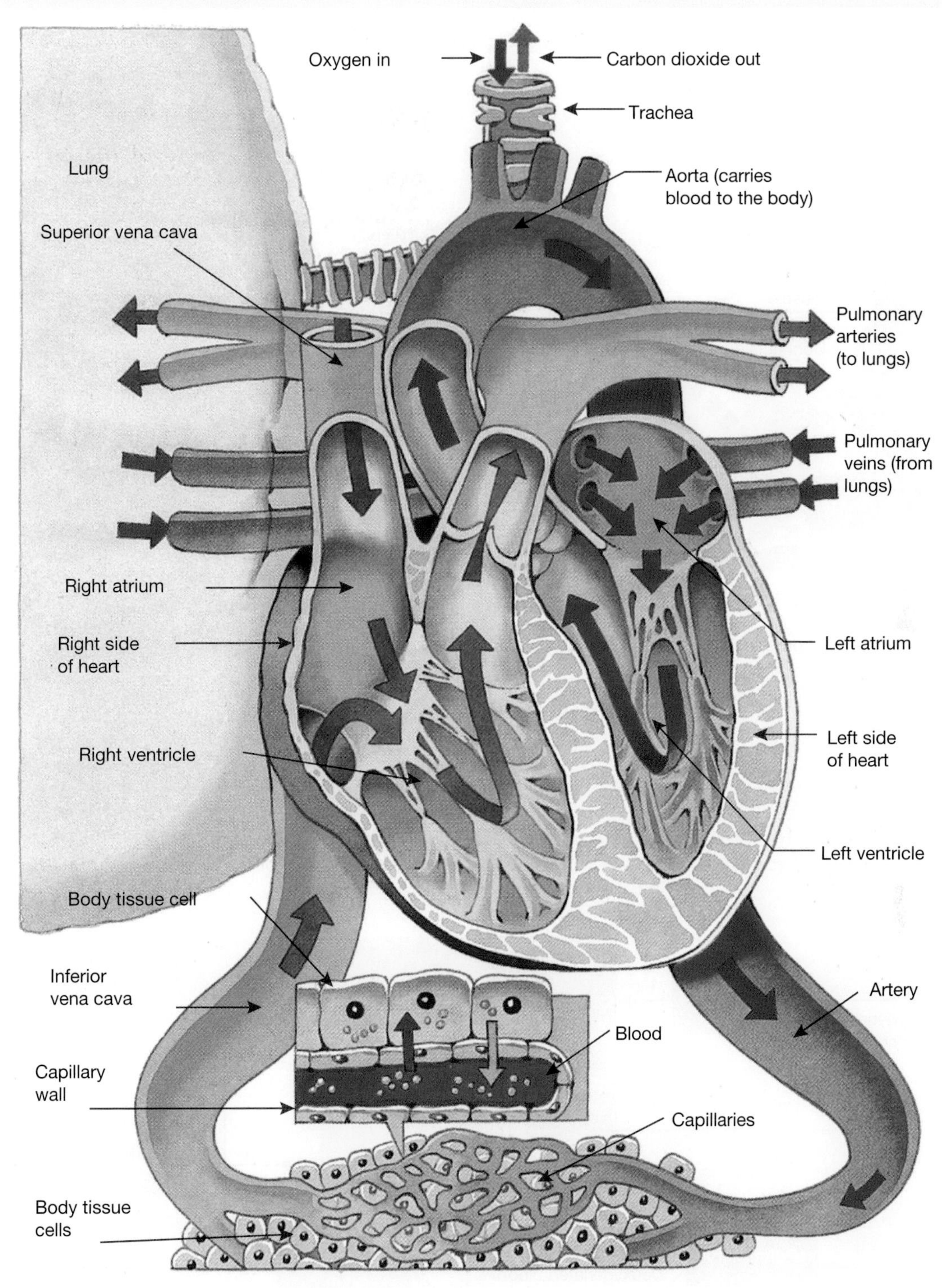

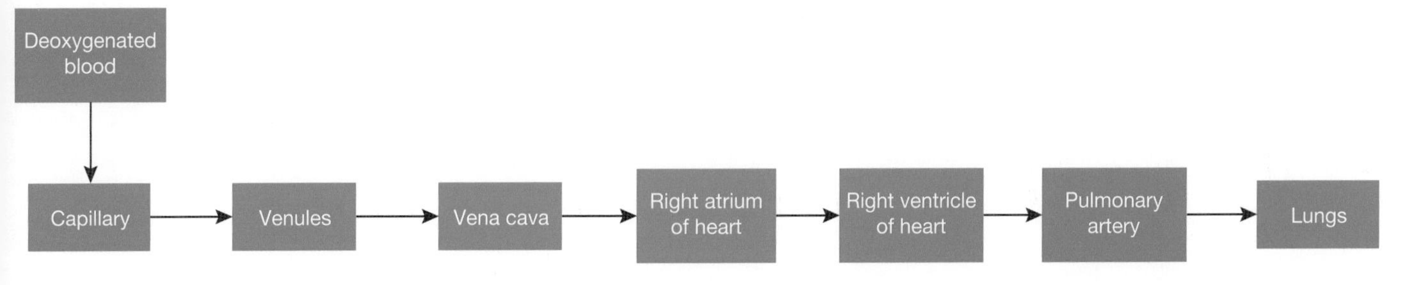

Your circulatory system consists of your heart, blood vessels and blood. Arteries, capillaries and veins are the major types of blood vessels through which your blood travels.

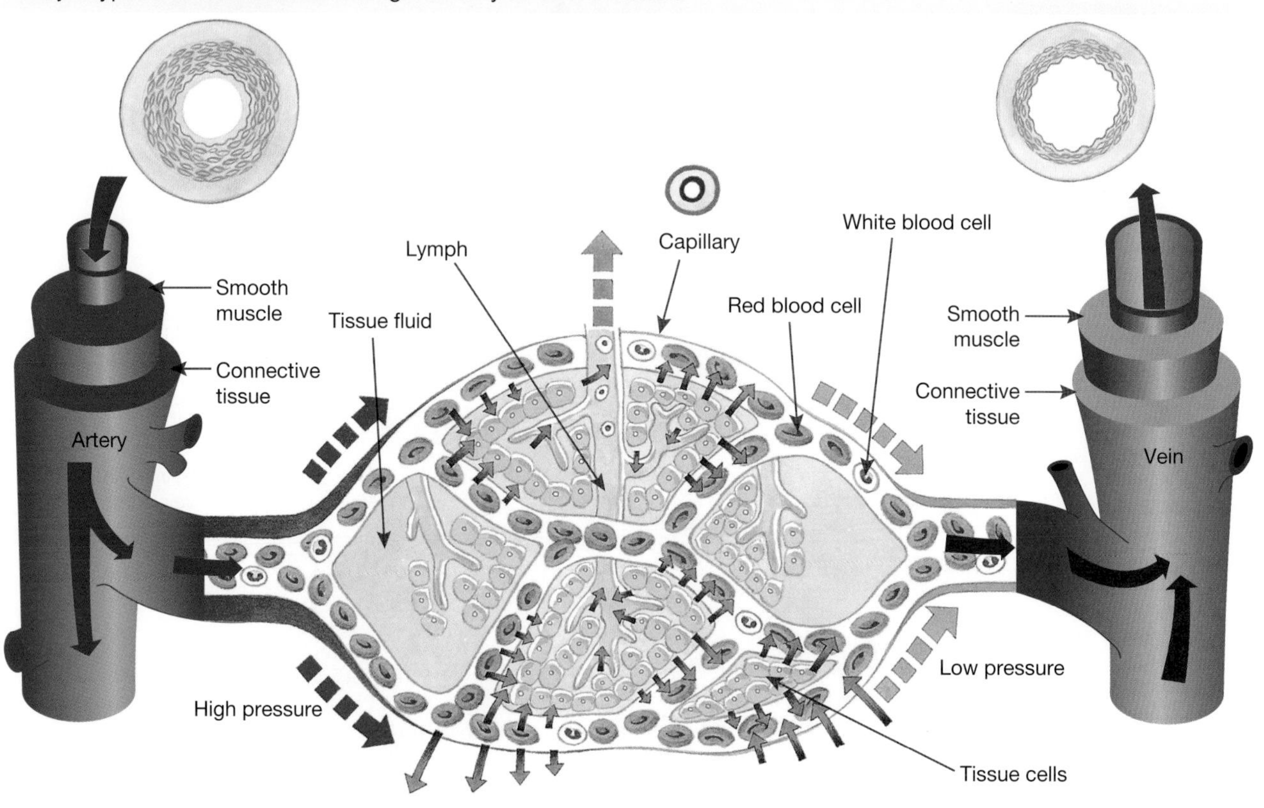

The heart is actually two pumps. One side pumps oxygenated blood and the other deoxygenated blood.

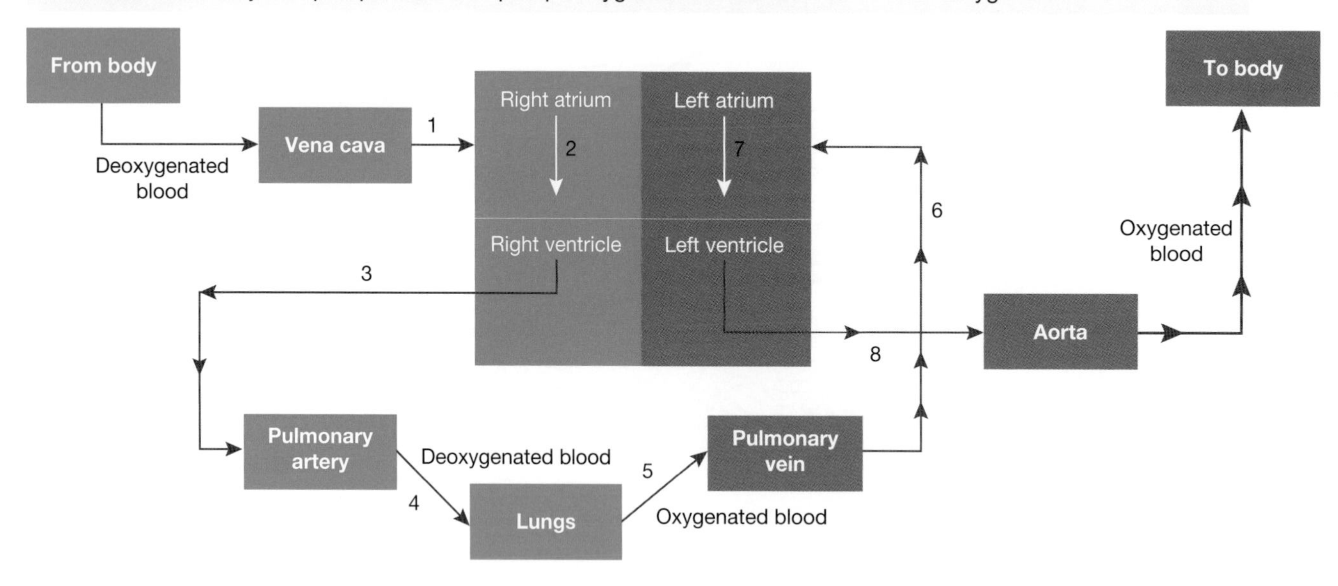

3.2.9 Putting it all together

Your body systems do not work in isolation. They work together to supply your cells with nutrients and to remove waste products that may otherwise be harmful. The transport of oxygen and carbon dioxide described here is merely an outline of the process. It is actually much more complex and is regulated by your nervous and endocrine system. Later in this chapter you will find out more about how these other systems are involved in keeping you alive.

Connected highways — the routes for blood circulations

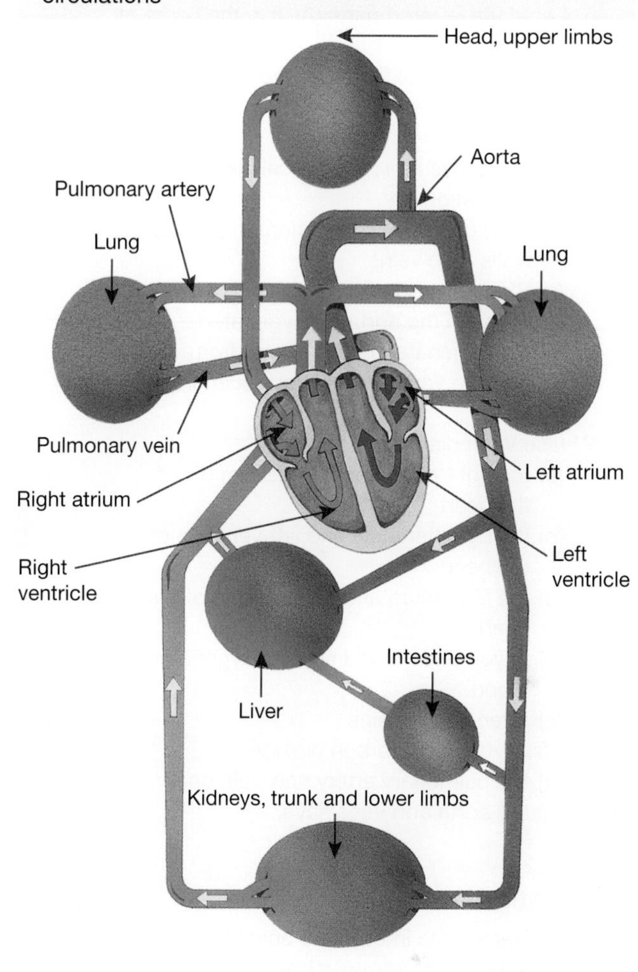

3.2 Exercises: Understanding and inquiring

To answer questions online and to receive **immediate feedback** and **sample responses** for every question, go to your learnON title at www.jacplus.com.au. *Note:* Question numbers may vary slightly.

Remember

1. State the word equation for cellular respiration.
2. Identify the molecule that the respiratory system and circulatory system work together to:
   (a) supply to your cells
   (b) remove from your cells.
3. Construct a flowchart to show the transport of:
   (a) oxygen from your nose to your alveoli
   (b) oxygen from your lungs to haemoglobin in your red blood cells
   (c) oxygenated blood from your lungs to your body cells
   (d) deoxygenated blood from your body cells to your lungs
   (e) carbon dioxide from your lungs to your nose.
4. Construct a diagram to show the interactions between your heart, body cells and blood vessels.

Think and discuss

5. Label the lettered parts (A–J) in the figure at right.
6. Identify which of the following statements are true and which are false. Justify your response.
   (a) Oxygen is a product of cellular respiration.
   (b) Arteries have thicker, more muscular walls than veins.
   (c) Blood travels to the heart in arteries.
   (d) Blood in the aorta is oxygenated.
   (e) Deoxygenated blood travels from your heart to your lungs in your pulmonary vein.
7. Use Venn diagrams to compare:
   (a) the right atrium and left atrium of the heart
   (b) the right ventricle and left ventricle of the heart
   (c) the left atrium and left ventricle of the heart
   (d) oxygenated blood and deoxygenated blood
   (e) arteries and veins
   (f) oxygen and carbon dioxide
   (g) the pulmonary artery and pulmonary vein
   (h) the aorta and vena cava.

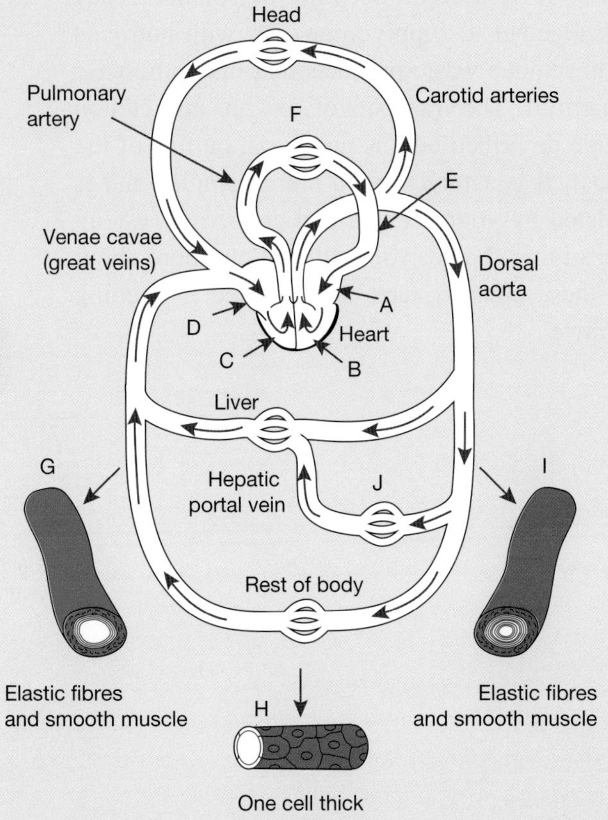

Investigate, think and create

8. (a) Search the internet for animations or simulations showing how the circulatory or respiratory systems function.
   (b) Select your favourite animation or simulation.
   (c) Construct a PMI chart that outlines what you liked about the animation, what you didn't like and how it could be improved.
   (d) Create your own multimedia version on the circulatory system and/or respiratory system.
   (e) Share your creation with the class. 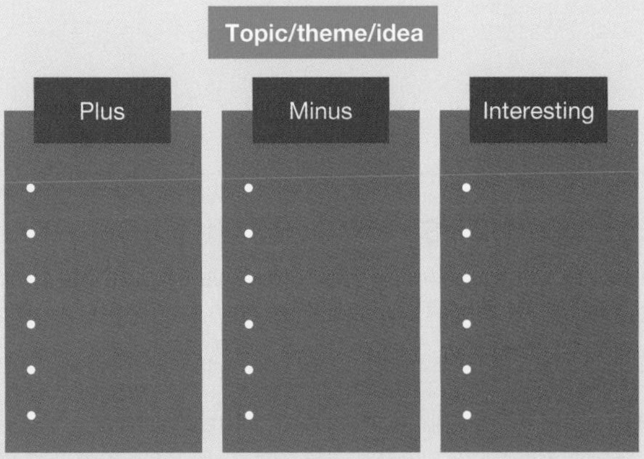
9. Find out more about the structure and function of either the circulatory or respiratory system and create a model that helps you to explain why the system is so important to survival.
10. Find examples of how developments in imaging technologies have improved our understanding of the functions and interactions of our body systems. Share your findings with others.
11. Investigate how technologies using electromagnetic radiation are used in medicine; for example, in the detection and treatment of cancer of the circulatory or respiratory system.
12. (a) Find examples of scientific research on either the circulatory or respiratory system.
    (b) Create a poster, PowerPoint presentation or podcast on the research that interests you most and present your findings to the class.

13. (a) Use the internet to identify problems relating to either the circulatory or respiratory system.
    (b) Select one of these problems and construct a model or animation to demonstrate its effect on normal body function.

In the capillaries, oxygen diffuses out of the blood and waste produced by cells diffuses into the bloodstream.

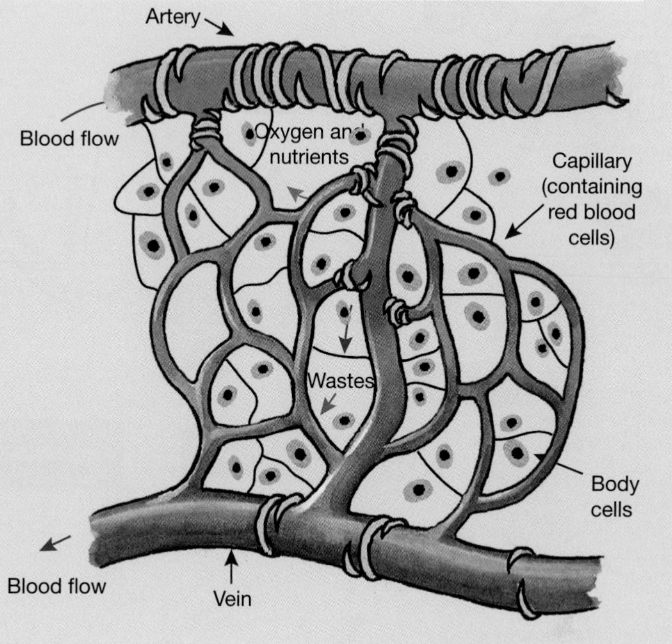

Complete this digital doc: Worksheet 3.1: Cells, tissues, organs and systems (doc-18864)

3.3 Essential intake

3.3.1 Essential nutrients

Feeling hungry? Tummy rumbling? You need to eat to provide your body with nutrients.

Nutrients are substances needed for energy, cell functioning and for your body's growth and repair. The five main groups of nutrients that your body needs to stay alive are:

- carbohydrates
- proteins
- lipids
- vitamins
- minerals.

All of these except minerals are called organic nutrients because they contain carbon, hydrogen and oxygen.

Carbohydrates and **lipids** are nutrients that provide you with an immediate source of energy and a back-up supply. While **proteins** can supply some energy, their key role is as bodybuilding compounds. They provide the raw materials required for cell growth and the repair of damaged and worn-out tissues. They are also involved in many other activities in your body; important chemicals such as enzymes and hormones are made of protein. Although vitamins have no energy value, they are needed in small amounts to keep you healthy and to speed up a variety of chemical reactions in your body.

Mind-mapping your essential nutrients

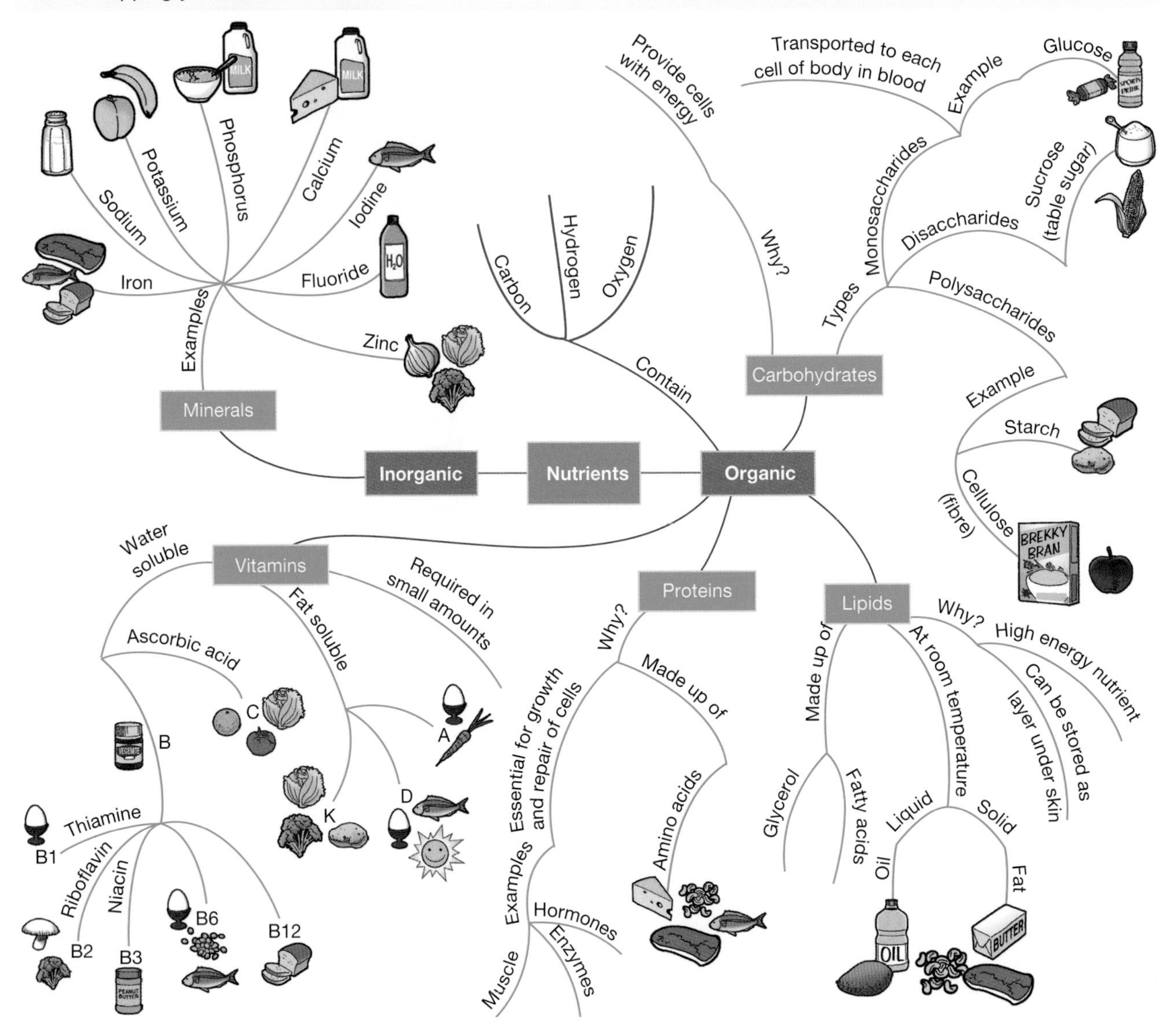

3.3.2 Feeding your emotions

Your body also needs the raw materials from nutrients to make neurotransmitters that can affect your emotions. For example, dopamine and norepinephrine are made up of three amino acids. These are tryptophan, tyrosine and phenylalanine. Tryptophan can be obtained from food sources such as cottage cheese, peanuts, red meat and brown rice, tyrosine from foods such as almonds, avocados, bananas and dairy products, and phenylalanine from meat, fish, eggs and soy products. Tryptophan is also important in the synthesis of another neurotransmitter called serotonin. For more information on these neurotransmitters see section 3.7.2.

3.3.3 Small but important

Even though **vitamins** and **minerals** are required in only small amounts, they are very important to your health. They are both needed to control chemical changes in your body. Your **endocrine system** and **nervous system** also require a number of these to be able to effectively function and maintain a healthy environment for your cells.

Vitamins
Vitamin
Why it is needed
Deficiency signs
A
Healthy skin and eyes; lining of digestive and respiratory systems; healthy bones and teeth
Mouth ulcers, poor night vision, frequent colds, flaky skin, dandruff
B1 (thiamine)
Release of energy from carbohydrates; helps heart work properly
Loss of appetite, indigestion, irritability, loss of muscle tone, poor concentration, poor memory
B2 (riboflavin)
Healthy skin and eyes, repairs damaged tissue; helps cells use oxygen
Sore tongue, sensitivity to bright lights, red burning eyes, cracked lips
B3 (niacin)
Healthy skin; helps release of energy from carbohydrates, fats and proteins
Lack of energy, dermatitis, diarrhoea, mental confusion
B6
Healthy teeth and gums, red blood cells and blood vessels
Depression, irritability, muscle tremors, dermatitis, reduced immunity
B12
Proper development of red and white blood cells
Poor hair condition, lack of energy, pale skin, shortness of breath
C (ascorbic acid)
Healthy bones and teeth; healthy tissues; healing of wounds
Frequent infections, bleeding gums, dry skin
D
Allows calcium and phosphorus to be used, to help bones and teeth grow and stay healthy
Joint pain or stiffness, backache, muscle cramps, tooth decay, hearing loss
E
Keeps body cells healthy
Easy bruising, slow wound healing, dry and dull skin and hair
K
Normal blood clotting
Impaired blood clotting
Minerals
Mineral
Why it is needed
Deficiency signs
Calcium
Healthy bones and teeth; blood clotting
Brittle bones, muscle cramps, insomnia, joint pain, arthritis, high blood pressure, tooth decay
Magnesium
Strengthens bones and teeth; promotes healthy muscles; important in nervous system
Muscle tremors, insomnia, depression, kidney stones, high blood pressure
Zinc
Healing wounds; proper growth; vision
Poor sense of taste, loss of appetite, slow wound healing, recurring infections, poor night vision
Selenium
Antioxidant that reduces risk of cancer; detoxifies heavy metals; strengthens effect of vitamin E
Signs of premature ageing, high blood pressure, frequent infections
Potassium
Proper working of nerves; keeps blood healthy
Fatigue, slow reflexes, muscle weakness, dry skin
Iron
Needed by red blood cells to carry oxygen to tissues
Anaemia, hair loss, sensitivity to cold

A lack of any of the 13 vitamins can cause disease. Diseases caused by a lack of vitamins are called **vitamin-deficiency diseases**. Diseases such as scurvy, rickets and beriberi have become less common as people have become more aware of the importance of vitamins. Deficiencies of minerals can also cause a number of significant problems.

Nutrients are needed for energy, cell functioning and body growth and repair.

3.3.4 Essential non-nutrients

Foods contain other important substances that are not nutrients. They are not used for energy or for growth and repair, but they are still essential to your health. Two of these substances are water and fibre.

3.3.5 Essential water

Did you realise that about two-thirds of your body is water? Water is another essential substance that you need to stay alive. Many of the chemical reactions that take place inside you use water. Your blood is 90 per cent water — the fluid part (plasma) is mostly water. Blood helps carry nutrients around your body and wastes away from it. You may be able to survive 40 days without food, but no more than 3 days without water.

How much water have you drunk today? Each day you lose water when breathing out (0.5 litres), sweating (0.5 litres) and urinating (1.5 litres). Have you replaced water that you have lost today? If you lose too much water, you may become **dehydrated**. A dry throat and mouth and dark-coloured urine are signs of mild dehydration. If you lose more than 20 per cent of your body's water volume, you could die!

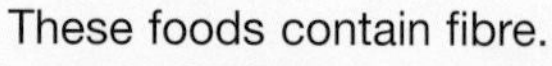

These foods contain fibre.

3.3.6 In one end and out the other

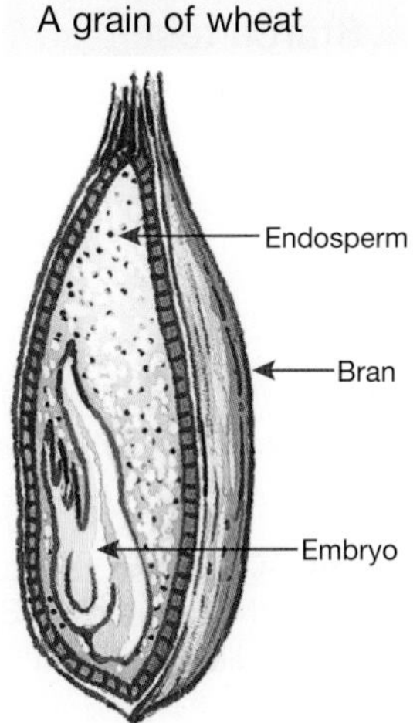

Fibre is found in the walls of plant cells. It is only partially broken down by your digestive system. Although it really does go 'in one end and out the other', it serves a very useful purpose and is an essential part of your diet. It provides bulk to your food, allowing it to move properly through your intestines. Without fibre, undigested food travels too slowly through the large intestine, losing too much water. The result is difficulty in releasing the solid food waste from the body, a condition called **constipation**. Lack of fibre in the diet can also lead to haemorrhoids (varicose veins around the anal passage, also known as piles), bowel cancer and several other diseases.

Fibre can be found only in foods that come from plants — foods such as fruits and vegetables, wholegrain breads and cereals, nuts and seeds.

Wholegrain products are higher in dietary fibre because they contain the outer covering, or bran, of the grain. When grains are highly processed, as they are in the production of white bread, white flour and many breakfast cereals, the bran is removed.

INVESTIGATION 3.1

Essential testing

AIM: To investigate the nutrients found in foods

Materials:

test-tube rack
4 test tubes
safety glasses
glucose solution
starch solution
gelatine solution
distilled water
iodine solution
test-tube holder
Benedict's solution
tongs
candle or Bunsen burner
matches
heatproof mat
0.01 M copper sulfate solution
1.00 M sodium hydroxide solution
food samples

Method and results

1. Copy and complete the table below for recording the test results.

Test results	Water	Glucose solution	Starch solution	Gelatine solution
Starch test				
Glucose test				
Protein test				

From your kitchen cupboards, select five foods to test. If they are solid, you may need to use a mortar and pestle to grind them into a 'mash' with a small amount of water before testing.

2. Predict which of your food samples will contain starch, glucose and/or protein.

Essential standards

- For each of the tests in this experiment, set up the four test tubes as shown below. After each test, clean the test tubes by rinsing with water. Make sure a fresh sample of each liquid is used for each test.

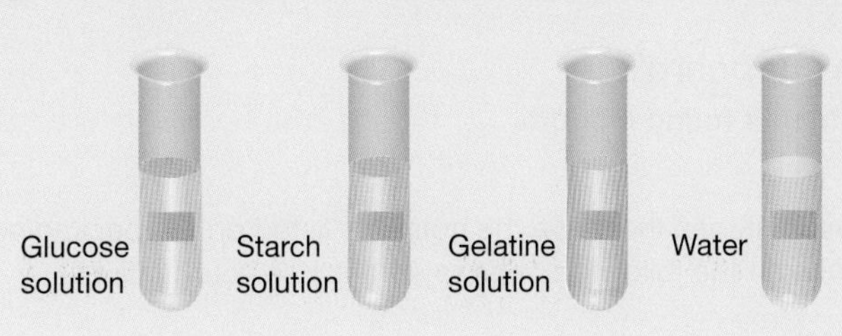

Starch test

3. Add two drops of iodine solution to each of the four test tubes. Observe any colour change and record the results.

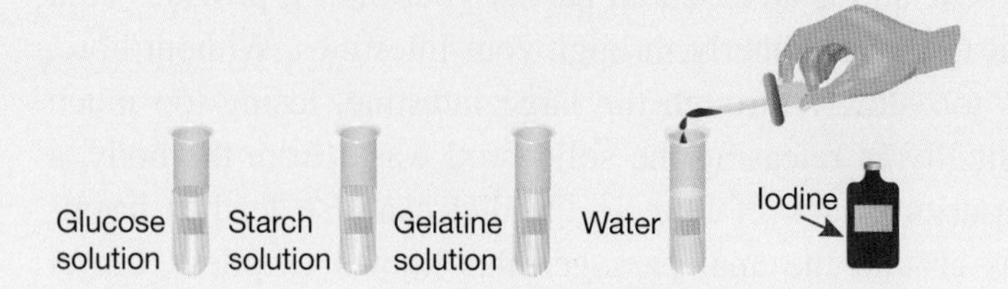

Glucose test

4. Add four drops of Benedict's solution to each of the four test tubes. Gently heat each test tube over the candle or Bunsen burner flame. Observe any colour change and record the results.

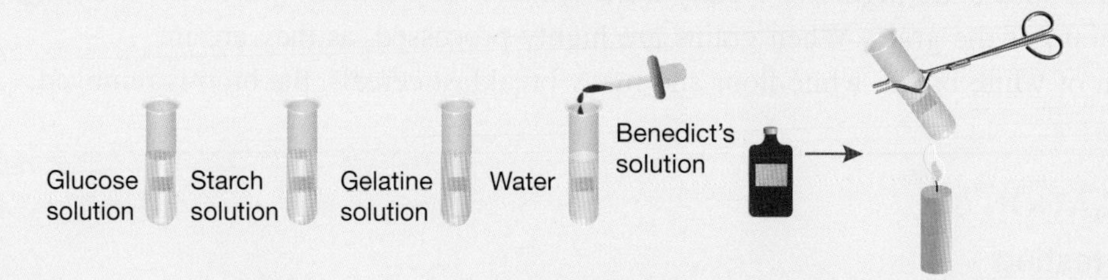

Protein test

5. Add ten drops of copper sulfate solution to each of the four test tubes. Then add five drops of sodium hydroxide solution to each test tube. Observe any colour change and record the results.

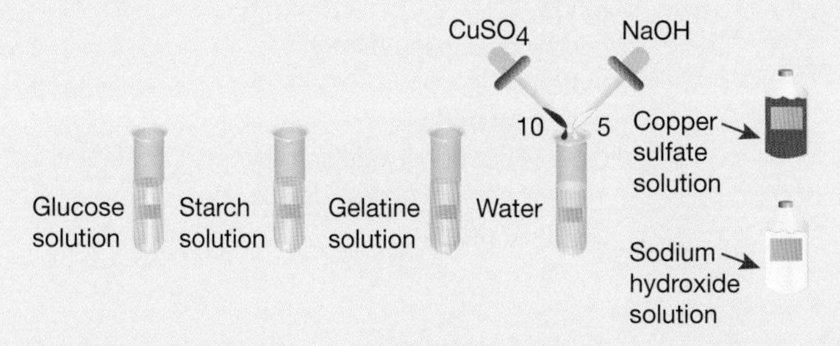

Essential food tests

6. Using the three tests above, investigate the food samples for the presence of starch, glucose and protein. (*Note*: Add *only* your food samples to these tests, *not* the glucose, starch or gelatine solutions.)

Discuss and explain

7. Suggest why you set up standard tests and added the same volumes of solutions to each test tube.
8. Which foods contain two or more of the nutrients tested for?
9. Were your predictions supported by your results?
10. Comment on your overall findings.
11. If you were to do the food testing again, suggest how you might improve the procedure.

INVESTIGATION 3.2

What's in your kitchen cupboard?

AIM: To investigate the nutrients found in foods

Method and results

- Find ten food items in your kitchen that have the nutrients listed on the packaging.

1. Draw up a table like the one on the following page to summarise your findings.

Discuss and explain

2. Which of the foods was highest in:
   (a) energy
   (b) protein
   (c) fibre
   (d) sodium?
3. Rank the foods in order from highest to lowest for:
   (a) fat
   (b) fibre
   (c) energy.
   Are your results what you expected? Why?
4. The recommended daily fibre intake is 30–40 g. On the basis of your findings, put together a meal of your packaged foods that would meet this requirement.
5. Draw a bar graph of your results for your foods and their fat content.
6. (a) Compare your results with those of two other students.
   (b) How were they similar and how were they different?
   (c) Select foods from your group to put together a meal. Using the nutrient tables, calculate the amount of each nutrient in your designed meal.
7. (a) Suggest three questions that you could research on the topics of nutrients or packaging labels.
   (b) Collate the questions from the whole class and select one of these questions to research.
   (c) Report your findings back to the class.

Name of food	Nutrients per 100 grams						
	Energy (kJ)	Protein (g)	Fat (g)	Total carbo-hydrate (g)	Sugar (g)	Dietary fibre (g)	Sodium (mg)
'Light'n 'tasty' cereal	1540	8.7	3.1	71.3	23.6	7.7	225
Potato chips	2130	8.5	31.9	46.1	1.8	3.0	518
Apricot jam	1140	6.4	0.1	66.6	59.4	–	17
Barbecue-flavoured 'Shapes' biscuits	2184	10.2	25.2	63.3	1.4	–	752
Multigrain corn thins	380	9.5	3.0	77.6	0.7	8.5	201

HOW ABOUT THAT!

Nutrient careers

Nutritionists and dietitians are examples of two careers with a focus on nutrition. These careers may involve communication of nutrition messages to individuals or to various groups within the community. These careers may be in private, public or community health, in the food industry or in various types of research.

Clinical nutritionists may have face-to-face consultations and discussions with their clients about dietary changes that may be required. While clinical nutritionists approach issues from a 'nutrient' perspective, dietitians may be working in a hospital or private practice to advise their clients about food and lifestyle changes.

Dietitians may also be involved in determining the appropriate food solution for patients who require a drip or nasogastric tube (a tube that goes through the nose and down into the stomach).

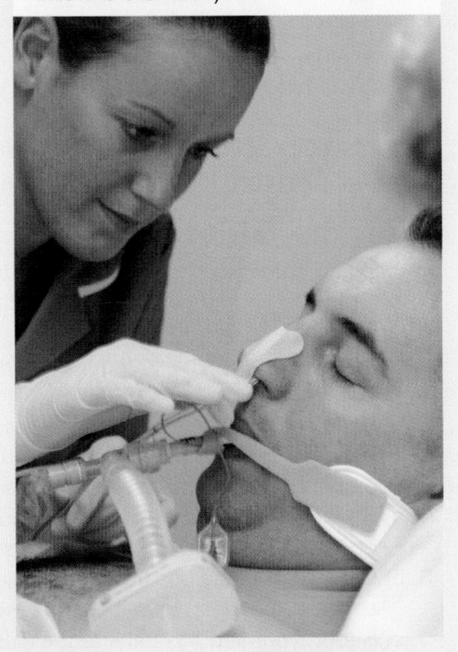

Dietitians provide advice to people diagnosed with diet-related diseases such as diabetes, coeliac disease, heart disease and certain types of cancers.

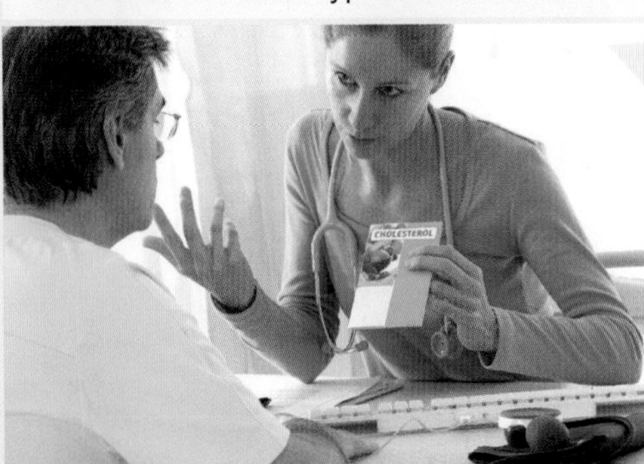

3.3 Exercises: Understanding and inquiring

To answer questions online and to receive **immediate feedback** and **sample responses** for every question, go to your learnON title at www.jacplus.com.au. *Note:* Question numbers may vary slightly.

Remember

1. Give two reasons why you need to eat.
2. (a) List the five main groups of nutrients.
   (b) Which of these are organic nutrients?
3. Identify shared features for the following pairs.
   (a) Carbohydrates and lipids
   (b) Cellulose and starch
   (c) Fats and oils
   (d) Iron and potassium
   (e) Hormones and enzymes
4. How do cells get the energy that they need?
5. (a) What are proteins made of?
   (b) Why are they important?
6. List some ways you can lose water.
7. Explain why it is important to drink water.
8. Describe the symptoms of dehydration.
9. What is fibre?
10. Why is it important to eat fibre even though the chemicals in it are not used by your body?
11. List the types of food you would recommend to a person lacking:
    (a) vitamin C
    (b) calcium
    (c) iron.

12. Explain why the following vitamins are important to your health.
    (a) A
    (b) C
    (c) K
13. Describe the deficiency signs of:
    (a) calcium, zinc and magnesium
    (b) vitamins A, B2 and D.
14. (a) Which amino acids are required for the synthesis of dopamine and norepinephrine? Name some foods that these are found in.
    (b) Which amino acid is important for the synthesis of serotonin? Name some foods that it can be found in.

Think and investigate

15. Suggest why pregnant women, children and adolescents need more protein than other adults.
16. Milk and other dairy products are well known as good sources of calcium. Which nutrients would be missing from the diet of someone whose food intake consisted mainly of dairy products?
17. Too much salt (sodium chloride) in your diet is not healthy. Why do we need salt at all?
18. Find out more about pregnancy and folate deficiencies.
19. Select a vitamin or mineral, find out details about it that interest you and then create a brochure to advertise it to others.
20. Suggest ways to encourage people to drink more water.
21. (a) An increased number of Australian women are being diagnosed as deficient in vitamin D. Suggest possible reasons for this situation.
    (b) List the symptoms associated with vitamin D deficiency.
    (c) Other than vitamin supplements, suggest ways in which this deficiency can be treated.
    (d) Design an investigation to test the effectiveness of your suggested treatments.
22. The addition of fluoride to our water supplies has caused much controversy. Find out and report on the key arguments for and against fluoridation.
23. Suggest how science can help individuals and communities make choices about their diet in terms of vitamins and minerals that they require.
24. Predict the relationship between darker skin and the amount of sunlight required for vitamin D production. Research and record information about the link, citing your references. Pose three questions that could be used to guide further research.
25. What do the results in the graph at right suggest?
26. Wholegrain products are high in fibre. Why do you think the word 'wholegrain' is used?
27. Some high-fibre breakfast cereals have more sugar added to them than some of the more highly processed breakfast cereals. Why do you think this is so?

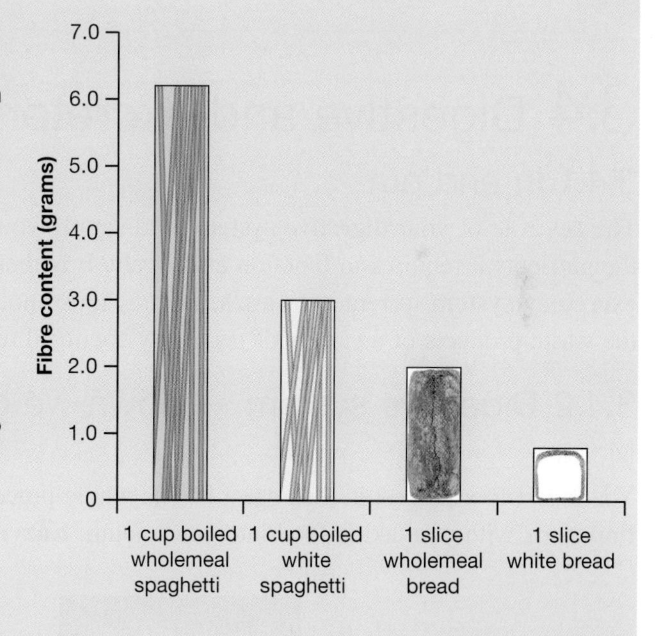

Investigate, report or create

28. Create your own vitamin and mineral learning tool to teach younger children the benefits of these nutrients to their health. This may be in the form of a song, play or poem, colourful flash cards with text and pictures, poster, brochure or booklet.
29. Construct your own vitamin or mineral wheel that is made up of two layers with a cut-out section in one area that allows the second layer to appear at the right time. One of the following formats should be used.
    - One layer with questions and the other with answers to the questions
    - One layer with types of foods and the other with vitamins or minerals found in those foods
    - One layer with types of vitamins or minerals and the other with diseases that can be caused by a deficiency in that vitamin or mineral and the types of foods in which it can be found

30. Prepare an advertisement to promote increasing fibre in people's diets. Your aim is to make high-fibre foods attractive to consumers. Your advertisement could be in the form of a poster, a dramatic performance, or a TV or radio commercial.
31. (a) In your team, brainstorm questions about proteins, lipids, vitamins and minerals.
    (b) Select one of these questions and suggest five further questions that you could use to find out more about it.
    (c) Use your questions to structure your research.
    (d) Organise your findings into a format that you will be able to share with others.
    (e) Report your findings back to your team or class.
32. A high-carbohydrate meal can increase your brain's tryptophan levels.
    (a) What effect might this have on your mood?
    (b) Which neurotransmitter is likely to be involved?
    (c) At what time of the day would it be a good idea to have such a meal? Why?
33. A high-protein meal can raise tyrosine levels in your blood and brain.
    (a) What effect might this have on your mood?
    (b) Which neurotransmitter(s) is/are likely to be involved?
    (c) At which time of the day would it be a good idea to have such a meal? Why?
    (d) If tyrosine is also needed to make active thyroid hormones, what may result if there are insufficient levels of this amino acid in your blood?

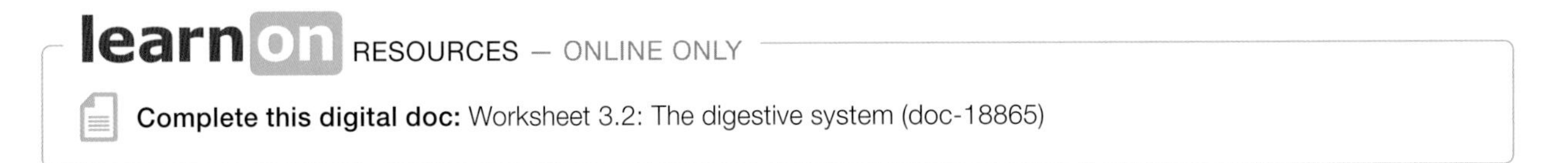

Complete this digital doc: Worksheet 3.2: The digestive system (doc-18865)

3.4 Digestive and excretory systems

3.4.1 In and out

The key role of your **digestive system** is to supply your body with the nutrients it requires to function effectively. It is then up to your **excretory system** to remove wastes, such as those not digested or the waste products of a variety of necessary chemical reactions.

3.4.2 Digestive system — down we go

Mouth

You ingest food, digest it, then egest it. The whole process of **digestion** starts with you taking food into your mouth. **Enzymes** (such as

The human digestive system

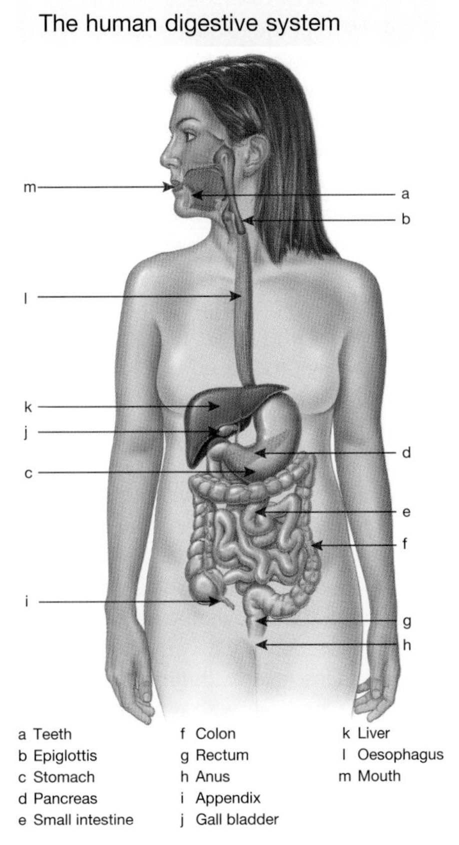

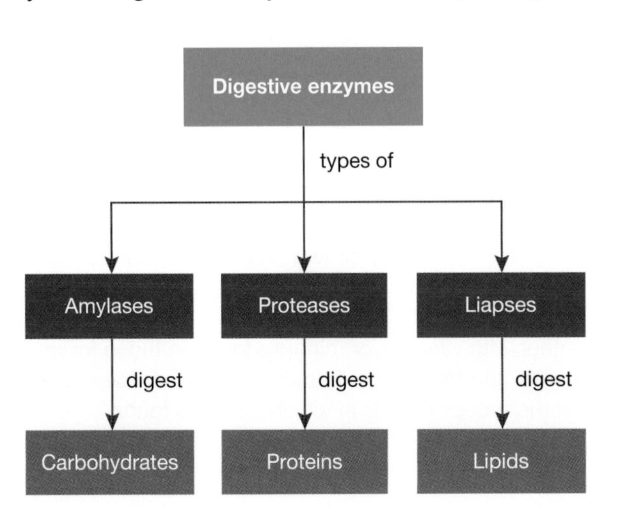

amylases) in your **saliva** are secreted by your **salivary glands** begin the process of **chemical digestion** of some of the carbohydrates. Your teeth physically break down the food in a process called **mechanical digestion**, then your tongue rolls the food into a slimy, slippery ball-shape called a **bolus**.

3.4.3 Oesophagus to stomach

The bolus is then pushed through your **oesophagus** by muscular contractions known as **peristalsis**. From here it is transported to your stomach for temporary storage and further digestion.

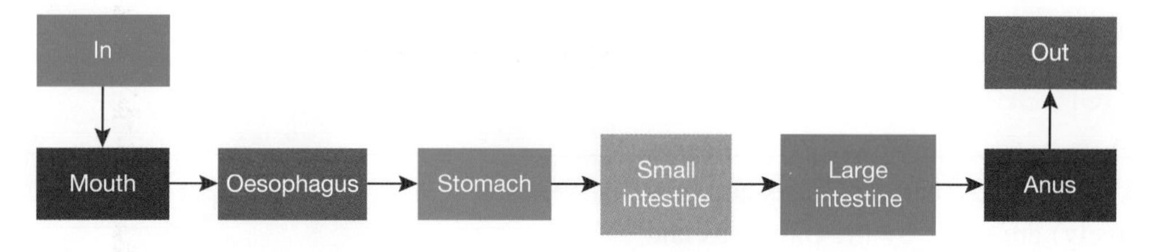

3.4.4 Stomach to small intestine

Once the food gets from your stomach to your **small intestine**, more enzymes (including amylases, proteases and lipases) turn it into molecules that can be absorbed into your body. The **absorption** of these nutrient molecules occurs through finger-shaped **villi** in the small intestine. Villi are shaped like fingers to maximise surface area to increase the efficiency of nutrients being absorbed into the surrounding capillaries. Once absorbed into the capillaries (of your circulatory system) these nutrients are transported to cells in the body need them.

3.4.5 Large intestine

On its way through the digestive tract (alimentary canal), undigested food moves from the small intestine to the **colon** of the **large intestine**. It is here that water and any other required essential nutrients still remaining in the food mass may be absorbed into your body. **Vitamin D** manufactured by bacteria living within this part of the digestive system is also absorbed. Any undigested food, such as the **cellulose** cell walls of plants (which we refer to as fibre) also accumulate here and add bulk to the undigested food mass.

The **rectum** is the final part of the large intestine and it is where faeces is stored before being excreted through the **anus** as waste.

The absorption of most nutrients into your body occurs in the ileum, the last section of the small intestine. The finger-like villi on its walls give it a large surface area that speeds up nutrient absorption. Many tiny blood vessels called capillaries transport the nutrients from the villi into your bloodstream. Undigested material continues on to the large intestine where water and vitamins may be removed, and then the remainder is pushed out through the anus as faeces.

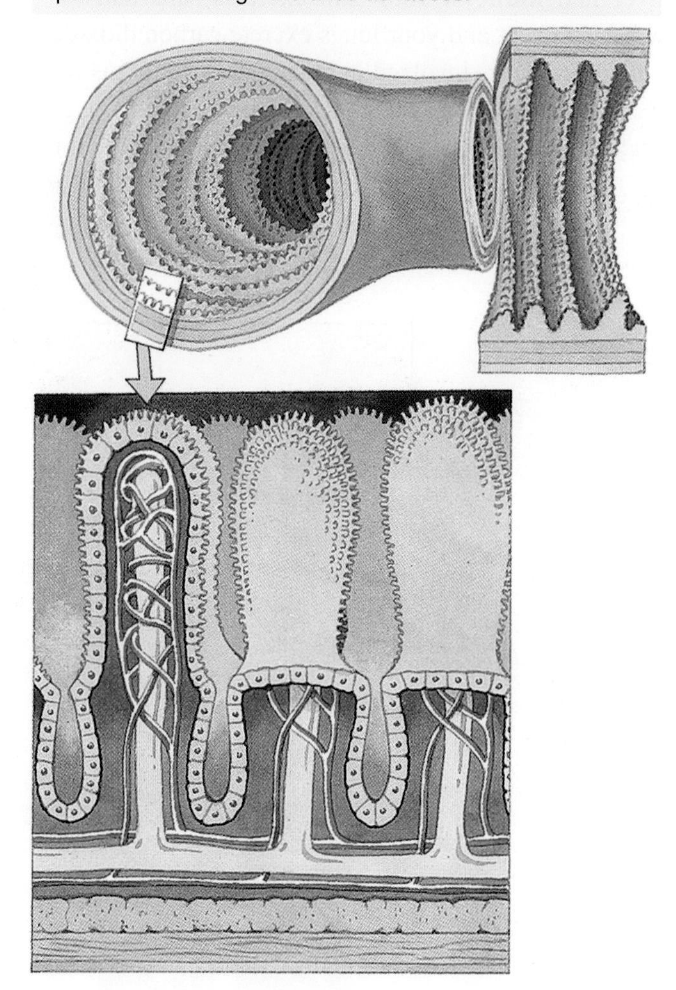

3.4.6 Liver

Your liver is an extremely important organ with many key roles. One of these is the production of **bile** which is transported to your gall bladder via the bile ducts to be stored until it is needed. Bile is transported from the **gall bladder** to the small intestine where it is involved in the mechanical digestion of lipids such as fats and oils.

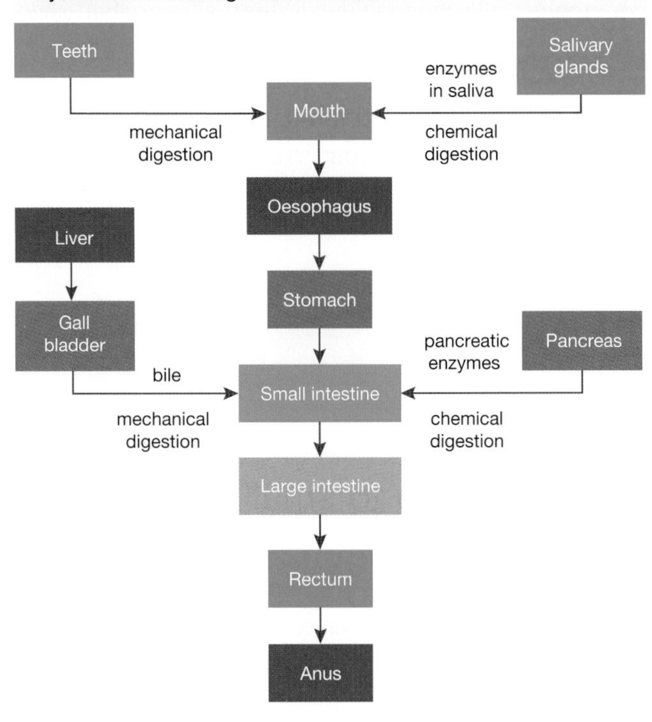

3.4.7 Pancreas

Enzymes such as **lipases**, **amylases** and **proteases** (which break down lipids, carbohydrates and proteins respectively) are made by the **pancreas** and secreted into the small intestine to chemically digest these components of food materials.

3.4.8 Excretory systems — out we go

Excretion is any process that gets rid of unwanted products or waste from the body. The main organs involved in human excretion are your **skin**, **lungs**, **liver** and **kidneys**. Your skin excretes salts and water as sweat and your lungs excrete carbon dioxide (produced by cellular respiration) when you breathe out. Your liver is involved in breaking down toxins for excretion and your kidneys are involved in excreting the unused waste products of chemical reactions (e.g. urea) and any other chemicals that may be in excess (including water) so that a balance within our blood is maintained.

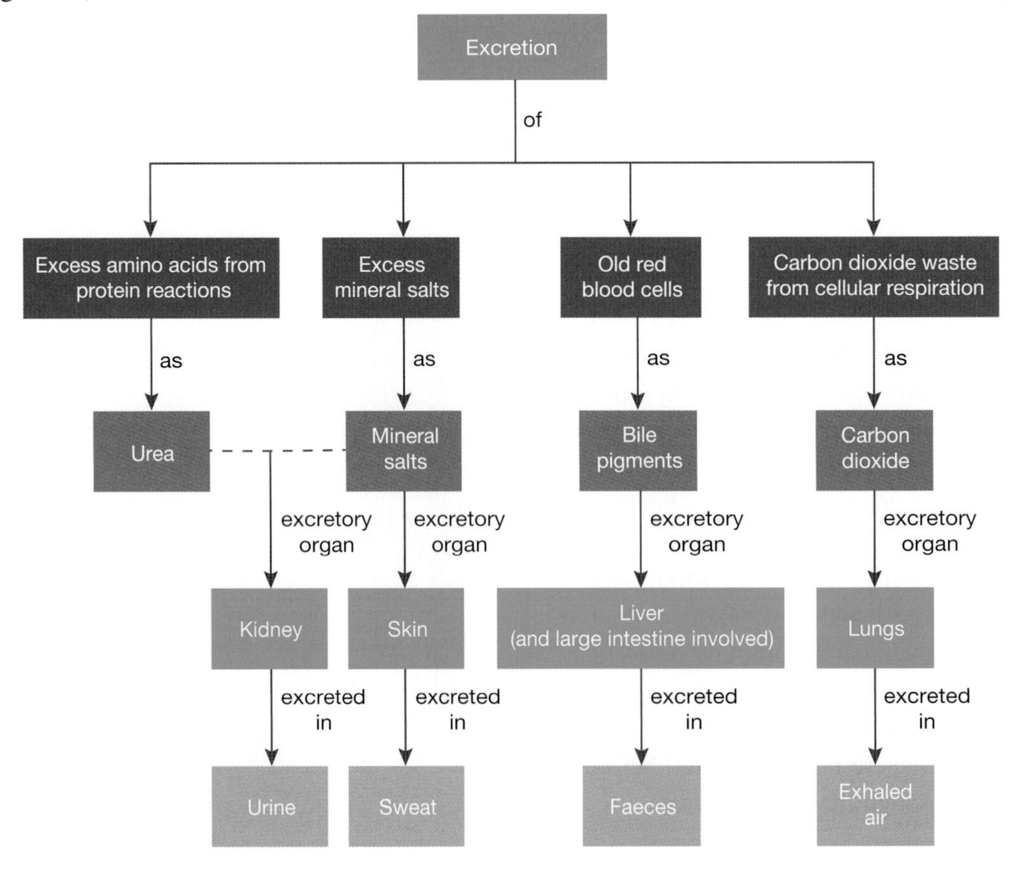

3.4.9 Liver

Over a litre of blood passes through your liver each minute. Your liver is like a chemical factory, with more than 500 different functions. Some of these include sorting, storing and changing digested food. It removes fats and oils from the blood and modifies them before they are sent to the body's fat deposits for storage. It also help get rid of excess protein, which can form toxic compounds dangerous to the body. The liver converts these waste products of protein reactions into urea, which travels in the blood to the kidneys for excretion. It also changes other dangerous or poisonous substances so that they are no long harmful to the body. Your liver is an organ that you cannot live without.

3.4.10 Kidneys

Your kidneys play an important role in filtering your blood and keeping the concentration of various chemicals and water within appropriate levels. Each of your kidneys is made up of about one million **nephrons**. These tiny structures filter your blood, removing waste products and chemicals that may be in excess. Chemicals that are needed by your body are reabsorbed into capillaries surrounding them. The fluid remaining in your nephrons at the end of its journey then travels through to your **bladder** via your **ureters** for temporary storage until it is released as **urine**.

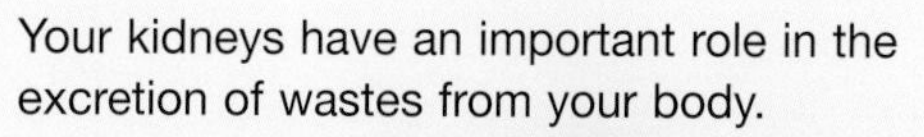

Your kidneys have an important role in the excretion of wastes from your body.

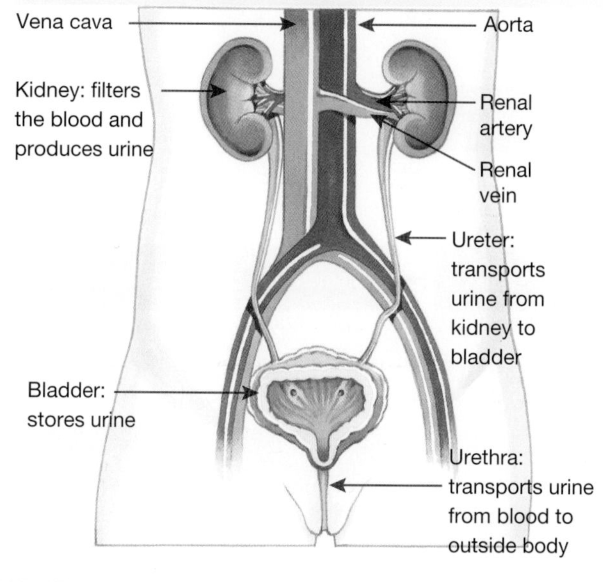

HOW ABOUT THAT!

Fish maintain their salt and water balance in different ways. Saltwater fish, such as snapper, drink sea water constantly and produce a small volume of urine. Freshwater fish, such as Murray cod, however, rarely drink, but make lots of urine.

Diagram of a nephron. Each of your kidneys is made up of about a million nephrons.

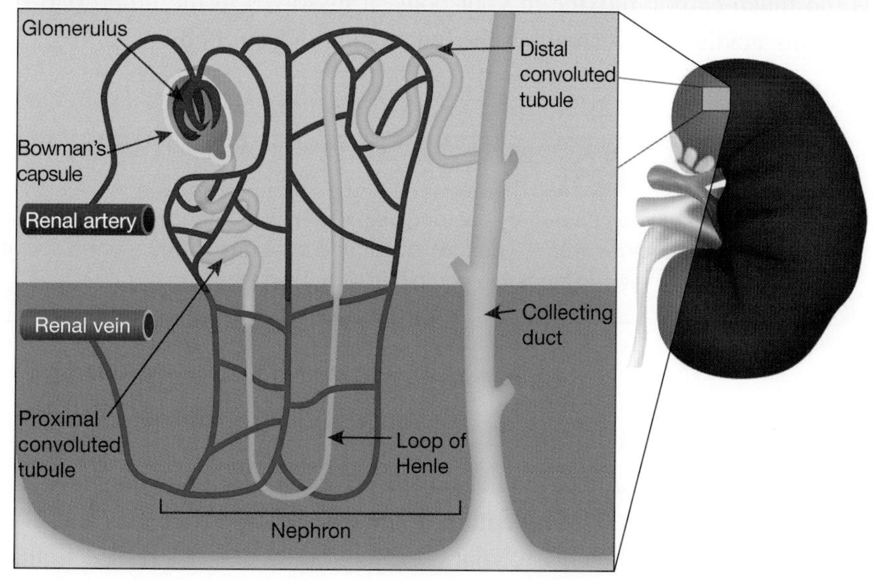

3.4.11 Blood, water and urine

Both blood and urine are mostly made up of water. Water is very important because it assists in the transport of nutrients within and between the cells of the body. It also helps the kidneys do their job because it dilutes toxic substances and absorbs waste products so they may be transported out of the body.

3.4.12 Too much or too little?

The concentration of substances in the blood is influenced by the amount of water in it. If you drink a lot of water, more will be absorbed from your large intestine, and the kidneys will produce a greater volume of dilute urine. If you do not consume enough liquid you will urinate less and produce more concentrated urine.

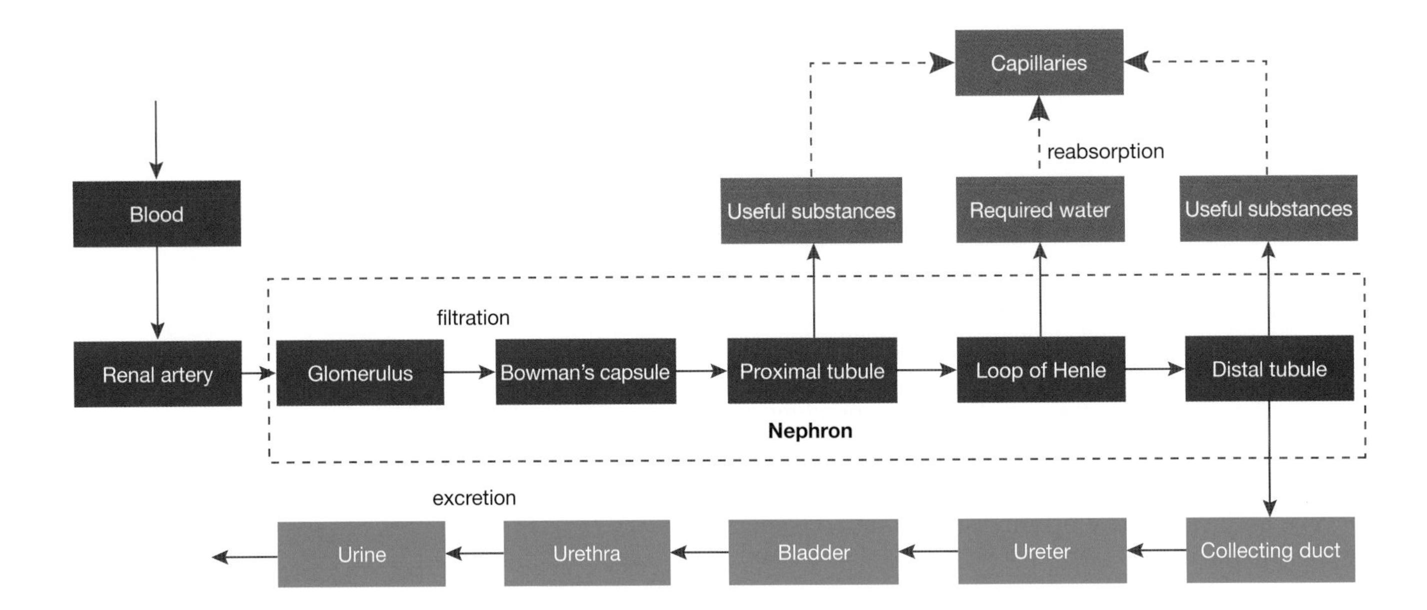

3.4.13 Lungs

Did you know that your body is more sensitive to changes in levels of carbon dioxide than oxygen? If there is too much carbon dioxide in your body, it dissolves in the liquid part of blood and forms an acid. The resulting acidic blood can affect the functioning of your body.

HOW ABOUT THAT!

The amount of oxygen carried by haemoglobin varies with altitude. At sea level, about 100 per cent of haemoglobin combines with oxygen. At an altitude of about 13 000 metres above sea level, however, only about 50–60 per cent of the haemoglobin combines with oxygen.

3.4.14 Blood and carbon dioxide

The amount of carbon dioxide in your blood influences your breathing rate. The level of carbon dioxide in the blood is detected by **receptors** in the walls of some arteries and in the brain. If the levels of carbon dioxide in your blood increase, your breathing rate will be increased so that carbon dioxide can be exhaled from your lungs and passed out of your body.

If you were to climb up high on a mountain, you would need time for your body to adjust. Initially you would feel tired and out of breath because you would be restricted by the limited amount of oxygen available to your cells. Your breathing and heart rate would increase in an effort to get more oxygen around your body. In time, your body would begin to produce more red blood cells and hence more haemoglobin. After this, your breathing and heart rate would return to normal.

3.4.15 Cellular respiration

Glucose is an example of a nutrient that may be released from digested food. It is absorbed in your small intestine and then taken by the capillaries to cells for use in **cellular respiration**. In this process the glucose is combined with oxygen, and is then broken down into carbon dioxide (a waste product that needs to be removed from the cell) and water. During this reaction energy, in the form of **ATP** (adenosine triphosphate), is also released. ATP provides the cells with the energy needed to perform many of its activities, and is essential to life.

Glucose + oxygen → carbon dioxide + water + energy (ATP)

This is an example of systems working together. Glucose is supplied via the digestive system and oxygen is supplied via the **respiratory system**. The **circulatory system** transports nutrients (such as glucose) and oxygen to your cells and removes wastes (such as carbon dioxide) from your cells. These wastes are then removed from your body by your excretory systems. Without a supply of glucose and oxygen, cellular respiration could not occur. Without removal of wastes, your cells may die. If you systems did not work together like they do, you would not be able to stay alive.

3.4 Exercises: Understanding and inquiring

To answer questions online and to receive **immediate feedback** and **sample responses** for every question, go to your learnON title at www.jacplus.com.au. *Note:* Question numbers may vary slightly.

Remember

1. Outline why the digestive system and the excretory system are important to the survival of your cells.
2. Identify examples of types of enzymes involved in the digestion of:
   (a) carbohydrates
   (b) proteins
   (c) lipids.
3. Explain why the villi in the small intestine are the shape that they are.
4. Describe how and where the nutrients are absorbed into your body from your digestive system.
5. Outline a way in which the liver is involved in digestion.
6. Identify the part of the digestive system in which water is absorbed into your body.
7. Construct a flowchart to show the route that undigested food material travels from your mouth to your anus.
8. Is cellulose digested? What happens to it?
9. Define the term *excretion*.
10. List examples of organs that are involved in human excretion.
11. Describe what happens when you drink a lot of water.
12. Describe one way in which excess salt is removed from your body.
13. Suggest reasons why you can't live without your liver.
14. Identify the name given to the:
    (a) tiny structures that make up the kidney
    (b) fluid that travels from your kidneys to your bladder for excretion.
15. Construct flowcharts to show the route travelled from the:
    (a) renal artery, through the nephron to the collecting duct
    (b) collecting duct to the urethra.

16. Suggest why a supply of water is important to your cells.
17. Is your body more sensitive to changes in carbon dioxide or oxygen levels? Explain.
18. Explain why mountain climbers sometimes find it difficult to breathe during a climb.
19. Explain how your cells obtain glucose and why it is important to survival.

Think and discuss

20. Use Venn diagrams to compare:
    (a) the digestive system and excretory system
    (b) the small intestine and large intestine
    (c) ingestion and egestion
    (d) proteases and lipases
    (e) cellulose and glucose
    (f) bile and enzymes
    (g) ureter and urethra
    (h) nephron and villi
    (i) the digestive system and respiratory system
    (j) the excretory system and circulatory system.

Analyse and evaluate

21. Use the table and the other information on these pages to answer the following questions.
    (a) Draw two bar graphs to show the quantity of water, proteins, glucose, salt and urea in blood and in urine.
    (b) Which substance is in the greatest quantity? Suggest a reason for this.
    (c) Which substances are found only in blood?
    (d) Which substances are found in urine in a greater quantity than in blood? Suggest a reason for this.
    (e) When would the amount of these substances in the urine become greater or less than in the blood?

	Quantity (%)	
Substance	In blood	In urine
Water	92	95
Proteins	7	0
Glucose	0.1	0
Chloride (salt)	0.37	0.6
Urea	0.03	2

Investigate, think and create

22. (a) Search the internet for animations or simulations that show how the excretory or digestive systems function.
    (b) Select your favourite animation or simulation.
    (c) Construct a PMI chart that outlines what you liked about the animation, what you didn't like and how it could be improved.
    (d) Create your own multimedia version on the circulatory system and/or respiratory system.
    (e) Share your creation with the class.
23. Find out more about the structure and function of either the digestive or excretory system and create a model that helps you to explain why the system is so important to our survival.
24. (a) Find examples of scientific research on either the digestive or excretory system.
    (b) Create a poster, PowerPoint presentation or podcast on the research that interests you most and present your findings to the class.
25. Imagine that you are a scientist working in a field related to the study of the circulatory or excretory system. Propose a relevant question or suggest a hypothesis for a scientific investigation and outline how you would design your investigation. Search the internet for relevant research or information.
26. Find examples of how developments in imaging technologies have improved our understanding of the functions and interactions of the digestive and excretory systems. Share your findings with your class.
27. Investigate how technologies using electromagnetic radiation are used in medicine; for example, in the detection and treatment of cancer of the digestive or excretory system.

28. There are often claims made in the media about particular products relating to the digestive and excretory systems. Examples include indigestion tablets, laxatives and weight loss tablets. Select one of these examples or one of your own to evaluate and/or test the claims being made in advertising or in the media.
29. Research and report on one of these conditions: urinary incontinence, kidney stones, dialysis, kidney transplants, cystitis.
30. Find out:
    (a) the differences between the urethra in human males and females
    (b) why pregnant women often need to urinate more frequently
    (c) how the prostate gland in males may affect urination in later life
    (d) which foods can change the colour or volume of urine
    (e) which tests use urine in the medical diagnosis of diseases.

Complete this digital doc: Worksheet 3.3: Removing waste from the blood (doc-18866)

3.5 Living warehouses

3.5.1 Living warehouses

It can be confusing trying to figure out what a healthy diet is when you are bombarded by so many different fad diets!

Many of these diets eliminate whole food groups and may put you at risk of developing a nutritional deficiency. Knowing how your body stores and uses energy — like a living warehouse — may help you to weigh up the risks and benefits of these 'wonder diets'.

To maintain a healthy weight, it is importatnt to balance your energy intake with the energy you use.

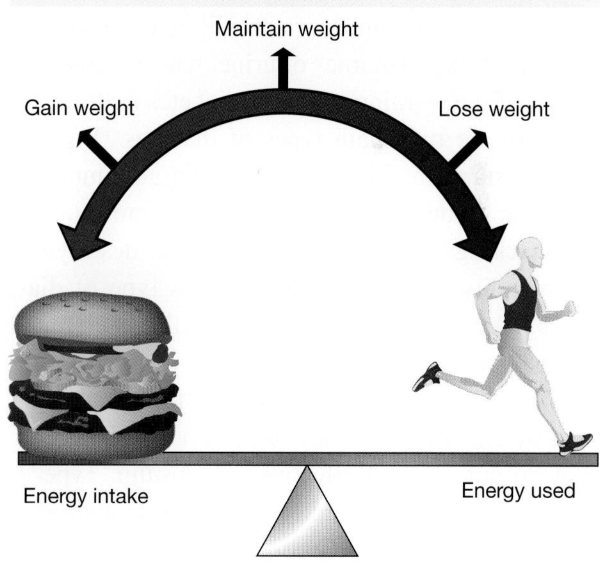

3.5.2 How much energy do you need?

To function effectively, your body needs energy. We gain energy from the foods that we eat. The amount of energy stored in this food is measured in kilojoules (kJ) or calories.

The amount of energy that you need depends on how big and active you are, how quickly you are growing and how fast your body uses it.

3.5.3 Balancing blood glucose

Your cells need **glucose** to use in the process of **cellular respiration** to make **ATP** (adenosine triphosphate) molecules. ATP is used by cells in reactions that require energy. This glucose is obtained from the food that you eat. Glucose molecules are transported in blood in your circulatory system to cells throughout your body.

If you have high levels of glucose in your blood, special cells in your **pancreas** detect this and release **insulin** into your bloodstream. Target cells in your muscles and liver receive this chemical message and glucose is taken out of the blood and converted into the storage polysaccharide **glycogen**. If the levels of blood glucose are too low, then another hormone, **glucagon** is released by the pancreas. Glucagon triggers the breaking down of glycogen into the monosaccharide glucose. This is how the glucose levels in the blood can be kept within a narrow range.

The hormones insulin and glucagon are secreted by the pancreas to control glucose levels in your blood.

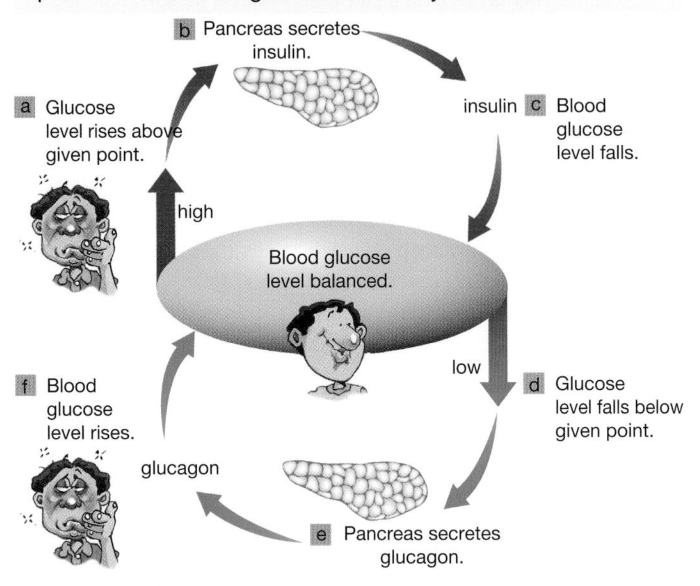

3.5.4 Diabetes

Diabetes mellitus is an endocrine disorder that is caused by a deficiency of insulin or a loss of response to insulin in target cells (such as those in liver and muscle tissue). This results in high blood glucose levels. Glucose levels can become so high that it is excreted by your kidneys and hence found in urine. Glucose in urine is one of the tests that are indicative of diabetes. The higher the glucose levels, the more water will be excreted with it. This results in the loss of large volumes of urine, which leads to persistent thirst; this is one of the warning signs for diabetes mellitus.

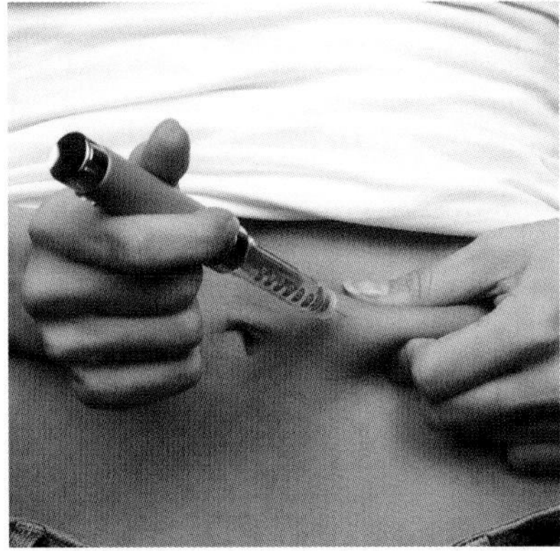

There are two main types of diabetes. **Type 1 diabetes mellitus** usually starts in childhood and is an autoimmune disorder. In this case, the immune system mounts an attack against cells in the pancreas, destroying their ability to produce insulin. This type of diabetes requires treatment with insulin injections. **Type 2 diabetes mellitus** usually starts later in life and is the most common form. It is characterised by either a deficiency of insulin or target cells that do not respond effectively to insulin. Type 2 diabetes has been linked to hereditary factors and obesity. It is usually controlled through exercise and diet.

Foods with a lower GI release energy more slowly and help you to feel full longer.

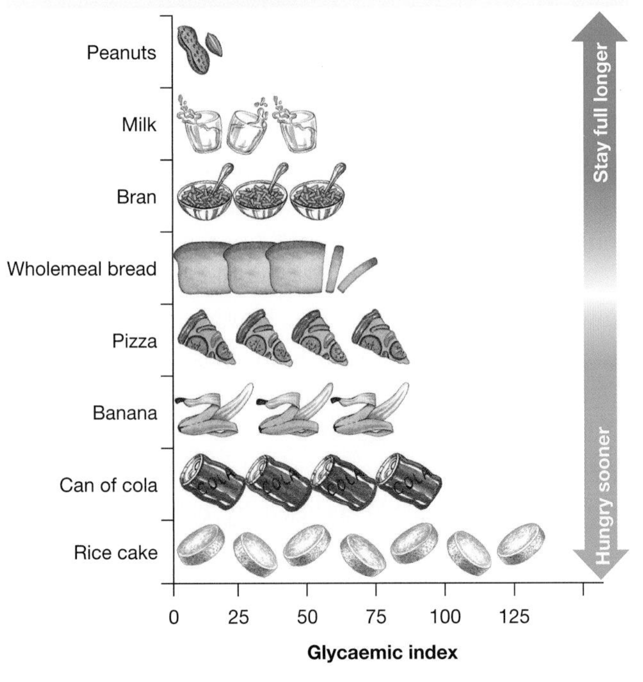

3.5.5 GI: high or low?

You may have noticed that some foods are labelled as 'low GI' or 'high GI'. This refers to the **glycaemic index** of the food. This is a measure of the time it takes for your blood sugar level to rise after you have eaten it. Foods that are considered to be low GI are digested more slowly than those that are high GI. This means that blood glucose levels will rise more slowly and over a longer period of time. This will mean that you will feel fuller for longer. High GI foods provide a short burst of glucose and you may start to feel hungry as your blood glucose levels drop.

3.5.6 Foods with a high GI

Foods such as white bread, rice and mashed potatoes contain starch and sugar, which are porous and have a high surface-to-volume ratio. This means that they can be digested easily by the enzyme amylase. Such foods have a high glycaemic index and can cause a sharp rise in blood sugar.

These foods are very good if you have been active and need to replenish energy stores quickly. The chart on the previous page and table below show some foods that will supply energy quickly and some that will help you to feel full longer.

Foods with a high GI, such as chocolate, cause a sharp rise in blood sugar. Foods with a low GI, such as nuts, result in a more moderate but longer lasting rise in blood sugar.

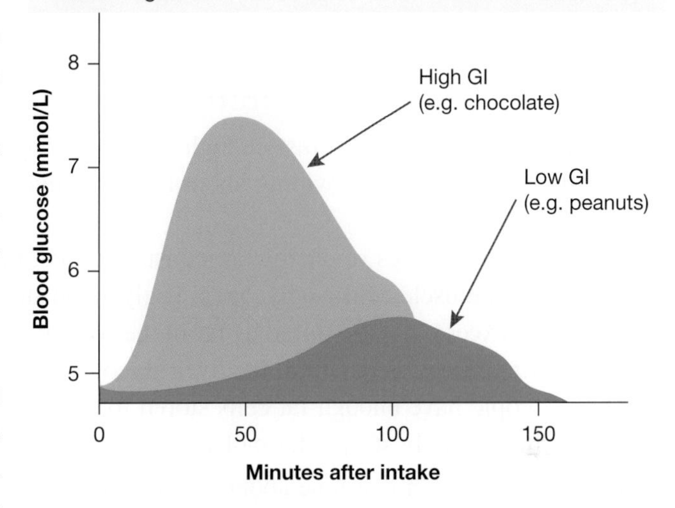

3.5.7 Foods with a low GI

Foods rich in fibre, such as wholemeal bread and thick pasta like spaghetti, are digested more slowly. This is because the more compact physical form of these foods makes it harder for the enzyme amylase to reach its substrate. These foods cause only a moderate change in your blood sugar level, so can help provide you with lasting energy throughout your day.

The table below indicates the glycaemic index of a range of foods. The graph in section 3.5.6 shows the energy spike and drop that occurs after eating high GI foods, and the more moderate, longer lasting rise in blood sugar level after eating low GI foods.

Glycaemic index	Extremely high	High	Moderately high	Moderately low	Low
Grains	Puffed rice Cornflakes White bread	Wholemeal bread Muesli Brown rice Porridge oats	Bran Rye bread White pasta Brown pasta	Tomato soup Lima beans	Barley
Fruit and vegetables	Parsnip Baked potato Carrot	Sweetcorn Mashed potato Boiled potato Apricots Bananas	Sweet potato Peas Baked beans Grapes Orange juice	Pears Apples Orange Apple juice	Red lentils Soybeans Peaches Plums
Sugar	Glucose Honey	Sucrose			
Snacks		Corn chips Chocolate Crackers Biscuits Low-fat ice-cream	Potato chips Sponge cake	Yoghurt High-fat ice-cream	Peanuts

HOW ABOUT THAT!

When food supplies are scarce, or during hibernation, an animal's ability to store energy reserves greatly assists its chances of survival. Penguins, for example, use their fat reserves to provide them with energy when required. Male emperor penguins are able to keep eggs warm for nine weeks at a time without any food. Animals that live in cold regions also use their fat storage ability to insulate themselves against the very cold weather conditions. Whales and seals have a thick layer of fat cells called blubber, which serves as an insulation layer in their cold, watery habitats.

3.5.8 Fats, feasts and famines

The ability of your ancestors to store high-energy molecules may be how you got to be here today. **Fats** are especially rich in energy, providing about twice as much energy as the equivalent amount of carbohydrate or protein.

When more kilojoules of energy are consumed than required, the body tends to store the excess energy in the liver and muscle cells as glycogen. If glycogen stores are full and the energy intake still exceeds that required, the excess may be stored as fat in the form of fat cells just beneath the skin.

When extra energy is required, the liver glycogen is used first, then the muscle glycogen and finally the fat. Most people have enough fat cells stored to provide energy for 3–7 weeks. The human body tends to hoard fat, immediately storing fat molecules obtained from food.

Most people should consume about 30–40 g of fat a day.

The amount of fat in your diet can have a more direct effect on weight gain than carbohydrates. Although fat hoarding can be a **liability** today, it may have increased the chances of survival of your hunting and gathering ancestors. Recent discoveries suggest that the regulation of fat storage may be controlled by a hormone called leptin and several genes inherited from your parents.

3.5.9 Banned! It's for your own good!

Imagine being told 'No treats for you! You will have spinach, capsicum and tomato on wholegrain bread and no butter!' Who tells you what to eat? Should you listen? Do others really care what you put into your mouth?

In 2006, the Victorian government decided to address the types of food that are available to school students. One of the reasons for this was the growing concern about the number of obese children in the state. Soft drinks containing sugar were the first to be on their no-go list. Do you think the government has the right to make such a decision? What is your opinion on this issue?

INVESTIGATION 3.3

Measuring the energy in food

AIM: To compare the amounts of energy stored in a range of foods

Materials:

small metal basket (used to fry food)
samples of small biscuits, potato chips, uncooked pasta, crouton or small piece of toast
safety glasses
thermometer
retort stand, bosshead and clamp
large test tube
Bunsen burner
measuring cylinder

CAUTION

Before starting this experiment, read all the steps below and make a list of the risks associated with this activity and how you plan to minimise these risks.

Method and results

1. Copy and complete the table below.

Food	Biscuit	Chip	Pasta	Crouton/toast
(a) Mass of food (g)				
(b) Volume of water (mL)				
(c) Initial temperature of water (°C)				
(d) Final temperature of water (°C)				
(e) Increase in temperature (= $d - c$)				
(f) Energy in food (J) (= $4.2 \times 30 \times e$)				
(g) Energy in food (kJ) (= $f \div 1000$)				
(h) Energy per gram of food (kJ/g) (= $g \div a$)				

- Use the clamp to attach the test tube to the retort stand.
- Measure 30 mL of water and pour it into the test tube.

2. Measure the temperature of the water.
3. Weigh the biscuit.
   - Place the small biscuit in the wire basket and set fire to it using the Bunsen burner. When the biscuit is alight, put the basket containing the biscuit underneath the test tube. The heat released from the burning biscuit will heat the water. Hold the basket under the test tube until the biscuit is completely burned. You can tell that the biscuit is completely burned if it is all black and will not re-ignite in the Bunsen burner flame.
4. Measure the temperature of the water again.
5. Calculate the amount of energy that was stored in the biscuit, using the following equation.

$$\text{Energy (in joules)} = 4.2 \times \text{volume of water (in mL)} \times \text{increase in temperature (in °C)}$$

6. Calculate the amount of energy per gram of food by dividing the amount of energy by the mass of the food.
7. Repeat the steps above using the other food samples.

Discuss and explain

8. Copy and complete the aim of this experiment: 'To compare the amount of __________ contained in a range of foods'.
9. Copy and complete the conclusion: 'The food that contained the most energy per gram was __________'.
10. Why was it necessary to calculate the amount of energy per gram of food?
11. Did all the heat from the burning food go into heating the water? Explain how this might have affected the validity of this experiment.

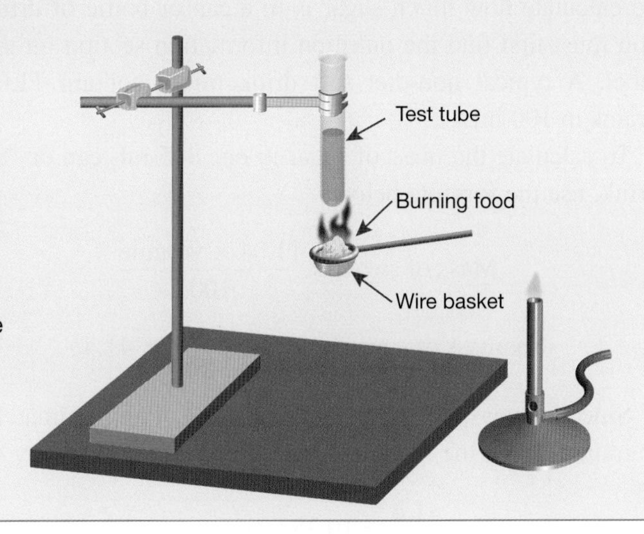

3.5.10 Childhood obesity on the rise

Jacqueline Freegard

Obese primary school children are showing signs of diseases normally only seen in overweight adults, research has revealed.

High levels of hyperinsulinism, fatty liver and other complications have been found in children as young as six. Obese and overweight children were also found to suffer from sleeping disorders, depression, bullying and muscle pain.

Dr Zoe McCallum from the Murdoch Childrens Research Institute said increasing numbers of overweight children were presenting with the early stages of serious diseases, including type 2 diabetes.

'We know there is increasing liver disease, increasing hyperinsulinism and increasing raised blood fats,' Dr McCallum said.

'And we know that there are children being diagnosed with type 2 diabetes at a younger age, which is traditionally seen in the adult population.'

Dr McCallum said the incidence of childhood obesity in Victoria has tripled in the past 15 years.

One-in-four Victorian children is overweight or obese, with 5 to 6 per cent classified as obese.

Australia now matched US rates of childhood obesity.

'We are following our American cousins and getting fatter faster than America,' she said.

'There are more obese children and they are carrying much more weight than they ever have.'

The study found overweight and obese children were unlikely to show symptoms of underlying diseases.

'These children on the whole, apart from the fact they are clearly carrying too much weight, will actually be quite healthy and may not suffer any ill effects from having these abnormal blood tests,' she said.

Dr McCallum said the study, from Perth's Princess Margaret Hospital, found parents were generally unaware of the problems associated with obesity.

'It won't be until 10 or 20 years later that there will be an impact,' she said.

'Children who have raised insulin, raised blood fats and elevated liver enzymes have hard evidence of future risks of diseases that do shorten life.'

Detecting the diseases early meant young children could be cured.

'The beauty of detecting it in kids is you can actually do something about it,' she said.

'If we slow the rate at which a child puts weight on then we can reverse some of those results.'

But she said even very young obese children suffer from the stigma associated with the disease.

Source: *Herald Sun*

3.5.11 How much sugar?

To calculate how much sugar is in a can or bottle of drink you must first find the nutrition information section on the label. A typical non-diet soft drink might contain 11.04 grams in 100 mL.

To calculate the mass of sugar in one 375 mL can of drink, use the formula below:

$$\text{Mass of sugar} = \frac{11.04 \times \text{volume}}{100}$$

$$\text{So, mass of sugar} = 11.04 \times \frac{375}{100} = 41.4g$$

Since one teaspoon of sugar has a mass of approximately 4 grams, divide the mass of sugar in one can of drink by 4.

$$\frac{41.4}{4} = 10.35 \text{ teaspoons}$$

Therefore, one can of soft drink might contain over 10 teaspoons of sugar.

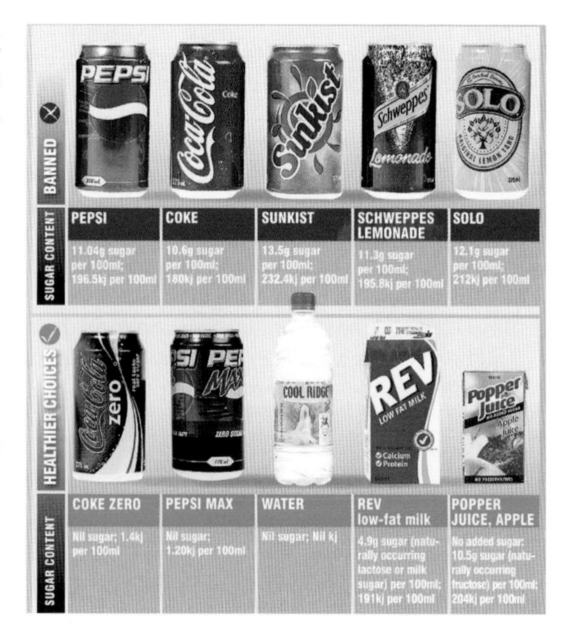

INVESTIGATION 3.4

Energy for living

AIM: To increase awareness of energy intake from food and energy output during different types of activities

Method and results

- Table 1 below provides an approximate amount of energy used for a variety of activities.

Activity	Approximate energy use (kJ) per hour
Sleeping	250
Very light: sitting, reading, watching television, driving	450
Light: leisurely walking, washing, shopping, light sport such as golf	950
Moderate: fast walking, heavy gardening, moderate sports such as cycling, tennis, dancing	1800
Heavy: vigorous work, sports such as swimming, running, basketball and football	3500

1. Construct a table (Table 2) with the headings shown below.
2. Record all the activities which you have been involved in over a 24-hour time period.
3. Complete the table by calculating the energy used for each activity.

Activity	Time spent on activity in hours or part of an hour	Energy use (kJ) per hour (from table above)	Total energy used (in kJ)

4. Calculate your total energy (kJ) used during the 24-hour period.
   - Select a two-course lunch and a drink from Table 3 shown below.
5. Calculate the energy (kJ) and fat (g) in your selected lunch.

Food	Energy (kJ)	Fat (g)
Pizza (two slices)	2060	20
Hamburger	1900	20
Salad sandwich	940	9
Chocolate eclair	1320	15
Fresh fruit salad	290	0.3
Apple pie with ice cream	2310	26
Medium cola	384	0
Strawberry thick shake	1230	15
Medium orange juice	530	0

Discussion and evaluation

6. Which activity used the most energy?
7. Subtract the total energy used (kJ) from the energy calculated in your lunch (kJ).
8. Based on the value calculated in question 7, how many kilojoules could you eat for breakfast and dinner to balance the rest of the energy used that was calculated in question 4? Comment on this amount.
9. Comment on the amount of fat (g) calculated in your selected meal.
10. Suggest why the values of energy used in daily activities are only approximate.
11. On the basis of your data, write a conclusion.
12. Identify the strengths and limitations of this investigation and suggest how it could be improved.
13. Propose a related research question that could be explored.
14. Suggest a hypothesis relevant to your research question.
15. Design an investigation that could test your hypothesis.

INVESTIGATION 3.5

Fizz and tell

AIM: To increase awareness of the amount of sugar in soft drinks and the amount consumed in a week, and to analyse data

Method and results

1. Survey the class to find out:
   (a) how much soft drink they consume in a week (in millilitres)
   (b) which types of soft drinks are consumed.
2. Present your results in a format that can be shared with others.
3. Comment on your results. Were they what you expected or were you surprised? Were there patterns? What other sorts of information would you like to know to further analyse the data?
4. Comment on whether your data support the following statement: 'Almost 80 per cent of teenagers consume soft drinks weekly, with 10 per cent drinking more than one litre per day.'

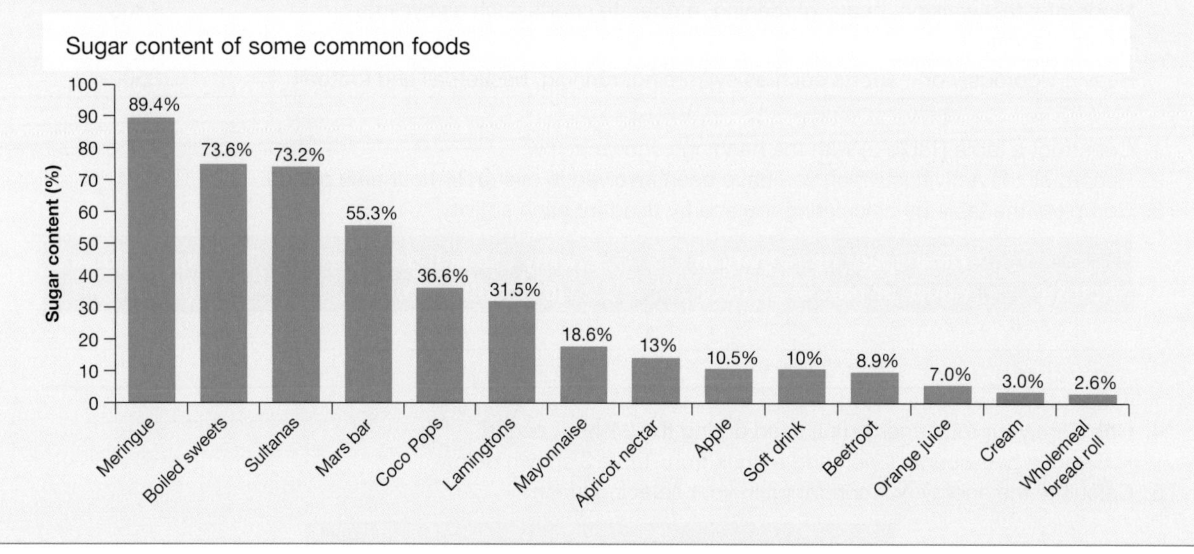

INVESTIGATION 3.6

More fizz and tell

AIM: To increase awareness of attitudes and opinions about the relationship between soft drink consumption and teenage obesity

Method and results

Consider the following statement:
'Sugar-loaded soft drinks should be banned from all Australian schools to reduce teenage obesity.'

1. Construct a PMI chart on the statement.
2. Do you agree with this statement?
3. In the classroom, construct a human graph to show people's opinions on the statement. Stand in positions to indicate your feelings about the statement. For example:
   Strongly disagree (0) — stand next to the left-hand wall
   Agree (2) — stand in the centre of the room
   Strongly agree (4) — stand next to the right-hand wall.
4. Have a discussion with students standing near you to find out the reasons for their opinion.
5. Listen to the discussions of students in other positions.
6. Construct a SWOT diagram to summarise what you have found out.
7. Record the results of the human graph.
8. (a) What was the most popular attitude? Suggest a reason for this.
   (b) What was the least popular attitude? Suggest a reason for this.
   (c) Do you think this attitude pattern is representative of other Australians your age? Explain.
9. On the basis of your discussions, have you changed your attitude since the start of this activity? If so, how is it different and why?

3.5 Exercises: Understanding and inquiring

To answer questions online and to receive **immediate feedback** and **sample responses** for every question, go to your learnON title at www.jacplus.com.au. *Note:* Question numbers may vary slightly.

Remember

1. Name the unit in which energy is often measured.
2. Explain why your cells need glucose.
3. Describe how your:
   (a) cells obtain glucose
   (b) blood glucose levels are kept within a narrow range.
4. Explain why a person with diabetes type 1 needs insulin injections.
5. Distinguish between:
   (a) high GI foods and low GI foods
   (b) diabetes type 1 and diabetes type 2
   (c) carbohydrate storage and fat storage.
6. Outline two ways in which fat storage assists the survival of animals.
7. Explain what happens when we eat more kilojoules than we use.
8. How can you eat a diet high in kilojoules and not put on weight?
9. Describe the relationship between:
   (a) insulin and glucagon
   (b) glycaemic index of foods and sugar levels.
10. When would it be a benefit to eat high GI foods?
11. List examples of:
    (a) high GI foods
    (b) low GI foods.

Using data and calculations

12. Use the table below that shows recommended energy intakes to answer the following questions.
    (a) Plot a graph to show how energy needs change with age. You will need to plot two lines: one for males and one for females. The age should be on the horizontal axis. (If a computer is available, you could use a spreadsheet.)
    (b) Suggest why females seem to need less energy.
    (c) Suggest why you need more energy as you approach your late teens.

	Recommended daily energy intake (kJ)		
Group	Age	Male	Female
Children	1	5000	4800
	5	7600	6800
	9	9000	7900
Adolescents	12	9800	8600
	13	10400	9000
	14	11200	9200
	15	11800	9300
Adults (height 190 cm)	18–30	12000	10600
	30–60	11400	9500
	over 60	9700	8800

Analyse and evaluate

13. (a) What types of drinks may be banned from Victorian state schools?
    (b) How much sugar do most non-diet soft drinks contain?
    (c) Calculate the mass of sugar in a two-litre bottle of Coke.
    (d) Calculate the number of teaspoons of sugar in a two-litre bottle of Coke.
    (e) Calculate and graph the amount of sugar in a 375 mL can or bottle of each of the drinks in the table at the beginning of this subtopic.

Think and discuss

14. Do you think that too much soft drink is being drunk by people your age? Should it be changed or monitored? What are some implications about the amount of soft drink consumed?
15. What are your opinions on the state government being able to dictate the types of foods that are available to children in schools?
16. Do you think the Victorian government's ban on soft drinks in schools will help reduce obesity in teenagers? Give reasons to support your opinion.
17. What other lifestyle habits should the government be involved in? How should they approach this? Provide reasons why you think they should be involved.

Think and investigate

18. Read the article *Childhood obesity on the rise* on page 144 and answer the following questions.
    (a) How much has the incidence of childhood obesity in Victoria increased in the last 15 years in the article?
    (b) Create a mind map on the problems that have been found in obese primary school children.
    (c) Brainstorm ways to reduce the incidence of obesity in Australia.
    (d) Since this article was written in 2005, find out if childhood obesity has increased over the last ten years. Support your response with information from at least three different sources. Share and discuss your findings with others.

Investigate and share

19. Is childhood obesity a real issue in Australia? Research various resources to gather as much relevant information as you can. On the basis of your research and personal beliefs, construct an argument to prepare for a debate with someone who has an opposing view.
20. Use the nutrition information in subtopic 3.3 and other sources to put together a menu for a day that has the recommended daily kJ for your age. Find out and report on diabetes research.

3.6 Myths, moods and foods

Science as a human endeavour

3.6.1 Folk legends

Knowledge often passes from one generation to the next through stories and tales. Some old-fashioned remedies have been passed on in this way. The truth of some of these folk legends may have altered or disappeared along the way, while others may have a sound basis.

For instance, is chicken soup good for fevers? Yes; but this is also true for many other protein-rich foods. Although your body produces about 2000 immune cells per second, many of these can be lost when you are feverish. The amino acids in proteins help you to reinforce and rebuild new immune cells and molecules.

There are many folk legends related to food, and some of these are shown in the bubble map at right.

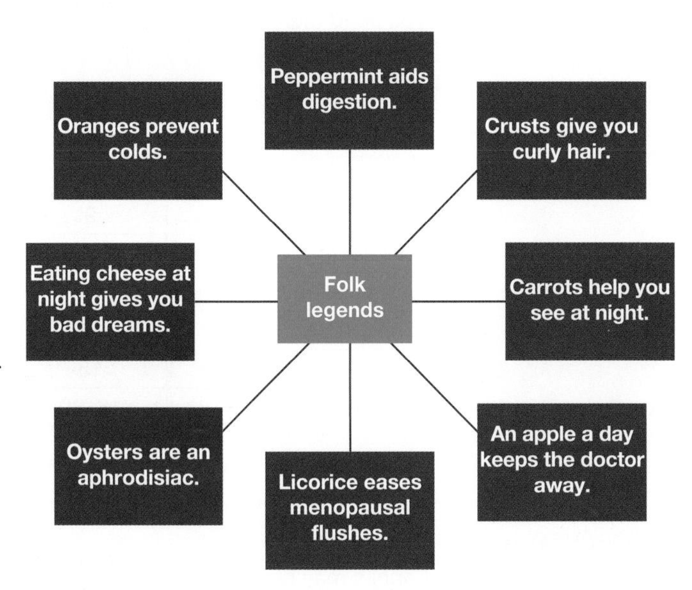

3.6.2 Diet to lift spirits

Fay Burstin

We know them as comfort foods — those warm hearty meals or rich treats — because the mere act of eating them makes us happy.

But research suggests that consuming the right foods could make us feel so good they could even relieve depression.

Two key nutrients in fish, nuts and beets have been found to work just as well as prescription antidepressants in preventing depression in laboratory rats.

Harvard University researchers in the US found omega-3 fatty acids and uridine, both linked to improved brain function, affected the rats' behaviour during a standard depression test.

Rats forced to swim in chilled water with no way to escape will normally become hopeless and float motionlessly. But when treated with antidepressants, they remain active for longer, searching for an escape.

A team led by neurobiologist William Carlezon at Harvard-affiliated McLean Hospital found rats whose diets were supplemented with high levels of omega-3 fatty acids for at least 30 days stayed active and focused on escape.

Similarly, the study published in Biological Psychiatry found rats injected with high levels of uridine were equally tenacious.

And combined doses of omega-3 oil and uridine were just as effective as three different antidepressants in prompting the rats to start swimming again, Dr Carlezon said. But they didn't see the same results in untreated rats.

Dr Carlezon speculated that the drugs and dietary supplements acted on brain cells' mitochondria, the power source that produces energy for cells.

'Imagine what happens if your brain does not have enough energy,' he said.

'Basically, we were giving the brain more fuel on which to run.'

Associate Professor Luis Vitetta, from Swinburne University's Graduate School of Integrative Medicine, said major medical advances had been made in recent years linking illnesses such as cancer and cardiovascular disease to diet.

Now, similar links were being drawn between nutrition and brain function disorders such as dementia, ADHD, depression and bipolar disorder, he said.

'We're starting to put the pieces of the puzzle together, based largely on why some cultures with certain diets suffer less from these disorders than others,' he said.

'Japan had one of the world's lowest rates of depression and we're beginning to think it's because they eat oily fish like salmon every day that's rich in omega-3 essential fatty acids.'

Dr Vitetta said at least 50 per cent of our brain was made up of essential fatty acids (EFAs).

But our brain can't manufacture EFAs itself so we need to get them from our diet.

Dr Vitetta said research showed anyone (or anything, including lab rats) fed omega-3 fatty acids performed better on brain function tests.

Studies show dyslexic children given an omega-3 dietary supplement can make two years' reading progress in six months and 70 per cent of kids diagnosed with ADHD no longer met the clinical criteria after four months of taking an EFA supplement.

But it's not just EFAs we need to lift our mood and brain power.

Dr Vitetta said good nutrition, including at least five or six portions of fresh fruit and vegetables a day, could ultimately have the same effect on the brain as antidepressant drugs.

'The vitamins and minerals in fresh fruit and vegetables are crucial for every bodily function, including the heart, the liver and the gastrointestinal system,' he said.

'When your body is working well, your weight is healthy and your skin looks good, all of which have a positive effect on your self-image'.

'And if you feel good about yourself, you're less likely to feel anxious and depressed, which is reflected in good mental health.'

Source: *Herald Sun*

3.6.3 Mood food

Ever heard of 'mood food' or comfort food? Do you crave particular foods when you are in a particular mood? Some foods don't just make you feel happy, but actually affect your brain. The article *Diet to lift spirits* discusses some recent research on the antidepressant properties of two key nutrients.

3.6.4 Seeing

Dr Lisa Smithers won the 2008 South Australian Young Investigators Award for her research on omega-3 oils, tuna oil and premature babies.

DHA (docosahexaenoic acid) is an omega-3 oil important for brain, nerve and eye tissue development. The highest concentration of omega-3 DHA in the human body is in the retina of the eye. Premature babies may have low levels of DHA and rely on milk to supply it to them after their birth.

Dr Lisa Smithers. Although they may look like jelly beans, these are omega-3-oil supplements.

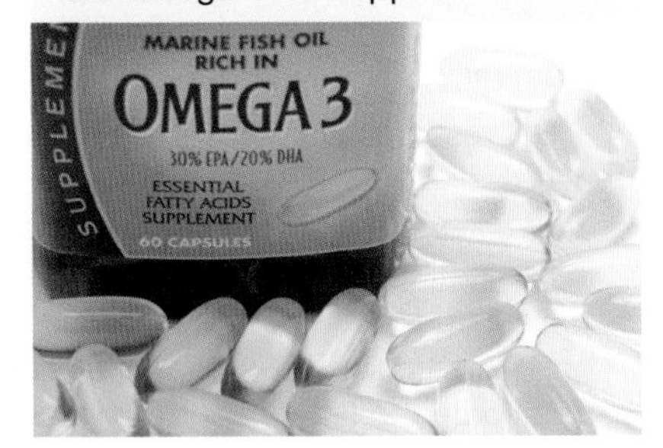

Dr Smithers's PhD research at the University of Adelaide involved a clinical trial. One group of breastfeeding mothers ingested tuna oil capsules with DHA. This raised the levels of DHA in the milk to four times higher than would normally be present. The DHA-enriched milk was provided until the premature babies reached their full-term date. The other group of mothers received placebo capsules that did not contain DHA.

Testing the babies at four months of age showed that those who were fed higher levels of DHA were able to visually detect a finer pattern than those who had not. This suggests that the addition of DHA to the milk assisted in their visual development.

3.6.5 Mood food

Dark chocolate (at least 70 per cent cocoa solids) contains catechins, strong antioxidants which enhance endorphins, the brain's natural feel-good chemicals, and increase libido.

Seafood and oily fish contain high levels of omega-3 essential fatty acids, nerve and brain cells' building blocks that will ultimately improve mood more than any other food.

Nuts and seeds, emu meat and other wild game also contain high levels of EFAs.

Chicken, turkey and legumes such as beans and lentils contain tryptophan, a protein converted into the brain chemical serotonin, usually low in people with depression.

Caffeine boosts mental alertness and concentration. But many regular tea and coffee drinkers confuse this effect with the unpleasant symptoms of caffeine withdrawal when they don't get their daily cuppa.

Carbohydrate cravings may be a subconscious attempt to raise levels of serotonin, as tryptophan is absorbed more quickly into the brain after eating carbohydrate 'comfort' food such as potatoes.

Junk food has high levels of sugar and animal fats, which send blood sugar and endorphin levels soaring, giving you an instant hit. But the effect is short-lived, quickly plunging blood sugar and energy levels downward, sending you into depression, so the overall effect is bad.

HOW ABOUT THAT!

In your great-grandparents' days, many children were given a daily dose of cod-liver oil to maintain good health. It turns out that your great-grandparents may have been right about the benefits of fish oil. Fish oil is rich in omega-3 fatty acids. These fatty acids are being investigated as a possible treatment for conditions including rheumatoid arthritis, depression, attention deficit disorder and heart disease.

A number of scientific studies have shown that omega-3 fatty acids affect behaviour and mood. For example, Bernard Gesch carried out an experiment involving British prison inmates. He gave half the people who had volunteered for his study a daily supplement that contained omega-3 fatty acids and other vitamins and minerals. The other prisoners were given a placebo (a tablet

Omega-3 fatty acids are found in oily fish (for example, tuna), some seeds and vegetable oils, and supplements.

that looked just like the supplement but did not contain fatty acids, vitamins or minerals). Over time, he found that the prisoners taking the supplement were involved in a lot fewer violent incidents. The prisoners taking the placebo showed no significant change in their behaviour.

3.6 Exercises: Understanding and inquiring

To answer questions online and to receive **immediate feedback** and **sample responses** for every question, go to your learnON title at www.jacplus.com.au. *Note:* Question numbers may vary slightly.

Remember

1. Read through the text entitled *Diet to lift spirits* and respond to the following questions.
   (a) Identify in which foods you would find the two key nutrients that act as antidepressants in depressed laboratory rats.
   (b) State the names of these antidepressant-type nutrients.
   (c) Identify which part of the brain cells Dr Carlezon suggested the drugs acted on.
   (d) List some links that were drawn between nutrition and brain function disorders.
   (e) Suggest why it is thought that Japan may have one of the world's lowest rates of depression.
   (f) What does EFA stand for?
   (g) State the percentage of our brain that is made up of essential fatty acids.
   (h) Describe the results of studies on dyslexic children given an omega-3 dietary supplement.
2. Suggest why you might get cravings for carbohydrates.
3. What are the benefits of ingesting caffeine?
4. Describe the effect of junk food on your endorphin and sugar levels.
5. Suggest why chicken and lentils might be good to eat when you are depressed.

Think and discuss

6. Construct a bubble map on the benefits of eating portions of fresh fruit and vegetables each day.
7. Suggest why coffee and tea drinkers may crave more each day.
8. With a partner, read through the article *Diet to lift spirits* in this subtopic. Outline the experiments performed on rats at Harvard University. How do you feel about this treatment of the rats? How do you think others may view these experiments and their outcomes?

Think, discuss and investigate

9. Select one of the folk legends from the bubble map at the beginning of this subtopic. Using one of the visual thinking tools from subtopic 1.10 to organise your thinking:
   - outline the history of the legend
   - make your own decision about the truth of the legend. What are your reasons for making this decision?
   - Share your findings with your partner, team or class.
10. Search for other folk legends that relate to food. Present your findings in the form of flash cards, with the legend on the front of the flash cards and the information on the back.
11. Research the following chemicals.
    (a) Omega-3 fatty acids
    (b) Endorphins
    (c) Uridine
    (d) Tryptophan
    (e) Catechins
    (f) Serotonin
    Summarise your findings in a mind map. With a partner, discuss your combined findings. Add any more relevant information to your mind map as you chat.
12. Investigate the history, manufacturing, composition and biological effects of dark chocolate.
13. Read through the text on the research by Dr Lisa Smithers and respond to the following questions.
    (a) Which of the two groups in her clinical trial were the control group? Why?
    (b) State the independent and dependent variables in her clinical trial.
    (c) Suggest which variables she would have needed to control.
    (d) In the clinical trial, some of the mothers were not breastfeeding. Find out or suggest how they could still be a part of the trial.

(e) Suggest how Dr Smithers may have decided which mothers received DHA and which did not. How would have you decided? Why?
(f) Discuss issues related to the decision of who gets the 'test drug/chemical' and who doesn't. If you had the choice, which group would you like to be in? Are there any other factors that may change your response? Discuss and explain.
(g) State what the findings of this research suggested.
(h) Suggest a myth that could result from this research.

14. (a) Formulate your own questions about one of the folk legends shown in the bubble map at the beginning of this subtopic.
(b) Research and report on relevant information or research on these.
(c) State a hypothesis.
(d) Design your own investigation.
(e) Suggest results that would support your hypothesis.
(f) Suggest limitations or difficulties that you may encounter if you were to actually conduct the investigation.

15. *Not all chocolate is created equal*. Suggest what this statement may mean and how it could relate to the myths and truths about the benefits of eating chocolate.

16. In a 2010 newspaper there was an article labelling some foods as superfoods. The table below provides some examples of these and the suggested implications of chemicals that they contain.
(a) Research the active chemical in each of these 'superfoods'.
(b) Find out whether there is any other scientific data to support:
(i) the suggestion that these foods are high in these chemicals
(ii) the implied effect of these chemicals on our health.
(c) Summarise your findings and discuss these with others in your class.
(d) Decide whether you think each of these foods deserves being labelled a superfood. Provide reasons for each decision.

Food		Super' property	Active chemical	Examples of other foods with high levels of this chemical
Watermelon		Sun protection	Lycopene	Red capsicums, tomatoes, green tea
Coriander		Anti-ageing	Beta-carotene and vitamin C	Berries, broccoli, carrots
Onions		Cancer fighting	Quercetin	Apples, oranges, parsley
Mussels		Metabolism	Selenium	Tuna, eggs, Brazil nuts
Black pepper		Antidepressant	Piperine	Salmon, dark chocolate, bananas

3.7 Drugs on your brain?

Science as a human endeavour

3.7.1 Drugs on your brain

Popping a pill or taking something that you shouldn't? Are you aware of the short- and long-term effects of your actions?

Introducing various chemicals into your body can have both beneficial and terrifying consequences. After all, we all need to eat and drink to obtain our nutrients. But there are some chemicals that can cause you great damage.

3.7.2 Passing the message

Neurotransmitters are key players in our memory, learning, mood, behaviour, sleep and pain perception. These chemicals pass a message from one neuron (pre-synaptic neuron) to another (post-synaptic neuron) across a gap between them called a synapse.

Although there are many different neurotransmitters, only one is used at each synapse. The type of neurotransmitter that is released at the synapse can be used to classify them into groups. For example, in your brain some synapses release acetylcholine, whereas others may release noradrenaline, dopamine or enkephalins. The effect that these neurotransmitters have depends on the type of receptor that is present on the membrane of the neuron that receives it. Once the message has been received, enzymes break the neurotransmitter down.

Neurotransmitters carry the message from one neuron to the next. They are stored in sacs called vesicles. When neurotransmitters are released from the vesicles of one neuron, they travel across the synapse to bind to specific receptors on the membrane of the next neuron.

3.7.3 Uppers and downers

Some drugs can affect your brain or personality by either increasing or decreasing transmission of messages across the synapse. These are collectively known as **psychoactive drugs**. These drugs can bind to the receptors, mimic the neurotransmitter or block the binding of the neurotransmitter to its receptor. Nicotine is an example of a drug that mimics the working of acetylcholine.

Some examples of **excitatory psychoactive drugs** include nicotine, caffeine, cocaine and amphetamines ('speed'). Many of these drugs come from natural sources. They all stimulate or increase the synaptic transmission. Like many other drugs of abuse, these stimulants activate your brain's reward circuit.

Excitatory psychoactive drugs can be thought of as **stimulants** or 'uppers', while **inhibitory psychoactive drugs** can be considered as **depressants** or 'downers'. As their name implies, they work by inhibiting or decreasing synaptic transmission. Barbiturates, benzodiazepines (such as Valium), alcohol and cannabis (marijuana) are examples of drugs that decrease the activity of your nervous system.

Examples of excitatory psychoactive drugs and inhibitory psychoactive drugs

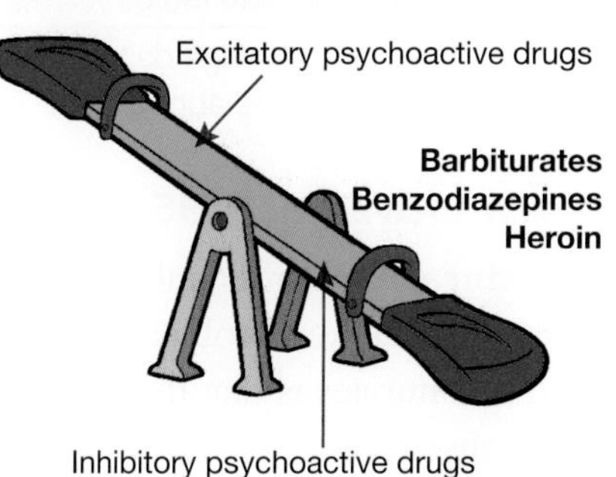

3.7.4 Caffeine

What do coffee, tea, cocoa, chocolate and some soft drinks have in common? They all contain **caffeine**. In moderate doses, this central nervous system stimulant can increase alertness, reduce fine motor coordination, and cause insomnia, headaches, nervousness and dizziness. In massive doses it is lethal.

One effect of caffeine is to interfere with adenosine at multiple sites in your brain, but this drug also acts on other parts of your body. It increases your heart rate and urine production.

3.7.5 Cocaine

Cocaine (coke, snow, crack, gold dust or rock) works by inhibiting or blocking the uptake of neurotransmitters — dopamine, norepinephrine or serotonin — in a synapse, prolonging effects within the central nervous system. This results in elevated heart rate and body temperature, increased alertness and movement, and dilation of pupils. High levels of norepinephrine may result in strokes, organ failure and heart attacks.

Large doses of cocaine can cause heart attacks, strokes, paranoia and hallucinations.

3.7.6 Amphetamines

Amphetamines (speed, ice, ecstasy, meth, pep pills or fast) are synthetic chemicals that affect levels of neurotransmitters — dopamine, norepinephrine or serotonin. Long-term use can result in insomnia, hallucinations, tremors, and violent and aggressive behaviour. Some amphetamines are **neurotoxic** and cause neuron death.

While the short-term effects may be a dry mouth, enlarged pupils, headaches and increased confidence, frequent use of amphetamines may result in psychosis.

3.7.7 Ecstasy

Ecstasy or MDMA is distributed in small tablets, capsules or powder form. Short-term effects include increased blood pressure, body temperature and heart rate. Larger doses can result in convulsions, vomiting and hallucinations. There is also a risk of heart attack or brain haemorrhage and swelling, and there is evidence that it causes long-term damage to the neurons in your brain.

Substance	Quantity of caffeine (mg)
Filter coffee (200 mL)	140
Instant coffee (200 mL)	80
Tea (200 mL)	80
Dark chocolate (30 g)	35
Typical cola (330 mL)	32
Milk chocolate (30 g)	15

An adult's average daily consumption of caffeine is about 280 mg. A fatal dose is about 10 g.

3.7.8 Barbiturates

Barbiturates are often taken to calm someone down and are used as sedatives. Sleeping pills are one such example. One key problem is that they may lead to tolerance and dependence. A key danger associated with barbiturates is that there is only a small difference between a dose that produces sedation and one that may cause death.

3.7.9 Marijuana

In 1964, the psychoactive ingredient in **marijuana** (also known as grass, pot, reefer or weed) was identified as a **THC** (delta-9 tetrahydrocannabinol). This chemical comes from a plant called *Cannabis sativa*. THC activates cannabinoid receptors in your brain located on neurons in your hippocampus (memory), cerebral cortex (concentration), sensory portions of your cerebral cortex (perception) and your cerebellum (movement). High doses of this drug may cause hallucinations, delusions, impaired memory and disorientation. As it is one of the world's most commonly used illegal drugs, there has been a great deal of research into how it works and the consequences of using it.

3.7.10 GHB

GHB (gamma hydroxybutyrate, sodium oxybate, also known as liquid E, fantasy or gamma-OH) is an odourless, colourless, salty liquid that acts as a depressant on your nervous system. One of the dangers of this drug is the difficulty of determining a safe dosage. Although a small amount may have a euphoric effect, more can lead to amnesia, respiratory difficulties, delirium, loss of consciousness and possibly death. Likewise, combining GHB with alcohol can also lead to deep unconsciousness and may cause coma or death. GHB also has the reputation of being used as a 'date-rape' drug.

3.7.11 Heroin

Diacetylmorphine or **heroin** (also known as smack, jive, horse or junk) is an illegal opiate drug that contains morphine as its active ingredient. Its source is the opium poppy, *Papaver somniferum*. **Opiates** stimulate a pleasure system in your brain that involves the neurotransmitter dopamine.

In 1973, scientists found neurons in the brain that have receptors for opiates. These are located in areas involved in pain, breathing and emotions. The discovery of these receptors led to further research about their purpose. Two years later, scientists discovered that the brain manufactures its own opiates known as **endorphins**. Although endorphins are always present in the brain, when you are in pain or stressed they are released in larger amounts.

3.7.12 Blood and alcohol

Unlike water, some drinks can have a negative effect on your health. One such drink is **alcohol**.

Alcohol is a depressant and can alter your mood, thinking and behaviour. Many parts of your body are affected by alcohol, as shown on the next page.

3.7.13 Alcohol and the digestive system

Alcohol is a substance that is directly absorbed into your bloodstream through your stomach and small intestine. It irritates your stomach and causes more stomach acid to be produced, which can result in painful heartburn and stomach ulcers. Alcohol is also linked to mouth, oesophagus, stomach and intestinal cancers. The part of the digestive system that is affected most is the liver. Alcohol can destroy liver cells and can cause fat to accumulate around the liver, resulting in a fatal condition known as cirrhosis.

3.7.14 Alcohol and the brain

Did you know that alcohol slows down your brain activity by interfering with your cerebellum? This may affect your coordination and perception and cause memory blackouts. When alcohol reaches your midbrain, your reflexes diminish, confusion and

(a) A healthy liver and (b) a liver from a person with cirrhosis caused by excess alcohol consumption. Alcohol is an example of a toxin broken down by the liver — excess consumption of alcohol can cause extra strain on liver tissue and damage to liver cells.

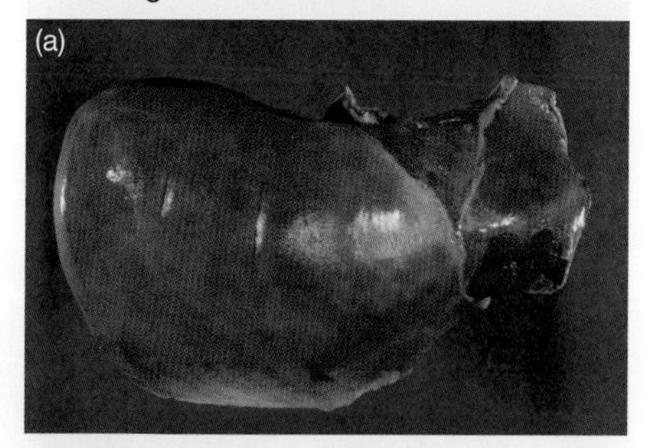

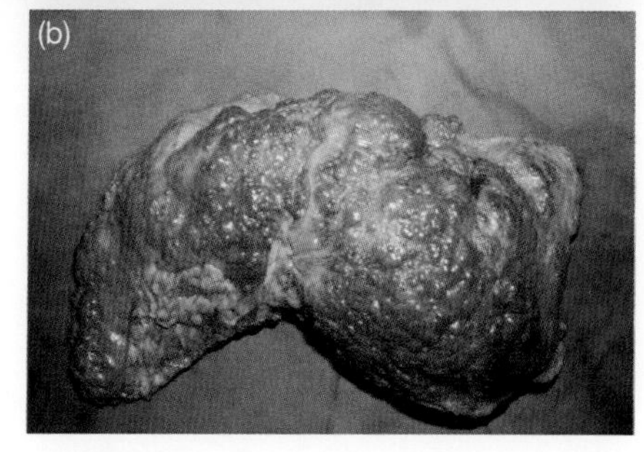

stupor follow, and then you may lapse into a coma. When the alcohol reaches your medulla, your heart rate may drop and your breathing may stop, possibly resulting in death.

> **HOW ABOUT THAT!**
> Australian scientists are currently in the race to develop 'smart' drugs through research on neurotransmitters. Smart drugs belong to a class of drugs called ampakines. These drugs work by boosting chemicals that allow information to flow from one part of the brain to another. Our scientists are also discovering neurotransmitters that were previously unknown, and are trying to find out about the cause and effects of imbalances of brain chemicals and drug addiction.

3.7.15 Some common questions

Does eating food stop you from getting drunk?

The rate at which alcohol is absorbed may be slowed by the presence of food in your stomach but it won't prevent you getting drunk or intoxicated.

How can you sober up more quickly?

Your liver works at a fixed rate. It will detoxify or clear about one standard drink each hour (see the standard drinks guide opposite). So, black coffee, cold showers, fresh air and vomiting won't speed up the process of getting rid of alcohol from your body.

Why do people who drink too much alcohol smell?

Although the liver breaks down about 90 per cent of the alcohol, the rest leaves the body in urine, sweat and breath.

Drinking coffee does not increase the speed at which the body processes alcohol.

Alcohol affects all parts of your body.

Nervous system
- Nerve damage
- Loss of sensation

Brain
- Loss of memory
- Stroke

Heart
- High blood pressure
- Heart disease

Throat
- Cancer

Stomach
- Bleeding
- Ulcers

Sexual organs
- Infertility
- Damaged fetus

Liver
- Cancer
- Cirrhosis
- Hepatitis

Intestine
- Cancer
- Ulcers

Bones
- Osteoporosis

Sexual organs
- Impotence
- Infertility
- Shrunken testicles

Should pregnant women drink alcohol?

During the first three months of pregnancy, alcohol interferes with the migration and organisation of brain cells. Heavy drinking during the next trimester, particularly between 10 and 20 weeks after conception, can have the biggest impact on the baby, leading to fetal alcohol syndrome (FAS). Drinking during the last trimester may affect the baby's hippocampus, which may reduce the child's future ability to encode visual and auditory information (reading and maths).

Heavy drinking during pregnancy can damage the baby's brain.

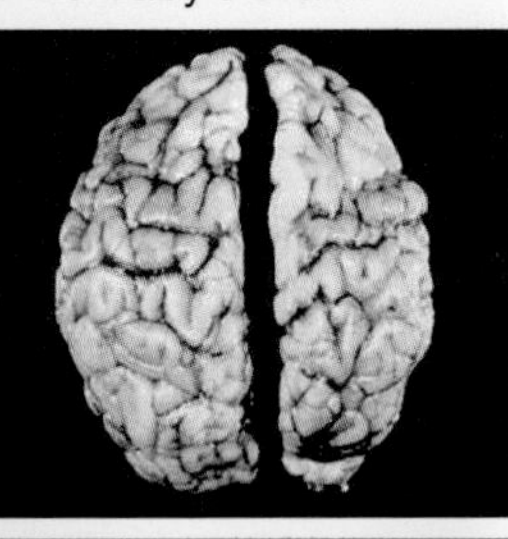

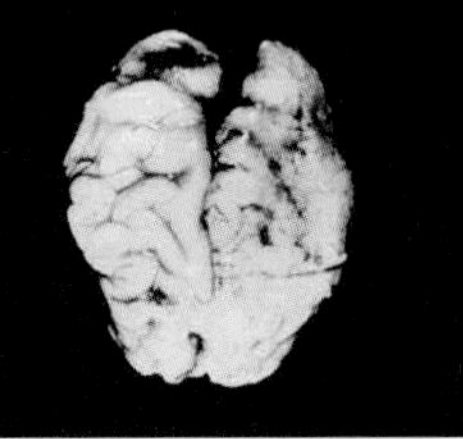

3.7.16 Australia and alcohol

Headlines in Australian news stories increasingly relate alcohol abuse to accidents or violence which result in injury or death. There are data to suggest that drinking at dangerous levels is increasing within our culture.

Over the last decade, there has also been an increase in the number of women drinking at risky or high levels. This has implications not just for the woman and those close to her, but potentially to the health of an unborn child.

While some Australians believe that they have a right to drink and eat whatever, whenever and however they wish, the government is not of the same belief. There are already restrictions on the amount of alcohol in your blood when you are driving and in a number of public places the consumption of alcohol is illegal. With increasing evidence of the dangers of alcohol not just to ourselves but also to others, where will the line be drawn and how will it be implemented?

A standard drink contains about 10 grams of alcohol. It takes the liver about an hour to break down the alcohol in one standard drink.

1.5
375 mL
Full-strength beer
4.9% alc/vol

0.8
375 mL
Light beer
2.7% alc/vol

1
285 mL
Middy full-strength beer
4.9% alc/vol

0.5
285 mL
Middy light beer
2.7% alc/vol

1.5
375 mL
Pre-mix spirits
5% alc/vol

1.5
300 mL
Alcoholic soda
5.5% alc/vol

1
30 mL
Spirit nip
40% alc/vol

0.9
60 mL
Port/sherry glass
18% alc/vol

1.5
170 mL
Sparkling wine/champagne
11.5% alc/vol

1.8
180 mL
Average restaurant serve of wine
12% alc/vol

This graph indicates that the proportion of adults in Australia drinking at a risky or high level is increasing. Suggest what the percentage of risky or high alcohol consumption may look like for 2015–16.

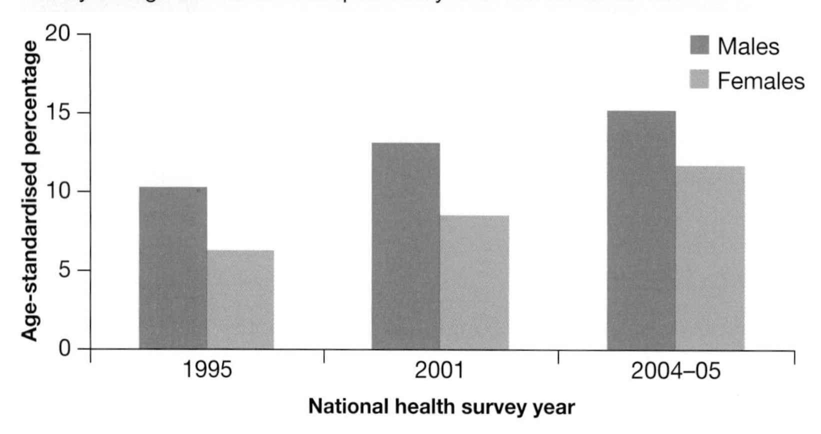

3.7 Exercises: Understanding and inquiring

To answer questions online and to receive **immediate feedback** and **sample responses** for every question, go to your learnON title at www.jacplus.com.au. *Note:* Question numbers may vary slightly.

Remember

1. State the name of the gap across which neurotransmitters pass.
2. Use a flowchart to show the links between a pre-synaptic neuron, a neurotransmitter and a post-synaptic neuron.
3. List three examples of neurotransmitters.
4. What do the vesicles in neurons contain?
5. What are psychoactive drugs?
6. What is the key difference between excitatory and inhibitory psychoactive drugs?
7. State other names for:
   (a) inhibitory psychoactive drugs
   (b) excitatory psychoactive drugs.
8. Construct a double bubble map to show the similarities and differences between excitatory and inhibitory psychoactive drugs.
9. Use a cluster map to show examples of the effects of the following drugs.
   (a) Caffeine
   (b) Cocaine
   (c) GHB
   (d) Heroin
   (e) Ecstasy
10. What is meant by the term neurotoxic?

Double bubble map

Think and evaluate

11. Construct a mind map to summarise the effects of alcohol on your body.
12. Which type of alcoholic drink in the standard drinks guide has the:
    (a) most alcohol
    (b) least alcohol?
13. How many standard drinks are there in three glasses of wine?
14. How many standard drinks are there in a 750 mL bottle of wine?

Investigate, think and create

15. Create a song that can be used to persuade people to reduce alcohol abuse.
16. Suggest ways in which young people can become more aware of alcohol abuse.
17. What is fetal alcohol syndrome? Find out about some other effects of alcohol on the developing fetus.
18. Construct graphs that show the different amounts of alcohol in different types of alcoholic drinks.
19. Find out how an alcohol breathalyser works. Construct a model.

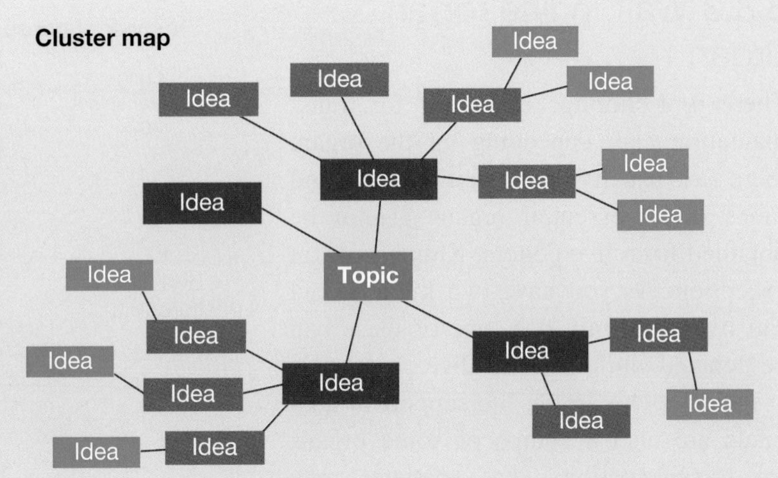

20. *In Australia, increased abuse of alcohol is directly linked to increased drownings, violence, accidents and death.*
    (a) Research various sources to see whether there is evidence to support this claim.
    (b) Using a matrix table, summarise your findings for and against the statement.
    (c) Discuss your findings with those of others, adding comments that you may have missed in your own research.
    (d) Do you agree with the statement? Explain.
    (e) In your team, construct a priority grid on 'Australia and alcohol'.
    (f) Share your team's grid with other teams and discuss similarities and differences.
21. Research and report back to your team, for discussion, on one of the following.
    - The caffeine content in a variety of foods, including different brands of coffee, tea, cola drinks, cocoa drinks and chocolates
    - The history of coffee
    - The symptoms of caffeine addiction
    - The effects and dangers of inhalants and methods of prevention
    - The connection between morphine, opium, codeine and heroin
22. The barbiturate sodium pentothal is also known as 'truth serum'. Find out how it works.

3.8 Wanted! Need an organ?

3.8.1 Wanted!

Organs within your body systems play an important role in keeping you healthy — and alive. But what if one of them fails? What if an organ with a critically important function could no longer do its job and needed to be replaced?

There are some organs that you just cannot live without. For example, if you didn't have a heart, what would pump the blood around your body? Without lungs, how could you obtain the oxygen that you need and remove the carbon dioxide that you don't?

3.8.2 Organ transplants

A solution to having a faulty organ is to replace it with another one that works. This may be achieved by transplanting a healthy organ from another person. In many cases, the source of the replacement organ is a recently deceased person. Organ transplantation presents a variety of medical challenges and raises a number of ethical issues.

3.8.3 Wait in line for an organ ...

There is a shortage of organs for transplantation and, depending on the organ, there are usually long waiting lists and times. Most essential organs cannot be obtained from live donors. Kidneys are an exception. As you have two kidneys and can live with only one, one of these can be donated while you are alive.

Patients who are on life support in hospitals are also a source of some organs for transplantation. These patients may have no brain function, or very limited, if any, chance of recovery. Another source of organs may be those harvested from people in car accidents. When applying for a driver's licence, some people register as a donor, so that their organs may be transplanted into others when they die.

Number of transplants performed in Australia in 2013

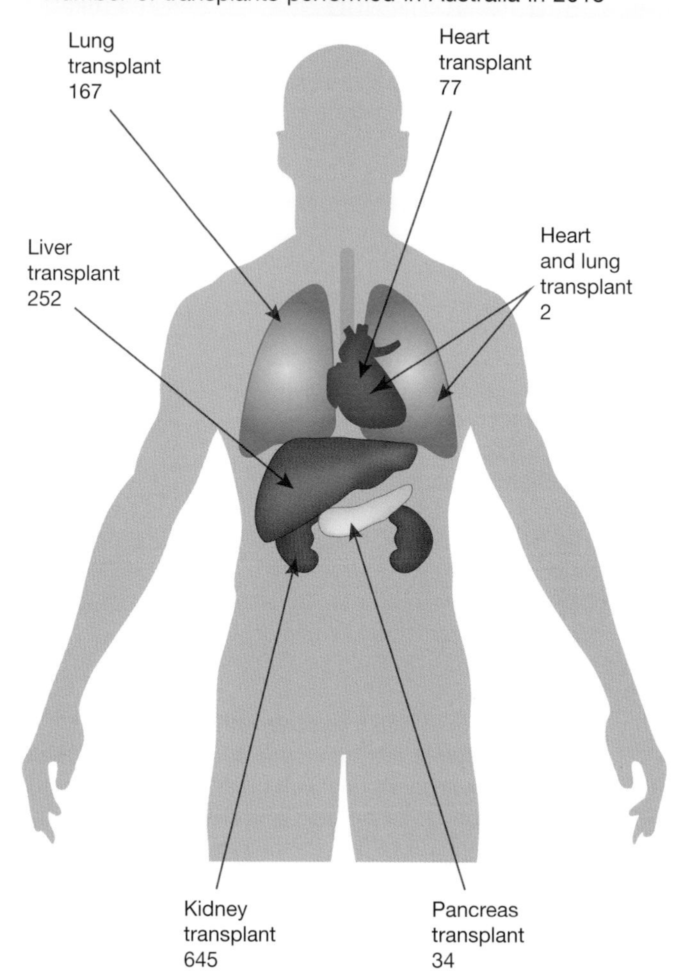

3.8.4 Is it a match?

Even when an organ becomes available for transplant, it needs to match the recipient's blood type and have a reduced chance of being rejected by the recipient's immune system. There needs to be a matching of special proteins called **antigens** between the donor and the recipient. Even so, the recipient's immune system will still attack the transplanted organ as a foreign 'non-self' invader, so drugs are required to suppress this response.

Cardiac researcher Doris Taylor has revived the dead. The process involved rinsing rat hearts with a detergent solution to strip the cells, until all that remained was a protein skeleton of translucent tissue — a 'ghost heart'. She then injected this scaffold with fresh heart cells from newborn rats and waited. Four days later, she saw little areas beginning to beat. After eight days, the whole heart was beating. Could this research lead to new transplant technologies for use in humans?

3.8.5 Growing body parts in labs

Scaffolded or printed?

Researchers are investigating the construction of skin, cartilage, heart valves, breast, ears and other body tissues in tissue-engineering laboratories. Some of these technologies involve injection of synthetic proteins to induce tissues to grow and change; some use scaffolding techniques; and some even use 3D printers to print out tissues such as blood vessel networks.

3.8.6 Stem cells

Stem cells are also being investigated as a possible solution to the shortage of donor organs. While stem cell research is showing great promise, it is also linked to considerable debate.

3.8.7 What are stem cells?

Stem cells are unspecialised cells that can reproduce themselves indefinitely. They have the ability to differentiate into many different and specialised cell types. Stem cells in a fertilised egg or zygote are **totipotent** — they have the ability to differentiate into *any* type of cell. The source of the stem cell determines the number of different types of cells that it can differentiate into.

The ability to differentiate into specific cell types makes stem cells invaluable in the treatment and possible cure of a variety of diseases. For example, they may be used to replace faulty, diseased or dead cells. The versatility of stem cells is what makes them very important.

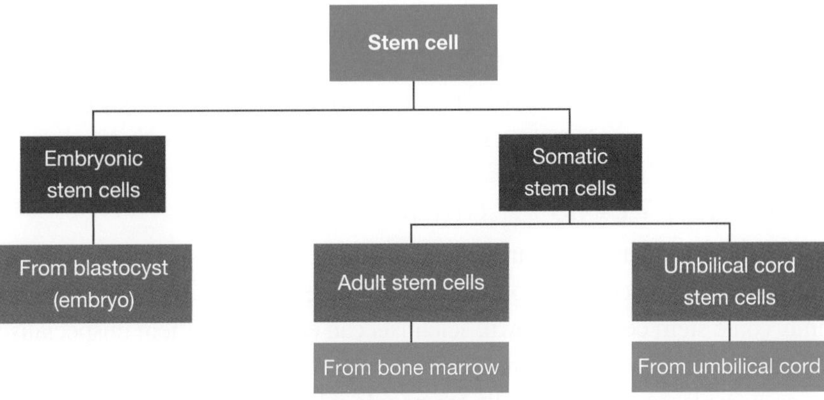

Stem cells can be divided into categories on the basis of their ability to produce different cell types.

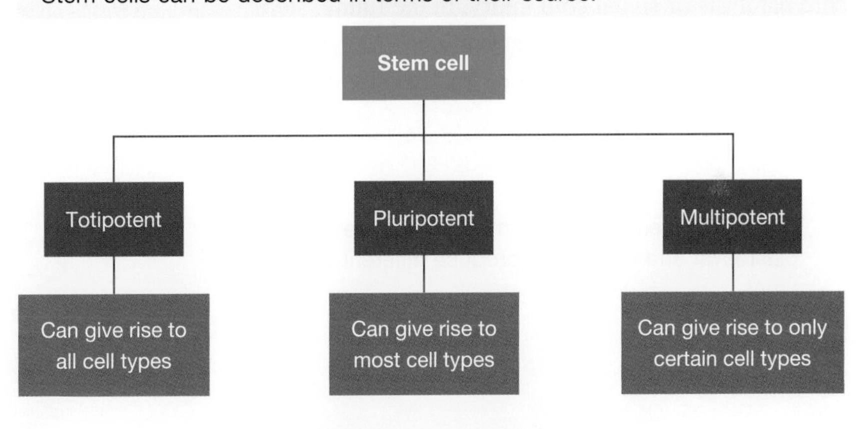

Stem cells can be described in terms of their source.

3.8.8 What are the sources of stem cells?

Embryonic stem cells can be obtained from the inner cell mass of a blastocyst. The blastocyst is the term used to describe the mass of cells formed at an early stage (5–7 days) of an embryo's development. Embryonic stem cells are **pluripotent** and can give rise to most cell types; for example, blood cells, skin cells, nerve cells and liver cells.

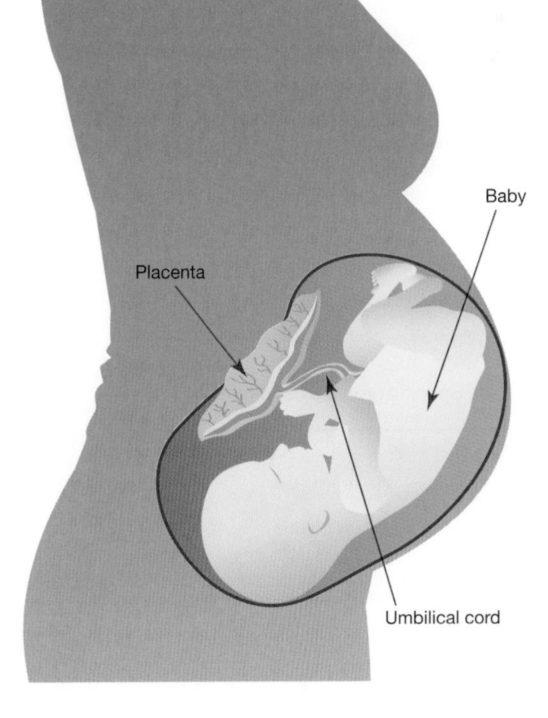

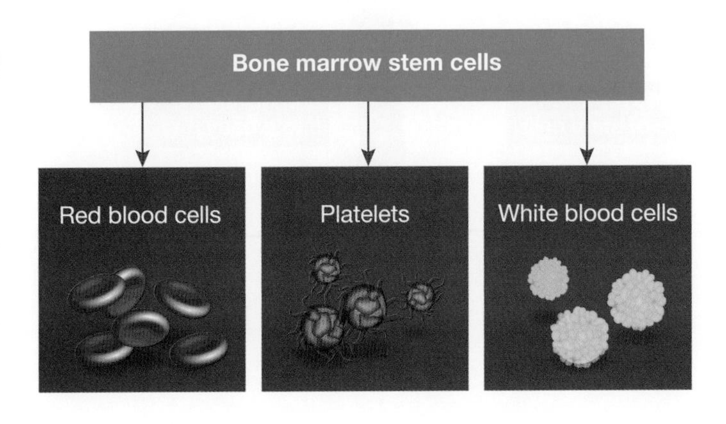

Somatic stem cells can be obtained from bone marrow and umbilical cord blood. Stem cells obtained from the bone marrow are often referred to as **adult stem cells**. These cells are **multipotent** and can develop into many kinds of blood cells.

The umbilical cord is the cord that connects the unborn baby to the placenta. This is how the baby gets nutrients and oxygen while it is still inside the mother's body. This cord contains stem cells that can develop into only a few types of cells, such as blood cells and cells useful in fighting disease. **Umbilical cord stem cells** can be taken from this cord when the baby is born.

3.8.9 Stem cells — made to order?

While the information in your genetic instructions tells your cells which types of cells they should become, scientists have also been able to modify the 'future' of some types of cells. By controlling the conditions in which embryonic stem cells are grown, scientists can either keep them unspecialised or encourage them to differentiate into a specific type of cell. This provides opportunities to grow replacement nerve cells for people who have damaged or diseased nerves. Imagine being able to cure paralysis or spinal cord injury. In the future, stem cells may also be used to treat and cure Alzheimer's disease, motor neurone disease, Parkinson's disease, diabetes and arthritis.

Embryonic stem cells are removed from the blastocyst and cultured to produce different cell types.

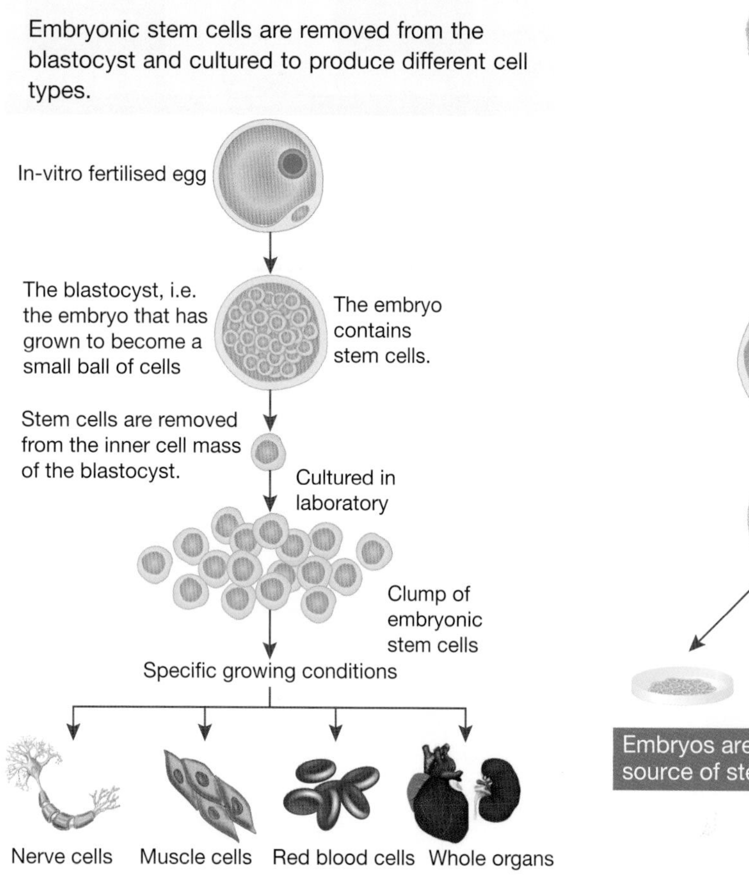

An embryo is the result of a sperm fertilising an egg. If this happens outside a woman's body, it is called in-vitro fertilisation.

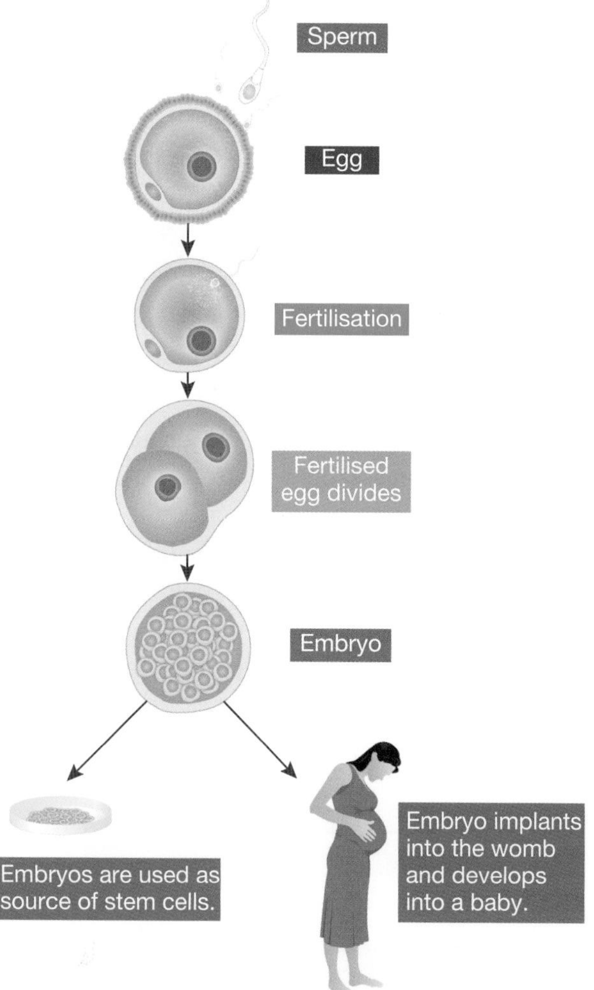

3.8.10 So what's the problem?

It is the source of embryonic stem cells that raises so many issues. Embryonic stem cells can be taken from spare human embryos that are left over from fertility treatments or from embryos that have been cloned in the laboratory. Some argue that this artificial creation of an embryo solely for the purpose of obtaining stem cells is unethical. There has also been concern about the fate of the embryo. In the process of obtaining stem cells, the embryo is destroyed.

Some parents have decided to have another child for the sole purpose of being able to provide stem cells for a child who is ill or has a disease. In this case, the blood from the umbilical cord or placenta is used as the source. Some suggest that this is not the 'right' reason to have a child and that they should not be considered to be a 'factory' for spare parts for their siblings.

3.8.11 Grow it for me!

There is also research into growing human organs in other animals, such as pigs. Once these 'pre-matched' organs have grown, they could then be transplanted into humans. While such a technology brings with it many benefits, it also is associated with a variety of issues and possible future consequences.

Open minds are crucial in the debate on stem cell research
Politicians who decide on issues involving complex science need to be thoroughly briefed . . .
Sydney Morning Herald

We must move ahead on stem cells
To maintain its world-leading reputation, Australia needs to change the law to allow the therapeutic cloning of cells.
The Age

Discovery offers hope on cancer
Melbourne researchers have taken a giant step towards understanding the genesis of breast cancer by inducing female mice to grow new mammary glands from stem cells.
The Age

3.8.12 Constructing synthetic replacement parts?

Synthetic materials can be used to construct some replacement body parts. Artificial joints made of plastic and titanium or ceramic can be used to replace a damaged joint. A variety of prosthetic limbs have also been developed that are suited to different types of activities.

Artificial hip joint

3.8.13 Radical bid to grow new bone

STEM CELL THERAPY A WORLD FIRST
Michelle Pountney

Health reporter
A MELBOURNE man is the first person in the world whose own stem cells are being used to try to mend a broken leg.
The cutting-edge stem-cell technology has helped Jamie Stevens, 21, back on his feet.

A motorcycle crash nine months ago left him with a severely broken left thigh bone. Part of the femur stuck through his leg, and other parts of the bone were missing.

The bone failed to heal and Mr Stevens's leg was held together by a large titanium plate.

Royal Melbourne Hospital orthopedics director Richard de Steiger decided Mr Stevens was the ideal first patient for a revolutionary stem-cell trial at the hospital.

About seven weeks ago, Mr de Steiger harvested bone marrow from Mr Stevens' pelvis. The adult stem cells were then separated from the other cells. A sub-group of stem cells called mesenchymal precursor cells — those that can transform into tissues including bone cartilage and heart — were isolated and grown.

Last week about 30 million of these cells were implanted into the 5 cm × 3 cm hole in Mr Stevens' thigh bone. The cells are expected to regenerate new bone and grow through the calcium phosphate.

Yesterday, just four days after surgery, Mr Stevens went home.

'It's good to be part of something that is on the brink,' he said. 'I wouldn't say I understand it. It's all pretty cool.'

It will be three to four months before the result of the operation is known.

'This is radical and the first procedure in the world to use a patient's own stem cells and make them turn into bone-forming cells that are the patient's own cells, to stimulate healing of a fracture,' Mr de Steiger said.

The cells are harvested, cultured and expanded using Australian biotechnology company Mesoblast's specialist adult stem-cell technology.

Mr de Steiger hopes that eventually the technique will be refined to a simple injection, avoiding further surgery.

Using a patient's own cells avoids the potential problem of the body's rejection of foreign cells.

The only other alternative to repair Mr Stevens' leg would have been a painful bone graft, taking a chunk of bone from his hip and plugging it into the hole in his thigh.

'The conventional way used over many years involves a large incision at the pelvis and taking out quite a large amount of bone in Jamie's case,' Mr de Steiger said.

'In this situation there is the risk of a separate incision, reported continuing pain from that incision, and separate infection risk at that site.'

Jamie Stevens prepares to leave Royal Melbourne Hospital. Pictured with his surgeon Richard de Steiger. If there are too many arguments about using someone else's stem cells, why not grow your own to mend, replace and renew?

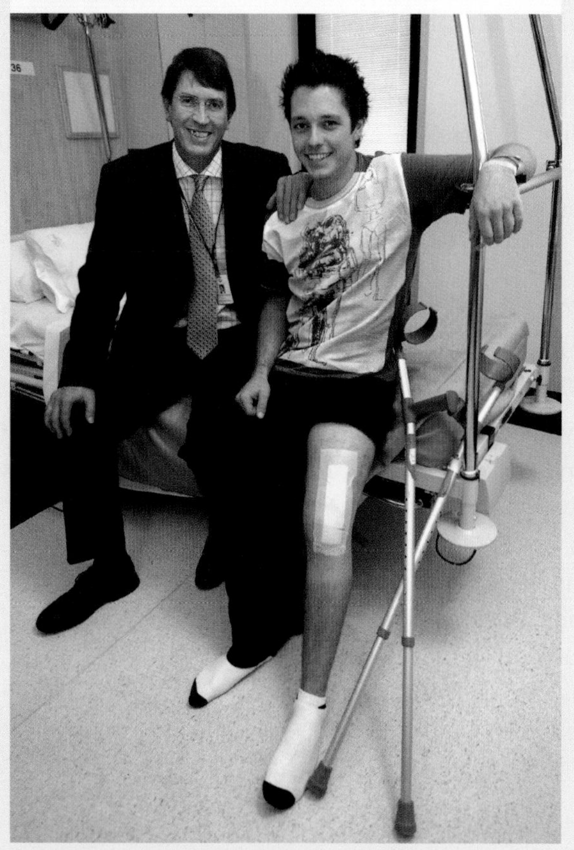

Mr de Steiger said orthopedic specialists at the Royal Melbourne hospital treated about 200 fractures of the long leg bones each year.

About 19 per cent become 'non-union' fractures that fail to heal; 10 of these patients will be recruited to be part of the year-long trial.

Mr Stevens said he was no more nervous about being the first patient in the world to have the procedure than he would have been having a graft.

'I think the benefit outweighs the old procedure, and being able to avoid having a big chunk of bone taken out of the hip . . . the recovery period of it was a lot quicker.'

After living in limbo for nine months, Mr Stevens said he was looking forward to resuming the life he enjoyed before his accident.

HOW ABOUT THAT!

Professor Alan Trounson is an Australian scientist who has spent a great part of his working life perfecting the technique for creating embryos outside the human body. He was part of the team that produced the first test-tube baby in Australia in 1980. He has also done a lot of work on embryonic stem cells. In 2000, his team showed that it was possible to produce nerve cells from embryonic stem cells, which meant that stem cells could potentially be used to cure diseases that have up to now been incurable. This has led to a surge of interest in the field of stem cell research.

In 2003, he was named the Australian Humanist of the Year. In 2007, he was appointed as the president of the California Institute for Regenerative Medicine, which specialises in stem cell research. It is the best-funded facility of its kind in the world, so Trounson will have the best facilities at his disposal to move stem cell research forwards.

Alan Trounson, an Australian scientist and one of the world's top stem cell research scientists

INVESTIGATION 3.7

What's your stance on organ transplants?

AIM: To increase awareness of attitudes and opinions on organ transplants

METHOD AND RESULTS

Consider the following statement:

'Should it be legal to buy transplant organs from either a live donor or from the family of a deceased donor in Australia?'

1. Construct a PMI chart for the statement.
2. Do you agree with the statement?
3. In the classroom, construct a human graph to show people's opinions about the statement. Stand in positions to indicate your feelings about the statement. For example:
   Strongly disagree (0): stand next to the left-hand wall
   Agree (2): stand in the centre of the room
   Strongly agree (4): stand next to the right-hand wall
4. Have a discussion with students standing near you to find out the reasons for their opinion.
5. Listen to the discussions of students in other positions.
6. Construct a SWOT diagram to summarise what you have found out.
7. Record the results of the human graph and examine them to answer the questions in your discussion.

Discuss and explain

8. (a) What was the most popular attitude? Suggest a reason for this.
   (b) What was the least popular attitude? Suggest a reason for this.
   (c) Do you think this attitude pattern is representative of other Australians your age? Explain.
9. On the basis of your discussion, have you changed your attitude since the start of this activity? If so, how is it different and why?

3.8 Exercises: Understanding and inquiring

To answer questions online and to receive **immediate feedback** and **sample responses** for every question, go to your learnON title at www.jacplus.com.au. *Note:* Question numbers may vary slightly.

Remember

1. What are stem cells?
2. Distinguish between the terms 'totipotent', 'pluripotent' and 'multipotent'.
3. Outline the importance of stem cells.
4. List sources of stem cells.
5. Outline issues regarding stem cell research.
6. Describe a scientific contribution made by the Australian scientist Professor Alan Trounson.

Investigate, share and discuss

7. Investigate some of the following questions.
   (a) Which inherited genetic diseases are potentially treatable with stem cells?
   (b) How many different kinds of adult stem cells exist and in which tissues can they be found?
   (c) Why have the adult stem cells remained undifferentiated?
   (d) What are the factors that stimulate adult stem cells to move to sites of injury or damage?
8. In your team, discuss the following questions to suggest a variety of perspectives.
   (a) Is it morally acceptable to produce and/or use living human embryos to obtain stem cells?
   (b) Each stem cell line comes from a single embryo. A single cell line allows hundreds of researchers to work on stem cells. Suggest and discuss the advantages and disadvantages of this.
   (c) If the use of human multipotent stem cells provides the ability to heal humans without having to kill another, how can this technology be bad?
   (d) Parents of a child with a genetic disease plan a sibling whose cells can be used to help the diseased child. Is it wrong for them to have another child for this reason?
9. Find out how stem cell research is regulated in Australia and in one other country. What are the similarities and differences of the regulations? Discuss the implications of this with your team-mates.
10. Investigate aspects of stem cell research and put together an argument for or against the research and its applications. Find a class member with the opposing view and present your key points to each other. Ask questions to probe any statements that you do not understand or would like to clarify. Construct a PMI to summarise your discussion.
11. Investigate and report on research at the Australian Stem Cell Centre.
12. (a) Use a bubble map or mind map to summarise the key points in the article *Stem cell therapy a world first*.
    (b) As a team, discuss the article and construct a PMI.
    (c) Formulate questions that may help you to develop an informed opinion.
    (d) Research your questions.
    (e) State your opinion on the use of stem cell therapies like the one in the article. Give reasons for your opinion.
    (f) Do you have the same opinion about other types of stem cell therapies? Explain.

Investigate, discuss and debate

13. Recent scientific and technological advances are associated with some very complex and difficult decisions, responsibilities and ethical issues.
    (a) On your own, score each of the statements below on a scale of 0–4, where 0 = strongly disagree and 4 = strongly agree.
       - Animals such as pigs should be used to grow organs for human transplants.
       - The creation of human embryos for stem cell research in Australia is acceptable.
       - Doctors should be allowed to harvest the organs of a deceased patient for organ transplants without the permission of the patient's relatives.
       - It should be compulsory for all Australians over the age of 18 to sign the donor register.
       - Smokers, heavy drinkers and drug users should be further down the organ transplant waiting list than those with a healthy lifestyle.
       - Close relatives of humans, such as monkeys and apes, should not be used as animals in scientific research testing the effectiveness of treatments against various diseases.

(b) Research two of these issues above. Construct a table with reasons FOR and AGAINST. Compare and discuss your table with others. Organise a class debate on one of the issues.
(c) For at least two of the statements, share your opinions by being involved in constructing a class 'opinionogram' such as in the *'What is your stance on organ transplants?'* investigation 3.7.
  (i) Suggest questions that could be used to probe students in different opinion zones.
  (ii) Share reasons for your opinion with students in other zones and listen to their reasons for their stance.
  (iii) Reflect on what you have heard from others. Decide if you want to change positions, and if so, change. Give a reason why you are changing.
  (iv) Have a member of the class record the number of students at each point of the scale.
  (v) Reflect on what you have learnt about the opinions and perspectives of others.
  (vi) Construct graphs showing the opinion scales for each statement and comment on any observed patterns.
  (vii) Construct a PMI on each statement based on opinions and statements made by others in the class.
  (viii) Select one of the statements and organise a class debate (different from the statement in (b)).

3.9 Systems working together

3.9.1 A team effort!

Multicellular organisms contain systems that perform particular jobs. The cells within these systems depend on each other and work together; they cannot survive independently of each other. Working together requires organisation, coordination and control.

Multicellular organisms show a pattern of organisation in which there is an order of complexity.

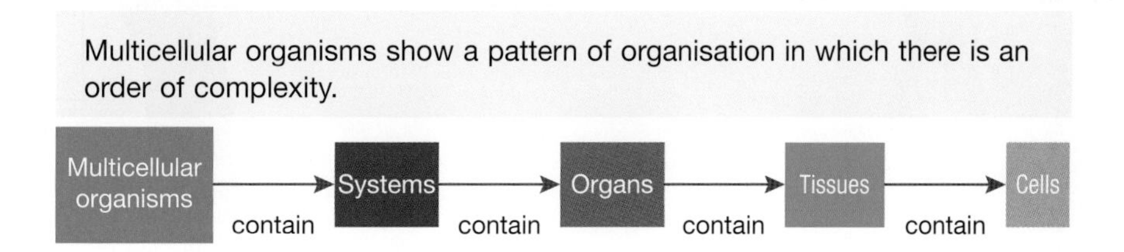

3.9.2 Control and coordination

Within your body, your nervous and endocrine systems play key roles in the coordination and control of the body systems. These systems also work together to keep your cells functioning, and they also provide a balanced internal environment that is essential to their survival.

Your **nervous system** is composed of the **central nervous system** (brain and spinal cord) and the **peripheral nervous system** (the nerves that connect the central nervous system to the rest of the body). Messages are sent as **electrical impulses** along neurons and then as a chemical message in the form of **neurotransmitters** across the **synapses** between them.

Your **endocrine system** is composed of **endocrine glands** that secrete chemical substances called **hormones** into the bloodstream. These chemical messages are transported throughout the circulatory system to specific target cells in which they bring about a specific response.

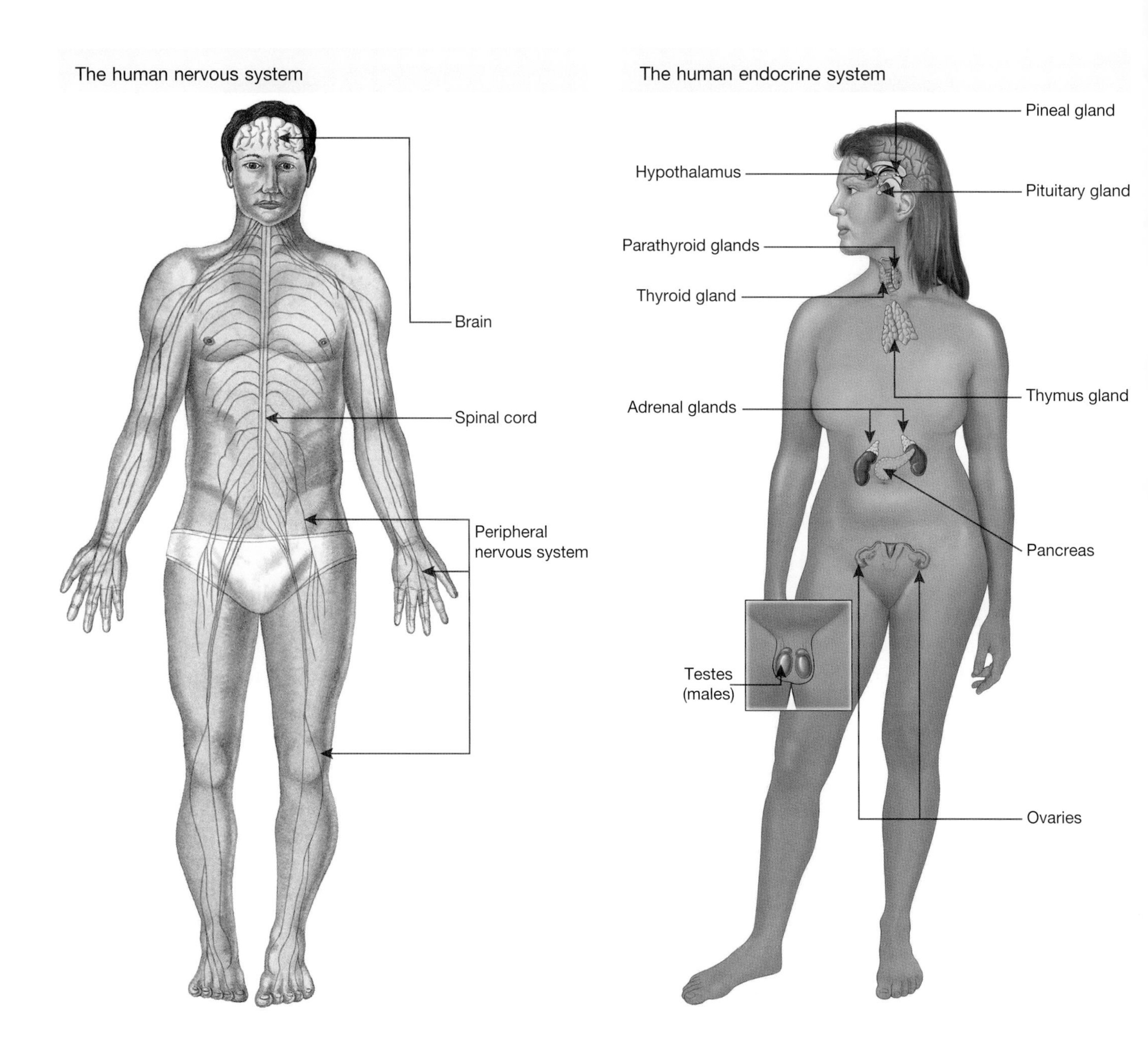

3.9.3 Feeding cells within systems

Cells within systems are alive. They require energy, oxygen and nutrients to function. They also need wastes that have been produced by the chemical reactions of life to be removed. The **digestive**, **respiratory**, **circulatory** and **excretory systems** work together to achieve this.

Your brain has a very important role in the control and coordination of other body systems. It is also involved in coordinating both the nervous system and the endocrine system.

The brain needs to be 'fed' and 'watered'. Other body systems work together to provide cells in your brain with what they need and what they don't.

3.9.4 Need gas?

Like other organs of your body, your brain needs oxygen and glucose to use in cellular respiration. Breathing helps you to feed your brain oxygen. Your **respiratory system** enables you to bring oxygen into your body through your nose and **trachea** to your **lungs**. Within your lungs, it passes through **bronchi** and **bronchioles** to finally reach your **alveoli**.

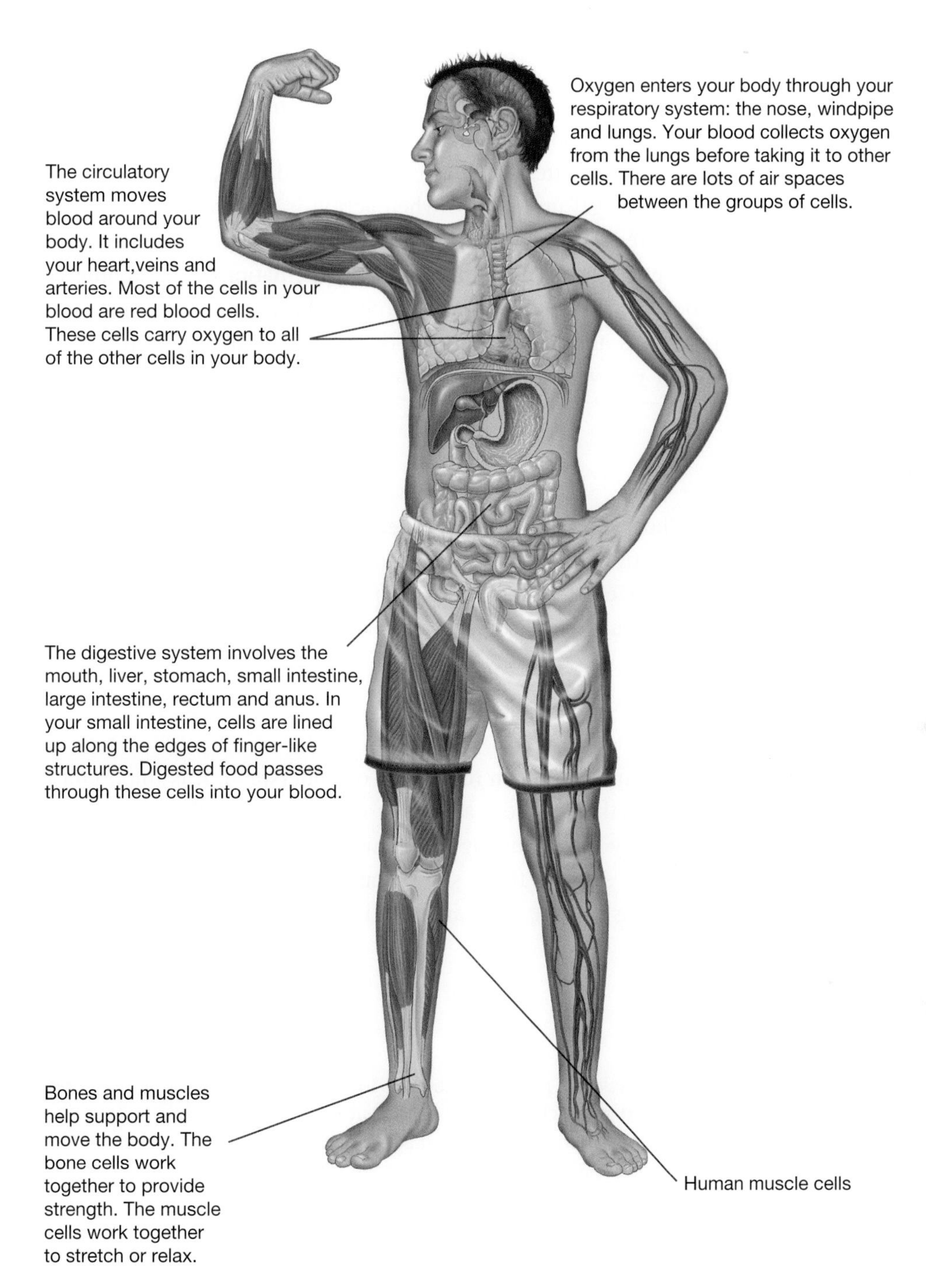

It is at the alveoli that oxygen enters the **capillaries** to be transported via your **circulatory system** to cells throughout your body, such as those in your brain. The circulatory system also transports carbon dioxide (a waste product of cellular respiration) back to your lungs. Carbon dioxide is then exhaled via the respiratory system. The rate at which these gases are exchanged is increased during exercise. The breathing rate is regulated by the brain to ensure that your cells get the oxygen they need, and that carbon dioxide is removed.

3.9.5 Sweet stuff

Eating helps feed your brain glucose. Some of this glucose may have been digested or broken down from a more complex form such as starch in potatoes or bread. This process of digestion occurs in the **digestive system** and is aided by biological catalysts called **enzymes**. Glucose is transported to your cells in your body by the **circulatory system**. The levels in your blood are kept within a narrow range. Your **endocrine system** uses hormones such as insulin and glucagon, which are released by your pancreas to maintain glucose blood levels.

3.9.6 Protein alert!

Your brain also needs nutrients such as amino acids to make **neurotransmitters**. The proteins eaten in food may need to be broken down into their component amino acids by your digestive system. Your brain uses the amino acid tyrosine to make neurotransmitters such as dopamine and **norepinephrine**.

Norepinephrine enables your body to 'get up and go' when action is required. When norepinephrine is released throughout your body, it increases the blood flow to your brain, which increases your alertness. This neurotransmitter is also very useful when you are doing maths calculations, maintaining your attention span and increasing your conscious awareness. Too much norepinephrine can make you feel 'hyper' and stressed; too little can cause drowsiness and make you feel 'out of it'. Foods with tyrosine include meat, fish, eggs, tofu and milk products. If you want to get your day going, a 'thinking breakfast' including any of these would be a great start!

3.9.7 Water me

Your brain can also get thirsty. When you feel thirsty, it means that the amount of water in your body has dropped and that salt concentration in your blood is increasing. Such an increase in blood salt levels can lead to fluids leaving your cells and moving into your bloodstream. This can result in an increase in your blood pressure and an increase in stress. Drinking water can reduce these effects within five minutes.

3.9.8 Telling the 'story'

No matter how clever you are or how much you can remember or understand, it is important to be able to communicate your ideas to others effectively. There are many different ways that you can communicate and share your understanding. One way of organising your thinking and coordinating your approach to a task is by using visual thinking tools such as those shown on the following page. These can help you to control and coordinate your thinking so that you can effectively communicate your understanding of scientific ideas.

HOW ABOUT THAT!

Bruno Annetta is a film-maker, graphic designer, actor and science teacher. Through his animated cartoons, Bruno communicates scientific ideas and shares his excitement about the wonders of science.

'I've always found it easier to remember information by doing something with it that was personal to me. The basic idea for the film *The Life of a Red Blood Cell* came out of me trying to remember the circulatory system when I first encountered it in Year 9 (Form 3). I created the animation in my head and later turned that idea into a film.'

'I believe that, through the use of animation, humour, music and storytelling, learning can be fun and easy. I have used science in the making of animated films. An understanding of light, lenses and different coloured filters is crucial in exposing film correctly. I believe that through combining the sciences with the arts, one has a richer and more fulfilling life.'

In *The Life of a Red Blood Cell*, we follow the red blood cell as it goes from the bone marrow, where it is formed, into the blood circulatory system and starts its journey.

You may find visual thinking tools useful in planning your story.

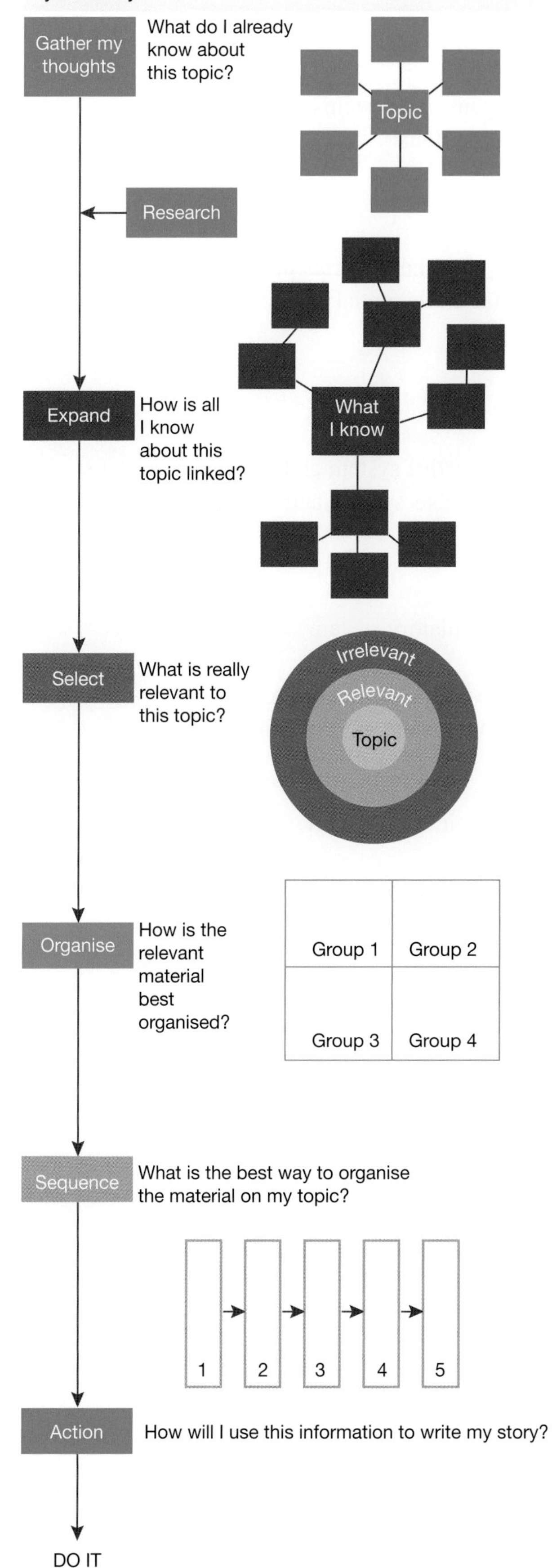

3.9 Exercises: Understanding and inquiring

To answer questions online and to receive **immediate feedback** and **sample responses** for every question, go to your learnON title at www.jacplus.com.au. *Note:* Question numbers may vary slightly.

Remember

1. Draw and label diagrams of the respiratory, circulatory, digestive and urinary (excretory) systems.
2. Compare and contrast the endocrine system and the nervous system.
3. Explain why your brain needs oxygen and glucose.
4. Describe how your brain cells obtain oxygen and glucose.
5. Describe how waste products are removed from cells such as those in the brain.
6. Explain why it is important for systems to work together.
7. Outline the relationship between proteins in food and neurotransmitters in the brain.
8. Describe some of the effects of norepinephrine.
9. Explain why it is important to drink water.

Think and discuss

10. Who am I? Identify the body part or organ that matches the function statement.
    (a) I am an organ that pumps blood around the body.
    (b) We are the top two chambers of the heart.
    (c) I am a blood vessel that takes blood to the heart.
    (d) I am an organ from which nutrients move from the digestive system into the circulatory system.
    (e) I am an organ from which oxygen diffuses into the bloodstream and from which carbon dioxide is removed.

Investigate, think, create and present

11. Select one of the research questions related to two different body systems below to investigate or identify your own. Use the coordinated thinking and planning figure at left to investigate this question and produce a summary of your findings that can be shared with others.
    - What evidence supports the link between obesity and diabetes?
    - My dad has diabetes; does that mean that I have an increased chance of getting it too?
    - What is the link between folate deficiency and neural tube defects?
    - What evidence supports the link between depression (Seasonal Affective Disorder) and vitamin D deficiency?
    - Can omega-3 fatty acids really affect behaviour and mood?
    - Why is a person who has coeliac disease more likely to be anaemic?
    - Why do heavy drinkers of alcohol have a greater risk of developing cirrhosis of the liver?

3.10 Review

3.10.1 Study checklist

Multicellular organisms

- explain why multicellular organisms need specialised organs and systems
- outline the relationship between cells, tissues, organs and systems

Digestive system

- identify nutrients that are essential for a healthy body
- describe the roles of carbohydrates, proteins, lipids, vitamins and minerals in the diet
- provide examples of two vitamins and two minerals that are essential for your health, and possible consequences of deficiencies
- explain why animals need to eat food
- label a diagram of the digestive system
- describe the role of enzymes in digestion
- outline the overall function and key components of the digestive system and circulatory system
- use flowcharts to describe how structures within the digestive system and circulatory system work together

Circulatory system

- outline the overall function and key components of the circulatory system
- describe how blood circulates around the body
- use flow charts to describe how structures within the digestive system work together
- describe how and why the circulatory system and digestive system work together

Respiratory system

- outline the overall function and key components of the respiratory system
- describe what happens at an alveolus
- use flowcharts to describe how structures within the respiratory system work together
- describe how and why the respiratory system and circulatory system work together

Excretory system

- outline the overall function and key components of the excretory system
- label a diagram of the urinary system
- describe the roles of the main organs of the excretory system
- describe how the excretory system and respiratory system work together

Nervous and endocrine systems

- outline the overall function and key components of the endocrine and nervous systems

Individual pathways

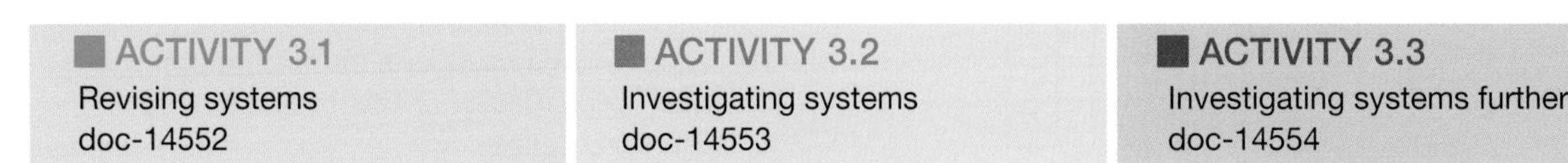

learn on ONLINE ONLY

3.10 Review 1: Looking back

To answer questions online and to receive **immediate feedback** and **sample responses** for every question, go to your learnON title at www.jacplus.com.au. *Note:* Question numbers may vary slightly.

Remember

1. Select the appropriate terms to complete the flow charts shown below:
   trachea, bronchi, arterioles, pulmonary vein, artery, left atrium, capillary, alveoli, body cells, left ventricle, capillary, vein, nose, bronchioles

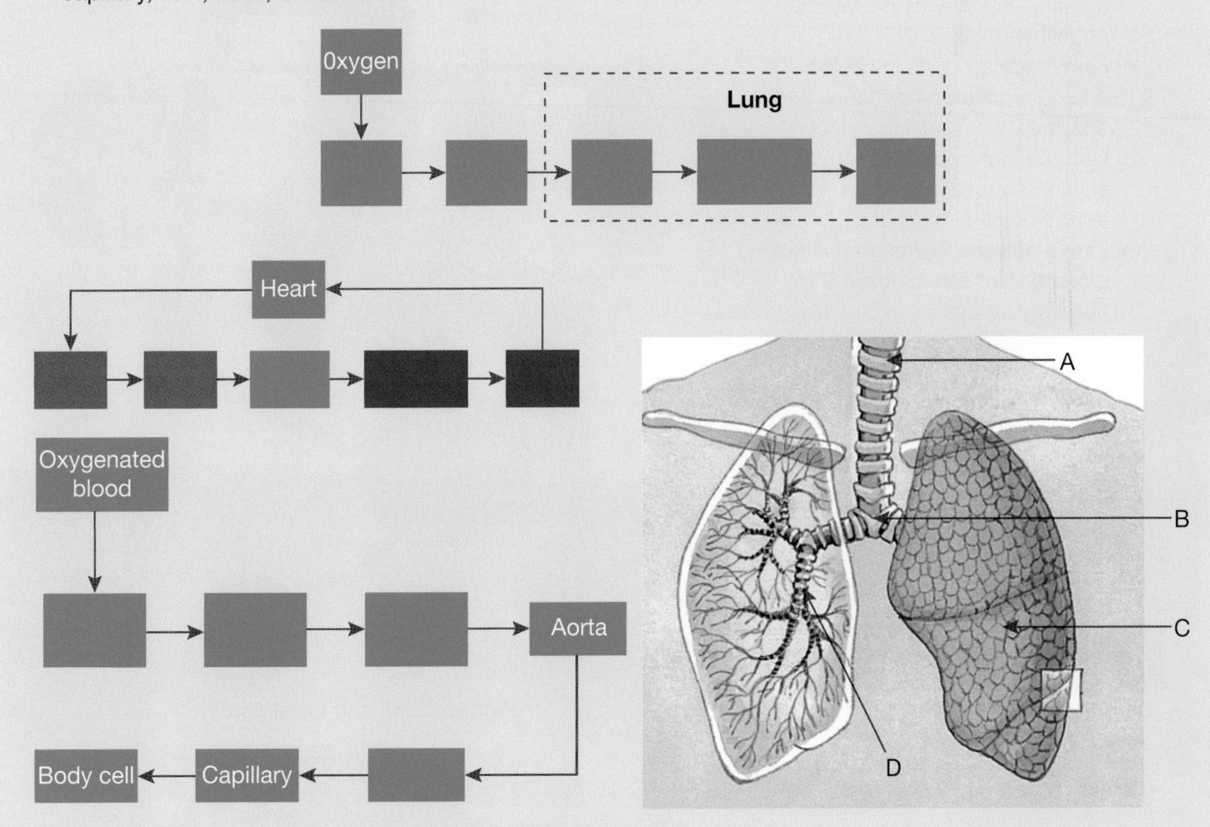

2. Label the diagrams of the respiratory system (above right) and the circulatory system (on the following page).
3. Identify whether the following statements are true or false. Justify your response.
   (a) Your circulatory and respiratory systems work together to provide your cells with carbon dioxide for cellular respiration.
   (b) Arteries transport blood to the heart whereas veins transport blood away from it.
   (c) Oxygen moves from the alveoli of your respiratory system into the red blood cells in the surrounding capillaries of your circulatory system.
   (d) The left atrium of the heart contains oxygenated blood.
   (e) Oxygenated blood travels from your lungs via the pulmonary vein to the right atrium of your heart.

4. Which am I? Match the chemical in the list below to the most appropriate description.
 potassium, vitamin A, fat, iron, protein, cellulose, starch
 (a) I am needed by red blood cells to carry oxygen to tissues.
 (b) I assist nerves in functioning and a deficiency may cause fatigue and slow reflexes.
 (c) I am made up of amino acids and can make up hormones and enzymes.
 (d) I can be stored under the skin and am solid at room temperature.
 (e) I am a polysaccharide that is not digested but can increase the fibre in your diet.
 (f) I am a fat-soluble vitamin that is needed for healthy lining of your digestive and respiratory systems.
 (g) I am a polysaccharide that can be digested and broken down into glucose.
5. Copy and complete the following table.

Nutrient	Why it is needed?	Examples of sources
Carbohydrate		
Protein		
Lipid		
Vitamin		
Mineral		

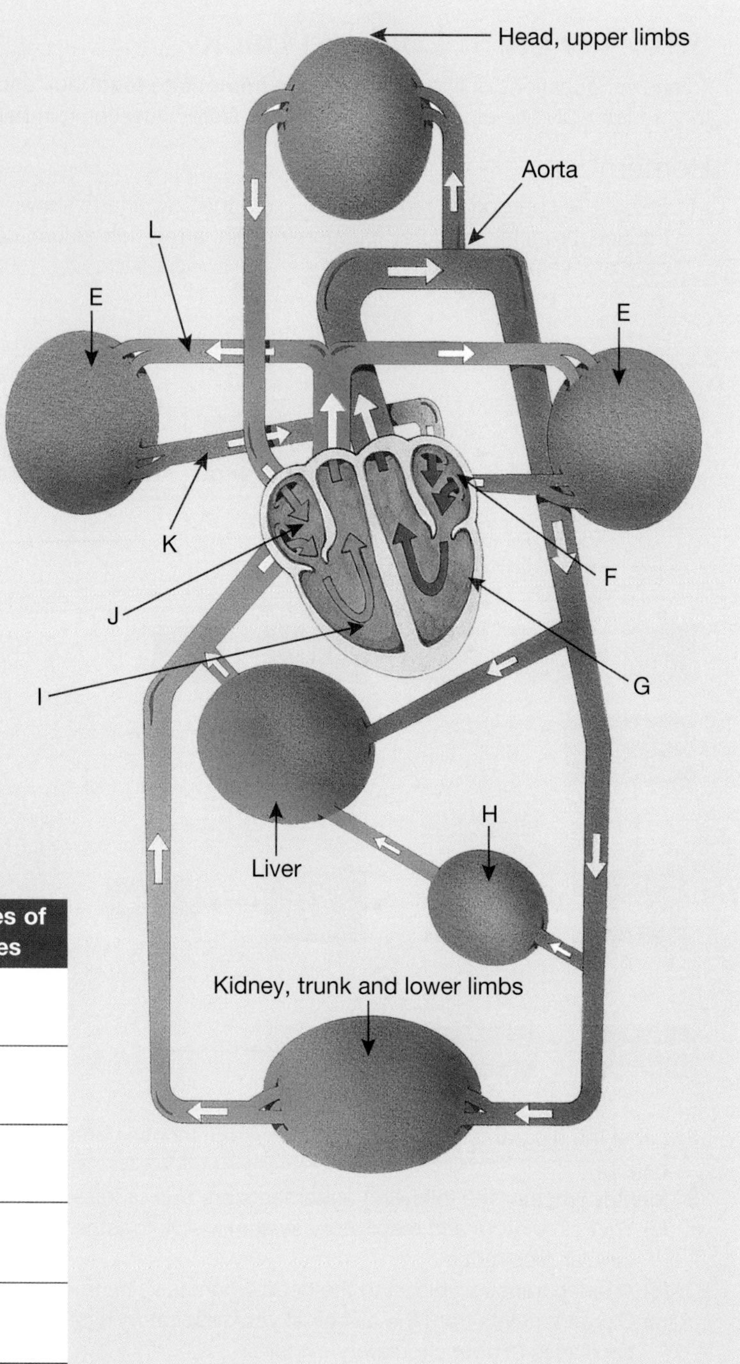

6. Select the appropriate terms to complete the flowchart at right:
   oesophagus, teeth, salivary glands, stomach, large intestine, mouth, anus, small intestine, rectum, liver, gall bladder
7. Label the diagrams of the digestive and excretory systems below.

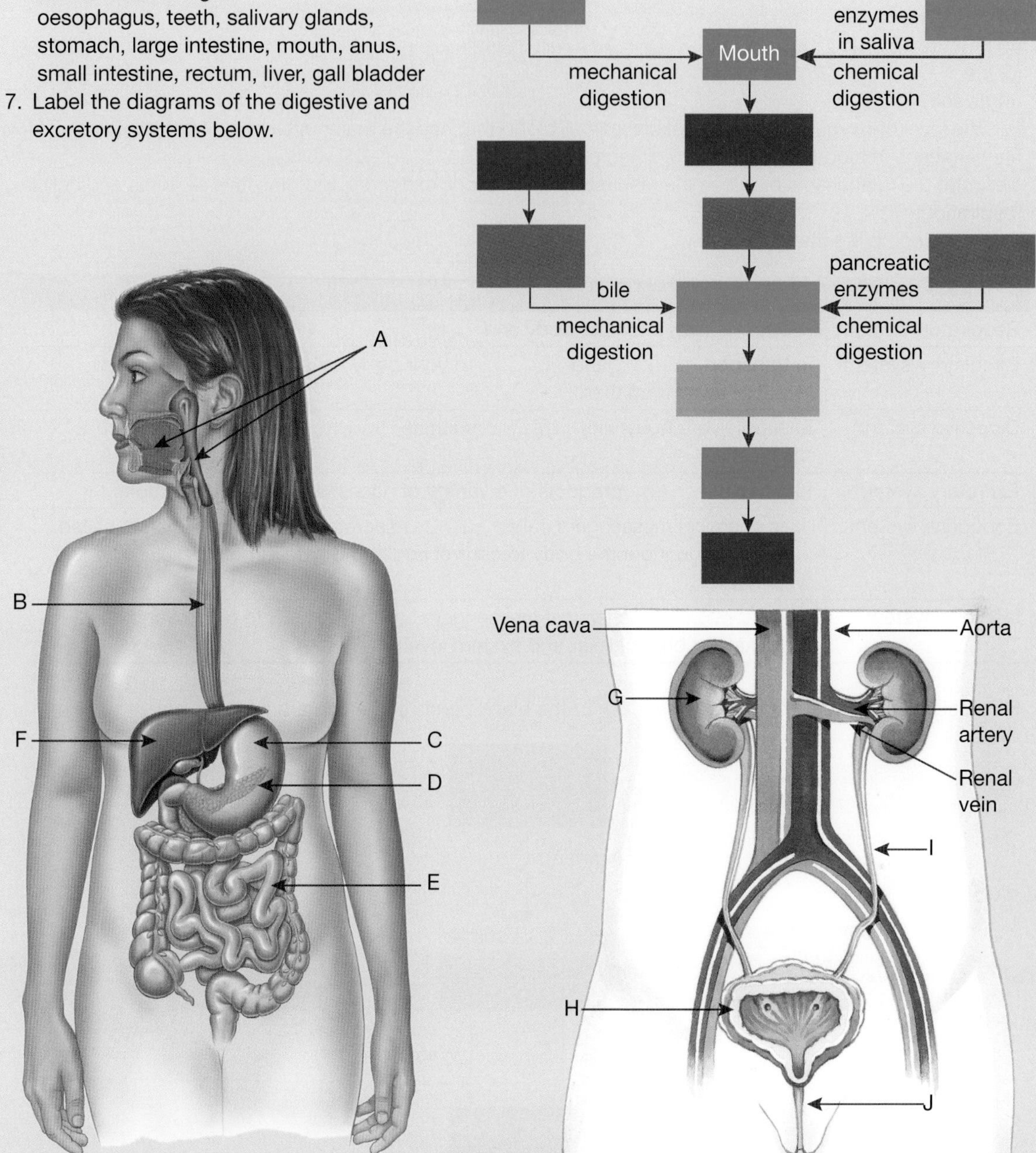

8. Match the organ up with the unwanted product or waste that it excretes.

Organ	Unwanted product or waste to be excreted
Kidney	Bile pigments from old red blood cells
Liver and large intestine	Carbon dioxide
Lungs	Urea

9. Explain how the shapes of the following structures well suit them to their function.
   (a) trachea
   (b) oesophagus
   (c) nephrons
   (d) villi
   (e) alveoli
10. Is it the level of oxygen or carbon dioxide in your blood that has the major influence on breathing rate? How are variations in blood concentrations detected?
11. Describe the relationship between the respiratory, circulatory, excretory and digestive systems and cellular respiration.
12. Copy and complete the table below.

Term	Function
Respiratory system	To get .... into your body and .... out
Circulatory system	To transport .... and .... to your body cells, and wastes such as .... away from them
Digestive system	To supply your body with .... such as .... so that it functions effectively
Excretory system	To remove .... products of a variety of necessary chemical reactions
Endocrine system	Uses chemical messengers called .... secreted from special glands called .... throughout the body to control and coordinate at both cellular and system level
Nervous system	Uses .... and chemical messengers called .... to control and coordinate at both cellular and system level

13. Suggest which body systems belong in each of the blank boxes in the figure below.

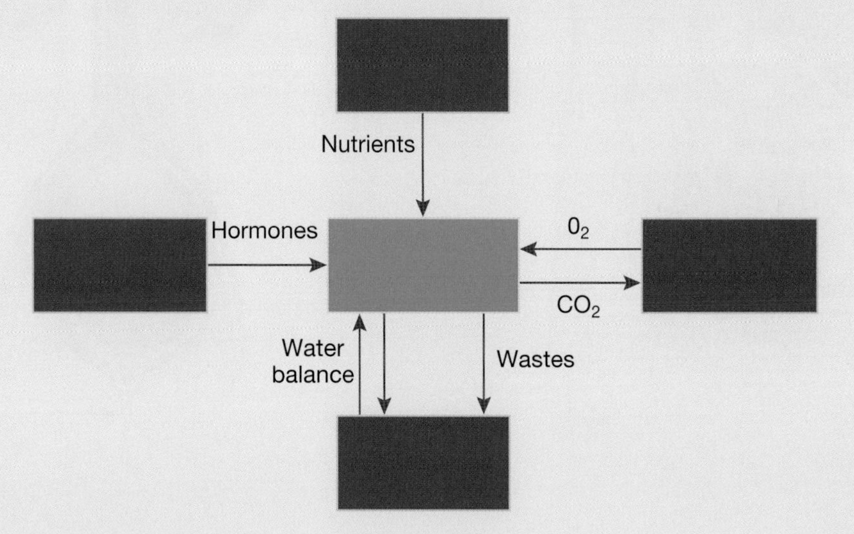

14. (a) In a team of four, brainstorm:
      (i) examples of situations and tactics that might be used to encourage or pressure young people to take drugs
      (ii) possible consequences of taking the drugs offered
      (iii) strategies (both verbal and non-verbal) that could be used to say 'No thanks!' or remove the pressured person from the situation.
    (b) Write a story or play that would help to provide young people with ideas on how to say 'no' or get out of difficult drug situations.
    (c) Present your play to the class.

15. List the following in order from highest to lowest alcohol content (for a volume of 180 mL): whisky, full-strength beer, white wine, port.
16. Classify the following as being (a) excitatory or (b) inhibitory psychoactive drugs.
    - Barbiturates
    - Nicotine
    - Heroin
    - Caffeine
    - Amphetamines
    - Benzodiazepines
    - Cocaine

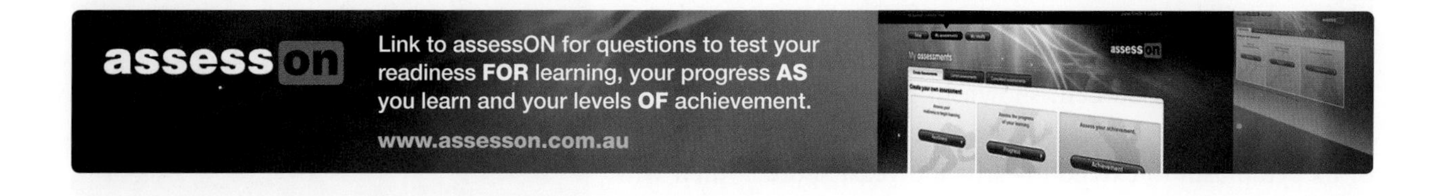

TOPIC 4 The body at war

4.1 Overview

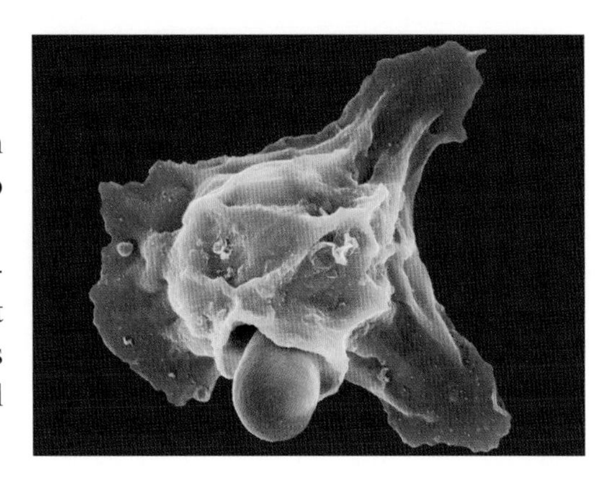

Lines of defence, soldier cells and chemical weapons all form a part of the amazing array of strategies used by our bodies to keep us healthy.

This coloured scanning electron micrograph shows a lymphocyte (white blood cell, shown in green) engulfing a yeast cell (shown as orange). The lymphocyte is using projections of its cytoplasm to extend towards the yeast spore, which will be swallowed up and digested.

4.1.1 Think about the body at war

- What is an infectious disease?
- What are the differences between viruses and bacteria?
- What is a parasite?
- What is the Black Death and how does it spread?
- Is immunisation necessary?
- What animal up to ten metres long can live inside a living human body?
- What does H5N1 have to do with birds?
- Is diabetes contagious?
- Why are anthrax, cholera, botulism and smallpox attractive to terrorists?

LEARNING SEQUENCE

Numerous **videos** and **interactivities** are embedded just where you need them, at the point of learning, in your learnON title at www.jacplus.com.au. They will help you to learn the content and concepts covered in this topic.

4.1.2 Food warnings

Throughout history, stories and rituals have been used to pass knowledge about food and nutrition from one generation to the next.

Think and investigate

1. Imagine that Goldilocks got sick after eating the porridge. Suggest some reasons why this may have happened.
2. The Three Bears did not cover their porridge while they went for their walk. Was this a good idea or not? Give reasons for your answer.
3. How long can porridge stay uncovered at room temperature before it is dangerous to eat? Find out the spoiling time of four other foods.
4. Baskets of food, along with jars of wine and oil, were found in Tutankhamen's tomb in Egypt in 1922. Other Egyptian tombs contained honey that was in a well-preserved state; when opened it retained some of its aroma. Today, most foods have a use-by date on the packaging. For three different foods, find out what might happen, and why, if you used it well after its use-by date.
5. Sometimes canned food is unsafe to eat. Find out why.
6. Find out what strategies humans have to survive eating lots of different foods, some of which may cause food poisoning.
7. Create your own fairytale to teach young children about poisonous or spoiled food. Present your story as a PowerPoint presentation, storybook, pantomime or puppet play.

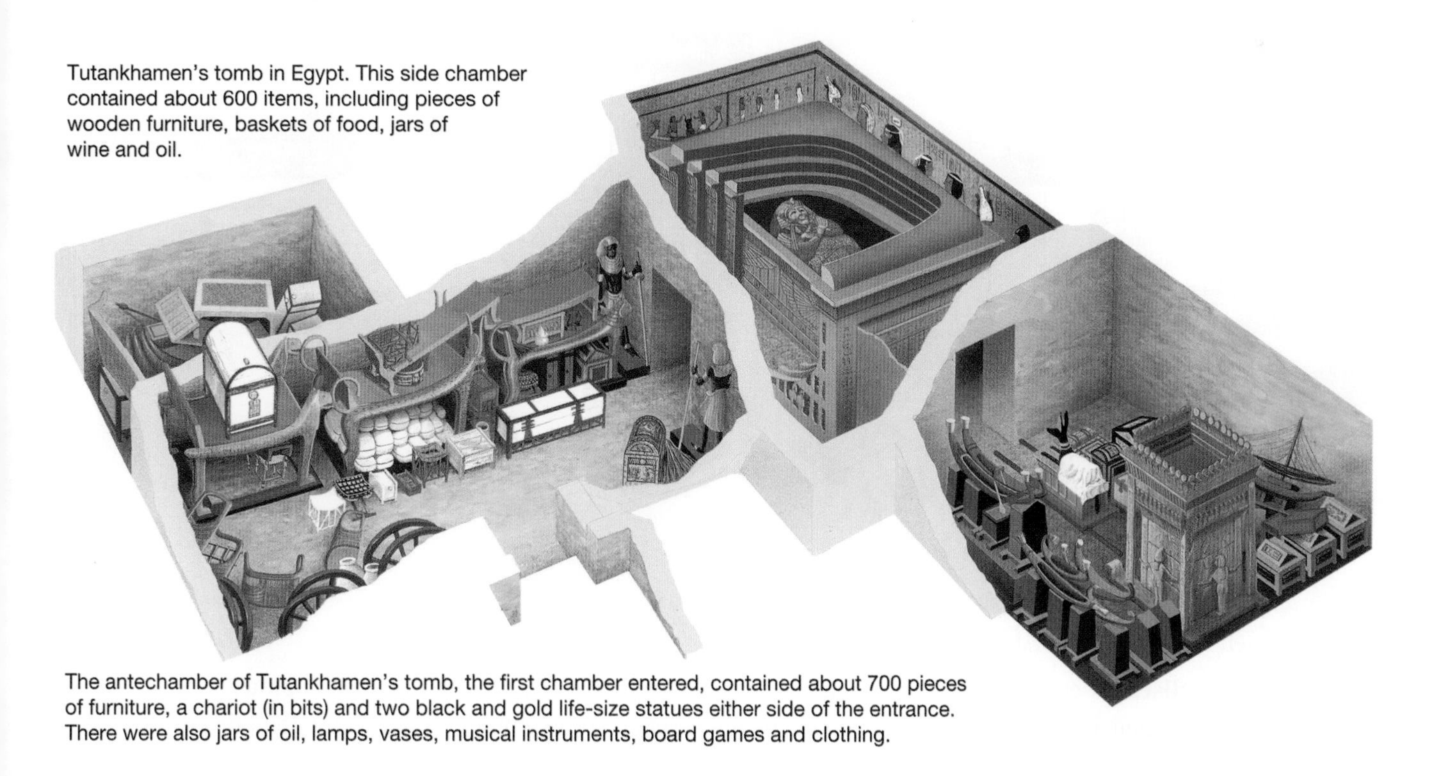

Tutankhamen's tomb in Egypt. This side chamber contained about 600 items, including pieces of wooden furniture, baskets of food, jars of wine and oil.

The antechamber of Tutankhamen's tomb, the first chamber entered, contained about 700 pieces of furniture, a chariot (in bits) and two black and gold life-size statues either side of the entrance. There were also jars of oil, lamps, vases, musical instruments, board games and clothing.

4.2 Catch us if you can

4.2.1 Catch us if you can

Something wrong? Not feeling well? You may have a disease! A **disease** can be defined as being any change that impairs the function of an individual in some way — it causes harm to the individual. Diseases can be classified as being infectious or non-infectious.

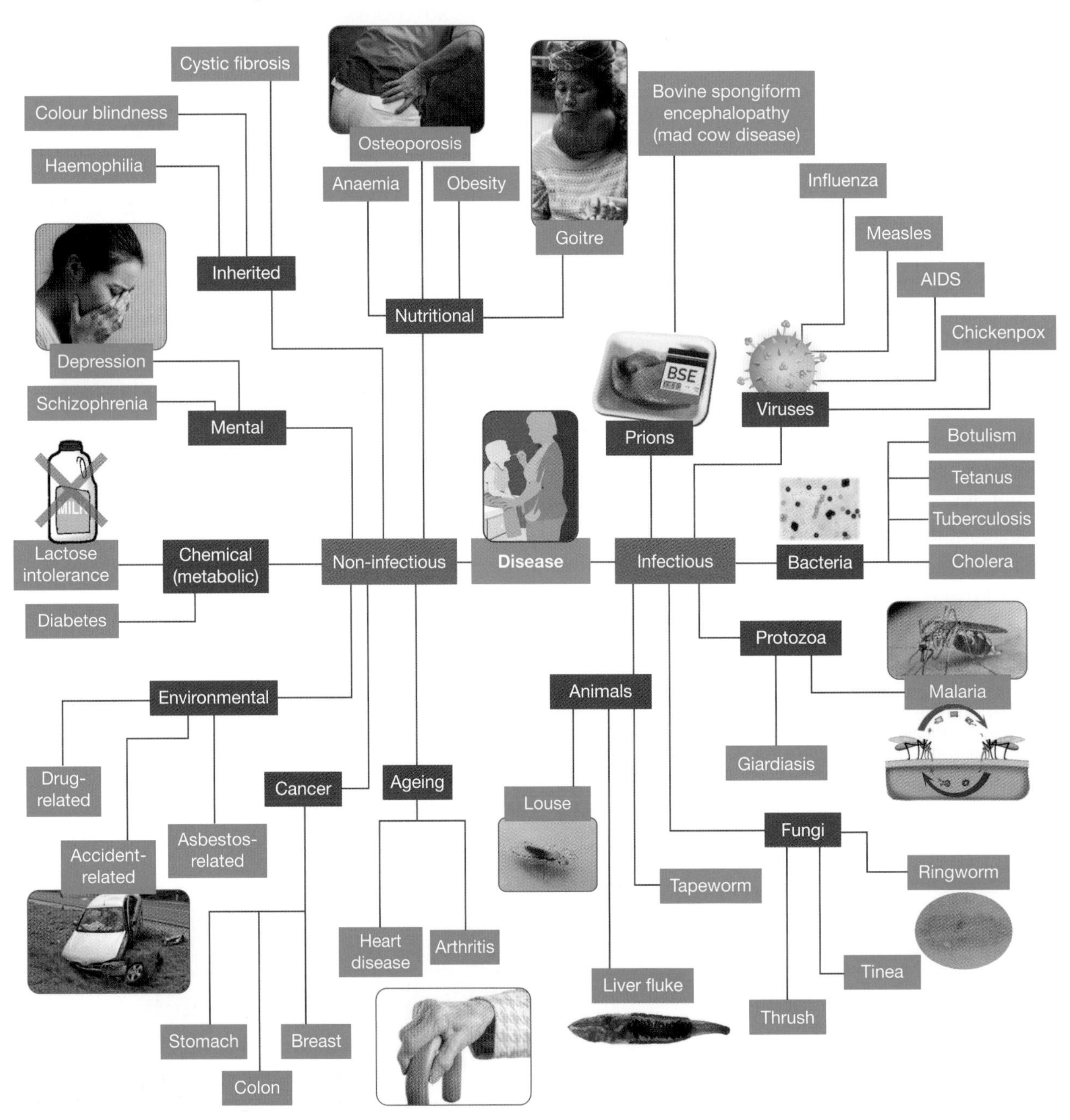

4.2.2 Can't catch us!

Non-infectious diseases cannot be spread from one person to another — they are not contagious (transferred from one organism to another). Obesity, rickets and scurvy are examples of non-infectious diseases that may be related to unbalanced diets or nutritional deficiencies. Inherited diseases such as haemophilia and cystic fibrosis and diseases related to exposure to particular poisons or drugs are also non-infectious.

Although viruses have been implicated in some cancers (for example, cancer of the cervix), most cancers are considered to be non-infectious diseases.

4.2.3 Can catch us!

Infectious diseases are diseases that are contagious and are caused by a **pathogen**. Tapeworms, head lice, liver flukes, fungi, protozoans and bacteria are examples of pathogens that are made up of cells and can be referred to as **cellular pathogens**. Some other pathogens, such as viruses, prions and viroids, are not made up of cells and for this reason are sometimes referred to as **non-cellular pathogens**.

4.2.4 Spreading it around

Keep out!

Preventing the spread of infectious diseases has been a challenge throughout history. The ancient Hebrews isolated those with disease by keeping them away from others or by sending them beyond the boundaries of the towns. In the Middle Ages, Mediterranean people refused to allow ships to dock for forty days if they carried sick people. The separation of sick people from healthy people to avoid infection was the beginning of **quarantine**. Unfortunately these methods were not enough to stop large outbreaks of disease.

WHAT DOES IT MEAN?

The word quarantine is derived from the Latin word *quadrāgintā*, meaning 'forty'.

The knowledge of how infectious diseases are transmitted is important if ways to control their spread are to be found. Some key ways in which pathogens may be transmitted include direct contact, vectors, contaminated objects or contaminated water supplies.

4.2.5 Direct contact

Some diseases are spread by direct contact. Touching others or being touched is one way in which pathogens can be directly transferred from one person to another.

Another way is via airborne droplets that are produced when you cough, sneeze or talk. These droplets may contain pathogenic bacteria or viruses and may land on objects or people around you, which may result in disease.

4.2.6 Vectors

Some diseases are spread by vectors. **Vectors** are organisms that carry the disease-causing pathogen between organisms — without being affected by the disease themselves. Mosquitoes, houseflies, rats and mice are examples of organisms that can act as vectors to spread disease.

4.2.7 Contaminated objects

While fungal diseases such as tinea and ringworm can be spread by direct physical contact, they may alsobe transmitted by towels or surfaces that have been contaminated with skin cells of an infected person.

Food poisoning is often caused by contamination of food (or food utensils) with particular types of pathogenic bacteria. This is why washing your hands is so important after going to the toilet and before touching food or being involved in food preparation.

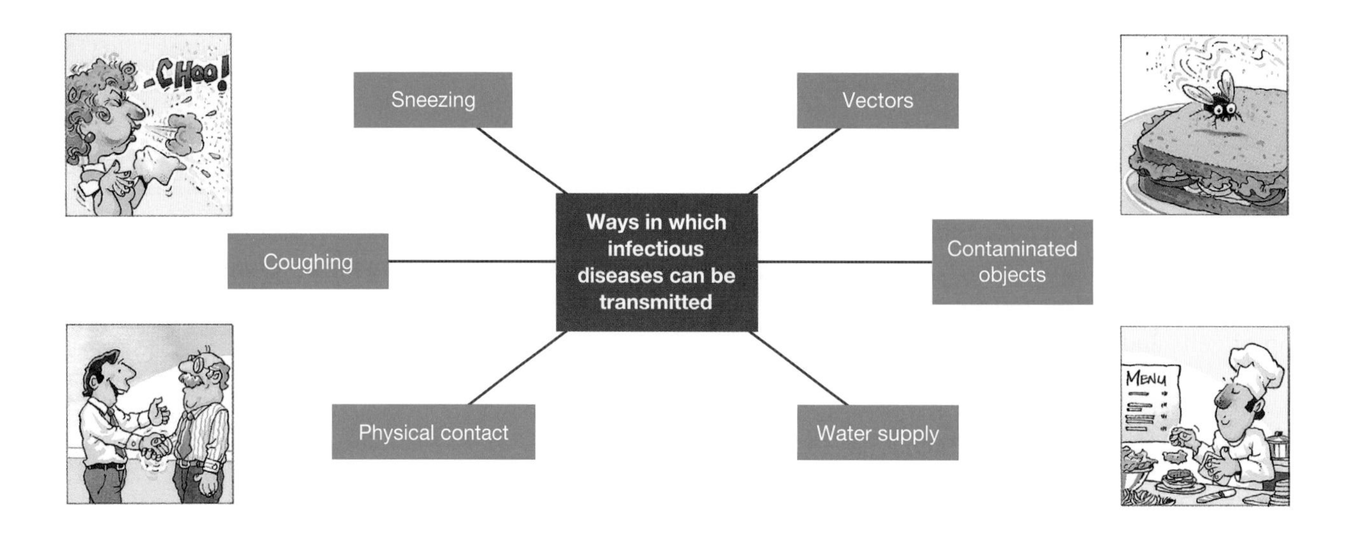

4.2.8 Contaminated water

Many pathogenic organisms live in water and are carried about in it. Our domestic water supplies are usually chemically treated to kill disease-causing micro-organisms within it. This may not be the case, however, with water drunk directly from water tanks, rivers or creeks. This water may need to be boiled before it is drunk.

During the summer months, the Environment Protection Authority (EPA) measures the levels of *Escherichia coli* (*E. coli*) bacteria in water in coastal beaches. The level of *E. coli* in the water is used as an indicator of levels of potentially pathogenic bacteria, as it is found in faeces.

4.2.9 Fighting the spread

There are a number of ways in which the spread of disease may be controlled. These include personal hygiene, care with food preparation, proper disposal of sewage and garbage, chemical control of vectors, chemical treatment of clothes, surfaces and water, pasteurisation of milk, public education programs, quarantine laws and the use of drugs such as antibiotics.

4.2 Exercises: Understanding and inquiring

To answer questions online and to receive **immediate feedback** and **sample responses** for every question, go to your learnON title at www.jacplus.com.au. *Note:* Question numbers may vary slightly.

Remember

1. Define the following terms: disease, non-infectious disease, infectious disease, pathogen, contagious, vector.
2. List three types of (a) non-infectious diseases and (b) infectious diseases.
3. Classify the following diseases as either non-infectious or infectious: ringworm, colon cancer, thrush, arthritis, cholera, diabetes, osteoporosis, malaria, measles, depression, anaemia, AIDS.
4. Distinguish between:
   (a) goitre and arthritis
   (b) haemophilia and anaemia
   (c) AIDS and malaria
   (d) tinea and chickenpox.

Think and create

5. On the next page is one flowchart that describes how disease can be spread by sneezing. Construct similar flow diagrams or cycles to show three different ways in which diseases can be spread.

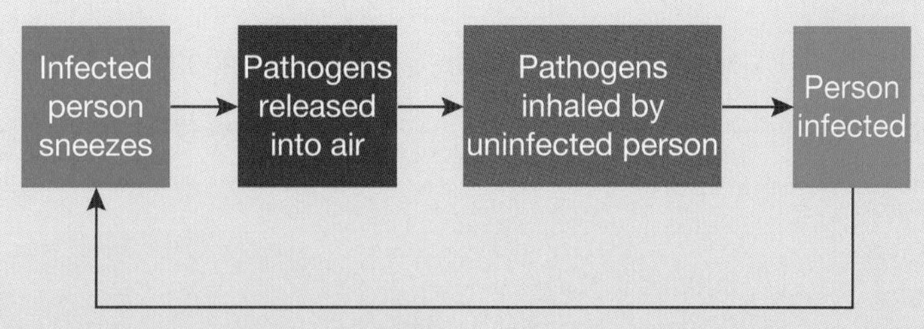

6. Create a poster or wall chart that can be used to serve one of the following purposes:
   (a) to detail the ways in which salmonella food poisoning could be avoided
   (b) for an advertising campaign designed to promote personal hygiene at home
   (c) to provide information about Louis Pasteur and the development of pasteurisation.

Think and investigate

7. Suggest why nutritional diseases are not classified as infectious diseases.
8. Until the middle of the twentieth century, infectious diseases killed many more people than non-infectious diseases. However, since about 1930, in the developed countries of Australia, North America and Europe, more people have died from non-infectious diseases. Account for this change.
9. The biggest killer of Australians in 1992 was heart disease. How might conditions of this kind, including heart attacks, be related to nutrition?
10. Find examples of at least six different types of diseases. Construct a matrix table with the headings shown below, and use a variety of resources to fill in answers for each disease.

Disease	Symptoms	Pathogen involved	Duration (how long it lasts)	Transmission (how it is spread)	Prevention or treatment	Interesting findings

11. How is a chemical or metabolic disease different from an environmental disease?
12. Can ageing be called a disease? What do you think? Explain your response.
13. Choose an antiseptic and a disinfectant found at school or at home and read the directions for use. Explain why they have different directions.
14. Two West Australian doctors, Barry Marshall and Robin Warren, isolated a *Helicobacter pylori* spiral bacillus from gastric biopsies and linked its presence to gastric ulcers. Find out more about this discovery and how it changed theories about ulcers and their treatment.
15. (a) Research and report on one of the following scientific careers: molecular parasitologist, microbiologist, virologist, bacteriologist, forensic microbiologist.
    (b) Use your new knowledge of this field of science to identify two problems that could be investigated.
16. Find out about quarantine laws in Australian states and other countries.
17. (a) Why did the levels of *Giardia* and *Cryptosporidium* become dangerous in Sydney water in 1998?
    (b) Find out how such pathogens are normally controlled.
    (c) What measures were taken by authorities to ensure the water was once again safe to drink?
18. In a team, find out more about bioterrorism and biological warfare. Each member is to select and research a disease that might be used as a biological weapon. Report to your team and discuss your findings. As a team, construct a visual fact sheet or web page to share your information with the rest of the class.

Investigate and create

19. (a) In a team, find out the cause, symptoms and methods of prevention of one of the following diseases: osteoporosis, schizophrenia, haemophilia, anaemia, arthritis, heart disease, lung cancer, skin cancer.
    (b) Report your findings back to your team or class in a PowerPoint presentation, visual thinking tool or poster.
    (c) In your team, discuss with others any ways in which the community or government may be involved in reducing the impact or frequency of these diseases.
20. Select a disease and research the structure and key features of its pathogen. Create a brochure for your pathogen with information about its life cycle and on the disease that it causes.

Analysing data

21. Study the pie chart on the next page. It shows the main causes of death worldwide in 2002.
    (a) What percentage of people died from infectious diseases:
        (i) worldwide
        (ii) in low-income African and Asian countries?
    (b) Why do you think there is such a large difference in the percentage of people who died from infectious disease between wealthier countries and poor countries?

(c) What percentage of children who died before the age of 5 died of infectious diseases? Why is this figure so high? (*Hint*: Think about the other main causes of death and who they are likely to affect.)

(d) Draw a column graph to represent the data shown in the pie chart.

(e) If the same data was collected for 2014 and a similar graph drawn, how do think the two graphs would differ? Give a reason for your answer.

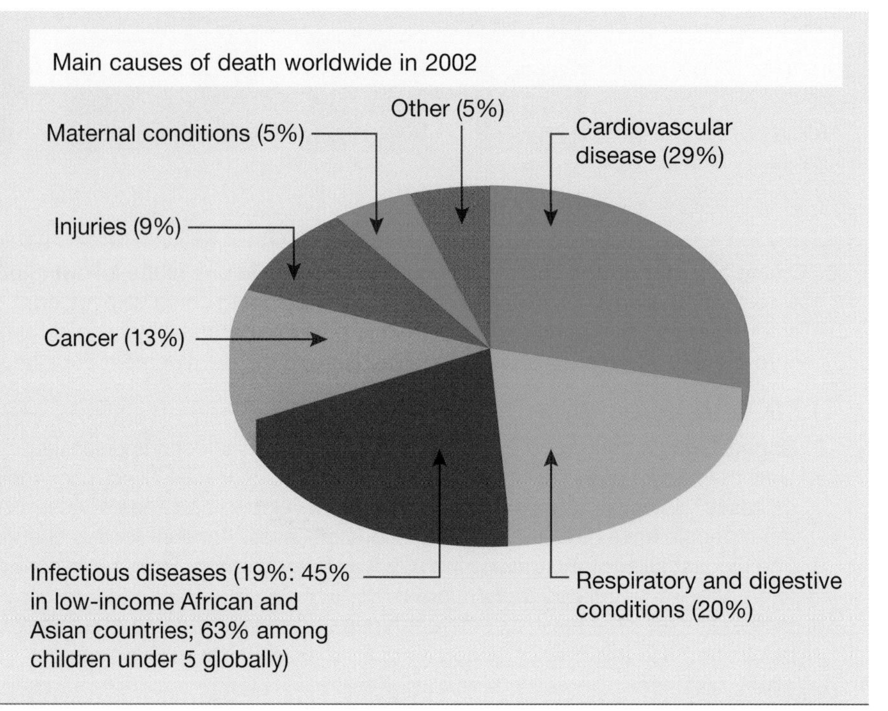

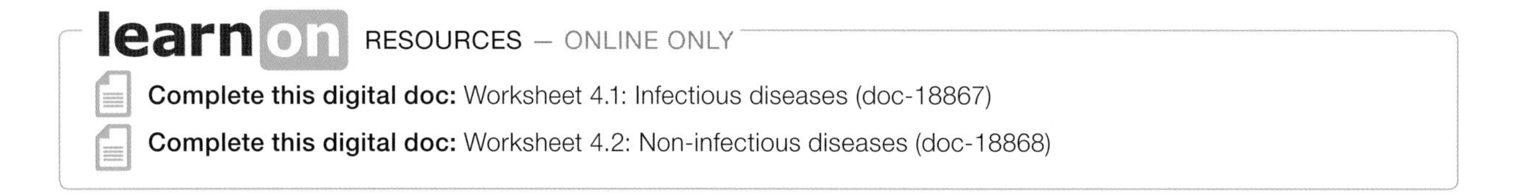

4.3 Invasion! Alien alert!

4.3.1 Invasion! Alien alert!

Feeling hungry? Need shelter? We do not live alone! We need others to survive. Sometimes others need to use us to be able to survive. Sometimes it isn't just about food and shelter — sometimes they even need our help to reproduce.

4.3.2 Parasites

Some relationships between organisms may provide one with resources, but not necessarily cause harm to the other. An example of this relationship is that involving a **parasite** and its host.

The organism that a parasite lives in or on is referred to as its **host**. The life cycle of parasites can involve one or more hosts. The **primary host** is the organism used for the adult stage and the **intermediate host** (or **secondary host**) is used for the larval stage.

Parasites can be classified on the basis of the part of your body in which they live. Parasites that live inside your body are called **endoparasites** and those that live outside your body are called **ectoparasites**.

The organism in which the parasite completes some part of its life cycle is referred to as its host.

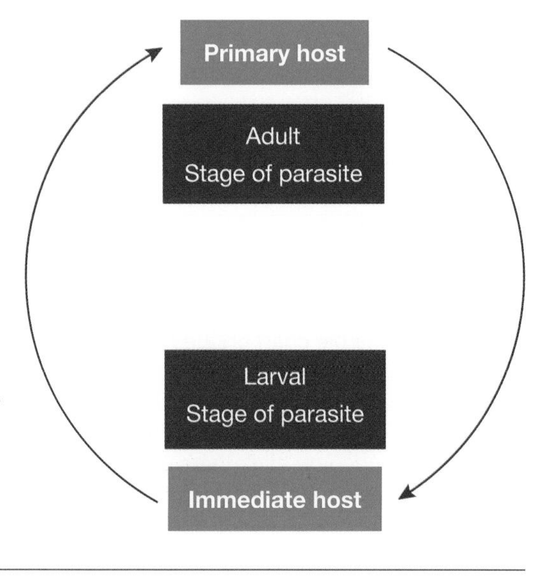

Some parasites can harm their hosts and cause disease; these parasites are also considered to be pathogens. Not all parasites, however, cause harm to their host. It's probably a very good idea if they don't, because they rely on their host for resources.

4.3.3 Pathogens

Infectious diseases are caused by pathogens. Pathogens may be cellular (made up of cells) or non-cellular. Disease-causing bacteria, protists, fungi and animals are examples of cellular pathogens. Viruses, prions and viroids are examples of non-cellular pathogens. Pathogens cause harm to their hosts (the organism that they infect).

4.3.4 Prions

Prions are non-cellular pathogens. The word *prion* is derived from the terms *protein* and *infection*. They are abnormal and infectious proteins that can convert your normal protein into prion protein. When cells containing prions burst, more of these infectious proteins are released to infect other cells. The bursting of these cells can also result in damage to the tissues of which they are a part.

Prions are thought to be responsible for degenerative neurological diseases. These diseases are also called **transmissible spongiform encephalopathies (TSE)**. The term *spongiform* is included because of the tiny holes that result from the bursting of infected cells, giving the brain a spongy appearance. Examples of these diseases include Kuru, Creutzfeldt-Jakob disease (CJD) and bovine spongiform encephalopathy (BSE).

BSE is commonly known as 'mad cow disease' because of the nervous or aggressive behaviour observed in infected cows. Hundreds of thousands of cattle were destroyed when it was discovered that humans could become infected with this disease by eating meat from infected cows.

4.3.5 Viruses

Viruses are another example of non- cellular pathogens. They consist of **DNA** or RNA enclosed within one or more protein coats. Viruses are so small that they can only be seen with very powerful electron microscopes.

Scientists debate whether viruses should be called living things as they are **obligate intracellular parasites**. This means that they need to infect a host cell before they can reproduce — they cannot do it on their own. As viruses cause damage to their host cell in the process, they are also classified as being pathogens. Examples of infectious diseases caused by viruses include warts, rubella, mumps, poliomyelitis, influenza, AIDS and the common cold.

Viruses come in many shapes and sizes.

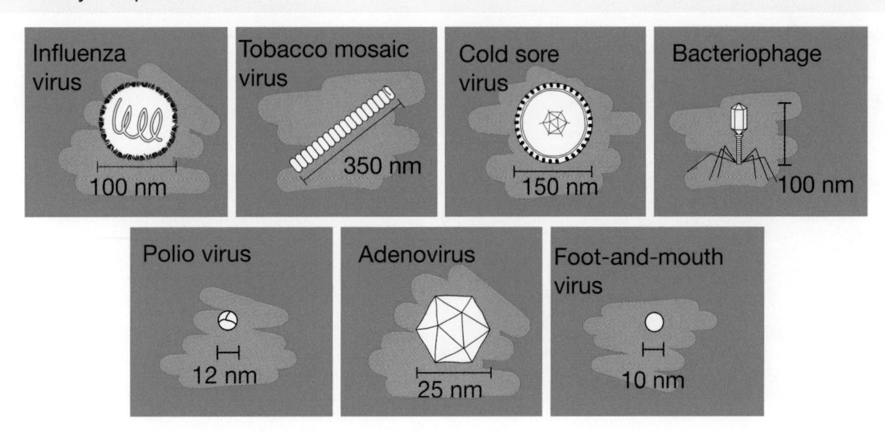

The influenza virus consists of RNA surrounded by protein and lipid layers. It is not cellular.

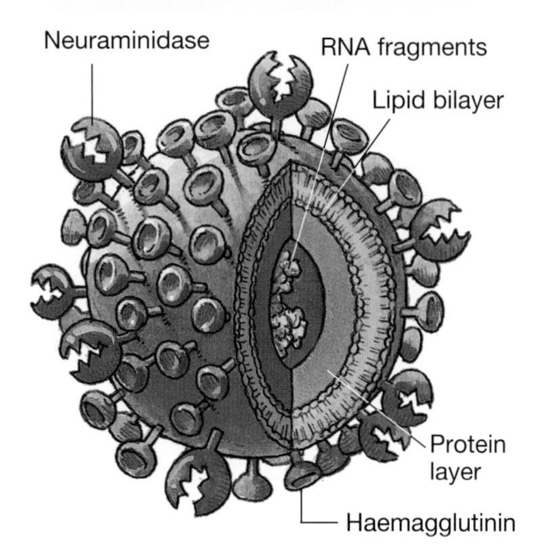

This cycle depicts how a virus is spread.

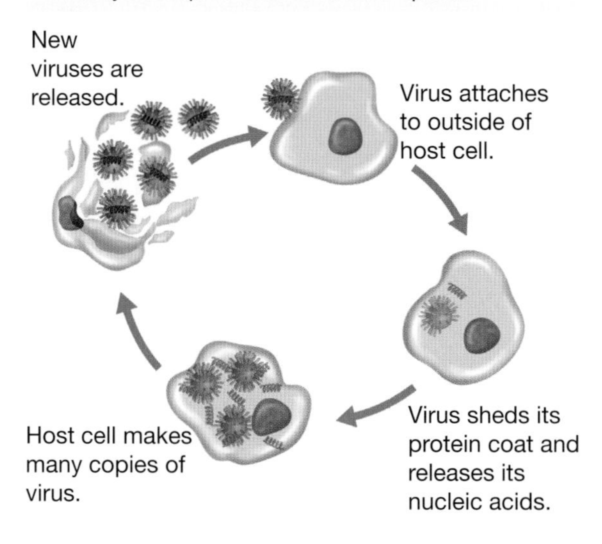

4.3.6 Bacteria

Disease-causing bacteria are cellular pathogens that consist of a single cell. They can be classified on the basis of their cell shape, the organisation of colonies of bacteria and the presence or absence of structures (such as a flagellum) or particular chemicals in their cell wall.

A spherical bacterium is referred to as **coccus** (for example, *Staphylococcus*), a rod-shaped bacterium as **bacillus** (for example, *Bacillus*) and a spiral-shaped bacterium as **spirochaete**. Their colonies can be described as being single, in pairs, in chains or clustered together.

Vibrio bacteria

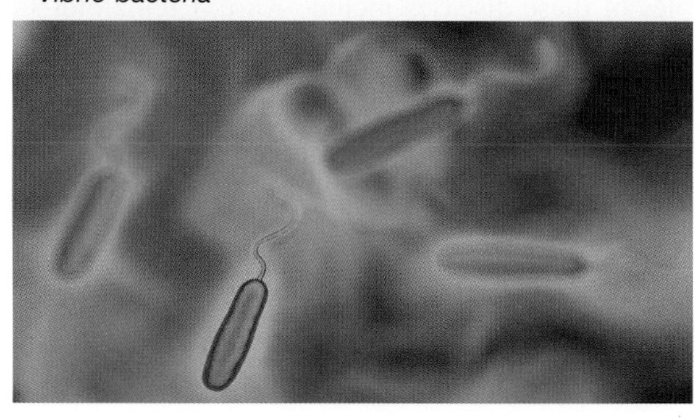

Examples of diseases caused by bacteria include strep (short for *Streptococcus*) throat, tetanus, pneumonia, food poisoning, gastroenteritis, cholera, gonorrhoea, leprosy, tetanus, scarlet fever, whooping cough, meningitis, typhoid and even pimples!

Streptococcus bacteria

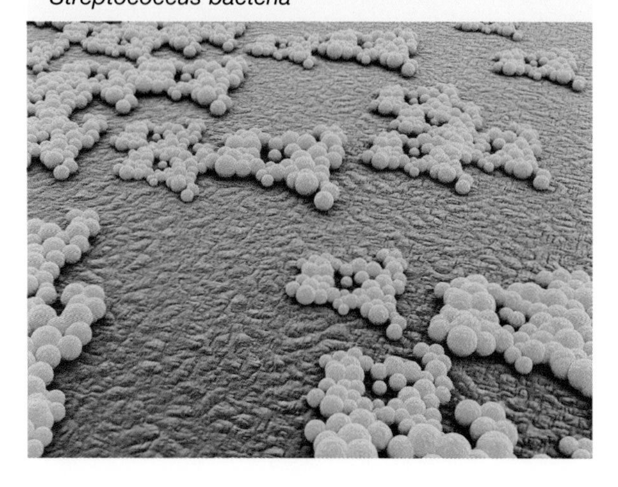

Some types of disease-causing bacteria

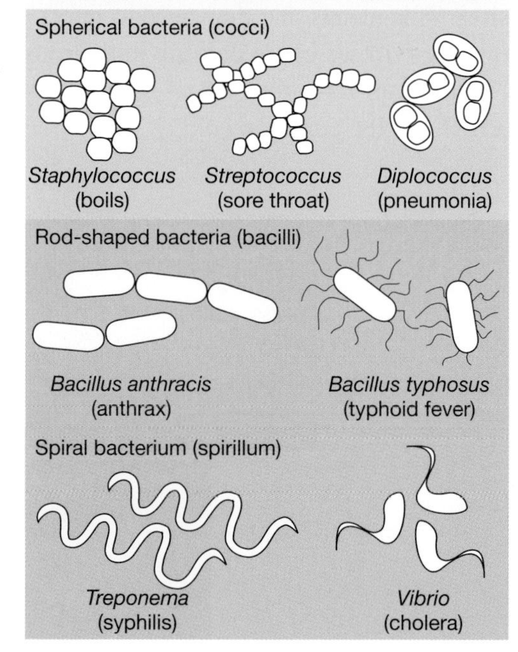

4.3.7 Protozoans

A number of infectious diseases are caused by parasitic protozoans. These single-celled organisms are usually found within their host's body. It is a good idea to know more about these diseases if you intend to go to tropical regions, where such diseases are more common. Examples of diseases caused by protozoans include malaria, amoebic dysentery and African sleeping sickness.

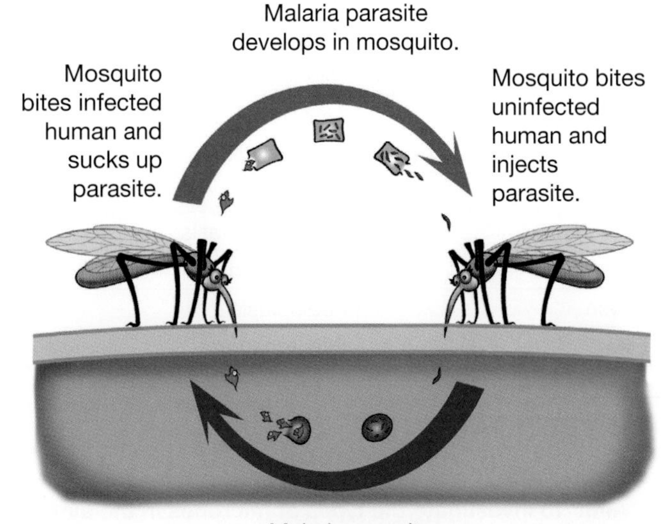

4.3.8 Fungi

Fungi belong to one of the biggest groups of organisms. They include some that are large, such as toadstools, and others that are microscopic, such as the **moulds** that grow on bread. Many fungi are parasites, feeding on living plants and animals, including humans. This often results in disease.

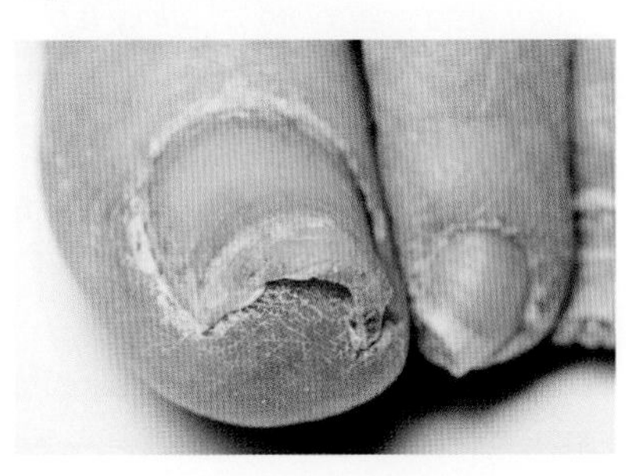

Common human diseases caused by fungi are tinea or athlete's foot, thrush and ringworm. Some fungi live in the mouth, the vagina and the digestive system at all times without causing harm. However, if resistance to disease is low, the fungi in these places can become active and cause problems such as thrush.

4.3.9 Worms and arthropods

Larger parasites include endoparasites such as tapeworms, roundworms and liver flukes, and ectoparasites such as ticks, fleas and lice.

Tapeworms are the largest of the parasites that feed on the human body and can be up to 10 metres long! They have hooks and suckers to keep a firm hold on your intestine. Tapeworms don't have to worry about finding a mate. When they are reproductively mature, their end segment, which is full of eggs, passes out with their host's faeces and on to its next host.

(a) The head of a tapeworm, magnified to 15 times its real size, showing its hooks and suckers (b) A pork tapeworm has a flat, ribbonlike body. It can infect humans.

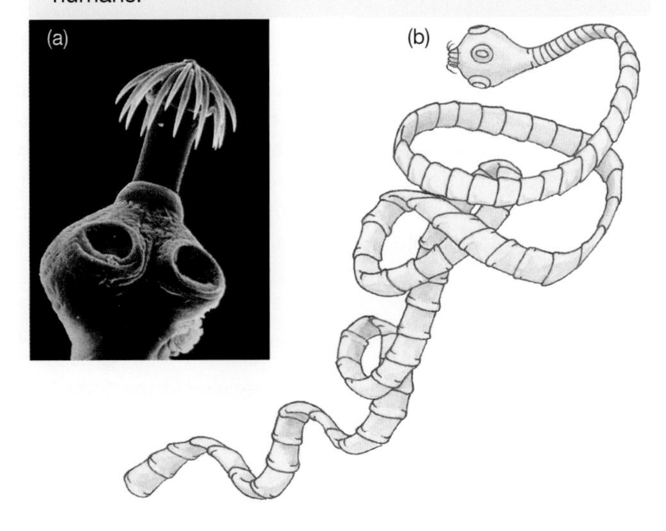

Did you get an itchy bottom at night when you were little? You probably had a roundworm infection such as threadworm or pinworm. Although these worms usually live in the large intestine, when ready to lay her eggs the female worm moves down to lay them on the moist, warm skin of your anus. The sticky material they are covered with irritates your skin so that you scratch it, picking up some eggs in your nails as you do. Better remember to wash your hands before you eat!

Some common parasite pathogens

Parasite	Condition caused	Source of infection
Amoeba	Amoebic dysentery	Contaminated food and drink
Malarial parasite	Malaria	Bite from infested mosquito
Tapeworm	Tapeworm	Raw or poorly cooked meats
Blood fluke	Schistosomiasis	Contaminated water
Tick	Skin infestation	Tick-infested areas
Louse	Pediculosis	Contact with human carrier, bedding, clothing
Flea	Skin irritation	Animal and human carriers

INVESTIGATION 4.1

Microbes

AIM: To investigate the types of microbes in the air of the laboratory

Materials:

prepared agar plate *sticky tape*

marking pen

Background

Agar is a jelly-like material made from seaweed. It provides a source of nutrients for microbes.

CAUTION

Do not open the tape seals after incubation.

Method and results

- Take the lid off the agar plate to expose the agar to the air in your laboratory for about five minutes.
- Seal the lid on the agar plate carefully, using the sticky tape.
- Give the plate to your teacher to incubate at about 35 °C for 2 days.

1. Check the agar plate after it has been incubated, and draw what you see on it.

- Give the unopened plates back to your teacher for proper disposal.

CAUTION

Do not open the tape seals.

Discuss and explain

2. Describe the general appearance, colour, size and shape of the groups or colonies on the agar plate.
3. What can you conclude about the air in your science laboratory?
4. Do you think that the air in other parts of your school would be different? Explain.
5. Discuss the risks that could be associated with the experiment and ways to reduce these risks.
6. Formulate your own question or hypothesis about microbial growth, and design an experiment that could be used to investigate it. Include an explanation of your choice of variables and required specific safety precautions.

4.3 Exercises: Understanding and inquiring

To answer questions online and to receive **immediate feedback** and **sample responses** for every question, go to your learnON title at www.jacplus.com.au. *Note:* Question numbers may vary slightly.

Remember

1. Outline the differences between:
   (a) pathogens, antigens and hosts
   (b) prions, viruses and bacteria
   (c) allergens and allergies.

2. Construct a cycle map to show how prions replicate.
3. Why do many biologists consider viruses to be non-living?
4. Describe two ways in which bacteria can
   (a) cause disease
   (b) be useful.
5. Outline the difference between a parasite, an endoparasite and an ectoparasite.
6. Suggest why it is a good idea to read up on protozoans before you travel to tropical climates.
7. Describe how mosquitoes are related to malaria.

Think and create

8. Why is the cell invaded by a virus called a host cell?
9. Compare the ways in which viruses and bacteria reproduce. Use a visual thinking tool to share your thinking.
10. The skin under your armpits contains over two million bacteria per square centimetre. Why don't they cause disease more often?
11. Construct a model or play to simulate how viruses infect and replicate.

Investigate

12. Explain how dental caries or tooth decay forms. Why is this related to a study of diseases?
13. After taking medicine or being ill, people are often advised to eat yoghurt. Find out what the benefits of this food might be. How is yoghurt made?

Investigate, report and create

14. When neurologist Stanley Prusiner suggested that a rogue protein was the cause of spongiform diseases such as Creutzfeldt-Jakob disease, it was considered so outrageous that he was ridiculed. Find out more about this scientist and his work and why his theory was eventually accepted.
15. Research the structure and life cycle of a pathogen. Create models or animations to show activities key to its survival.
16. Investigate the link between Kuru and cannibalism.
17. Research and report on one of these scientists specialising in research into disease.
    - Dr Carleton Gajdusek — transmissible degenerative disease of the brain
    - Professor Staley Prusiner — discovery of prions as a new biological principle of infection
    - Associate Professor Robin Gasser — parasitic infections
    - Professor Colin Masters — Alzheimer's disease and neurodegenerative disorders

4.4 Medieval medicine

Science as a human endeavour

4.4.1 Hippocrates and body humors

Before the invention of the microscope the causes of many infectious diseases were not only invisible, but also beyond our imagination. Without awareness of cells, other theories were developed to explain what we saw and what could not be seen.

Around 430 BC, the Plague of Athens killed one-third of the population of Athens. About thirty years later, a person who would have a considerable effect on our understanding of disease was born. His name was Hippocrates.

Hippocrates (c. 460–377 BC) was a Greek doctor who believed that everything was created from four elements: water, earth, air and fire. He also believed that linked to these elements were four humors within the human body. These were blood, yellow bile (choler), black bile (melancholy) and phlegm. He thought that these humors not only shaped a person's character but also, if unbalanced, could cause disease.

Hippocrates and his disciples looked for natural causes of disease. They based their medical practice on reason and experiment and used diet and medication to restore the body's balance of humors. Hippocrates also established the rules and principles that were followed by medieval doctors and are still followed by doctors today. The **Hippocratic Oath** requires doctors to take care of the ill and not do them harm.

Hippocrates (c. 460–377 BC) had a considerable impact on our understanding of disease. He believed that the body contained four humors: blood, yellow bile (choler), black bile (melancholy) and phlegm.

4.4.2 Galen and anatomy

Claudius Galen (c. AD 129–199), a Greek physician who lived in Rome, was influenced by Hippocrates and developed his theories further. When he was in his late twenties (about 157 AD), Galen was appointed physician to gladiators in his home town of Pergamum. Within about ten years he became the personal physician to Emperor Marcus Aurelius in Rome.

Galen believed that all diseases were caused by an imbalance in the elements or their associated body humors and that all cures must be based on correcting the imbalance. He also believed in the importance of anatomical knowledge. Galen wrote hundreds of books about human anatomy, surgery and herbal medicines — books that were to be used by doctors for the next thousand years.

It was thought that body humors not only shaped a person's character but, if they were unbalanced, could cause disease.

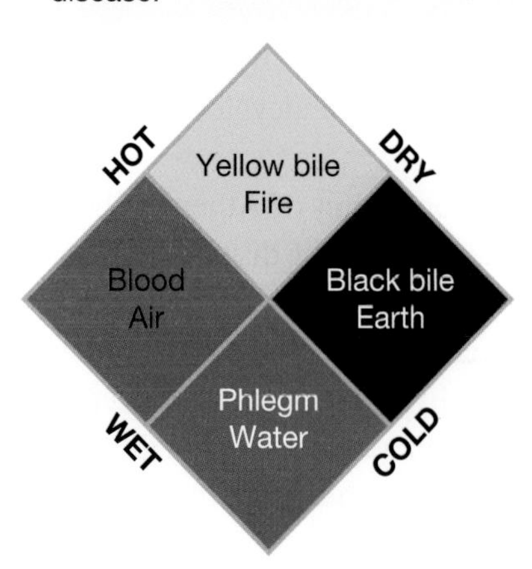

Claudius Galen (c. AD 129–199) was influenced by Hippocrates. His published books were used to train doctors for over a thousand years.

4.4.3 The Middle Ages — times of change

Beginning with the collapse of the Roman Empire, the Middle Ages (AD 500–1500) were a time in which Europe changed dramatically. This was a time of growing populations, developing technology, increased trade and new ideas. It was also a time of hardship, deadly disease and wars in which only about 50 per cent of children reached the age of 15; of those surviving, many died in their twenties and thirties.

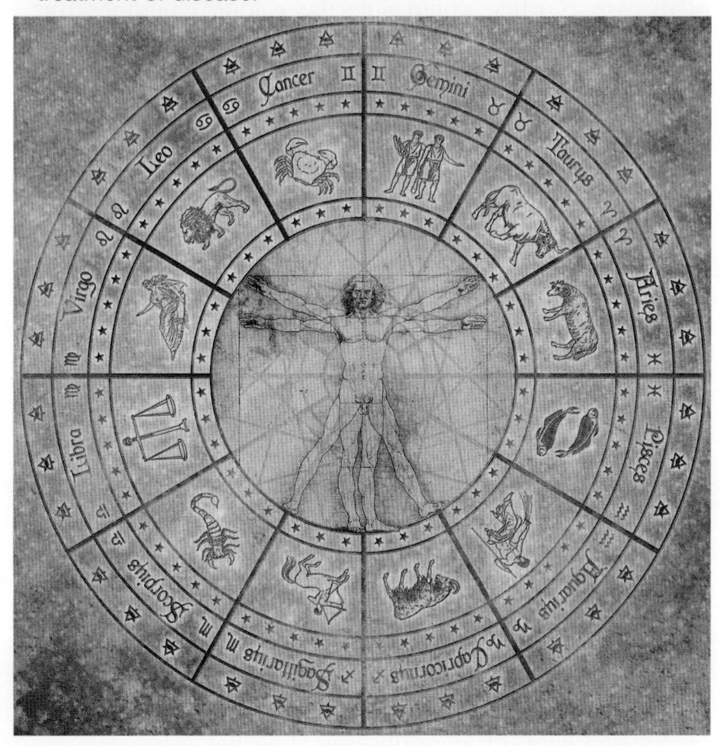

In medieval times, astronomy was linked to body humors and treatment of disease.

4.4.4 Causes of disease

Medieval doctors were influenced by the ideas of Hippocrates and also linked each of the four body humors to the stars and planets. Medieval villagers, however, relied on their own practical knowledge and traditional superstitions to explain causes of diseases, and used natural substances to create potions. There were also those who believed that evil spirits, curses or mysterious magical powers may be the cause of disease.

As the Christian Church grew more powerful, old superstitions were banned and traditional healers were controlled. Church leaders spread their own view of the cause of illness — God's punishment for sins. The church also offered several different methods of spiritual healing, including prayers, charms, relics and pilgrimages.

Leprosy was a common disease in medieval Europe. This disease destroys skin, muscle and bones and was thought to be punishment from God. Not only were lepers treated with fear and loathing, but they were also not allowed to marry, had to carry a warning bell and were often driven out of villages.

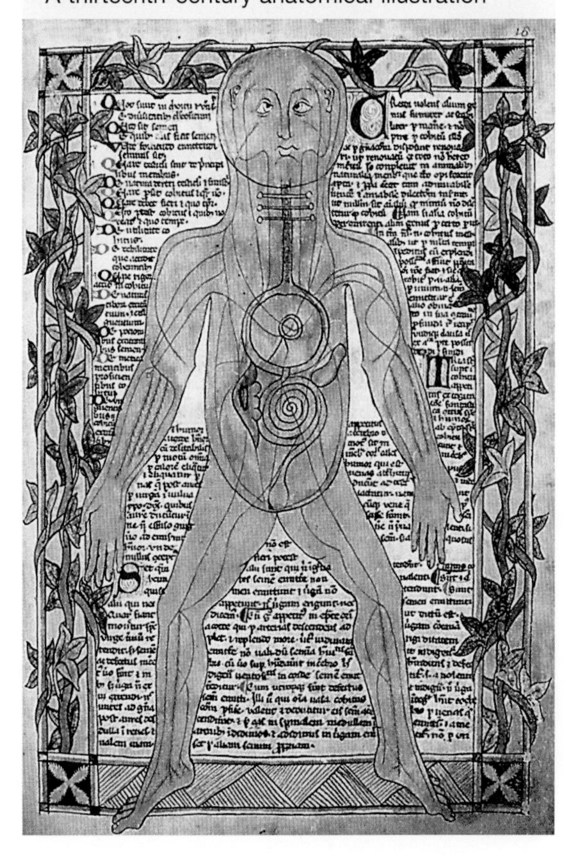
A thirteenth-century anatomical illustration

4.4.5 Medieval medicines

Medieval physicians advised their patients how to live healthy lifestyles, and used their training in mathematics and astronomy to map out healthy and sickly times. They also worked closely with apothecaries, who sold medicinal plants. Sometimes these plants and their knowledge of the four humors, astronomy, chemical sciences and religion were used to create medicines.

Medieval women made many of their families' medicines. They used seeds, stems and leaves of herbs, trees and flowers. Some medicines even used animals or animal products.

4.4.6 Testing the waters

Many medieval physicians considered that testing 'the waters' (urine) was an effective method in the diagnosis of disease. Scribes would note the colour, cloudiness or sediment of the urine and charts were used to match these features to particular diseases. Sometimes blood samples were collected, which may have been tasted to detect a diagnostic sweetness or bitterness.

A fifteenth-century diagram showing possible colours of urine to help doctors diagnose diseases

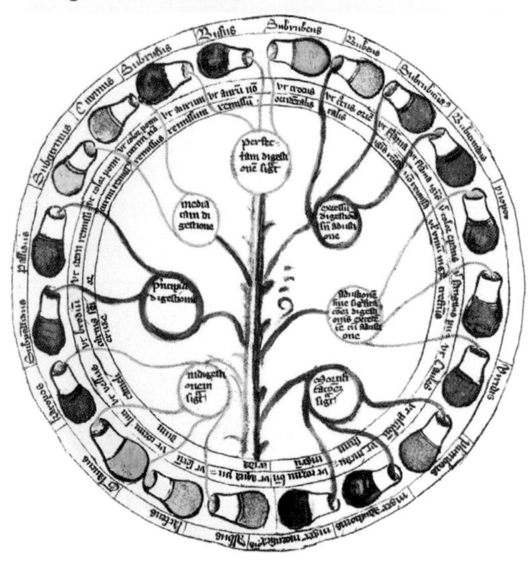

4.4.7 Balancing humors to treat disease

Medieval barbers not only cut and shaved men's faces but also performed minor surgery, such as removing rotten teeth and bloodletting. The red and white striped pole often associated with barbers was a symbol that they let (released) blood, with the white stripes representing the bandages over the cuts.

Not only barbers were involved in bloodletting — this widespread medieval treatment was also performed by doctors and surgeons and was meant to improve the balance of humors within the body. Medical texts of the time showed which veins to cut to release each humor and cure different illnesses. Leeches were also applied to the skin to suck out poisons or bad blood from wounds.

Surgeons also used cupping and cauterising to treat disease. Cupping involved placing hot metal glasses or cups on a patient's cut skin, in the belief that poisons would be released from the body into the cup. To cauterise wounds or help heal internal disorders, surgeons would burn the tissues with red-hot irons or boiling oil.

Bloodletting (top) and leeches (bottom) were used to treat disease during medieval times.

4.4.8 Making sense of disease

Our knowledge and understanding of our world is shaped by what we can sense about it. While some of the ideas of those living in early Greece and medieval times may seem silly or strange to us, they made sense of their world with the tools that were available to them at the time.

The development of new technologies that have enhanced our senses have changed the way in which we see the world. These technologies, that were unavailable to early Greek and medieval humans, provide us with an awareness of our world and the opportunity to explore it that was unavailable to them. Although we shape technology to meet our needs, we are also shaped by it.

4.4 Exercises: Understanding and inquiring

To answer questions online and to receive **immediate feedback** and **sample responses** for every question, go to your learnON title at www.jacplus.com.au. *Note:* Question numbers may vary slightly.

Remember

1. Who was Hippocrates, and why is he important to our understanding of disease?
2. List the four humors of which the body was thought to be made up.
3. State Galen's belief about the cause of disease.
4. Suggest how Galen was able to influence doctors, hundreds of years after his death.
5. What did medieval people think caused disease?
6. What were many medieval medicines made up of?
7. Describe how medieval doctors used body humors to:
   (a) diagnose disease
   (b) treat disease.

Investigate, think and discuss

8. Prior to Hippocrates' theories of disease, what was the belief that early Greeks had about disease? Suggest reasons for these beliefs.
9. Suggest why Galen's books were used for a thousand years to train doctors, rather than new books being written and used.
10. If you had lived during the Middle Ages, what connections do you think you would have made about the cause and effects of disease? Why?
11. Using body chemistry as a key to unlocking health and disease is an idea that has been around for a very long time. Use information on these pages and other resources to investigate this idea. Do you agree with it? Provide reasons for your response.
12. Find out more about the life and scientific contributions of either Hippocrates or Galen. Present your findings in a 'This is your life' multimedia, brochure or poster format.
13. Research the Hippocratic Oath. What is it? Do all doctors take it? Discuss any issues that are related to it.
14. Suggest how scientific theories can be influenced by the period in which you live.
15. Many medieval remedies treated the symptoms of the disease rather than the cause. Find examples of at least four different medieval remedies for disease, state their claims of healing and evaluate their effectiveness based on our current medical knowledge. Display your findings as an advertisement or pamphlet, or in a multimedia format.
16. Paracelsus (c. 1493–1541) was the pioneer of alchemy in medicine; he challenged some of the Greek and Roman medical ideas and suggested specific remedies for specific diseases. Find out more about his life, theories and contributions.
17. Although Galen's texts were used for over a thousand years, some of his ideas were incorrect. Some of this inaccuracy was due to the fact that the ideas were based on his observations of animals rather than humans.
   (a) Find examples of Galen's inaccuracies.
   (b) Over a thousand years later another physician, Vesalius (1514–1564), questioned all previous medical theories, dissected human bodies (against the laws of the Church), recorded his detailed observations and published his work — over time, his corrections to some of Galen's theories were accepted. Find out what these corrections were and why Vesalius is so well respected even now.
18. Phlebotomy (bloodletting) is still used in the treatment of some diseases. Some of the diseases that are treated in this way include haemochromatosis, polycythemia vera and porphyria cutanea tarda. Find out more about these diseases and why phlebotomy is used to treat them.
19. Phlebotomy is also used in blood transfusions and diagnosis of disease. Find out more about what a working day would be like as a phlebotomist.
20. While much of our recorded history of science relates to that of Europe, scientific discoveries were being made in other parts of the world. An example is that of the ancient Arabic scientists. The Arabic Golden Age spanned from AD 750–1258. Some of their discoveries later spread through Europe. Select, investigate and report on one of the following ancient Arabic scientists: Ibn Al-Nafis, Ibn-Sina, Al-Zahrawi or Al-Razi.

4.5 Zooming in — micro tales

Science as a human endeavour

4.5.1 Zooming in — micro tales

Zooming in ... The invention of the microscope opened a whole new world to explore and led to changes in how we saw and thought about not only the world, but also ourselves!

Hooke's *Micrographia* (1665) and van Leeuwenhoek's *Arcana naturae* (1695)

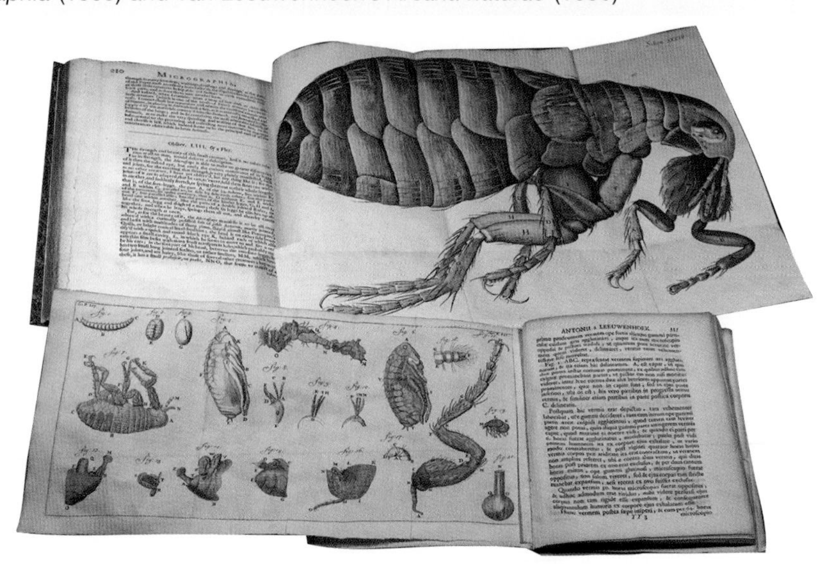

4.5.2 Hooke and cells

The microscope was invented in 1609, opening up a whole new world of discovery. Scientists began to develop new ideas rather than relying on those in Greek and Roman texts. In 1665, a bubonic plague epidemic in London killed 75 000 people. It was also during this year that an English physicist, Robert Hooke (1635–1703) observed a sliver of cork under the light microscope and noted a pattern of tiny regular holes that he called **cells**. Later that year, Hooke published his book of microscopic drawings called *Micrographica.* His recorded observations led to many discoveries in related fields.

Robert Hooke observed a sliver of cork under his microscope and noted cells.

4.5.3 Leeuwenhoek and cells

In 1674, Anton van Leeuwenhoek (1632–1723) observed 'animacules' in lake water through a ground glass lens. Although this marked the beginning of the formal study of microbiology, little progress was made for over a century. A possible reason for this was because few could equal his skill in grinding lenses to the accuracy required for simple microscopes. It was not until the mid-19th century that technological advances in optics led to the production of compound microscopes that did not produce distorted images.

Anton van Leeuwenhoek

4.5.4 Jenner and vaccination

In 1718, Lady Mary Wortley Montagu (1689–1762) introduced **variolation** to England from Turkey to help fight against smallpox. Almost sixty years later in 1774, King Louis XV of France, like many others, died from smallpox. In 1762 Austrian physician Marcus Anton von Plenciz (1705–1786) suggested that infectious diseases were caused by living organisms and that there was a specific organism for each disease. In 1796 Edward Jenner (1749–1823), influenced by the idea of variolation, used cowpox virus to develop a smallpox vaccine. By 1853, Jenner's **vaccination** against smallpox was made compulsory in England. The development and use of many more vaccines was to follow.

Edward Jenner developed the smallpox vaccine, saving many from scarring and death.

Schleiden and Schwann — cell theory

In 1838, Matthias Schleiden suggested that plants were made up of cells and Theodor Schwann recognised that animals were also composed of cells. This led to the establishment of the cell theory — that all living things are made up of cells.

Sketches of cells made by Theodor Schwann

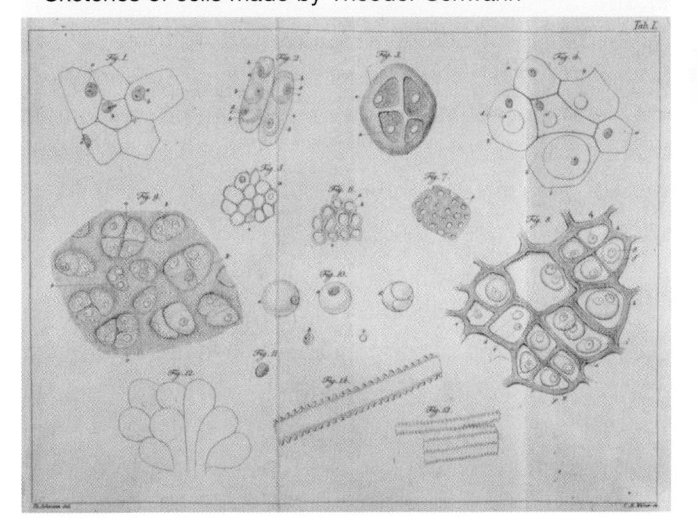

Theodor Schwann and Matthias Schleiden

4.5.5 Pasteur and Koch — germ theory

Between 1857 and 1880, Louis Pasteur (1822–1895) performed a series of experiments that disproved the doctrine of spontaneous generation — the notion that life could arise out of non-living matter. Until Pasteur's work, it was thought that microbes were produced only when substances went rotten. Pasteur showed that microbes were around all the time and could cause disease. He also introduced vaccines for fowl cholera, anthrax and rabies that were made from altered or weakened strains of viruses and bacteria.

In 1867, Joseph Lister published a study associating micro-organisms with infection. This led to the use of disinfectants during surgery, reducing post-operative infections and death. In this same year, Robert Koch (1843–1910) established the role of bacteria in anthrax and formulated postulates that could be used to confirm whether the cause of an infection was viral or bacterial. Eight years later, in 1875, his postulates were used for the first time to demonstrate that anthrax was caused by *Bacillus anthracis*. This validated the germ theory of disease.

4.5.6 Virchow — body humors out of favour

In 1858, Rudolf Ludwig Carl Virchow (1821–1902) argued that all cells arose from pre-existing cells and that the cell, rather than body humors, was the ultimate locus of all disease. His paper *Cellular Pathology* established the field of cellular pathology, linking cells and disease.

With the availability of microscopes providing more detailed observations, scientists continued to make many more discoveries. With these observations, new theories were generated. More pathogens were identified as being the cause of infectious diseases and vaccines for specific diseases were developed.

Rudolf Virchow argued that cells rather than body humors were the locus of all disease.

4.5.7 Knowledge of cells leads to discovery of our immune response

In 1884, Elie Metchnikoff (1845–1916) discovered **white blood cells** that showed antibacterial activity and called them **phagocytes**. He then formulated the theory of **phagocytosis** and developed the cellular theory of vaccination. In 1891, Paul Ehrlich (1854–1915) proposed that **antibodies** were involved in **immunity**. In 1949, Australian Macfarlane Burnet began research that led to the **clonal selection theory** and, in 1961, Noel Warner established the physiological differences between **cellular** and **humoral immune responses**. In 1974, another Australian, Peter Doherty, together with Rolf Zinkernagel, discovered the basis of identifying self and non-self that is necessary for immunity. This was just the beginning of many new discoveries to be made regarding how we fight disease.

4.5.8 Knowledge — a powerful weapon against disease

The development of new technologies has enabled us to expand our senses and magnify the world around and within us. With these new observations came many new discoveries, prompting new ways of thinking and many new theories to identify the causes of disease and how the diseases could be prevented, treated or cured. Often the drive for these discoveries was the devastation and despair associated with the effects of disease within the society in which these scientists lived.

In 1909 Paul Ehrlich introduced the idea of 'magic bullets' — chemicals that could destroy bacteria without harming the host. In 1928 Alexander Fleming (1881–1955) discovered the **antibiotic** penicillin — opening the era of 'wonder drugs'. In 1941, Australian Howard Florey (1881–1955) effectively showed that penicillin killed *Strepococcus* bacteria and persuaded companies to manufacture the antibiotic, saving millions of lives.

New vaccines have been developed against many diseases, saving lives and reducing suffering. Understanding of our body systems and how we fight disease led to new discoveries and technologies. New technologies would often give rise to many other new technologies.

In 1959, Sydney Brenner and Robert Horne developed a method for studying viruses at the molecular level using the electron microscope. Such technologies enhanced our senses and enabled us to observe and be aware of our environment in a way we could never previously have imagined. What future discoveries will new technologies allow us to make? Will our descendants consider our current theories in the same way that we now consider those of Hippocrates, Galen and those who lived in medieval times?

4.5 Exercises: Understanding and inquiring

To answer questions online and to receive **immediate feedback** and **sample responses** for every question, go to your learnON title at www.jacplus.com.au. *Note:* Question numbers may vary slightly.

Remember

1. State a contribution made by each of the following scientists to our knowledge of disease.
   (a) Robert Hooke
   (b) Anton van Leeuwenhoek
   (c) Edward Jenner
   (d) Matthias Schleiden and Theodor Schwann
   (e) Louis Pasteur
   (f) Joseph Lister
   (g) Robert Koch
   (h) Rudolf Virchow
   (i) Elie Metchnikov
   (j) Paul Ehrlich
   (k) Frank Macfarlane Burnet
   (l) Peter Doherty
   (m) Alexander Fleming
   (n) Howard Florey
   (o) Sydney Brenner and Robert Horne

Investigate, think and discuss

2. Construct a timeline that shows:
   (a) how our knowledge of disease has changed over time. Include examples of theories or discoveries. Comment on any connections between these.
   (b) the life dates of scientists mentioned in this subtopic. Comment on any patterns, overlaps or possible influences between these scientists.
   (c) connections between outbreaks of diseases and discoveries related to disease.
3. Select a scientist discussed in this subtopic. Identify the year in which they were the same age as you currently are. Find out what their life may have been like at this time and key events that they may have been influenced by. Construct a paper or electronic diary that describes a week of their life at your age.
4. Find out more about the life and scientific contributions of one of the scientists mentioned in this subtopic. Present your findings in a 'This is your life' multimedia, brochure or poster format.
5. Find out more about the contributions of one of the following physicians to our knowledge or understanding of disease.
   (a) Samuel Hahnemann
   (b) Selman Abraham Waksman
   (c) Hans Christian Gram
   (d) Jonas Edward Salk
   (e) Rodney Porter

6. Describe how scientific arguments can be used to make decisions regarding personal and community issues.
7. The table below shows some examples of diseases and when the pathogen that caused it was identified. In some cases, the identification was made independently by different researchers around the same time. Select one of the diseases in the table and find out about and report on:
   (a) the pathogen
   (b) the symptoms, effects, treatment and prevention of the disease
   (c) the story behind the discovery of the disease and identification of its pathogen.

Date discovered	Disease	Pathogen	Discoverer
1879	Gonorrhoea	*Neiserria gonorrhoeoe*	Albert Nisser
1881	Pneumonia	*Streptococcus pneumoniae*	Louis Pasteur, George Sternberg
1883	Diphtheria	*Corynebacterium diphtheriae*	Edwin Theodore Klebs, Frederich Loeffler
1883	Cholera	*Vibrio cholerae*	Robert Koch

8. Rational drug design has not only opened up opportunities to help us in our fight against disease, but also enabled scientists from different fields to collaboratively work together. Find out more about this new technology, the types of science involved and examples of its application.
9. Our new technologies have enabled us to observe and think at molecular levels. This has led to the development of nanotechnologies. Some of these technologies are being applied in the treatment of disease. Find out about how nanoparticles and nanotechnologies are being used to deliver drugs to cells.
10. Some disease research has involved using human subjects without their consent. Other research has put the subject at great risk. Find out about examples of either of these types of research and the consequences or related issues.
11. Find out more about the Nuremberg code and the reason for it being developed. Do you agree with these reasons? Justify your response.
    (a) Find out about Australian research and our contribution to knowledge and understanding of disease. Include information on the research methods used.
    (b) Report your findings in part (a) as a newspaper article, podcast or video interview.
    (c) Identify a related problem that could be investigated.

4.6 Zooming in — nano news

Science as a human endeavour

4.6.1 Zooming in — nano news

Just when we thought we knew it all, the development of nanotechnologies has enabled us to see even smaller! With our new vision came another world to explore, another journey of discovery, excitement and questions … and possible new ways to protect ourselves against disease.

On the flu virus's surface there are only two types of proteins: haemagglutinin and neuraminidase.

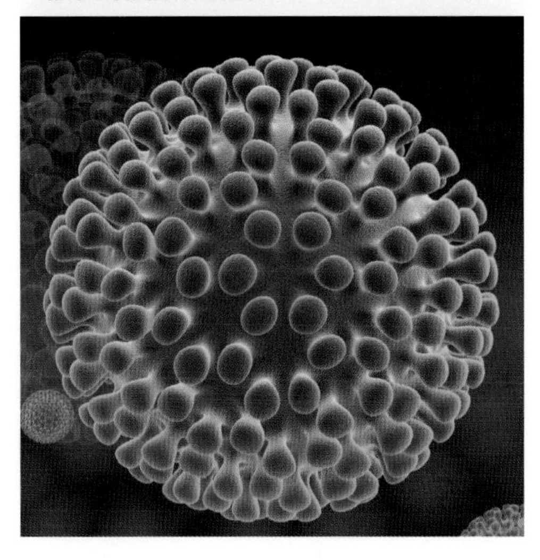

4.6.2 From tiny to tinier

When we talk about cells, we usually talk in terms of **micrometres** (a millionth of a metre, or $\frac{1}{1000000}$ m). When we talk about nanotechnology, we need to think in nanometres. A nanometre is one billionth of a metre, or $\frac{1}{1000000000}$ m.

To understand nanotechnology and nanoscience, you need to learn to think very, very small.

4.6.3 Tiny but powerful!

While nanotechnology is about the very small, its implications and potential application are enormous. The development of this technology has given us not only super-smart and super-strong materials and medicines, but also the possibility of creating other technologies that are currently not possible for mainstream production.

4.6.4 The race is on

We are in a time of biological revolution. Every day more secrets are unlocked, and with each discovery more questions are generated. One of the current questions for scientists to investigate involves the mysteries of viruses that are a current or potential threat to members of our species or species important to us.

4.6.5 Making models

Scientists often use models to help them visualise the features of their studies. They may even add colour to different parts to emphasise features or chemical composition. These models can be representations of the types and arrangements of atoms or molecules in a virus. Aside from helping scientists visualise their shapes, models also enable them to predict how viruses may interact with other molecules. This may provide clues as to what we can do or create to protect us from attack.

4.6.6 Nanobots

Some suggest that, in the future, nanomedicine may bring about the eradication of disease. This major accomplishment would require combining nanotechnology with biotechnology.

Nanotechnology may make possible the creation and use of materials and devices working at the level of molecules and atoms. Imagine minuscule machinery that could be injected to perform surgery on your cells from the inside. They could be programmed to seek out and destroy invaders such as bacteria, protozoans or viruses, or even cancerous cells. Heart attacks due to the blockage of your arteries may also be a thing of the past. These nanobots may be able to cruise through your bloodstream to clear plaque from your artery walls before it has a chance to build up. Could these nanobots also be programmed to stop us from growing old?

This figure of the avian flu H5N1 virus is an example of a model that scientists may use. Can you suggest the differences between this virus and the swine flu H1N1 virus?

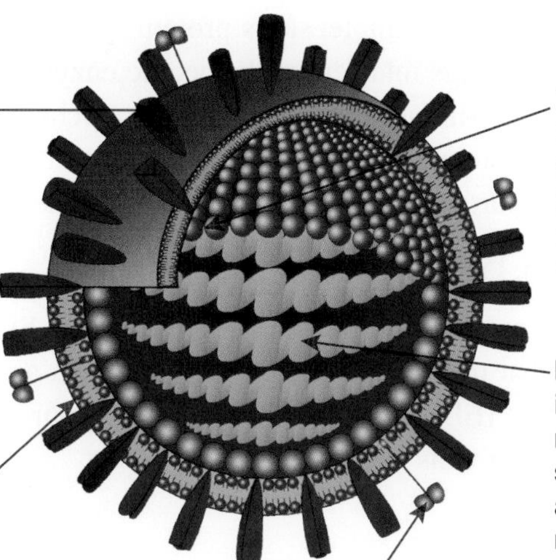

4.6.7 Golden nanoparticles kill brain parasite!

Toxoplasmosis gondii is a parasite that causes cysts in the brain of about a third of the people it infects. Michael Cortie, an Australian scientist, has developed a technique that involves the use of gold nanospheres (about 20 nm in diameter) that are coated with an antibody that selectively attaches to the parasite. When a laser is applied, the nanospheres heat up and kill the parasite. This is groundbreaking research and may have further applications related to other parasites.

This electron micrograph shows a cyst from a human brain that has been infected with the parasite *Toxoplasmosis gondii*.

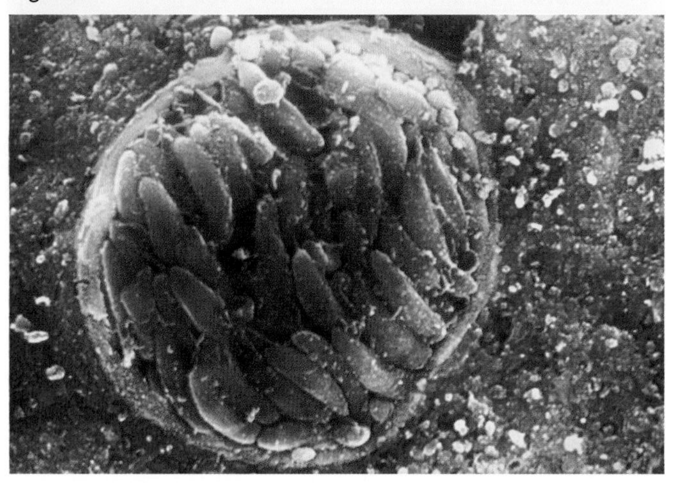

4.6.8 Delving deeper

Tiny human-made **nanoparticles** (about 0.1–100 nanometres) are small enough to pass through a cell membrane. They are currently being developed to deliver drugs directly to cancer cells. The basic structure of nanoparticles is called a **dendrimer**. Attached to these cancer-fighting dendrimers are methotrexate, folic acid and a fluorescent stain. Methotrexate is a drug that kills the cancer cells and the fluorescent stain allows monitoring of the process. Folic acid acts as the bait to attract the cancer cells. This vitamin is essential in cell reproduction and, as cancer cells are actively multiplying, they have a high need for it. When they accept the nanoparticle, the methotrexate poisons the cell, killing it. These dendrimers are then removed from the bloodstream as they pass through the kidneys.

4.6.9 Rational drug design

Viruses enter the cells of their host to use the cell's machinery to replicate themselves. This makes it difficult to develop drugs to kill them without killing their host's cells as well. Knowledge of the structure of the virus and how it replicates has provided scientists with information enabling them to design and develop drugs that can be used to reduce infection.

Some antiviral drugs inhibit DNA or protein synthesis and hence interfere with the replication of the virus within the cell. Interferon, for example, stops protein synthesis, and idoxuridine interferes with DNA synthesis. Some other antiviral drugs interfere with specific enzymes that are important to the virus.

Relenza is an example of an anti-influenza drug that was researched by Australian CSIRO scientists. Relenza binds to the binding site of neuraminidase, one of the proteins on the protein envelope of the influenza virus. This action prevents new viral particles from being able to leave the infected cell to infect other cells.

4.6.10 Using loaded bugs to target drugs?

Jennifer MacDiarmid and Himanshu Brahmbhatt at EnGeneIC in Sydney have been involved in the development of a new technique that uses fragments of bacteria (such as *Salmonella enteric* and *E. coli*) called 'EnGeneIC Delivery Vehicles' (EDVs) to carry drugs to tumour cells. Once these little biorobots have unloaded their cargo, the tumour cells are destroyed.

4.6.11 No boundaries

While some see nanotechnology as the technological saviour of the twenty-first century, others are concerned it has a dark side and are watching warily. Nanotechnology operates at the scale of atoms and molecules. It is fundamentally different from current technologies in that it builds from the bottom up. The underlying principle of nanotechnology is both disturbing and mesmerising — if you can control and rearrange atoms, you can literally create anything! What are our boundaries and responsibilities, or don't we need or want any?

4.6 Exercises: Understanding and inquiring

To answer questions online and to receive **immediate feedback** and **sample responses** for every question, go to your learnON title at www.jacplus.com.au. *Note:* Question numbers may vary slightly.

Investigate and create

1. (a) Research one of the current or possible future applications of nanotechnology.
   - Medicines, e.g. drug delivery
   - Nanobots
   - Artificial cells or body parts
   - Cosmetics

   (b) Construct a model that could be used to communicate information about the use of your nanotechnology application.
2. There is increasing concern about the safety of nanotechnology. Some of these claims are frightening.
   (a) Find out more about issues surrounding this exciting and powerful new field. Take special note of the types of language being used and the experts or authorities being quoted.
   (b) Discuss your findings and then construct a PMI chart on nanotechnology.
   (c) Create a campaign that would communicate your point of view on nanotechnology.
   (d) Organise a class debate on the applications of nanotechnology.
3. Find out the names and features of at least four different viruses that cause human diseases. In teams of four, write a story that involves these viruses and then create puppets or toys for each type of virus to act out the story.

Investigate, share and discuss

4. Find out more about types of research outlined in this subtopic. Share your findings with others.
5. Find out more about the research on H5N1 and how it may be applied in the future.
6. The success of the flu virus is due to tricks that it has evolved to dodge or hijack the defence mechanisms of the cells it invades. Find examples of how it achieves this.
7. (a) What is a pandemic?
   (b) How is it different from an epidemic?
   (c) Find out more about two pandemics in human history and write up your findings as a news report, diary entry, web page or PowerPoint presentation.
8. (a) What is known about the structure of the human immunodeficiency virus (HIV) that causes AIDS?
   (b) How may some of this information be used in the future?
9. (a) Select a virus that is responsible for a human disease and find out what is known about its structure.
   (b) Make a model of the virus. Use labels to provide details of the different parts, describing what they are and what they do.
10. Virologists are involved in the study of viruses. Find out more about what these scientists do. Write a science fiction story that includes a virologist as the key character.
11. Suggest dangers that are associated with the study of viruses. Do you think that research on viruses should be allowed? Give reasons for your opinion.
12. What are the differences between viruses, viroids and prions? Give examples of a disease that can be caused by each.

4.7 Outbreak

4.7.1 Shaped by disease

Disease has shaped our human history. It could be argued that we are who we are because of and in spite of disease.

learnON RESOURCES — ONLINE ONLY

Watch this eLesson: Understanding HIV (eles-0125)

4.7.2 Local or global?

Throughout history there have been records of **plagues** — contagious diseases that have spread rapidly through a population and resulted in high death rates. There are also other terms used to describe the spread of disease. **Epidemics** occur when many people in a particular area have the disease in a relatively short time and **pandemics** are diseases that occur worldwide.

4.7.3 The Black Death — bubonic plague

The Plague of Justinian in the sixth century was one of the first recorded pandemics. It is thought to have been the result of **bubonic plague**. Of all of the plagues throughout history, the bubonic plague (known as **Black Death** in Europe) has been the most widespread and feared. Its name is due to the presence of black sores on the skins of victims. The cause of the disease is the bacteria *Yersinia pestis*. These bacteria were transmitted by fleas that had bitten an infected rat and then bitten a human, infecting the human with the disease.

First recorded in the north-eastern Chinese province of Hopei in 1334, it is thought that bubonic plague was responsible for the death of about 90 per cent of its population (about 2 million people). By 1348, bubonic plague had reached Europe. Within five years, an outbreak of this disease had resulted in the death of almost one-third of Europe's population. After this time, plague visited England another six times before the end of the century.

Nearly all those infected died within three days of their first symptoms appearing. Lack of medical knowledge and great fear resulted in the development of a diverse range of methods being used to fight the condition. Some people tried special diets or were cut or bled in the hope that the disease would leave their bodies with their bodily fluids. Others (flagellants) whipped themselves to show their love of God, hoping to be forgiven their sins and spared the disease. Most importantly, bodily wastes and the bodies and clothes of those infected with the disease were burned in deep pits. In some areas, improved public sanitation resulted from these outbreaks.

The last recorded epidemic of the Black Death was around 1670. A victim of its own success, it had killed so many so quickly that those remaining had either immunity or genetic resistance. While it could still infect, its hosts were able to fight back and destroy it. Its demise paved the way for another disease, smallpox, to take over as the number one infectious disease.

Plague doctors wore protective clothing that included a long beak filled with antiseptic substances.

Some medieval people believed that self-flagellation would protect them against bubonic plague.

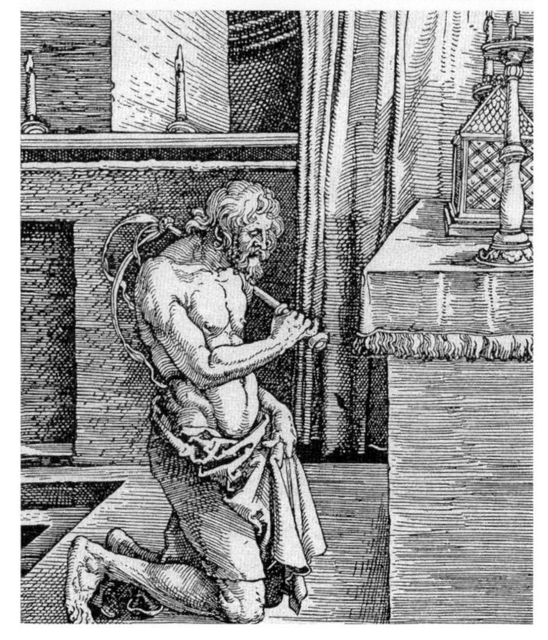

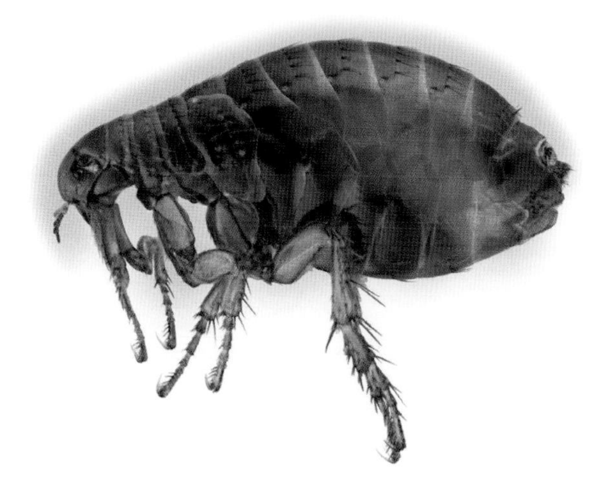

4.7.4 Crossing boundaries

Recent years have seen not only the discovery of new infectious agents, but also the emergence of some of our old infectious enemies. Some of these new diseases are crossing the species barrier and are now infecting species that they previously did not affect. Increasing resistance of many pathogens to antibiotics or vaccines has also raised concerns about the potential for sudden outbreaks of infectious diseases around the world.

Some of the new diseases and pathogens that have been identified or crossed the species barrier over the last few decades include Lyme disease, rabies, henipavirus, bovine spongiform encephalopathy (mad cow disease), Legionnaire's disease, HIV, Marburg virus, hantavirus, SARS, H5N1 and Ebola virus.

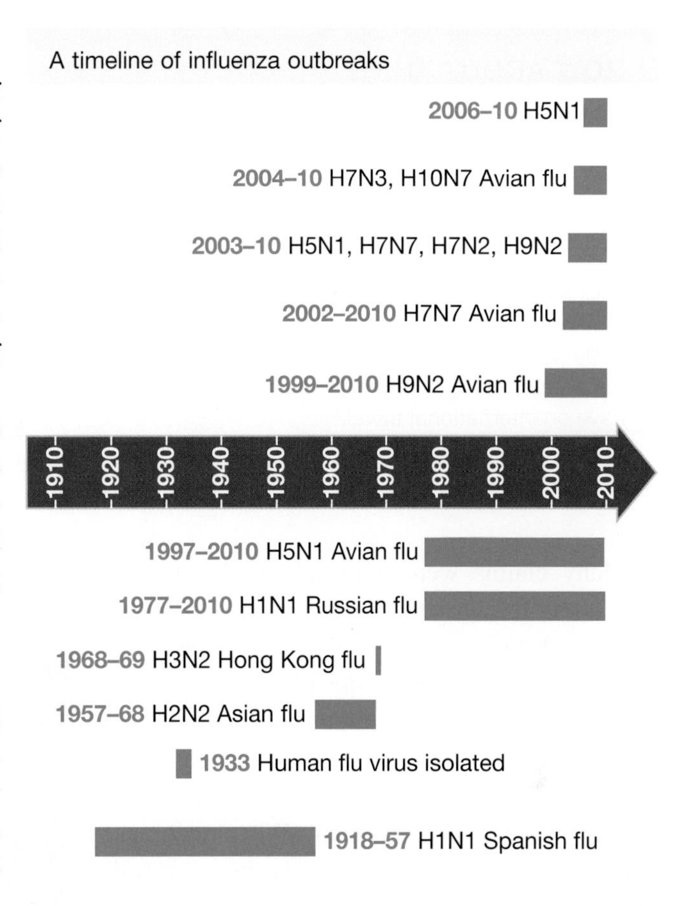

4.7.5 Influenza

Throughout history, there have been numerous outbreaks of influenza. The influenza virus constantly evolves, and pandemics happen every few decades when the flu virus gets new **surface proteins** that people have little immunity to, generally because they come from an animal strain.

By the end of 1918, more than 25 million people had died from a virulent strain of **Spanish influenza** (H1N1). In 1919, The Health Organization of the League of Nations was established with the aim of preventing and controlling disease around the world.

The **Asian influenza** (H2N2) pandemic followed in 1957, followed by a series of others over the next decades. **Avian influenza** (H5N1) made its debut in 1997 in a form that was highly contagious among birds and also infected humans. Since that time, it has devastated East Asian poultry industries. By 2006, a particular strain of H5N1 had been transmitted to humans and had caused a number of fatalities. H5N1 was dangerous because its H5 surface protein was totally new to humans — this is why it has killed more than half of the people who have been infected with it.

4.7.6 Swine flu

In 2009, there was a **swine flu** (H1N1) pandemic. This strain of influenza contained a mixture of genes from the swine flu, human flu and avian flu viruses. It was of particular concern because it was thought that this new strain may have surface proteins that the human immune system may not recognise.

The media was full of headlines expressing fear and concern. This caused global panic and a rush to develop vaccines or treatments for swine flu. Antiviral drugs (such as Tamiflu) that had not been rigorously tested against this new strain of flu were mass produced and supplied to doctors. In Australia, the families of those infected with the disease were quarantined — advised to stay at home, rather than go to work or school. Doctors were advised to have separate areas for those possibly infected (or send them to other surgeries), keeping them away from the general public.

FLU HYSTERIA IF FLU CRISIS HITS

Schools, restaurants, theatres and gyms could be shut and AFL matches cancelled or played in empty stadiums if swine flu grips Australia.

Source: *Herald Sun*, 1 May 2009.

HOW ABOUT THAT!

Increased travel between continents brought new knowledge and discoveries. It also brought death. At the turn of the century, around 1500 expeditions by Columbus and other explorers brought venereal diseases, smallpox and influenza to areas that had no prior history of them. This resulted in the deaths of millions of native people, who had no prior exposure to enable them to develop immunity. In some areas, up to 95 per cent of the native population died. If a severe pandemic was to occur, what effect do you think it could have on international travel?

Many claims were made about the dangers and possible consequences of swine flu. There was a rush to create policies and procedures that could be followed if some of thc predictions became reality. The community took an interest in the disease and began asking questions. How was it transmitted? What were the symptoms? Could infection be prevented? What treatments were there and how could they be accessed? How many would die? Who would die? Was this a taste of living in the past? What technologies could we use or develop to defend ourselves against this new infectious threat?

The swine flu (H1N1) virus contains a mixture of genes from the swine flu virus, human flu virus and avian flu virus.

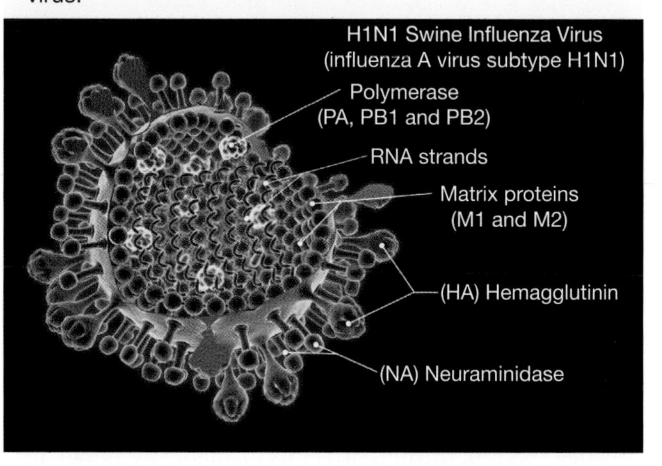

During a pandemic, would your school gym or assembly area become a ward for those infected?

4.7 Exercises: Understanding and inquiring

To answer questions online and to receive **immediate feedback** and **sample responses** for every question, go to your learnON title at www.jacplus.com.au. *Note:* Question numbers may vary slightly.

Remember

1. Define the term *plague* and give an example of one.
2. State the difference between an epidemic and a pandemic.
3. Suggest why bubonic plague is often referred to as the Black Death.
4. Identify the name of the pathogen that causes bubonic plague.
5. Describe three ways in which medieval people tried to defend themselves against bubonic plague.
6. Construct a flowchart to show the relationship between the bubonic plague pathogen, fleas and rats.
7. Suggest why the last recorded epidemic of bubonic plague was around 1670.
8. Provide three examples of new infectious diseases that have been identified over the last few decades.
9. State the names of three different types of influenza.
10. State which Asian industry was most affected by the outbreak of avian influenza.
11. In which year was there a swine flu pandemic, and which antiviral drug was used to try to defend us against it?
12. Suggest a relationship between international travel and pandemics.

Investigate, think and discuss

13. Doing what science tells us invokes the weight of scientific opinion. Citing scientific evidence as reason can at times lead to different interpretations of truth. How was the data collected and analysed? Were there biases in the methods used to obtain and interpret the results? With this in mind, investigate and discuss aspects of disease that have recently been cited in the media.
14. Humans who eat BSE-infected meat can contract Creutzfeldt-Jakob disease. Discuss issues that may result from Australia lifting the blanket ban on beef imports from countries that have recorded cases of bovine spongiform encephalopathy (BSE).
15. Find out more about one of the following Australian disease outbreaks: Legionnaire's disease, severe acute respiratory syndrome (SARS), swine flu, whooping cough (pertussis).
    (a) Present your findings in a brochure or multimedia advertisement to inform others of key points about the disease.
    (b) Create a model or animation to show how this disease is transmitted and how it causes disease.
16. Research the history, cause, symptoms and effects, treatments and relevant issues for one of the following diseases: H5N1 avian flu, anthrax, leprosy, Ebola, hepatitis C, SARS, tuberculosis, HIV/AIDS, syphilis, gonorrhoea, salmonella, cholera, bubonic plague, rabies.
17. Find out more about the sixth-century Plague of Justinian and its importance in history. Imagine you are living in this period; write a letter to a friend living outside Europe to describe what life was like during the time of this plague.
18. *Yersinia pestis* is named in honour of French bacteriologist Alexander Yersin, who successfully isolated the bacteria in Hong Kong in 1894. Find out more about his work and create a virtual Twitter or Facebook page that he may have created if he were alive today.
19. Find out more about the relationship between fleas (*Xenopsylla cheopis*) that fed on infected black rats (*Rattus rattus*) and the transmission of bubonic plague.
20. In 2001, the entire genome (genetic map) for *Yersinia pestis* was mapped. Find out more about this discovery.
21. *Yersinia pestis* was originally a harmless bacterium that lived in the stomachs of rats. About 1500 years ago it mutated in such a way that it was able to enter the rat's bloodstream. Find out what changes occurred that enabled it to do this.
22. Find out what happens once a flea transmits the bubonic plague pathogen into humans. Show your findings in a flowchart.
23. In 1999 in Malaysia more than 100 people died from pig-borne Nipah virus, and over a million pigs were slaughtered to prevent the spread of the disease. Find out more about this disease.
24. Find out more about Typhoid Mary and her involvement with the transmission of typhoid.
25. *Clostridium botulinum* is the pathogen responsible for a type of food poisoning called botulism. Find out more about how this bacteria is transmitted, how it causes paralysis and how chances of infection can be reduced.
26. Find out what cholera and diphtheria have in common and how they are different.

27. Find out more about swine flu and present your information in a creative format to share with others. Consider the following points in your presentation.
    - What is a pandemic?
    - What causes swine flu?
    - Why all the fuss about swine flu?
    - What are the symptoms of swine flu?
    - How can we prevent or treat swine flu?
    - Do we need policies or to take any action to prepare for a possible future pandemic?

4.8 Putting up defences

4.8.1 Antigens — you don't belong here!

Pathogens possess specific chemicals that are recognised as being non-self or foreign to your body. These non-self chemicals, referred to as **antigens**, trigger your immune response.

Your first line of defence aims to stop pathogens from entering your body.

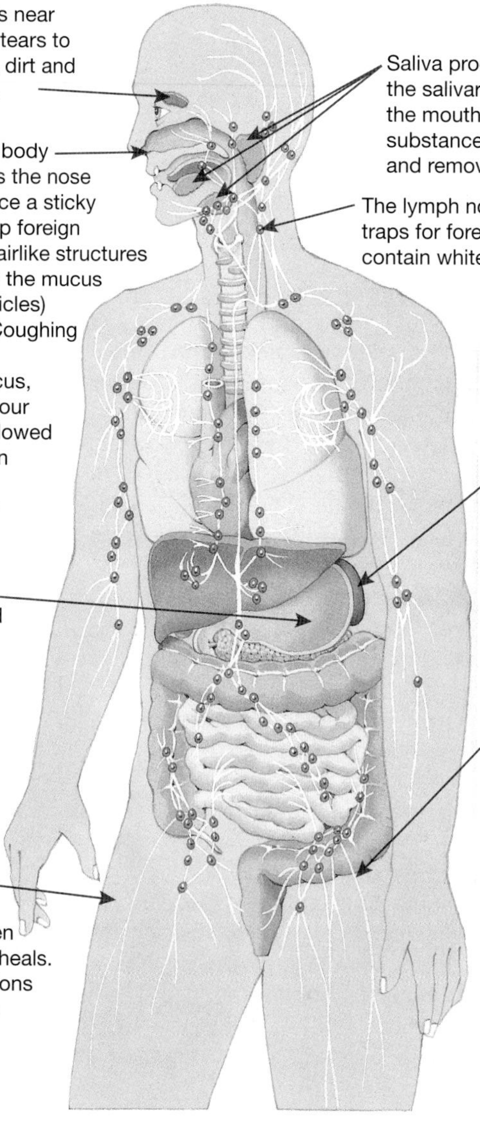

Lachrymal glands near the eye produce tears to wash away dust, dirt and foreign particles.

Saliva produced by the salivary glands in the mouth contains substances to help resist and remove microbes.

The linings of the body openings, such as the nose and throat, produce a sticky mucus to help trap foreign particles. Small hairlike structures called cilia sweep the mucus (and trapped particles) out of the body. Coughing is another way of removing the mucus, bringing it up to your mouth to be swallowed so that any foreign material can be destroyed in your stomach.

The lymph nodes are filters or traps for foreign particles and contain white blood cells.

The spleen is part of the lymphatic system, which helps filter and remove foreign particles from the lymph fluid.

The stomach produces an acid that kills many microbes before they reach the intestines.

The lymphatic system carries white blood cells that destroy foreign particles.

The skin is a surface barrier to most diseases. It is waterproof and if it is cut, the hole is sealed by a blood clot that then forms a scab to help protect you while it heals. It is also dry and slightly acidic — conditions that prevent the growth of many bacteria and fungi.

4.8.2 Lines of defence

Pathogens can cause disease, preventing or stopping your body from working well. A healthy body helps you to defend yourself against infectious disease by setting up natural barriers, or lines of defence. The first and second lines of defence are described as being non-specific. They fight the same way for all infections, regardless of whether they have encountered them before. The third line of defence is specific. It fights differently for different types of invaders and may react differently if it has been exposed to them before.

4.8.3 The first line of defence

Your body's first line of defence is designed to prevent the entry of invading pathogens. Some of these defences are physical barriers (such as skin, coughing, sneezing, cilia and nasal hairs) and others are chemical barriers (body fluids such as saliva, tears, stomach acid and acidic vaginal mucus).

Phagocytes (such as some types of white blood cells) engulf and destroy materials. This action is called phagocytosis.

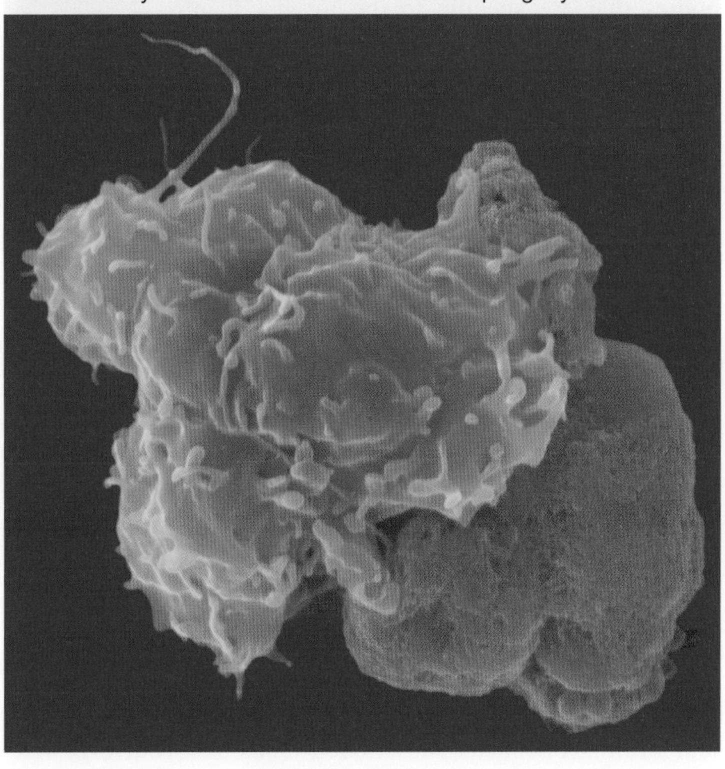

4.8.4 The second line of defence

If pathogens manage to get through your first line of defence, the second line of defence comes into play. If you have had a cut that became infected you may have noticed that the area became red, warm and swollen (inflamed). The redness (caused by the increased blood flow to the area) and **inflammation** are signs that your second line of defence has been triggered.

Special types of white blood cells, phagocytes, that engulf and destroy pathogens (and other foreign material) move to the site of the infection. This action of engulfing and destroying materials is called phagocytosis.

Your second line of defence, (b) and (c), is active once the pathogen has breached your first line of defence, your skin (a).

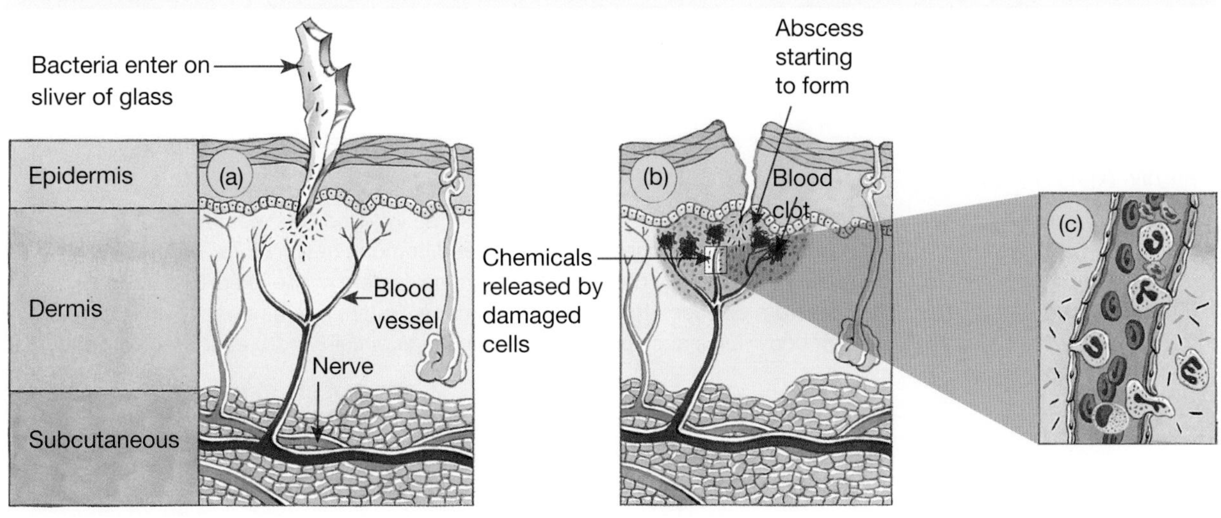

4.8.5 The third line of defence

Have you ever felt swollen glands in your neck when you had an infection? These glands are part of a network of fine tubes running throughout your body called your **lymphatic system**. Your lymphatic system contains lymph vessels, lymph nodes, lymph and white blood cells. Some of these white blood cells are lymphocytes.

4.8.6 Lymphocytes

Lymphocytes are involved in your specific immune response. When triggered by infection, your **B lymphocytes** divide into **plasma cells**. These cells produce chemicals called antibodies that are specific to the invader's antigens. These antibodies assist in the destruction of the invading pathogen. Your **T lymphocytes** fight at a cellular level. These cells not only attack foreign invading cells, but may also attack your own cells that have been invaded. By destroying these infected cells, they also destroy the cause of infection and reduce the chance that it will be spread to other cells.

Your lymphatic system is involved in your third line of defence.

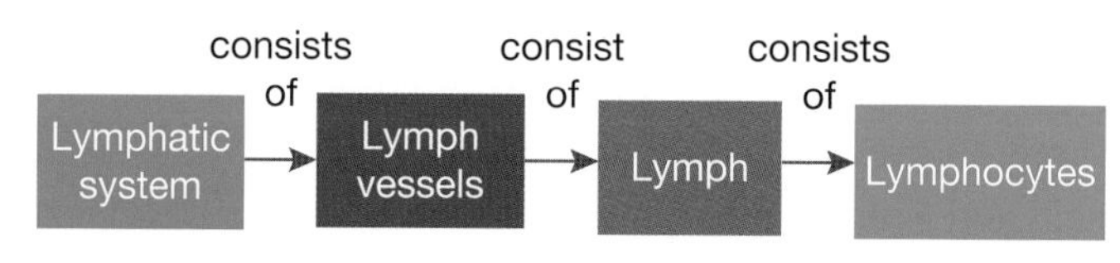

The actions of lymphocytes can assist phagocytes in their duties. For example, some T lymphocytes produce substances that can attract or activate phagocytes. Antibodies (produced by B lymphocytes) can bind to antigens, causing pathogens to clump together. This clumping makes it easier for the phagocytes to engulf them.

Your immune system can be so effective that you can be infected with a pathogen but not develop any symptoms. Lymphocytes can form **memory cells**, so that next time you encounter the same type of invader your immune response can be faster and stronger. Sometimes it is so fast and strong that, even though you may be infected with the pathogen, you may not show any symptoms of the disease that it could cause.

4.8.7 Systems working together

Defence against disease is another example of how your systems work together. Your respiratory system's lining of mucus and ciliated tubes and your digestive system's enzymes and stomach acids help your fight against invaders. White blood cells produced in your bone marrow include those that will become phagocytes and lymphocytes. These defending cells will be circulated throughout your body in your circulatory system and lymphatic system to areas of infection where they perform their task of destroying invaders. The remnants of these invaders are then excreted from your body via your excretory system.

HOW ABOUT THAT!

A type of T lymphocyte called the *helper T lymphocyte* (helper T cell) can be infected by the human immunodeficiency virus (HIV). This is the virus that causes AIDS (acquired immune deficiency syndrome). HIV destroys the helper T cells, and in doing so gradually damages the immune system of the infected person — this is why people with AIDS often die from diseases that a healthy immune system could normally defend itself from. HIV can be transmitted through body fluids such as blood, semen, vaginal fluid and breast milk. Currently there is no known cure.

Your blood is involved in your body's defence against disease.

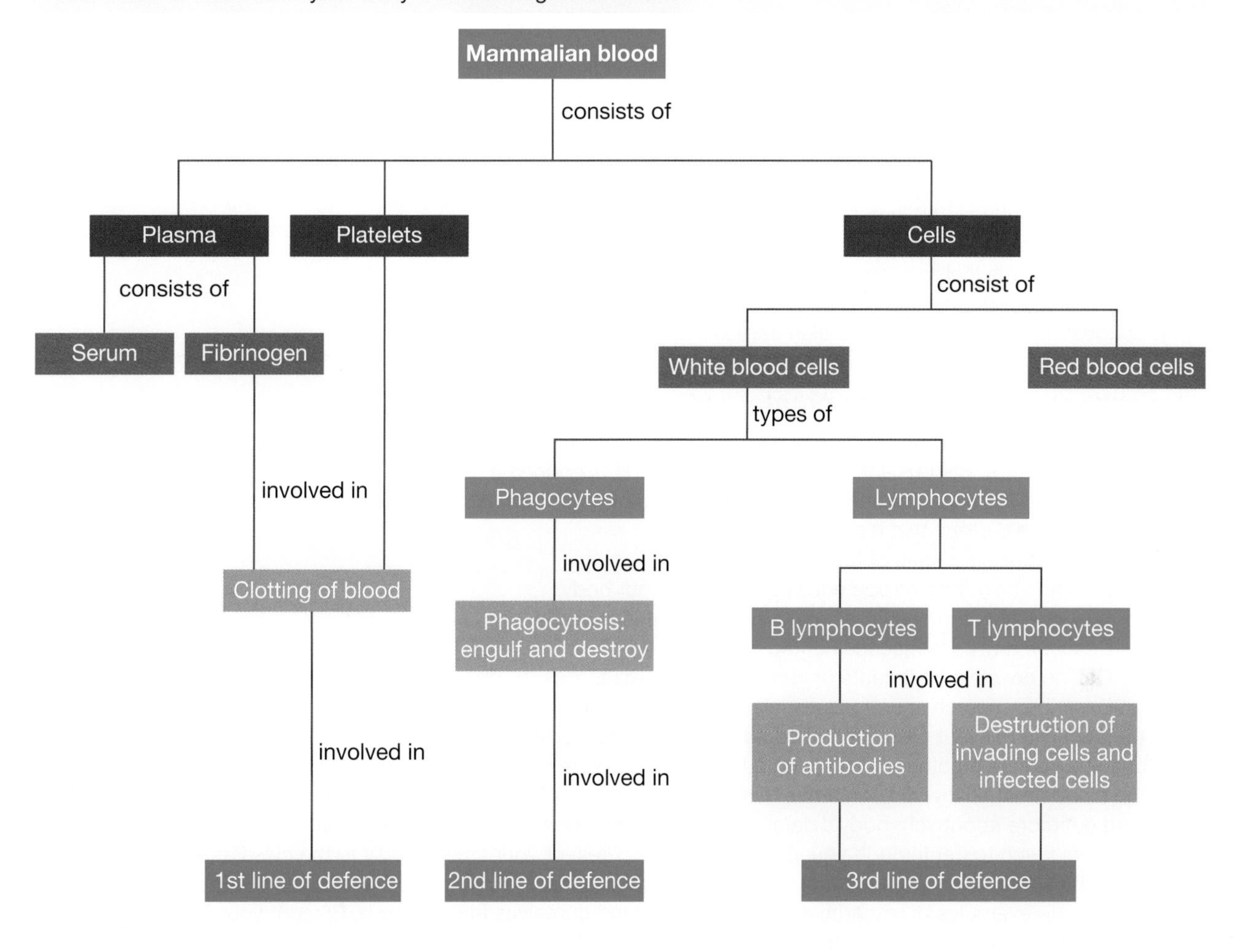

4.8 Exercises: Understanding and inquiring

To answer questions online and to receive **immediate feedback** and **sample responses** for every question, go to your learnON title at www.jacplus.com.au. *Note:* Question numbers may vary slightly.

Remember

1. (a) How many lines of defence are there?
   (b) Describe the key differences between these lines of defence.
2. Suggest how cilia, mucus, coughing and stomach acids can work together to help defend you against pathogens.
3. Suggest why bacteria and fungi find it difficult to grow on skin.
4. Provide two examples of (a) physical barriers and (b) chemical barriers involved in the first line of defence.
5. Use labelled flowcharts to show the relationship between:
   (a) pathogens, phagocytes and phagocytosis
   (b) lymphocytes, lymph, lymphatic system and lymph vessels
   (c) antigens, pathogens, antibodies, lymphocytes, phagocytes.
6. Suggest how you can be infected with a pathogen but not show any symptoms.
7. Outline how your blood is involved in each line of defence against disease.
8. Construct connecting flowcharts (or a mind map) to show how systems of your body can work together to fight disease.

Think and discuss

9. Use Venn diagrams to compare the following.
   (a) First and second lines of defence
   (b) Second and third lines of defence
   (c) Physical and chemical barriers in the first line of defence
   (d) Inflammation and phagocytosis
   (e) Phagocytes and lymphocytes
   (f) T lymphocytes and B lymphocytes
10. Use relations diagrams to show why:
    (a) some glands feel swollen when there is an infection in your body
    (b) the area around an infected cut becomes red and inflamed (swollen).
11. Suggest three ways in which foreign particles might be able to enter your body.

Investigate, imagine and create

12. Find out more about how your body responds to the invasion of a specific invading micro-organism, either a bacterium (e.g. *Bacillus anthracis*, *Bacillus typhosus*, *Treponema*, *Vibrio*) or a protoctistan (e.g. malaria, amoebic dysentery). Research and report on the following points about your selected microbe.
    (a) Name the disease that it causes.
    (b) Outline the symptoms of the disease.
    (c) Describe how the microbe enters and infects your body.
    (d) Describe how the body defends itself against this microbe.
    (e) Describe any consequences of the disease.
    (f) Outline possible treatments or cures for the disease.
    (g) Describe research and/or issues related to this microbe or the disease that it causes.
    (h) Identify a relevant problem that could be investigated further.
13. Imagine you are a pathogen attacking a human body. Write an imaginative paragraph about your invasion of the bloodstream. How did you arrive there? What lines of defence did you encounter along the way?
14. Find out more about your body's defence systems. In a team or on your own, create a story or play about invaders trying to get through your body's defences. Present your story or play to the class as a puppet play, animation or movie.
15. Investigate examples of research about how we defend ourselves from disease. Present your findings as a documentary or newspaper report.
16. Find out more about research relating to HIV and AIDS. Present your findings as a poster or in multimedia format.
17. Research examples of animations or simulations that show how your body responds to the threat of invading pathogens. Select the best ones and have a 'Body at war movie' class session. Construct a PMI chart for each that can be collated into a class chart.

4.9 Choosing immunisation

Science as a human endeavour

4.9.1 Choosing immunisation

Does it make sense, does it have meaning? In your learning, memory and behaviour (and your self-concept), sense and meaning are key elements. If you or a close relative or friend are infected by a disease — it has meaning. Does it make sense if there is nothing that anyone can do to help them survive?

learnon RESOURCES — ONLINE ONLY

Watch this eLesson: Immunisation in Australia (eles-0126)

4.9.2 A small start, a giant leap!

Smallpox

Observations that, once infected, a survivor of a disease often did not catch that disease again must have been made throughout history. A long time before vaccination had been created in England, the Chinese used this observation as a basis for a process called variolation.

In the case of smallpox, variolation involved transferring material from the lesions of those infected with smallpox to healthy individuals. The transference was achieved by inserting infected material under the skin or inhaling the infected powder. The relative success of this process in reducing mortality and morbility rates resulted in its spread to other countries.

It was an English aristocrat and writer, Lady Mary Wortley Montagu (1689–1762) who was responsible for bringing variolation to England from Turkey around 1721. She had been scarred by this disease herself and had also lost close relatives to it. Although variolation was used by some of the aristocracy (including the royal family), it was not until 1797 that Edward Jenner (1749–1823) refined this method into a process we now call **vaccination**. Jenner's vaccination method was able to be used by wider populations and occasionally its use was enforced. By 1980, because of the use of vaccination, the World Health Organization (WHO) was able to announce the elimination of smallpox from our planet.

Smallpox leaves the sufferer with scarred skin. In 1980 The WHO announced that smallpox had been eliminated.

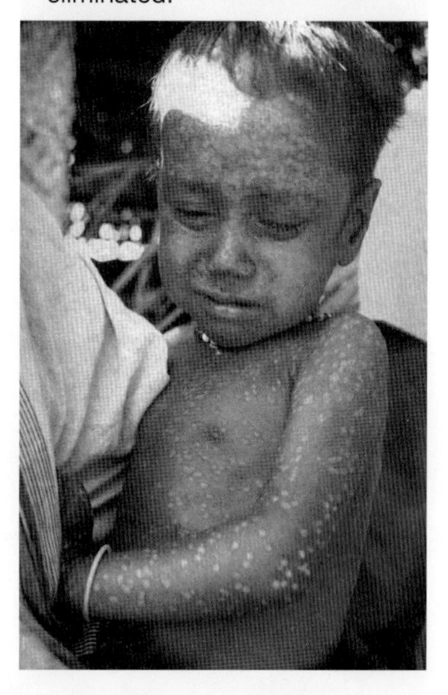

HOW ABOUT THAT!

In 1796, Edward Jenner found a safe way of developing immunity to the disease smallpox. He noticed that people who had contracted cowpox, a much less serious disease, did not seem to ever develop smallpox. Jenner took some pus from an infected cow and deliberately gave a person cowpox. Some time later he exposed this person to smallpox, but the person never showed signs of the illness. Jenner had successfully produced an immunity to smallpox. He called the method *vaccination*, from the Latin word for cow.

Lady Mary Wortley Montagu

4.9.3 Polio

Poliomyelitis (polio) is an infectious disease caused by the *Picornaviridae* virus. This disease is highly infectious and consequences can include complete recovery, limb and chest muscle paralysis, or death.

A vaccine for polio was developed by Jonas Salk in 1955 using a dead virus. This vaccine, however, required a booster shot about every three years and occasionally a live virus contaminated the vaccine. One batch in 1955 infected 44 children with polio — this resulted in some fear within the population about its use. In 1956, American doctor Albert Sabin announced that his oral live virus polio vaccine was ready for mass testing. Public mistrust in the safety of a vaccine using a live virus resulted in Sabin using Soviet (Russian) school children in his large population tests. His tests indicated that this vaccine was not only safer, but also more effective, providing lifelong immunity — and it was cherry-flavoured and could be taken by mouth! By 1961, Sabin's oral polio vaccine was adopted as the standard in America. In 1966, Australia also introduced this oral vaccine.

Dr Albert Sabin and his vaccine against polio

People suffering with polio may become paralysed.

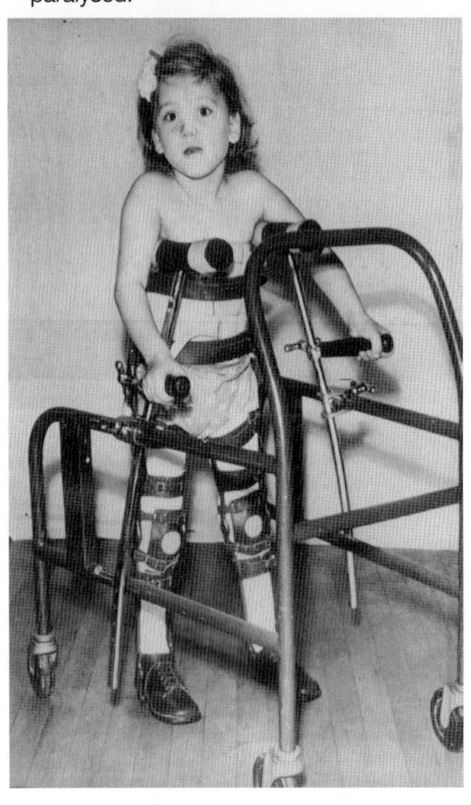

Patients with polio may be incapable of breathing unassisted. In some cases patients needed to be in an 'iron lung', a machine which provided artificial respiration.

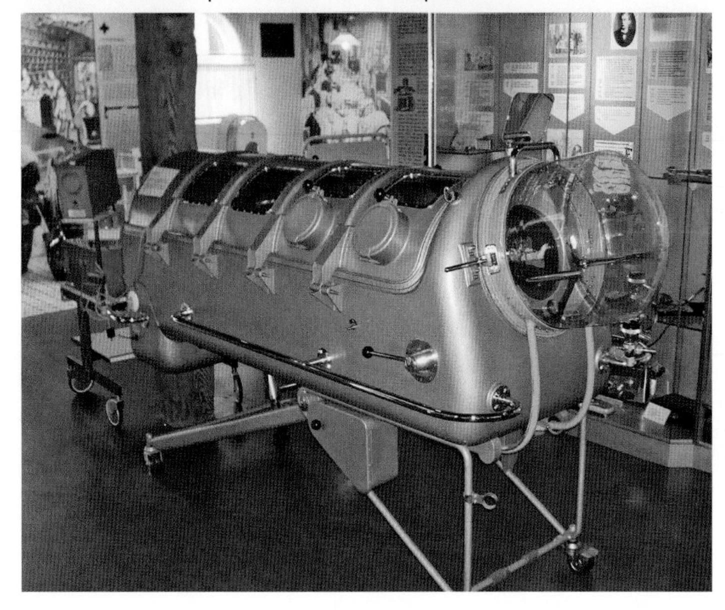

4.9.4 Cervical cancer

Cervical cancer is responsible for the deaths of more than 300 Australian women each year. A significant risk factor associated with this cancer is the common human papillomavirus (HPV). It is for this reason that every couple of years most Australian women have the **Pap test**, which is used to detect abnormal cervical cells that can lead to cancer.

A vaccine against the papillomavirus was developed by Professor Ian Frazer from the University of Queensland's Centre for Immunology and Cancer Research. He was recognised as Australian of the Year in 2006 for his involvement in this development. This vaccine may assist in the prevention of cervical cancer in more than half a million women worldwide each year.

Ian Frazer developed the HPV vaccine which may assist our fight against cervical cancer.

4.9.5 What is immunity?

Immunity is resistance to a particular disease-causing pathogen. A person who is immune does not develop the disease.

If a person is exposed to the antigen of a particular pathogen, or non-self material, they may make specific antibodies against it. The next time they encounter that antigen, their response may be so fast and effective that they can resist infection.

The development of one type of immunity involves the use of a vaccine. Vaccination or **immunisation** is the giving of the vaccine to produce a type of immunity called artificial immunity.

On the second exposure to an antigen, the immune system is able to start producing antibodies more rapidly and in greater amounts.

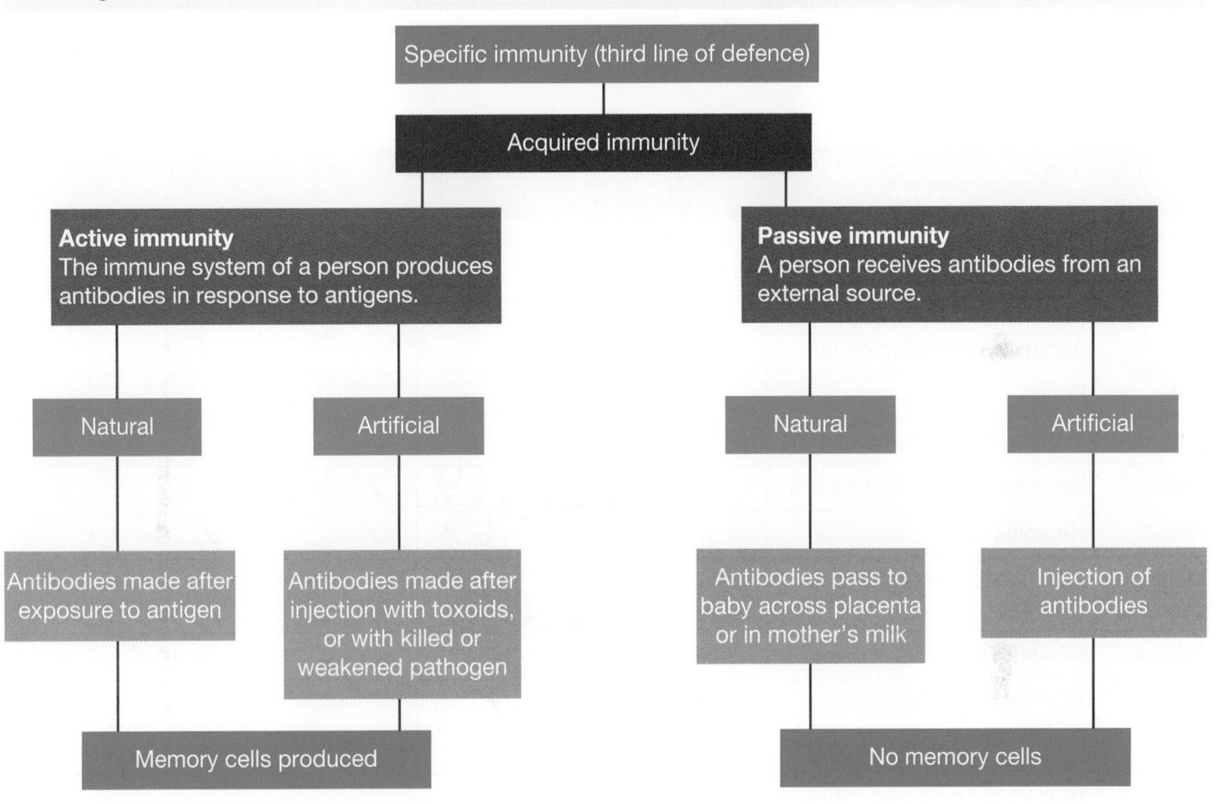

4.9.6 Active or passive?

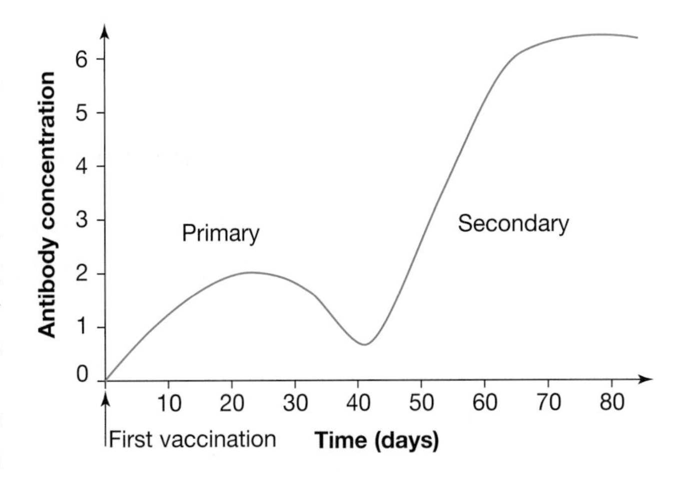

If your body makes antibodies to a specific antigen, this is described as **active immunity**. Your body has memory cells that remember the antigen and you can make more identical antibodies very quickly. You could also gain artificial (or induced) active immunity by producing antibodies after being injected with a toxoid or a killed or treated pathogen that contains the antigen.

If you receive antibodies from an outside source, this is called **passive immunity**. In this case, you don't have memory cells for this infection so, if you were exposed to it again, your body would react as it did the first time. You could get passive immunity from your mother's milk, across the placenta or through an injection of antibodies.

4.9.7 With just one jab!

Vaccinations have been developed by scientists against many diseases and are available to the majority of Australians. This has meant that many children have not had to experience some of the horrors experienced by previous generations. Community health programs ensure that children are vaccinated to protect them against infectious diseases such as tetanus, rubella, mumps, diphtheria, poliomyelitis and whooping cough. Many of these diseases have now been controlled so are rarely seen in Australia.

4.9.8 Don't jab me!

Alarmingly, there is an increasingly low child immunisation rate in some areas in Australia. This has resulted in the government taking steps to boost the numbers of children immunised. Should they do this and, if so, why? Some people question the safety of such vaccinations. Are there no disadvantages to immunisation? Do we know what may go wrong? How and on whom should we test vaccines?

Vaccine	Diseases vaccinated against	Ages								
		Birth	2 months	4 months	6 months	12 months	4 years	10–13 years	12–13 years	15–17 years
DTP	Diphtheria, tetanus and pertussis (whooping cough)		✓	✓	✓	✓	✓			✓
IPV	Polio		✓	✓	✓		✓			
Hib	*Haemophilus influenzae* type b		✓	✓	✓	✓				
MMR	Measles, mumps and rubella					✓	✓			
Hep B	Hepatitis B	✓	✓	or ✓	or ✓	✓		✓		
Meningo–coccal C	Meningococcal C					✓				
7vPCV	Pneumococcal		✓	✓	✓					
VZV	Varicella (Chicken pox)							✓		
HPV	Human papillomavirus								✓	

Examples of vaccines in the National Immunisation Program Schedule (from 2007)

4.9.9 Guinea pigs

Vaccines are not the only chemicals that require testing. Many cosmetics advertise their products as not having been tested on animals. What does this mean? If they haven't been tested, how do we know that they are safe for us to use? Should we use only humans for testing and, if so, who should we use? Should we pay to test them on people? Should we draft or randomly select members of the population for testing? These are some controversial views for you to think about.

HOW ABOUT THAT!

Scientists are currently looking at the possibility of inserting genetically modified viruses into the DNA of plants such as bananas and soybeans. These genetically altered plants may be used as a source of vaccines for injections or simply to be ingested as an oral vaccine. Such vaccines would be cheap to mass-produce and easy to distribute. What are the implications for developing countries? How could this new technology affect food chains in ecosystems?

4.9 Exercises: Understanding and inquiring

To answer questions online and to receive **immediate feedback** and **sample responses** for every question, go to your learnON title at www.jacplus.com.au. *Note:* Question numbers may vary slightly.

Remember

1. Describe a way in which each of the following people contributed to the fight against disease.
   (a) Lady Montagu
   (b) Edward Jenner
   (c) Jonas Salk
   (d) Albert Sabin
2. Distinguish between the following:
   (a) immunity and immunisation
   (b) antigen and antibody
   (c) active immunity and passive immunity
   (d) natural passive immunity and artificial passive immunity
   (e) natural active immunity and artificial active immunity.

Think and investigate

3. (a) Research the cause and effects of the diseases that Australian children are currently vaccinated against, as listed in the table in this subtopic.
   (b) Research reasons parents may not vaccinate their children against any of these diseases. Include examples of claims that are being made and comment on whether these claims are justified.
   (c) Find out the possible consequences to others if children are not vaccinated against these diseases.
   (d) Do you think that vaccination against these diseases should be compulsory? Provide reasons for your opinion.
   (e) Investigate and report on current research related to one of these diseases.
   (f) Provide examples of how the media has influenced or can influence opinions about whether children should be vaccinated. Are there any claims made in these examples that are not scientifically justified? If so, what are they?
   (g) Hold a class debate about whether vaccinations should be compulsory in Australia.
   (h) Write an article for a local newspaper or school newsletter about the importance of vaccination.
   (i) What is the HPV vaccine? Find out more about it and how it may affect you.
   (j) Find out more about the history of vaccination and how various vaccines have been trialled on humans. How do you think vaccines should be trialled? Justify your opinions.
   (k) If there is a pandemic but not enough vaccine for everybody, who should it be given to? Justify your response and share it with others.
   (l) Find out more about the relationship between immunity and babies being breastfed.
   (m) Do you think that immunisation should be compulsory in Australia? Justify your reasons.
   (n) In 1952, the World Health Organization (WHO) set up the Global Influenza Surveillance Network. Find out more about the structure of this network and what it hopes to achieve.
   (o) In the early 1900s, diphtheria caused more deaths in Australia than any other infectious disease. Find out more about this disease and the effect of vaccination on it.
4. Australian scientists Sir Gustav Nossal and Professor Peter Doherty have made significant contributions to science. Find out more about their work and how it relates to the study of viruses.
5. Find out why we currently have no vaccine for the common cold. Comment on whether you think there may be one in the future. Discuss this with your team.
6. Find out more about the use of nanotechnology in the delivery of drugs.
7. Nicotine and alcohol drug abuse can kill millions of people each year. Could a new generation of vaccines stop the explosion of this 'epidemic'? If it were possible to develop vaccines that could target physical addictions would you use them? Trials have already begun on such vaccines. Find out more, share and discuss it with others and react to what you have found out.

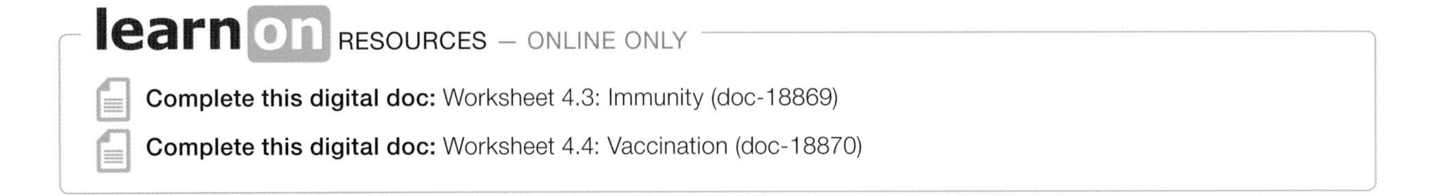

4.10 Travel bugs

4.10.1 Travel bugs

If you are planning an overseas trip, it's recommended that you research the conditions in your holiday destination carefully. Otherwise you may bring back more than you expect!

4.10.2 Pass the toilet paper!

The most common illness suffered by overseas travellers is diarrhoea. While this may cause a little discomfort in the short term, it may be lethal if it continues for a long time. It is responsible for the deaths of almost five million children in tropical regions each year. There are no vaccines to protect you against it, but you can reduce your risk of getting it by following a few simple precautions. These include avoiding locally made ice and ice-cream, lettuce, salads and uncooked foods that may have been washed with contaminated water or handled unhygienically. Only bottled or boiled water may be safe to drink.

4.10.3 A quick jab before you go

Before travelling outside Australia, it is a good idea to check the vaccination requirements for entry into your destination. Vaccines are currently available for some strains of hepatitis, typhoid, yellow fever, Japanese encephalitis, cholera, influenza, rabies and bacterial meningitis.

If you are travelling to a region where malaria is a problem, you are advised to begin a course of antimalarial tablets one week before leaving. This preventative action should be continued for at least a month after your return.

4.10.4 All about malaria

How can I catch malaria?

You catch malaria by being bitten by a female *Anopheles* mosquito that has been infected by the *Plasmodium* parasite. The parasite moves into the salivary glands of the mosquito and is passed into your bloodstream when it bites you.

How do you know if you have malaria?

Some people infected with malaria show no symptoms, while most have high fevers, aches, pains, shivering and night sweats. Fatigue, low blood cell counts and yellowing of the skin and whites of eyes (caused by jaundice) may also result. Severe complications include cerebral malaria, anaemia and kidney failure, and can often result in death.

What causes malarial night sweats?

Once inside your body, malaria parasites grow and multiply first in your liver cells and then in your red blood cells. Successive broods of malaria parasites grow inside your cells until your red blood cells burst open and are destroyed. The new malaria parasites (or merozoites) seek other cells to infect and destroy. This causes night sweats.

Malaria transmission cycle

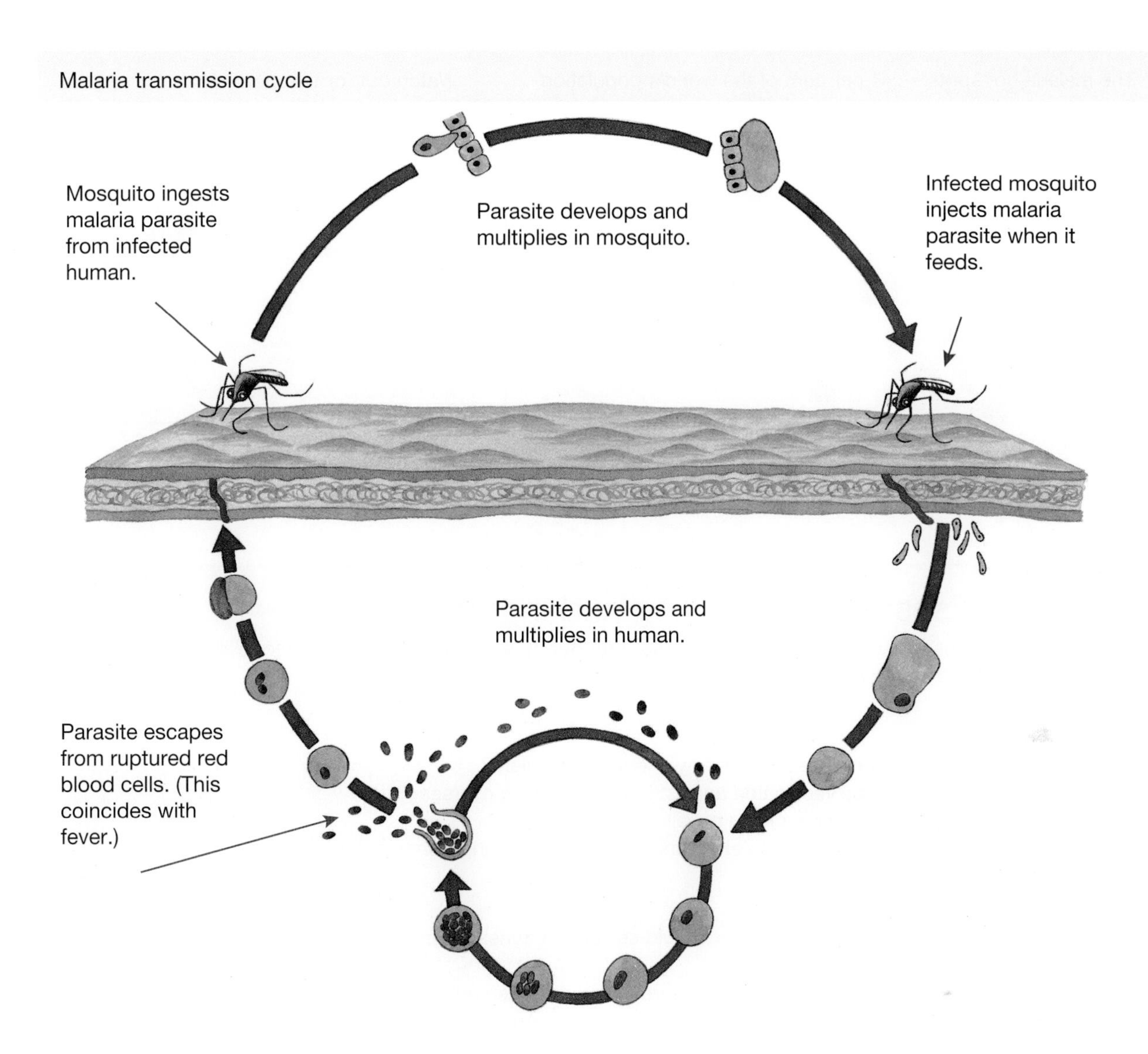

How dangerous is malaria?

Malaria kills over one million people each year. It is one of the most serious public health problems worldwide. It is also a leading cause of death and disease in many developing countries, in which pregnant women and young children are most affected. An infected mother can transmit the malaria parasite to her unborn child through the placenta.

What's new in malaria research?

Mosquitoes, the vectors for malaria, are increasingly resistant to many of the available pesticides. In 2005, British researchers published their findings on two types of fungi that can kill malaria-causing mosquitoes. Their investigations in this field may help reduce the number of malaria victims.

In Australia, teams led by Professor Alan Cowman at the Walter and Eliza Hall Institute of Medical Research have studied how the malaria parasite uses genetic trickery to evade our immune systems. Their research may lead to the development of drugs that disrupt the malaria parasite's ability to disguise itself. This will increase the chance of detecting and destroying it.

Watch out for the mozzies!

Mosquitoes are not only vectors for the malaria parasite, but can also transmit elephantiasis, dengue fever, yellow fever and Japanese encephalitis.

The malaria hot spots — 41 per cent of the world's population live in areas where malaria is transmitted.

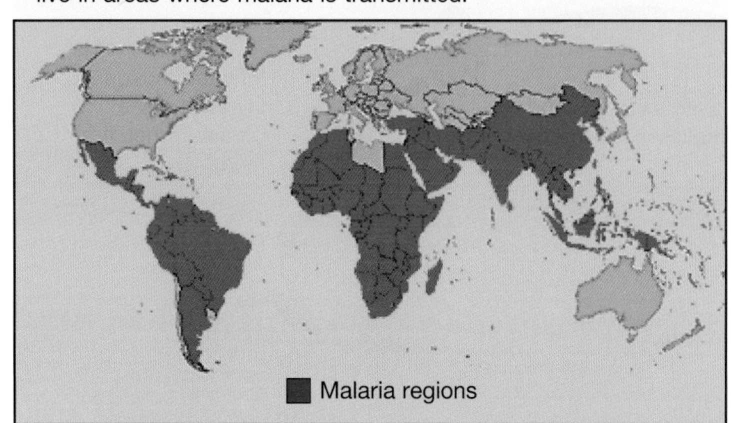

Watch out for the mozzies!

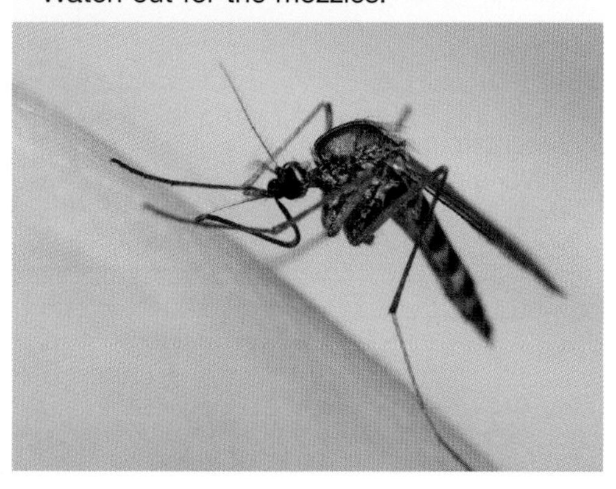

4.10 Exercises: Understanding and inquiring

To answer questions online and to receive **immediate feedback** and **sample responses** for every question, go to your learnON title at www.jacplus.com.au. *Note:* Question numbers may vary slightly.

Remember

1. What is the most common illness suffered by overseas travellers?
2. Which diseases can you be vaccinated against before travelling overseas?
3. Suggest practical methods for avoiding traveller's diarrhoea.
4. Construct a cycle map, relations diagram or mind map to summarise what you have learned about malaria.

Investigate and discuss

5. (a) In a group of four, each team member should select two questions from the following list so that a total of eight questions will be answered.
   (b) Use any resources available to you to research them and then report your findings back to your team.
   (c) As a team, construct a brochure that outlines advice for Australians travelling overseas.
      (i) Describe how mosquitoes are affected by the malaria parasite.
      (ii) Explain why only the female mosquito transmits the malaria parasite.
      (iii) Of the 430 known species of *Anopheles*, how many transmit malaria? Name three.
      (iv) Find out the difference between the gametocyte stage and the sporozoite stage of the malaria parasite. Draw a diagram to summarise your findings.
      (v) Explain how travelling at particular times of the year can affect your chances of getting malaria.
      (vi) In which hours of the day are you at a greater risk of catching malaria? Explain why.
      (vii) Find out six things that you can do to reduce your chances of getting malaria.
      (viii) There have been four Nobel prizes awarded for work on malaria. Who were they awarded to, when and for what?
      (ix) Suggest why travelling to a particular country can reduce your chances of being able to donate blood.
      (x) Malaria, or a disease resembling malaria, was recorded more than 4000 years ago. Research the history of malaria and use a timeline to share your results with your team.
      (xi) Quinine, atebrin, chloroquine, Fansidar, Malarone and artemisinins are examples of drugs used to combat malaria. Find out about their history, effectiveness, similarities and differences.
      (xii) Antimalarial drugs have been found to have reduced effectiveness over time. Explain how this has happened.

4.11 Our noble Nobels

4.11.1 Our noble Nobels

Australian scientists have made significant contributions to disease control and to the quality of life that we enjoy today. Sir Howard Florey, Sir Frank Macfarlane Burnet and Professor Peter Doherty each won a Nobel Prize in Medicine.

One hundred years ago, many children died from both infectious diseases and bacterial infections. A small scratch was sometimes enough to allow deadly bacteria to enter the body and cause swelling, the formation of pus and severe pain. Children born today can avoid the harsh consequences of bacterial infections.

4.11.2 Marvellous mould

Howard Florey was born in Adelaide in South Australia in 1898. He was a keen student who loved sport and chemistry. He studied medicine at the University of Adelaide where he won a Rhodes scholarship to Oxford University, England. While in England he led the team who finally extracted **penicillin** in 1940. In 1945 he shared his Nobel Prize with Alexander Fleming and Ernst Chain. In speaking of his discovery, he modestly stated, 'All we did was to do some experiments and have the luck to hit on a substance with astonishing properties.'

Howard Florey

Penicillin was so successful in saving lives that population control became an issue for medical researchers. Florey later worked on contraception research. In honour of his contribution to medicine, he was knighted in 1944. His likeness appeared on an Australian $50 banknote and a suburb of Canberra was named after him.

4.11.3 Miracle cure-all

Penicillin is an antibiotic and is a chemical made by the mould (fungus) *Penicillium*. If you leave oranges for too long in the fruit bowl, you will sometimes find them growing a greenish mould. This is *Penicillium*. **Antibiotics** destroy bacteria, and they are widely used to treat diseases caused by bacteria.

Today, we take antibiotics to avoid the harsh consequences of bacterial infections. These photographs of a young patient in 1942 show how serious an infection can be. After being treated with penicillin, the patient's condition improved and she recovered fully.

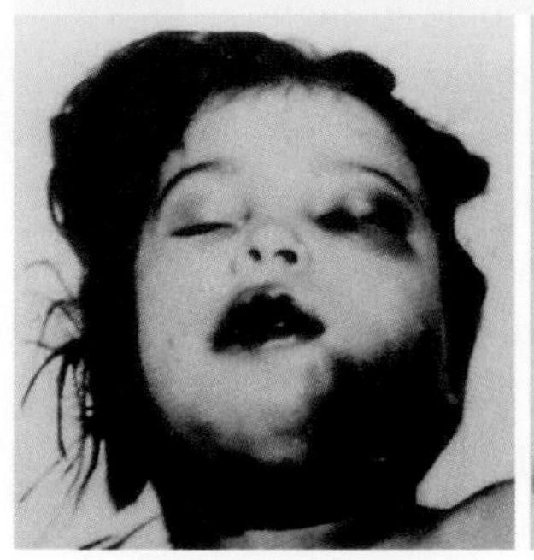

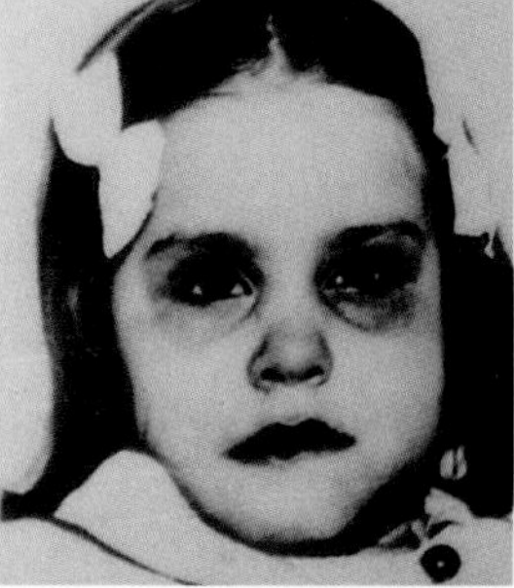

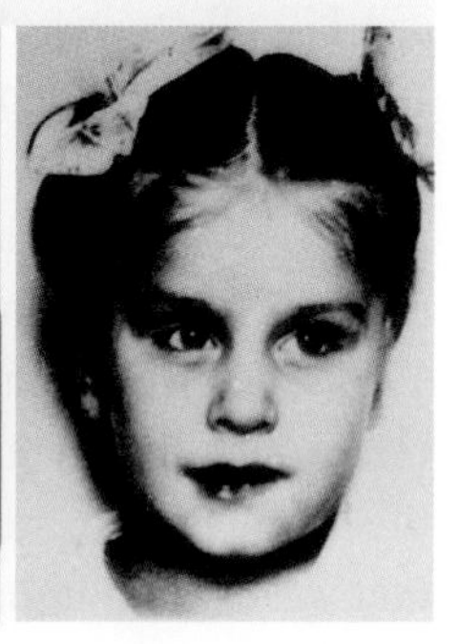

In the human bloodstream, penicillin works by stopping bacteria from forming cell walls as they try to divide. Natural penicillin must be given by injection as otherwise it is destroyed by stomach acid. Some people are allergic to penicillin, but luckily there are now several different antibiotics to choose from. There are few people in the community who have not taken antibiotics at some time in their lives.

4.11.4 The father of immunology

Frank Macfarlane Burnet, known as 'Mac', was born in Traralgon, Victoria in 1899 and died in 1985. As a boy, he loved science and spent hours exploring the bush near his home searching for beetles. Charles Darwin was his hero. After graduating from the University of Melbourne as a medical researcher, he started work at the Walter and Eliza Hall Institute (WEHI) in Melbourne. He then worked in England for many years, returning to Australia in 1944 to become director of the WEHI. He was knighted in 1951 and received his Nobel Prize in 1960. In 1961 he was named Australian of the Year, and four years later he was elected President of the Australian Academy of Science.

Immunology, the science that deals with protection from diseases, was Mac's specialty and he spent most of his career studying viruses. His doctorate thesis was on the **phage**, a type of virus that infects and kills bacteria. Scientists of the time thought there was only one species of phage. Mac showed that there are, in fact, several species.

In 1928, there was public hysteria against vaccination when 12 children died after receiving their diphtheria injections. Mac was part of a team that investigated this tragedy. His experiments showed that contamination of the vaccine caused the deaths, rather than the vaccine itself. This no doubt saved many further lives as people regained their confidence in vaccination.

Mac demonstrates his method of growing viruses by injecting them into eggs to a class of American postgraduate students.

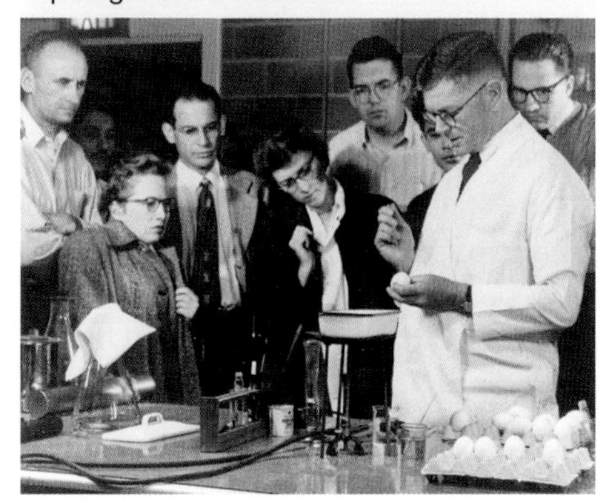

4.11.5 Influenza strains

While in England, Mac worked on the human influenza (flu) virus and developed a successful method of growing high concentrations of the virus using fertilised chickens' eggs. This work led to the development of an influenza vaccine. Mac determined that there were several different strains of influenza. This meant a new vaccine had to be developed each year once the particular strain of influenza had been identified. His work laid the foundation for the discovery by Dr Peter Coleman from CSIRO that all influenza viruses had a common part. Researchers then focused on ways to attack this common part and were able to produce drugs that can kill all strains of influenza virus. Now, people in high-risk categories are encouraged to be vaccinated each autumn to avoid contracting the disease.

In these photographs, bacteria were grown in penicillin for 30 minutes. The bacteria grow longer as shown at (b), but eventually rupture (d), unable to divide due to the influence of the penicillin.

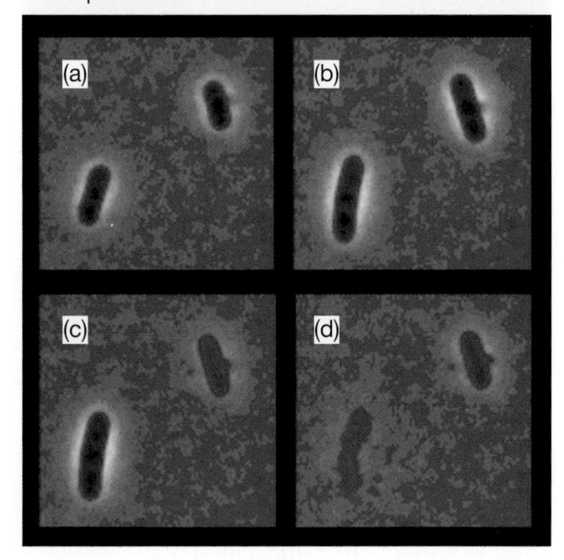

Mac was so dedicated to his work that he was willing to risk his life to show others what he knew. In the early 1950s, CSIRO released the myxomatosis virus so it would infect and reduce the rabbit population in Australia. At the same time, there was an outbreak of encephalitis that made hundreds of people sick. The public started to blame myxomatosis. Mac knew how the myxoma virus worked and that it could not affect humans. He set up an experiment where he and two colleagues, Professor Frank Fenner and Dr Ian Clunies Ross, injected themselves with live myxoma virus. When it was shown that their health was not affected, the panic died down.

4.11.6 Matching body parts

Mac's work inspired other scientists, contributing to our ability to perform transplants. Mac believed that the body learns about immunity at an early age. He suggested that if you could put cells from another body into a fetus at the right time, the fetus would learn not to reject such cells later in life.

Dr Peter Medawar and his team of scientists used this idea when they injected donor tissue from a mouse into the embryo of another mouse. When the mouse was born, the team grafted skin from the donor mouse onto the newborn mouse. No rejection occurred. Now scientists know that they must match the genes carefully when they are looking for possible transplant organs. They use a close genetic match between recipients and donor organs, together with drugs that deaden the immune system, to perform successful transplantations. Today organs including heart, lung, kidney, cornea, bone marrow, skin and pancreas may be transplanted, extending the lives of many people. Immunology is still an important area of scientific research.

4.11.7 Killer cells

Professor Peter Doherty was born in Brisbane in 1940. He received a veterinary science degree from the University of Queensland and a graduate medical degree from the University of Edinburgh. He shared his Nobel Prize in 1996 with Rolf Zinkernagel when they described the way the **immune system** recognises virus-infected cells. In 1997 Peter Doherty was named Australian of the Year. Doherty and Zinkernagel worked at the John Curtin School of Medical Research in Canberra from 1973 to 1975.

Professor Peter Doherty

The immune system uses special white blood cells called T lymphocytes, or **T cells**, to protect an organism from infection by eliminating invading microbes. T cells have to be smart enough to avoid damaging their own organism. They need a recognition system so that they can identify the parts they must destroy and those they must protect. The body also needs to know when to activate them.

Doherty and Zinkernagel studied mice to learn how their immune systems (particularly their T cells) protect them against the virus that causes meningitis. They discovered that mice can make killer T cells that protect them. However, when these T cells were placed in a test tube with infected cells from another mouse, they did not work. Doherty and Zinkernagel developed a model to explain why this happened. They said that each T cell carries a marker that allows it to recognise the cell of the organism it is protecting, as well as the antigen of the invading microbe. At the spot where the antigen attaches itself to the host, the T cell can make a matched fit and destroy the antigen. It works like two interlocking pieces of a jigsaw puzzle.

When your body is exposed to a microbe, it develops T cells that give it immunity. If there are enough of the right type of T cells, these can eliminate the microbes faster than they can reproduce and you remain well. Your body keeps some of these T cells as immunity against future attacks from the same microbe.

This work has had a major impact on our understanding of organ transplantation and vaccines. Scientists now realise they must try to match both tissue and immune system types for a successful transplantation.

4.11.8 From Pasteur to penicillin

Understanding and finding cures for infectious diseases has been a long process involving the efforts of many scientists around the world. Some of the key researchers in the discovery and development of penicillin, and their ideas and breakthroughs, are listed in the tables below and on the next page. If it were not for their contributions, we may not have the antibiotic medicines that we take for granted today.

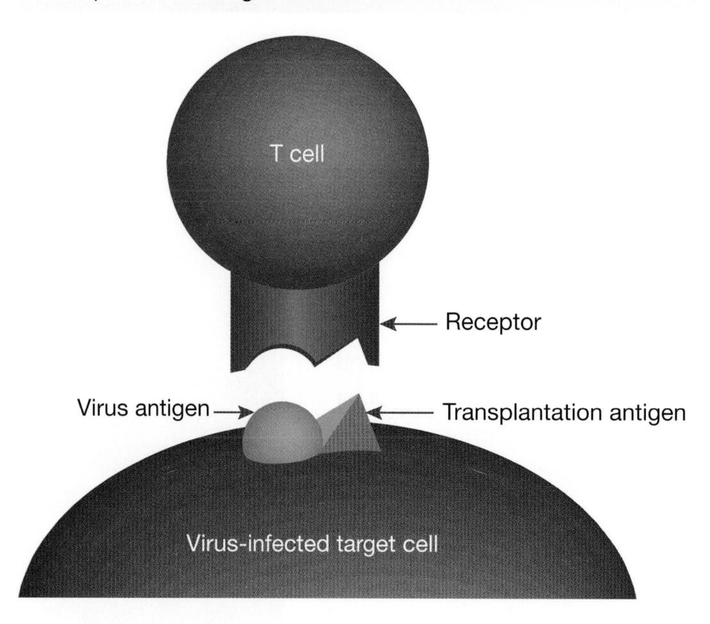

Australian Nobel Prize–winning scientists

Year of Nobel prize	Scientist	Contribution to our understanding of disease
1945	Howard Florey (1898–1968)	Isolation and manufacture of penicillin and discovery of its curative effect in various infectious diseases
1960	Frank Macfarlane Burnet (1899–1985)	Discovery of acquired immunological tolerance
1996	Peter Doherty (1940–)	Discoveries about the specificity of the cell-mediated immune defence
2005	Barry Marshall (1951–) and Robin Warren (1937–)	Discovery of the involvement of the *Helicobacter pylori* bacterium in stomach ulcers and gastritis

Other notable Nobel Prize–winning scientists

Scientist	Field	Contribution to our understanding of disease
Louis Pasteur (1822–1895)	French chemist	Discovered that infectious diseases are spread by bacteria. Observed that mould stopped the spread of anthrax.
Joseph Lister (1827–1912)	British surgeon	Noted that samples of urine contaminated with mould prevented bacterial growth.
Alexander Fleming (1881–1955)	Scottish bacteriologist	In 1928, while studying the influenza virus, Fleming went on holiday and left several discarded Petri dishes on his bench. He had been using them to grow bacteria in nutrient jelly. When he returned, he noticed that where some of the mould had fallen, the bacteria had been killed. He called this substance penicillin but was unable to extract it and did not pursue it further.

4.11 Exercises: Understanding and inquiring

To answer questions online and to receive **immediate feedback** and **sample responses** for every question, go to your learnON title at www.jacplus.com.au. *Note:* Question numbers may vary slightly.

Remember

1. Sir Howard Florey, Sir Frank Macfarlane Burnet and Professor Peter Doherty are Australians who have received a Nobel Prize.
   (a) Which Nobel Prize did each win and in what year?
   (b) In which area of science did each one specialise?
2. What would have happened to you if you had a bacterial infection in the time before penicillin was discovered?
3. Explain how penicillin works.
4. How do T cells protect us from disease?
5. Who discovered that infectious disease was spread by bacteria?

Think

6. What precautions must we all take to ensure that antibiotics remain useful to us?
7. Find out more about the important discoveries made by Sir Frank Macfarlane Burnet and Sir Howard Florey. Present your findings as a flowchart.

Investigate

8. Find out how scientists made moulds for their research and then investigate how to produce your own moulds.
9. Use the internet to find out more about Nobel Prize winners.

4.12 Cycle maps and relations diagrams

4.12.1 Cycle maps and relations diagrams

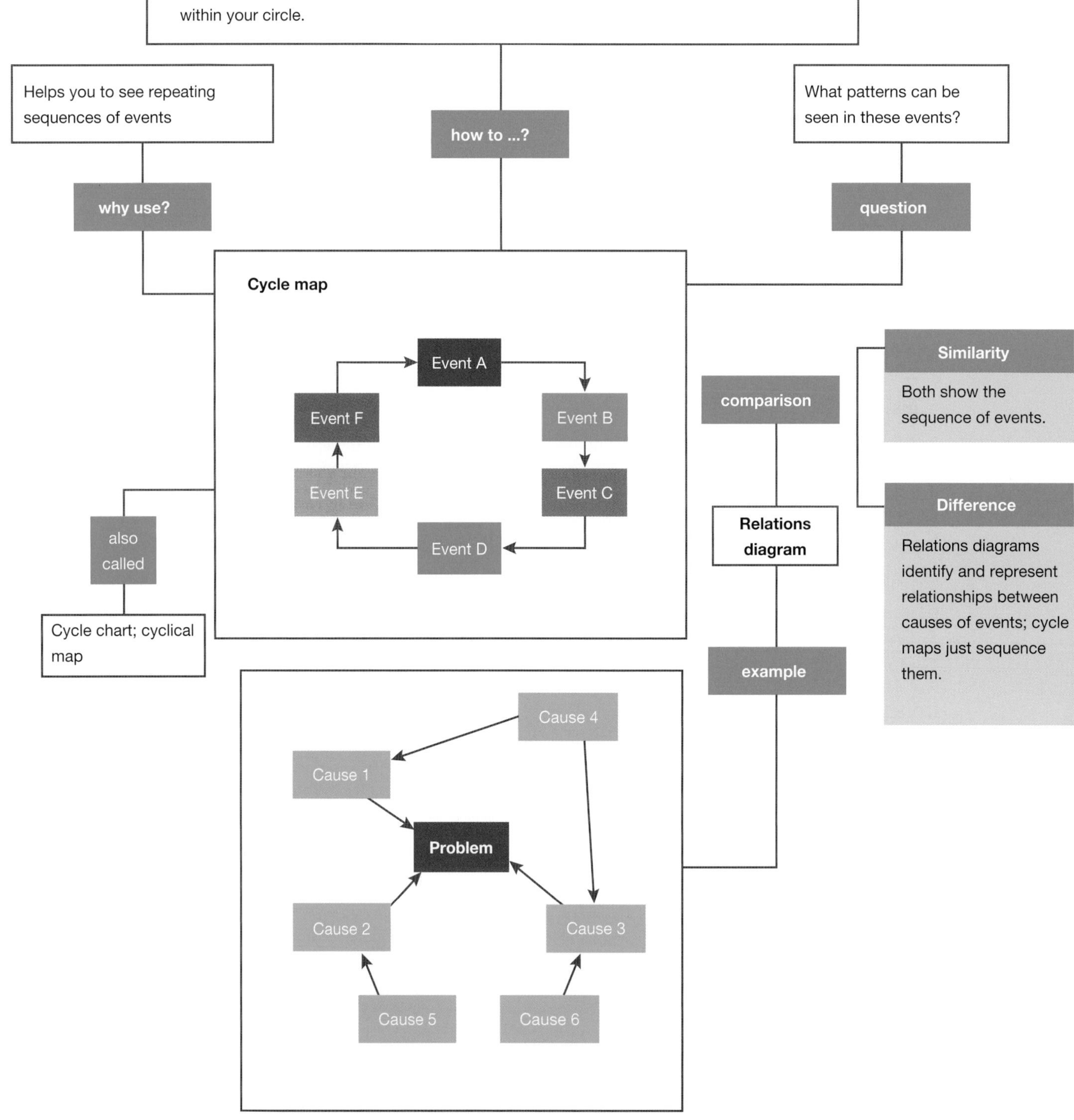

4.13 Review

4.13.1 Study checklist

Cause of disease

- compare infectious and non-infectious disease
- define the following terms: pathogen, parasite, cellular pathogen, non-cellular pathogen, host, primary host, secondary host, endoparasite, ectoparasite, plague, epidemic, pandemic
- describe ways in which diseases can be transmitted
- recall examples of disease caused by prions, viruses, bacteria, fungi, protozoa and animal parasites

History of disease

- describe how ideas about disease transmission and treatment have changed from medieval times to the present as technology and knowledge have developed
- describe how technology has changed the way in which we view disease

Defence against disease

- distinguish between the first, second and third lines of defence against disease
- describe the role of the skin, mucous membranes, chemical barriers and other components of the first line of defence against disease in the human body
- outline how inflammation, fever and phagocytosis assist in the maintenance of health
- explain how specific immunity against a particular pathogen is acquired
- distinguish between specific and non-specific defence against disease
- distinguish between antibodies and antigens
- state the relationship between the lymphatic system, lymph, lymph vessels and lymphocytes
- distinguish between T lymphocytes and B lymphocytes
- compare and contrast active and passive immunity

Human endeavour and disease

- evaluate issues relevant to vaccination
- recognise aspects of science, engineering and technology within careers associated with disease
- comment on the use of nanotechnology in medicine
- consider how the values and needs of contemporary society can influence the focus of scientific research

Individual pathways

ACTIVITY 4.1	ACTIVITY 4.2	ACTIVITY 4.3
Revising the body at war doc-8440	Investigating the body at war doc-8441	Investigating the body at war further doc-8442

learnON ONLINE ONLY

4.13 Review 1: Looking back

To answer questions online and to receive **immediate feedback** and **sample responses** for every question, go to your learnON title at www.jacplus.com.au. *Note:* Question numbers may vary slightly.

1. Construct your own summary mind maps or concept maps on the following topics, using the terms suggested below (as well as any others that may be relevant).
   (a) Infectious disease: contagious, infected, pathogen, cellular pathogens, non-cellular pathogens, quarantine, direct contact, vectors, contaminated objects, contaminated water, sneezing, coughing, physical contact, antibiotics, personal hygiene, tapeworms, head lice, fungi, protozoans, bacteria, viruses, prions

(b) Pathogens and parasites: parasite, host, primary host, intermediate host, endoparasite, ectoparasite, pathogen, non-cellular pathogen, cellular pathogen, prions, Kuru, mad cow disease, viruses, obligate intracellular parasites, mumps, AIDS, warts, influenza, bacteria, coccus, bacillus, *Streptococcus*, cholera, pneumonia, typhoid, whooping cough, Gram stain, protozoans, malaria, amoebic dysentery, fungi, tinea, ringworm, thrush, worms and arthropods, tapeworm, liver fluke
(c) Putting up defences: lines of defence, first line of defence, second line of defence, third line of defence, antigen, non-self, specific, non-specific, physical barriers, chemical barriers, inflammation, phagocytosis, phagocytes, white blood cells, inflammation, cilia, skin, acid, enzymes, nasal hairs, sneezing, coughing, lymphocytes, B lymphocytes, plasma cells, antibodies, T lymphocytes, lymphatic system, lymph, lymph vessels, memory cells
(d) Immunity: vaccine, vaccination, immunisation, active immunity, passive immunity, artificial immunity, natural immunity, antibodies, active natural immunity, active artificial immunity, passive natural immunity, passive artificial immunity

2. Anthrax, cholera, botulism, smallpox and Ebola are examples of diseases with potential as biological weapons. Although these infectious agents have been used as weapons for centuries, advances in technology have accelerated their threat to large populations.
   (a) Find out more about one of these diseases.
   (b) In your group, discuss and identify specific problems or threats relevant to your disease.
   (c) Discuss what you would do to minimise the impact of a possible bioterrorist threat using this disease.
   (d) Design equipment and protective clothing that may help protect you from this disease, and make a model of these.
   (e) Create a training video showing the procedures you would use in the event of such an attack.
   (f) Write a story that describes the history and consequences of an imaginary bioterrorist attack.
3. Write a story about how non-self material passes through the first line of defence and then how the second and third lines of defence work together against attack. Use the diagram on the right to help develop some of the characters in your story.

T helper cell
activates
B plasma cell
activates
Antibody
attacks
Macrophage (phagocytic white blood cells)
attacks
Non-self cell, agent or material
Antigen non-self
activates

4. Imagine that you are employed as a member of a team with the responsibility of public health in your community. Either on your own or in your team, design a plan of attack for each of the scenarios below, including how you would communicate to others what you were doing and why, and how you would implement it.
   (a) About a third of the parents in your community refuse to get their children immunised against whooping cough.
   (b) Five families living in different areas of your community have severe food poisoning.
   (c) A child in one of your schools (who has just returned from an overseas holiday) has been diagnosed with swine flu.
   (d) One of your colleagues has broken out in a red rash, but refuses to go home.
5. Use the figure on the right as a framework to construct a summary of this chapter.
6. Design an experiment that would show which disinfectants and antiseptics are most effective against the growth of bacteria in your kitchen.
7. Select your favourite scientist from the past. Find out more about them and then write and perform a speech that they might have given to their colleagues or scientific community. If possible, make a recording of your speech to share with others.

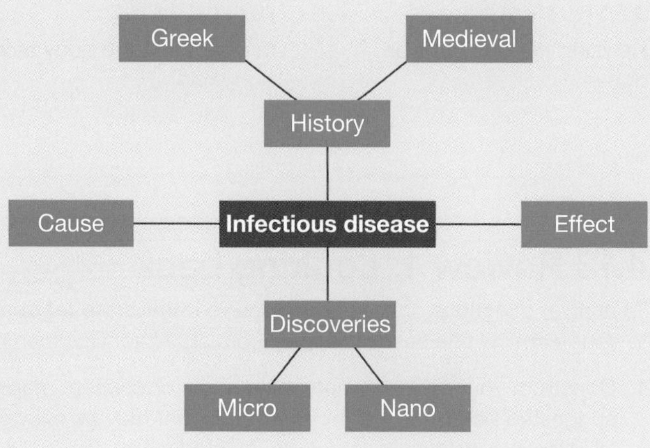

8. Claims that the MMR vaccine resulted in autism in some children led to a measles epidemic. Follow this story and discuss your opinion with others. Organise a class debate about whether there is evidence to support this claim.
9. Examine the cluster map below. Find out where the words in the top box fit in the map. Then create your own Venn diagrams, relations diagrams, cycle maps, concept maps or mind maps of what you have learned in this chapter.

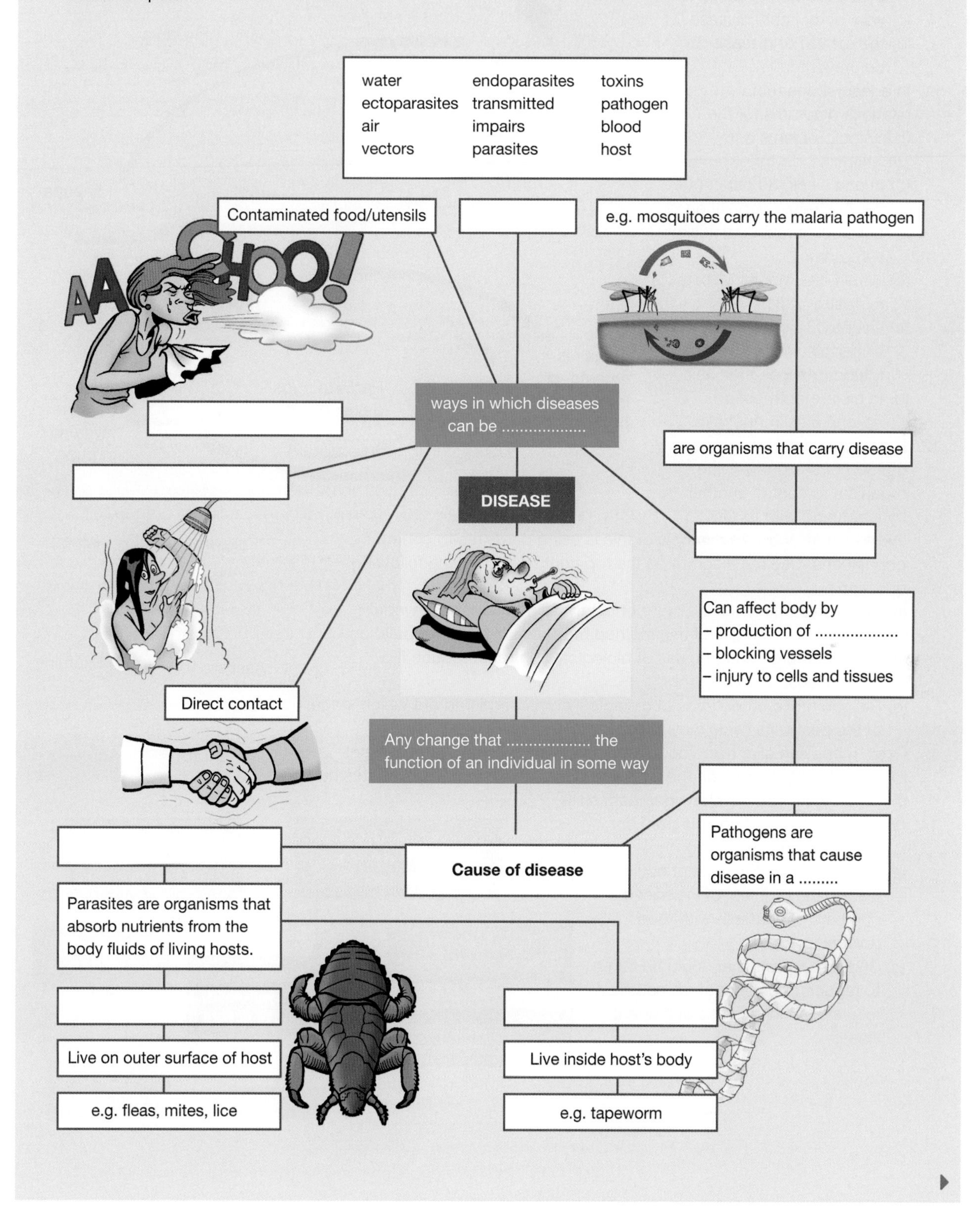

10. Use a cycle map or a relations diagram to outline how:
    (a) infection can be spread within a community
    (b) you can become infected by a pathogen or parasite
    (c) your body fights disease
    (d) the spread of disease can be prevented.
11. In a team, construct relations diagrams for the following problems and situations.
    (a) In one week, 80 per cent of the Australian population develops a rash and flulike symptoms.
    (b) Within one month, all cats in Australia die.
    (c) Over a 48-hour period, all people within 10 km of Melbourne lose their hair.
    (d) In a six-month period, no-one dies in Australia.
12. Biological control is a method of using one living organism to control another by interfering with its life cycle in some way. An example of this is using parasites to control fly populations. Use the diagram on the top right to answer the following questions.
    (a) At which stage in the life cycle of the fly do the parasites invade?
    (b) Suggest how the use of this method may control the fly population.
    (c) Find out more about the use of biological control to reduce fly populations.
    (d) Research two other types of biological control and find out which stage of the pest's life cycle they affect.
13. Cycle maps can also help you to organise your tasks and project time. Examine the cycle map on the right and discuss with your team-mates how this could help you develop your team skills.
14. (a) Observe the relations diagram on the right.
    (b) Suggest other causes that could lead to the problem of increased numbers of students with food poisoning.
    (c) Suggest actions that could be taken to reduce or prevent these causes resulting in an increase of student illness.

Parasites that use flies as hosts can be used to help control fly populations.

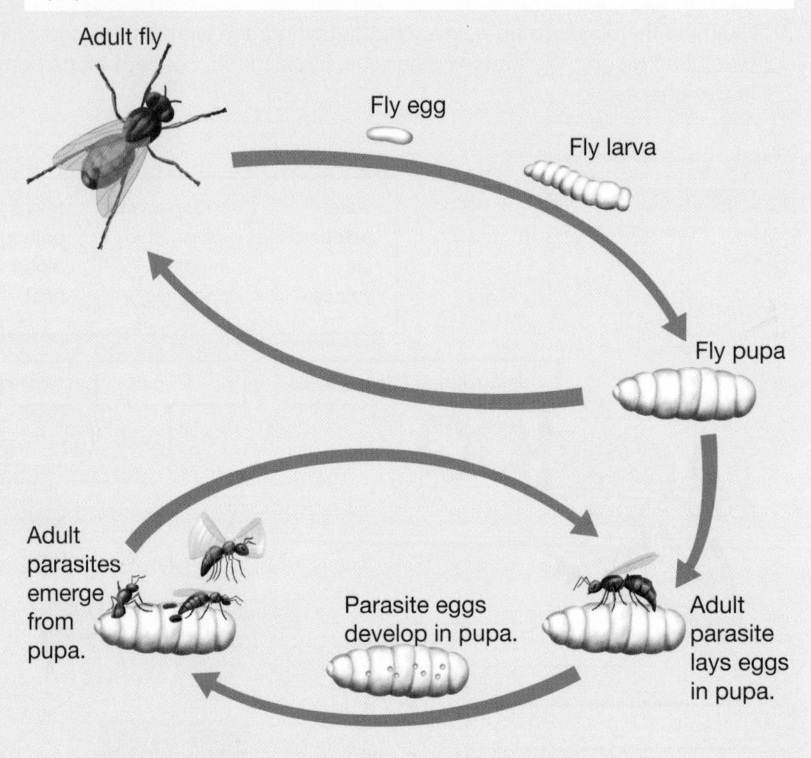

Team meeting cycle map

Collect → Examine → Agree → Perform → Assess → Collect

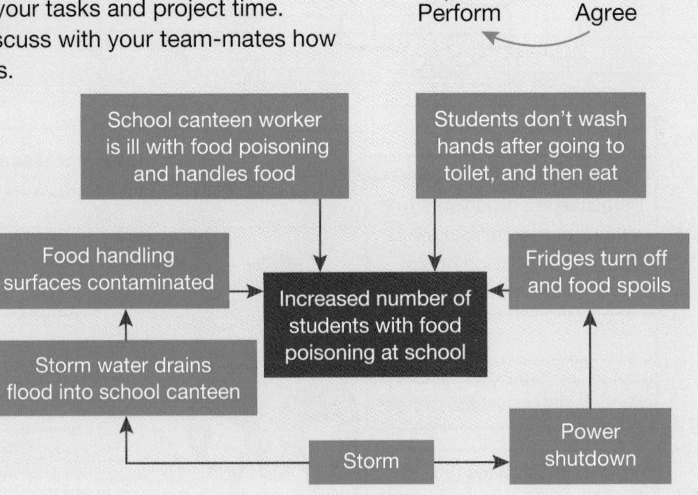

15. Read the paragraph below on Gram stains and bacteria. Summarise the information as either a cycle map or a relations diagram.

HOW ABOUT THAT!

In 1884, Joachim Gram (a Danish bacteriologist) developed the Gram stain. This stain divides bacteria into two groups on the basis of the chemical composition of their cell wall. Gram-positive bacteria take up the purple colour of the stain, whereas Gram-negative bacteria don't take up the stain, and therefore stain pink. This information can be used to determine which antibiotics would be most effective in killing them. Gram-positive bacteria are generally more susceptible to penicillin and sulphonamide drugs and Gram-negative bacteria are more susceptible to other types of antibiotics such as streptomycin and tetracycline.

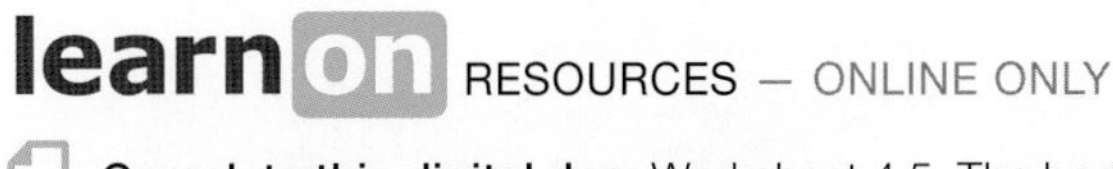

Complete this digital doc: Worksheet 4.5: The body at war: Puzzle (doc-18871)

Complete this digital doc: Worksheet 4.6: The body at war: Summary (doc-18872)

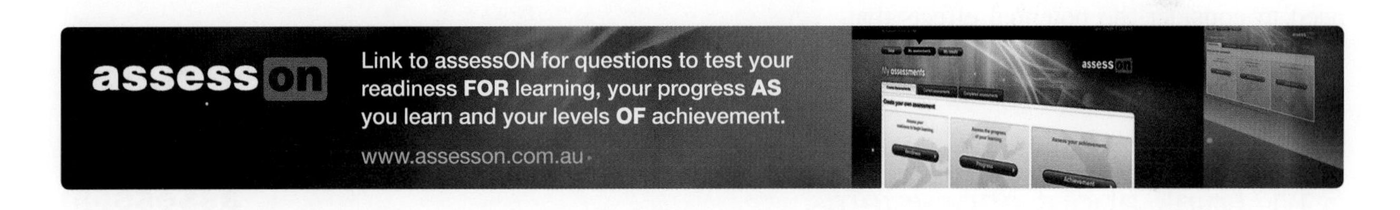

TOPIC 5
Ecosystems — flow of energy and matter

5.1 Overview

When we think about ecosystems, we need to think both big and small. We need to consider the recycling of atoms between organisms and within their environment and the flow of energy through living organisms and its changes from one form to another. We need to appreciate the relationships between organisms, and between organisms and their environment. We also need to consider the potential effects that these relationships have, not only on individual organisms and their environment, but also on our planet.

5.1.1 Think about ecosystems

assess**on**

- How can stomata help pull water up a plant?
- Is being green essential for photosynthesis?
- Why do some cells have more mitochondria than others?
- What's the difference between nitrifying and denitrifying bacteria?
- Why do energy pyramids always have the same basic shape?

LEARNING SEQUENCE

Numerous **videos** and **interactivities** are embedded just where you need them, at the point of learning, in your learnON title at www.jacplus.com.au. They will help you to learn the content and concepts covered in this topic.

5.1.2 Water

Organisms need water to survive. The good news is that water cycles through ecosystems. The bad news is that, at times, the amount of water available can be too great (as in the case of floods) or too little (as in the case of drought).

Some species have adapted to these conditions and possess adaptations that increase their chances of survival. Other organisms are not so fortunate and severe conditions of too much or too little water can result in their death. If too many of a particular type of organism die, then the decrease in their population size can have implications not only for other members of their food web, but also for other biotic and abiotic factors within their ecosystem.

Think, investigate and create

Carefully examine the water cycle and the 2013 Australian rainfall figures. Australia is considered to be one of the driest continents on Earth, yet there is a variety of ecosystems within it. Use a range of resources to answer the following essential question: How can *where* an organism lives affect *how* it lives? Present your findings as a set of models or in a creative multimedia format.

Not all of Australia received the same amount of rainfall in 2013. Climatic conditions differ across our continent.

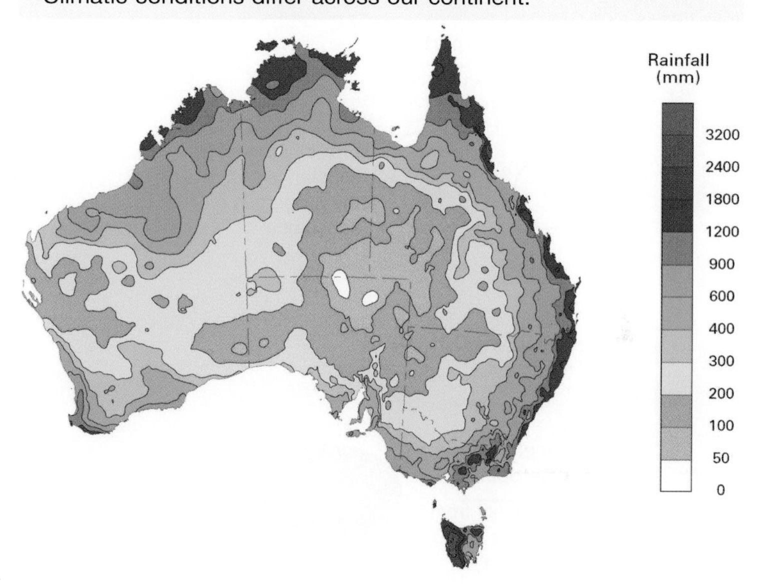

Water cycles through ecosystems. The wider the arrow, the greater the amount of water that moves through that part of the cycle. Is the movement of water by transpiration greater or less than the movement by evaporation from the ocean surface?

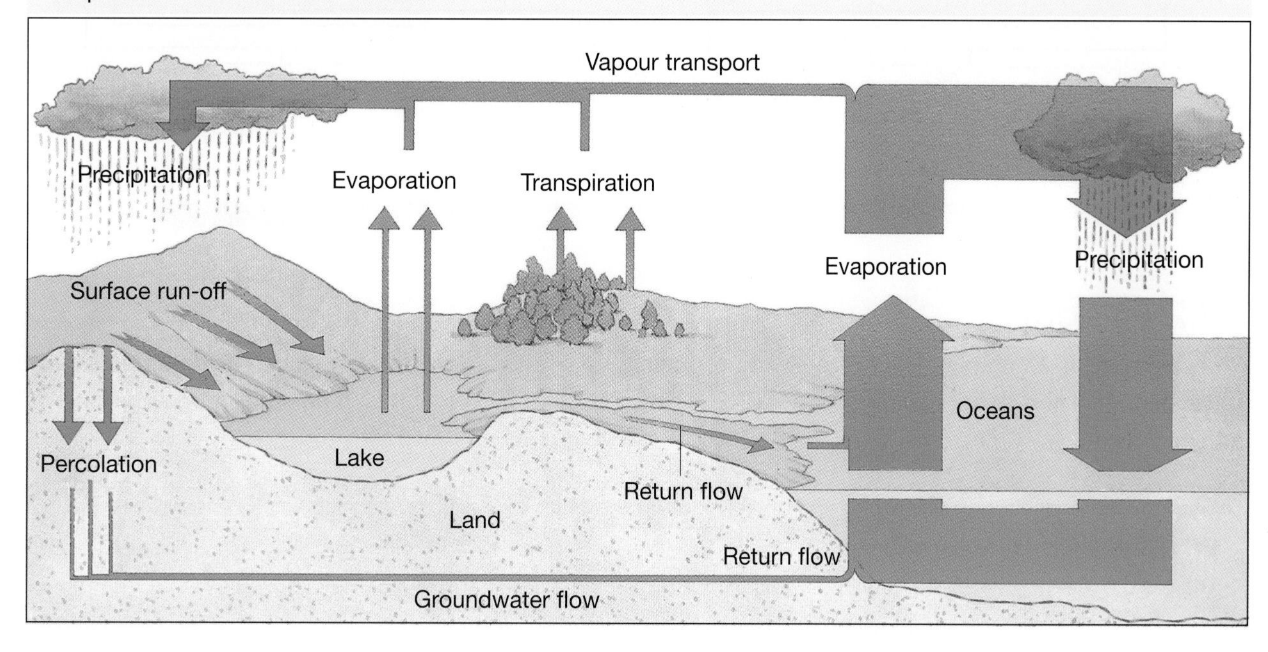

5.2 Systems: Ecosystems

5.2.1 Living together

You are a multicellular **organism** of the **species** *Homo sapiens*. When you are with others of your species in the same area at a particular time, you belong to a **population**. When the population you are part of is living with populations of other species, then collectively you could be described as a **community**. Communities of organisms living together interact with each other and their environment to make up an ecosystem.

Levels of biological organisation. As each level increases, structural complexity increases and unique phenomena may emerge.

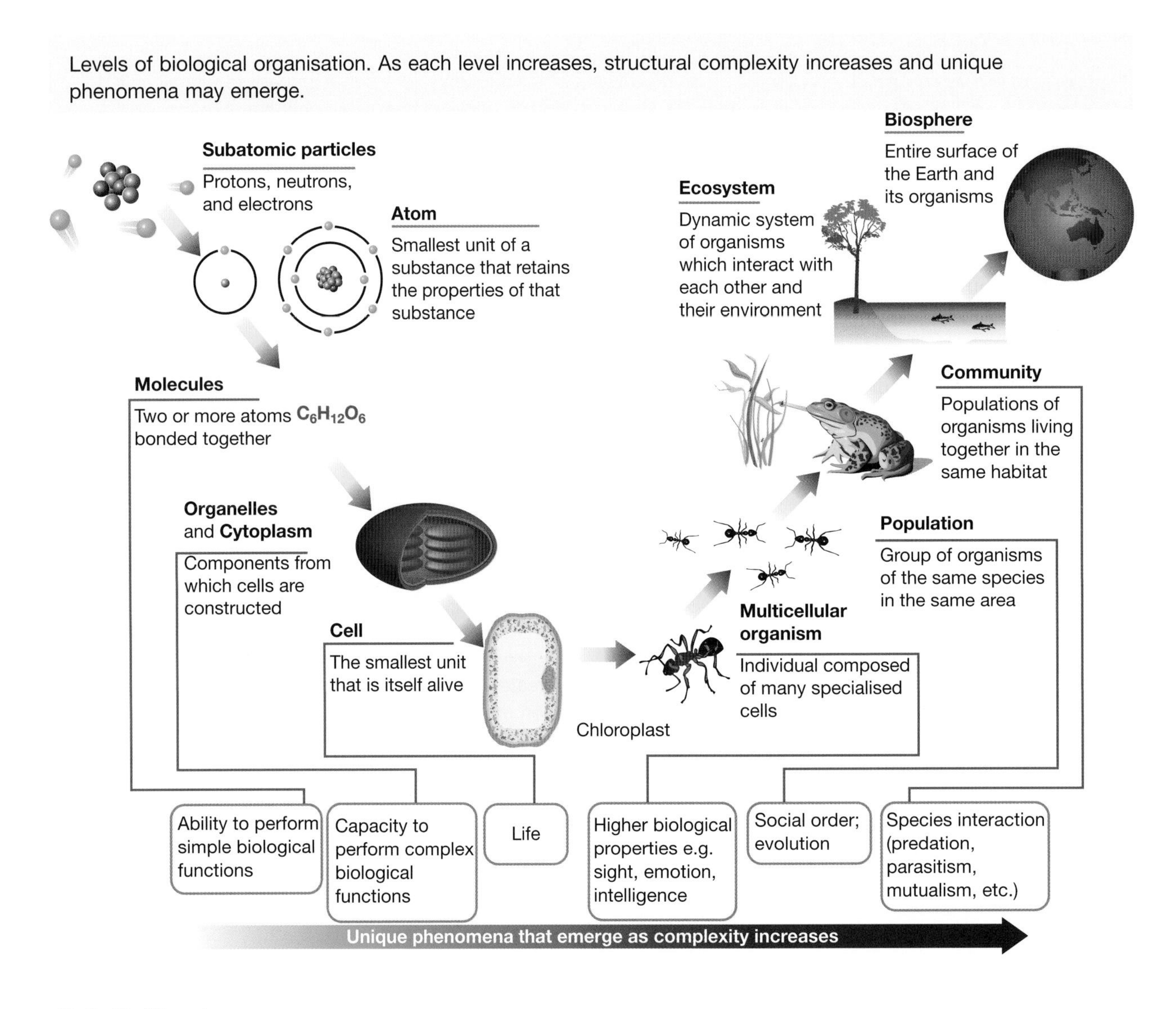

5.2.2 Ecology

An **ecosystem** is a complex level of organisation made up of living (biotic) parts, such as communities of organisms, and non-living (abiotic) parts, such as the physical surroundings. The study of ecosystems is known as **ecology**.

WHAT DOES IT MEAN?

The word *ecology* comes from the Greek terms *oikos*, meaning ‘home’, and *logos*, meaning ‘study’.

5.2.3 How do you get your food?

The members of every community within an ecosystem can be identified as being either a producer (autotroph), consumer (heterotroph) or decomposer. The feeding relationships between these groups can be shown in food chains or food webs (see section 5.7.4).

While **producers** are responsible for capturing light energy and using this energy to convert inorganic materials into organic matter, **decomposers** break down organic matter into inorganic materials (such as mineral ions) that can be recycled within ecosystems by plants.

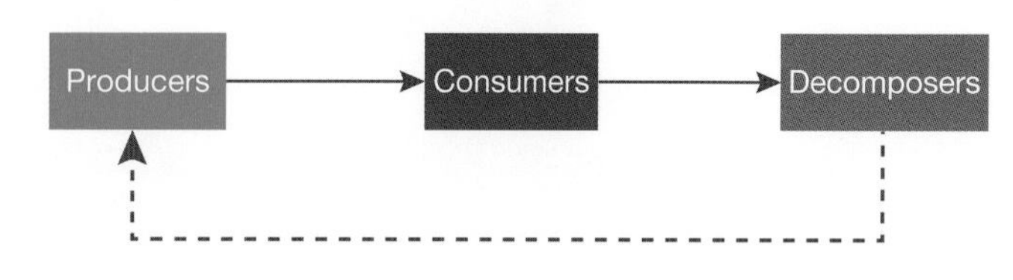

5.2.4 Producers

Producers within ecosystems are essential as they are at the base of the food chain. Plants are examples of producers. They use the process of **photosynthesis** to capture light energy and use it to convert simple inorganic substances (carbon dioxide and water) into organic substances (glucose). Since plants are able to convert glucose into other essential organic substances and do not need to feed on other organisms, they are often referred to as **autotrophs** ('self-feeders').

Plants also release oxygen gas as a waste product of photosynthesis. This molecule is essential for a type of cellular respiration called aerobic respiration — a process essential to the survival of the majority of organisms on our planet.

5.2.5 Consumers

As animals are unable to make their own food, they are called **heterotrophs** ('other-feeders'), and because they obtain their nutrition from consuming or eating other organisms, they are called **consumers**. Consumers are divided into different types on the basis of their food source and how they obtain it.

Herbivores eat plants and are often described as being **primary consumers** because they are the first consumers in a food chain. **Carnivores** eat other animals and are described as secondary or tertiary consumers in food chains or webs. Humans are examples of **omnivores**, which means we eat both plants and animals.

Another group of consumers releases enzymes to break down the organic matter in rotting leaves, dung and decaying animal remains, and then absorbs the products that have been externally digested. This group of consumers is known as **detritivores**, and they include earthworms, dung beetles and crabs.

The koala is a herbivore.

The Tasmanian devil is a carnivore.

5.2.6 Decomposers

While producers convert inorganic materials into organic matter, decomposers convert organic matter into inorganic materials. This is an example of how matter can be recycled within ecosystems so that they remain sustainable.

Fungi and bacteria are common examples of decomposers within ecosystems. These heterotrophs obtain their energy and nutrients from dead organic matter. As they feed, they chemically break down the organic matter into simple inorganic forms or mineral nutrients. Their wastes are then returned to the environment to be recycled by producer organisms.

5.2.7 Interactions between species

Species exist in an ecosystem within a specific **ecological niche**. The niche of a species includes its habitat (where it lives within the ecosystem), its nutrition (how it obtains its food) and its relationships (interactions with other species within the ecosystem).

Interactions within an ecosystem may be between members of the same species or between members of different species. Examples of types of interactions include **competition**, predator–prey relationships and symbiotic relationships such as **parasitism**, **mutualism** and **commensalism**.

5.2.8 Competition

Organisms with a similar niche within an ecosystem will compete where their needs overlap. Competition between members of different species for the same resource (e.g. food or shelter) is referred to as **interspecific competition**. Competition for resources between members of the same species (e.g. mates) is referred to as **intraspecific competition**.

5.2.9 Predator–prey relationship

In a **predator–prey relationship**, one species kills and eats another species. The predator does the killing and eating and the prey is the food source. Examples of predator–prey relationships include those between eagles and rabbits, fish and coral polyps, spiders and flies, and snakes and mice. How many others can you think of?

5.2.10 Herbivore–plant relationship

Plants cannot run away from herbivores! How then can they protect themselves against being eaten? Some plants protect themselves by using physical structures such as thorns, spines and stinging hairs; others use chemicals that are distasteful, dangerous or poisonous.

5.2.11 Symbiotic relationships

Symbiotic relationships are those in which the organisms living together depend on each other. Examples of symbiotic relationships include parasitism, mutualism and commensalism.

Interaction	Species 1	Species 2
Parasitism	✓(Parasite)	× (Host)
Mutualism	✓	✓
Commensalism	✓	**0**

✓ = benefits by the association; × = harmed by the association; 0 = no harm or benefit

5.2.12 Parasite–host

Parasites are organisms that live in or on a host, from which they obtain food, shelter and other requirements. Although the host may be harmed in this interaction, it is not usually killed. Some parasites are considered to be pathogens, as they can cause disease. This means that the functioning of their host is in some way impaired or damaged.

Parasites living on the host are called ectoparasites (e.g. fungi, fleas, ticks, leeches and some species of lamprey [see photo at right]). An example of an ectoparasite is the fungus that causes tinea or athlete's foot. The fungus secretes enzymes that externally digest the skin that it is attached to. It then absorbs the broken-down nutrients. This causes your skin to break and become red and itchy.

Parasites living inside their host are called endoparasites (e.g. flatworms such as Echinococcus granulosus or roundworms such as Ascaris lumbricoides). Tapeworms (see photo at right) are an example of an endoparasite. Their heads have suckers (and sometimes hooks) to firmly attach themselves to the walls of their host's intestine. They do not need a digestive system themselves as they live off the digested food within the intestine. Tapeworms vary in length from 1 cm to 10 cm. As each tapeworm contains both male and female sex organs, they don't need a mate to reproduce.

5.2.13 Parasitoids

A new group of consumers has been suggested, called parasitoids ('-oid' means '-like'). These organisms are halfway between predators and parasites. While they act like parasites, they kill their hosts within a very short period. Examples of organisms that may be classified as parasitoids are mainly wasps (see photo at right) and flies. The female parasitoid lays her egg(s) in the body of the host; when the eggs hatch, they eat the host from the inside. The host is killed when vital organs have been eaten. This relationship has applications in horticulture as a potential biological control method for pests feeding on crops.

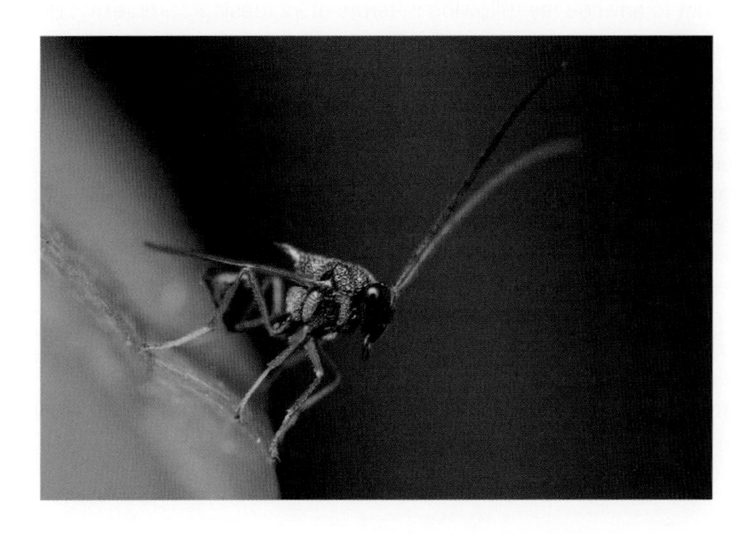

5.2.14 Mutualism

An interaction between organisms of two different species in which they both benefit is called mutualism. In many cases, neither species can survive under natural conditions without the other. Tiny protozoans found in the intestines of termites help them to digest wood. These organisms are dependent on each other for their survival. Another example is that of lichen (see photo at right), which is often found growing on rocks or tree trunks. Lichen is made up of a fungus and an alga living together. The alga uses light from the sun to make glucose and the fungus uses this as food. The fungus shelters the alga so that it does not get too hot or dry out.

5.2.15 Commensalism

An example of commensalism is found between remora fish and sharks. Remora fish are often found swimming beneath sharks and benefit by being able to feed on leftover scraps; the sharks are not harmed but receive no benefit. The organism that benefits is referred to as the commensal and the other is sometimes referred to as the host. Clownfish and sea anemones are another example. While the clownfish (*Amphiprion melanopus*) lives among the tentacles of the sea anemone, it is unaffected by their stinging cells and benefits from shelter and any available food scraps.

5.2 Exercises: Understanding and inquiring

To answer questions online and to receive **immediate feedback** and **sample responses** for every question, go to your learnON title at www.jacplus.com.au. *Note:* Question numbers may vary slightly.

Remember

1. Use Venn diagrams to compare the following relationships.
   (a) Commensalism and mutualism
   (b) Parasitism and commensalism
   (c) Predator–prey and parasite–host
2. State the name of the species to which you belong.
3. Outline the relationship between species, organisms, populations, communities and an ecosystem.
4. Define the term *ecology*.
5. Construct a flowchart that shows the relationship between producers, consumers and decomposers.
6. Construct a continuum to arrange the following in terms of increasing complexity: biosphere, cell, population, molecules, organisms.
7. Explain why producers are essential to ecosystems.
8. Construct a Venn diagram to compare autotrophs and heterotrophs.
9. Distinguish between herbivores, carnivores, omnivores and detritivores.
10. Identify a type of organism that you may find in
    (a) a temperate marine kelp forest ecosystem
    (b) a temperate closed forest ecosystem
    (c) an Antarctic marine ecosystem.
11. Distinguish between producers and decomposers.
12. Identify two common examples of decomposers.
13. Define the term *ecological niche*.
14. Distinguish between *interspecific competition* and *intraspecific competition*.
15. Construct a table to summarise the similarities and differences between parasitism, mutualism and commensalism.

Investigate, think and discuss

16. (a) List three examples of predators and then match them to their prey.
    (b) Suggest structural, physiological and behavioural features that may assist:
        (i) predators in obtaining food (e.g. webs, teeth, senses, behaviour)
        (ii) prey in avoiding being eaten (e.g. camouflage, mimicry, behaviour, chemicals).
17. In the interaction between a clownfish and a sea anemone, which is benefited?
18. Use a flowchart to describe how a parasite obtains its food.
19. Suggest why a parasite does not normally kill its host.
20. Use a visual thinking tool to show the difference between a commensal and a parasite.
21. Is a mammalian embryo a parasite? Explain your answer.
22. Parasite–host relationships also exist within the plant kingdom. The two main types of these relationships are holo-parasitism (in which the parasite is totally dependent on the host for food) and hemi-parasitism (in which the parasite obtains some of its nutrients from the host but can make some itself). Some plant species belonging to the genus *Rafflesia* are examples of holo-parasites, and many Australian species of mistletoe are hemi-parasites. Research and report on:
    (a) *Rafflesia* parasites and their host *Tetrastigma*
    (b) pollinators for *Rafflesia* flowers
    (c) one of the following hemi-parasites:
        (i) sheoak mistletoe (*Amyema cambagei*) and its host *Casuarina cunninghamiana*
        (ii) paperbark mistletoe (*Amyema gaudichaudii*) and host *Melaleuca decora*.
23. Decide whether the following relationships are examples of parasitism, commensalism or mutualism.
    (a) A dog with a tapeworm in its intestine, absorbing the digested food
    (b) Egrets staying near cows and feeding on the insects they stir up
    (c) Harmless bacteria *Escherichia coli* living in human intestines
    (d) Root nodules of clover contain bacteria — the clover benefits, but can survive without the bacteria; the bacteria don't live anywhere else
    (e) A fungal disease on human skin, such as ringworm
24. The koala and the bacteria that live in its gut have a symbiotic relationship. Find out how each of the organisms benefits from this relationship.
25. Choose one of the following parasites: malaria parasites, tapeworms, ticks, insects that make galls in trees, blight-causing bacteria. Explain how it infests its host and how it affects its host.
26. Find out about the symptoms and treatment of some common parasites in humans.
27. Some clovers (Trifolium) produce cyanide. Find out how this may protect them against being eaten.
28. Find examples of ways that Australian plants try and protect themselves from being eaten by herbivores.
29. Use internet research to identify three problems that can be investigated about interactions between organisms.
30. Construct a model that simulates interactions between at least four different types of organisms.

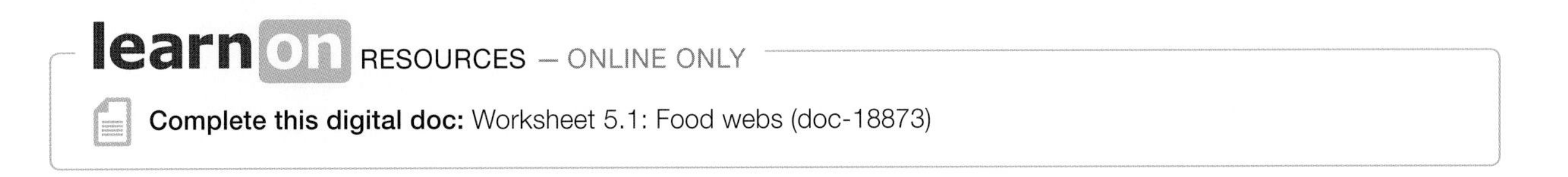

5.3 Mapping ecosystems

5.3.1 Mapping ecosystems

Are you at home? What does that mean to your survival? A habitat is the name given to the place where an organism lives.

It needs to be convenient and provide conditions that are comfortable to the functioning of cells and life processes of its inhabitants. The match between the environmental conditions and the needs of organisms is responsible for the **distribution** and **density** of species within it.

An **ecosystem** may contain many habitats. It is made up of living or **biotic factors** (such as other organisms) and non-living or **abiotic factors** (such as water, temperature, light and pH) that interact with each other.

5.3.2 Tolerance — the key to survival

Each species has a **tolerance range** for each abiotic factor. The **optimum range** within the tolerance range is the one in which a species functions best. Measuring the abiotic factors in a habitat can provide information on the abiotic requirements for a particular organism in that habitat. Can you think of features that organisms possess to increase their chances of survival in some habitats more than in others?

5.3.3 How many and where?

Investigation of an ecosystem involves studying how different species in it interact. To do this, you need to:
1. identify the organisms living in the ecosystem by using keys and field guides
2. determine the number or density of different species in the particular area. This indicates the biological diversity (**biodiversity**) within the ecosystem.
3. determine the distribution of the different species or where they are located.

5.3.4 Sampling an ecosystem

Sampling methods are used to determine the density and distribution of various populations and communities within the ecosystem. **Transects** are very useful when the environmental conditions vary along the

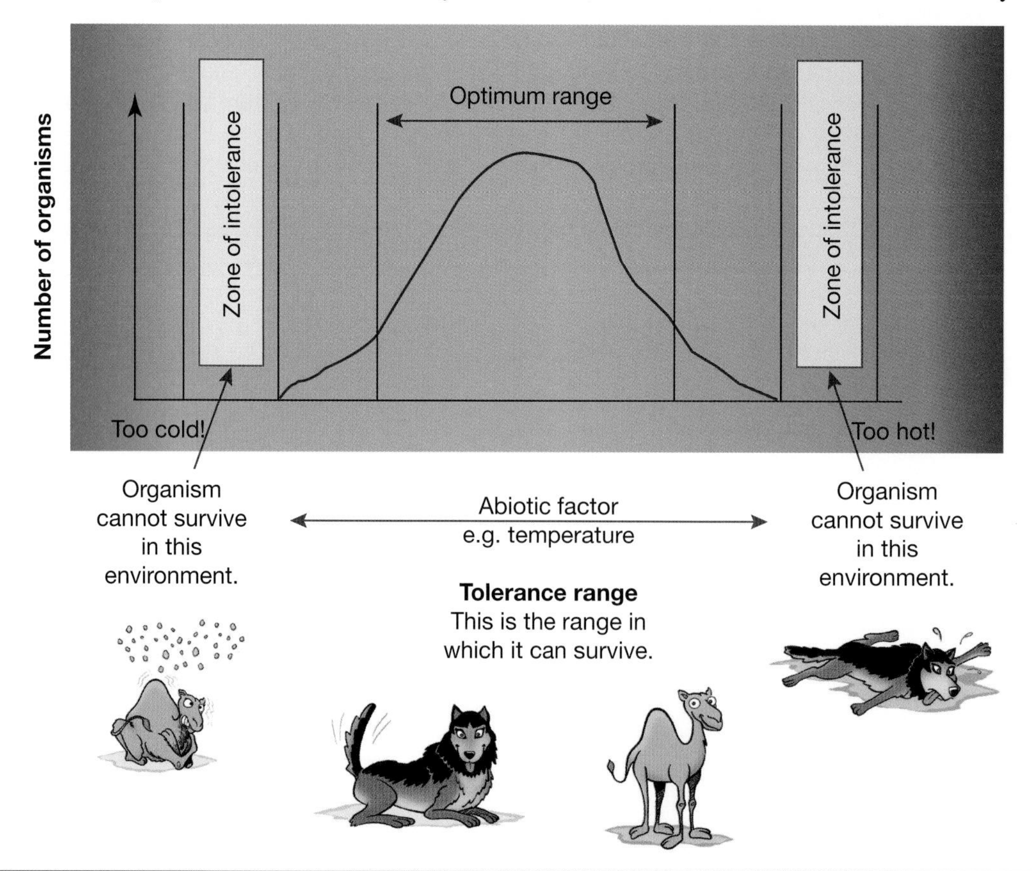

sample under investigation. **Quadrats** can be used to estimate the distribution and abundance of organisms that are stationary or do not move very much. The **mark, release and recapture** sampling method is used to determine the abundance of mobile species.

5.3.5 Life in a square

A quadrat is just a sampling area (often 1 square metre) in which the number of organisms is counted and recorded. When organisms are counted in a number of quadrats, this is usually considered to be representative of the total area under investigation. The average density of the total area can be estimated using the equation shown:

Line transects provide information on the distribution of a species in a community.

Estimated average density

$$= \frac{\text{total no.of individuals counted}}{\text{no.of quadrats} \times \text{area of each quadrat}}$$

For example, if the total number of individuals counted = 100, number of quadrats = 4, area of each quadrat = 1 m^2, then estimated average density

$$= \frac{100 \text{ individuals}}{4 \times 1 \text{ m}^2} = 25 \text{ individuals/m}^2$$

INVESTIGATION 5.1

Using quadrats

AIM: To estimate the abundance of eucalypts in two different environments

Background

- The maps of environments A and B show each eucalyptus tree as a cross.

Materials:

maps of environments A and B (provided by your teacher) transparent sheet

Method, results and discussion

1. Copy and complete the table below.

	Number of eucalypts	
Quadrat number	**Environment A**	**Environment B**
1		
2		
3		
4		
5		
Average		

2. Measure the length and width of each map and calculate the area of each using the following equation.

$$\text{Area} = \text{length} \times \text{width}$$

3. Make a quadrat by cutting a 3 cm × 3 cm square out of transparent sheet. Calculate the area of the quadrat.
4. Close your eyes and drop the quadrat anywhere on the map. Count how many eucalypts (crosses) are inside the quadrat. Repeat four more times. Do this for both maps.
5. Estimate the abundance of eucalypts in each environment using the equation shown in the previous box.
6. Ask your teacher for the actual abundance of eucalypts in each environment. Compare your estimate with the actual abundance.
7. Suggest what you could have done to make your estimate more reliable.

INVESTIGATION 5.2

Measuring abiotic factors

AIM: To investigate some abiotic factors in two different environments

Background

In this investigation you will measure some abiotic factors for environments A and B. The soil samples were collected from these environments. The water samples were collected from rivers that run through each environment.

Materials:

water samples A and B and soil samples A and B (provided by your teacher)
thermometer
dropper bottle of universal indicator solution
universal indicator colour chart
dropper bottle of silver nitrate solution (0.1 mol/L)
calcium sulfate powder
test tubes

Method and results

1. Review the table at end of this "Method and results".
2. Copy the table below and add your abiotic factor results for each environment.

Abiotitc factor	Environment A	Environment B
Water temperature (°C)		
Soil temperature (°C)		
Water pH		
Soil pH		
Water salinity		

3. Construct column graphs that show the abiotic factor results for each environment.

Abiotic factor	Equipment used and/or method
Water temperature (°C)	Thermometer
Soil temperature (°C)	Thermometer
Water pH	Pour 5 mL of tap water sample A into a test tube. Add 3 drops of universal indicator. Compare the colour of the water with the colour chart and record the pH of the water sample. Repeat using water sample B.
Soil pH	Place a small sample of soil A onto a watchglass. The soil should be slightly moist. If the soil is very dry, add a few drops of distilled water. Sprinkle some calcium sulfate over the soil. Add some drops of universal indicator over the calcium sulfate powder. Compare the colour of the powder with the colour chart and record the pH of the soil. Repeat using soil sample B.

Abiotic factor	Equipment used and/or method
Water salinity	Pour 5 mL of water sample A into a test tube. Add 3 drops of silver nitrate solution. Note whether the sample remains clear, becomes slightly cloudy or turns completely white/grey. Use the salinity table below to work out the salinity of the water sample. Repeat using water sample B.

Description	Salinity
Clear	Nil
Slightly cloudy	Low
Completely white/grey	High

Discuss and explain

4. pH is a measure of the alkalinity or acidity of a substance. A pH of more than 7 is considered to be alkaline whereas a pH below 7 is considered to be acidic.
   (a) Which water sample was most acidic?
   (b) Which soil sample was most acidic?
5. Identify any trends or patterns in your results. Suggest reasons for these patterns.
6. Which of the tests in this investigation were qualitative and which were quantitative?
7. In which way were the variables controlled in this investigation?
8. Are your temperature results accurate for each environment? Explain.
9. Suggest two ways in which you could improve this investigation.
10. Design an investigation that could examine abiotic factors in your local school environment.

INVESTIGATION 5.3

The capture–recapture method

AIM: To estimate population size using the capture–recapture method

Materials:

a large beaker
red and yellow beads (substitute other colours if needed)

Method, results and discussion

1. In your notebook, draw a table similar to the one shown below with enough room for 10 trials and the average.

Trial	Number of untagged fish (red beads)	Number of tagged fish (yellow beads)
1		
2		
3		
4		
5		
6		
7		
8		
9		
10		
Average		

- Place about 200 red beads in the large beaker (you do not need to count them exactly at this stage). These represent goldfish living in a pond.
- Catch 25 of the goldfish and tag them (replace 25 of the red beads with yellow beads).
- Mix the beads thoroughly.

2. With eyes closed, one student should randomly select 20 beads from the beaker. These are the recaptured goldfish. Count how many fish are tagged (yellow beads) and untagged (red beads), and enter the numbers in the table.
3. Return the beads to the beaker and mix thoroughly. Repeat the above step a further nine times.
4. Calculate the average number of tagged and untagged fish per capture.
5. Calculate the total number of fish using the equation:

$$\text{population size} = \frac{n_1 \times n_2}{n_3}$$

in which n_1 = number caught and initially marked,
n_2 = total number recaptured
n_3 = number of marked individuals recaptured.
6. Count how many beads were actually in the beaker and compare the actual number to the number you calculated using the capture–recapture method.
7. List any source of errors in this experiment.
8. Explain why this method can only be used for animals that move around. Why can't it be used to estimate the number of trees in a forest, for example?

INVESTIGATION 5.4

Biotic and abiotic factors

AIM: To measure biotic and abiotic factors in different areas in an environment

CAUTION

Be sun-safe!

Materials:

access to a natural area in your school grounds or bushland near your school
a data logger with temperature probe and light probe or a thermometer and hand-held light sensor
wet–dry thermometer (or humidity probe for data logger)
wind vane
soil humidity probe (optional)
calcium sulfate powder
water in a small wash bottle
Petri dish
universal indicator
tring
tape measure or trundle wheel
sunhat and sunscreen

Method and results

- Break up into groups. Each group will need to study a different area of the environment. Try to choose areas that are different (e.g. sunny and shady areas, or near paths and away from paths).

Abiotic factor	Materials used/method	Measurement
Temperature		
Air humidity	Wet–dry thermometer	
Light intensity		
Soil humidity	Soil humidity probe (if available)	
Water pH	Refer to Investigation 5.2	
Soil pH	Refer to Investigation 5.2	
Water salinity	Refer to Investigation 5.2	

Part A: Abiotic factors

1. Copy the table in previous page into your notebook. Fill in the missing materials of equipment in the second column.
2. Use the equipment available at your school to measure the abiotic factors listed in the table. Complete the third column of the table.
3. When you are back in the classroom, construct a table or spreadsheet to enter the results collected from each group. Calculate the average reading for each abiotic factor measured.
4. Choose one of the abiotic factors you measured and construct a column graph showing the reading for each location studied.

Part B: Biotic factors

- Use a trundle wheel to measure the length and width of the total area you are studying. If the area is too large to measure you may be able to estimate the surface area using a map.
- Use the tape measure and string to cordon off an area 1 m by 1 m. This is your quadrat.

5. List all the different species you can see inside your quadrat. If you do not know their names, describe or draw them.
6. Decide which plant(s) you will count; you may wish to count clovers, for example. Count how many of this type of plant(s) are in your 1 m × 1 m square.
7. Estimate the total number of each plant(s) counted using the equation below:

$$\text{Total number} = \frac{\text{average number per quadrat} \times \text{total area}}{\text{area of quadrat}}$$

8. Identify any trends in the results you obtained in your abiotic factor observation. For example, how did the results for sunny areas compare with those for shady areas?
9. Some organisms living in your quadrat cannot be seen. Give some examples. Why are these organisms very important?
10. Compare the class results for parts A and B. Identify any trends in the results. Is there a relationship between any of the abiotic factors and the type of organisms found?

5.3 Exercises: Understanding and inquiring

To answer questions online and to receive **immediate feedback** and **sample responses** for every question, go to your learnON title at www.jacplus.com.au. *Note:* Question numbers may vary slightly.

Remember

1. Outline the difference between density and distribution.
2. Define each of the terms below and provide an example.
   - Habitat
   - Ecosystem
   - Abiotic factor
3. Recall the type of information that is provided by:
   (a) quadrats
   (b) transects.

Think

4. Describe the difference between a habitat and an ecosystem.
5. (a) List five biotic factors that are part of the ecosystem in which you live.
   (b) List five abiotic factors that are a part of your ecosystem.
6. Suggest ways in which a freshwater habitat may vary from a marine habitat. Relate these differences to the differences in features of organisms located in each habitat.
7. Suggest the difference between the terms *environmental factors* and *environmental conditions*.

Using data

8. The location of five different types of trees in the two quadrats below is indicated by the five different symbols.
   (a) Count and record the number of trees in each quadrat.
   (b) Count and record the number of the different species in each quadrat.
   (c) Which quadrat provides the greatest variety of habitat types for wildlife? Give reasons for your response.
   (d) Suggest why the rainforest species in both quadrats are located most densely near the creek.

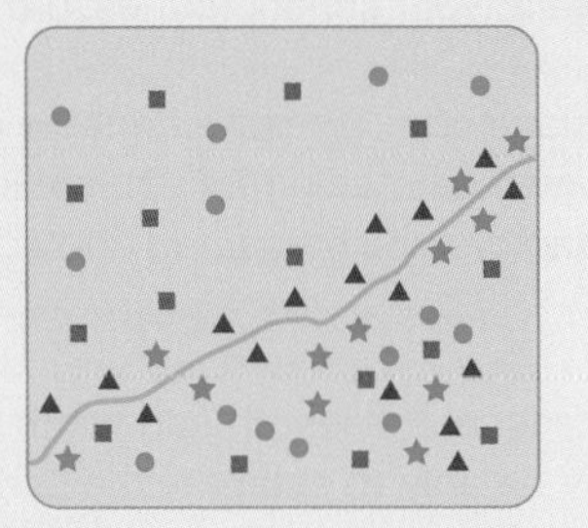

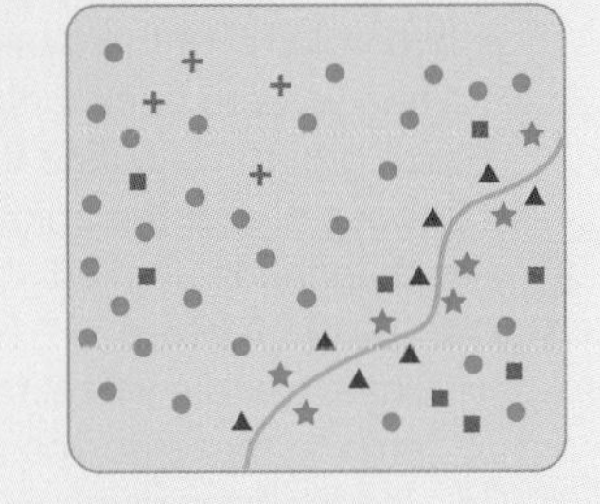

Key

★ Myrtlebeech, a rainforest tree
▲ Sassafras, a rainforest tree
■ Mountain ash, smooth-barked eucalypt
● Blackwattle tree
+ Messmate, rough-barked eucalypt
— Creek

9. (a) Carefully observe the diagram below. Describe the patterns along the rock platform to the sea for each of the abiotic factors measured.
   (b) Suggest the features that organisms living at these locations would need to possess.
      (i) Location A
      (ii) Location D
      (iii) Location F

Physical factors — salinity, temperature and dissolved oxygen at low tide on a rock platform. *Source*: Biozone International (Year 11 Biology 1996 Student Resource and Activity Manual).

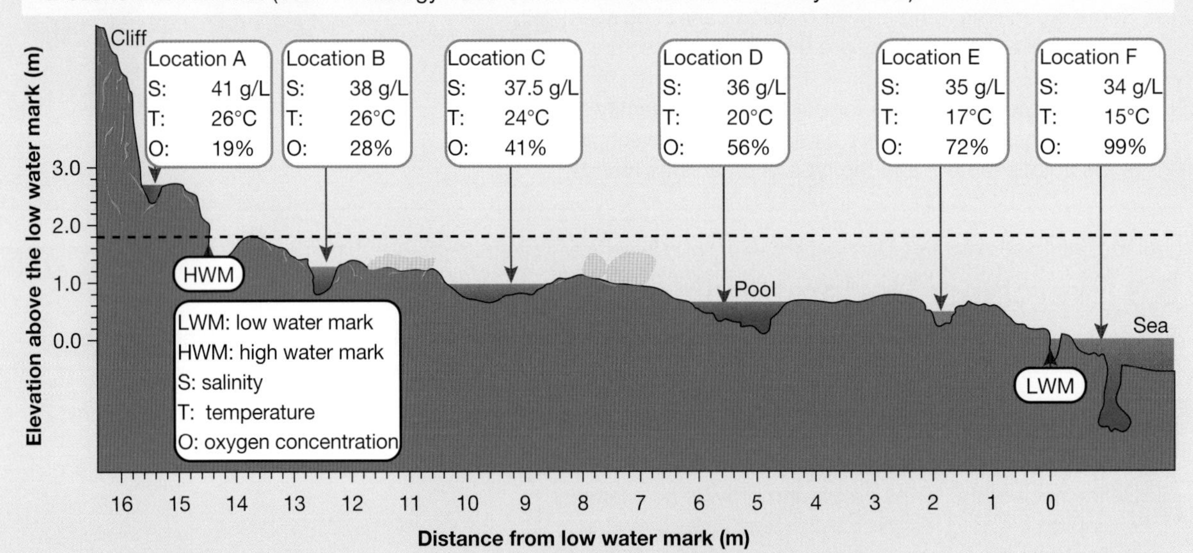

10. Suggest why wombats living in different areas have different features. Suggest how these differences may increase their chances of survival. Suggest reasons for the difference between their previous and recent habitat ranges.

Investigate, think and discuss

11. Find out more about research that the CSIRO, the Bureau of Meteorology or other Australian research institutions are involved in that is related to the climate and ecosystems. Share your findings with others. Suggest a question that you would like to investigate if you were involved in that field of research.

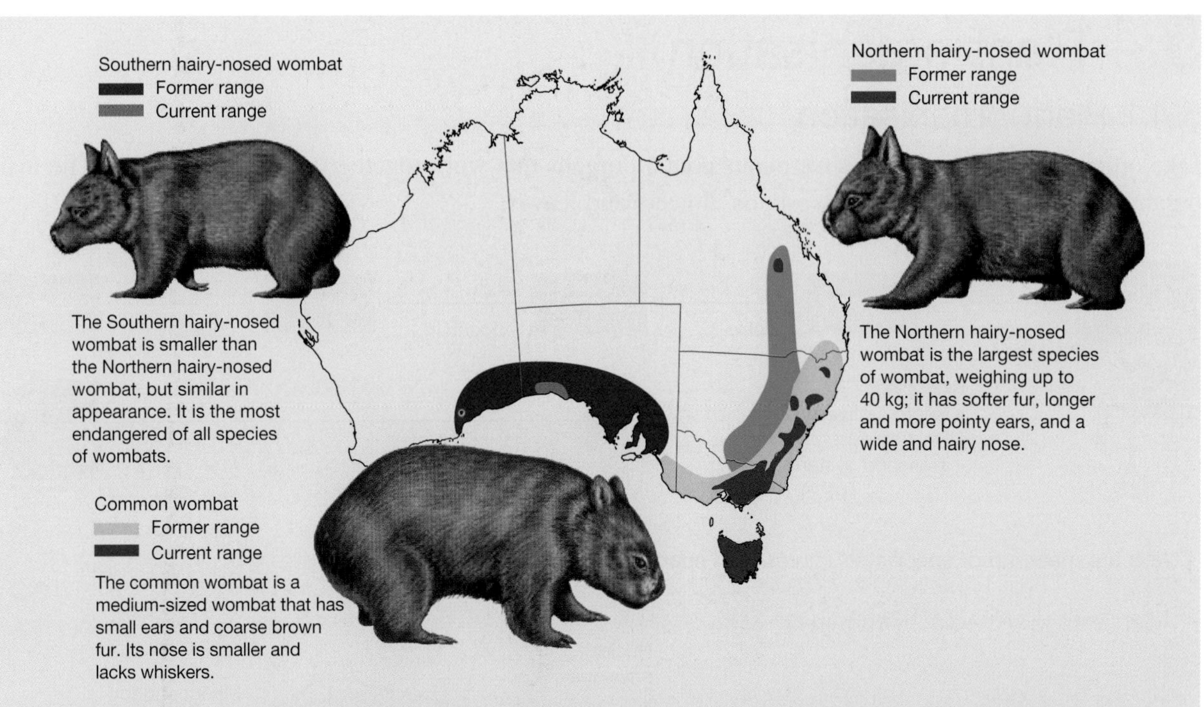

12. Light intensity, water availability, temperature and dissolved oxygen levels are examples of abiotic factors that may be limiting factors in determining which organisms can survive within a particular habitat. Consider the details in the table below and suggest responses for each of the blank cells.

Abiotic factor	Abiotic factor description	Example of habitat	Features of organism that could survive
Light intensity	Low	Floor of tropical rainforest	
Water availability	Low	Desert	
Temperature	Very high		
Dissolved oxygen levels	Low		

13. Select a particular Australian organism and then research its habitat range over the last 100 years. Construct graphs and labelled diagrams to share your findings.
14. Find out how radio or satellite tracking techniques are used to track animals whose habitats range over large areas.

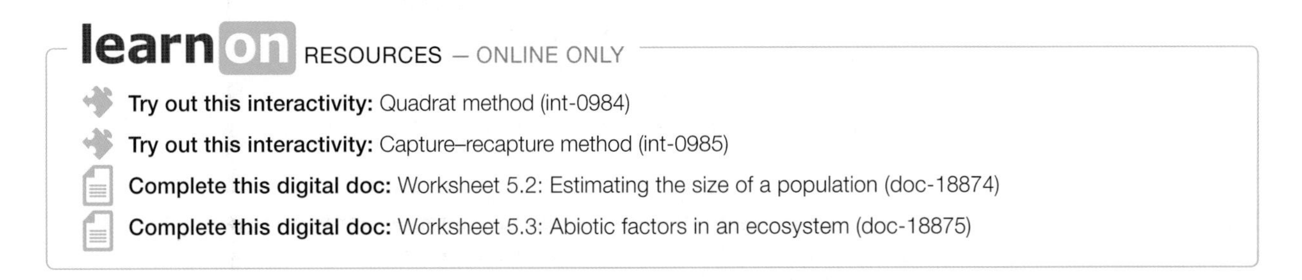

Try out this interactivity: Quadrat method (int-0984)

Try out this interactivity: Capture–recapture method (int-0985)

Complete this digital doc: Worksheet 5.2: Estimating the size of a population (doc-18874)

Complete this digital doc: Worksheet 5.3: Abiotic factors in an ecosystem (doc-18875)

5.4 Plant organisation

5.4.1 Plant organisation

Like other multicellular organisms, plants contain organs that work together to keep them alive. The main organs in vascular plants are roots, stems, flowers and leaves.

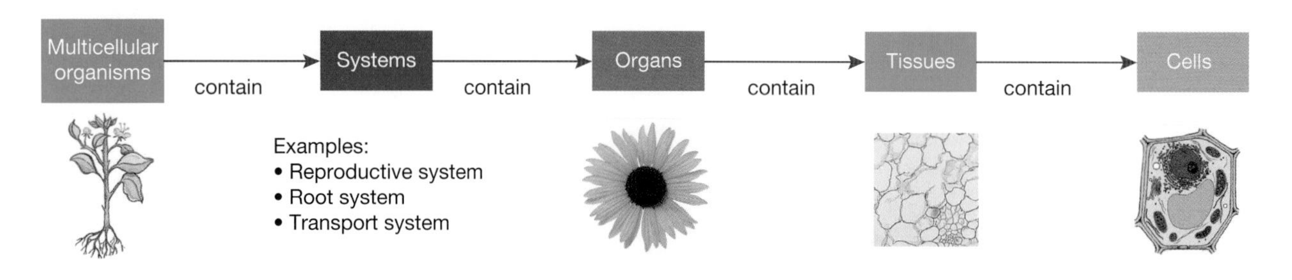

Cells are made up of specific structures with specific functions.

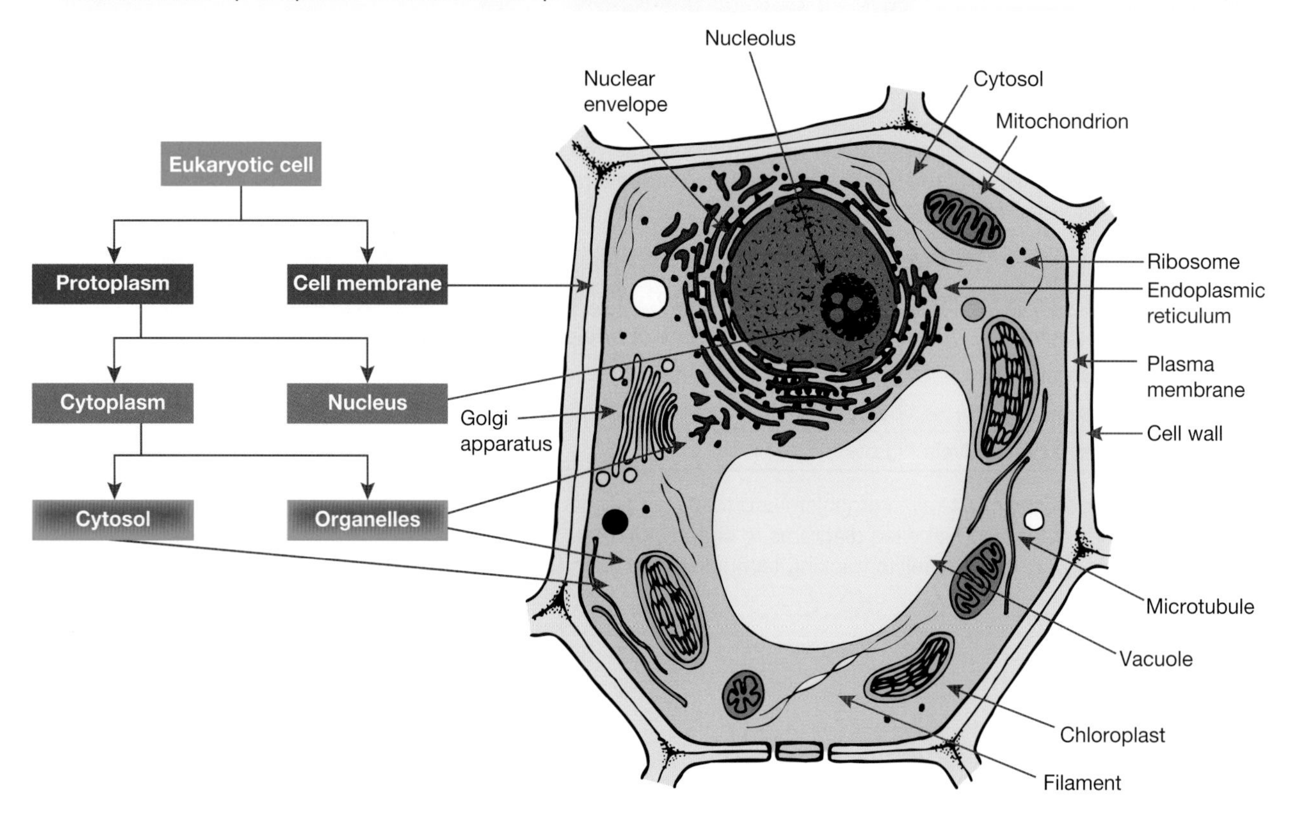

5.4.2 Roots: hairy roots

Roots both anchor plants and help them to obtain oxygen, water and mineral salts from the soil. **Root hairs** found on the outermost layer of the smallest roots can greatly assist this process by increasing the amount of surface area available for absorption. These long cells act like thousands of tiny fingers reaching into the soil for water and soluble salts.

5.4.3 Stems: transport tubes

Plants have a transport system made up of many thin tubes which carry liquids around the plant. The two main types of tubes in vascular plants are the **phloem** and **xylem vessels**. These tubes are located together in groups called **vascular bundles**.

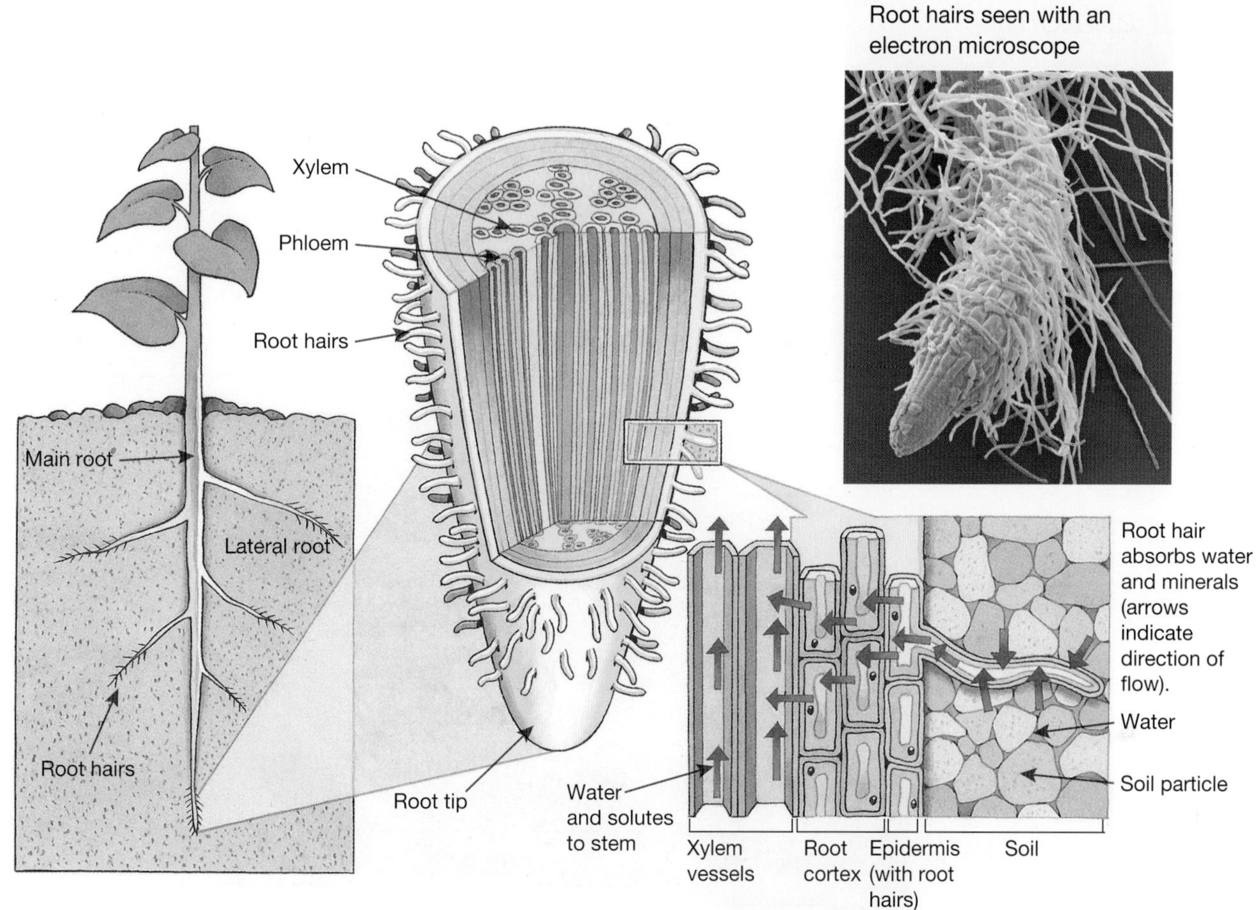

5.4.4 Translocation

Organic substances are transported up and down the plant in the phloem. This is called **translocation**. The two main types of organic molecules transported are nitrogenous **compounds** (for example, amino acids) and soluble carbohydrates (for example, sucrose).

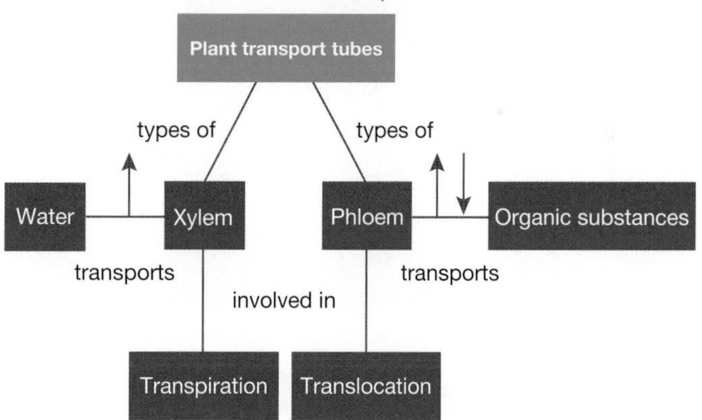

5.4.5 Transpiration

The transport of water up from the roots of the plant, through the xylem and out through the stomata as water vapour is called the **transpiration stream**. As this water vapour moves from the plant, suction is created that pulls water up through the xylem vessels from the roots. The loss of water vapour from the leaves (through their stomata) is called **transpiration**.

The strong, thick walls of the xylem vessels are also important in helping to hold up and support the plant. The trunks of trees are mostly made of xylem. Did you know that the stringiness of celery is due to its xylem tissues?

The photograph shows how the vascular tissue of a dicot (buttercup) appears when viewed under an electron microscope.

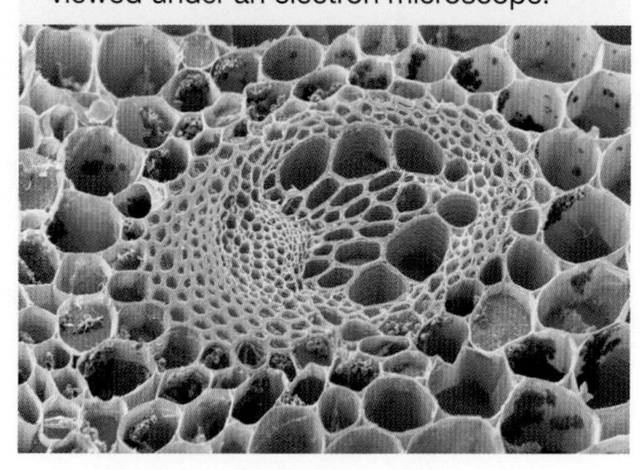

Leaf anatomy

Transverse section through a portion of a leaf showing xylem and phloem tissue.

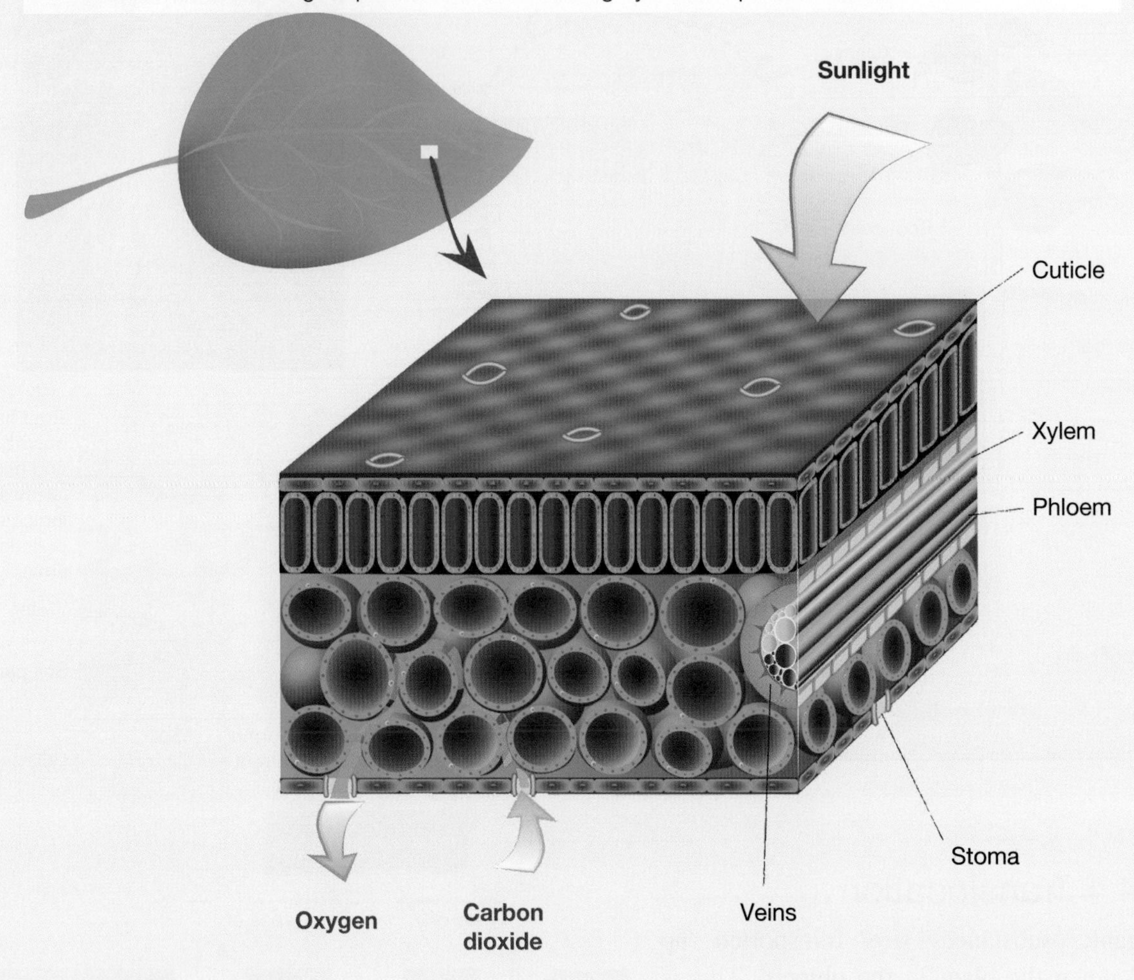

Diagrams of typical cross-sections of the stem of (a) a young dicot and (b) a monocot

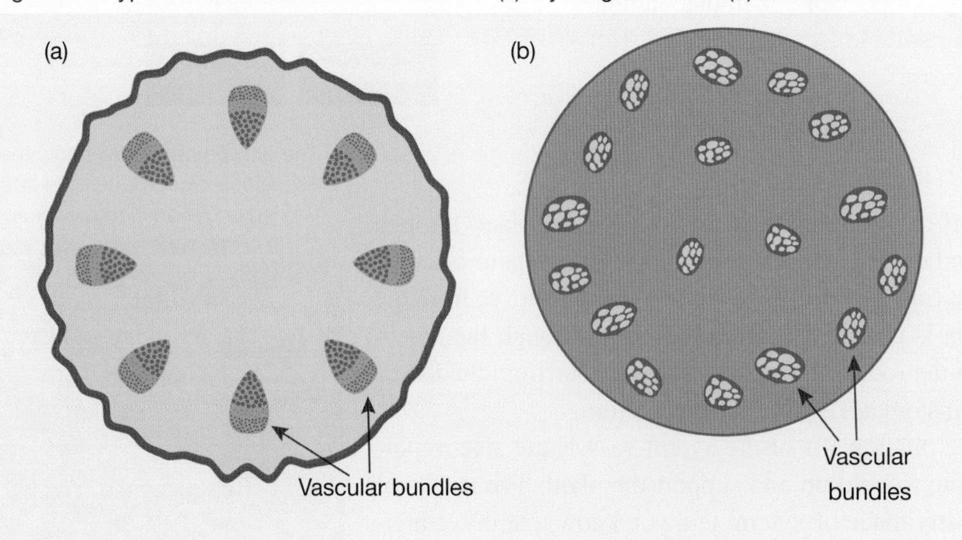

5.4.6 Leaves: chloroplasts

A plant leaf is an organ that consists of tissues such as epithelium, vascular tissue and parenchyma tissue. The structure of cells within the tissues and the organelles that they contain can vary depending on the function of the cell. Leaf cells, for example, contain **organelles** called **chloroplasts**. Chloroplasts contain **chlorophyll**, a green pigment that is involved in capturing or absorbing **light energy**. The synthesis of **glucose** also occurs in the chloroplast.

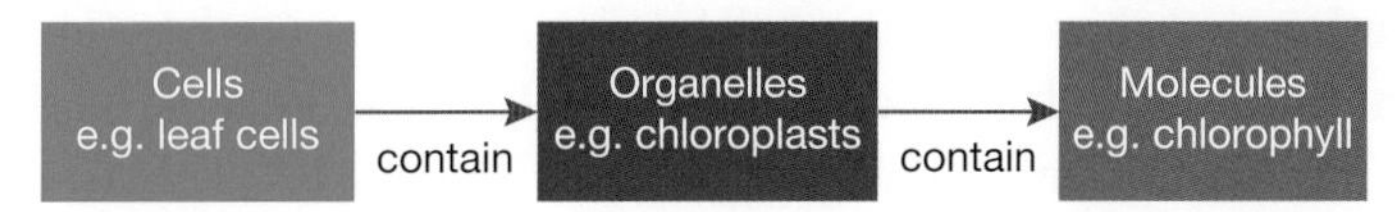

5.4.7 Flaccid or turgid

Plants need water to survive. If not enough water is available or too much water is lost, the plant may wilt. When this occurs, water has moved out of the cell vacuoles and the cells have become **flaccid**. The firmness in petals and leaves is due to their cells being **turgid**.

5.4.8 Leaves: stomata

The exchange of gases such as oxygen, carbon dioxide and water vapour between the atmosphere and plant cells occurs through tiny pores called **stomata**. These are most frequently located on the underside of leaves. Evaporation of water from the stomata in the leaves helps pull water up the plant.

Surface view of leaf showing distribution of stomata. Note epidermal cells, guard cells and their chloroplasts, and stomal pores.

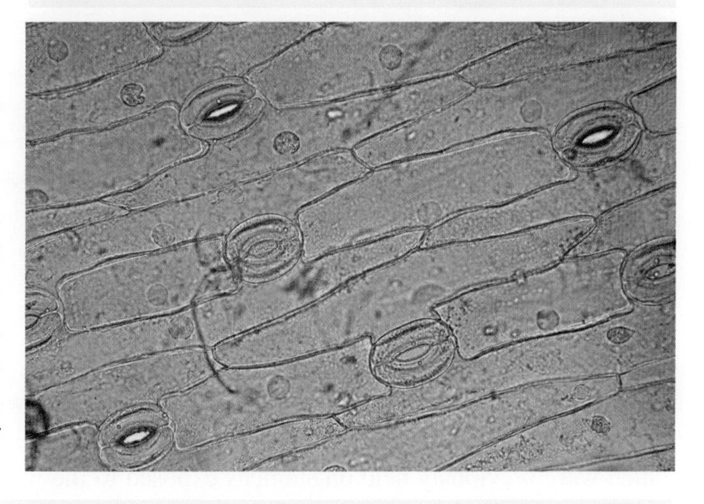

5.4.9 Guarding the pathway

Guard cells that surround each **stoma** enable the hole to open and close, depending on the plant's needs. When the plant has plenty of water, water moves into the guard cells, making them turgid. This stretches them lengthways, opening the pore between them (the stoma). If water is in short supply, the guard cells lose water and become flaccid. This causes them to collapse towards each other, closing the pore. In this way, the guard cells help to control the amount of water lost by the plant.

Inner and outer walls of guard cells may be of different thicknesses. A thinner outer wall can stretch more than a thicker inner wall. Microfibrils in guard cells also influence the extent to which walls of guard cells can stretch. As the outer walls of guard cells stretch, the stoma (pore) opens.

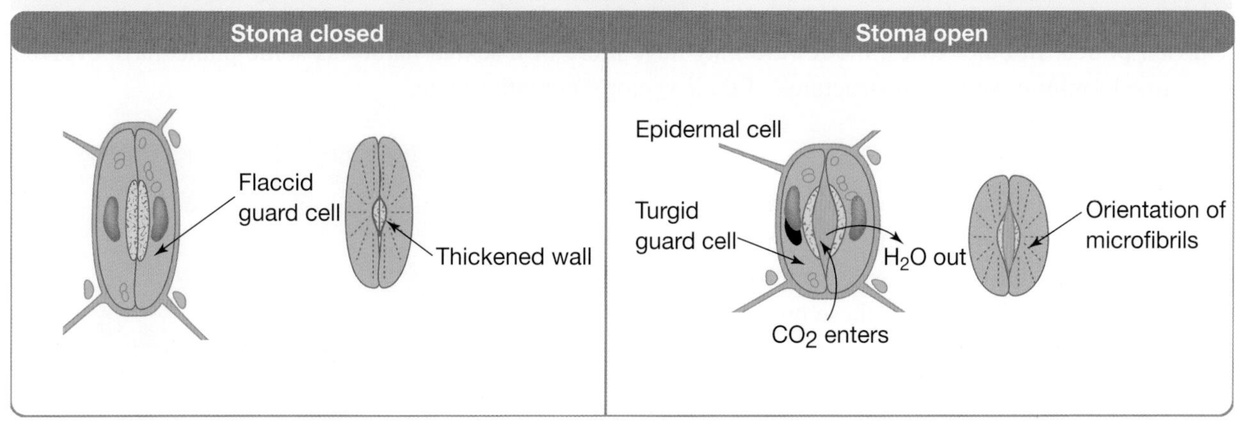

5.4.10 Flowers

Flowers make up the reproductive structures of some plants. Within the flower there are structures that produce sex cells or **gametes**. **Anthers** produce **pollen grains** (sperm) and **ovaries** produce **ova**.

5.4.11 Pollination

Before the gametes can fuse together (**fertilisation**) to make a new plant they need to find each other. First contact, or **pollination**, is achieved by the pollen grains landing on the **stigma**. Some plants pollinate themselves (**self-pollination**) and others require **cross-pollination**; that is, pollination involving others. Cross-pollination may involve not only other plants of the same species, but sometimes assistance from other species.

Parts of a typical flower. Note the presence of pollen-producing and egg-producing organs in the same flower.

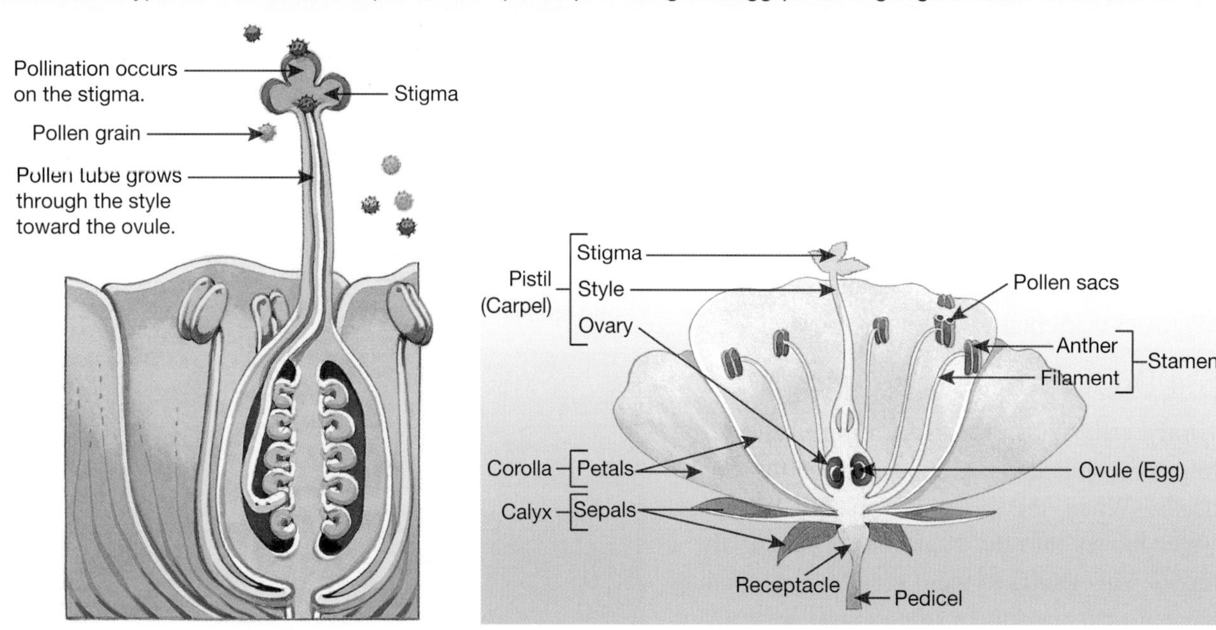

5.4.12 Wind-pollinated

Pollen is transferred between some plants by the wind (**wind pollination**). The flowers of plants that use this type of pollination are usually not brightly coloured and have a feathery stigma to catch pollen grains which were previously held on stamens exposed to the wind.

5.4.13 Animals as vectors

The flowers of plants that use animals as **vectors** to carry their pollen between plants are often brightly coloured and may reward the animal with food. In some cases, the reward may be sugar-rich nectar or protein-rich pollen. In other cases, the reward may have more of a sexual nature as some plants have evolved over time to mimic the sexual structures of their vector's potential mate.

5.4.14 Insect-pollinated

Flowers that are pollinated by insects (**insect pollination**) are often blue, purple or yellow (colours that insects can see), possess a landing platform, have an enticing scent or odour and contain nectaries offering a food supply to these hungry **pollinators**. When the insects visit their next sweet treat of nectar on another flower, they transport pollen from their previous visit to the stigma of their new meal provider.

5.4.15 Bird-pollinated

Flowers that are pollinated by birds are often red, orange or yellow (colours recognised by hungry birds as food) and possess petals in a tubular shape, with nectaries usually inside the base of the flower. As in insect pollination, the birds carry pollen from one meal to their next.

INVESTIGATION 5.5

Stem transport systems

AIM: To observe how water moves up celery stems

Materials:

celery stick (stem and leaves)
knife
two 250 mL beakers
water
blue food colouring
red food colouring
hand lens

Method, results and discussion

- Slice the celery along the middle to about halfway up the stem.
- Fill two beakers with 250 mL of water. Colour one blue and the other red with the food colouring.
- Place the celery so that each side of the celery is in a separate beaker.
- Leave for 24 hours and then observe the celery.
- Cut the celery stick across the stem.
- Use the hand lens to look at the inside of the stem.

1. Look at where the water has travelled in the celery. Draw a diagram to show your observations.
2. Draw a diagram to show what you can see when you cut across the stem.
3. Where is the differently coloured water found in the stem?
4. Where are the different colours found in the leaves?
5. Draw a diagram of the whole celery stick and trace the path of the water through each side to the leaves.
6. How could you turn a white carnation blue? Try it.

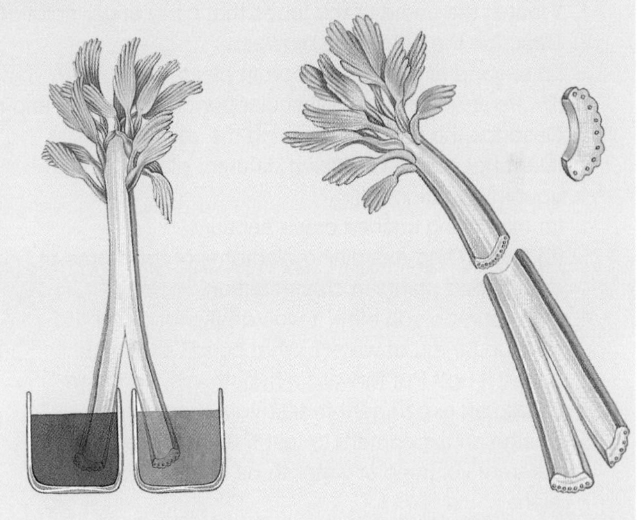

Evaluation

7. Did you encounter any difficulties or problems in this investigation?
8. Suggest ways in which you could improve your investigation if you were to repeat it.
9. Suggest a hypothesis that you could test using similar equipment.

INVESTIGATION 5.6

Observing leaf epidermal cells

AIM: To observe stomata and guard cells in leaf epidermal cells

Materials:

leaf
microscope
clear sticky tape
microscope slide

Background

You can make a slide of leaf epidermal cells with sticky tape.

Method and results

- Put some sticky tape over a section of the underside of a leaf.
- Press the sticky tape firmly onto the leaf.
- Tear the tape off. Some of the lining cells should come off with the sticky tape.
- Press the tape, sticky side down, onto a microscope slide.
- View the sticky tape under the microscope.
- Try to find a pair of guard cells and one of the stomata.

1. Is the stoma (the opening) open or closed?
2. Make a drawing of a group of cells, including the guard cells. Include as much detail in your drawing as possible.
3. Label the guard cells and stomata.
4. Date your drawing and give it a title. Write down the magnification used.

Discuss and explain

5. Summarise your observations.
6. Suggest a hypothesis that you could test using similar equipment.
7. Propose a research question that could be investigated.

5.4 Exercises: Understanding and inquiring

To answer questions online and to receive **immediate feedback** and **sample responses** for every question, go to your learnON title at www.jacplus.com.au. *Note:* Question numbers may vary slightly.

Remember

1. What is the name of the tubes that carry sugar solution around the plant?
2. Describe the difference between:
   (a) sugar and water transport in plants
   (b) the arrangement of vascular bundles in dicots and monocots.
3. Describe the patterns in which the vascular tissue is arranged in the stems of different plants. Obtain your information by:
   (a) examining stained cross-sections
   (b) finding and examining diagrams of the stems of different plants in cross-section.
4. How long do you think it would take for a plant to take up 50 mL of water? What conditions might speed it up? Put forward a hypothesis, and then design an experiment to test your hypothesis.
5. Design an experiment to test the time taken for different volumes of water to be taken up by the plant.
6. On which part of the plant are stomata usually found? Can you suggest why?
7. Describe how the guard cells assist the plant in controlling water loss.
8. Suggest why plant roots have small hairs.
9. Label the figure on the right using the following labels: cuticle, vascular bundle, water loss through stomatal pore, xylem, chloroplast, upper epidermis.

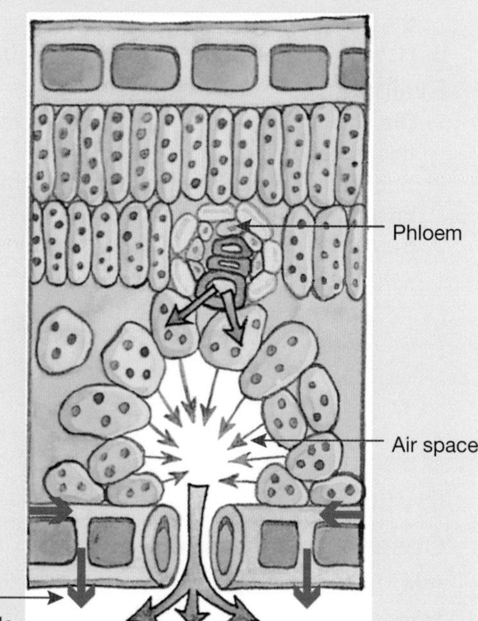

Investigate

10. Some plants have special features that help them reduce water loss. Some leaves have a thick, waxy layer (cuticle). Others have a hairy surface or sunken stomata. Plants that are able to tolerate extremely dry environments are called *xerophytes*. Find out some ways in which plants in dry environments, such as deserts, reduce their water loss. Present your information on a poster or as a model.
11. Design an experiment to measure the amount of water lost through the leaves of a plant.
12. Place a plastic bag over the leaves of plants growing in the school grounds. Seal the bag and record the amount of water collected over 24 hours. What conclusions can you draw from your results?
13. Find out more about the use of paid domestic honeybees in Australia to pollinate crops, and issues related to feral honey bees incidental pollination.
14. Some plants are described as being grown hydroponically. Investigate what this means and reasons for it.
15. (a) Suggest why flowers pollinated by birds are generally red, orange or yellow, not strongly scented and tubular in shape.
    (b) Use the internet and other resources to see if your suggestions were correct. Report back on your findings.
    (c) Investigate and report on Australian research into bird-pollination of plants.
    (d) On the basis of your research, propose two relevant questions that could be investigated and collate these with questions from other students in the class.
    (e) From the class question databank produced in part (d), select a question and research possible answers, sharing your findings with others.
16. Suggest why flowers pollinated by insects are blue, yellow or purple, can be scented and contain nectaries at the base. Report your findings.
17. There have been claims that bee populations are declining around the world and that their pollination of flowers may be becoming unsynchronised with the life cycle of the plants. Some suggest that the culprit is climate change. Investigate these claims and construct a PMI chart that shows support for the claim, support against the claim and interesting points relevant to the claim. What is your opinion? Do you agree or disagree with the claim? Justify your response.

Create

18. In a group, write and then act out a play or simulation of the way water moves through a plant.
19. Write a story about a group of water molecules that travels from the soil, through a plant and then into the atmosphere as water vapour.

5.5 Photosynthesis

5.5.1 Solar powered

Did you know that life on Earth is solar powered? The source of energy in all ecosystems on Earth is sunlight. Plants play a very important role in catching some of this energy and converting it into a form that both they and other organisms can use.

5.5.2 Why are plants called producers?

Photosynthetic organisms such as plants, algae and phytoplankton are called **producers** or **autotrophs** because they can produce and use their own food. They use light energy to make complex, energy-rich organic substances from simpler inorganic substances (such as carbon dioxide and water).

This process of capturing light energy and its conversion into chemical energy is called **photosynthesis** because it involves using light energy to synthesise glucose. Once it is in this chemical form, it can be used as food, stored as starch or converted into other organic compounds.

Solar energy coming into Earth has various fates. What percentage of this energy is immediately reflected out? In what form is the incoming energy? In what form is the major outgoing energy?

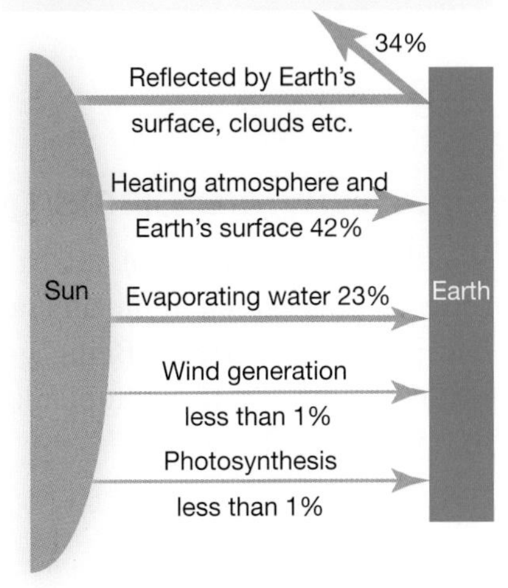

5.5.3 Photosynthesis

The light energy captured by chlorophyll provides energy to split water (H_2O) molecules into oxygen and hydrogen. The oxygen is released as oxygen (O_2) gas into the atmosphere through the stomata. The hydrogen combines with carbon dioxide (CO_2) obtained through stomata from the atmosphere to make glucose ($C_6H_{12}O_6$).

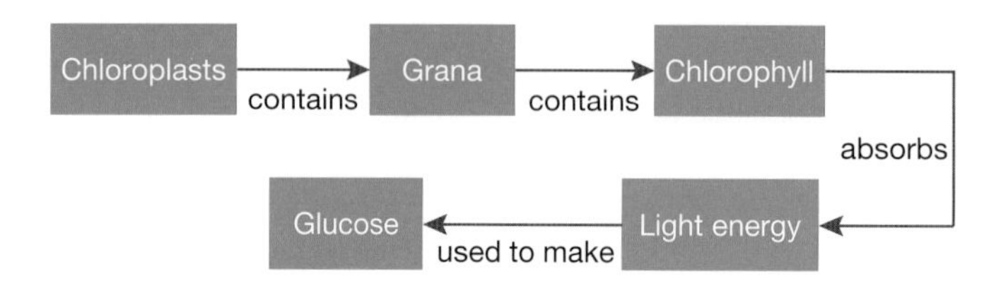

An overall chemical reaction for photosynthesis can be written as:

$$\text{carbon dioxide} + \text{water} \xrightarrow[\text{chlorophyll}]{\text{visible light energy}} \text{glucose} + \text{oxygen} + \text{water}$$

It can also be represented in chemical symbols as:

$$6CO_2 + 12H_2O \xrightarrow[\text{chlorophyll}]{\text{visible light energy}} C_6H_{12}O_6 + 6O_2 + 6H_2O$$

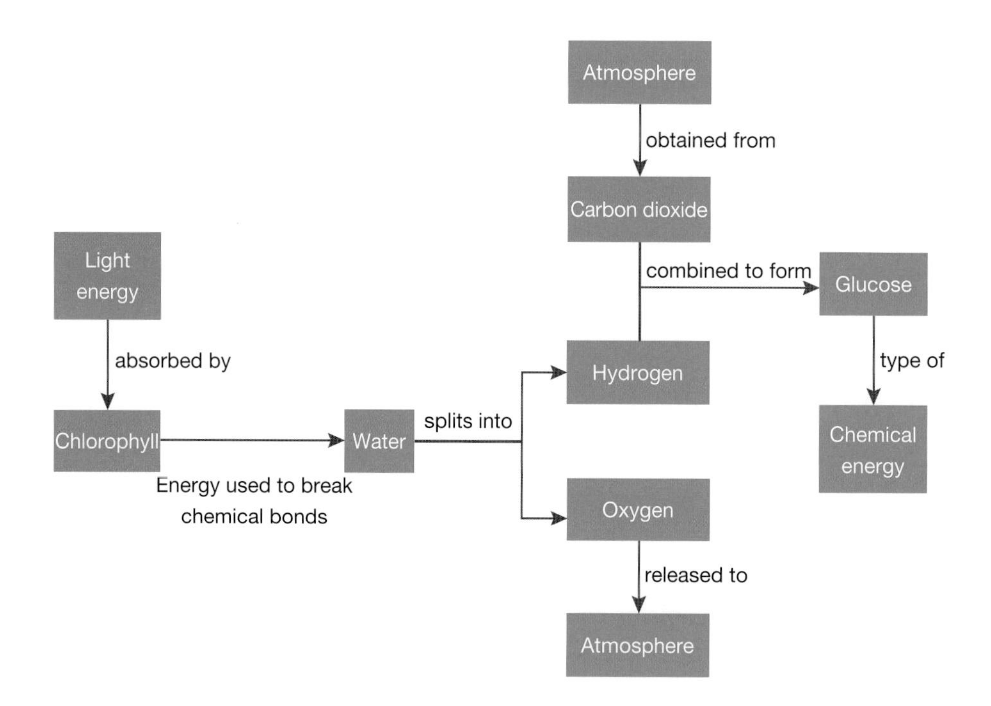

5.5.4 Why are plants green?

Visible light consists of all of the colours of the rainbow! Of the whole spectrum, chlorophyll reflects only green light and absorbs other wavelengths of light (colours). It is for this reason that plants look green. Being green, however, is not essential to be able to photosynthesise. Some plants — algae and phytoplankton, for example — may contain light-capturing pigments that are red, yellow or brown.

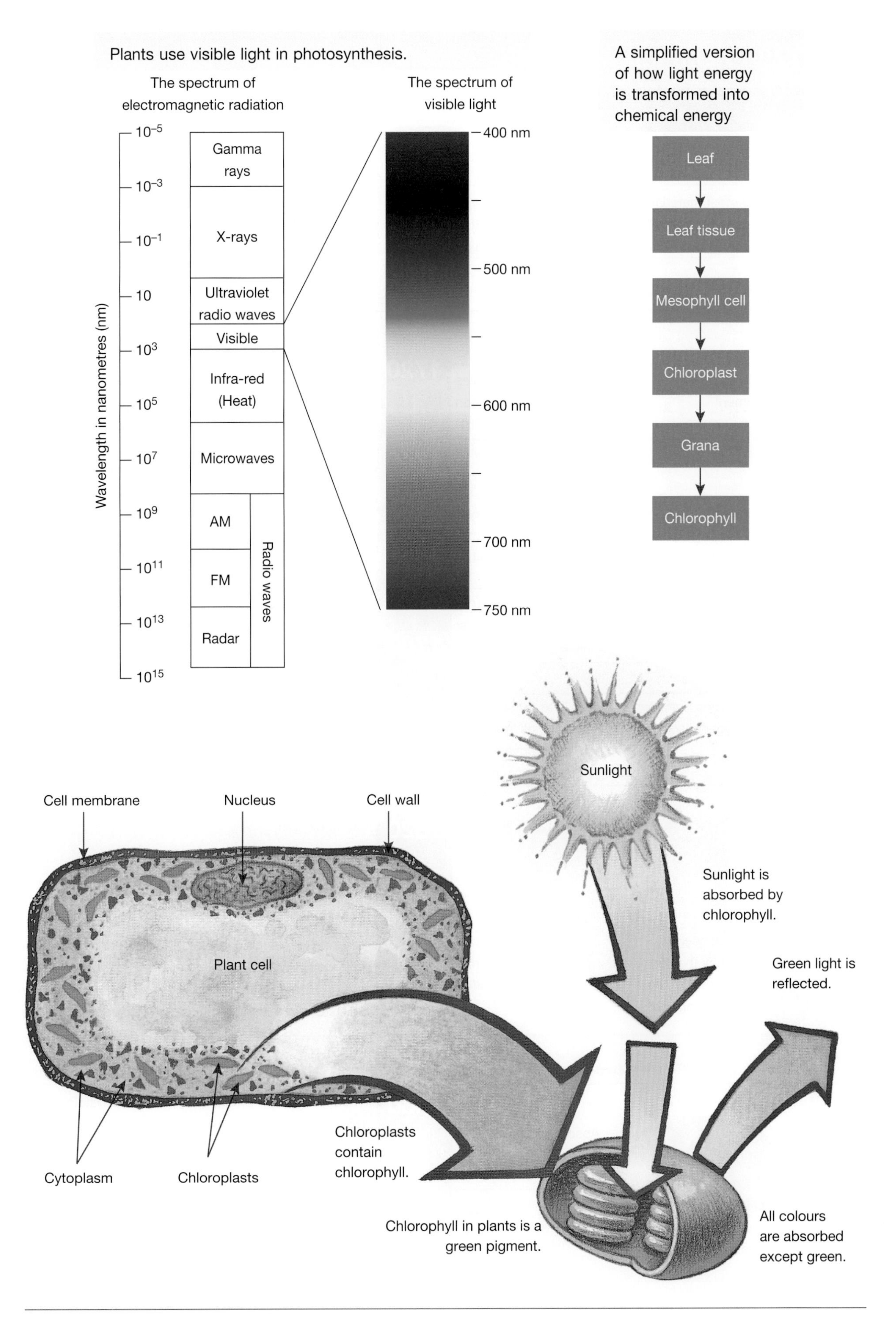
Plants use visible light in photosynthesis.
The spectrum of electromagnetic radiation
The spectrum of visible light
Wavelength in nanometres (nm)
10–5
10–3
10–1
10
10^3
10^5
10^7
10^9
10^11
10^13
10^15
Gamma rays
X-rays
Ultraviolet radio waves
Visible
Infra-red (Heat)
Microwaves
AM
FM
Radar
Radio waves
400 nm
500 nm
600 nm
700 nm
750 nm
A simplified version of how light energy is transformed into chemical energy
Leaf
Leaf tissue
Mesophyll cell
Chloroplast
Grana
Chlorophyll
Sunlight
Cell membrane
Nucleus
Cell wall
Plant cell
Sunlight is absorbed by chlorophyll.
Green light is reflected.
Cytoplasm
Chloroplasts
Chloroplasts contain chlorophyll.
Chlorophyll in plants is a green pigment.
All colours are absorbed except green.

HOW ABOUT THAT!

Discovery journal of photosynthesis

1450 — Nicholas of Cusa, German cardinal (1401–1464): Proposed idea that weight gained by plants is from water, not earth.

Jan Baptista van Helmont, Dutch physician (1577–1644): Demonstrated most material in a plant's body does not come from soil; suggested it comes from water.

1700

1720 — Stephen Hales, British physiologist/clergyman (1677–1761): Suggested plants get some nourishment from air.

1740

Joseph Priestley, British chemist/clergyman (1733–1804): Showed that plants could 'restore' air injured (by respiration).

Priestley's experiment

Burning candle floating on cork

Candle goes out.

Add green plant.

Later the candle can burn again.

Mouse with green plant survives.

Mouse alone dies.

1760

Jan Ingenhousz, Dutch physician (1730–1799): Showed that plants need sunlight to restore 'injured air' and that only the green parts do this; all parts of plants 'injure' air (i.e. respire).

1780

Various European chemists (late 18th century): Oxygen discovered and identified as 'restored' air, carbon dioxide discovered and identified as the 'injured' air.

1800 — Jean Senebier, Swiss minister (1742–1809): Plants use carbon dioxide dissolved in water as food.

Leaves in water without carbon dioxide give off no oxygen.

Leaves in water with carbon dioxide give off oxygen.

Oxygen-enriched air

1820

1840 — Julius von Sachs, German botanist (1832–1897): Discovered plant respiration, and that chlorophyll is found in chloroplasts; showed starch grains form during photosynthesis, and plants take in and use minerals from the soil; showed minerals were required for making chlorophyll.

Maize seedling held by the cork, with roots in the culture solution

1860

1880 — Theodor Wilhelm Engelmann, German physiologist (1843–1909) Showed that oxygen was produced by chloroplasts; and that red and blue light are the most important wavelengths for photosynthesis.

Light

Reflected light

Absorbed light

Chloroplast

Transmitted light

1900

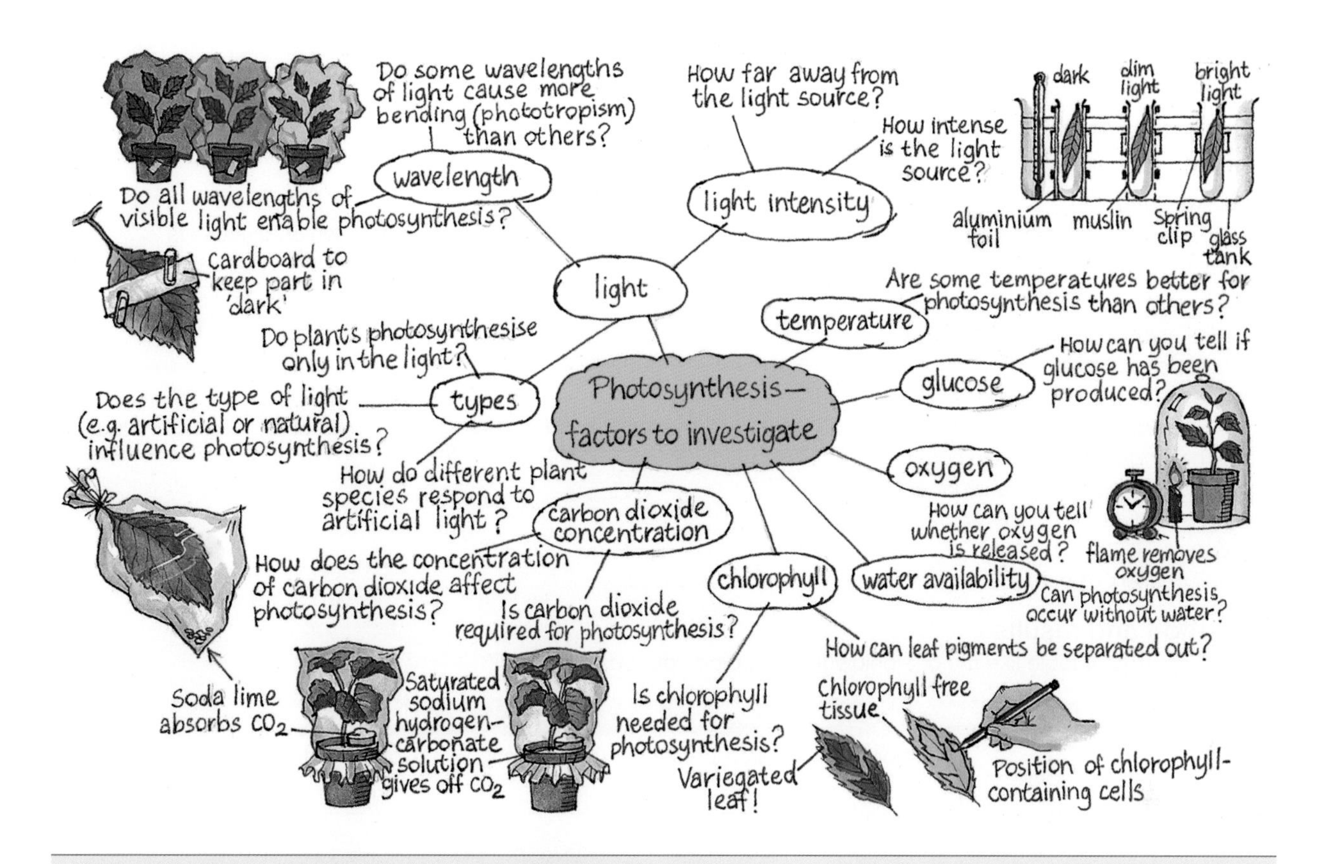

INVESTIGATION 5.7

Looking at chloroplasts under a light microscope

AIM: To observe chloroplasts under a light microscope

Materials:

tweezers
water
dilute iodine solution
moss or spirogyra
light microscope, slides, coverslips

Method and results

- Using tweezers, carefully remove a leaf from a moss plant or take a small piece of spirogyra.
- Place the leaf in a drop of water on a microscope slide and cover it with a coverslip.
- Use a light microscope to observe the leaf.
- Put a drop of dilute iodine solution under the coverslip. (Iodine stains starch a blue–black colour.)
- Using the microscope, examine the leaf again.

Discuss and explain

1. Draw what you see.
2. Label any chloroplasts that are present.
3. Describe the colour of the chloroplasts.
4. What gives chloroplasts their colour?
5. (a) Did the iodine stain any part of the leaf a dark colour?
   (b) If so, what does this suggest?
6. What conclusions can you make about chloroplasts?
7. Identify the strengths of this investigation.
8. Suggest improvements to the design of this investigation.
9. Suggest a hypothesis that could be investigated using similar equipment. (You may use internet research to identify relevant problems to investigate.)
10. Design an experiment to test your hypothesis. Include an explanation for your choice and treatment of variables.
11. Share and discuss your suggested hypothesis and experimental design with others and make any refinements to improve it.

INVESTIGATION 5.8

Detecting starch and glucose in leaves

AIM: To detect glucose and starch in plant leaves

Background

Glucose can be detected with a chemically sensitive paper. The polysaccharide starch, which glucose is converted into for storage, is detected by iodine.

Materials:

iodine solution in a dropper bottle
1% starch solution
white tile or blotting paper
leaves from seedlings or plants of one type (geranium, hydrangea, lettuce, spinach or silverbeet cuttings are good)
glucose indicator strip with colour chart
1% glucose solution in a dropper bottle
mortar and pestle
sand
small beakers or petri dishes for testing different substances.

Method and results

1. Construct a table like the one below for recording your observations.

Item tested	Iodine test		Glucose test	
	Colour	Starch present?	Colour	Concentration of glucose

Testing leaves for starch

- To observe the effect of iodine solution on starch, place a few drops of starch solution on a piece of blotting paper or a white tile. Add a few drops of iodine.
- Soften two or three leaves by dipping them with tongs into hot water for 10 seconds.
- Repeat the test with the softened leaves. Keep one leaf aside that is not tested with iodine to compare it with the leaves that you test.

2. Record the colour observed and the presence of starch in your table.

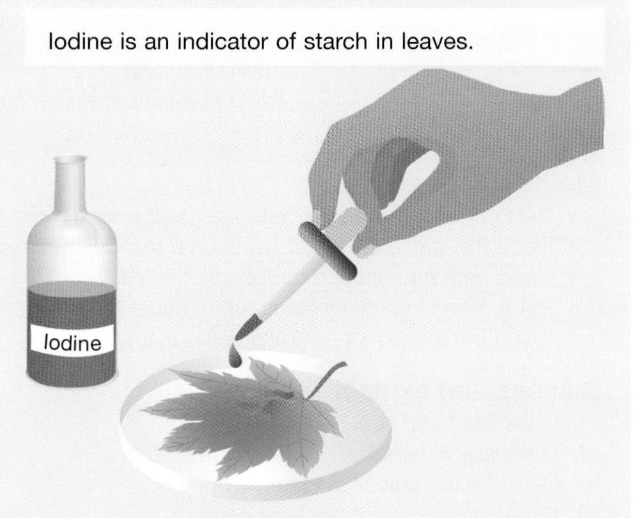

Iodine is an indicator of starch in leaves.

Testing leaves for glucose

- To observe the effect of glucose on the glucose indicator strip, place a drop of glucose solution on the end of the strip on a white tile.
- Use the chart of colours to determine the concentration of the glucose.
- Using the mortar and pestle, grind some fresh leaves with a little water and a sprinkle of sand.
- Allow a strip of glucose indicator paper to soak up the liquid.

3. Record the colour and glucose concentration in your table.

Discuss and explain

4. Describe the effect of the iodine on the starch solution.
5. Describe the effect of the glucose solution on the indicator strip.
6. What do your results suggest about the way energy is stored in leaves?
7. Why was sand added to the mixture in the mortar?

8. The sand does not affect the result on the indicator strip. How could you show this?
9. Identify the strengths of this investigation.
10. Suggest improvements to the design of this investigation.
11. Suggest a hypothesis that could be investigated using similar equipment. (You may use internet research to identify relevant problems to investigate.)
12. Design an experiment to test your hypothesis. Include an explanation for your choice and treatment of variables.
13. Share and discuss your suggested hypothesis and experimental design with others and make any relevant refinements to improve it.

INVESTIGATION 5.9

Out of the light

AIM: To investigate differences in a plant's production of starch when light is removed

CAUTION

Ethanol is flammable. Do not place it near a naked flame.

Materials:
pot plant that has been kept in the dark for a few days
several strips of aluminium foil
scissors and sticky tape
hotplate
500 mL beaker of boiling water
test tube of ethanol
forceps
iodine solution and dropping pipette
Petri dish
watchglass with a small sample of potato starch

Make sure that the aluminium strips are secured, and that you do not damage the leaf.

Method and results

- Fix aluminium strips to one leaf of a plant as shown in the figure at right. Make sure that both sides of the leaf are covered by the strip and that you do not damage the leaf.
- Leave the plant in the light for 3 days.
- Remove the leaf from the plant and take off the foil.
- Dip the leaf into boiling water for 10 seconds, then place it in a test tube of ethanol.
- Stand the test tube in the beaker of hot water and leave for 10 minutes. This treatment will remove the chlorophyll.

1. While the leaf is in the ethanol, test a small sample of potato starch on a watchglass with the iodine solution. Note any colour change.

- Remove the leaf from the ethanol with the forceps and dip it into the hot water in the beaker again to remove any excess ethanol.

2. Place the leaf into a Petri dish and cover with iodine solution. Note any colour change and where on the leaf any such change occurred.

Discuss and explain

3. Glucose is produced during photosynthesis and is then converted to starch and stored. Did your test show any differences in starch production between the sections of leaf exposed to the light and the sections kept in the dark?
4. Which variable has been investigated in this experiment?
5. Why was the plant kept in the dark for a few days prior to the experiment?
6. What inferences (suggested explanations) can you make from your observations?
7. What is the control in this experiment?

Evaluate

8. Identify the strengths of this investigation.
9. Suggest improvements to the design of this investigation.
10. Suggest a hypothesis that could be investigated using similar equipment. (You may use internet research to identify relevant problems to investigate.)
11. Design an experiment to test your hypothesis. Include an explanation for your choice and treatment of variables.
12. Share and discuss your suggested hypothesis and experimental design with others and make any suitable refinements to improve it.

5.5 Exercises: Understanding and inquiring

To answer questions online and to receive **immediate feedback** and **sample responses** for every question, go to your learnON title at www.jacplus.com.au. *Note:* Question numbers may vary slightly

Remember

1. Identify the source of energy for all ecosystems on Earth.
2. (a) Name the green pigment that can capture light energy.
   (b) Name the structure in which you would find this pigment in a plant.
3. Recall the word equation for photosynthesis.
4. Is being green essential for photosynthesis? Explain.
5. Identify an example of an autotroph and an example of a heterotroph.
6. Use a Venn diagram to compare producers and consumers.
7. Identify the scientist who:
   (a) suggested that plants get some nourishment from air
   (b) proposed that plants use carbon dioxide dissolved in water as food
   (c) discovered that chlorophyll was found in chloroplasts
   (d) showed that oxygen was produced in chloroplasts.

Think and discuss

8. How can you test whether photosynthesis has occurred in all parts of a leaf?
9. If you were testing a leaf for carbon dioxide and enclosed it in a plastic bag with soda lime in it, why would you also put the control plant without soda lime in a plastic bag?
10. Starch found in a leaf is used as evidence of photosynthesis in the leaf. Where else might the starch have come from?
11. If you were measuring the effect of light intensity, why would you also need a thermometer?
12. Name the chemical produced in photosynthesis that contains chemical energy for the plant.
13. Explain how it can be said that photosynthesis is the reverse reaction of aerobic respiration.
14. During a 24-hour period of day and night, when will a plant be respiring and when will it be photosynthesising?
15. Give two examples of movements that plants make.
16. Apart from the production of food, how are plants important to life on Earth?

Investigate

17. Refer to the *Discovery journal of photosynthesis* in this subtopic to answer the following questions. Which scientists made discoveries and when did they make them? Were their ideas accepted immediately? Which ideas did their new ideas replace?
18. Select one of the questions about photosynthesis experiments in the diagram above Investigation 5.7. Design (and if possible perform and report on) an experiment to try to find the answer or more information about it.

19. The diagram at right shows a chloroplast and a mitochondrion in a plant cell.
    (a) Which energy conversion takes place in the chloroplast?
    (b) The arrows on the diagram at right show the flow of energy and substances into and out of the cell. Choose words from the box below that are represented on the diagram by the letters A–H.

Water	Oxygen	Sun
Carbon dioxide	Heat energy	Chemical energy
Light energy	Glucose	

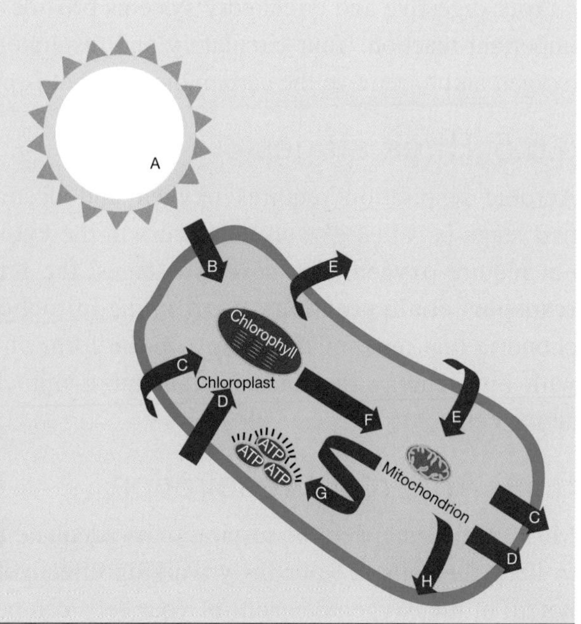

20. If someone said to you: 'If all photosynthesis on Earth stopped, humans would eventually become extinct', would you agree or disagree? Justify your answer.
21. Investigate and report on the hottest creatures on Earth that live in deep-sea hydrothermal vents. In your report, describe the creatures and how they survive.

5.6 Cellular respiration

5.6.1 Taking what we need

Unlike plants, animals cannot convert light energy into chemical energy. Our energy and nutritional demands are met by taking in or consuming other organisms. That is why we and other organisms with this need are called consumers or heterotrophs.

When a consumer eats another organism, not all of the chemical energy is used to form new tissues or stored for later use. Humans can store some of the unused energy as glycogen in their liver or as **fat** in fatty tissue beneath the skin. Some of the chemical energy is also converted into other forms; for example, some of it is released as heat.

5.6.2 Cells are busy places

The conversion of chemical energy and the growth and repair of cells result from thousands of chemical reactions that occur in your body. These reactions are known collectively as **metabolism** and occur in all living things.

5.6.3 Cells need energy

All living things respire. **Cellular respiration** is the name given to a series of chemical reactions in cells that transforms the chemical energy in food into **adenosine triphosphate (ATP)** — this is a form of energy that the cells can use. The energy in ATP can later be released and used to power many different chemical reactions in the cells.

5.6.4 Aerobic respiration

Aerobic respiration is the process that involves the breaking down of glucose so that energy is released in a form that your cells can then use. The overall equation for cellular respiration is shown below:

$$\text{Glucose} + \text{oxygen} \xrightarrow{\text{enzymes}} \text{carbon dioxide} + \text{water} + \text{energy}$$

$$C_6H_{12}O_6 + 6O_2 \xrightarrow{\text{enzymes}} 6CO_2 + 6H_2O + \text{energy}$$

Your digestive and circulatory systems provide your cells with the glucose that is required for this very important reaction. Your circulatory and respiratory systems also work together to supply your cells with oxygen and to remove the carbon dioxide that is produced as a waste product.

5.6.5 Three stages

Aerobic respiration requires oxygen and occurs in three stages. The first stage is called **glycolysis**, occurs in the **cytosol** of the cell and does not require oxygen. The next two stages, the **Krebs cycle** and **electron transport chain reactions**, occur in the **mitochondria**. It is in the mitochondria that most of the energy, in the form of ATP, is produced. Cells with high energy demands contain more mitochondria than other less active cells.

Mitochondria (shown in red) are the powerhouses of your cells.

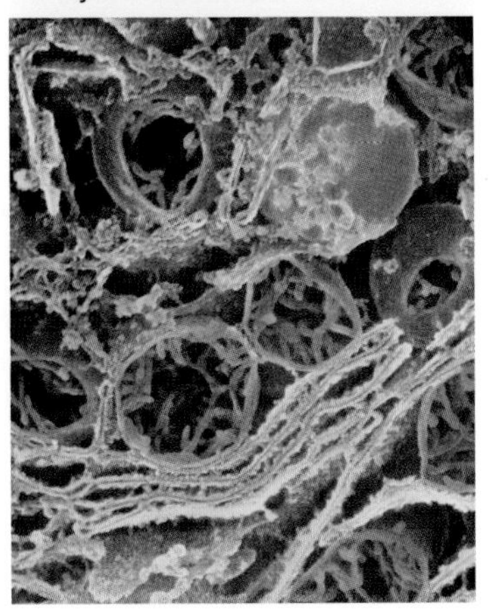

5.6.6 Anaerobic respiration

Most of the time, aerobic respiration is adequate to supply enough energy to keep the cells in your body working effectively. Sometimes, however, not all of the oxygen demands of your cells can be met.

Your muscle cells have the ability to respire for a short time without oxygen using **anaerobic respiration**. Using this reaction, glucose that has been stored in your muscle cells is converted into **lactic acid**.

5.6.7 What's your end?

While the end products of aerobic respiration are always the same, the end products of anaerobic respiration can be different, depending on the organism. Humans and other animals produce lactic acid (or **lactate**), whereas plants and yeasts produce **ethanol** and carbon dioxide.

Although less energy is produced in anaerobic respiration (2 ATP) than in aerobic respiration (36–38 ATP), it is produced at a faster rate. This is very helpful when a quick burst of energy is needed for a short time.

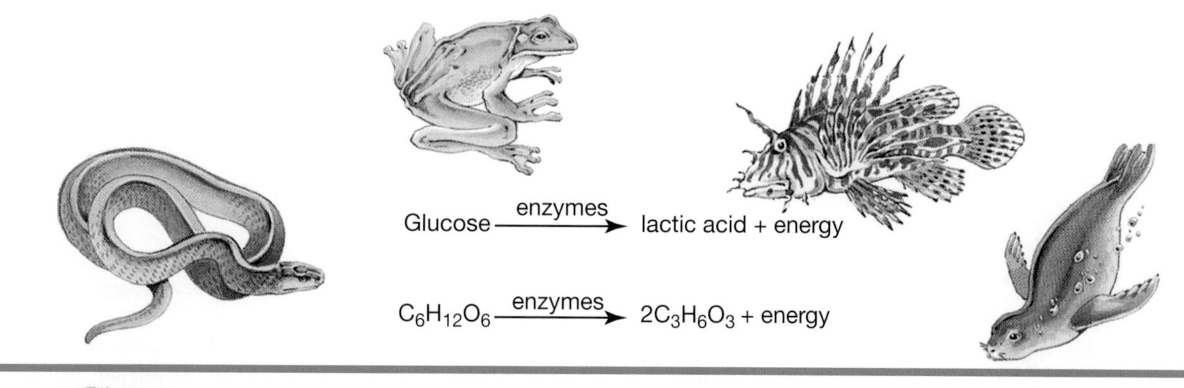

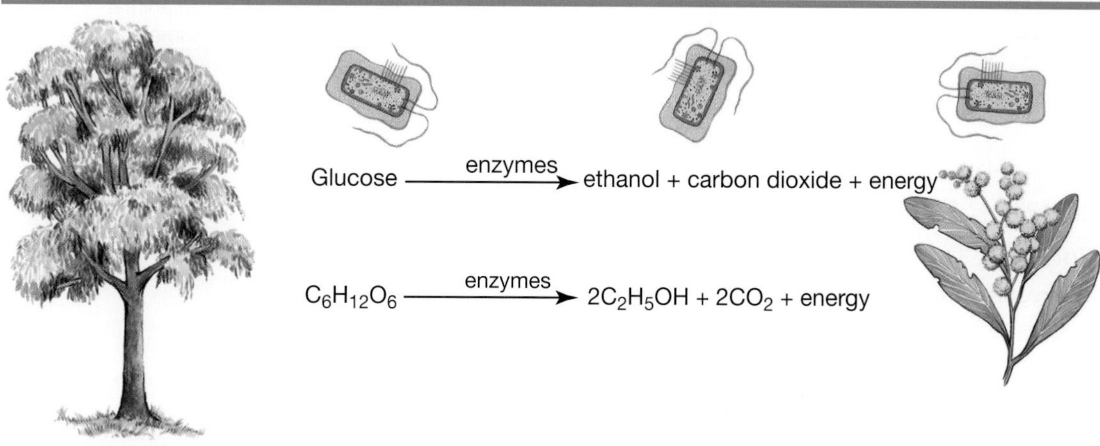

Features of aerobic and anaerobic respiration

Question	Aerobic respiration	Anaerobic respiration
Is oxygen required?	Yes	No
What is glucose broken down into?	Carbon dioxide + water + energy	Lactic acid or ethanol + carbon dioxide + energy
For how long can the reaction occur?	Indefinitely	A short time only
Is the energy transfer efficient?	Yes	No
How fast is ATP production?	Slow	Fast
How many molecules of ATP are produced in each reaction?	36	2
What are the end products?	All organisms: carbon dioxide and water	Animals: lactic acid Plants and yeasts: ethanol and carbon dioxide
About how much energy is released per gram of glucose?	16 kJ	1 kJ

HOW ABOUT THAT!

Some micro-organisms respire only anaerobically, and can die in the presence of oxygen. These are referred to as **obligate anaerobes**. The bacterium *Clostridium tetani* is an example of an organism that can thrive only in the total absence of oxygen. At the site of a wound, tissue necrosis (death) provides a locally anaerobic environment in which these bacteria can grow. This bacterium releases toxins which cause tetanus (or lockjaw), a painful condition in which muscles remain contracted.

How did we find out about the link between respiration and oxygen? Robert Boyle (1627–1691) performed experiments that showed that something in air was needed to keep a candle burning and an animal alive. Joseph Priestley (1733–1804) took Boyle's experiment a step further and showed that plants produced a substance that achieved this.

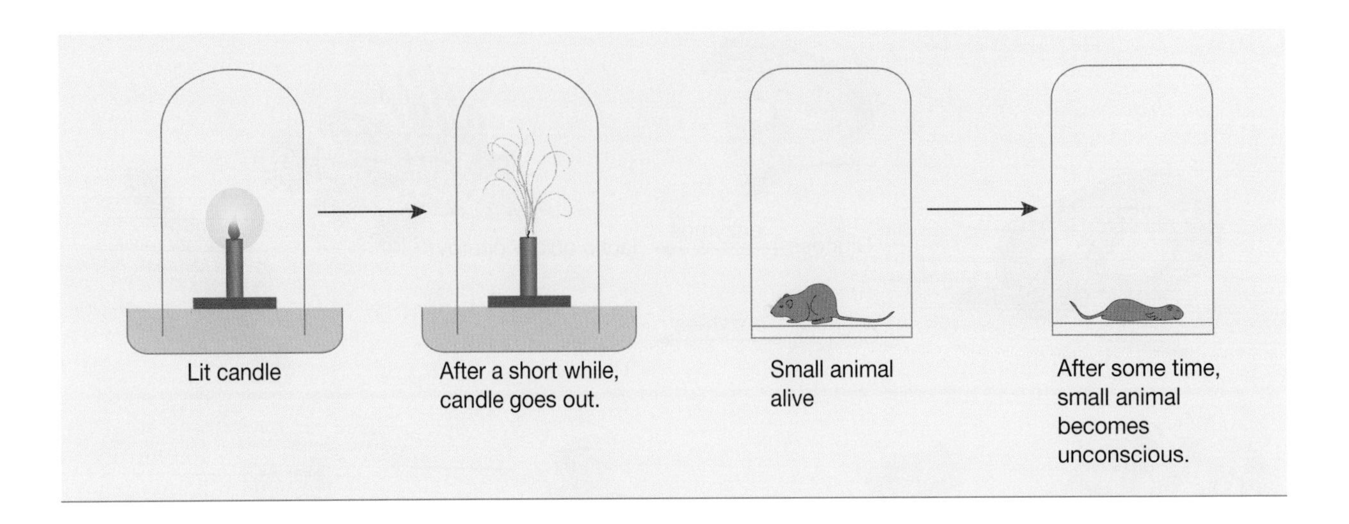

INVESTIGATION 5.10

Fermenting fun

AIM: To observe the process of fermentation

Materials:

safety glasses and laboratory coat
large side-arm test tube
rubber stopper to fit side-arm test tube
length of rubber hosing (25 cm) fitted with a short end of glass tubing
test-tube rack
½ spatula full of yeast
½ spatula full of sugar
20 mL of warm water
test tube half filled with water, with a few drops of bromothymol blue added
dropping bottle of bromothymol blue indicator

Method and results

- Set up the apparatus as shown in the diagram below.
- Place the yeast, sugar and 20 mL of water in the side-arm test tube.
- Seal the top of the side-arm test tube with the rubber stopper and connect the rubber hosing between the two test tubes.
- Leave the apparatus set up in a warm place overnight.

1. Next day observe the colour of the bromothymol blue indicator and note the odour in the yeast test tube.

Discuss and explain

2. List the changes that occurred in the yeast test tube.
3. Bromothymol blue changes from blue to green to yellow in the presence of carbon dioxide. What does the colour change you observed suggest?
4. Describe the odour in the yeast test tube.

Experimental set-up for the fermentation of yeast

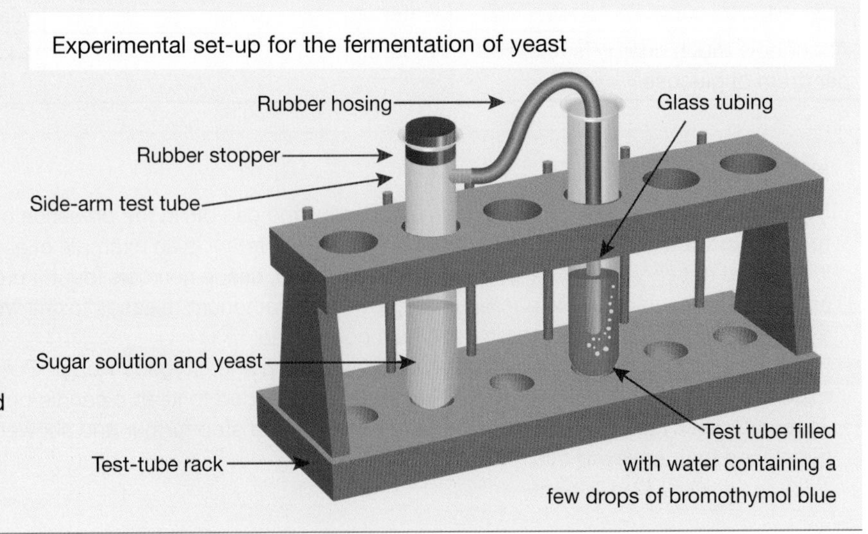

5.6 Exercises: Understanding and inquiring

To answer questions online and to receive **immediate feedback** and **sample responses** for every question, go to your learnON title at www.jacplus.com.au. *Note:* Question numbers may vary slightly.

Remember

1. Suggest why humans are referred to as *consumers* or *heterotrophs*.
2. Describe what happens to chemical energy that is stored in your body.
3. Define the term *metabolism*.
4. Which organisms respire?
5. What does ATP stand for?
6. Outline the importance of ATP.
7. Write a word equation for aerobic respiration.
8. Write the chemical equation for aerobic respiration.
9. Outline the purpose of aerobic respiration.
10. Name the three stages of aerobic respiration.
11. State the location within the cell where each stage of aerobic respiration occurs.
12. Copy and complete the flowchart at right to illustrate the differences between aerobic and anaerobic respiration 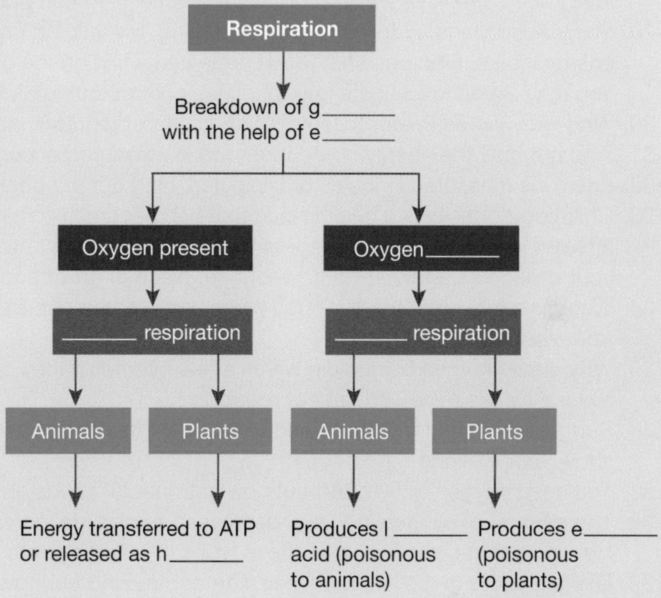
13. Suggest a link(s) between cellular respiration and the:
    (a) digestive system
    (b) circulatory system
    (c) respiratory system.
14. Write the word equation for anaerobic respiration in:
    (a) animals
    (b) plants, yeasts and bacteria.
15. Use a Venn diagram to compare aerobic and anaerobic respiration.

Respiration
Breakdown of g______
with the help of e______
Oxygen present
Oxygen______
______ respiration
______ respiration
Animals
Plants
Animals
Plants
Energy transferred to ATP or released as h______
Produces l______ acid (poisonous to animals)
Produces e______ (poisonous to plants)

Investigate, think and discuss

16. (a) What are the advantages of anaerobic respiration to a person who is a short-distance swimmer?
    (b) What are the disadvantages of anaerobic respiration to the body?
    (c) Which type of respiration, aerobic or anaerobic, is more likely to occur in each of the following activities?
        (i) Ten sit-ups
        (ii) A leisurely walk
        (iii) Lifting a very heavy steel bar above your head
        (iv) Watching TV from the sofa
        (v) A 30-metre sprint to catch the dog
17. The muscles of a sprinter respire anaerobically throughout a race.
    (a) Suggest how the sprinter could compete in a 100 m race without breathing.
    (b) Why does the sprinter need to pant to get extra oxygen at the end of the race?
    (c) Where does the extra oxygen enter the body while panting is used for breathing?
18. Intestinal tapeworms excrete lactic acid directly into their host's gut. Suggest the advantage to the tapeworm of this behaviour.
19. *Clostridium botulinum* is an anaerobe that cannot survive in the presence of oxygen. This microbe causes a potentially lethal form of food poisoning called botulism. Suggest why this bacterium may be a problem in canned foods.

20. Suggest why one type of cellular respiration releases more energy than the other.
21. Find out the methods used to make white wine, red wine, beer and Swiss cheese. Report your findings in the form of a recipe book.
22. How can anaerobic respiration be involved in the pickling of foods such as cabbage, pickles and olives?
23. What is an energy transformation?
24. Does metabolism occur in plants?
25. What can living energy converters do that non-living energy converters cannot?
26. How is the human body similar to a moving motor vehicle?
27. Why is it correct to say that your body burns food?
28. Suggest possible answers to the question below.
    What am I?
    You have to fill me with fuel regularly. The fuel is burned inside me in a chemical reaction in which chemical energy is transformed into mechanical energy. I get very hot when fuel is burned. The faster I go, the more fuel I need. I release waste products that smell and can pollute the environment.
29. Many Australians try to lose mass by dieting. Investigate one of the different types of diets used to reduce energy intake. Find out what foods are eaten while on the diet and explain how it can result in mass loss. You may like to investigate low-fat diets, joule-counting diets or high-fibre diets.
30. Who was James Prescott Joule? Write a short biography about him.
31. Find out how the energy content of food is measured accurately.
32. Energy is measured in joules or kilojoules. Find out the origin of the name given to this unit.
33. Childhood obesity is a health issue that has become prominent recently in Australian news. Search the internet, journals and newspapers to create a collage of headings and key points about this topic. Organise your collage as a PowerPoint, webpage, piece of art or poster.
34. Water is released from your body when you breathe out. Describe two other ways in which water is released from your body.
35. Why would human beings die within a few minutes if they could not breathe, even though they can live for several weeks if they do not eat any food?
36. Suggest the effect of activity and exercise on the rate of aerobic respiration. Use a visual thinking tool to show your thinking.
37. You need oxygen to live; find out how astronauts' needs are met.
38. The amount of oxygen in the air decreases as altitude increases when you go high up in the mountains. How do people who live up in these types of environments get enough oxygen?
39. Oxygen levels in the sea decrease with depth. Find out how deep-sea divers get their oxygen.
40. Prepare a PowerPoint presentation that summarises the information in this subtopic. Present your summary to the class.
41. Find out more about mitochondria and then make a model of a mitochondrion.

5.7 Sustainable ecosystems

5.7.1 Sustainable ecosystems

Communities within ecosystems are made up of populations. Interactions between these populations and their environment enable matter to be recycled and energy to flow through the ecosystem.

5.7.2 Energy flows through ecosystems

Light energy is captured by producers and converted into chemical energy using the process of photosynthesis. Some of this energy is used by the producers themselves, some is released into the atmosphere and some is passed on through food chains to consumers. Energy flows through ecosystems.

Cellular respiration is a process that all living organisms use to convert energy into a form that their cells can use. Glucose and oxygen (the products of photosynthesis) are used in cellular respiration.

Photosynthesis is the process involved in capturing light energy and converting it into chemical energy. Respiration is the process that uses chemical energy produced from photosynthesis for energy to live.

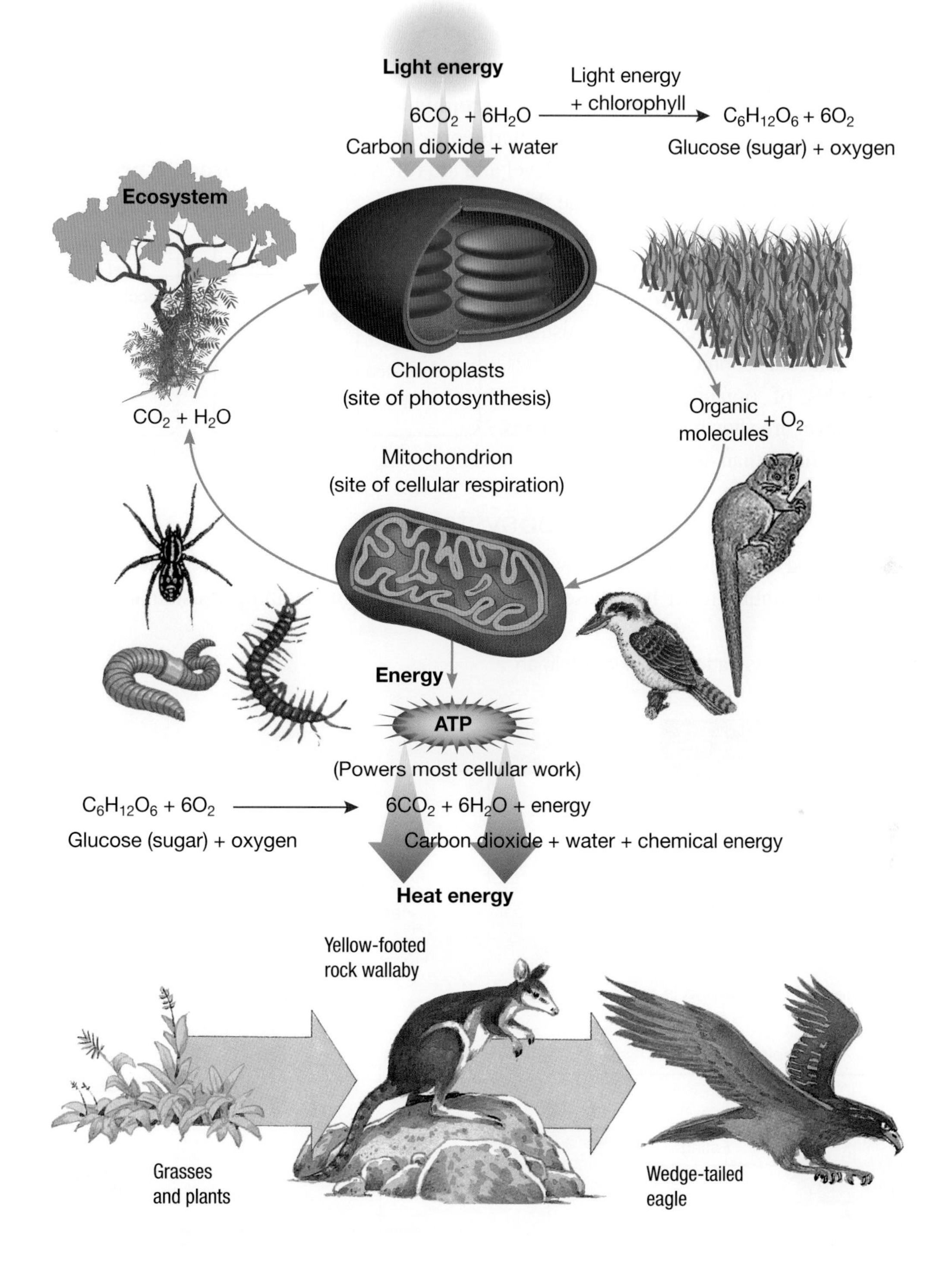

5.7.3 Living in the dark

How can ecosystems exist on the ocean floor, where there is no light for photosynthesis? Rather than being photosynthetic, some organisms are **chemosynthetic**. They use energy released from chemical reactions (rather than light) to produce organic molecules. Examples of these non-photosynthetic producers are autotrophic bacteria such as *Thiobacillus* spp.

5.7.4 Food chains and food webs

Feeding relationships in ecosystems can be described as **food chains** and **food webs**. Food chains show the direction of the flow of energy. Interconnecting or linked food chains make up a food web.

5.7.5 Trophic levels and orders

Within a food chain, each feeding level is called a **trophic level**.

Producers make up the first trophic level and herbivores the next. As herbivores are the first consumers in the food chain, they are at the second trophic level and are referred to as **first-order** or **primary consumers**. Consumers that eat the herbivores are at the third trophic level and are called **second-order** or **secondary consumers**. As only about 10 per cent of the chemical energy is passed from one trophic level to the next, most food chains do not usually contain more than four trophic levels.

Some examples of organisms that could be present at each level are shown at right. Organisms can appear within more than one trophic level.

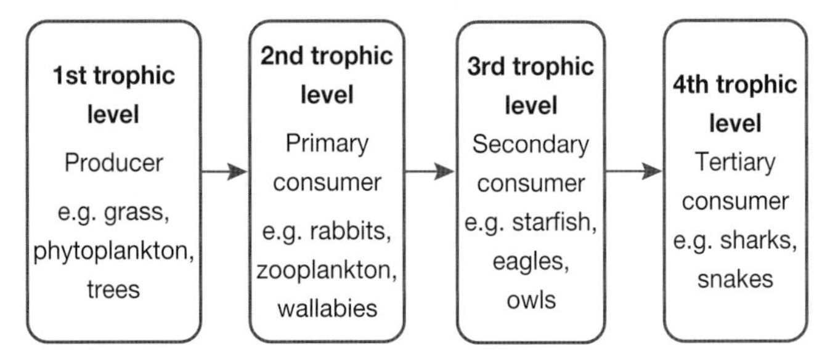

5.7.6 Matter cycles through ecosystems

Food chains and food webs also describe how matter can be recycled through an ecosystem. Carefully observe each of the following figures to see how these relationships assist in maintaining a sustainable ecosystem.

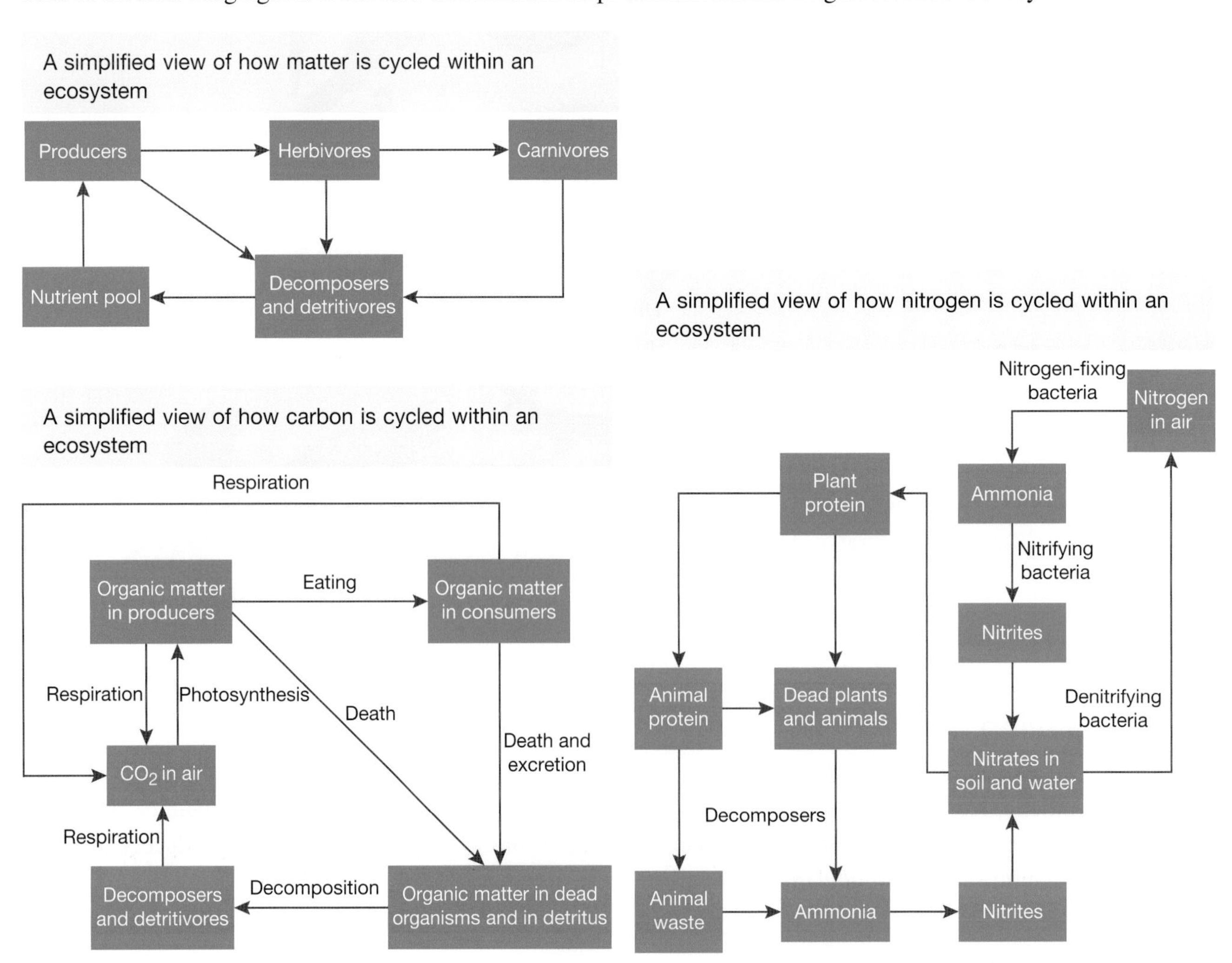

A simplified view of how phosphorus is cycled within an ecosystem

Plant protein → Animal protein
Animal protein → Waste and dead animals
Waste and dead animals → Dissolved phosphate in soil and water
Dissolved phosphate in soil and water → Plant protein
Phosphate in rocks → (Weathering, Erosion) → Dissolved phosphate in soil and water
Shallow sea sediments → (Run-off from rivers) → Dissolved phosphate in soil and water
Shallow sea sediments → Deep sea sediments
Deep sea sediments → (May be trapped for millions of years) → Phosphate in rocks

5.7.7 Ecological pyramids

Ecological pyramids can provide a model that can be used to describe various aspects of an ecosystem. They can show the flow of energy, the recycling of matter through an ecosystem or the numbers of organisms and the relationships between them.

These pyramids are constructed by stacking boxes that represent feeding (or trophic) levels within a particular ecosystem. The size of the box indicates the number or amount of the feature being considered.

5.7.8 Energy pyramids

An **energy pyramid** for a food chain as described below would show a larger box at the bottom and smaller boxes as you move up the food chain. Energy pyramids always have this basic shape, because only some of the energy captured by producers is converted into chemical energy. Of the energy captured, only about 10 per cent is passed on through each feeding level, with about 90 per cent of the energy being transferred to the environment as heat or waste.

This decrease in energy along the food chain is one reason that the numbers of levels in food chains are limited. There is also a limit to the number of organisms that can exist at each level of the food chain. Energy pyramids show that, as you move up the food chain or web, there is less food energy to go around and therefore fewer of each type of organism.

An energy pyramid — only about 10 per cent of the food energy received at each level is passed through to the next; the other 90 per cent is transferred to the environment.

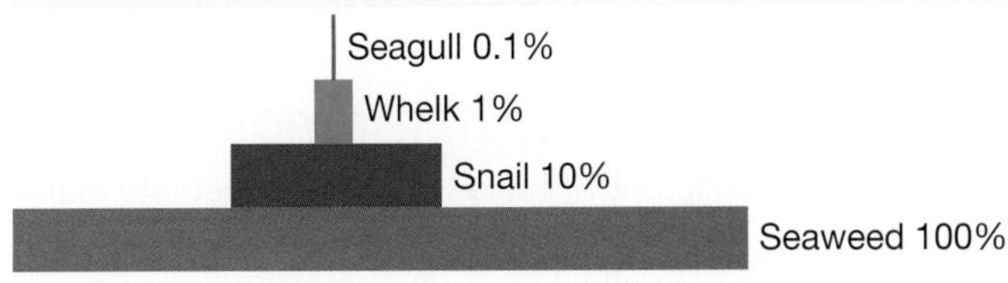

5.7.9 Pyramids of numbers and biomass

A **pyramid of numbers**, as the name suggests, indicates the population or numbers of organisms at each trophic level in the food chain. A **pyramid of biomass** shows the dry mass of the organisms at each trophic level. These pyramids can appear as different shapes due to reproduction rates or mass differences between the organisms.

A pyramid of biomass shows the dry mass of the organisms at each trophic level.

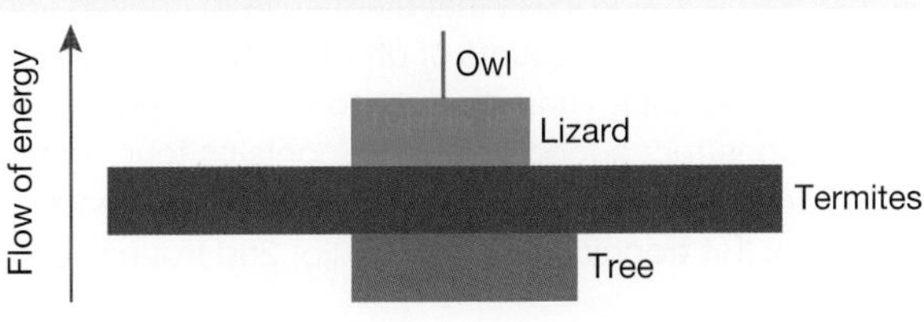

5.7.10 Population growth

The rate at which a population can grow is determined by its **birth rate** minus its **death rate**. The size of the population is also influenced by **immigration** (the number of individuals moving into an area) and **emigration** (the number of individuals leaving an area). It is also influenced by available resources, predators and disease. The overall growth rate can be calculated by the formula:

population growth = (births + immigration) – (deaths + emigration)

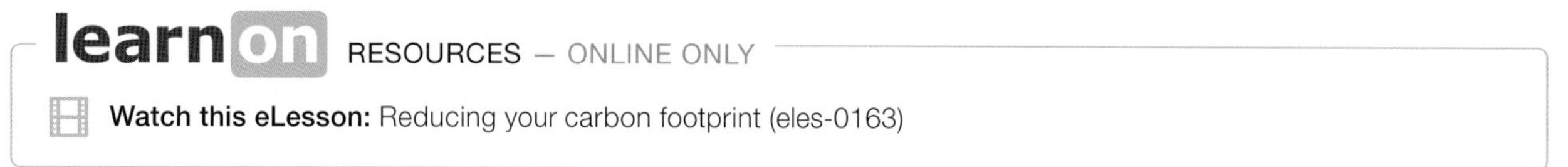

5.7.11 Growth without limits

If you were to provide a population with plenty of food, lack of predators and disease, it would grow rapidly. A bacterium, for example, divides every 20 minutes. Under favourable conditions, that single bacterium would produce a population of 1 048 578 individuals within 7 hours! Graphing this population growth would result in a J-shaped growth pattern known as **exponential growth**.

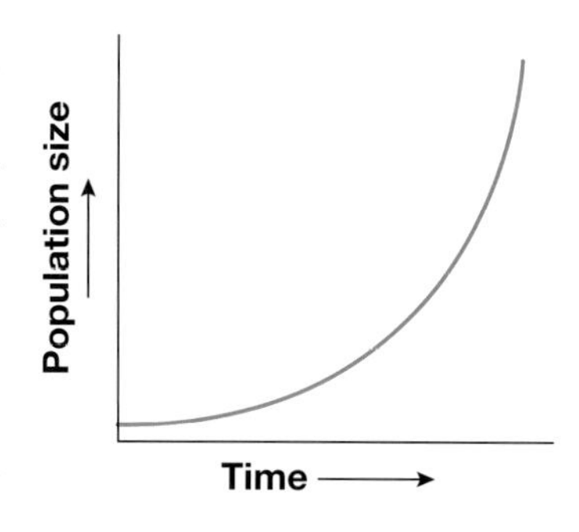

Organisms need particular resources in order to survive. Some of these resources will be in limited supply, and organisms will need to compete with other organisms to get what they need. An ecosystem has limited resources and can carry only a particular number of organisms. This is called its carrying capacity, and is what causes a population to plateau when it reaches a particular size.

5.7.12 Carrying capacity

Populations, however, have only a limited amount of resources and if you were to graph their growth it would look more like an S-shaped, or **sigmoid**, graph. Eventually the population growth would be zero (overall). When this occurs the population is described as having reached a **steady state**, **plateau phase** or **equilibrium**. When the birth and death rates balance each other out, a point of **zero population growth** is reached. A population in its plateau phase contains the maximum number of individuals that its particular environment can carry — it has reached its **carrying capacity**.

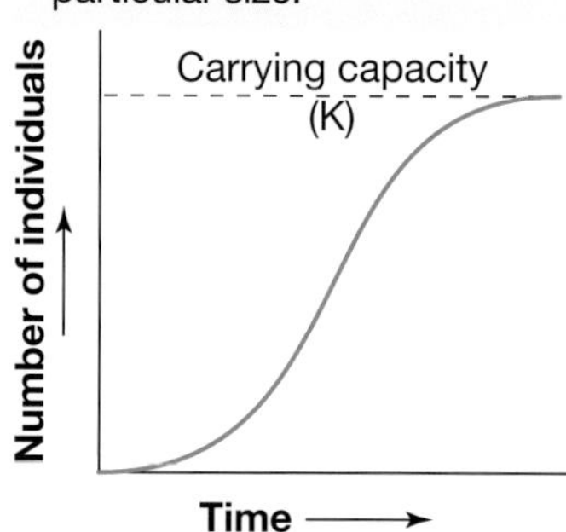

5.7 Exercises: Understanding and inquiring

To answer questions online and to receive **immediate feedback** and **sample responses** for every question, go to your learnON title at www.jacplus.com.au. *Note:* Question numbers may vary slightly.

Remember

1. Rank these terms in order of complexity:
   - ecosystem
   - population
   - community.
2. Name the process that plants use to convert light energy into chemical energy.
3. Identify the products of photosynthesis that are used as inputs for cellular respiration.
4. Construct a Venn diagram to compare photosynthesis and chemosynthesis.
5. Construct a food chain that contains four organisms.
6. Identify an organism that you may find at each of the following trophic levels:
   (a) 1st trophic level (b) 2nd trophic level (c) 3rd trophic level (d) 4th trophic level.

7. Suggest why food chains rarely contain more than four trophic levels.
8. Construct linked flowcharts to show a simplified view of how:
   (a) matter is cycled within an ecosystem
   (b) carbon is cycled within an ecosystem
   (c) nitrogen is cycled within an ecosystem
   (d) phosphorus is cycled within an ecosystem.
9. Describe how ecological pyramids are used to describe various aspects of an ecosystem.
10. State a formula that could be used to calculate an increase in a population.
11. List three factors that may result in:
    (a) an increased population size
    (b) a decreased population size.
12. Describe the difference between sigmoid and exponential growth patterns.
13. Outline the relevance of carrying capacity to population growth.

Think, discuss and investigate

14. Look at the graph at right showing the relationship between photosynthesis and respiration.
    (a) Which line of the graph represents the rate of photosynthesis?
    (b) When would the rate of respiration be greater than the rate of photosynthesis?
    (c) State three similarities between respiration and photosynthesis.
15. Suggest where the following labels fit on the figure below right: producers, decomposers, second-order consumer, eaten, first-order consumer, eaten, death and wastes, respiration.
16. Suggest where the following labels fit on the figure below left: producers, decomposers, second-order consumer, heat, first-order consumer, death and wastes, nutrients in soil, energy from sunlight.

This graph shows the rates of photosynthesis and respiration for a plant over a 24-hour period.

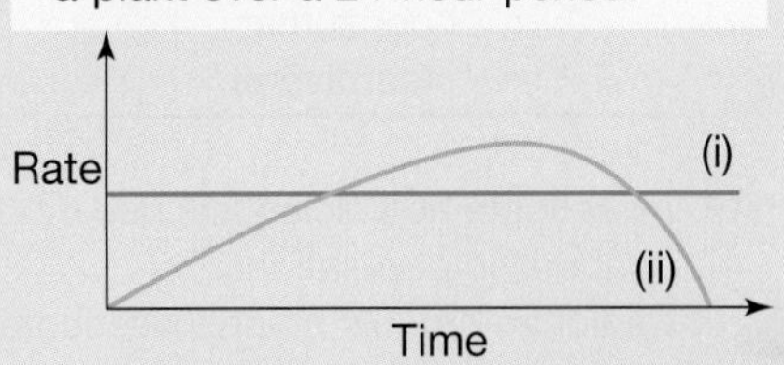

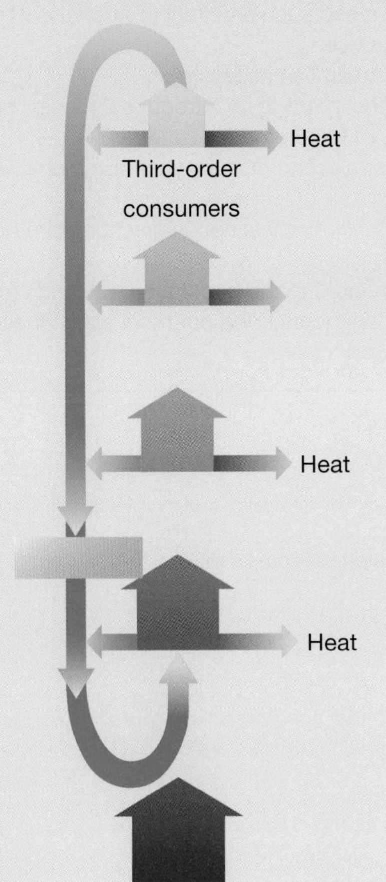

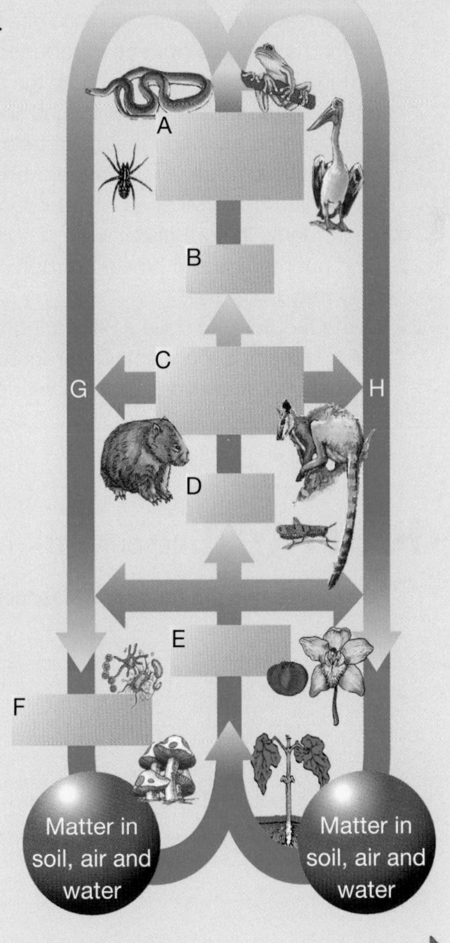

17. Carefully observe the predator–prey graphs at right.
    (a) Comment on any patterns observed.
    (b) Suggest reasons for the observed pattern.
18. Complete the following table.

Trophic level	Organism	Food source
First	Producer	Convert inorganic substances into organic matter using sunlight energy and the process of photosynthesis
	Primary consumer (herbivore)	Plants or other producers
Third		
	Tertiary consumer (carnivore)	

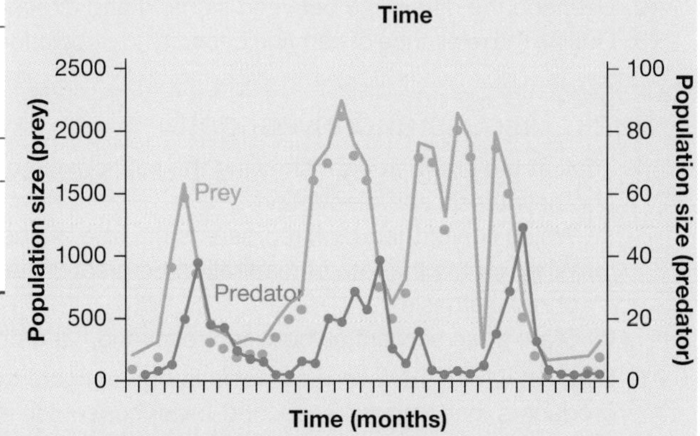

19. Investigate how scientists use models to predict changes in populations.
20. Select an organism and investigate the effect of seasonal changes on its population size.
21. Suggest why it is (a) necessary for energy to flow through ecosystems and (b) essential for matter to cycle through ecosystems.
22. Introduced species can have an effect on the feeding relationships within food webs.
    (a) Find examples of three introduced species that have had such an effect in Australia.
    (b) Find out when, why and how the species were introduced.
    (c) Investigate the effect of the introduced species on Australian ecosystems.
    (d) Research methods that have been used to reduce the effect of an introduced species or control its population size, and comment on the effectiveness of the methods.
    (e) Construct a PMI chart on introduced species in Australia and discuss your comments with other students in your class.
23. Find out more about how autotrophic bacteria (such as *Thiobacillus* spp.) use chemosynthesis and where they may be found.
24. Some factors that have an impact on populations are related to the population size and are referred to as being density dependent. Other factors are density independent. Find out more about these two types of factors and share your findings with other students in your class.

RESOURCES — ONLINE ONLY

Complete this digital doc: Worksheet 5.4: Cycling of materials (doc-18876)

5.8 Changes in populations

5.8.1 Changes in populations

The effect of a change in size of a particular population on an ecosystem can be predicted by observing feeding relationships in food webs. If, for example, the number of the producers is reduced, it will not just be herbivores that are affected, but also the animals that eat them, which in turn affects the organisms that either eat them or are eaten by them.

If a new species is introduced it may compete for the resources of another species, leading to a reduction in that population. This may have implications for organisms that either eat or are eaten by the affected population.

5.8.2 Artificial ecosystems

Humans have created artificial ecosystems to maximise the production of their own food supplies and resources. The purpose of agriculture is to turn as much of the sun's light energy as possible into chemical energy in particular crops or pasture plants for animals. In order to achieve this, humans have attempted to control populations of other organisms. This has led to interference in food webs and hence the ecosystems that contain them.

5.8.3 Taking away

To make room for crop plants, land has often been cleared of forests. The local habitats of many organisms have been destroyed. Organisms that may compete for resources or in some way potentially lower crop yields are considered to be pests and are also removed, or their populations killed or controlled.

Crops are often monocultures, consisting of only one species of plant. At the end of each growing season the crops are harvested, processed and removed from the ecosystem. There is little natural decomposition of dead material and exposed soil may be blown away by the wind. Valuable nutrients are lost. Such activities have led to the destruction of many natural ecosystems.

5.8.4 Giving back

Fertilisers are added in an attempt to replace some of the lost nutrients. Some of these may end up in waterways, adding large quantities of nitrogen and phosphorus to the water. This can lead to algal blooms or **eutrophication**, which may result in the death of organisms within the ecosystem.

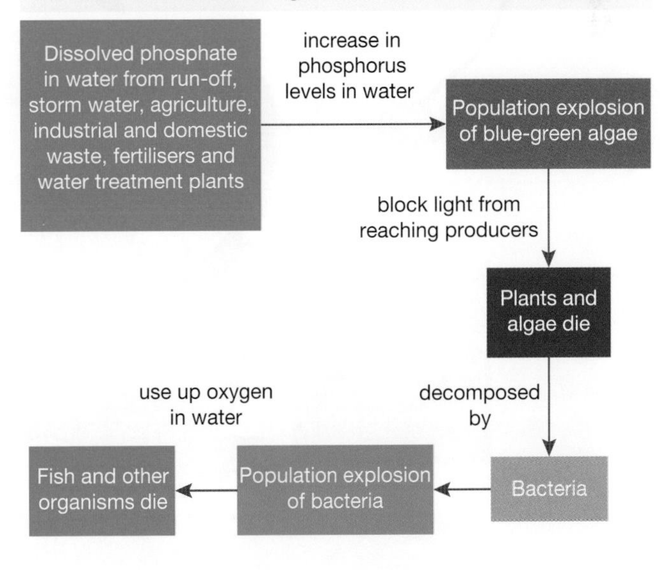

5.8.5 Controlling pests

Organisms that compete for resources or potentially lower the yield of the plant crop being grown are considered to be pests that need to be controlled. Pest control may be carried out using chemical or biological control.

5.8.6 Biological control

Biological control of unwanted organisms can exploit naturally existing ecological relationships. Predators or competitors may be used to kill or reduce numbers of the pests or somehow disrupt the pests' reproductive cycle. A disease, for example, might be used to kill the unwanted organism without harming other species. While some cases of biological control have proven to be successful, others (such as the introduction of cane toads and prickly pear plants) have caused a variety of new problems.

5.8.7 Chemical control

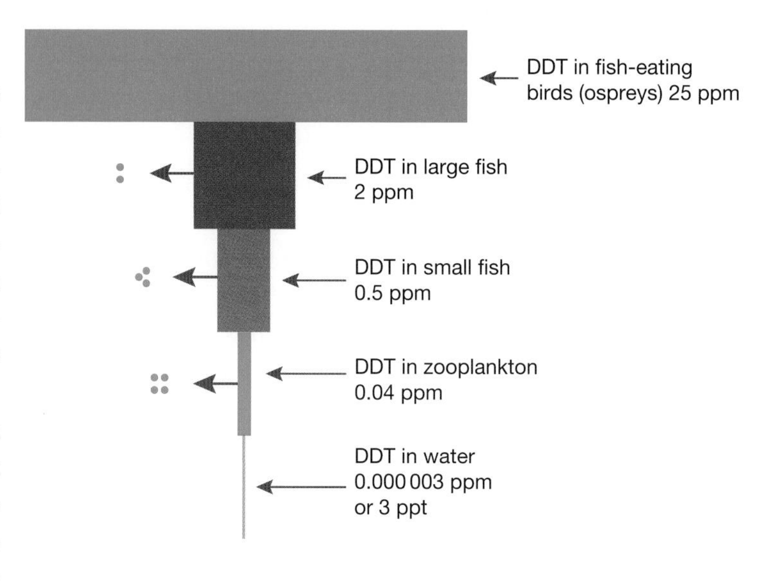

Chemical methods of control include the use of pesticides such as **insecticides**, **fungicides**, **herbicides** or **fumigants**. Herbicides kill plants other than the planted crop so that they do not compete for nutrients and water in the soil, and light from the sun. Insecticides are used to kill organisms that compete with humans for the food crops.

Although pesticides are still used in agriculture, their effectiveness on target pest species often decreases. Other species may also be affected within the ecosystem and the food webs in which the target species belongs. In some cases, concentrations of non-biodegradable pesticides (such as DDT) can be magnified along the food chain by a process described as **bioaccumulation** or **biological magnification**.

5.8.8 Introduced species

An **introduced species** is one that has been released into an ecosystem in which it does not occur naturally. The food webs in ecosystems are very delicate and can easily be unbalanced, especially when new organisms are introduced. These introduced organisms compete with other animals for food, provide predators with a new source of prey, or may act as predators themselves.

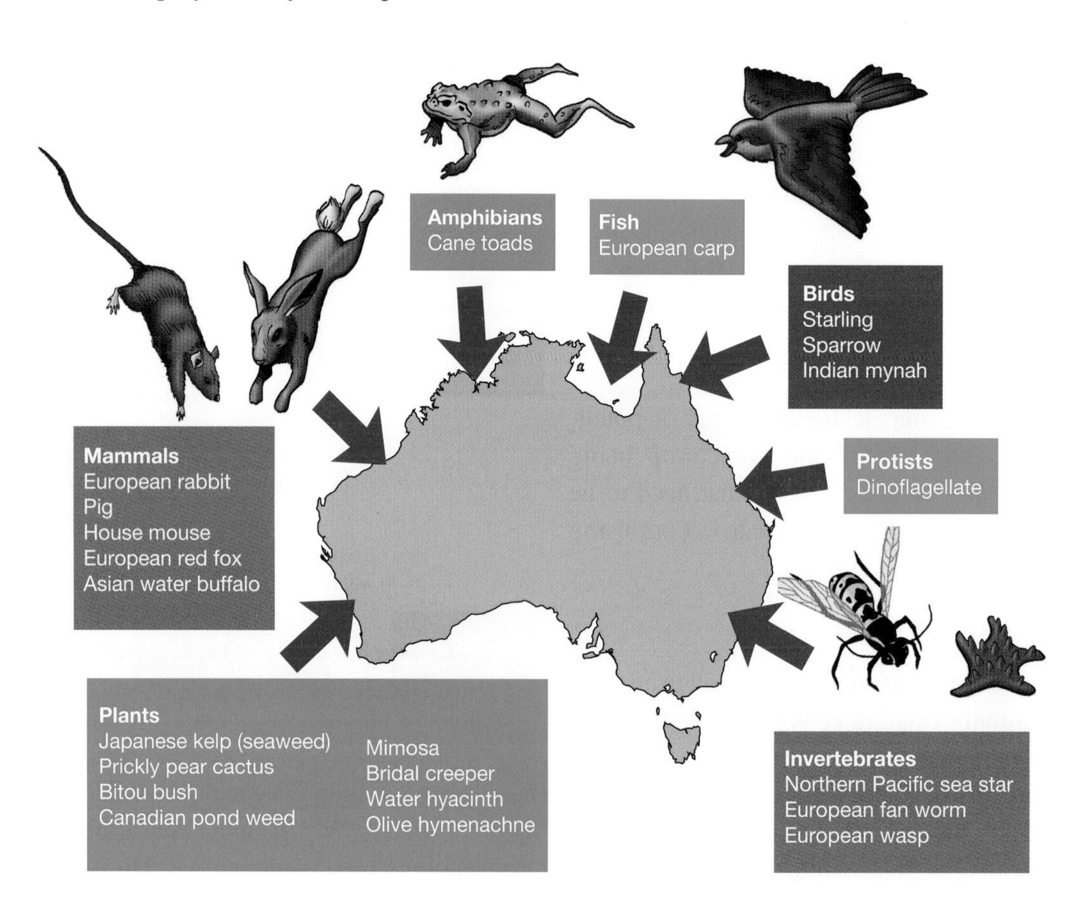

5.8.9 Cane toads

Ecological impact:

- Occupying water habitats so that native tadpoles cannot live there
- Killing fish that eat the tadpoles and other animals that eat the adult toads
- Eating our natural wildlife including frogs, small lizards, birds, fish and insects.
- Poisonous; fatal to animals that eat them

5.8.10 Northern pacific sea star

Ecological impact:

- Potential great harm to our marine ecosystem and to marine industries
- Threatening biodiversity and shellfish aquaculture in south-eastern Tasmania and Port Phillip Bay
- It is a voracious predator. Some of our native marine species, such as scallops and abalone, don't recognise it as a predator, so do not try to escape it.
- No natural predators or competitors to keep the population under control

5.8.11 Rabbits

Ecological impact:

- Competing for food with the native animals such as kangaroos, wallabies, wombats and bandicoots
- Disrupting food webs and unbalancing ecosystems
- Building extensive underground warrens
- Stripping most of the vegetation in their area, causing another problem — erosion. Without plant roots to hold the soil, wind and rain carry the soil into creeks, rivers and lakes, causing further problems for the organisms that lived there.

RESOURCES — ONLINE ONLY

Watch this eLesson: Native rats fighting for their habitat (eles-1083)

HOW ABOUT THAT!

Dr Susan Wijffels, CSIRO oceanographer and leader of the IMOS bluewater and climate node, is involved in research that investigates the impact of ocean ecosystems and the oceans' role in the carbon cycle. She is currently investigating the use of floating sensors, underwater gliders and satellite tags on marine animals. The tagging of these marine animals will provide information about when and where they feed and the types of salinity, pressure and temperatures that they feed in.

Dr Susan Wijffels with a robotic float that acts like an underwater weather balloon

Cameras mounted on seals provide information from the depths of the sea.

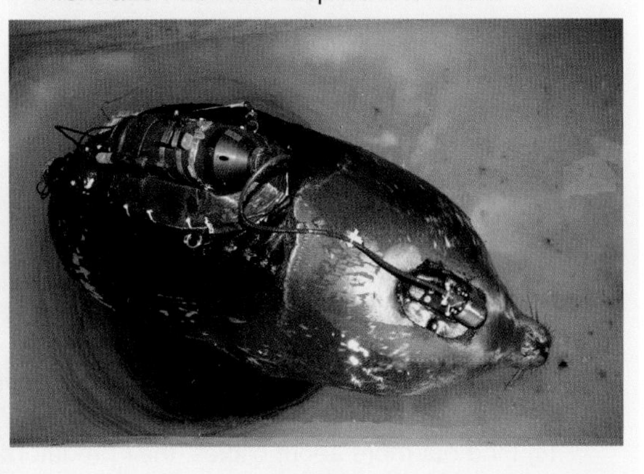

HOW ABOUT THAT!

Researchers are working on an immunocontraception method that aims to block conception in rabbits to control their numbers. This method will use a virus that has been modified to contain genetic material that codes for the production of a protein essential for reproduction. When infected with this modified virus, (it is hoped that) the female rabbit produces the protein and her immune system responds by producing antibodies against it. These antibodies should then attack her eggs, blocking conception.

5.8 Exercises: Understanding and inquiring

To answer questions online and to receive **immediate feedback** and **sample responses** for every question, go to your learnON title at www.jacplus.com.au. *Note:* Question numbers may vary slightly.

Analyse, think and investigate

1. Carefully examine the eutrophication flowchart at the beginning of this subtopic.
   (a) Suggest how dissolved phosphate in waterways can be linked to algal blooms.
   (b) Do you think that suffocation is an appropriate description of the effect of eutrophication?
   (c) Find examples of algal blooms in Australia. Investigate:
      (i) the cause of the algal bloom
      (ii) consequences or effects of the algal bloom on the local ecosystem
      (iii) how algal blooms can be treated or prevented.

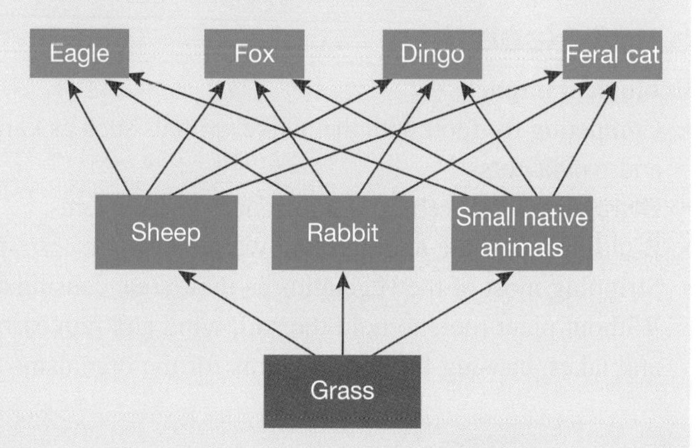

2. Observe the figure of the food web at right and then answer the following.
   (a) Suggest possible consequences of a decrease in the population size of:
      (i) rabbits
      (ii) feral cats
      (iii) grass.
   (b) Suggest possible consequences of an increase in the population size of:
      (i) grass
      (ii) sheep
      (iii) eagles.
3. Habitat loss and introduced species have led to a decline in reptile populations. Find out more about the effects of habitat loss and introduction of cats, foxes and cane toads on Australian reptile populations and the ecosystems in which they live.
4. Although populations of greater bilbies once ranged over most of mainland Australia, predation by introduced species such as cats and foxes has eliminated them from most of their former habitats.
   (a) Carefully observe the figure at right and describe the change in bilby distribution pattern over the recorded times in Queensland.
   (b) From the figure, estimate the percentage population values in Queensland for each year and construct a graph to show your data.
   (c) Find out more about bilbies, their predators and ways in which bilbies try to avoid their predators.

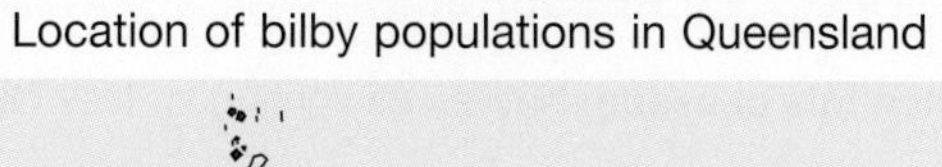
Location of bilby populations in Queensland

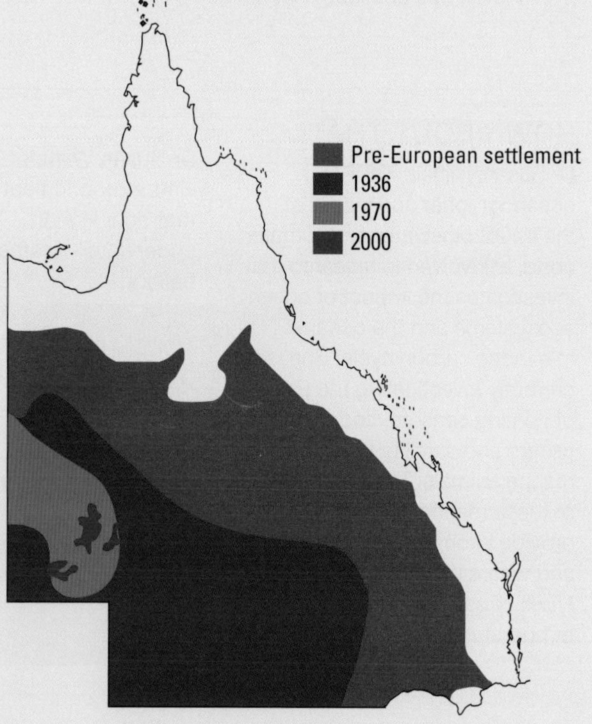

(d) Find out more about the introduction of either cats or foxes to Australia, their effect on Australian ecosystems, and Australian research projects that are investigating ways in which to reduce their impact.

(e) Find out more about Australian research into protection of bilby populations.

(f) Suggest what you think should be done to protect the greater bilby from becoming extinct. Outline how this could possibly be achieved.

5. The lesser bilby, *Macrotis leucura*, is now extinct. Find out more about this species and possible reasons for its extinction. Knowing what we know now, suggest how this could have been avoided.
6. Carefully examine features of the planigale and wallaby, and the location of their distribution in Australia in the figure on the right. Find out more about each of these organisms and their habitats and relate their features to their suitability in these environments. Research the food webs in which they are linked and discuss possible (or real) implications of human activity to their survival.

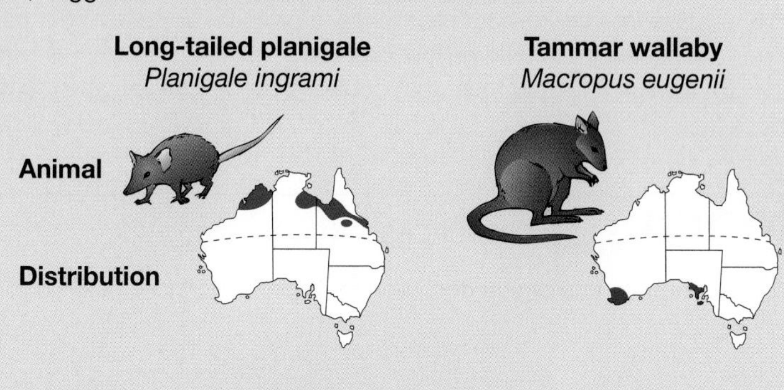

7. Carefully examine the graph on the right showing crown-of-thorns starfish populations recorded in Great Barrier Reef surveys between 1986 and 2004. Suggest why scientists measure and record the size of populations of organisms. Find out more about the food webs that this starfish is linked to. What type of food does it eat? Which organisms eat it? Suggest possible implications of the changes in crown-of-thorns starfish populations for other members of its food web. Research issues relating to the crown-of-thorns starfish and the effect it has had on Australian marine ecosystems.

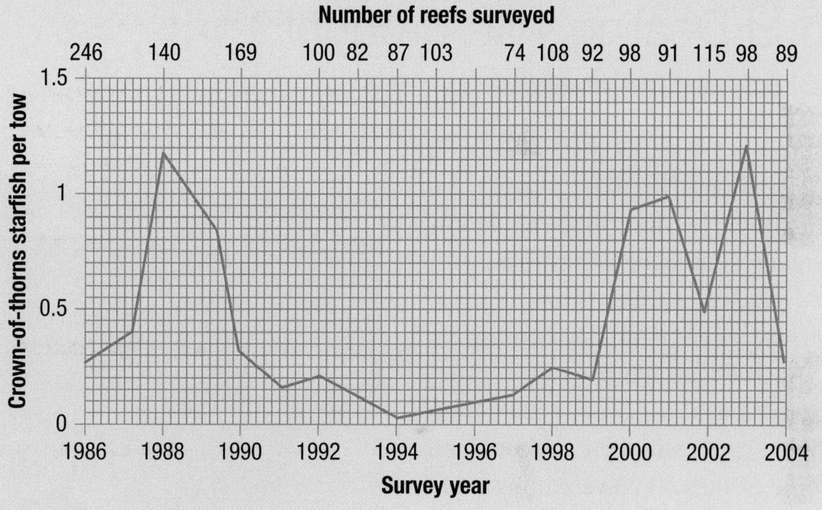

Investigate and discuss

8. Investigate ways in which human activities have affected either the nitrogen cycle or the carbon cycle, leading to an imbalance within an ecosystem.
9. Investigate the introduction of rabbits into Australia and research their effect on our ecosystems. Identify and research examples of methods that have been used to control the population growth of rabbits. Comment on the effectiveness of these methods.
10. Suggest a definition for the term sustainable agriculture. Find out whether your suggestion was correct and find out more about ways in which this form of agriculture differs from unsustainable agriculture.
11. Find out more about critically endangered and endangered species and ecological communities that are affected by rabbits.
12. Research one of the following introduced marine species to find out why it is considered an environmental pest.
    - Green shore crab (*Carcinus maenas*)
    - Pacific oyster (*Crassostrea gigas*)
    - European fan worm (*Sabella spallanzanii*)
    - Japanese goby (*Tridentiger trigonocephalus*)

13. Find examples of non-indigenous fish or marine vegetation that have been introduced in each of the following ways.
    - Ship fouling, ballast waters or dry ballast
    - Stock enhancement, mariculture, or biological control
    - Wave action or ocean currents
14. Find out more about Susan Wijffels and her research on ocean ecosystems.
15. Find out more about oceanographers and then write a short story about one of their adventures.
16. (a) Use internet research to identify a problem or issue about pests, population control or introduced species that could be investigated.
    (b) Design an experiment that could be used to investigate the problem.

Complete this digital doc: Worksheet 5.5: Introduced pests (doc-18877)

5.9 Dealing with drought

Science as a human endeavour

5.9.1 Dealing with drought

Approximately 80 percent of Australia is described as having arid or semi-arid conditions. How can Australian plants and animals survive under such dry conditions?

Australian animals have some clever ways to cope with limited water supplies in their environments. Some of these involve putting reproduction on hold, while others produce extremely concentrated urine, and yet others possess water collection structures.

Common name: Thorny devil. Scientific name: Moloch horridus

The thorny devil (*Moloch horridus)* collects dew overnight on the large spines covering its body. Moisture eventually collects in a system of tiny grooves or channels running between their scales. These channels help direct the collected water towards the lizard's mouth where a gulping action takes in the water and quenches the animal's thirst.

The spinifex hopping mouse (*Notomys alexis*) also has a few useful tricks. These small, nocturnal animals can survive without drinking water, and produce extremely concentrated urine. The figure on next page shows some ways in which they are well adapted to surviving arid conditions.

An outline of how *Notomys alexis*, or the spinifex hopping mouse, achieves a water balance. For survival, water inputs must balance water outputs.

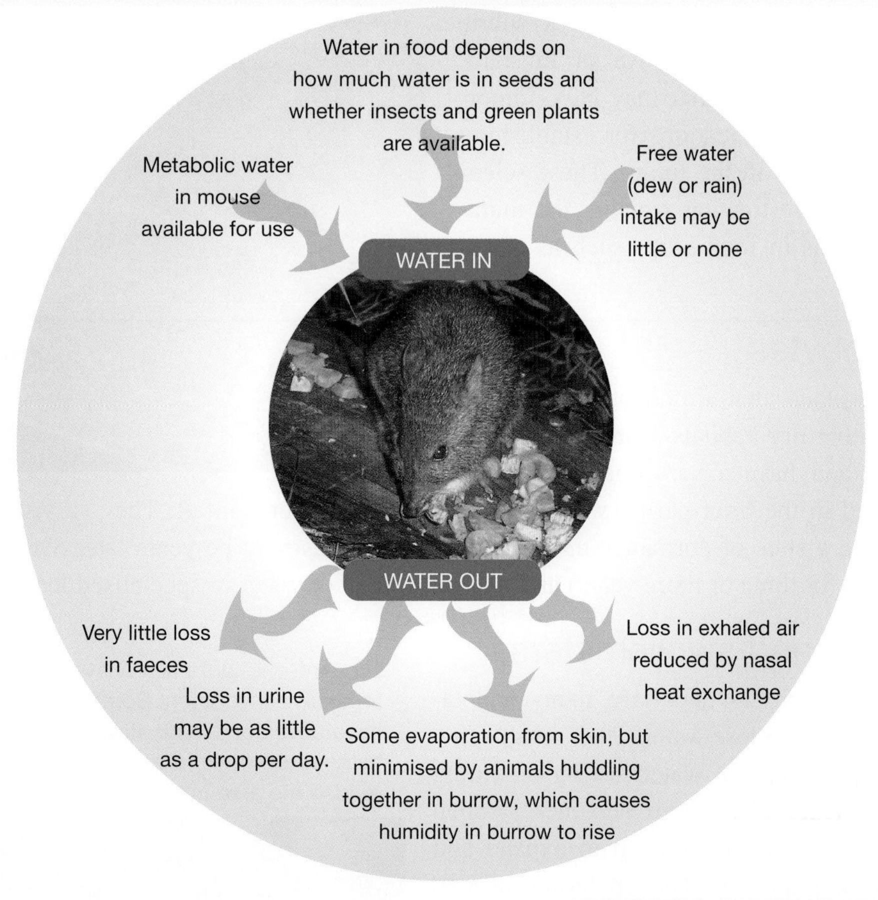

5.9.2 Finding food in the dark

Not only is water scarce, but temperatures in our Australian outback can be very high during the day. Many Australian animals are **nocturnal** — that is, they are active only at night. Nocturnal animals have adaptations that help them find their food in the dark.

Snakes such as pythons, for example, possess heat-sensitive pits in their lower jaw that contain **thermoreceptors**, allowing the location of their warm-blooded prey to be detected.

Bilbies are also well adapted to sense their food in the dark by using their sharp hearing and long, sensitive nose and whiskers to detect their food.

Common name: Central carpet python.
Scientific name: *Morelia bredli*

Common name: Greater bilby.
Scientific name: *Macrotis lagotis*

5.9.3 Hot and thirsty

Even though our native plants are unique and some have strategies to cope with our continent's harsh conditions, global warming and scarcity of water are a threat to their survival. Scientists around the world are seeking solutions to our current and future problems. One of these problems involves ways in which we can help plants survive conditions associated with droughts.

5.9.4 Hot plants of the future

In 2009, a team of scientists at the Australian National University (ANU) discovered a subtle mutation in *Arabidopsis* (a relative of mustard, cabbage and canola plants) that may have important and far-reaching implications for establishing drought-resistant plants in the future. These scientists are currently investigating whether the mutation has applications in food crops such as wheat and rice.

The plant *Arabidopsis* in which the drought-resistant mutation was discovered is a relative of mustard, cabbage and canola plants.

5.9.5 Xerophytes

Xerophytes are plants that are adapted to survive in deserts and other dry habitats. Some xerophytes are **ephemeral** and have a very short life cycle that is completed in the brief period when water is available after rainfall. They survive periods without water by entering a state of dormancy until the next rains. This may be years later. Other xerophytes are **perennial** (living for three or more years) and rely on storage of water in specialised leaves, stems or roots.

5.9.6 Spines and swollen stems

Most cacti are xerophytes and have many adaptations to store rather than lose water. Their small spiny leaves reduce the amount of water lost by providing a small surface area. Their stem becomes swollen after rainfall, with pleats allowing it to expand and contract in volume quickly. The epidermis around the stem has a thick waxy cuticle and contains stomata, which usually open during the night (to collect carbon dioxide required for photosynthesis) rather than during the day when water can be lost through them.

Drought-resistant plants have developed strategies to avoid dehydration. Some plants, such as cacti, have developed thick, fleshy, water-storing leaves, hairy or reflective foliage and small leaves to reduce the area from which water can be lost.

5.9.7 Can our Aussie plants survive?

But what if we can't wait for research and investigations to produce these drought-resistant plants … which plants can we plant and grow now? What can we do to help them survive? How do they survive, when others would die? Observe the figures of Australian plants on the next page and try to see the structures that may assist them in drought tolerance.

5.9.8 Resistant or tolerant?

Our native plants are unique and have developed some strategies to cope with the harsh conditions that global warming and scarcity of water threaten them with. Although drought resistant and drought tolerant are often used as the same term, they are not. **Drought tolerant** means that the plant can tolerate a period of time without water. Plants that are **drought resistant** can store their water and live for long periods of time without water. Many of our Australian plants that live in water-limited environments would be classified as being drought tolerant.

5.9.9 Drought-tolerant features

Just because a plant is an Australian native doesn't mean that it is drought tolerant. It may have evolved to be better suited to high rainfall zones or cool mountain forests. Many drought-resistant plants already grow in areas where water is scarce. Examples of drought-tolerant adaptations to look for include:

- small narrow leaves
- grey or silver foliage
- furry texture
- water-retaining succulent (juicy) leaves or stems
- modified or absent leaves.

Adaptations that help eucalypt trees survive in a dry environment

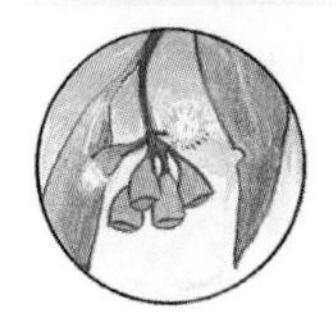

Grevillea 'Robyn Gordon' is a drought-tolerant plant named after the eldest daughter of David Gordon (a keen **botanist** and founder of Myall Park Botanic Garden).

The leaves of many eucalypts hang vertically to reduce their exposure to the sun, and pendant branches move with the wind to create 'holes' in the soil under the tree which may later collect water.

5.9 Exercises: Understanding and inquiring

To answer questions online and to receive **immediate feedback** and **sample responses** for every question, go to your learnON title at www.jacplus.com.au. *Note:* Question numbers may vary slightly.

Remember

1. Describe two ways in which Australian organisms can increase their chances of survival in arid or semi-arid environments.
2. Identify the common names of the following organisms: *Moloch horridus*, *Notomys alexis*, *Macrotis lagotis*, *Morelia bredli*.
3. Suggest why the recent discovery of the mutation in *Arabidopisis* is important.
4. Define the term xerophyte.
5. Describe ways in which xerophytes can increase their chances of survival.
6. Describe how cacti can survive in areas with limited water.
7. Describe features of eucalypts that help them to survive hot, dry conditions.

8. Outline the difference between drought resistant and drought tolerant.
9. List adaptations to look for that suggest a plant may be drought tolerant.
10. Who was *Grevillea* 'Robyn Gordon' named after?

Investigate, think and discuss

11. Kangaroos and wallabies can have joeys at different stages of development by producing different types of milk. Find out more about reproductive strategies that can increase the chances of surviving in arid environments for these animals.
12. Find answers to the following questions and then suggest your own question for research.
    (a) Why do kangaroos lick their forearms?
    (b) How did magnetic termite mounds get their name?
    (c) Why do spinifex hopping mice huddle together during the day?
13. The Queensland bottle tree (*Brachychiton rupestris*) has an ability to store a significant amount of water. Find out more about how it reacts to drought and how it is used by indigenous populations.
14. Identify two examples of endangered species that live in the Australian outback. Investigate and report on the effect of introduced species on these endangered species. Outline research or projects that are currently underway to help control populations of any of the species involved.
15. *Even during drought conditions, kangaroos need to be killed so that our kangaroo industry meets its requirements for meat and skins.*
    (a) Research information relevant to this quote and summarise your findings into a SWOT analysis.
    (b) Discuss your findings with others, adding any more details to your SWOT figure.
    (c) Consider and discuss the various perspectives relevant to this quote.
    (d) Organise a class debate that considers at least four different perspectives.
    (e) Outline your opinion on the quote. Include reasons for your opinion.

Investigate and create

16. Identify an example of a plant and an animal that lives in an arid or semi-arid Australian environment. Investigate the effect of seasonal changes on each of these species. Construct a seasonal calendar that includes graphs and diagrams to display your findings.
17. Investigate ways in which human activity is having an impact on ecosystems in regions affected by drought. Discuss your findings with others. Construct a PMI chart to summarise your findings.
18. Investigate how models can be used to predict changes in populations due to drought. Construct a graphical model that could be used to show changes in kangaroo or rabbit populations within a particular ecosystem over time.
19. Research feeding relationships between organisms living in arid or semi-arid Australian environments. Summarise your findings into a food web.
20. Design and create your own virtual drought-tolerant garden. Include captions or labels to identify the plants and describe how they can survive on limited water supplies.
21. Visit a nursery and make notes on the information about drought tolerance on the plant labels. Design and create your own labels.
22. (a) Find out more about various types of research into drought-tolerant and drought-resistant plants in Australia
    (b) Formulate your own questions to help you identify your own hypothesis to test.
    (c) Design and conduct your own investigation.
    (d) Report your findings as a class presentation and a scientific journal article.
23. Organisms within an ecosystem can also be affected by too much water. Research the effects of floods on Australian ecosystems.
24. Find out more about one of the following Australian plants and construct a model that shows features that increase its chances of survival: *Hardenbergia violacea*, *Dianella revoluta*, *Anigozanthos flavidus*, *Callistemon viminalis*.
25. Create your own multimedia presentation about Australia's climate, geography, and plants and animals that can survive in arid conditions.
26. Create your own advertisement promoting Australia's unique climate, plants and animals.

5.10 Fire proof?

5.10.1 Fire proof?

Natural disasters are not uncommon. They happen all over the world. Extremes of droughts, fire, flood, lightning, landslides, earthquakes, tornadoes, hurricanes and tsunamis are examples of natural disasters.

While their effects can be devastating to people, natural disasters can also have a great impact on ecosystems. In some situations, the species living within these ecosystems have developed strategies to survive such natural disasters.

Many Australian plants have adaptations to help them survive bushfires.

5.10.2 Bushfires

Since the last ice age, bushfires have been a natural part of Australia's unique ecosystem. Over time, this has led to natural selection of various adaptations in its inhabitants. The PMI chart below shows pluses, minuses and interesting points about fire and our Australian ecosystem.

Plus	Minus	Interesting
Clears the undergrowth	Slows down growth of mature trees	Aboriginal communities have traditionally used fire to manage land.
Helps regeneration of plants which produce seeds in hard or waxy husks	Damages timber, reducing value of trees	Eucalypt species have many features that help them to survive bushfires.
Releases stored nutrients, providing good conditions for new growth	Causes property damage if out of control	Without fire, open eucalypt forests would disappear, replaced by more dense forests containing trees that do not require fire to flower and produce seed.
Makes soil structure finer and easier for seeds to grow in	Makes soil finer and more easily washed or blown away by water or wind	Some animals burrow underground to escape a fire.
Destroys growth inhibitors in adult trees and allows young plants to thrive	Kills or drives off many animals, fungi and insects, and threatens other organisms in the food web with starvation	Some animals survive bushfire but die because of the reduced food supply in the period after the fire.
Allows sunlight to reach the soil, helping young plants to grow	Kills decomposers, which are then lost to the life cycle	Lightning and arson cause many of the large bushfires that result in death and property damage.

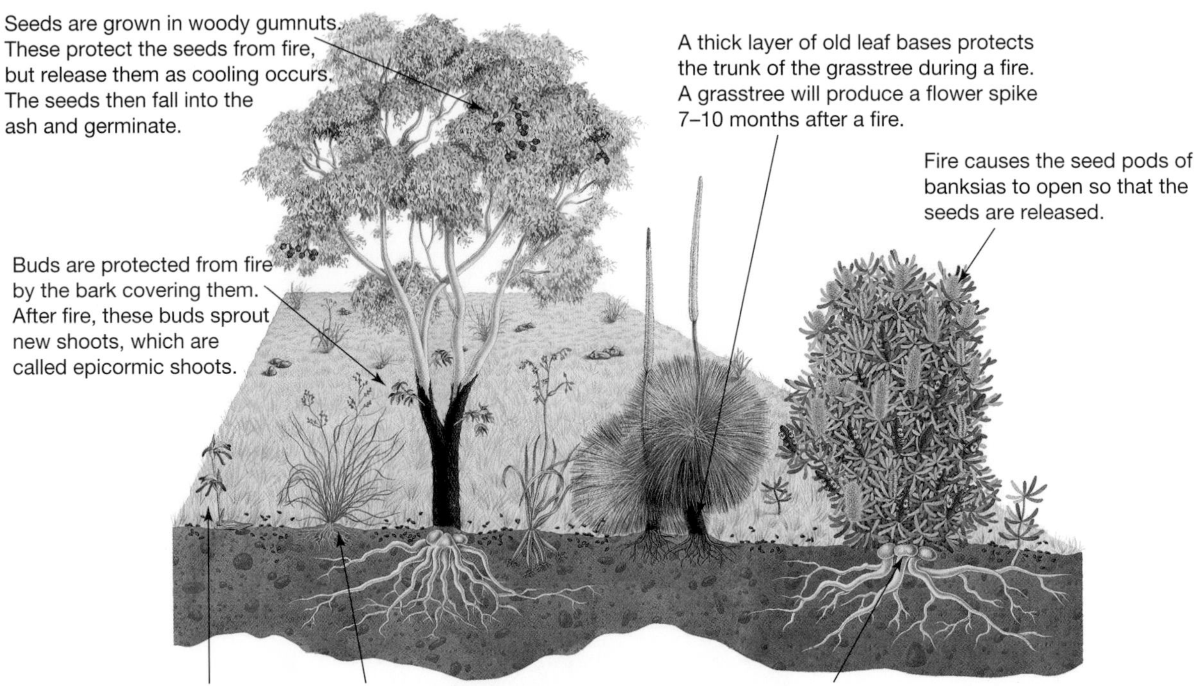

5.10.3 Sunburnt country

I love a sunburnt country,
A land of sweeping plains,
Of ragged mountain ranges,
Of droughts and flooding rains,
I love her far horizons,
I love her jewel sea,
Her beauty and her terror —
The wide brown land for me!

So wrote Australian poet Dorothea Mackellar in her poem *My Country*. Some arid regions of Australia have highly erratic rainfall with long drought and flooding rains. These climatic conditions have a considerable effect on the plants and animals within these ecosystems. They also have implications for our farmers, who may not share this poet's love of our sometimes challenging climatic conditions!

Big or small?

<table>
<tr><th>Small controlled fire</th><th>Large bushfire</th></tr>
<tr><td>Some leaf litter destroyed; many insects and decomposers survive on the ground.</td><td>All leaf litter destroyed; no insects and decomposers survive on the ground.</td></tr>
<tr><td colspan="2">Soil releases stored nutrients. Ash provides many minerals and fine texture. This helps seeds to germinate and new plants to grow.</td></tr>
<tr><td>Heat and smoke cause some plants to release seed and some seeds to germinate.</td><td>Heat and smoke cause most plants to release seed and many seeds to germinate.</td></tr>
<tr><td>Many unburnt patches where grasses and shrubs survive; animals can find food and shelter here.</td><td>Few unburnt patches; no food or shelter is left for animals.</td></tr>
<tr><td>Many animals survive and can stay in the area. Food is still available.</td><td>Many animals are killed, or must move to another habitat. No food is available.</td></tr>
<tr><td>Fallen branches and logs survive to provide shelter for animals.</td><td>No fallen branches or logs survive so there is less shelter.</td></tr>
</table>

INVESTIGATION 5.11

Germinating seeds with fire

AIM: To investigate the effect of heat on germination of acacia and hakea or banksia seeds

Background

Many seeds need fire to germinate. It could be the smoke, heat or the chemicals in ash that cause the seeds to germinate.

CAUTION

- Make sure you are supervised by an adult.
- Burn seed pods only in a safe area.
- Do not do this activity on a hot windy day or a day of total fire ban.
- Have a bucket of water or a fire extinguisher ready.
- Pods stay hot for some time after burning. Give them time to cool before touching them.

Materials:

hakea or banksia seed pods — unopened

newspaper

bucket of water

seedling mix

acacia seeds (silver or black wattle work well)

matches

seedling trays

oven

Method and results

Part A

- Collect unopened banksia or hakea seed pods from trees in your local area.
- Wrap the seed pods in newspaper and burn them in a safe area. (Alternatively, heat the pods in an oven.)

1. Observe the seed pods after burning.

- Collect the seeds and plant them in the seedling trays. Care for them until they are large enough to plant in the garden.

Discuss and explain

Part B

- Divide the acacia seeds into two equal piles. Record the number of seeds in each pile.
- Plant one pile of seeds in a seedling tray.
- Heat the second pile in the oven.
- Plant these seeds in a separate seedling tray. Sprinkle some ash over the seedling tray.
- Keep the trays moist. Wait for the seeds to germinate. This could take many days.

2. Count the number of seedlings that have germinated in each tray. Compare class results.

- Look after your seedlings and, when large enough, plant them in a garden.

Discuss and explain

3. Describe the effect that heat or fire had on the pods.
4. Identify which group of seeds germinated most effectively.
5. Suggest what caused one group to germinate more than the other.
6. Explain how this is similar to the effect that fire would have on the seeds.
7. Suggest how opening the seed pods in response to heat may help the plants to grow at the right time.
8. Outline the strengths and limitations of this investigation, and suggest how this investigation could be improved.
9. (a) Suggest your own investigation question about germination and fire.
   (b) Design an experiment to investigate your research question.

5.10.4 A fiery start

It is believed that Indigenous Australians first arrived in Australia over 40 000 years ago and successfully managed the land. They used very different hunting and gathering practices from those of Europeans. While Aboriginal people did have significant impact on the Australian environment, their lifestyle was sustainable and allowed resources to renew.

Some of the ways in which Indigenous people cared for their land included:

- moving from place to place rather than staying in the same location. This ensured that the plants and animals they fed on had a chance to replenish.
- eating a wide variety of food so that no single food source was depleted
- leaving enough seeds to ensure that plants could regenerate
- leaving some eggs in a nest when collecting
- not hunting young animals or the mothers of young animals
- not allowing particular members of a group to eat certain foods. This ensured that a wide variety of food was eaten and that 'taboo foods' were not depleted.
- leaving the land to recover for a period of time after harvesting a crop, such as bananas. This allowed time for the crop to regenerate and nutrients to return to the soil.

5.10.5 Using fire

One way that early Indigenous Australians affected the environment significantly was through their use of fire. Fire was used for hunting. Setting fire to grassland revealed the hiding places of goannas, and possums could be smoked out of hollows in trees. Fire was also used to clear land. The grass that grew back after the fire attracted grazing animals, which could be hunted more easily.

Aborigines Hunting Kangaroos by Joseph Lycett, c. 1817. Early Indigenous Australians used fire to increase grassland areas, providing grazing land for kangaroos which could then be hunted. This image of Indigenous land management and hunting was painted by Joseph Lycett when he was a convict in Newcastle, New South Wales.

Over time, some species of plants that were sensitive to fire became extinct whereas the plants with adaptations that allowed them to survive a fire or regenerate rapidly after a fire became more common. Adaptations are features that help an organism survive in its environment. Some modern-day species such as the banksia are not just well adapted to frequent bushfires; they actually need to be exposed to the high temperatures of a fire for their seeds to germinate.

5.10.6 Reducing the impact of bushfires

Large wild fires such as occurred in Victoria in 2009 can have devastating consequences including loss of lives and damage to property. They can also impact on ecosystems. As the bush burns, animals become victims of the flames or must flee, and habitats are destroyed. One way of reducing the frequency and severity of wild fires is through regular back-burning. This involves deliberately setting fire to vegetation when temperatures are low and the winds are calm to minimise the chance of the fire spreading out of control. Controlled burning removes highly flammable vegetation that acts as fuel for bushfires.

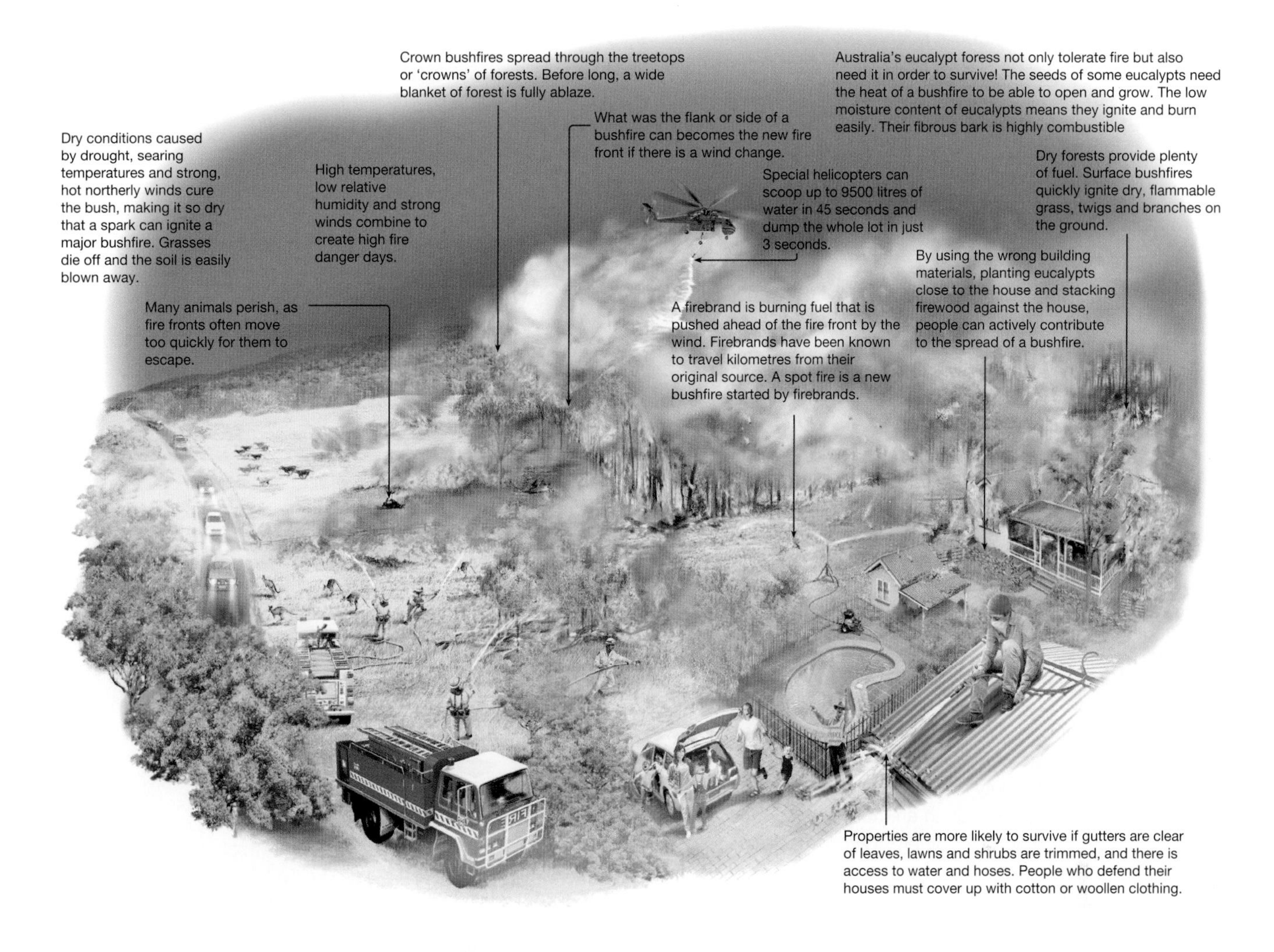

5.10 Exercises: Understanding and inquiring

To answer questions online and to receive **immediate feedback** and **sample responses** for every question, go to your learnON title at www.jacplus.com.au. *Note:* Question numbers may vary slightly.

Remember

1. Use a single bubble map to show examples of natural disasters.
2. Use a cluster map to show how some Australian plants have adapted to surviving a fire.

Think

3. Select a visual thinking tool (see subtopic 1.10) and use it to help organise your thinking on fires, drought, tsunamis, floods, earthquakes or tornadoes.
4. Convert the information in the PMI chart at the beginning of this subtopic into a different type of visual thinking tool.

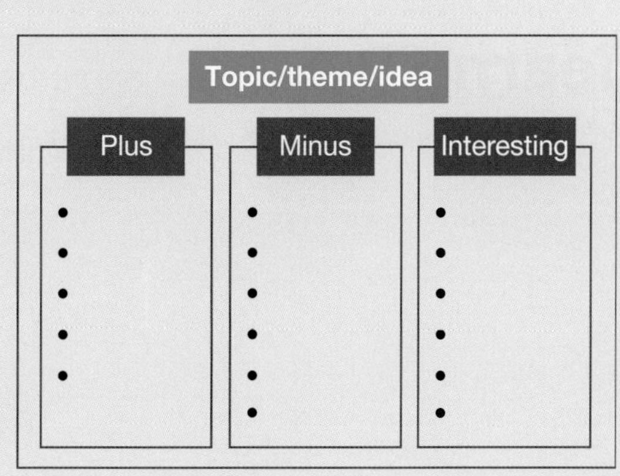

Investigate

5. Investigate drought, tsunamis, floods, earthquakes or tornadoes, and present your findings in a PMI chart.
6. Find out more about how one of the following plants is affected by fire: mountain ash (*Eucalyptus regnans*), wattle (*Acacia*), grasstree, banksia, orchid.
7. Find out how bushfires, drought or floods affect farmers in Australia. Construct a poem, song, play or video to share your findings.
8. Many of Australia's plant and animal species have adapted to survive conditions in their environments. Investigate the adaptations of a particular species. Construct a model, poster or PowerPoint presentation to illustrate how these adaptations increase its chance of survival.
9. Organise a class or team debate on whether controlled burning should be legal.
10. (a) Find out what the regulations or local guidelines are for bushfire safety in terms of the bush ecosystem.
    (b) In a team, construct a SWOT (Strengths, Weaknesses, Threats and Opportunities) analysis or PMI chart to evaluate these regulations or guidelines.
    (c) As a team, create your own improved version of these regulations or guidelines.
11. Research the uses of fire by early Indigenous Australians.
12. Find out why the bushfire seasons are different across our continent.
13. In the extremely intense 2009 Victorian bushfires, many bush animals died. Suggest ways in which five different types of animals could survive a less intense fire.
14. Find out what to do if you are caught in a bushfire. Create an emergency survival information card that people could use if they were caught in a bushfire.

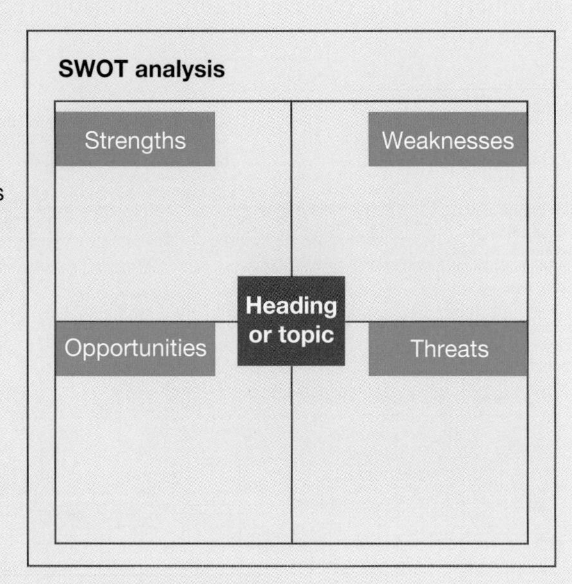

Bushfire seasons throughout Australia

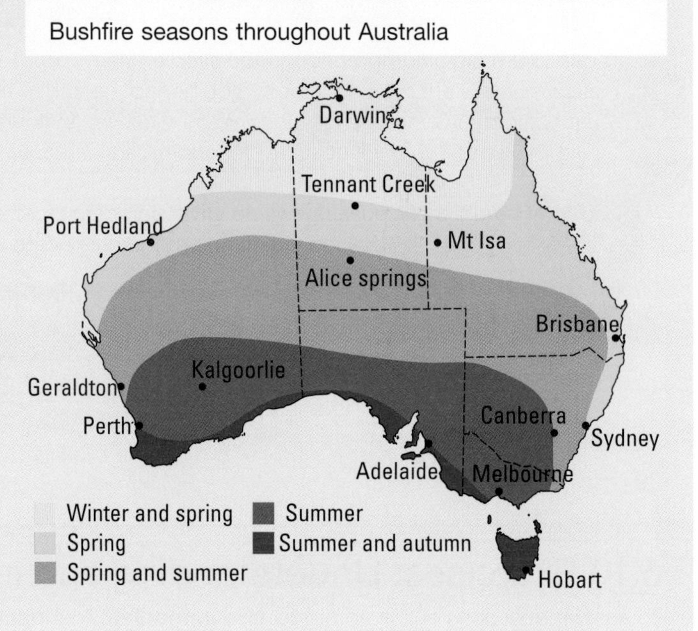

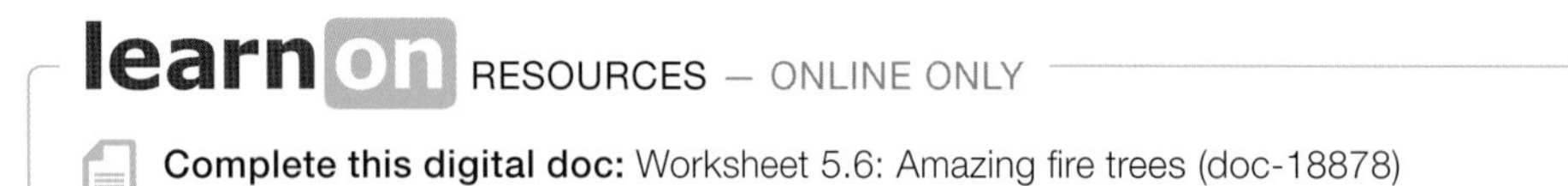

Complete this digital doc: Worksheet 5.6: Amazing fire trees (doc-18878)

5.11 SWOT analyses and fishbone diagrams

5.11.1 SWOT analyses and fishbone diagrams

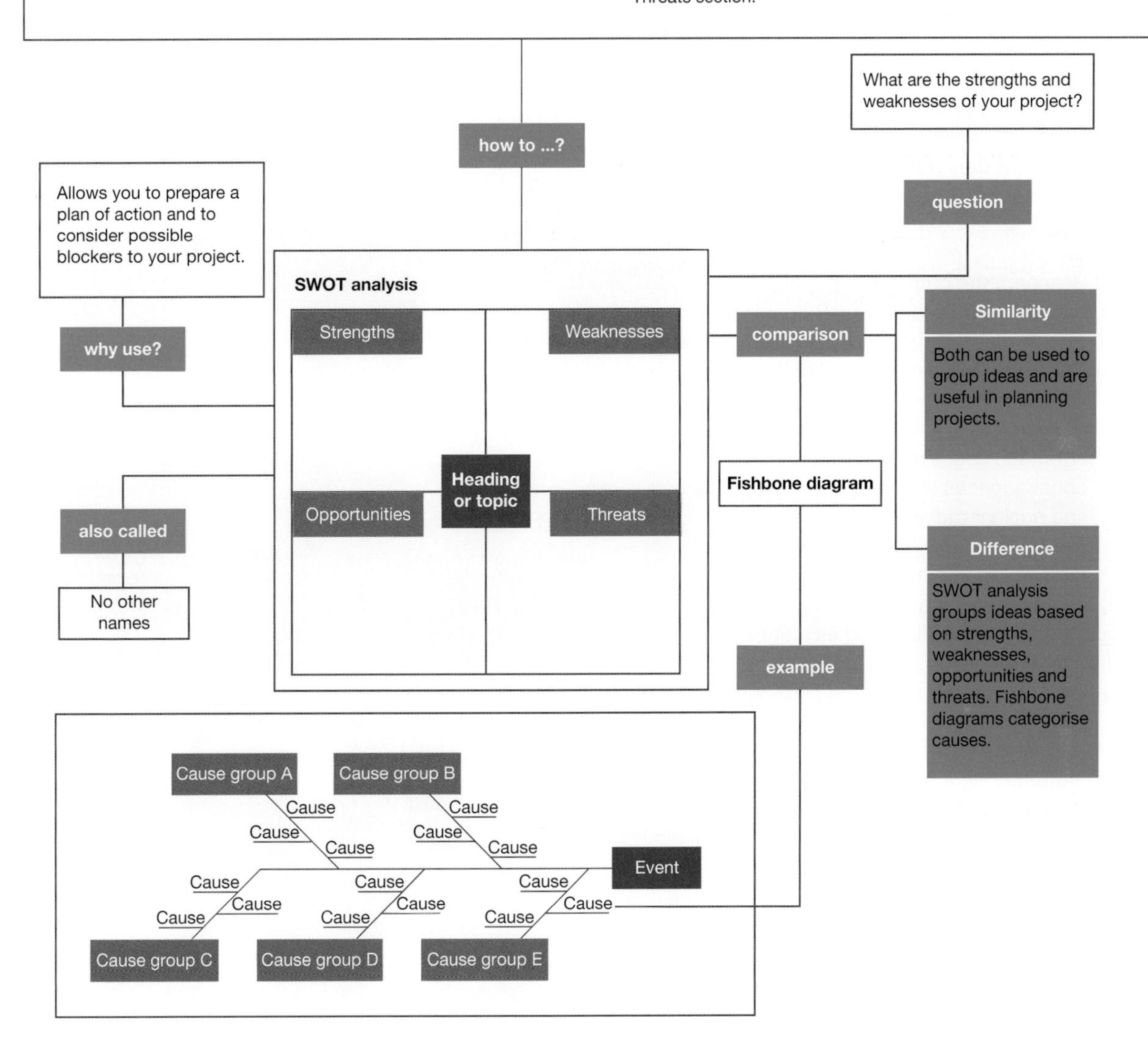

5.11 Exercises: Understanding and inquiring

To answer questions online and to receive **immediate feedback** and **sample responses** for every question, go to your learnON title at www.jacplus.com.au. *Note:* Question numbers may vary slightly.

1. Examine the map on the next page showing the 2012 Australian climate highlights for each Australian state or territory.
   (a) Research possible consequences of these highlights to populations living within ecosystems in each Australian state or territory.
   (b) Construct an overall SWOT analysis on the consequences of the 2012 Australian climate highlights.
   (c) Select a consequence and construct a fishbone diagram to categorise possible causes.
   (d) Research and report on current research on floods and their effect on Australian ecosystems.

2. Earthquakes and tsunamis in Japan in 2011 caused considerable destruction of habitats and damage to ecosystems. They also damaged a number of nuclear power plants, causing radiation to leak into the surrounding areas. In the media frenzy, varied reports, claims and statements were made. Not only did these cause panic, but they also suggested that there was a need to educate the population about different types of radiation and its possible effect on humans.
   (a) Locate examples of media reports about this disaster from the early stages to the more recent. On a timeline, collate examples of the headlines about this environmental disaster.
   (b) Research the damage to populations, habitats and ecosystems. Summarise your findings on your timeline from part (a).
   (c) Comment on any observed patterns or confusion.
   (d) Research relevant scientific knowledge about the cause and possible consequences of earthquakes, tsunamis and nuclear radiation.
   (e) Research possible connections between food webs and nuclear radiation.
3. Examine the maps at the right showing the occurrence of bushfires and incidence of drought in Australia.
   (a) Describe patterns in the occurrence of bushfires.
   (b) Suggest reasons for the observed bushfire pattern.
   (c) Research possible causes and consequences of bushfires in Australian ecosystems.
   (d) Construct an overall SWOT analysis on the consequences of bushfires to populations within Australian ecosystems.
   (e) Construct a fishbone diagram to categorise possible causes of bushfires.
   (f) Describe patterns in the incidence of drought.
   (g) Suggest reasons for the observed drought incidence pattern.
   (h) Research possible causes and consequences of drought in Australian ecosystems.
   (i) Construct an overall SWOT analysis on the consequences of drought to populations within Australian ecosystems.
   (j) Construct a fishbone diagram to categorise possible causes of drought.

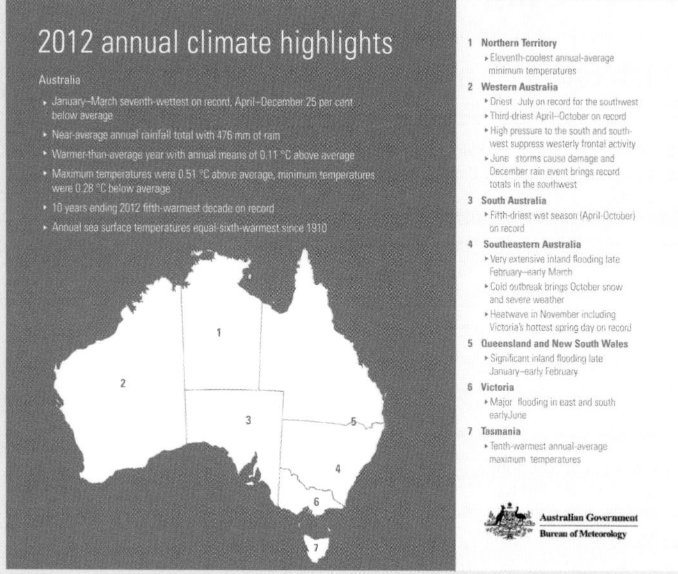

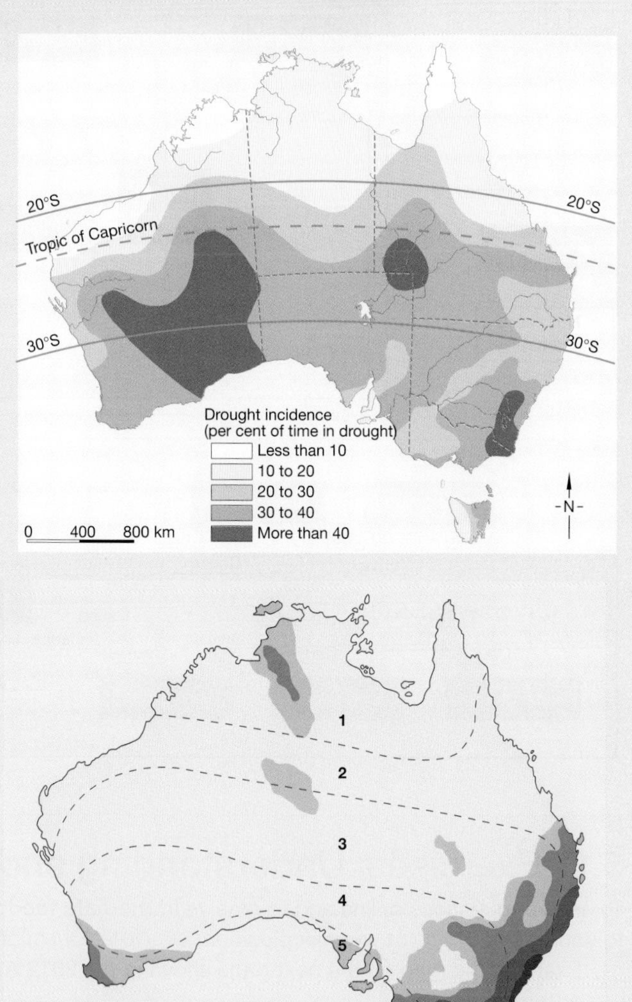

learn on

5.12 Project: Blast off!

Scenario

By 2050, there will be 10 billion people on our small world and the human population will have expanded far beyond a number that the Earth's resources are capable of supporting. If we are to survive into the next century, we will need to find other worlds for settlement and resources. Exoplanets are planets that are found orbiting stars far from our own. So far, 560 exoplanets have been discovered, some of which could be similar enough to Earth for us to colonise. Of course, these star systems are very far away and the people who travel out into space to set up the new colonies will be on board their spaceships for very long periods of time.

The Australian Space Exploration Agency has been formed with the specific aim of sending a crew of 80 colonists to the exoplanet XY2305 — a world that is very similar to Earth — to form the basis of a much larger future settlement. The spacecraft engines that are presently available to the Agency are capable of getting the colonists there with a total journey time of twenty years. As part of the spacecraft design team at the ASEA, you will need to design a spacecraft that will meet all of the survival needs of the crew during their long journey.

The design brief for the spacecraft has the following specifications:

- As the spacecraft will be built in Earth's orbit and will not need to land on the planet at the other end, it does not have to have an aerodynamic shape. It can also be as large as you need; however, keep in mind that the best use should be made of the interior space so that it is easily negotiated by the colonists.
- Apart from an initial intake of supplies, all food, water and oxygen for the journey will need to be grown, recycled or produced on board the ship itself.
- If the ship is to have artificial gravity, the design must include a method of generating this gravitational field.
- Facilities need to be provided for research, sleeping, recreation and exercise.
- There will be equal numbers of male and female colonists who will be aged between 20 and 30 years of age.

Your task

Your group has been given a project brief to design a spacecraft that will be able to provide life support for 80 colonists for their twenty-year journey to the exoplanet XY2305. You will present your final design to the Administration of the ASEA in the form of (i) a PowerPoint demonstration and (ii) a labelled model of your spacecraft.

In your presentation, you will need to consider, among other things:

- the types of activities that an average crew member would be involved in and for how long, in each normal 24-hour period of time onboard the spaceship
- the amount of food, water and oxygen that each person will need every day to perform these activities
- how carbon dioxide would be converted into oxygen
- how water will be produced/recycled
- the different types of waste that will be produced and how these wastes will be managed /recycled.

5.13 Review

5.13.1 Study checklist

Features of ecosystems

- identify examples of biotic and abiotic factors in ecosystems
- measure a range of abiotic and biotic factors in an ecosystem
- use the quadrat method and the capture–recapture method to estimate the abundance of a particular species in an ecosystem

Relationships within ecosystems

- describe the following interactions between species: parasitism, mutualism and commensalism
- construct food chains and food webs
- describe the role of decomposers and detritivores in ecosystems

Energy and materials in ecosystems

- describe the flow of energy through an ecosystem
- identify three types of ecological pyramids
- interpret ecological pyramids
- extract information from food webs
- use cycle diagrams to describe the carbon and nitrogen cycle
- outline the role of chlorophyll in photosynthesis
- recall word and symbol equations for photosynthesis and cellular respiration
- describe the role of photosynthesis and respiration in ecosystems

Human impact on ecosystems

- describe ways in which humans have affected on ecosystem
- identify an introduced species and describe its impact on other populations within an ecosystem

Natural events

- describe the effects of bushfires, floods and droughts on Australian ecosystems
- describe examples of adaptations of animals and plants to Australian ecosystems Find out more about various types of research into drought-tolerant and drought-resistant plants in AustraliaCaution

Individual pathways

ACTIVITY 5.1	ACTIVITY 5.2	ACTIVITY 5.3
Revising ecosystems	Investigating ecosystems	Investigating ecosystems further
doc-8443	doc-8444	doc-8445

learnon ONLINE ONLY

5.13 Review 1: Looking back

Blast off into the future

What will life be like in your future? What sorts of challenges will you face? How can you best prepare yourself for a future that you have not yet seen? What skills will you need to survive?

You might be better able to not only survive, but also thrive, in an unknown future if you can:

- effectively observe, analyse and evaluate your environment
- generate relevant questions to explore perspectives and clarify problems

- selectively apply a range of creative thinking strategies to tackle challenges
- get along with others and respect their differences, viewpoints and ideas
- effectively communicate with others
- take responsible risks and be innovative and willing to tackle challenges
- manage your own learning and reflect on what and how you learn.

One way to develop these skills is to work as a team to solve a challenge or problem. Use the Blast off! situation shown below to find out more about living in a sustainable future and to present some possible solutions to the problem. It is important to realise that sustainability is not just about your environment and individual physical requirements; you also need to consider the social, political and emotional aspects.

Blast off!

Earth is about to explode! You have been selected as one of the humans to be sent into space. Your spacecraft is on a journey that will take 300 years to reach its destination — a planet that your descendants will colonise if they manage to reach it.

Blasting off

This activity works best in teams of four. It can be a brief experience or a more intensive journey, depending on how deeply you delve into each of the questions. You can use the material in the rest of this chapter to help you.

You may also decide to share and rotate various team roles throughout your journey (such as various combinations of project manager, recorder/scribe, devil's advocate, time keeper, researcher, resource manager, celebrator, presenter).

Before you begin, brainstorm ideas for the name of your spaceship. As a team, agree on one of them.

1. Pre-blast thinking

Life needs

(a) Use a table like the one below to list the things you need to survive and suggest how these needs or requirements are met on Earth.

What I need and how I get it

Life requirement	How the life requirement is met on Earth

(b) Compare your list and your suggestions with those of others in the class.

2. Who's going?

In your team, brainstorm and discuss the following points. (Visual thinking tools may be useful to record summaries of your discussions.)

(a) How many people will be going on the journey? Investigate models and theories of population growth to justify your response.

(b) What are their ages and jobs? Construct a table with the headings as shown below to list details of the members of your spacecraft community.

Give reasons why you would include these people and not others.

- How many males and females will you include?
- Why did you select individuals of these ages?
- Why did you select the various occupations of journey members?

Who will be on your spaceship and why?

Name (use your imagination)	Gender	Age	Occupation	Reason why individual selected

3. Fuelling up

(a) Write out a plan of the types of activities that a crew member may be involved in, and for how long, throughout a 24-hour period on your spaceship.
(b) Use your activity plan to help you estimate the amount of energy that would be needed by this crew member in a 24-hour period.
(c) Calculate the average daily amount of energy used by all of the travellers in your spaceship.

4. What's to eat?

(a) As well as energy, your crew will need nutrients such as proteins, carbohydrates, lipids, vitamins and minerals. Individually design a daily menu that would meet the requirements for a particular member of your crew, giving reasons for your selections. Share and discuss your menus with others in your team.
(b) Discuss some ways that these nutrients could be available throughout your 300-year journey.

Daily menu
Name: ....
Occupation: ....
Particular nutritional requirements: ....

Breakfast	Lunch	Dinner	Snacks

5. Home alone

(a) Which types of organisms (and how many) will you take on your journey?
(b) Suggest some examples of food chains and food webs that could exist within your spaceship.

6. Gassed up

(a) Estimate the amount of oxygen that you inhale in an average day. Multiply this by the number of people on your spaceship to get an estimate of the total amount that would be inhaled each day.
(b) Suggest why this oxygen is required.
(c) Estimate the amount of carbon dioxide that you exhale in a average day. Multiply this by the number of people on your spaceship to get an estimate of the total amount that would be exhaled each day.
(d) Discuss with your team some ways in which carbon dioxide may be converted into oxygen aboard your spacecraft. Summarise your discussion in the form of an annotated diagram.

7. Recycle me

(a) On board your spacecraft, different types of waste will be produced. Suggest examples of these wastes and discuss how they could be managed.

(b) Water is also essential for life. Suggest some ways that water could be recycled on your spaceship.

8. Sustainability

(a) Discuss how the idea of an ecological footprint may relate to your spaceship design.

(b) Will the number of people on your spaceship remain the same throughout your journey? Discuss the implications of your response with your team.

(c) Compare your spaceship ecosystem with an artificial ecosystem such as the Biosphere 2 project.

9. Blast off!

(a) As a team, construct a model of your spacecraft. Include labels or a brochure to outline key features and details.

(b) Individually, write a week's diary entry describing life 300 years after leaving Earth. Share your diary entries with each other and with other teams.

(c) Present a summary of your findings for the entire activity using hyperlinked electronic text, PowerPoint, web pages or another creative format.

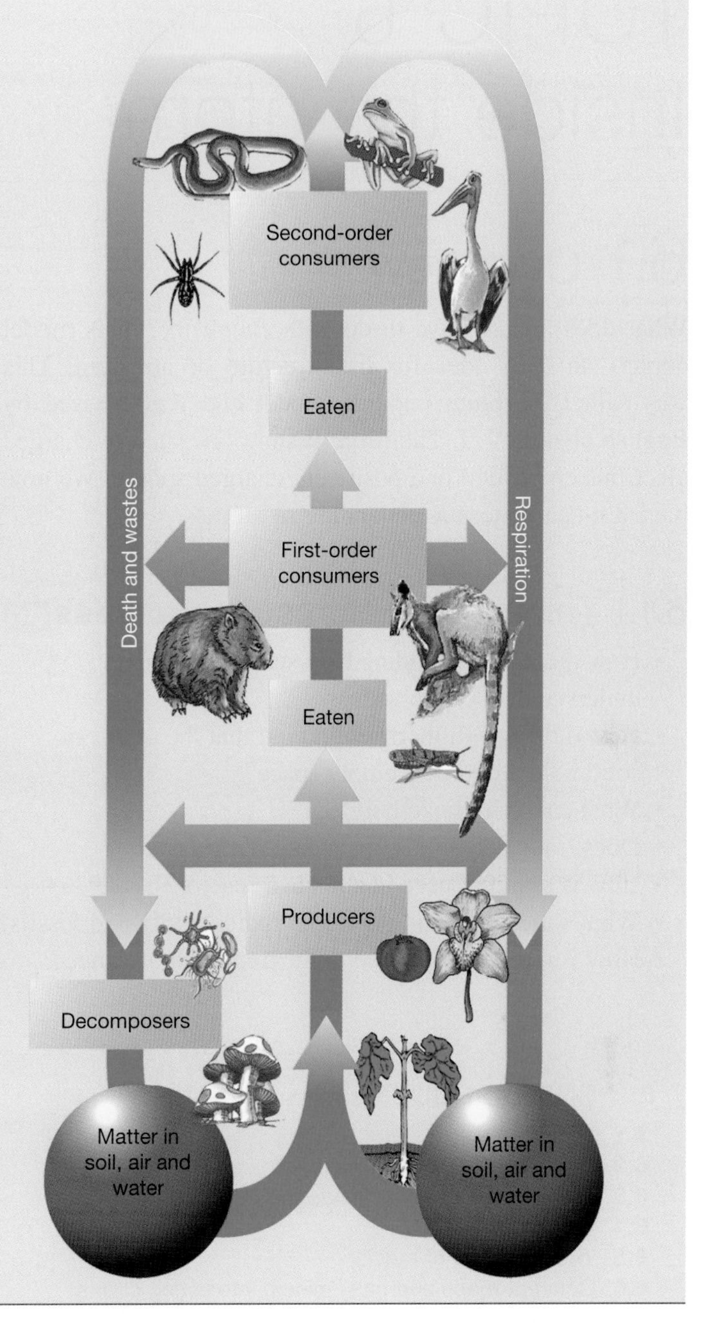

RESOURCES — ONLINE ONLY

Complete this digital doc: Worksheet 5.7: Ecosystems: Puzzles (doc-18879)

Complete this digital doc: Worksheet 5.8: Ecosystems: Summary (doc-18880)

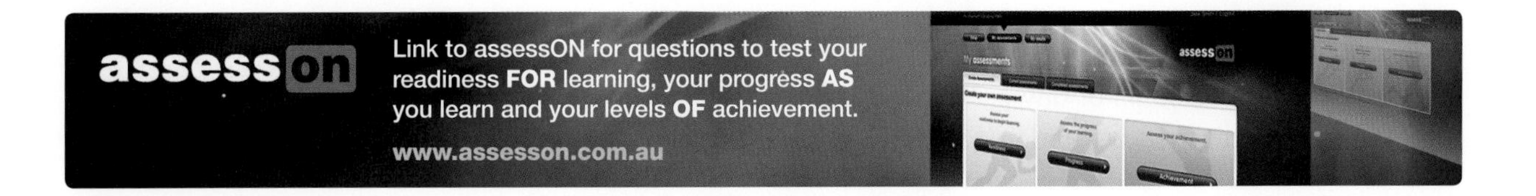

TOPIC 6
Inside the atom

6.1 Overview

What does a cake have to do with chemistry? This model depicts an early idea for the structure of an atom. This was called the plum pudding model and was devised by English chemist J. J. Thomson. It shows negatively charged electrons embedded in a positively charged sphere. We now have a much better understanding of atoms.

6.1.1 Think about atoms **assessON**

- How did a plum pudding help scientists gain an understanding of atoms?
- How did Lord Rutherford find out that the atoms in solid gold are mostly empty space?
- What causes radioactivity?
- Does 'radioactive' always mean 'dangerous'?
- How is uranium used in a nuclear reactor?
- What's the connection between radioactivity and fossils?
- How is radioactivity used in the treatment of cancer?

LEARNING SEQUENCE

Numerous **videos** and **interactivities** are embedded just where you need them, at the point of learning, in your learnON title at www.jacplus.com.au. They will help you to learn the content and concepts covered in this topic.

6.1.2 Do you have the inside information?

All substances are made up of tiny particles. You probably already know quite a lot about the particles inside substances. This knowledge is the first step in your quest to find out why substances behave the way they do.

Your quest

Answer the questions below to find out how much you already know about the inside story on substances.

1. The substances around you and inside you can be placed into three groups — elements, compounds and mixtures.
   (a) Which one of these groups contains substances that are made up of only one type of atom?
   (b) Which one of these groups is the least likely to be found naturally in the Earth's crust?
   (c) What is the difference between a compound and a mixture?
   (d) Arrange the substances listed below into the three groups of substances to complete the affinity diagram below.

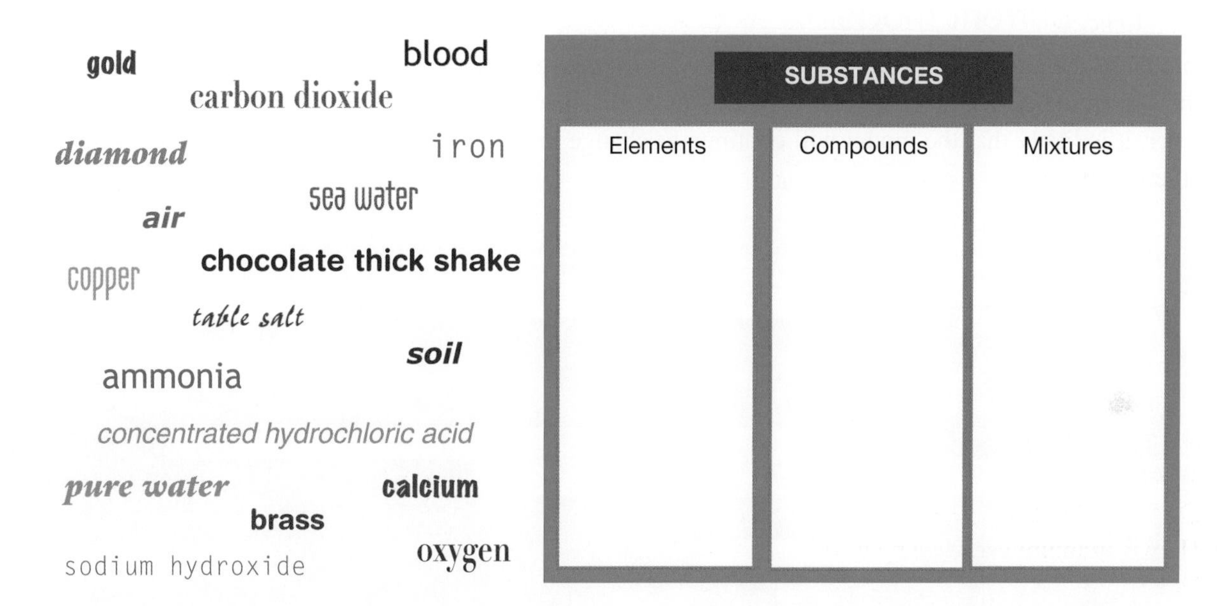

2. Elements, compounds and mixtures are made up of tiny particles called atoms and molecules.
   (a) How is a molecule different from an atom?
   (b) List two elements that can be made up of molecules.
   (c) List two compounds that are made up of molecules.
   (d) Name one compound that is not made up of molecules.
3. Name three particles found inside an atom.
4. Which of the diagrams below represents:
   (a) an atom of an element
   (b) a molecule of an element
   (c) a molecule of a compound?

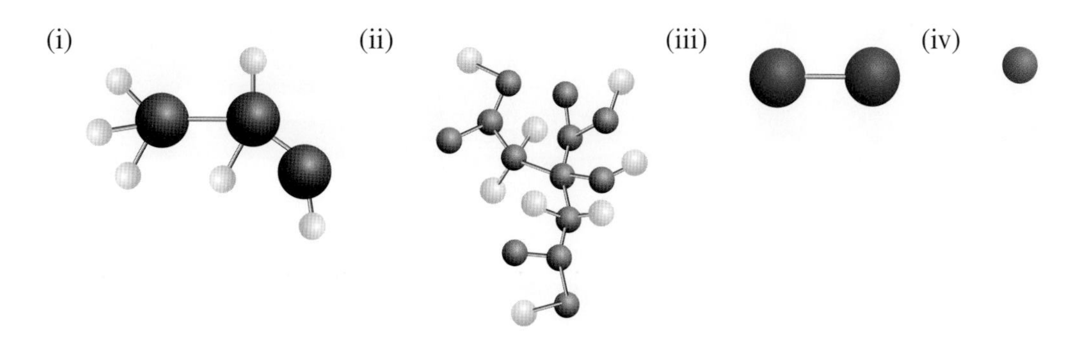

6.2 Chemical building blocks

6.2.1 The mystery of matter

Most of our knowledge about the 'building blocks' of matter that we call atoms is less than 100 years old. But the idea that matter was made up of atoms was first suggested about 2500 years ago by the great philosopher and teacher Democritus. Since then, various theories and models of the atom have been accepted, rejected and modified. The flowchart below shows some of the important developments in our knowledge of the atom.

6.2.2 The current model

The model of the atom accepted today consists of a tiny, dense **nucleus**, made up of **protons** and **neutrons**, which is surrounded by **electrons**. It provides us with an explanation of many observable phenomena. However, it is likely that the model will continue to change or become more detailed as scientists continue with their research.

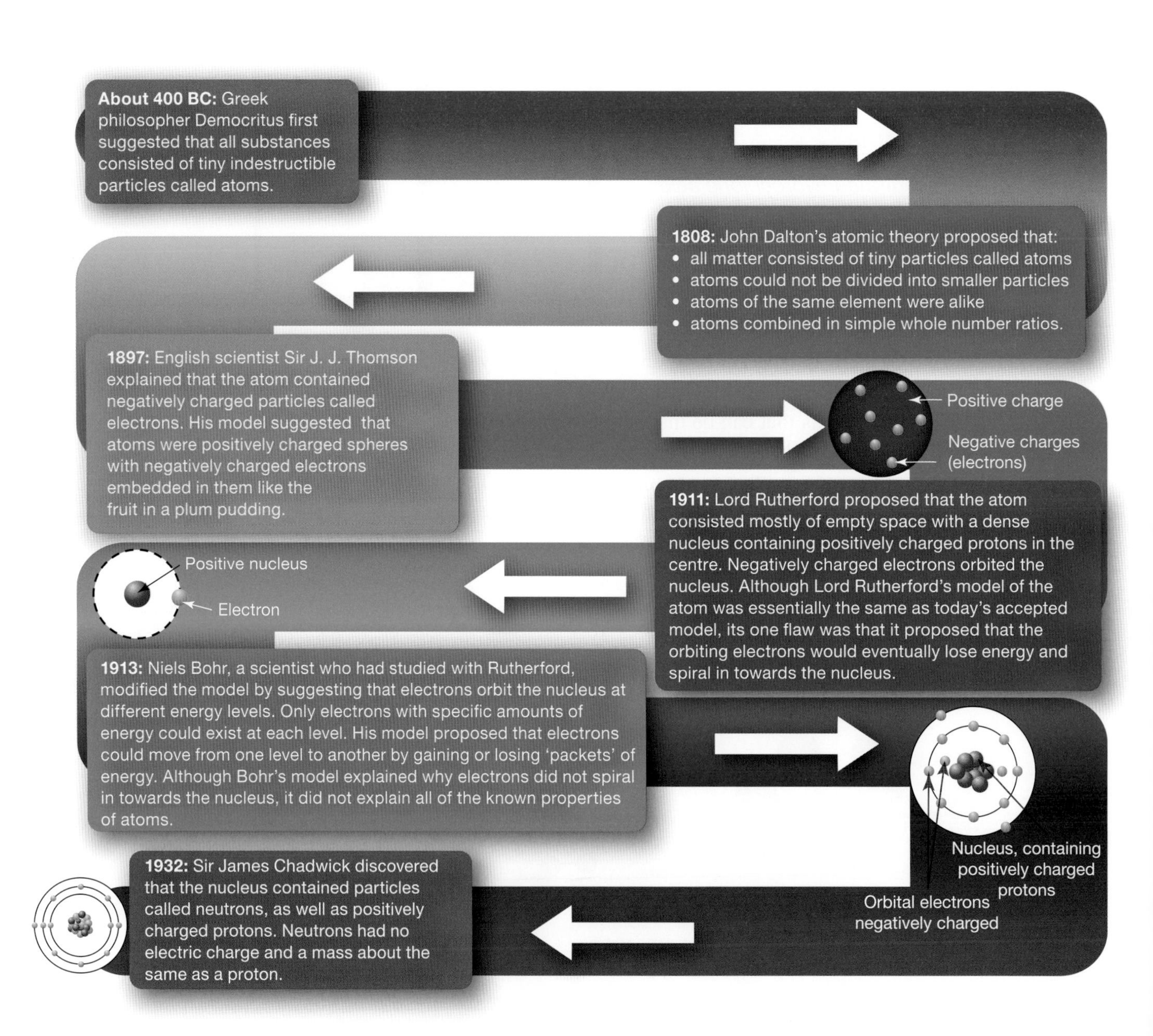

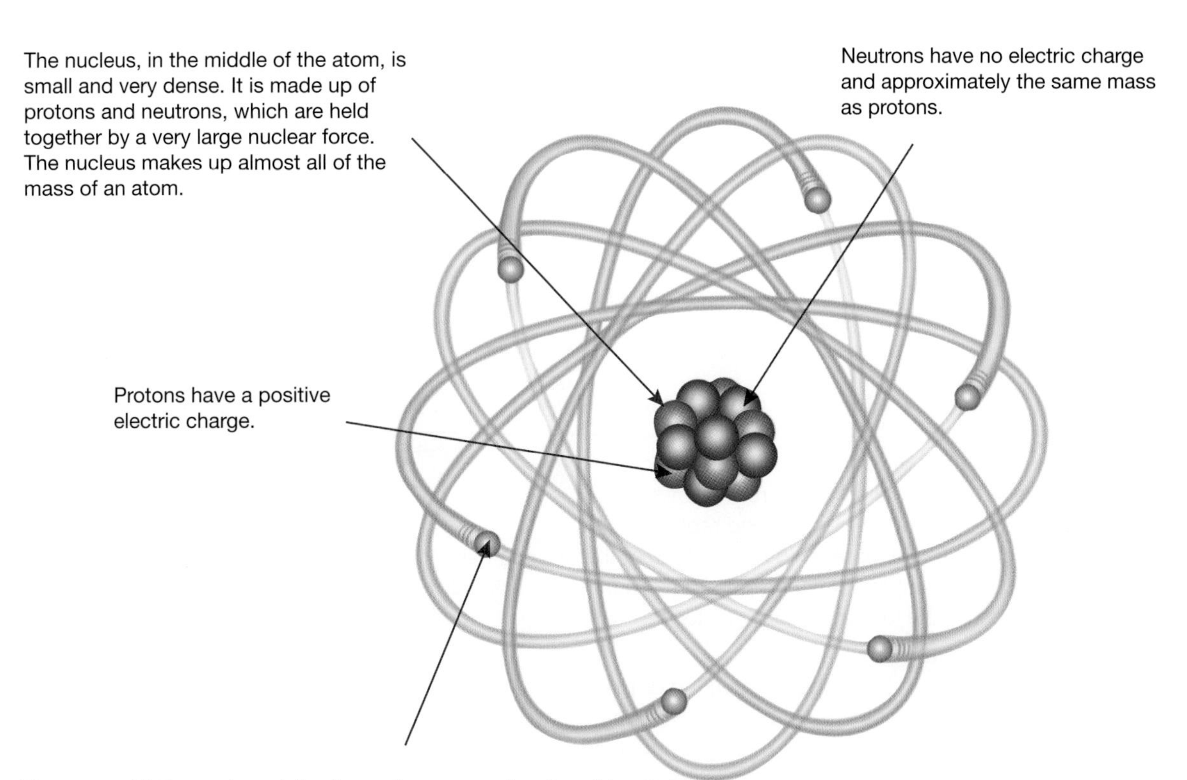

Electrons orbit the nucleus, following paths commonly referred to as electron clouds. Electrons have a negative electric charge equal in size to the positive charge of a proton, and a tiny mass, about $^{1}/1800$ of the mass of a proton or neutron. The number of electrons in an atom is equal to the number of protons in its nucleus. This means that an atom is electrically neutral.

HOW ABOUT THAT!

Even the largest atoms are less than one billionth of a metre across. That's a millionth of a millimetre and about 1/20 000 of the diameter of the finest of human hairs.

HOW ABOUT THAT!

Lord Rutherford's model of the atom was based on experiments in which he fired tiny positive alpha particles at very thin sheets of gold foil. Most of the particles went straight through the gold foil and very few were reflected back. He explained that the few particles that were reflected back were repelled by a very small, positively charged nucleus in the atoms of the gold. Most of the alpha particles, he said, continued through the foil because each gold atom consists mainly of empty space. Lord Rutherford said later that his observations were about as credible as if you had fired a 16-inch shell at a piece of tissue paper and it had come back and hit you!

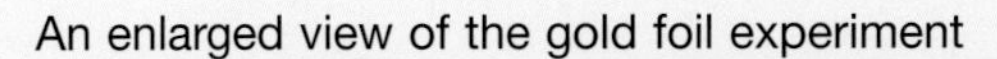
An enlarged view of the gold foil experiment

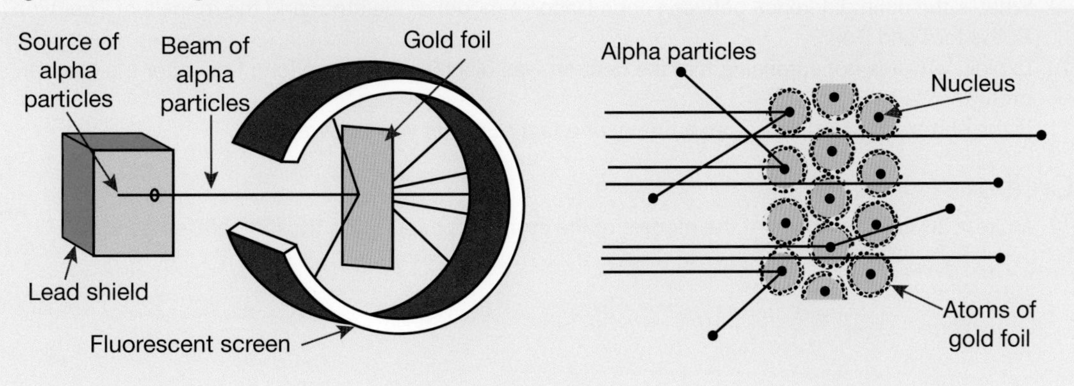

INVESTIGATION 6.1

Locate the nucleus

AIM: To model Rutherford's experiment

Materials:

a hardcover book of at least A4 size
5 plastic soft drink bottle lids
a 10 mm diameter ball bearing or 12 mm diameter marble

A model of Rutherford's experiment

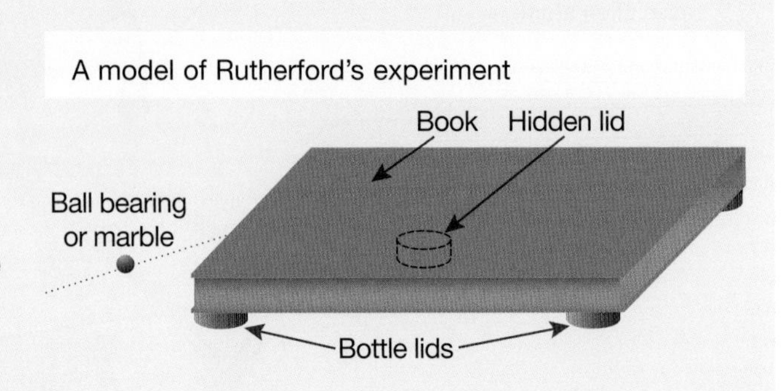

Method and results

- Support the book on a benchtop using a bottle lid under each corner.
- Have one member of your group lift the book, place the fifth bottle lid somewhere in the area surrounded by the other four lids and replace the book. The fifth lid represents the nucleus of the atom in this model.
- After the other members of your group turn around, they take turns to roll the ball bearing or marble under the book to find the location of the 'nucleus'.

Discuss and explain

1. Record the number of times the ball bearing or marble is rolled before striking the 'nucleus' for the first time.
2. Comment on how difficult it is to locate the 'nucleus' in this model.
3. What is represented in this model of Rutherford's experiment by:
   (a) the area under the book in the area surrounded by the four lids?
   (b) the ball bearing or marble?
4. Comment on the weaknesses of this model of Rutherford's experiment.

6.2 Exercises: Understanding and inquiring

To answer questions online and to receive **immediate feedback** and **sample responses** for every question, go to your learnON title at www.jacplus.com.au. *Note:* Question numbers may vary slightly.

Remember

1. Describe the 'plum pudding' model of the atom proposed by J. J. Thomson.
2. Why did most of Rutherford's alpha articles go through the thin sheets of gold foil?
3. What was the main weakness of the Rutherford model of the atom?
4. Describe, with the aid of a labelled diagram, the modern view of the structure of the atom.
5. Name the three important particles that make up an atom.
6. Where is most of the atom's mass located?
7. How are protons different from neutrons? How are they similar?
8. Describe the differences between protons and electrons.

Think

9. What is the main difference between John Dalton's model of the atom and the models of Thomson, Rutherford and Bohr?
10. Explain why it is not surprising that the neutron was discovered quite a long time after the electron and proton.
11. Is the current model of the atom a theory or a fact? Explain your answer.

Create

12. Make a 3D version of one of the models of the atom proposed since the time of Democritus.

Investigate

13. Find out more about one of the following scientists and describe their contributions to our knowledge about the structure of the atom. In your report you need to include:
    - full name, place of birth, date of birth and death
    - a brief description of the type of work the scientist did in his/her lifetime
    - their contribution to our understanding of the structure of the atom
    - the technology available to the scientist that enabled him/her to make the discovery
    - a description of how relevant the scientist's theory is to today's understanding of the structure of the atom.

    Choose from: John Dalton, Sir William Ramsay, Marie Curie, J. J. Thomson, Henry Moseley, Max Planck, Eugen Goldstein, Lord Rutherford, Frederick Soddy, Sir James Chadwick, Niels Bohr, Louis de Broglie.
14. Today we know of many particles smaller than neutrons and protons. Find out more about some of these particles and how they are investigated.

learnon RESOURCES — ONLINE ONLY

Watch this eLesson: The atom (eles-1775)

Complete this digital doc: Worksheet 6.1: Chemical building blocks (doc-18881)

Complete this digital doc: Worksheet 6.2: How big is an atom? (doc-18882)

6.3 Stability and change: Inside the nucleus

6.3.1 The core of the atom

At the centre of every atom is a tiny, solid core called the nucleus. Within the nucleus, protons and neutrons are usually held together by incredibly strong forces. Some of the mysteries of radioactivity can be unravelled by taking a closer look inside the nucleus.

6.3.2 Neutrons and isotopes

All atoms of a particular element have the same number of protons. However, often the number of neutrons in atoms of the same element is different. Such atoms have the same atomic numbers but different mass numbers. Atoms of the same element with different mass numbers are called **isotopes**. Most elements exist as two or more isotopes. These isotopes all have the same chemical properties, but slightly different masses.

Hydrogen, for example, has three isotopes. Each of the three isotopes has one proton. However, the different isotopes have 0, 1 or 2 neutrons respectively.

Naming isotopes

In symbols, isotopes are represented as ${}^{A}_{Z}E$, where:
A = the mass number; the sum of the number of neutrons and number of protons in the nucleus
Z = the atomic number; the number of protons in the nucleus
E = the symbol of the element.

In words, isotopes are described by using the element name and the mass number. For example, the isotope of hydrogen that has two neutrons has a mass number of 3 and an atomic number of 1. It is therefore represented as ${}^{3}_{1}H$ or, in words, hydrogen-3.

The three isotopes of hydrogen. Hydrogen-2 and hydrogen-3 are also known as deuterium and tritium respectively.

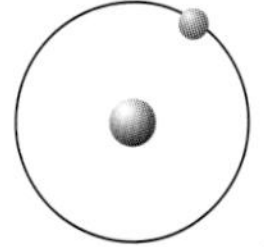

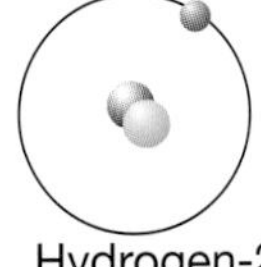

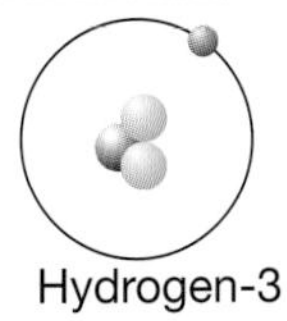

WHAT DOES IT MEAN?

The word *isotope* is derived from the Greek words *isos*, meaning 'equal', and *topos*, meaning 'place'. It came about because even though each isotope of the same element had different numbers of neutrons and therefore different weights, they occupied the same place on the periodic table of the elements.

Stable or unstable

In most atoms, the protons and neutrons found in the nucleus are held together very strongly. The nuclei of these atoms are said to be **stable**. However, in some atoms the neutrons and protons in the nucleus are not held together strongly. These nuclei are **unstable**. Consequently, some isotopes of elements are stable and some are unstable. Isotopes that are unstable decay to form other elements. These isotopes are said to be radioactive and are called radioactive isotopes, or **radioisotopes**. For example, two isotopes of carbon, carbon-12 and carbon-14, have identical chemical properties. However, the nucleus of carbon-14 is not stable and disintegrates naturally. Carbon-12 is a stable isotope while carbon-14 is a radioactive isotope.

Element	Symbol	Number of protons	Number of neutrons	Stable or radioactive?
Carbon-12	$^{12}_{6}C$	6	6	Stable
Carbon-14	$^{14}_{6}C$	6	8	Radioactive
Uranium-235	$^{235}_{92}U$	92	143	Radioactive
Uranium-238	$^{238}_{92}U$	92	146	Stable

6.3.3 Natural and artificial radioactivity

Natural radioactivity is radioactivity emitted from matter without energy being supplied to atoms. There are about 50 isotopes that emit radioactivity naturally. They exist in the air, in water, in living things and in the ground. Most radioactive isotopes (about 2000 in total) are made radioactive artificially by bombarding their atoms with sub-atomic particles like protons and neutrons.

6.3.4 Three of a kind

The energy emitted by radioactive substances is called **nuclear radiation** because it comes from the nucleus. Lord Rutherford showed that there were three different types of nuclear radiation: **alpha particles**, **beta particles** and **gamma rays**.

Alpha particles

Alpha particles are helium nuclei that contain two protons and two neutrons. Alpha particles are positively charged. They cannot travel easily through materials and can be stopped by a sheet of paper or human skin. They pose little hazard to the external body but can cause serious damage if breathed in, eaten or injected. The symbol for alpha particles is α.

Alpha decay: a nucleus ejects a helium nucleus.

Parent nucleus

Daughter nucleus

Alpha particle

Beta particles

Beta particles are the same size and mass as electrons, can have a negative or positive electric charge and can travel at speeds as high as 99 per cent of the speed of light. Beta particles can penetrate human skin and damage living tissue, but they can not penetrate thin layers of plastic, wood or aluminium. The symbol for beta particles is β.

Beta decay: a nucleus ejects an electron.

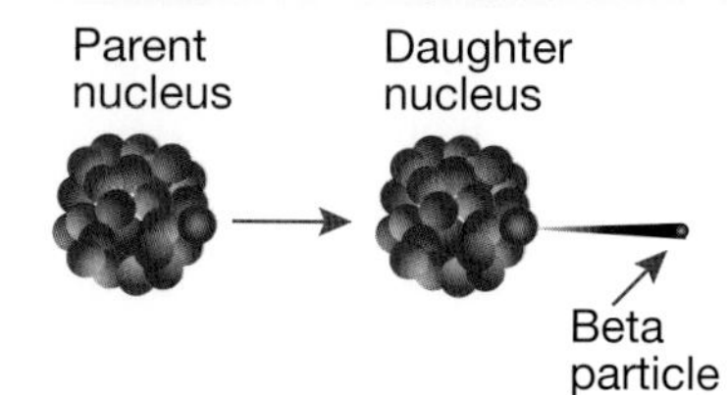

Gamma rays

Gamma rays are not particles, but bursts of energy released after alpha or beta particles are emitted. Gamma rays travel at the speed of light and are highly penetrating. They can cause serious and permanent damage to living tissue and can be stopped only by a thick shield of lead or concrete. The symbol given to gamma rays is γ.

Gamma decay: an excited nucleus ejects a photon.

Parent nucleus

Daughter nucleus

Gamma ray

The different penetrating powers of alpha (α), beta (β) and gamma (γ) radiation

α particles — absorbed in a few centimetres of air, or by a piece of paper or layer of dead skin

γ rays — barely affected by air; absorbed in many centimetres of lead

β particles — absorbed in about 100 cm of air, or a few centimetres of wood

Paper

Wood

Lead

Concrete

6.3.5 The lives and half-lives of radioisotopes

The nuclei of different radioactive substances decay at different rates. Some radioisotopes decay in a few seconds, while others take thousands of years. The time taken for half of all the nuclei in a sample of a radioisotope to disintegrate or decay is known as its **half-life**.

There are three naturally occurring isotopes of uranium; uranium-238, uranium-235 and uranium-234. Each of the isotopes spontaneously disintegrates or decays, producing alpha particles and gamma rays. Each isotope has its own half-life; that is, the time taken for the concentration to fall to half its initial value. Half-lives can vary from microseconds to billions of years. The half-lives of each of the uranium isotopes are more than a billion years.

A graph showing the radioactive decay of strontium-90, which has a half-life of 28 years

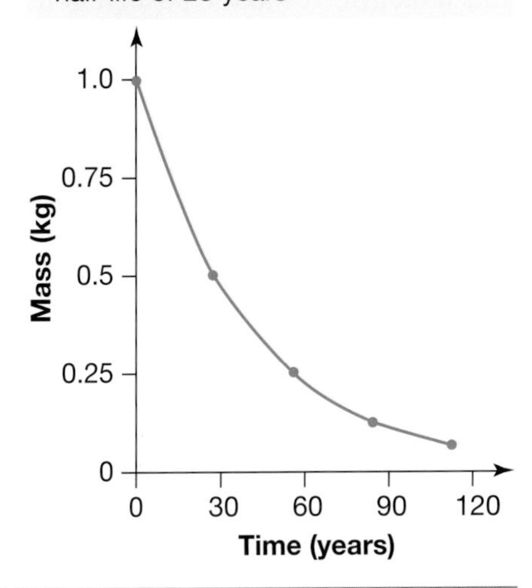

Number of half-lives	Fraction remaining
1	$\frac{1}{2}$
2	$\frac{1}{4}$
3	$\frac{1}{8}$
4	$\frac{1}{16}$

6.3.6 In the background

We are all exposed to background radioactivity every day. Fortunately it is quite safe. Most of it comes from naturally occurring radioactive elements in the Earth's atmosphere and crust. A smaller amount comes from outer space in the form of **cosmic radiation**, mostly in the form of high energy protons emitted by stars, including the sun. The word cosmic comes from the Greek word *kosmos*, meaning 'universe'. The Earth's atmosphere protects us from the dangers of cosmic radiation. There are even small amounts of radio-isotopes in the human body, including hydrogen-3 (tritium), carbon-14 and potassium-40.

HOW ABOUT THAT!

Radioactivity was discovered by accident. French physicist Henri Becquerel discovered radioactivity while investigating the fluorescence of uranium salts in 1896. When he developed a photographic plate that had been left in a drawer near his benchtop, he found that it had been fogged up by radiation from the uranium salts.

This effect of radioactivity is now used in a protective device worn by people who work with radioactive materials. The 'fogging' of the film in this device measures the amount of radioactivity they have been exposed to.

Becquerel was the first scientist to report the effects of radioactivity on living tissue. He suffered from burns on his skin as a result of carrying a small quantity of the element radium in his pocket.

6.3.7 Smoke alarms

Inside a smoke alarm, in the ionisation chamber, there are two plates that are oppositely charged. There is also a tiny amount of americium-241, which has a half-life of 432 years. Americium-241 atoms emit alpha radiation. The alpha particles knock electrons off the molecules in the air. This creates positive particles and free electrons. The positive particles are attracted to the negative plate, and the electrons are attracted to the positive plate. A small current is set up.

When smoke particles are drawn into the smoke alarm, they attach themselves to the positive ions, making them neutral and disrupting the current. This change is sensed by the detector and the siren sounds.

What happens in the ionisation chamber?

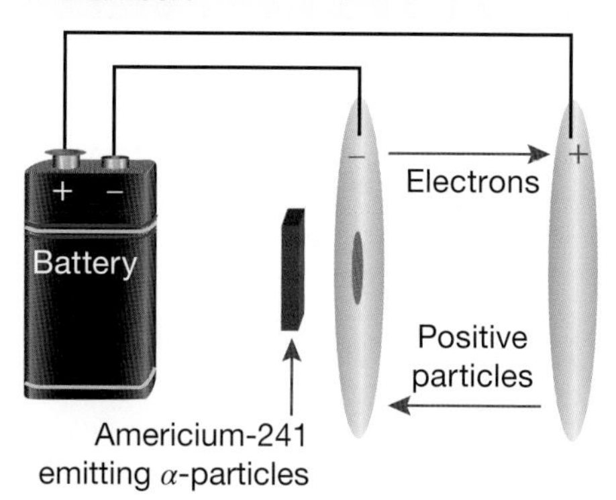

6.3 Exercises: Understanding and inquiring

To answer questions online and to receive **immediate feedback** and **sample responses** for every question, go to your learnON title at www.jacplus.com.au. *Note:* Question numbers may vary slightly.

Remember

1. How are isotopes of the same element different from each other?
2. In the symbol $^{A}_{Z}E$, what is represented by
   (a) the letter A
   (b) the letter Z?

3. Why are the isotopes of some elements radioactive?
4. Write down the type of nuclear radiation described by the following statements.
   (a) A radioactive particle that has the same size and mass as an electron
   (b) A radioactive particle that is made up of two protons and two neutrons
   (c) The type of radiation that can penetrate the human body and can be stopped only by a thick shield of lead or concrete
   (d) A radioactive particle that can travel almost at the speed of light
5. What electric charge is carried by an alpha particle?
6. How are we protected from cosmic radiation from outer space?

Using data

7. A scientist wished to determine the type of radiation emitted by a radioisotope. She had three materials (paper, plastic and lead) and an instrument called a Geiger counter, which detects nuclear radiation. She covered the radioisotope with each of the three materials and measured the radiation that passed through each material. The results of her experiment are shown in the table below.

Results of radioactivity experiment

Material	Effect on Geiger counter readings
Paper	No effect on readings
Plastic	Readings fell by two-thirds
Lead	Large fall in readings

   What type of nuclear radiation does this radioisotope emit? Explain your answer.

Think

8. About 0.01 per cent of the potassium in your body is the radioisotope $^{40}_{19}K$.
   (a) How many protons and neutrons are in each atom of this radioisotope?
   (b) The stable nuclei of potassium atoms have one less neutron than the nuclei of potassium's unstable radioisotope. Write down the complete symbol for the stable isotope of potassium.
9. Are the atoms $^{230}_{93}X$ and $^{239}_{94}Y$ isotopes of the same element? Explain.
10. An atom of uranium-238 ($^{234}_{90}U$) decays by emitting a single radioactive particle. The atom formed as a result of the decay is thorium-234 ($^{234}_{90}Th$). What type of radioactive particle is emitted? Explain how you got your answer.
11. The half-life of an isotope of tritium is 4500 days. How many days will it take an amount of tritium to fall to a quarter of its initial mass?

Investigate

12. Find out which radioactive gas in the atmosphere is responsible for most of the background radiation we are exposed to on Earth.

Analyse and evaluate

13. The graph at right shows the decay of a radioisotope over four minutes.
    (a) What is the half-life of this isotope?
    (b) How many radioactive particles would be left after five minutes?
    (c) When the decay takes place in a sealed container, helium gas is collected. Name one type of radiation produced in the decay.

Graph of radioactive decay of an isotope

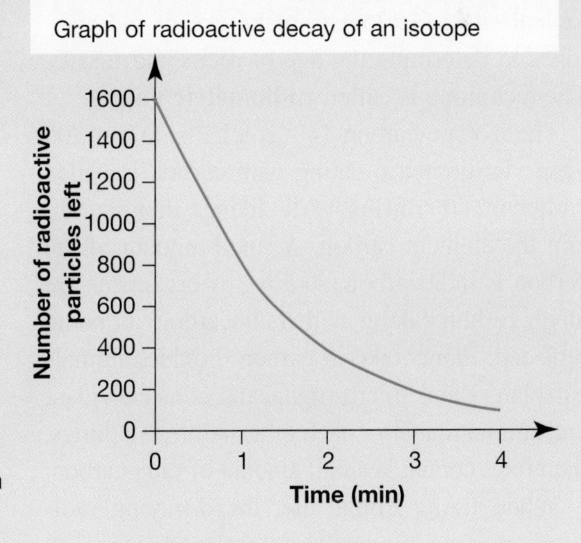

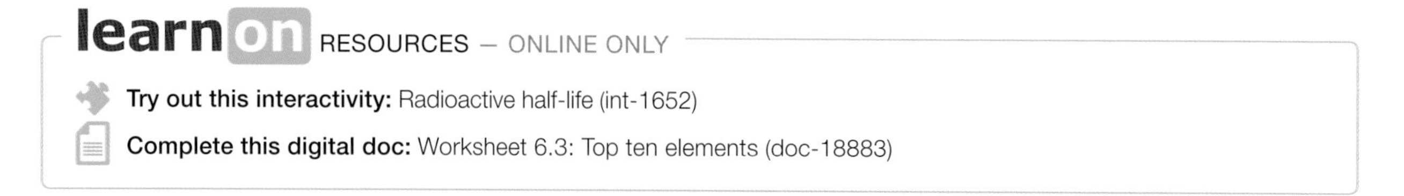

6.4 Using radioactivity

Science as a human endeavour

6.4.1 Radioisotopes

In 1903, Marie Curie, her husband Pierre, and Henri Becquerel were awarded the Nobel Prize in Physics for their discovery of radioactivity and their work on uranium. Little did they know that their discoveries and investigations would change the course of history.

They could not have imagined that their work would lead to the development of nuclear weapons capable of killing millions of people, nuclear power plants that generate electricity, and radioactive isotopes that can be used to treat cancers and detect life-threatening illnesses.

Radioisotopes are used in industry and research, and also have medical applications. They can be used as radioactive 'tracers' to follow the movement of substances through liquids (for example, sediment movement in rivers and the movement of substances in the blood). Radioactive isotopes are also used in smoke detectors, soil analysis, pollution testing, measuring the thickness of objects and criminology.

6.4.2 Radiometric dating

Naturally occurring radioisotopes can be used to calculate the age of samples from archaeological sites. Geologists make use of radioisotopes to determine the age of rocks and fossils. The technique is called **radiometric dating**.

The isotope carbon-14 has a half-life of 5700 years. Radiometric dating with carbon is called **radiocarbon dating**. All living things contain the element carbon. A small amount of the carbon is radiocarbon. As long as organisms are alive, carbon (along with radiocarbon) is being replaced. Plants take in carbon dioxide, animals eat plants, and micro-organisms consume plant and animal matter or each other. All living things, therefore, contain a small amount of radiocarbon.

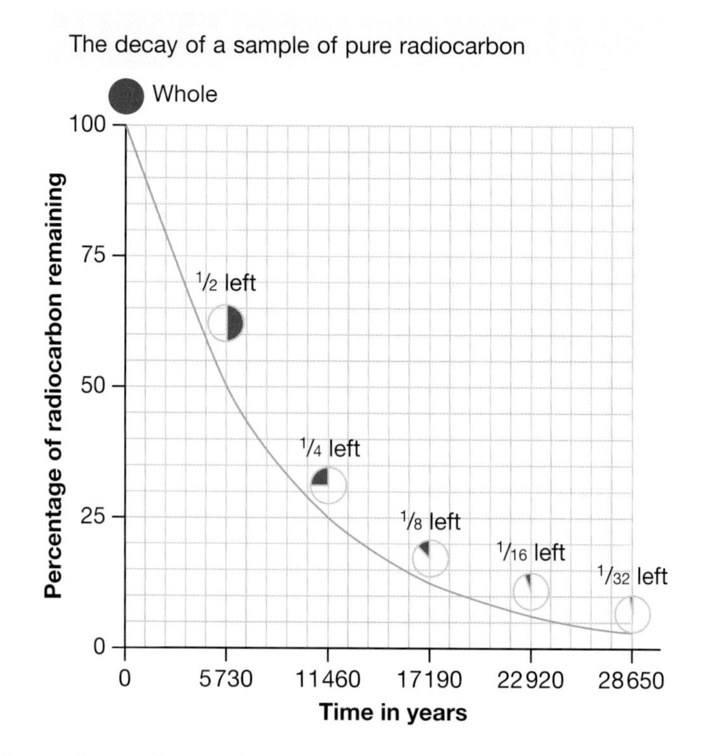

When living things die, the decaying radiocarbon is no longer being replaced. Since all fossils were once living, their age can be determined by measuring the amount of radiocarbon remaining. After 5700 years, only half of the usual amount of radiocarbon will be left. A graph can be used to estimate the age of a fossil. After about 50 000 years, the amount of radiocarbon becomes too small to measure accurately.

All rocks contain small amounts of radioactive elements such as uranium and potassium. The age of older rocks, and the fossils within them, can be determined by using radioactive elements with longer half-lives.

6.4.3 Radioisotopes and nuclear power

The radioactive properties of uranium are used in the generation of electricity in **nuclear reactors**. Australia is one of several countries that have large high-grade deposits of uranium. Uranium is converted to uranium dioxide and then sealed in rods, called **fuel rods**. The uranium undergoes a **fission** reaction in the reactor when neutrons are fired at the radioactive uranium. This causes the uranium nuclei to split and form two new elements, releasing neutrons, radiation and heat in the process. This heat energy is used to heat water to produce steam, which is used to turn the turbines that generate the electricity.

Fast breeders

In some countries fast breeder reactors use the artificial radioisotope plutonium-239 as a fuel. Plutonium-239 is made by bombarding uranium-238 with fast moving neutrons (that's why the term 'fast breeder' is used). The plutonium-239 produced is also used to produce nuclear weapons.

Nuclear waste

The used fuel rods in a nuclear reactor are radioactive and contain a mixture of radioisotopes. Some of the waste radioisotopes have half-lives of only minutes, while others have half-lives of thousands of years. These waste products are currently sealed in steel containers or glass blocks and stored in power stations or buried deep at sea or underground away from groundwater. There is, however, still no permanent solution to the problem of disposing of nuclear waste.

It has been suggested that nuclear waste should be sent by rocket to the sun or into outer space. However, the risk of a rocket carrying nuclear waste exploding before leaving the Earth's atmosphere makes that solution very risky.

A worker inspecting output at a nuclear power plant

6.4.4 Radiotherapy in the treatment of cancer

Radiotherapy is the use of radioisotopes, or other radiation such as X-rays, to kill cancer cells or prevent them from multiplying. It can be targeted at a small area so that surrounding tissue is not damaged. Radiotherapy is often used along with other treatments such as surgery or chemotherapy.

An example of a nuclear fission reaction

Nuclear equation: $^{235}_{92}U + ^{1}_{0}n \longrightarrow ^{141}_{56}Ba + ^{92}_{36}Kr + 3^{1}_{0}n$

Radiation can be directed at the cancer by a machine like the one at right. This method is known as **external radiotherapy**. The other method, known as **internal radiotherapy** or **brachytherapy**, involves placing radioisotopes inside the body at or near the site of the cancer. In some cases both methods are used. The type of treatment depends on the type of cancer, its size and its location as well as the general health of the patient.

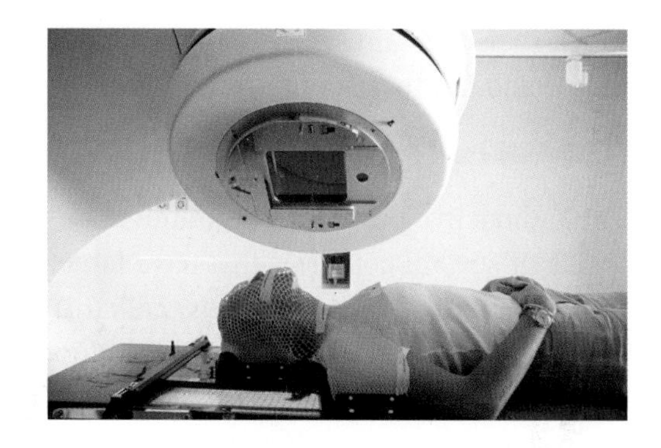

6.4.5 Radioisotopes in the diagnosis of disease

Radioactive substances may be inserted into the body to detect or identify the cause of disease. The radiation produced by the substance while it is in the part of the body under investigation is measured to diagnose the problem.

Some radioisotopes can be used to obtain images of parts of the body. The gamma rays emitted by these radioisotopes are used to produce the images. PET (positron emission tomography) scans use cameras surrounding the patient to detect gamma rays coming from radioisotopes injected into the body.

A PET image of the human brain

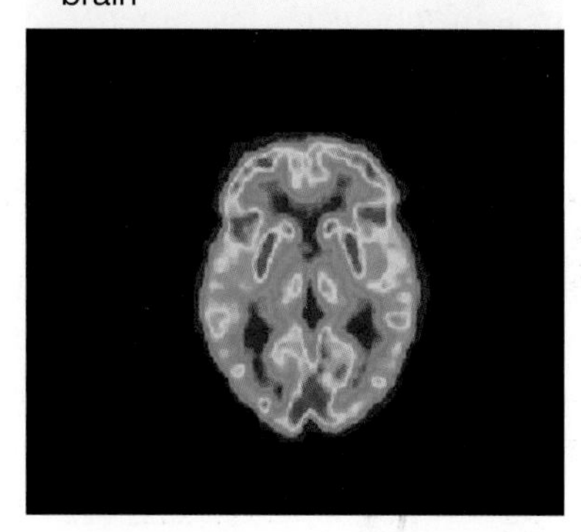

WHAT DOES IT MEAN?

The word *fission* comes from the Latin word *fissio*, meaning 'to split'.

Some of the radioisotopes used in the treatment and diagnosis of disease

Radioisotope	Use	Half-life
Phosphorus-32	Treatment of leukaemia	14.3 days
Cobalt-60	Used in radiotherapy for treating cancer	5 years
Barium-137	Diagnosis of digestive illnesses	2.6 minutes
Iodine-123	Monitoring of thyroid and adrenal glands, and assessment of damage caused by strokes	13 hours
Iodine-131	Diagnosis and treatment of thyroid problems	8 days
Iron-59	Measurement of blood flow and volume	46 days
Thallium-201	Detection of damaged heart muscles	3 days

RESOURCES — ONLINE ONLY

Watch this eLesson: Nuclear medicine (eles-1084)

6.4.6 Preserving food

If you've ever suffered from food poisoning you will understand why it is necessary to keep food from spoiling. Food in sealed containers can be preserved by exposing it to gamma radiation. The radiation kills the micro-organisms in the food and keeps it from spoiling. However, there has been much controversy about the safety of food that has been treated in this way.

INVESTIGATION 6.2

Radioactive decay

AIM: To investigate the decay of a radioisotope

Materials:

graph paper

Method and results

1. The half-life of the radioisotope iodine-131 is 8 days. Calculate the amount of iodine-131 left after 8, 16, 24, 32, 40, 48, 56, 64, 72 and 80 days if 100 g is given to a patient to treat a thyroid problem. Present your information in a table.
2. Draw a graph showing how the radioisotope decays. Make the horizontal axis represent time and the vertical axis represent the amount of radioisotope left.

Discuss and explain

3. What fraction of the iodine-131 is left after:
   (a) 8 days
   (b) 16 days
   (c) 24 days
   (d) 80 days?
4. Why is it difficult to store radioisotopes with short half-lives?

6.4 Exercises: Understanding and inquiring

To answer questions online and to receive **immediate feedback** and **sample responses** for every question, go to your learnON title at www.jacplus.com.au. *Note:* Question numbers may vary slightly.

Remember

1. What is the name of the nuclear reaction that takes place in nuclear power stations?
2. Describe three uses of radioactive elements.
3. What is radiotherapy and how does it prevent the spread of cancer through the body?
4. How is internal radiotherapy different from external radiotherapy?
5. How do radioisotopes used in food preservation stop food from spoiling?

Think

6. Is iodine-131 a more stable radioisotope than barium-137? Explain.
7. The use of barium-137 in the diagnosis of digestive illnesses involves the patient drinking it in a syrup. What property of barium-137 makes its use quite safe?

Imagine

8. It was Marie Curie who invented the word 'radioactivity' to describe the disintegration of the nucleus. What would Marie Curie think if she were still alive today and could see both the good and bad effects of radioactivity? Would she be proud? Would she be disappointed? Would she be angry? Imagine that you are Curie and write a letter explaining your feelings.

Investigate

9. Research the topic 'nuclear reactors' and find out:
   (a) what they are built from
   (b) what fuel rods and control rods are
   (c) what type of nuclear reaction occurs in the reactor
   (d) how the reactor is kept cool
   (e) how electricity is generated
   (f) what kinds of safety features are used.
10. Radiotherapy is an effective method of treating cancer. However, it has a number of side effects. Find out what the side effects are.

Analyse and evaluate

11. Use the graph at the beginning of subtopic 6.4 to answer the following questions.
   (a) Parts of the skeleton of a large animal are found buried in sand dunes. The amount of radioactive carbon-14 in the bones is about one-eighth of that found in the skeletons of living animals. How long ago did the animal probably die (to the nearest thousand years)?
   (b) Approximately what percentage of the original amount of radioactive carbon-14 would you expect to find in:
      (i) an Aboriginal spear 11 000 years old
      (ii) a skull 23 000 years old, found in a cave?

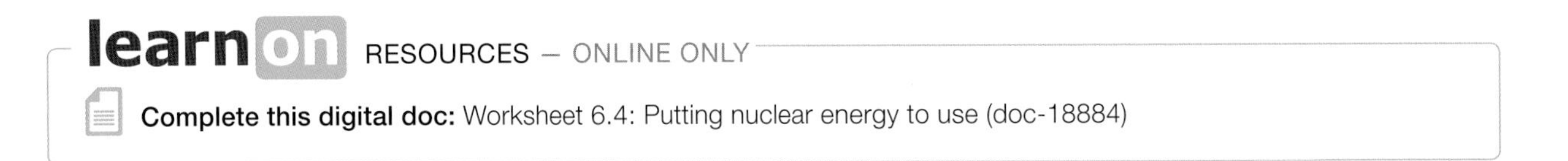

6.5 The dark side of radiation

6.5.1 The dark side of radioactivity

While nuclear radiation has many uses that are beneficial to society as a whole, there is no doubt that it is very much a two-edged sword. For every person whose life has been saved by radiotherapy or a smoke detector, there is someone who remembers the toll taken by Chernobyl, Fukushima, Hiroshima and Nagasaki.

Nuclear radiation can have a devastating effect on living things. Exposure to large doses of radiation can cause immediate effects such as nausea, headaches, vomiting and diarrhoea. Nuclear radiation damages living cells, and too much exposure can lead to diseases such as leukaemia, cancer and immune system collapse later in life. It can also damage the reproductive mechanisms in cells, including DNA, leading to birth defects in the offspring of exposed organisms. It is a sad irony that Marie Curie herself died of leukaemia at the age of 67. Her illness was almost certainly caused by her constant exposure to radioactivity.

6.5.2 When reactors go wrong

Like any other piece of complex technology, a nuclear reactor can work safely only if its many individual systems are functioning smoothly and efficiently. They must be well-maintained and well-managed by highly trained personnel. Unfortunately, in many cases the flaws of a nuclear reactor's design are not spotted until it is too late.

Chernobyl 1986

Reactor 4 was an old design that used graphite moderators, used water as a coolant and had no radiation containment shields around the reactor cores. On 25 April 1986, Reactor 4 at Chernobyl was scheduled to be shut down for routine maintenance. Due to a series of operational errors, nearly all of the control rods were withdrawn from the core to compensate for a power loss. This caused the reactor to become rapidly unstable and fission started to occur too quickly. While an attempt was made to fully insert all of the control rods (absorbing all of the neutrons in the core and stopping the fission reaction), a reaction with the graphite tips of the control rods suddenly caused an uncontrollable power surge in the reactor. In 4 seconds, the power rocketed up to 100 times its normal value and the reactor core reached 5000 °C (about the same temperature as the surface of the sun), causing some of the fuel rods to rupture. The hot fuel particles hit the cooler water and caused a steam explosion that destroyed the reactor core. The graphite core caught fire and, because it had no containment shield, some of the vaporised radioactive fuel went into the atmosphere.

While only two people were killed in the original explosion, three others died during the night and fifty emergency workers died from acute radiation poisoning. Since the accident, the rate of thyroid cancer in children has been ten times higher in the region around Chernobyl and, of the 600 000 people contaminated by radiation, 4000 have died from long-term cancers.

Pripyat in the Ukraine was home to 50 000 people, most of whom had jobs at Chernobyl. When Reactor 4 of the Chernobyl nuclear power plant exploded, the town was abandoned. Now, nature is starting to reclaim it despite the remaining radiation.

The remains of Chernobyl's Reactor 4 after the explosion

Fukushima 2011

The Fukushima Daiichi nuclear disaster was caused by a series of unlucky events occurring one after another. On 11 March 2011 a massive earthquake occurred off the coast of Honshu (the main island of Japan) leaving the Fukushima nuclear reactor complex relatively unharmed but reliant on its back-up generators. Unfortunately, the earthquake caused a tsunami that struck the coast of Honshu less than an hour later, killing more than 19 000 people and destroying over 1 000 000 buildings. The reactors at Fukushima Daiichi were flooded by the 15 m high tsunami, disabling 12 of the 13 back-up generators as well as the heat exchangers that released waste heat into the sea. Without power, the circulation of water coolant around the reactor cores ceased, causing them to become so hot that much of the coolant water was boiled off. The heat became high enough to melt the fuel rods in reactors 1, 2 and 3 (this is referred to as a **melt-down**). A reaction between the cladding of the melted fuel rods and the remaining coolant water produced hydrogen gas that exploded when mixed with the air. This threw

Map showing the amount of radiation absorbed per hour at ground level around Fukushima six weeks after the melt-down

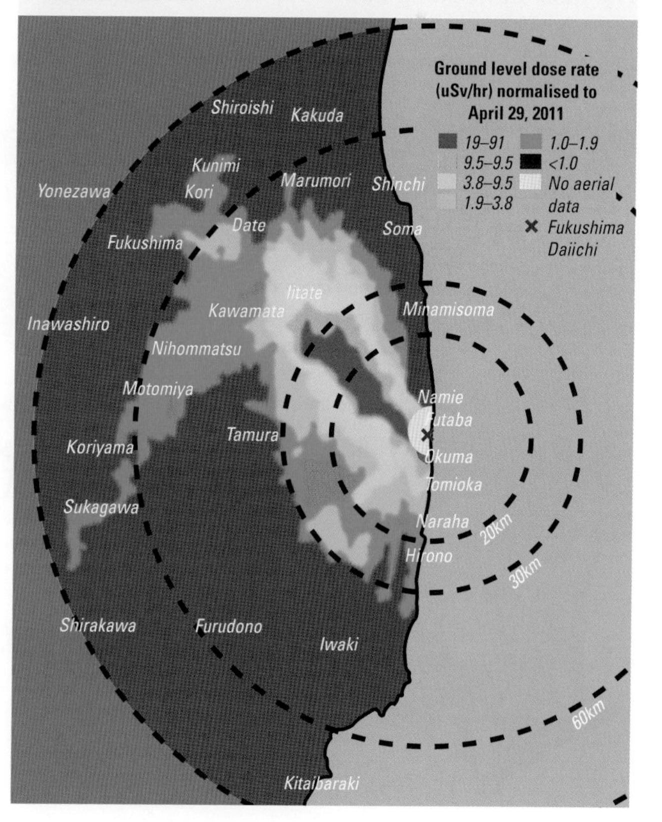

nuclear material up into the atmosphere. More than 160 000 people had to be evacuated from the area for fear of radiation. While three employees at the Daiichi plants were killed directly by the earthquake and tsunami, there were no fatalities from the nuclear accident.

6.5.3 Nuclear weapons

There are approximately 20 000 nuclear weapons in the world today, enough to destroy our planet many times over and effectively obliterate life from its face.

Effects of nuclear weapons

When nuclear weapons are detonated, enormous amounts of heat and radiation spread out from the centre of the blast (known as **ground zero**) in what is called a **thermal flash**. This radiation forms a fireball which generates the distinctive mushroom cloud associated with nuclear weapons. The fireball from the Hiroshima bomb formed a fireball 7 km across. At locations close to ground zero, most substances were melted or burned and organic matter (including people) was vaporised. People up to 50 km away received serious burns and those who looked directly at the flash were blinded.

Atomic bomb destruction, Hiroshima, Japan. Around 90 per cent of the buildings were destroyed, with only a few concrete-reinforced buildings surviving. Some 70 000 people died instantly, with tens of thousands more dying in the aftermath.

After the initial blast, the vaporisation of particles close to the blast causes an implosion of air from further out. When these inrushing air particles collide, they cause a high pressure shock wave to spread outwards at speeds of up to 3000 km/h. This shock wave causes the destruction of buildings, blowing them outwards from the centre of the blast.

The blast also releases large amounts of radiation in the form of gamma rays which can burn out electrical and electronic systems including computer networks and power grids, and even disrupt the electrical systems that control cars, planes and weaponry. This burst of energy is called an **electromagnetic pulse**. The most devastating effects for survivors are due to radiation exposure.

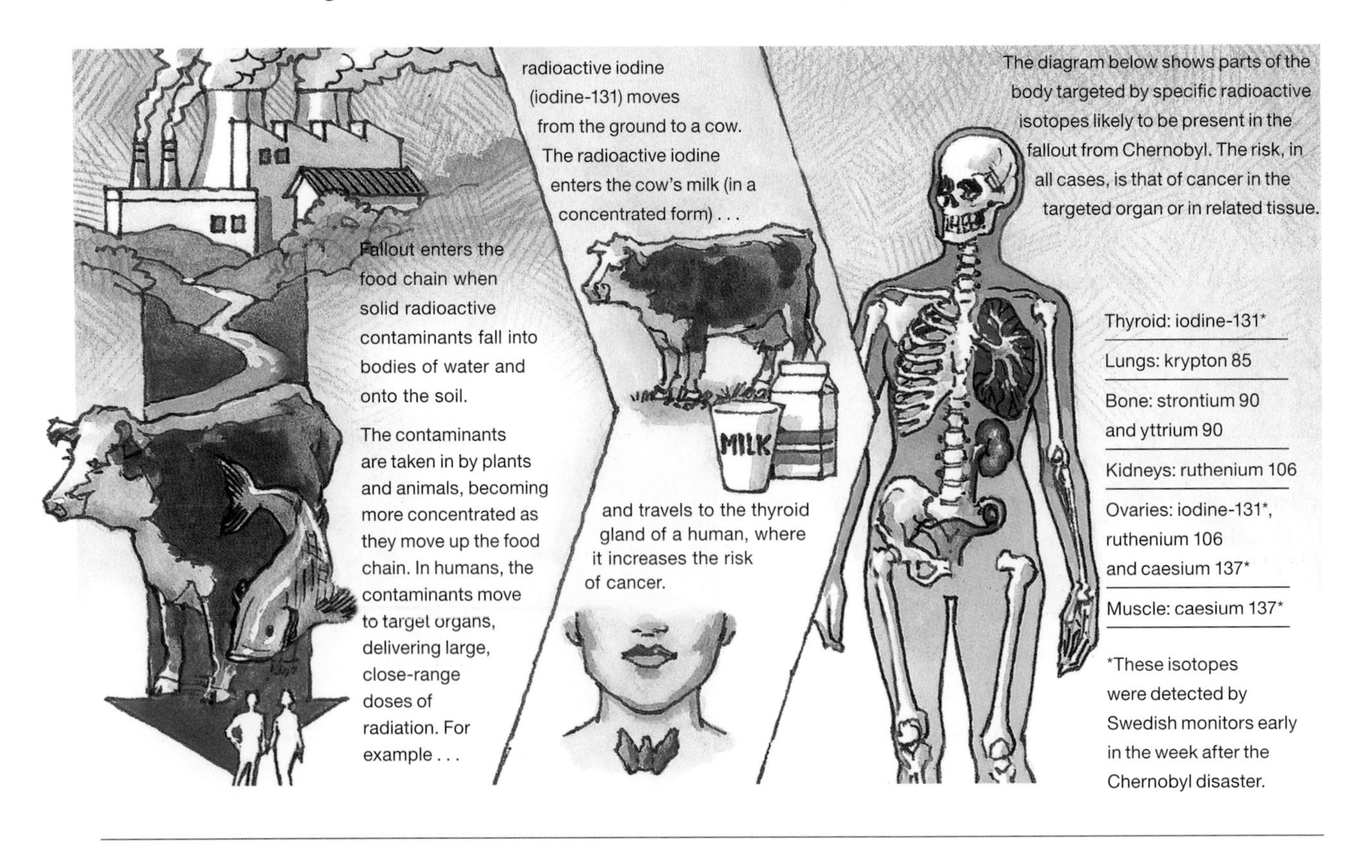

The radioactive nuclei formed during the nuclear reactions as well as tonnes of irradiated dust are blasted high into the atmosphere during detonation and the formation of the mushroom cloud. In the weeks following the nuclear explosion, these come back down to Earth as **nuclear fallout**. This radioactive fallout increases the background radiation for many years where it comes down, so people in the fallout zones are exposed to higher radiation levels with damaging effects.

6.5 Exercises: Understanding and inquiring

To answer questions online and to receive **immediate feedback** and **sample responses** for every question, go to your learnON title at www.jacplus.com.au. *Note:* Question numbers may vary slightly.

Remember

1. List the effects of exposure to large doses of nuclear radiation to humans.
2. Describe radioactive fallout.
3. Explain how the Chernobyl nuclear accident occurred.
4. What caused the melt-down in reactors 1, 2 and 3 at the power station in Fukushima?
5. Define the following terms: (a) melt-down (b) thermal flush (c) electromagnetic pulse (d) ground zero
6. Describe the short-term and long-term effects of an atomic explosion.

Think

7. Why did the incidence of leukaemia increase among young children rather than adults after Chernobyl?
8. Explain why nuclear energy is described by some as 'a blessing and a curse'.
9. One of the problems that led to the disaster at the Chernobyl nuclear reactor was due to the fact that the control rods could not be inserted into the reactor. Why would this have been a problem?

Investigate

10. Find out how a Geiger counter is able to measure the amount of radiation in a location.
11. Create a report on the accident at Chernobyl, Fukushima or Three Mile Island, explaining
    (a) how the accident affected the workers at the power plant and the surrounding towns and villages
    (b) the attempts made to reduce or control the damage caused by the radiation
    (c) the long-term effects of the accident.
12. Suppose you have been asked to write a report to discuss the following proposal: The use of radioactive elements should be banned in Australia. Give both sides of the argument, but present a conclusion for or against the proposal. You should search the internet using keywords such as uranium, radiation, mining, nuclear and waste to find useful sites.

Atomic bomb damage of Hiroshima

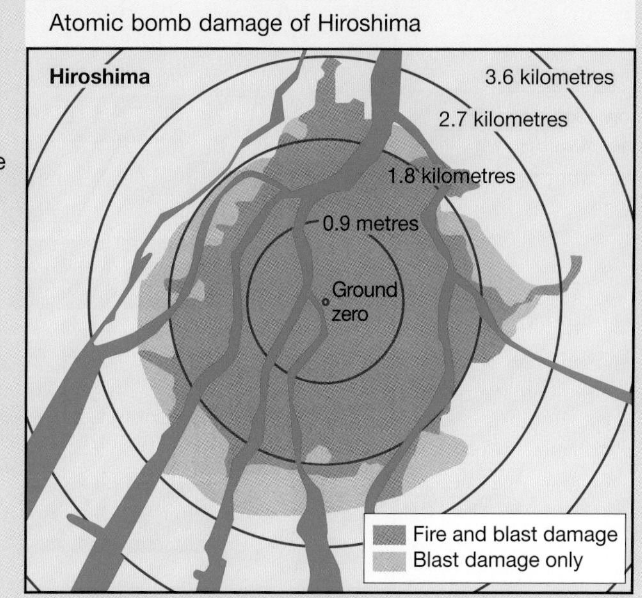

Using data

13. The map and table shown at right indicate the distribution of deaths and injuries caused by the Hiroshima bombing in 1945.
    (a) Use this information to determine:
        (i) original population of Hiroshima before the bombing
        (ii) number of people killed who were within 1 km of ground zero.
    (b) As you would expect, the number of people killed gets smaller the further from ground zero that they were located. What explanations can you give that the percentage wounded doesn't follow the same pattern?

Distance from ground zero (km)	Killed	Injured	Population
0–1.0	26 700 (86%)	3 000 (10%)	31 200
1.0–2.5	39 600 (27%)	53 000 (37%)	144 800
2.5–5.0	1 700 (2%)	20 000 (25%)	80 300
Total	**68 000 (27%)**	**76 000 (30%)**	**256 300**

6.6 Concept maps and plus, minus, interesting charts

6.6.1 Concept maps and plus, minus, interesting charts

1. On small pieces of paper, write down all the ideas you can think of about a particular topic.
2. Select the most important ideas and arrange them under your topic. Link these main ideas to your topic and write the relationship along the link.
3. Choose ideas related to your main ideas and arrange them in order of importance under your main ideas, adding links and relationships.
4. When you have placed all of your ideas, try to find links between the branches and write in the relationships.

how to ...?

why use?

To show what you understand about a particular topic

question

How can I explain this topic to someone else?
What do I understand about this particular topic?

also called

Knowledge map; concept web

Concept map

Topic
Link
Link
Main idea
Main idea
Link
Main idea
Link
Link
Link
First-level idea
First-level idea
First-level idea
First-level idea
First-level idea
Second-level idea
Link
Link
Link
Link
Second-level idea
Second-level idea
Link
Link
Second-level idea
Second-level idea
Second-level idea
Link
Third-level idea

comparison

Similarity

PMI charts help you consider the pros and cons of a decision; concept maps help you to see the relationships between ideas or concepts.

Difference

Concept maps show the links between ideas; PMI charts group ideas into various perspectives.

Plus, minus, interesting

example

PMI chart

Topic/theme/idea		
Plus	Minus	Interesting
•	•	•
•	•	•
•	•	•
•	•	•
•	•	•
•	•	•

6.6 Exercises: Understanding and inquiring

To answer questions online and to receive **immediate feedback** and **sample responses** for every question, go to your learnON title at www.jacplus.com.au. *Note:* Question numbers may vary slightly.

1. A concept map can be used to illustrate some of the important ideas associated with the atom and the links between the ideas.
   (a) Copy the concept map below into your workbook and complete it by adding the links between the ideas.

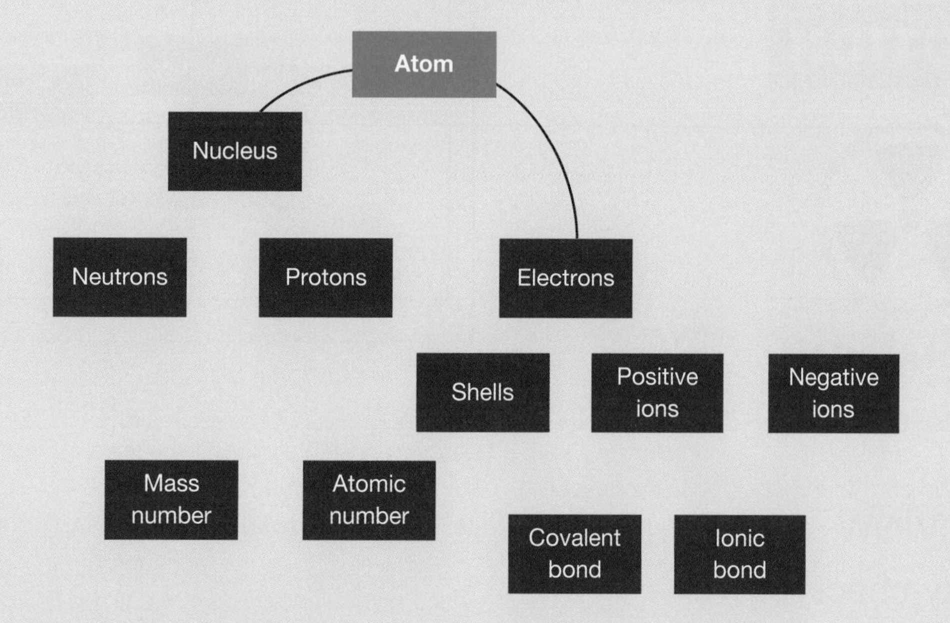

   (b) Construct your own concept map to show how ideas about what is inside substances are linked. Begin by working in a group to brainstorm the main ideas of the topic.
2. Construct a concept map of ideas associated with radioactivity.
3. Create a PMI chart on radioactivity, using the diagram below as a starting point.

Radioactive materials are classified as dangerous goods. The international symbol for radioactivity must be displayed wherever they are used.

Radioactivity		
Plus	Minus	Interesting
• Can be used to treat cancer • • • • • •	• Radioactive isotopes can be used in nuclear weapons • • • •	• Radioactivity was discovered by accident • • • • •

4. A SWOT analysis, like a PMI chart, is a visual tool that helps you think about different viewpoints related to an issue or topic. Work in a small group to perform a SWOT analysis to represent the positive and negative aspects of one of the following issues.
   (a) Nuclear power for Australia
   (b) Exporting uranium to other countries

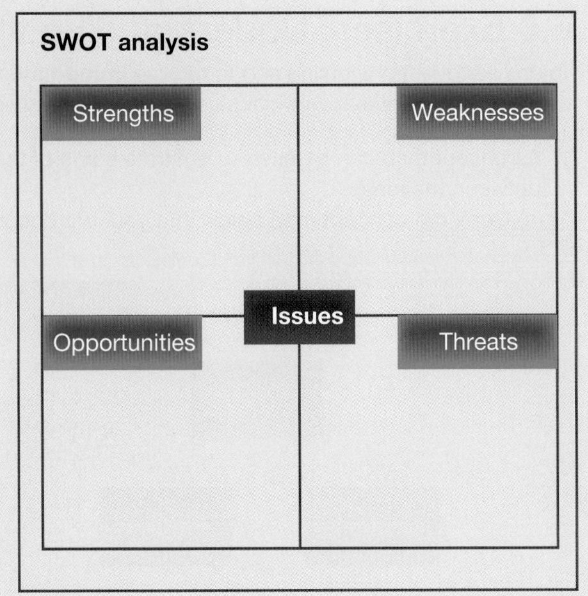

6.7 Review

6.7.1 Study checklist

Structure of the atom

- describe and model the main features of the currently accepted model of the atom
- identify the nucleus, protons, neutrons and electrons in a simple illustration of an atom
- compare the mass and charge of protons, neutrons and electrons

Radioactivity

- associate different isotopes of elements with the number of neutrons in the nucleus
- explain why, in terms of the stability of the nucleus, some isotopes are radioactive while others are not
- represent isotopes correctly in both symbols and words
- describe the characteristics of alpha, beta and gamma radiation, including penetrating power
- identify the main sources of background radiation
- define the half-life of radioisotopes
- explain how the known half-life of some radioisotopes can be used to determine the age of rocks, fossils and ancient artifacts
- describe the use of nuclear fission reactions in nuclear reactors

Science as a human endeavour

- investigate the historical development of models of the structure of the atom
- investigate the contribution of scientists such as Henri Becquerel, Marie and Pierre Curie, and Lord Rutherford to development of the model of the structure of the atom and radioactivity
- describe the impact of the discovery of radioactivity and the subsequent development of nuclear technology on the course of history
- explain how radioisotopes are used in nuclear reactors, radiometric dating, the treatment of cancer, medical diagnosis and food preservation
- examine the risks associated with radioactivity, nuclear power stations and nuclear weapons achivement.

Individual pathways

ACTIVITY 6.1 Revising the atom doc-8446	ACTIVITY 6.2 Investigating the atom doc-8447	ACTIVITY 6.3 Investigating the atom further doc-8448

learnON ONLINE ONLY

6.7 Review 1: Looking back

To answer questions online and to receive **immediate feedback** and **sample responses** for every question, go to your learnON title at www.jacplus.com.au. *Note:* Question numbers may vary slightly.

Remember

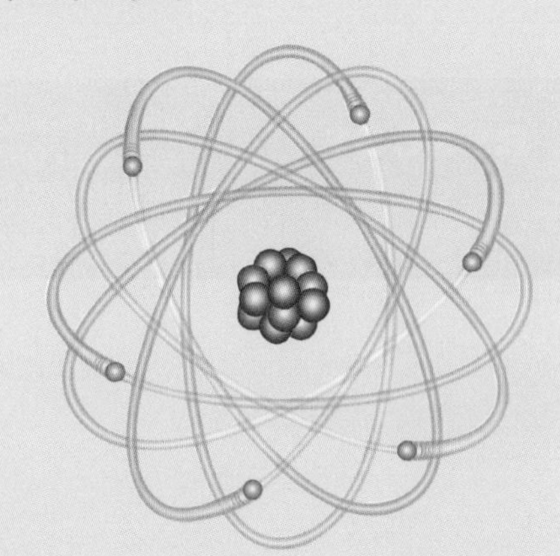

1. The diagram at right represents a model of a neutral atom.
   (a) Which two types of particles make up the nucleus of the atom?
   (b) Which particles are shown orbiting the nucleus in the atom?
   (c) What features of atoms are not very well represented by this particular model?
   (d) To which element does this atom belong?
2. Which of the particles in the neutral atom has:
   (a) a negative electric charge
   (b) a positive electric charge
   (c) no electric charge
   (d) the smallest mass?
3. Describe the contributions of the following scientists to our understanding of the structure of the atom.
   (a) J. J. Thomson
   (b) Lord Rutherford
   (c) Niels Bohr
4. The hydrogen atom exists as three different isotopes.
   (a) How are the atoms of each isotope different from the others?
   (b) Identify two features of the hydrogen atom that are the same for each of the three isotopes.
5. Alpha particles are helium nuclei containing two protons and two neutrons.
   (a) What is the electric charge of an alpha particle?
   (b) How do the mass and size of an alpha particle compare with the mass and size of a beta particle?
   (c) Suggest why alpha particles are easily stopped by human skin while beta particles are not.
   (d) Which type of radiation from the nucleus is more penetrating than either alpha or beta particles?
6. Which type of nuclear radiation travels at the speed of light?
7. Where does most of the natural background radiation that we experience every day come from?
8. Radioisotopes have many uses.
   (a) What property of radioisotopes makes them useful?
   (b) Describe some of the beneficial uses of radioisotopes.
   (c) Some radioisotopes are considered highly dangerous even after thousands of years. Why?
9. Two isotopes of the element carbon found naturally on Earth are carbon-12 and carbon-14.
   (a) How is every atom of carbon-14 different from every atom of carbon-12?
   (b) What features and properties do carbon-14 and carbon-12 have in common?
   (c) Which of the two carbon isotopes is stable?

10. The half-life of strontium-90 is 28 years. If a 400 gram sample of strontium-90 was left to decay, how many grams of the sample would be left after:
    (a) 28 years
    (b) 56 years
    (c) 84 years?
11. Estimate the half-life of the isotope whose decay is shown in the graph at right.
12. Explain how it is possible to use carbon-14 to estimate the age of the remains of a dead plant embedded in a rock.
13. Imagine that a nuclear power station has been proposed 50 kilometres north of Melbourne's city centre. Outline arguments for and against the proposal.
14. Investigate how and where radioactive waste is stored in Australia.

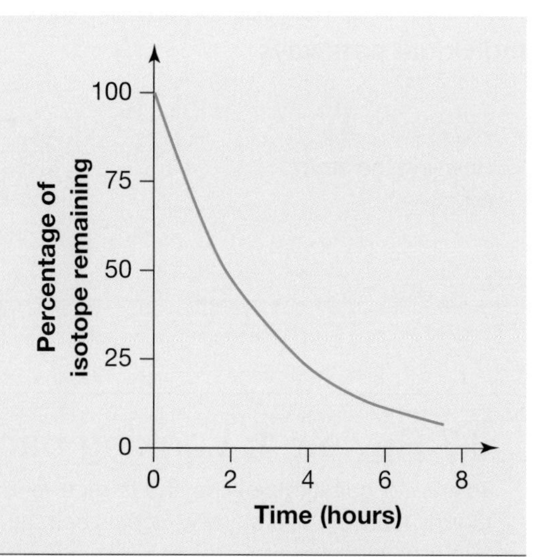

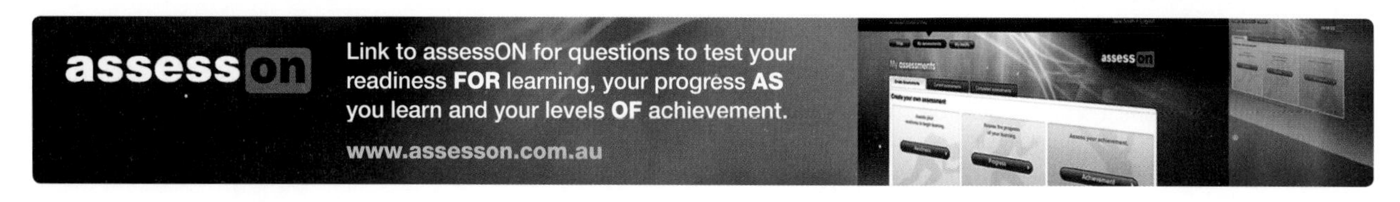

TOPIC 7
Chemical reactions

7.1 Overview

Every single living thing on Earth depends on chemical reactions — from the largest mammal, the blue whale, right down to the smallest insects. In plants, chemical reactions transform carbon dioxide and water into sugars and other nutrients such as proteins and starch. The burning of fuels to generate electricity, operate industry and transportation, and keep our homes at a comfortable temperature is a chemical reaction, as is the formation of crystals in a retort. So what are chemical reactions?

7.1.1 Think about chemical reactions

assesson

- How do atoms behave during chemical reactions?
- In chemical reactions, what is conserved other than energy?
- How does an icepack go cold without containing anything cold?
- What makes an airbag inflate during a car accident?
- How can you stop your stomach from burning and rumbling?
- What causes tooth decay?
- What is pickling and why is it done?
- Why does acid rain dissolve statues?
- What is the active ingredient of petrol?
- Which combustion reaction takes place in your own body?

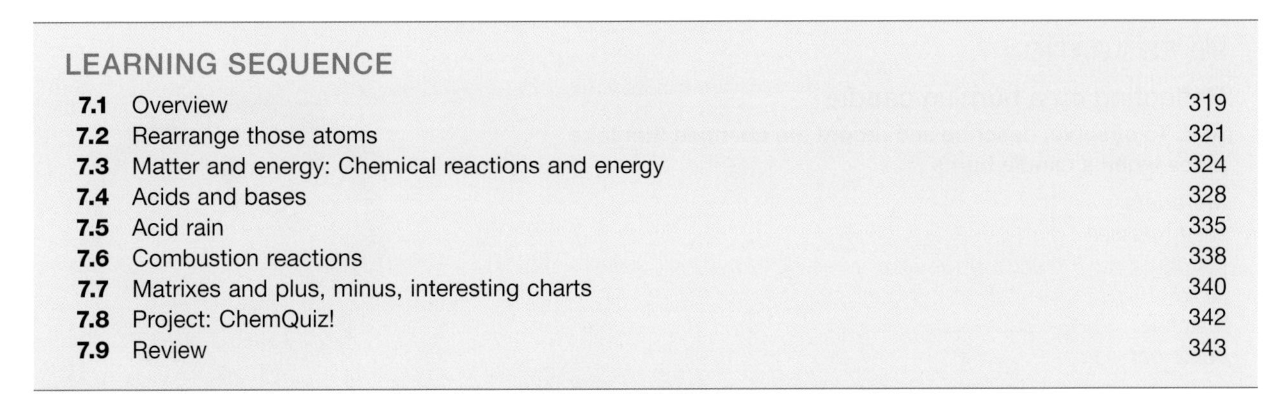

LEARNING SEQUENCE

Numerous **videos** and **interactivities** are embedded just where you need them, at the point of learning, in your learnON title at www.jacplus.com.au. They will help you to learn the content and concepts covered in this topic.

7.1.2 The chemistry of eating

WHAT AM I?

- You have to regularly provide me with fuel.
- The fuel burns in a chemical reaction known as combustion.
- The chemical energy stored in my fuel is transformed into mechanical energy and heat.
- The faster I go, the more fuel I need.
- I release waste products that smell and pollute the environment.

Your quest

Preparing, eating and digesting food all involve chemical reactions, many of which you already know about. Answer the following questions to find out what you already know about these important chemical reactions.

1. All of the food that we eat — including meat — begins with the growth of plants.
   (a) What is the name of the chemical reaction that produces the glucose that plants produce?
   (b) Which form of energy is necessary to allow this chemical reaction to take place?
2. The baking of bread makes use of a chemical reaction involving yeast and sugar. The same type of reaction is used in brewing to produce alcohol.
   (a) What is the name of this chemical reaction?
   (b) One of the products of this chemical reaction causes bread to rise while it is being baked. What is the name of this product? (*Hint:* It's a gas.)
3. The chemical digestion of food occurs when chemicals in your body react with the food.
   (a) What name is given to the chemicals that speed up these chemical reactions?
   (b) Much of the food that you eat is broken down to glucose, which takes part in a chemical reaction that occurs in every single cell of your body. What is the name of this chemical reaction, which releases useful energy?
4. Overeating can make your stomach produce too much acid.
   (a) Which type of substance is contained in the products that can be taken to reduce the discomfort and pain caused by the extra acid?
   (b) What is the name of the chemical reaction that provides you with relief from the effects of the extra acid?

Evidence of chemical reactions

INVESTIGATION 7.1

Reflecting on a burning candle

AIM: To observe, describe and record the changes that take place when a candle burns

Materials:

safety glasses
candle
jar lid
matches
heatproof mat

Method and results

- Place the tea light on a heatproof mat.
- Light the candle.
- Observe the candle burning and answer the following questions.

Discuss and explain

1. Describe three physical changes that take place while the candle burns.
2. Make a list of as much evidence as you can that a chemical reaction has taken place.

CAUTION

Do not touch the candle or flame and do not smell the vapour directly. You can fan the vapour towards your nose with your hand.

7.2 Rearrange those atoms

7.2.1 Chemical reaction

A cake rising in an oven, a bath bomb fizzing in a full bathtub, and an old car getting rusty — what do they have in common? They are all evidence of chemical reactions.

Chemical reactions take place when the bonds between atoms are broken and new bonds are formed, creating a new arrangement of atoms and at least one new substance. As the new substance is formed, observable changes take place — a change in temperature or colour, the formation of a visible gas or new solid, or perhaps even just an odour.

7.2.2 Reactants and products

The new substances that are formed during a chemical reaction are called the **products**. The original substances are called the **reactants**. For example, when hydrogen gas is added to oxygen gas and ignited, the new substance water is formed. The reactants are hydrogen and oxygen. The product is water. The bonds between the hydrogen atoms and oxygen atoms are broken and new bonds are formed between oxygen and hydrogen, as shown at right.

Notice that the hydrogen and oxygen atoms that were present in the reactants are also present in the product. There is no gain or loss of atoms. They have simply been rearranged.

This reaction can be represented by a word equation as shown: oxygen + hydrogen → water.

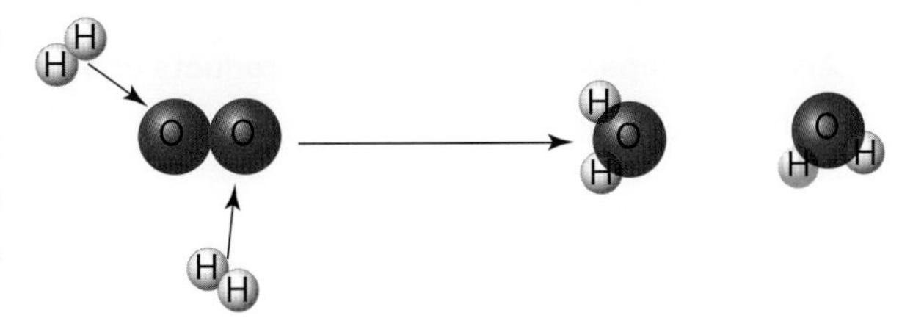

7.2.3 A burning question

The idea of atoms rearranging themselves may seem obvious now, but two hundred years ago it was not. It was thought, for example, that when a candle burned the wax simply vanished. In other words, it was thought that matter could disappear.

In the eighteenth century, French nobleman Antoine-Laurent Lavoisier showed that although a candle seems to disappear as it burns, there is as much mass present after it has completely burned as there was before. The apparent loss of mass was caused by gases moving into the atmosphere. Lavoisier's ideas led to the development of the **Law of Conservation of Mass**, which states that matter can be neither created nor destroyed during a chemical reaction. The diagram and word equation above is a simple representation of the Law of Conservation of Mass.

Lavoisier also provided evidence for the **Law of Constant Proportions**, which states that a compound, no matter how it is formed, always contains the same relative amounts of each element. For example, carbon dioxide (CO_2) always contains the same relative amounts of carbon and oxygen (about 27 per cent of the mass is made up of carbon). It does not matter whether the carbon dioxide forms from the reaction of sherbet in your mouth or from the reaction in the engine of a car, this proportion is fixed because every molecule of CO_2 is formed by the bonding of one carbon atom with two oxygen atoms. This law helped to shape our understanding of the way atoms bond together. In fact, after his unfortunate execution during the French Revolution, Lavoisier became known as the Father of Modern Chemistry.

7.2.4 Starting the ball rolling

Simply placing two chemicals together does not always mean they will react. For example, hydrogen and oxygen react violently, yet a mixture of these two gases can be stored indefinitely if kept cool in a secure container. Energy must be supplied to start the reaction. Sometimes only a small amount of energy is needed to start (or initiate) the reaction. Heat transferred from the surroundings may be enough.

Energy may also be supplied by an electric current, a beam of light or a Bunsen burner flame. This energy is needed to begin the process of breaking the bonds in the reactants, which allows the atoms to rearrange and form new bonds in the products.

In this case, the word equation is modified to show the word 'heat' written over the reaction arrow

$$\text{hydrogen} + \text{oxygen} \xrightarrow{\text{heat}} \text{water}$$

INVESTIGATION 7.2

Conserve that mass!

AIM: To compare the mass of the products of a chemical reaction with the mass of its reactants

Materials:

safety glasses
250 mL conical flask
4 Alka-Seltzer tablets
1 balloon
Matches
an electronic balance
100 mL measuring cylinder
water

Method and results

CAUTION

Wear safety glasses.

- Place the conical flask on the balance and pour in 100 mL of water.

1. Place two tablets alongside the conical flask and record the total mass.
   - Remove the flask from the balance and drop the tablets into the water.

2. When the reaction is complete, weigh the flask and record the mass.
   - Rinse out the flask thoroughly and again add 100 mL of water.
   - Place two tablets inside the balloon. You may need to break the tablets into pieces to do this.
   - Stretch the neck of the balloon over the conical flask, being careful not to drop the tablets into the water. The balloon should be flopped over, resting against the side of the flask.
3. Place the conical flask and balloon onto the balance and record the total mass.
   - Lift up the top of the balloon and drop the tablets into the water in the conical flask.
4. When the reaction is complete, weigh the flask and record the mass. **Do not remove the balloon**.
5. After you have recorded the mass, remove the balloon. Light a match and test the gas in the conical flask. Record your observations.
6. Describe what happened during the reaction.

Discuss and explain

7. Which gas do you think filled the balloon?
8. Comment on your results of the total mass before and after each reaction. Explain your answer.
9. Why do you think it took a long time for the Law of Conservation of Mass to be developed?

7.2 Exercises: Understanding and inquiring

To answer questions online and to receive **immediate feedback** and **sample responses** for every question, go to your learnON title at www.jacplus.com.au. *Note:* Question numbers may vary slightly.

Remember

1. What name is given to chemicals that:
   (a) react in a reaction
   (b) are formed in a chemical reaction?
2. What happens to the atoms in substances that take part in chemical reactions?
3. State the Law of Conservation of Mass and explain in your own words what it means.
4. State the Law of Constant Proportions and explain how it applies to carbon dioxide.
5. Energy can be required to start a reaction. List three possible sources of this energy.

Think

6. A piece of paper is weighed on an accurate balance and then burned, leaving a pile of ashes. The ashes are collected and weighed on the same balance.
   (a) Would you expect the mass of the ashes to be the same as the mass of the paper before it was burned?
   (b) Explain your answer in terms of the products produced.
7. Explain why, when a piece of steel wool burns, the mass of the blackened material is greater than the original mass of the steel wool.
8. A chemical reaction is described by the following word equation:

   sodium sulfate + barium chloride ⟶ barium sulfate + sodium

   Identify the second product.

Investigate

9. Find out more about Antoine-Laurent Lavoisier, his work and why he lost his head during the French Revolution.

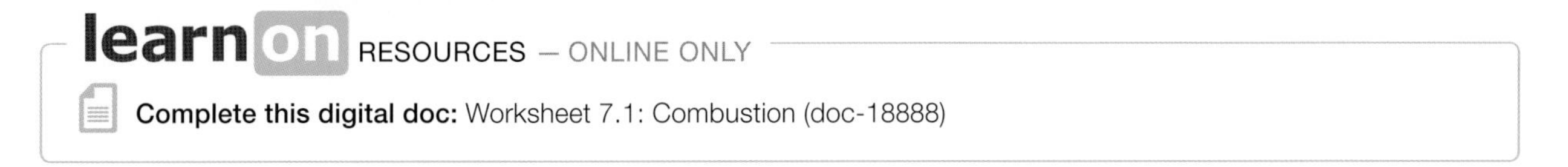

7.3 Matter and energy: Chemical reactions and energy

The energy that keeps this motorcycle moving comes from an exothermic chemical reaction.

7.3.1 Energy and chemical reactions

When fuels such as petrol are burned in motor vehicles, energy is released and used to keep the vehicle in motion. Burning is a chemical reaction in which fuel reacts with oxygen, producing carbon dioxide, water and several other products.

The energy released comes from the rearrangement of atoms. There is less energy stored in the chemical bonds in the products than there was in the reactants. Chemical reactions that release energy are called **exothermic** reactions.

Chemical reactions in which energy is absorbed from the surroundings are called **endothermic** reactions. There is more energy 'stored' in the chemical bonds of the products than there was in the reactants.

Whether energy is absorted or released during a chemical reaction can be observed by comparing the temperature of the substance before the reaction with their temperature after the reaction.

WHAT DOES IT MEAN?

The words exothermic and endothermic come from the Greek words *exo*, meaning 'out', *endo*, meaning 'in', and *therme*, meaning 'heat'.

7.3.2 Hot stuff

Portable hand warmers, commonly used by skiers and campers, become hot when shaken due to an exothermic chemical reaction in which energy is released to the surroundings. One type of hand warmer contains iron, water, salt and sawdust. When the contents of the packet are shaken quickly, the powdered iron reacts with oxygen to form iron oxide. During this chemical reaction, some of the **chemical energy** of the substances is transformed into heat energy which is transferred to the hands, increasing their temperature. We can show this chemical reaction with a word equation:

iron + oxygen ⟶ iron oxide

7.3.3. As cold as ice

Athletes use instant icepacks to treat injuries. The icepack may consist of a plastic bag containing ammonium nitrate or ammonium chloride powder and an inner bag of water. Squeezing the bag breaks the weaker inner bag and immediately causes the powder to dissolve in the water. The **chemical process** that takes place absorbs energy from the injured area, thus lowering its temperature. It is therefore an endothermic chemical process.

We can describe this chemical process with a word equation:

ammonium chloride + water ⟶ ammonium chloride in solution

INVESTIGATION 7.3

Exothermic and endothermic processes

PART 1 Is for Teacher Demonstration Only

AIM: To investigate some exothermic and endothermic processes

Materials:

safety glasses
bench mat
4 large test tubes and test-tube rack
10 mL measuring cylinder
Balance
thermometer (–10 °C to 110 °C)
stirring rod
magnesium ribbon
sandpaper
0.5 mol/L hydrochloric acid
lithium chloride
sodium thiosulfate
potassium chloride

Method and results

Construct a table like the one below in which to record the temperature changes as the five chemical processes described take place.

Exothermic and endothermic processes

Chemical process	Initial temperature (°C)	Final temperature (°C)	Change in temperature (°C)

Part 1: Magnesium in hydrochloric acid

- Pour 10 mL of 0.5 mol/L hydrochloric acid into a test tube in a test-tube rack. Place a thermometer in the test tube and allow it to come to a constant temperature. Record the temperature of the solution.
- Clean a 10 cm piece of magnesium ribbon using the sandpaper until it is shiny on both sides. Coil the magnesium ribbon and place it into the test tube of hydrochloric acid.
- Observe the temperature of the solution as the magnesium reacts with the hydrochloric acid.

1. Record the final temperature of this solution.

Part 2: Lithium chloride in water

- Pour 10 mL of water into a test tube in a test-tube rack. Place a thermometer in the water in the test tube and allow it to come to a constant temperature.

2. Record the temperature of the water.

- Use a balance to weigh 2 g of lithium chloride, add it to the water in the test tube and stir gently.
- Observe the temperature of the solution as the lithium chloride dissolves in the water.

3. Record the final temperature of this solution.

Parts 1 and 3 of this experiment can be demonstrated with the aid of a data logger.

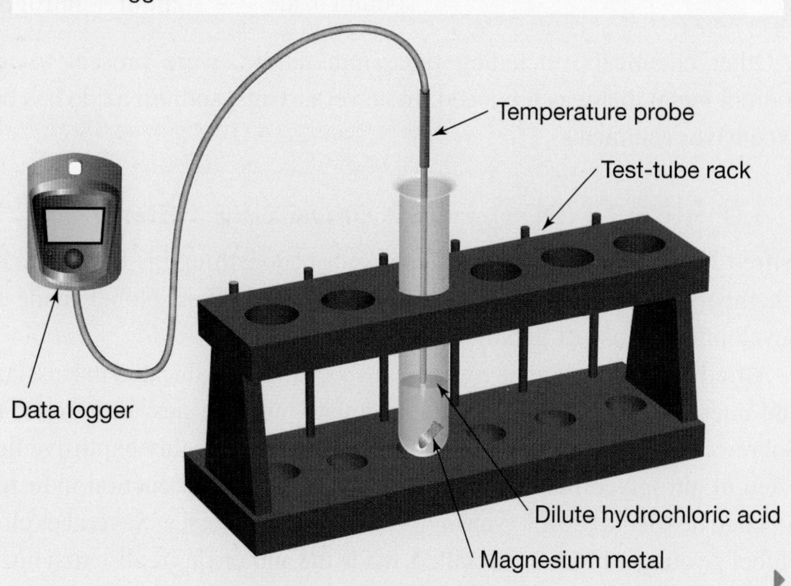

Part 3: Sodium thiosulfate in water

- Repeat part 2 from the previous page using 2 g of sodium thiosulfate instead of lithium chloride.

Part 4: Potassium chloride in water

- Repeat part 2 from the previous page using 2 g of potassium chloride instead of lithium chloride.

4. Complete the table by calculating the change in temperature resulting from each process. Use + or – signs to indicate whether the temperature decreased or increased.

Discuss and explain

5. Which processes were exothermic and which were endothermic?
6. Which one or more of the chemical processes above was a chemical reaction? How do you know?

Although the chemical bonds are broken in this process, it is not a chemical reaction because a new substance is not formed.

7.3.4 Airbags

Airbags have saved many people from death or serious injury in car accidents. When an airbag inflates, it creates a cushion between the occupant's body and the windscreen, dashboard and other parts of the inside of the car. Airbags, which are made from nylon, may be concealed in the steering wheel, dashboard, doors or seats.

The rapid inflation of an airbag is the result of an explosive exothermic chemical reaction. The reaction is triggered by an electronic device in the car that detects any sudden change in speed or direction of the car. The bag fills with a harmless gas. When the occupants move forwards or sideways into the bag, they push the gas out of the airbag through tiny holes in the nylon. The airbag is usually totally deflated by the time the car comes to rest.

Airbags inflate as a result of an explosive chemical reaction.

One of the chemical reactions commonly used in airbags produces a massive burst of nitrogen gas. In older airbags, the nitrogen. was released when the toxic chemical sodium azide (NaN_3) decomposed:

$$\text{sodium azide} \longrightarrow \text{sodium} + \text{nitrogen gas}$$

Other chemicals, including potassium nitrate, were present to react with the potentially dangerous sodium metal that was produced. In newer airbags, sodium azide has been replaced with less toxic (and less expensive) chemicals.

7.3.5 Alfred Nobel — an explosive career

Alfred Nobel is probably most famous for bequeathing his fortune to establish the Nobel Prizes in Physics, Chemistry, Medicine, Literature and Peace. However, Nobel made his fortune inventing **dynamite** and developing the use of explosives in the 1860s.

Alfred Nobel was born in Sweden in 1833. He was educated in Russia. Nobel was fluent in several languages and interested in literature, poetry, chemistry and physics. In Paris he met a young Italian chemist, Ascanio Sobrero, who had earlier invented **nitroglycerine**, a highly explosive liquid. Alfred Nobel became very interested in nitroglycerine and saw its potential in the construction industry. When he returned to Stockholm in Sweden he tried to develop nitroglycerine as an explosive. Several explosions, including one in 1864 in which Nobel's younger brother was killed, made the authorities realise that nitroglycerine was extremely dangerous.

Alfred Nobel had to move his laboratory out of Stockholm's city limits and onto a barge anchored on a nearby lake. He was determined to make nitroglycerine safe to work with. He discovered that mixing nitroglycerine with silica would turn the liquid into a paste that could be shaped into rods suitable for inserting into drilling holes. In 1866 he patented this material under the name dynamite.

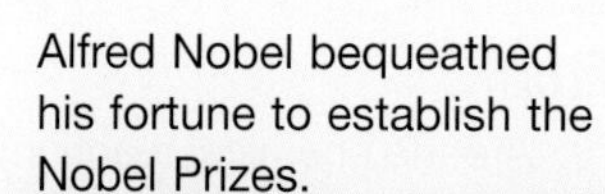

Alfred Nobel bequeathed his fortune to establish the Nobel Prizes.

Dynamite is mainly used in the mining and construction industries. Huge areas of rock can be broken apart because the chemical reaction involved in dynamite's explosion releases large amounts of energy and gas, which can exert great pressure. Explosives can release enough energy to cause a small earthquake.

The invention of dynamite could not have come at a better time than the middle of the nineteenth century. New mines were being opened to supply coal for heating and steam engines, iron and other building materials. Railways were being laid all over the world and passes had to be blasted through the mountains. Over the years, Alfred Nobel set up factories and laboratories in more than 20 countries.

Alfred Nobel died in 1896 and when his will was opened it came as a surprise that the interest earned by his $9 million fortune was to be used for the establishment of the Nobel Prizes. The prizes were to be awarded 'for the good of humanity' in the fields of chemistry, physics, physiology or medicine, literature and peace.

Explosion: ... the act of exploding; a violent expansion or bursting with noise, as of gunpowder or a boiler ...
(*The Macquarie Dictionary*)
... loud noise caused by this ... rapid or sudden increase ...
(*The Australian Pocket Oxford Dictionary*)

7.3 Exercises: Understanding and inquiring

To answer questions online and to receive **immediate feedback** and **sample responses** for every question, go to your learnON title at www.jacplus.com.au. *Note:* Question numbers may vary slightly.

Remember

1. How are exothermic reactions different from endothermic reactions?
2. In a chemical reaction in which energy is absorbed from the surroundings, where does the extra energy go?
3. Explain why the chemical process that takes place in an icepack containing ammonium chloride is not a chemical reaction.
4. Explain how an airbag works.
5. Write a word equation to describe one chemical reaction that occurs to inflate an airbag.
6. What was Alfred Nobel's most famous invention?
7. Describe how an explosive is able to split large volumes of rock.

Think

8. Are the chemical reactions described below exothermic or endothermic?
   (a) Dilute hydrochloric acid is added to dilute sodium hydroxide in a test tube. They react to produce sodium chloride and water. After the reaction, the test tube feels very warm.
   (b) As garden compost decomposes, the compost heap gets warmer.
   (c) Barium hydroxide and ammonium thiosulfate solutions are mixed and the temperature drops enough to freeze water.
9. Instant hot compresses are used by athletes to warm torn muscles. They relieve pain and speed up the healing process. The hot compresses contain calcium chloride powder and an inner bag of water. When the inner bag bursts, the calcium chloride dissolves in the water and releases energy.
   (a) Is the chemical process that takes place in the compress endothermic or exothermic?
   (b) How does the energy stored in the chemical bonds of the product compare with the energy stored in the chemical bonds of the calcium chloride and water?
   (c) Write a word equation to describe this chemical process.
10. Are explosions endothermic or exothermic reactions? Explain your answer.
11. Why do you think that Alfred Nobel donated his entire fortune to reward those who worked for the 'good of humanity'?
12. In exothermic chemical reactions, energy is released. Why is energy not included in the chemical equations that describe the reactions?

Investigate

13. Find the names of some Australians who have been awarded the Nobel Prize. Choose one Australian scientist who has won the Nobel Prize and write a short biography about him or her. Include in your biography information on when they were awarded the Nobel Prize and the work that they did to receive such a prestigious award.
14. Use a yearbook or the internet to find out who won the most recent Nobel Prizes for Chemistry, Physics and Medicine. Write a short biography about one of the laureates. (The winners of Nobel Prizes are referred to as laureates. The Nobel Prizes are announced in October of each year.)

7.4 Acids and bases

7.4.1 Acids and bases

Chemical reactions involving acids and bases play an important role in our lives. They occur in the kitchen, in the laundry, in the garden, in swimming pools and even inside the body.

Acids

Acids are **corrosive** substances. That means they react with solid substances, 'eating' them away. Acids have a sour taste — in fact, the word 'acid' comes from a Latin word meaning sour. Some acids, such as the sulfuric acid used in car batteries, are dangerously corrosive. The acids in ant stings and bee stings cause pain. Others, such as the acids in fruits and vinegar, are safe — even pleasant — to taste.

WHAT DOES IT MEAN?

The word *acid* comes from the Latin word *acidus*, meaning 'sour'.

Bases

Bases have a bitter taste and feel slippery or soapy to touch. Some bases are very corrosive, especially caustic soda (sodium hydroxide). Caustic soda will break down fat, hair and vegetable matter and is the main ingredient in drain cleaners. Other bases are used in soap, shampoo, toothpaste, dishwashing liquid and cloudy ammonia as cleaning agents. Bases that can be dissolved in water are called **alkalis**. Some common acids and bases are listed in the tables on the following page.

7.4.2 Giving an indication

Acid–base **indicators** are substances that can be used to tell whether a substance is an acid or a base. The indicators react with acids and bases, producing different colours in each. Two commonly used indicators are litmus, which turns red in an acid and blue in a base, and bromothymol blue, which turns yellow when added to an acid and a bluish-purple when added to a base.

The pH scale

You can describe how acidic or basic a substance is by using the numbers on the **pH scale**. The pH scale ranges from 0 to 14. Low pH numbers (less than pH 7) mean that substances are acidic. High pH numbers (more than pH 7) mean that substances are basic. If a substance has a pH of 7, it is said to be neutral — neither acidic nor basic. This is shown on the pH scale below.

Acids and bases can be graded from strong to weak. For example, a strong acid has a very low pH (pH 0 or 1) and a strong base has a very high pH (pH 13 or 14). The pH of a substance can be measured using a pH meter or a special indicator called **universal indicator**. Universal indicator is a mixture of indicators and it changes colour as the strength of an acid or base changes. The colour range of universal indicator is shown on the next page.

The pH values of some common substances

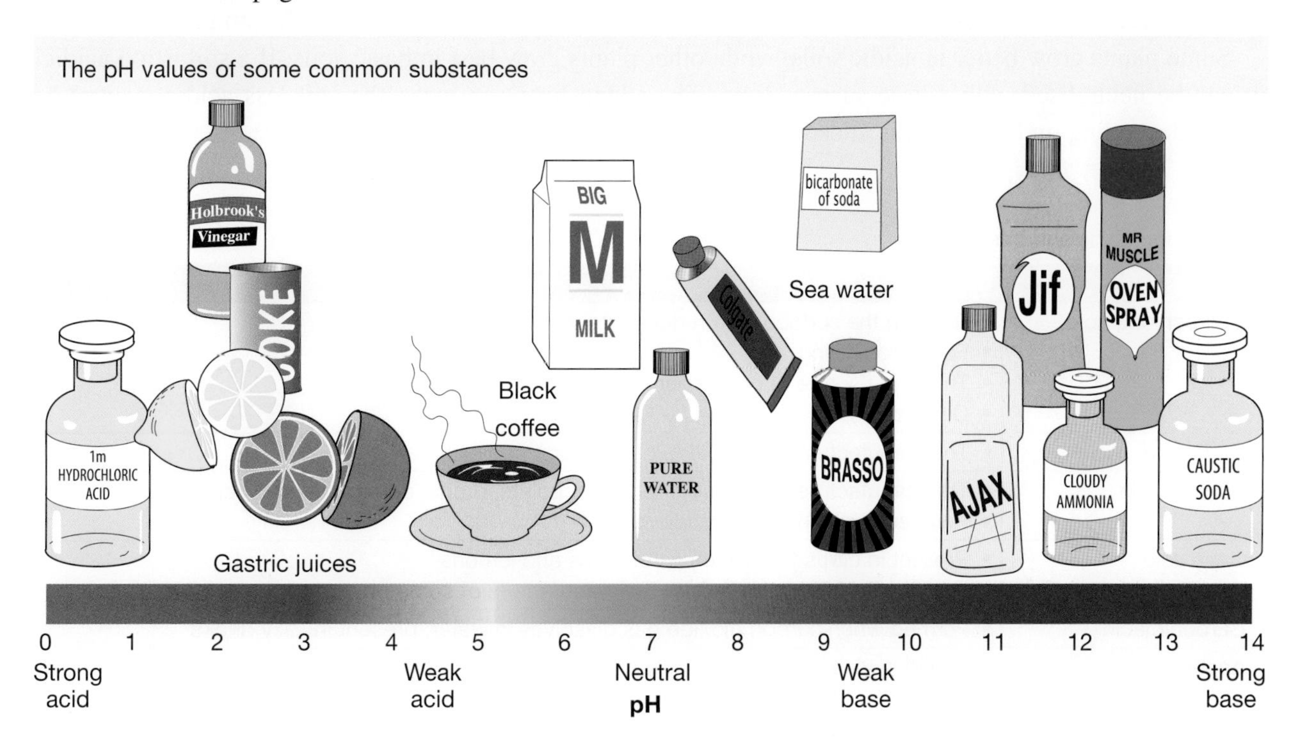

7.4.3 Neutralisation

When an acid and a base react with each other, the products include water and a salt. Such a reaction is called a **neutralisation** reaction. These reactions can be very useful. They can relieve pain caused by indigestion or the stings from wasps, bees and ants. They can be used to change the pH of soil to make it more suitable for growing particular plants. Neutralisation reactions are also used in cooking and to keep swimming pools and spas clean.

To neutralise means to stop something from having an effect. To stop the properties of acids from having an effect, a base can be added to it. Similarly, to stop a base from having an effect, an acid can be added. So, the pain caused by the acidic sting of an ant can be relieved by adding a weak base, such as sodium bicarbonate (baking soda). The pain caused by the base in the sting of a wasp can be relieved by adding a weak acid such as vinegar.

The word equation for a neutralisation is:

$$\text{acid} + \text{base} \longrightarrow \text{salt} + \text{water}$$

Sometimes, a gas is produced as well as a salt and water. For example, when hydrochloric acid is neutralised with sodium hydroxide, the products are water and the salt sodium chloride. When hydrochloric acid is neutralised with sodium bicarbonate, the products are the salt sodium chloride, water and carbon dioxide gas.

The colour range of universal indicator. It is pink in strong acid (pH 1), blue in strong base (pH 14) and green in neutral solutions (pH 7).

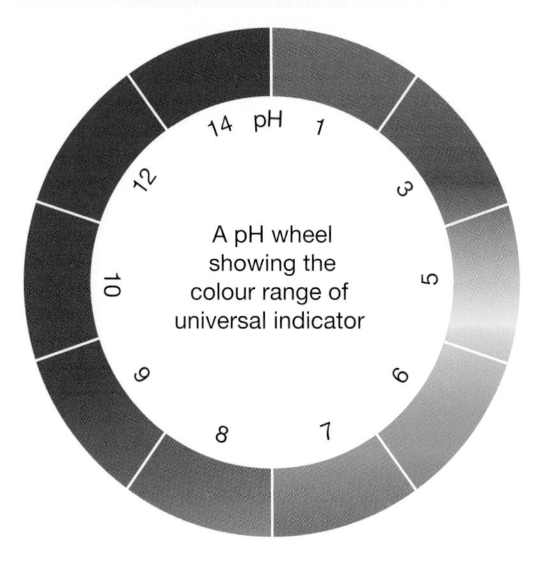

Neutralisation in the garden

Neutralisation reactions are used in many situations around the home. A sting from an ant or a bee is very painful as it contains an acid — formic acid. This can be neutralised by a base such as soap. A wasp sting is painful because it contains a base and can be treated by applying an acid such as vinegar. It is important to know what has bitten you so that the correct substance can be used to neutralise the sting.

Some plants grow better in acidic soils, while other plants grow best in basic soils. If a soil is too acidic, it can be neutralised with a base such as lime. The added lime can make the soil less acidic, neutral or basic, depending on how much is added.

Common acids and bases

Acid	Uses
Hydrochloric acid	• To clean the surface of iron during its manufacture • Food processing • The manufacture of other chemicals • Oil recovery
Nitric acid	• The manufacture of fertilisers, dyes, drugs and explosives
Sulfuric acid	• The manufacture of fertilisers, plastics, paints, drugs, detergents and paper • Petroleum refining and metallurgy
Citric acid	• Present in citrus fruits such as oranges and lemons • Used in the food industry and the manufacture of some pharmaceuticals
Carbonic acid	• Formed when carbon dioxide gas dissolves in water; present in fizzy drinks
Acetic acid	• Found in vinegar • The production of other chemicals, including aspirin

Base	Uses
Sodium hydroxide (caustic soda)	• The manufacture of soap • As a cleaning agent
Ammonia	• The manufacture of fertilisers and in cleaning agents
Sodium bicarbonate	• To make cakes rise when they cook

If the soil is too basic, ammonium sulfate can be added to the soil. This reacts with the soil to produce an acid, which helps to neutralise the bases in the soil. These neutralisation reactions in your garden can help your plants to grow by providing soil with the most suitable pH.

7.4.4 Indigestion

The hydrochloric acid in your stomach helps to break down the food you eat. It is a very strong acid, with a pH of less than 1.5. But if you eat too quickly, or eat too much of the wrong food, the contents of your stomach become even more acidic. You feel a burning sensation because of the corrosive properties of the acid.

To relieve the pain of indigestion, you can take antacid tablets. The active ingredients in antacid tablets are weak bases such as aluminium hydroxide, magnesium carbonate and magnesium hydroxide, which neutralise the acid. The cause of the relief you experience can be described by chemical word equations such as:

hydrochloric acid + aluminium hydroxide ⟶ aluminium chloride + water

hydrochloric acid + magnesium hydroxide ⟶ magnesium chloride + water

hydrochloric acid + magnesium carbonate ⟶ magnesium chloride + water + carbon dioxide

One product of this last reaction is carbon dioxide gas. You burp to get the gas out of your stomach.

7.4.5 Acids and bases in the kitchen

Some foods, such as pickles, chutney and tomato sauce, last a long time without refrigeration because they contain acids that prevent the growth of micro-organisms that would cause them to spoil. Others, such as onions and beetroot, are preserved by storing them in vinegar, which is also known as acetic acid. This process of preserving food is called **pickling**.

The base sodium bicarbonate is more commonly known as baking soda. When it reacts with an acid, the products are a salt, water and carbon dioxide. Self-raising flour is a mixture of an acid and baking soda. When water or milk is added to self-raising flour, the acid and base react together. The carbon dioxide produced causes the mixture to rise when it is heated.

Two ingredients in pancakes are buttermilk (an acid) and baking soda. When the two are mixed, a salt, water and carbon dioxide are produced. The bubbles of carbon dioxide get larger as the mixture is heated, causing the mixture to rise.

7.4.6 Swim safely

When chlorine is added to a swimming pool, it reacts with the water to produce hypochlorous acid. This acid kills bacteria and algae, keeping the pool water safe for swimming. All the chemicals in a swimming pool, when combined, need to have a pH in the range of 7.2–7.8 for a clean, hygienic pool and safe swimming.

If the pH falls below 7.2, the micro-organisms will still be killed but the swimmers will get red and stinging eyes, and the water may become corrosive and damage pool fittings. A base such as sodium carbonate (soda ash) or sodium bicarbonate (bicarbonate of soda) would have to be added to neutralise the excess acid.

If the pH rises above 7.8, bacteria and algae will grow and the water will be unfit for swimming. To reduce the pH, an acid such as sodium hydrogen sulfate would have to be added to neutralise the excess base.

7.4.7 Corrosive acids

Acids are corrosive. They can dissolve metals, eat away marble statues, destroy the enamel of your teeth and kill bacteria.

Because acids are corrosive, they can be very harmful. Strong acids can burn your skin and eat away clothes. If an acid is spilt on the floor, a basic powder, such as sodium bicarbonate, should be used to neutralise the acid. All spills in the science lab should be reported to your teacher.

Acid can destroy the enamel on your teeth. Teeth are protected by a 2 mm thick layer of enamel made of hydroxyapatite. After a meal, bacteria in the mouth break down some of the food to produce acids such as acetic acid and lactic acid. Food with a high sugar content produces the most acid. The acids produced by the bacteria can dissolve the enamel coating of the tooth. Once this protective coating is destroyed, the bacteria can get inside the tooth and cause tooth decay. The best way to prevent this chemical reaction between tooth enamel and acid from happening is to clean and floss your teeth after every meal and avoid eating sugary foods.

HOW ABOUT THAT!

The fizzy sensation that you get when you eat sherbet is due to an acid–base reaction. The sherbet consists of sodium bicarbonate and citric acid. Both of these substances are in powdered form in the sherbet and do not react with each other. When they dissolve in the saliva of your mouth a reaction takes place, producing carbon dioxide gas and hence the fizzing.

Acids and metals

When an acid reacts with a metal, the products are a salt and hydrogen gas. The word equation for an acid–metal chemical reaction is:

$$\text{acid} + \text{metal} \longrightarrow \text{salt} + \text{hydrogen}$$

For example, when sulfuric acid (hydrogen sulfate) reacts with copper, the products are copper sulfate and hydrogen. The chemical word equation for this reaction is:

$$\text{sulfuric acid} + \text{copper} \longrightarrow \text{copper sulfate} + \text{hydrogen}$$

INVESTIGATION 7.4

Antacids in action

AIM: To investigate the neutralising action of an antacid

Materials:

Petri dish
electronic balance
spatula
antacid powder
0.1 mol/L hydrochloric acid
250 mL conical flask
100 mL measuring cylinder
methyl orange indicator
white tile or white paper

Method and results

1. Measure and record the mass of the Petri dish.
2. Add a small amount of antacid powder to the dish and record the mass of the antacid and Petri dish.
3. Calculate the mass of the powder.
   - Add 50 mL of the dilute hydrochloric acid to the 250 mL flask.
   - Add 3 drops of methyl orange indicator.
   - Place the flask mixture on the white tile (or paper) and use the spatula to slowly add antacid from the Petri dish bit by bit. Swirl the flask to mix. Stop adding antacid when the colour changes from red to orange.
4. Measure and record the mass of the Petri dish and its contents (the unused antacid).
5. What was the mass of the antacid powder?
6. What colour change occurs when the methyl orange indicator is in the acid?
7. By subtraction, calculate the mass of antacid used to neutralise 50 mL of dilute hydrochloric acid.

Discuss and explain

8. How does your result agree with other groups in your class? Suggest reasons for the similarities or differences between your results.
9. Use your results to calculate how much antacid you would need to neutralise 500 mL of dilute hydrochloric acid.

INVESTIGATION 7.5

Reaction of acids with metals

AIM: To investigate the chemical reactions of an acid with a range of metals

Materials:

safety glasses
bench mat
test tubes and test-tube rack
pieces of metal such as copper, iron, zinc, magnesium, aluminium
dropping bottle of 2 mol/L hydrochloric acid solution
rubber stopper
matches

Method and results

When an acid reacts with a metal, a salt is formed and hydrogen gas is given off. You can test for hydrogen gas by holding a lighted match at the mouth of the test tube. If the gas is hydrogen, it will explode and make a 'pop' sound.

CAUTION

Do not push the stopper into the test tube firmly. Just hold it in the top of the test tube for a few seconds.

- Place a small piece of one of the metals in a test tube.
- Add the acid to the test tube to a depth of 1 cm.
- Observe the chemical reaction.
- Test for hydrogen gas by holding a rubber stopper over the end of the test tube for a few seconds and then placing a lighted match at the mouth of the test tube.

1. Record your observations.
2. Repeat the test with other metals.

Discuss and explain

3. When zinc metal reacts with hydrochloric acid, zinc chloride and hydrogen gas are formed. Write a word equation for this reaction.
4. When the lighted match produces a 'pop', the hydrogen gas is reacting with the oxygen in the air to form water. You may have noticed the water form at the top of the test tube after you performed the match test. Write a word equation for this chemical reaction.

7.4 Exercises: Understanding and inquiring

To answer questions online and to receive **immediate feedback** and **sample responses** for every question, go to your learnON title at www.jacplus.com.au. *Note:* Question numbers may vary slightly.

Remember

1. Use a two-column table to describe the properties of acids and bases.
2. What common property do some acids and bases have when they come into contact with solid substances?
3. Describe the difference between a base and an alkali.
4. Which type of substance has a pH value:
   (a) less than 7
   (b) more than 7
   (c) equal to 7?
5. Explain why the chemical reaction between an acid and a base is called neutralisation.
6. Which substance is produced in all neutralisation reactions?
7. Explain how self-raising flour helps cakes rise.
8. Which acid is present in your stomach to help you digest food?

9. How does an antacid relieve the pain of indigestion?
10. Why does soap relieve the pain of an ant sting?
11. Why do foods that are high in sugar cause so much tooth decay?
12. Identify two products of every chemical reaction between an acid and a metal.

Using data

13. A pH meter is used to measure the pH of 5 different substances. The results are as shown in the table at right:
    (a) Which substance is most likely to be:
        (i) orange juice
        (ii) milk?
    (b) Which substance could be:
        (i) a weak base
        (ii) pure water
        (iii) vinegar
        (iv) a strong base?
    (c) Which two of the substances would you expect to be the most corrosive?

Substance	pH value
A	6.0
B	12.0
C	3.0
D	7.0
E	8.0

Think

14. Write word equations for the reactions between:
    (a) hydrochloric acid and sodium hydroxide
    (b) hydrochloric acid and sodium bicarbonate
    (c) sulfuric acid (hydrogen sulfate) and sodium hydroxide.
15. When you add buttermilk (an acid) to baking soda (a base) in a mixing bowl, does the pH increase or decrease? Explain your answer.
16. Antacid tablets contain a base, which neutralises the excess acid in your stomach and relieves the pain. When you take an antacid tablet, would you expect the pH value in your stomach to increase or decrease? Explain your answer.
17. A stinging-nettle plant may contain an acid that is injected into your skin when you touch it. Describe how you could show that the plant does contain an acid.
18. Write a word equation to describe the chemical reaction between hydrochloric acid and calcium carbonate.
19. Why is it that the acids in the food and drink you consume do not damage your stomach?
20. When you add lime to a soil that is too acidic, are you increasing or decreasing the pH?

Investigate

21. Find the websites of two antacid products such as Gaviscon®, Mylanta®, Eno® or Alka-Seltzer®.
    (a) Research and report on:
        (i) the ingredients of the product or products
        (ii) the claims made about each of their antacid product or products
        (iii) advice and warnings
        (iv) side effects.
    (b) Find a medical site that provides information about antacids, including side effects.
    Comment on the adequacy of the advice on each of the two companies' websites.
22. On the packet of one brand of baking soda, there is a claim that you can deodorise your entire house by sprinkling it on your carpets and leaving it for a few minutes before vacuuming.
    (a) Suggest how the baking soda could have this deodorising effect.
    (b) Investigate other claims made on baking soda packaging and design some experiments to test one or more of the claims.

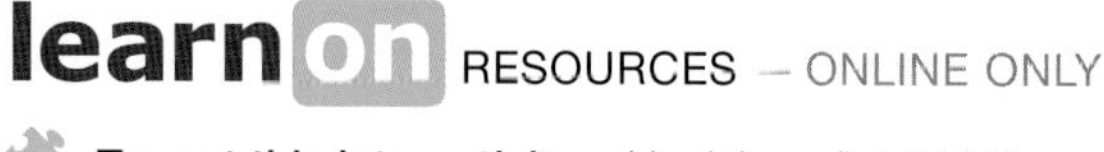

RESOURCES — ONLINE ONLY

Try out this interactivity: pH rainbow (int-0101)

Complete this digital doc: Worksheet 7.3: Acids and bases (doc-18890)

7.5 Acid rain

Science as a human endeavour

7.5.1 What causes acid rain?

Every year, **acid rain** causes hundreds of millions of dollars worth of damage to buildings and statues.

The photographs at right show the damage that has been caused to a statue over sixty years. Forests, crops and lakes are also affected by acid rain which is blown in from industrial areas.

These photographs were taken in 1908 (left) and in 1969 (right). You can see the damaging effects of acid rain on this statue.

Rain is normally slightly acidic. As clouds form and rain falls, the water reacts with carbon dioxide in the atmosphere to form very weak carbonic acid. If concentrations of sulfur dioxide and nitrogen oxide are high, these gases react with the water in the atmosphere to produce sulfuric, nitric and other acids. When this rain falls, it is far more acidic than it would normally be and is known as acid rain. If the acid rain falls as snow, acid snow can build up on mountains. When this snow melts, huge amounts of acid are released in a short period.

Where do the gases come from?

Most of the gases that cause acid rain come from the burning of fossil fuels (natural gas, oil and coal) in industry, power stations, the home and cars. North America and Europe have a greater problem with acid rain because of the use of coal with a higher sulfur content than Australian coal. The sulfur dioxide released by volcanoes also contributes to acid rain.

7.5.2 Damage caused by acid rain

Acid rain damages the cells on the surface of leaves and affects the flow of water through plants. It also makes plants more likely to be damaged by frosts, fungi and diseases. The acid rain collects in streams, rivers and lakes, making the waterways more acidic. A healthy lake has a pH of about 6.5 and fish, plants and insects can live in it. Acid rain causes the pH of the lake to fall. Some aquatic plants and animals cannot tolerate these acidic conditions and die. It is not only the acidic water that can kill the aquatic life. Acid rain reacts with soil, releasing minerals, which may contain elements such as aluminium. The aluminium is washed into the streams, rivers and lakes and poisons the aquatic plants and animals.

When acid rain eats into buildings and statues, it is reacting with calcium carbonate in the marble or limestone.

$$\text{calcium carbonate} + \text{acid rain} \longrightarrow \text{gypsum} + \text{water} + \text{carbon dioxide}$$

The gypsum formed by acid rain on a statue is a powdery dust (calcium sulfate), which is washed away by the rain. As this chemical reaction continues, the statue is slowly eaten away.

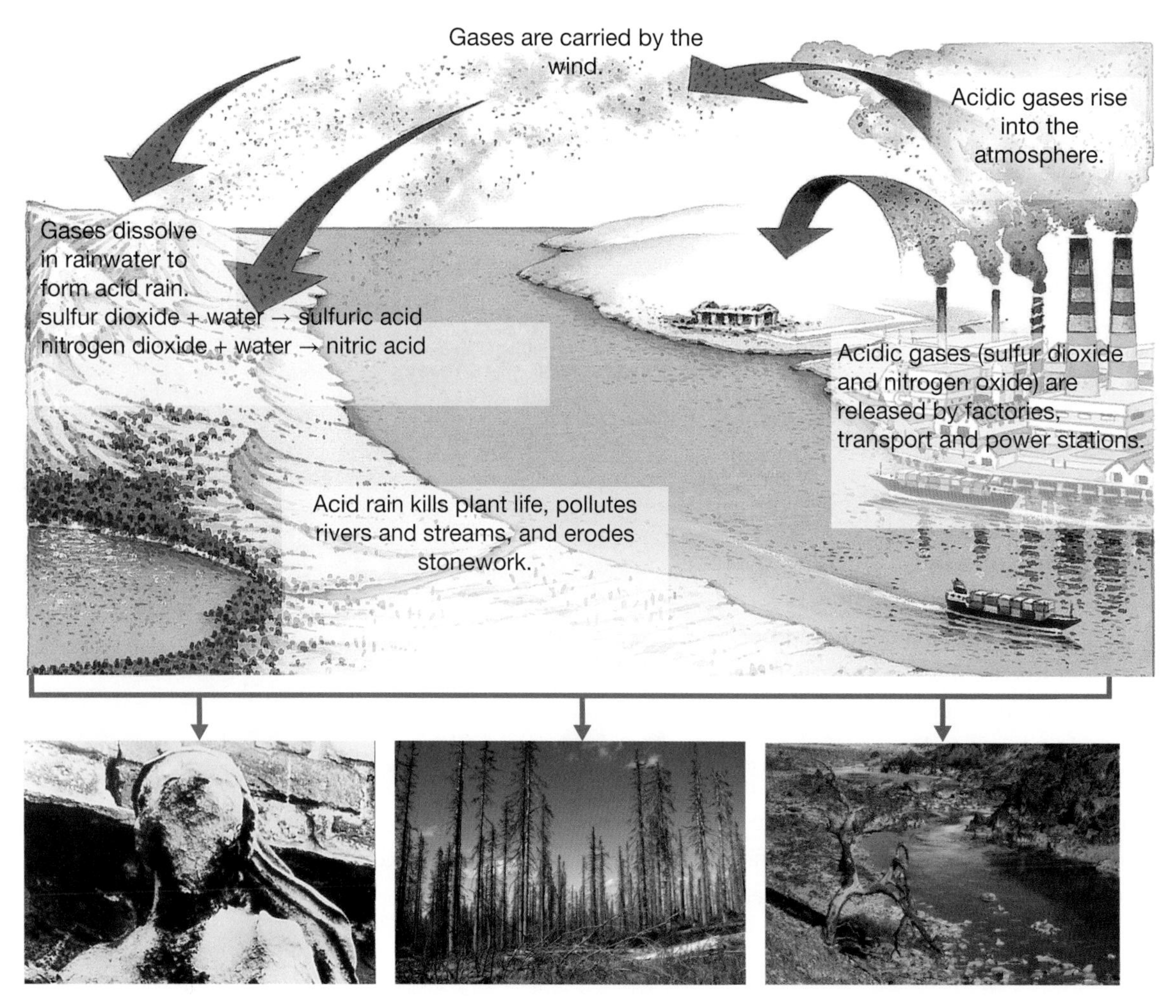

When acid rain eats into buildings and statues, it is reaching with calcium carbonate in the marble or limestone.
Calcium carbonate + acid rain → gypsum + water + carbon dioxide

Acid rain damages the cells on the surface of leaves and affects the flow of water through plants. It also makes plants more likely to be damaged by frosts, fungi and diseases. In northern Europe, entire forests have died as a result of acid rain.

Acid rain collects in streams, rivers and lakes, making the water more acidic. Acid rain causes the pH lakes to fall. Some aquatic plants and animals cannot tolerate these acidic conditions and die.

7.5.3 Solving the problem

The problem of acid rain and all the damage that it causes can be solved only by reducing the release of acidic gases into the air. Some ways of doing this include:

- looking for alternative ways of producing electricity
- encouraging people to use public transport or to car pool.

INVESTIGATION 7.6

Investigating acid rain

AIM: To investigate the effect of pH of acidic water on the growth of seeds

Materials:

empty milk cartons
potting soil

distilled water
measuring cylinder
vinegar (or 0.1 mol/L hydrochloric acid solution)
seeds (e.g. lucerne, peas, cress, beans)
universal indicator

Method and results

- Cut the milk cartons so that they are about 10 cm high. These will make suitable containers for growing the seeds, 5 seeds per container.
- Test the effect of water with different pH values on the growth of the seeds. To ensure that your tests are fair, you will need to keep everything the same in your experiment, except the one thing that you are varying. In this case you are varying the level of acidity (pH) of the water that you are putting on the plants.

Discuss and explain

1. Prepare a report on your investigation. This could be a written report, a video, a wall chart or an oral presentation.

7.5 Exercises: Understanding and inquiring

To answer questions online and to receive **immediate feedback** and **sample responses** for every question, go to your learnON title at www.jacplus.com.au. *Note:* Question numbers may vary slightly.

Remember

1. What is acid rain and how is it caused?
2. Why is rain slightly acidic even without air pollution?
3. Describe two different ways in which acid rain can harm the plants and animals in streams and lakes.
4. Complete this word equation: acid rain + calcium carbonate →

Think

5. Motor vehicles make a large contribution to the acid rain problem. Most of them use fuel that releases acidic nitrogen oxides when it is burned. Write an account of some ways in which motor vehicle pollution could be reduced over the next thirty years.

Create

6. Write a newspaper article about the devastation caused by acid rain.
7. Design a wall chart that would explain how acid rain is formed and the damage that it can cause.

Imagine

8. Imagine that you live near a factory or power station that is producing acidic gases and causing harm to the environment. You wish to be elected onto the local government board to try to stop this problem. Write a speech that you could give at an election meeting.

Investigate

9. Use the library to find out which countries are most affected by acid rain.
10. Find out some of the ways that damage caused by acid rain could be stopped or at least reduced.

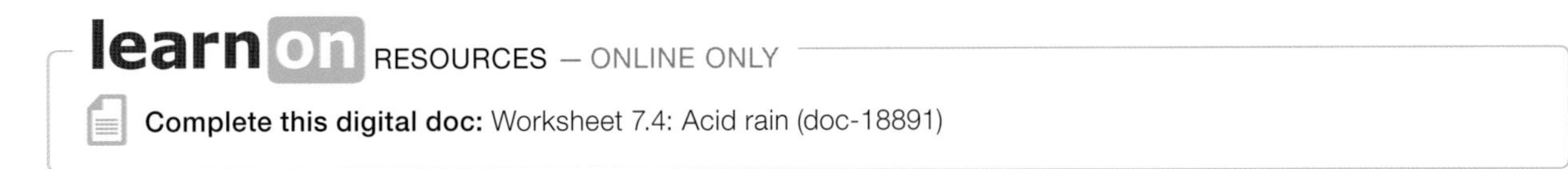

7.6 Combustion reactions

7.6.1 Combustion reactions

Some of the most spectacular chemical reactions to watch, including fireworks and the launching of spacecraft, are **combustion** reactions.

Combustion reactions are those in which a substance reacts with oxygen and heat is released. Burning is a combustion reaction that produces a flame. The substance that reacts with oxygen in a combustion reaction is called a **fuel**.

7.6.2 Fossil fuels

Fossil fuels such as natural gas, petrol and coal have formed from the remains of living things. They are compounds of hydrogen and carbon called **hydrocarbons**. The products of the combustion of fossil fuels always include carbon dioxide and water. Because of impurities in fossil fuels, these are not the only products of their combustion. In some cases various dangerous gases, including carbon monoxide, are also produced.

Cooking with gas

The **natural gas** used in gas stoves and ovens contains methane, a colourless, odourless and highly flammable gas. Natural gas formed millions of years ago from the remains of plants and animals and became trapped under rock. Its lack of colour and odour makes it very dangerous if there is a leak, so gas suppliers add chemicals that do have an odour so that the methane can be detected in the event of a leak or if the gas is accidentally left switched on. Methane reacts with oxygen, producing carbon dioxide and water, and it burns with a blue flame. The heat needed to start the reaction is provided by a match, lighter or spark. The chemical word equation for the combustion of methane is:

$$\text{methane} + \text{oxygen} \longrightarrow \text{carbon dioxide} + \text{water}$$

Motoring along

The fuel used in most Australian cars is liquid **octane**. This is the major component of petrol, usually between 85 per cent and 95 per cent — other fuels make up the remainder. Octane is obtained from **crude oil** which, like natural gas, is formed from the remains of marine plants and animals that died million of years ago. The vapour of liquid octane reacts with oxygen, producing carbon dioxide and water. The reaction is started in each cylinder of a car by a spark from a spark plug. Most of the energy released during the reaction is used to turn the wheels of the car. The chemical word equation for the combustion of octane is:

> **WHAT DOES IT MEAN?**
>
> The word *combustion* comes from the Latin word *comburere*, meaning 'to burn'.

$$\text{octane} + \text{oxygen} \longrightarrow \text{carbon dioxide} + \text{water}$$

Taking off

The fuel used in jet aircraft is **kerosene**, which is obtained from crude oil. Like the octane in cars, the vapour of this fossil fuel reacts with oxygen. An electrical spark is used to start the reaction. The chemical word equation for the combustion of kerosene in a jet engine is:

$$\text{kerosene} + \text{oxygen} \longrightarrow \text{carbon dioxide} + \text{water}$$

Generating electricity

Fossil fuels such as coal and natural gas are burned in power stations to generate electricity. The energy released during the combustion reaction is used to heat water to produce steam. The steam turns the blades of giant turbines, transforming its energy into electrical energy. The chemical word equation for the combustion of coal is:

$$\text{coal} + \text{oxygen} \longrightarrow \text{carbon dioxide} + \text{water}$$

7.6.3 Essential combustion

A chemical reaction called **respiration** takes place in every cell of your body. Respiration is a slow combustion reaction. The energy required by your body is released when the fuel, glucose from your digested food, reacts with oxygen from the air that you breathe. The products of respiration are carbon dioxide and water. The chemical word equation for respiration is:

$$\text{glucose} + \text{oxygen} \longrightarrow \text{carbon dioxide} + \text{water}$$

7.6.4 Blasting off

The energy to launch spacecraft is provided by a combustion reaction. The main rocket engines are fuelled by hydrogen, which reacts with oxygen in an exothermic reaction that releases enough energy to lift more than two million kilograms off the ground towards outer space. The only product of the reaction is water. The chemical word equation for the combustion of hydrogen is:

$$\text{hydrogen} + \text{oxygen} \longrightarrow \text{water}$$

Rocket engines are fuelled by a combustion reaction with hydrogen.

7.6.5 Oxidation reactions

Combustion reactions are examples of **oxidation** reactions. However, strangely enough, not all oxidation reactions involve oxygen. Oxidation is now defined as the transfer of electrons from a reactant. That is what happens to fuels when they are burned in oxygen. The reaction between copper and a silver nitrate solution is an example of an oxidation reaction that does not involve oxygen. Copper is oxidised when electrons are removed from copper atoms during the reaction that produces silver metal.

7.6 Exercises: Understanding and inquiring

To answer questions online and to receive **immediate feedback** and **sample responses** for every question, go to your learnON title at www.jacplus.com.au. *Note:* Question numbers may vary slightly.

Remember

1. What characteristics do all combustion reactions have in common?
2. How are fossil fuels different from other types of fuel?
3. How is each of the following combustion reactions started?
   (a) The burning of natural gas
   (b) The combustion of octane in a car
4. What are the products of all combustion reactions in which fossil fuels are burned?
5. What is the fuel in the combustion reaction known as respiration?

Think

6. Describe at least two effects on the environment of the combustion of fossil fuels.
7. Hydrogen and oxygen are cooled to extremely low temperatures so that they can be stored as liquids in the fuel tanks of rockets. Why is water, the product of the reaction, produced as a gas?
8. Respiration is the chemical reaction that takes place in every cell of your body. State two reasons it is classified as a combustion reaction.
9. Write a word equation for an oxidation reaction that does not involve oxygen.

Investigate

10. Find out how kerosene and octane are extracted from crude oil.
11. Find out why catalytic converters are used in cars and which chemical reactions take place within them.

12. Oil companies often make claims that their petrol is cleaner, more economical and provides superior performance than that of their competitors. Use the internet to investigate the following questions.
    (a) How do the oil companies go about improving their fuel products?
    (b) How do the oil companies try to convince consumers that their claims are correct?
    (c) What do you think? Is there a difference between the same fuel products made by different companies or are they all the same? Use your research to back up your opinion.

Create

13. Create a poster that shows how the burning of coal is used to generate electricity. Include the chemical equation for the combustion of coal on your poster. Also include information about where the reactants come from and what happens to the products.

7.7 Matrixes and plus, minus, interesting charts

7.7.1 Matrixes and plus, minus, interesting charts

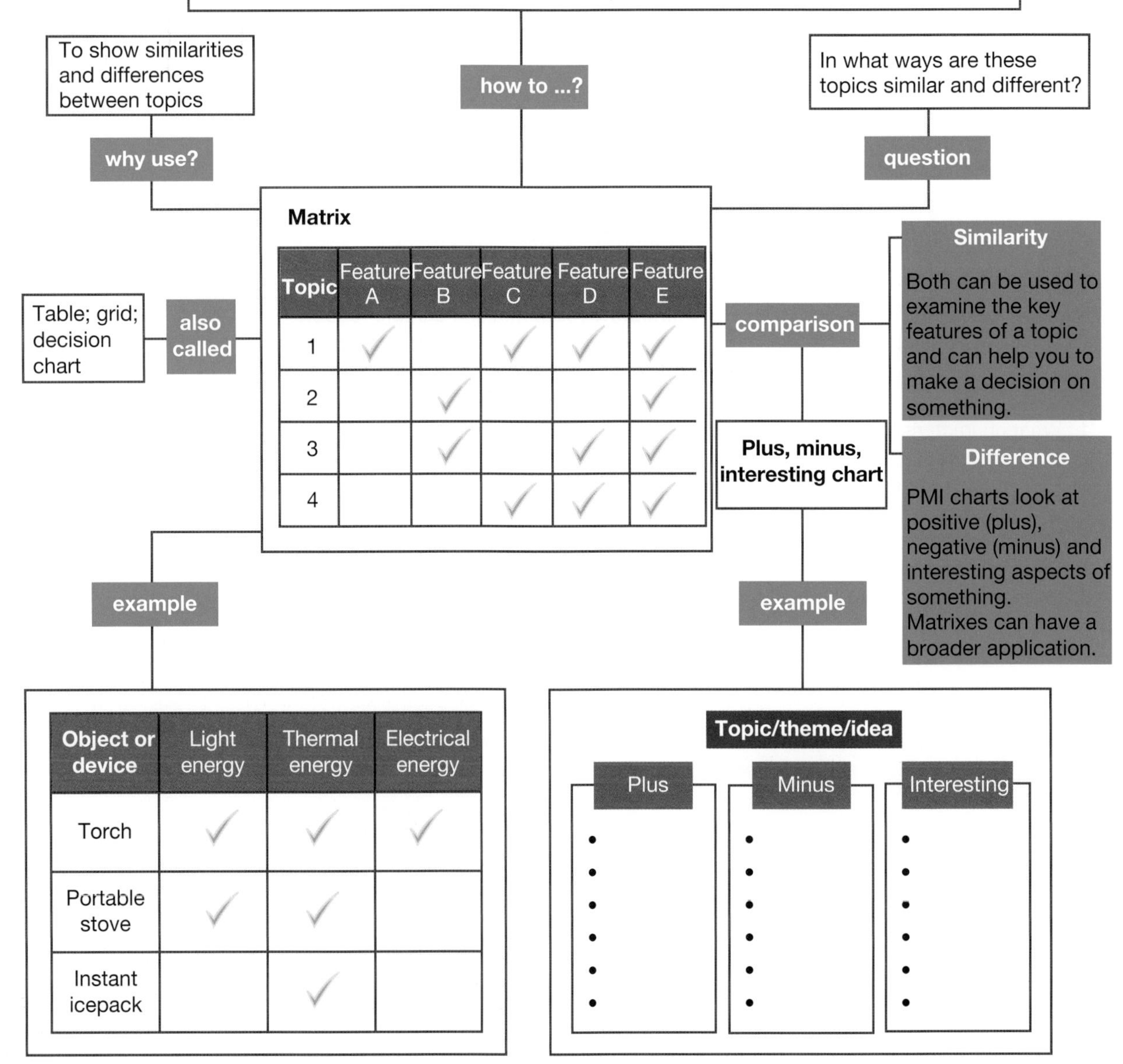

7.7 Exercises: Understanding and inquiring

To answer questions online and to receive **immediate feedback** and **sample responses** for every question, go to your learnON title at www.jacplus.com.au. *Note:* Question numbers may vary slightly.

Think and create

1. Copy and complete the matrix below to show which type of chemical reaction each statement refers to..

Statement	Endothermic reactions	Exothermic reactions	Neutralisation reactions	Combustion reactions
Chemical bonds are always broken.				
New chemical bonds are formed.				
Energy is released to the surroundings.				
Energy is absorbed from the surroundings.				
The Law of Conservation of Mass applies.				
A salt is always produced.				
Oxygen is always a reactant.				
One reactant is always an acid.				
Respiration in living cells is an example.				
Takes place to inflate a car airbag.				
A new substance is produced.				

2. Create your own PMI chart on chemical explosions, using the diagram below as a starting point.

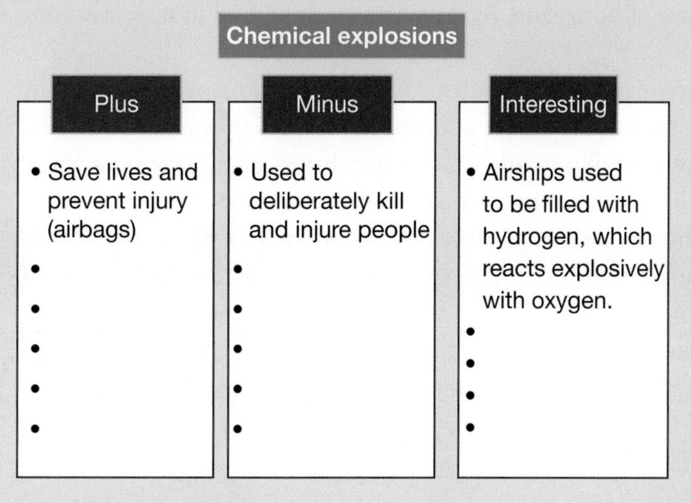

3. Create a PMI chart on the use of fossil fuels.

HOW ABOUT THAT!

In 1937, a hydrogen-filled airship called the Hindenburg exploded violently while docking at a refuelling tower. Until recently, it was believed that hydrogen was the cause of the disaster — hydrogen and oxygen react explosively when ignited by a spark. However, scientists now claim that it was, in fact, the flammable aluminium-coated skin of the airship and a stray spark that were to blame.

learn on

7.8 Project: ChemQuiz!

Scenario

You only have to have a glance at any page of your TV guide to see that Australians young and old love a good quiz show. Whether it's *Hot Seat, Jeopardy, Spit it Out or It's Academic*, programs with a quiz show format rate consistently well. While the idea of watching someone answer questions seems like an odd form of entertainment, psychologists theorise that their popularity arises from a combination of a desire to learn new information and a form of competition — after all, who hasn't watched a quiz show and yelled the answers at the screen? In recent educational studies, the use of quiz game formats as a teaching tool in the classroom is gaining support.

The Brain Mine is a company that specialises in educational resources for use in Science classrooms. On the basis of these educational studies of quiz games, they have decided that they would like to add a computer-based chemistry quiz show that teachers could purchase and run in their classrooms as a fun and effective way of improving student knowledge. As product developers at The Brain Mine, it is up to you and your team to make this happen! You and your team are going to develop *ChemQuiz*, a chemistry-based quiz show in which the class teacher will act as the show host, groups of students will be the contestants and the questions (which pop up on a computer screen so that the contestants can see them) are based on chemistry skills.

Your task

Using PowerPoint, you will create a series of question screens for a quiz show that should run for about ten minutes. For each question screen, the show host must be able to reveal the correct response after a contestant has given their answer. The question screens should be entertaining and eye-catching, and should also be easily readable by the contestants and the show host (who will read the questions out as they appear).

You will need to give a demonstration of your *ChemQuiz* show with one of your group acting as the show host (the role that would normally be taken by the teacher). The show host will need to explain the rules of the quiz show at the start. The contestants will be your fellow students (preferably not those in your group, who will already know the answers!).

7.9 Review

7.9.1 Study checklist

Reactants and products

- identify reactants and products in chemical reactions
- describe the rearrangement of atoms of the reactants during a chemical reaction
- describe chemical reactions using word equations
- state the Law of Conservation of Mass and the Law of Constant Proportions

Energy in chemical reactions

- recognise that many chemical reactions must be initiated by an input of energy
- distinguish between endothermic and exothermic reactions

Acids and bases

- describe the properties of acids and bases
- distinguish between acids and bases
- describe a variety of examples of the use of acids and bases in the home, garden and industry
- investigate neutralisation reactions and describe examples of their everyday use
- investigate chemical reactions of acids with metals

Combustion reactions

- recognise the role of oxygen in combustion reactions
- describe a variety of examples of the use of combustion reactions
- compare combustion reactions with other oxidation reactions
- identify respiration as a combustion reaction
- describe respiration using a word equation

Science as a human endeavour

- describe some applications of endothermic and exothermic reactions in everyday life and athletics
- describe the role of Alfred Nobel in the development of explosives and the awarding of the Nobel Prizes
- investigate the causes and effects of acid rain
- investigate methods of preventing acid rain
- investigate the effect of the products of combustion reactions on the environment
- evaluate claims made about chemical products

Individual pathways

ACTIVITY 7.1
Revising chemical reactions
doc-8449

ACTIVITY 7.2
Investigating chemical reactions
doc-8450

ACTIVITY 7.3
Investigating chemical reactions further
doc-8451

learnon ONLINE ONLY

7.9 Review 1: Looking back

To answer questions online and to receive **immediate feedback** and **sample responses** for every question, go to your learnON title at www.jacplus.com.au. *Note:* Question numbers may vary slightly.

1. A particular chemical reaction can be described by the word equation:

   hydrochloric acid + magnesium carbonate → magnesium chloride + water + carbon dioxide.

   (a) What are the reactants in this chemical reaction?
   (b) List the products of the reaction.
   (c) From which compound did the atoms present in the carbon dioxide come?
   (d) This is an exothermic reaction. Which substances have more energy stored in chemical bonds — the reactants or the products?
2. List four examples of observable evidence that a chemical reaction has taken place.
3. Use the Law of Conservation of Mass to explain why it is incorrect to say that when a candle burns it disappears.
4. State the Law of Constant Proportions.
5. When hydrogen reacts with oxygen in a rocket engine, a huge amount of energy is released.

   $$\text{hydrogen} + \text{oxygen} \xrightarrow{\text{heat}} \text{water}$$

   (a) Why does the word 'heat' appear above the arrow in the chemical word equation?
   (b) Is this reaction endothermic or exothermic? Explain how you reached your answer.
6. When an instant icepack is squeezed to activate it to treat an injury, ammonium chloride dissolves in water, producing a solution of ammonium chloride. As the ammonium chloride dissolves, energy is absorbed from the injured area, causing it to become very cold.
   (a) Is the production of the ammonium chloride an endothermic or exothermic process?
   (b) Explain why this process is not a chemical reaction, even though chemical bonds have been broken.
7. Are the chemical reactions that convert the chemical energy stored in your muscles into other forms of energy endothermic or exothermic? How do you know?
8. Use a three-column table to sort the substances listed below into acids, bases and salts.

caustic soda	gastric juices
hydrogen chloride	antacid tablets
sodium bicarbonate	copper nitrate
sodium chloride	magnesium chloride
ammonia	fizzy drinks
lemon juice	sodium sulfate

9. The liquids in the bottles below are labelled with their pH. Which of the bottles is most likely to contain:
   (a) distilled water
   (b) a strong acid
   (c) vinegar
   (d) bathroom surface cleaner?

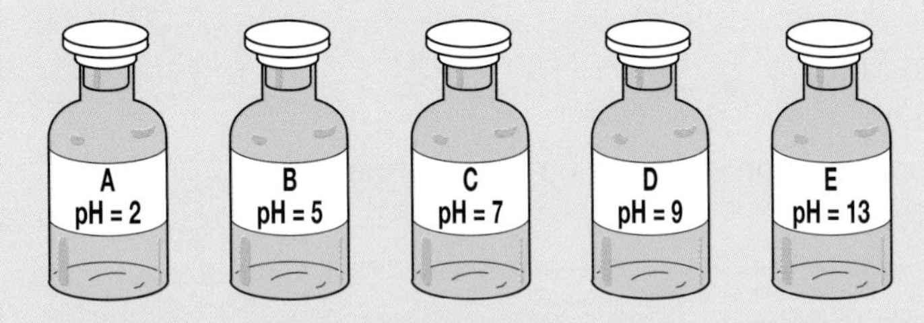

10. Predict the salts that would result from the neutralisation reaction between:
    (a) magnesium oxide and hydrochloric acid
    (b) copper (II) oxide and sulfuric acid
    (c) sodium hydroxide and acetic acid
    (d) sodium oxide and nitric acid.

11. If the water in a swimming pool has a pH that is too high for hygienic and safe swimming, which type of pool chemical should be added — an acid or a base?
12. What is pickling?
13. Complete the following chemical word equation:

acid + metal → ....

14. Outline the cause of acid rain and explain how it affects the natural environment and man-made structures.
15. There are at least two reactants in every combustion reaction. One of them is called a fuel.
    (a) With what substance does the fuel react?
    (b) Identify one product of every combustion reaction.
    (c) One product of combustion reactions is not a chemical. What is it?
16. Identify two chemical products of combustion reactions in which fossil fuels are burned.
17. Identify the main reactant in each of the following fuels in combustion reactions.
    (a) Natural gas
    (b) Petrol
    (c) Jet aircraft fuel
18. One combustion reaction takes place in every cell of your body.
    (a) What is the name of this combustion reaction?
    (b) Identify the reactants in the reaction.
    (c) Identify two chemical products of the reaction.
19. The diagram below shows the organisms that are normally found in a particular lake and the pH of water that they are able to tolerate.

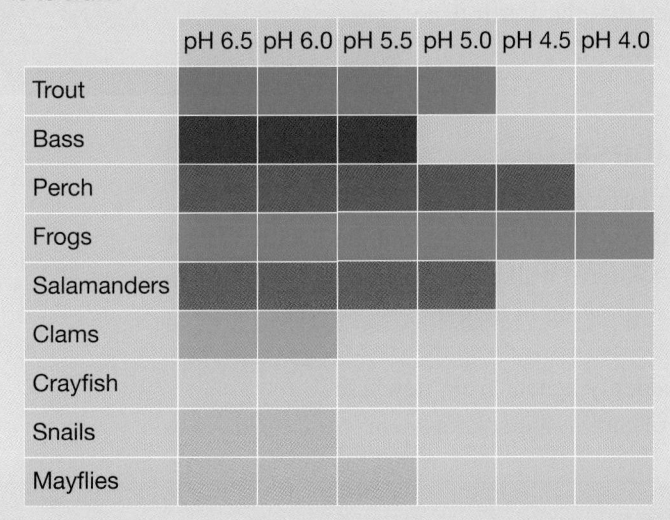

    (a) Which of these species would start to die first if the lake water started to increase in acidity?
    (b) Which of the species is the most tolerant of high acid levels in the lake? Explain.
    (c) Which species would remain if the acidity of the lake water increased until it had a pH of 5.0?

Complete this digital doc: Worksheet 7.5: Chemical reactions: Summary (doc-18893)

TOPIC 8
The dynamic Earth

8.1 Overview

The ground beneath you seems still. It might even seem dull. But first appearances can be deceiving. In fact, the Earth's crust is not still — it is constantly moving and changing. Nor is it dull — deep beneath the surface is a layer of red-hot molten rock, which sometimes bursts through and creates a volcano like the one shown (Eyjafjallajökull, Iceland, which erupted in 2010). Volcanoes and earthquakes provide spectacular evidence that the Earth is a dynamic, ever-changing planet.

assess on

8.1.1 Think about these

- How can something as large as a continent move?
- Why do volcanoes make a 'ring of fire' around the Pacific Ocean?
- How could Captain Cook have walked to Australia 250 million years ago?
- Why are the Himalayas growing in height?
- What causes tsunamis?
- How can a volcano suddenly appear from nowhere?
- Where is the largest volcano in the solar system?

LEARNING SEQUENCE

Numerous **videos** and **interactivities** are embedded just where you need them, at the point of learning, in your learnON title at www.jacplus.com.au. They will help you to learn the content and concepts covered in this topic.

8.1.2 Journey to the centre of the Earth ...

'Descend into the crater of Yokul of Sneffels, which the shade of Scataris caresses before the Kalends of July, audacious traveller, and you will reach the centre of the Earth. I did it.'

So wrote Jules Verne in his science fiction novel *Journey to the centre of the Earth*, which was published in 1864. The novel describes a fascinating journey by the adventurous Professor Lidenbrock, his nephew Axel and their guide Hans to the centre of the Earth. Their quest begins with a descent into the crater of the extinct volcano Snæfellsjökull in Iceland.

Create

1. Write your own science fiction short story about an attempted journey to the centre of the Earth. Before starting, think about what you would really expect to find beneath the surface and what sort of vehicle you would need to travel in.

Think

2. What would you expect to find at the very centre of the Earth?
3. Think about the substances found close to the surface of the Earth — close enough to the surface to be able to reach with drills and tunnels. Make a list of substances that are:
   (a) used to provide energy for heating, transport and industry
   (b) used for building and other construction
   (c) exceptionally valuable
   (d) able to find their way naturally to the surface.
4. The Earth's crust (the outer solid layer) is constantly changing. Make a list of events that occur because of these changes.

8.2 The Earth's crust

Science as a human endeavour

8.2.1 Beneath the crust

The Earth's crust is the very thin, hard, outer layer of our planet. To get an idea of how thin the Earth's crust is, compare it to a medium-sized apple. Imagine that the apple is the Earth. The crust would be as thin as the skin of the apple. Two questions have intrigued geologists for more than a hundred years: What lies beneath the crust? Why is the crust moving?

Comparing the Earth's crust to the skin of an apple makes it easier to understand how thin it is.

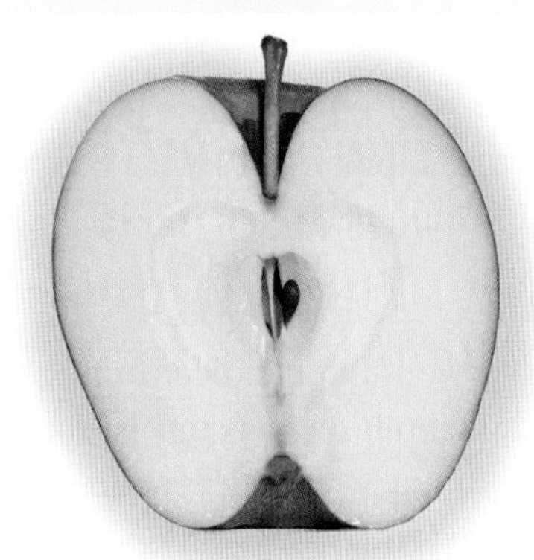

Questions about what is beneath the Earth's surface have inspired curiosity and imaginative writing — such as Jules Verne's novels. The idea of drilling through to or even travelling to the centre of the Earth is appealing. There could be no better way to find out what is down there. But the deepest man-made holes in the Earth have been drilled to only about 15 km of the 6370 km distance to the centre. Other methods had to be found to find out what is beneath the surface of the Earth.

Scientists use data from earthquakes to find out what lies inside the Earth. Earthquakes produce waves, known as **seismic waves**, that transfer energy through the crust. It is the energy of these waves that causes destruction at the surface. Seismic waves travel at different speeds and behave differently as they pass through different substances below the crust. By analysing the behaviour of seismic waves, scientists have been able to identify the state and chemical composition of the substances inside the Earth.

8.2.2 Moving continents

The shrinking theory

Geologists of the 1800s believed that, as the Earth cooled, the crust began to shrink and wrinkle. They believed that the continents were the high parts of the wrinkles and that oceans covered the lower

The structure of the Earth

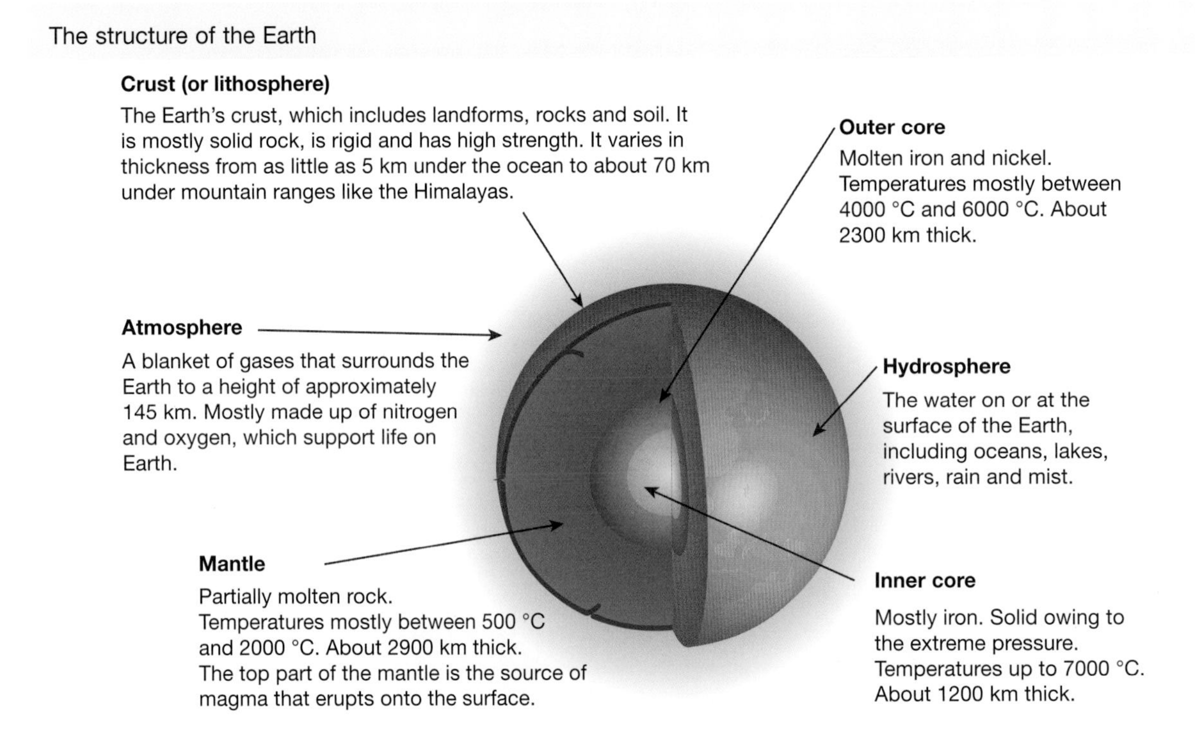

parts. During the late 1800s and early 1900s evidence was found that showed that the continents were moving. At first, the geologists thought that the movement was caused by the wrinkles moving relative to each other.

The continental drift theory

In 1912, a German meteorologist named Alfred Wegener proposed a new theory. He had noticed that the present day continents looked as though they would fit together very much like a jigsaw puzzle. Wegener suggested that the continents were floating, or drifting, on a denser material underneath. He proposed that the continents were breaking apart and rejoining in a process that he called **continental drift**.

A supercontinent

Wegener also believed that, at one time, all of the continents were joined together in a single 'supercontinent' that he called **Pangaea**. This belief was supported by the discovery of fossils of the same land animals on different continents. Pangaea was surrounded by a vast sea called **Panthalassa**.

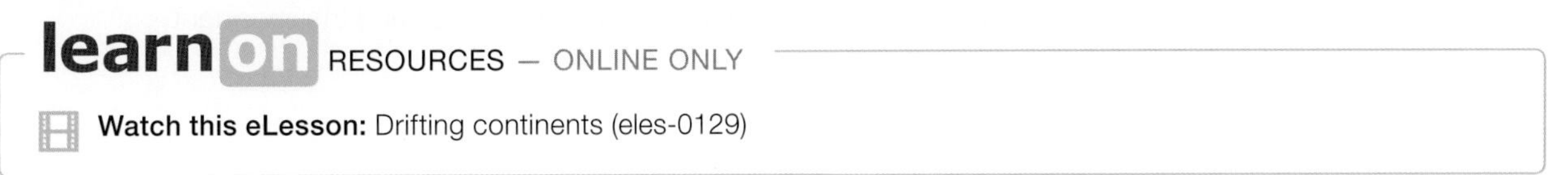

INVESTIGATION 8.1

Continental drift

AIM: To create a simple model to demonstrate continental drift

Materials:
enlarged copy of the map, scissors

Method

- Cut out the continents from the enlarged copy of the map right.
- Examine the distribution of fossils on each continent.
- Rearrange the continents into one supercontinent by matching the distribution of fossils.

Discuss and explain

1. How do you think the distribution of fossils helps to prove Wegener's theory of continental drift?
2. Suggest at least one other way in which the continents can be put together.

Distribution of a selection of fossils of ancient organisms

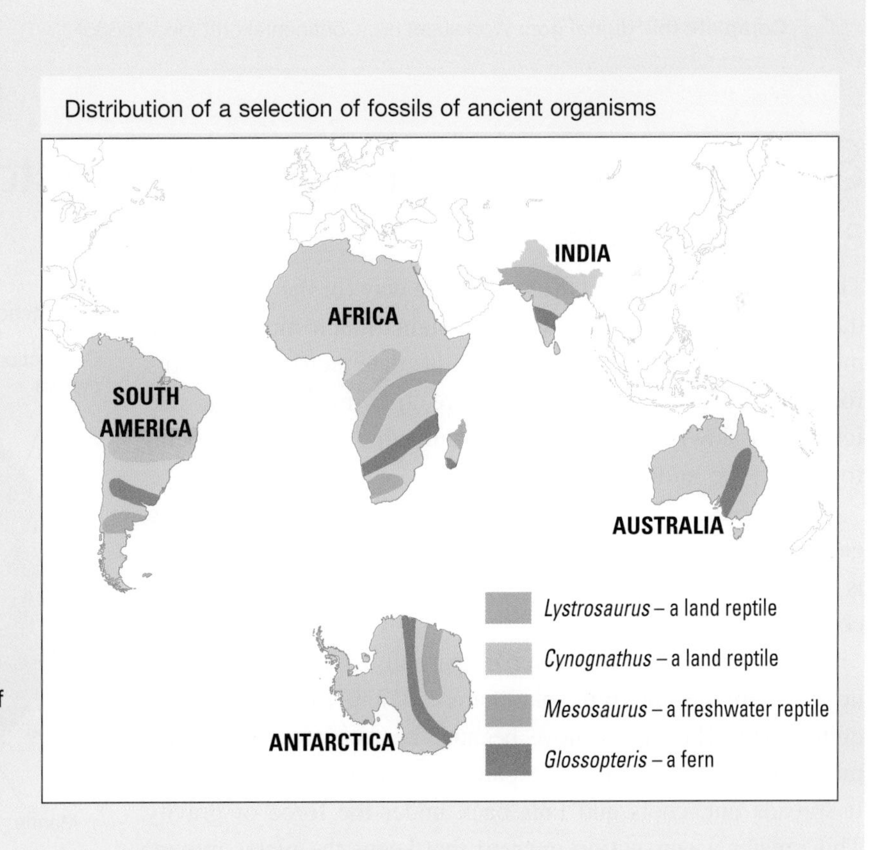

8.2 Exercises: Understanding and inquiring

To answer questions online and to receive **immediate feedback** and **sample responses** for every question, go to your learnON title at www.jacplus.com.au. *Note:* Question numbers may vary slightly.

Remember

1. Copy and complete the table at right.
2. Even though the inner core is hotter than the molten outer core, it is solid. Explain why this is the case.
3. Outline how the shrinking theory, popular during the 1880s, explained the existence and movement of the continents.
4. Describe two observations that provided evidence for Wegener's theory of continental drift.
5. What were Pangaea and Panthalassa?

Layer	Description
Atmosphere	
Hydrosphere	
Crust	
Mantle	
Outer core	
Inner core	

Think

6. According to Wegener's theory of continental drift, upon which layer of the Earth are the continents floating?

Create

7. Create a poster of the Earth that shows the four main layers beneath the surface and the important characteristics of each layer.

RESOURCES — ONLINE ONLY

Complete this digital doc: Worksheet 8.1: Continental drift (doc-18885)

8.3 Stability and change: Plate tectonics

8.3.1 Plates on the move

The theory of continental drift paved the way for the more recent theory of **plate tectonics**, which explains much more than the movement of continents. The theory of plate tectonics explains, for example, why the Himalayas are growing in height, why Iceland is slowly splitting in two and why new rock is being formed in the middle of the ocean.

With the use of technology such as sonar and satellite imaging, geologists have been able to demonstrate that the Earth's crust is divided into approximately 30 plates, not just the separate continents.

The plates move on a layer of partially molten rock in the upper **mantle**. Some of the plates are very large, while others are quite small. The plates move because heat causes the partially molten rock in the mantle to expand and rise towards the surface. It spreads out, cools and falls back under the force of gravity. This creates a **convection current** that keeps the plates moving slowly.

Convection currents in the mantle

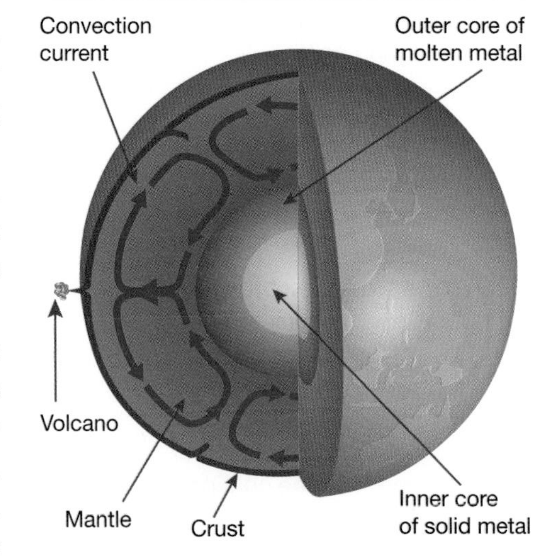

The plates can consist of two types of crust. The continents are made up of **continental crust**, which is between 30 km and 70 km thick. The plates beneath the oceans consist of **oceanic crust**. Oceanic crust is much thinner than continental crust and has an average thickness of about 6 km. It is also a little denser than continental crust due to differences in its chemical composition.

A simplified map showing the major tectonic plates that make up the Earth's crust. The arrows show the direction of plate movement.

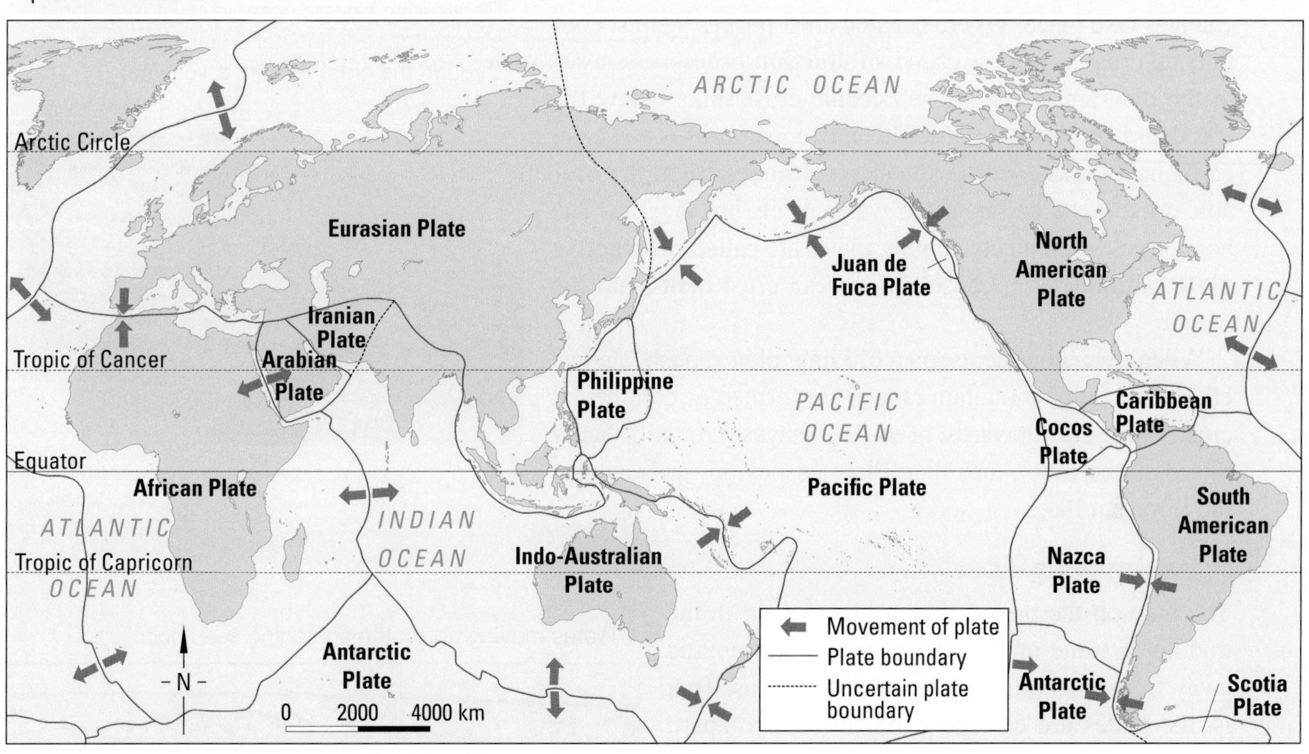

HOW ABOUT THAT!

The majority of the world's active volcanoes lie along the edges of the Pacific Plate. They form a circle around the Pacific Ocean known as the Ring of Fire.

The Ring of Fire

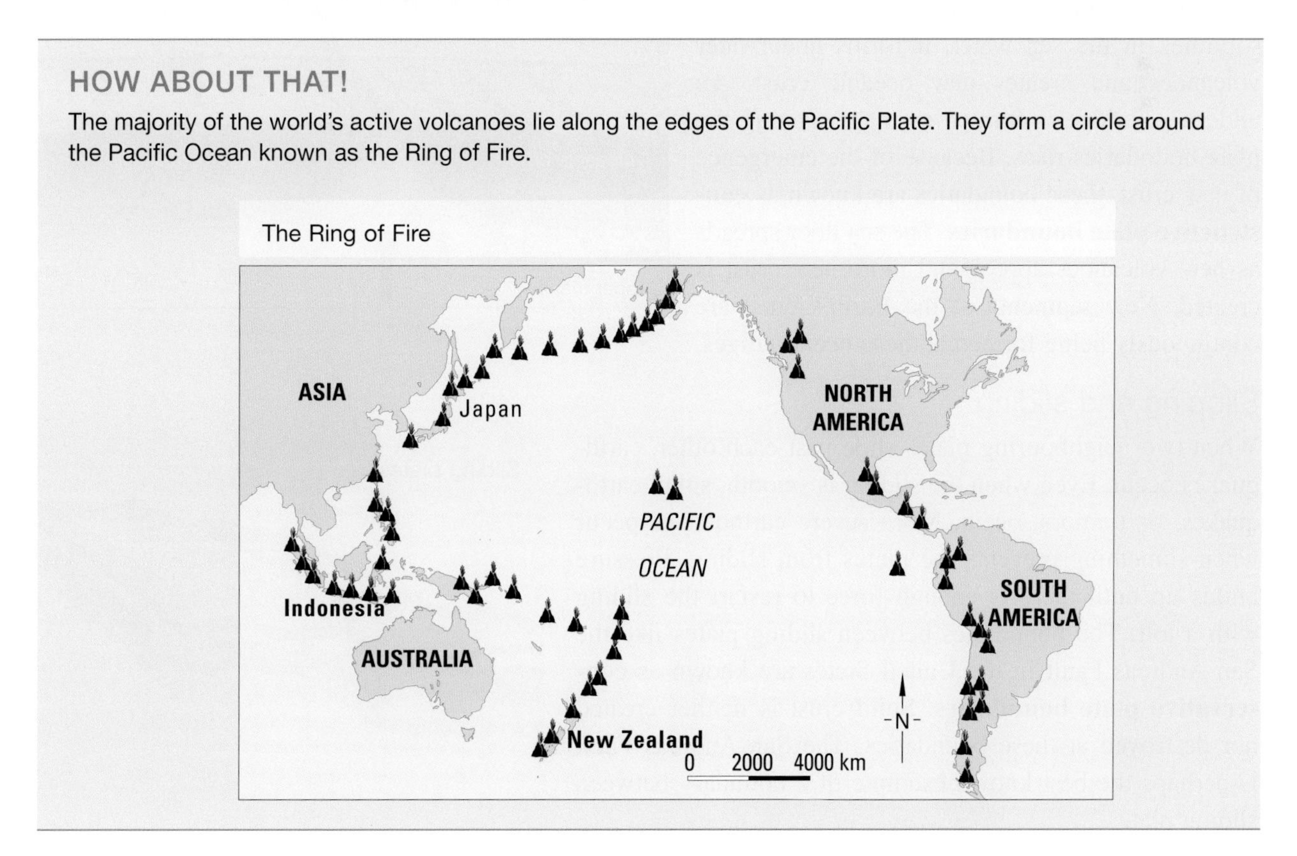

The plates move slowly (usually just a few centimetres in a year), and can slide past each other, push against each other or move away from each other.

The map on the previous page shows the location of some of the major plates and **boundaries**. The location of some of the boundaries is still not certain. These are shown on the map by dotted lines.

Converging plates

When two plates push against each other, two separate changes take place. Where oceanic crust pushes against continental crust, a process called **subduction** occurs, as shown in the diagram at right. The oceanic crust sinks below the less dense continental crust. This movement causes powerful earthquakes and creates explosive volcanoes when the oceanic crust melts and cold sea water meets hot magma. The boundaries between converging plates are called **destructive plate boundaries** because solid ocean crust melts into the mantle.

When two continents on colliding plates push against each other, huge mountain ranges are formed as continental crust crumples upwards. The Himalayas are still being raised as the Indo-Australian Plate pushes upwards against the Eurasian Plate.

Converging plates: oceanic crust melts as it sinks under continental crust. Magma bursts through the crust to form a volcano.

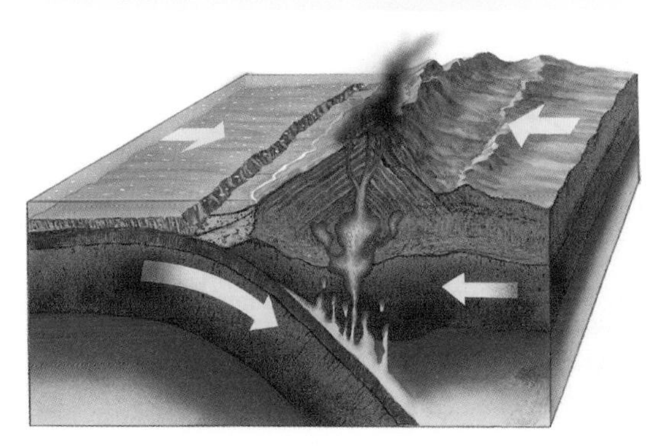

Plates moving apart

The map on the previous page shows that, in the middle of some major oceans, plates are moving away from each other in opposite directions. That is, the plates are diverging. As they move apart, magma from the mantle rises. As it cools and solidifies in the sea water, it forms underwater volcanoes and creates new oceanic crust. An underwater ridge is formed as magma along the plate boundaries rises. Because of the emergence of new crust, these boundaries are known as **constructive plate boundaries**. The sea floor spreads as new volcanoes appear and more new crust is created. New segments of the Earth's crust are continuously being formed at these **ocean ridges**.

Plates moving apart: the spreading sea floor

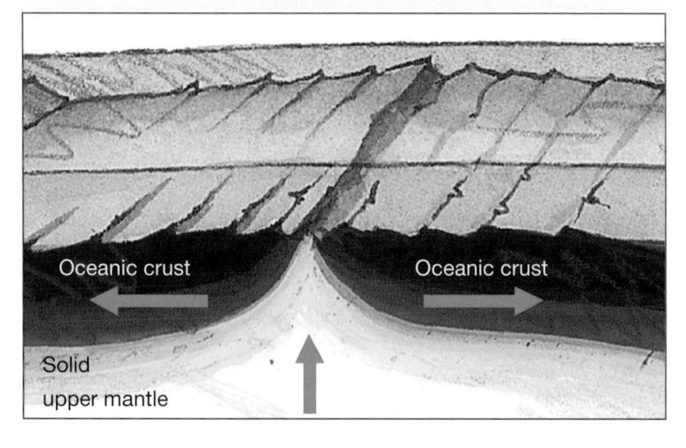

Slipping and sliding

When two neighbouring plates slide past each other, earthquakes occur. Even when the sliding is smooth, small earthquakes, or tremors, occur. More severe earthquakes occur when something prevents the plates from sliding. Pressure builds up until there is enough force to restart the sliding with a jolt. The boundaries between sliding plates like the San Andreas Fault in the United States are known as **conservative plate boundaries**. Solid crust is neither created nor destroyed at these boundaries. The San Andreas Fault is perhaps the best known example of a boundary between sliding plates.

Sliding plates

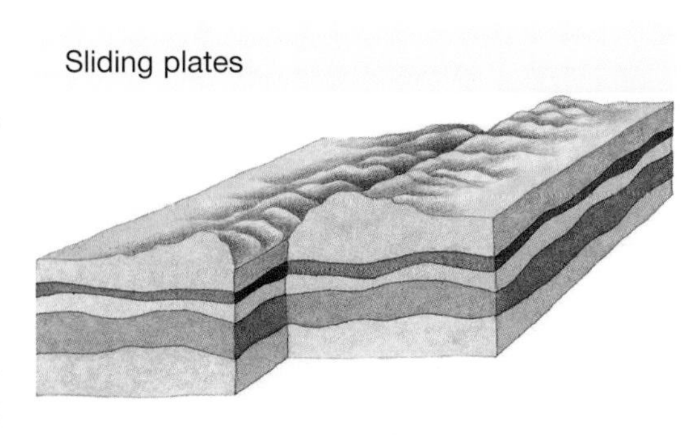

WHAT DOES IT MEAN?

The word *tectonic* is derived from the Greek word *tektonikos*, meaning 'builder'.

8.3.2 Magma recycled

While new oceanic crust is formed from cooling lava at ocean ridges, old oceanic crust is pushed downwards at subduction zones, eventually melting to form magma. This slow and continuing natural process of 'recycling' old crust and producing new crust takes place over millions of years.

8.3.3 Further evidence

Strong evidence for the theory of plate tectonics has been provided by the location of volcanoes and earthquakes, growing mountain ranges, spreading ocean ridges and the movement of the continents. However there is further evidence:

- Two-hundred-million-year-old fossils of the same land animals have been found in all of the southern continents. As these animals could not swim from one continent to another, this is evidence for the theory of continental drift and therefore supports the theory of plate tectonics.
- The rocks further away from the mid-lines of ocean ridges are older than those closer to the centre. This supports the idea that new rock is being formed in the middle of ocean ridges, continuously pushing the older rock aside.

8.3.4 The continental jigsaw

The theory of plate tectonics enabled a more complete reconstruction of the movement of continents proposed by the continental drift theory. Geologists now believe that about 200 million years ago the supercontinent Pangaea broke up into two smaller continents called **Laurasia** and **Gondwanaland**. The continents of Africa, South America, Antarctica and Australia were all part of Gondwanaland.

Australia on the move

Since Australia began to separate from Antarctica about 65 million years ago, it has slowly moved northward. Its climate has changed — from cold, to cool and wet, to warm and humid, to the hot and dry conditions that most of the continent experiences today. The movement of the tectonic plates is continuous and they are still moving today, taking the continents with them. Australia will continue to move north at the rate of a few centimetres each year.

A stable continent

Australia is geologically stable because it is near the centre of a tectonic plate, well clear of the boundaries. Volcanic activity and severe earthquakes are unlikely. The extreme age of Australia's rock is due to its distance from subduction zones, where new rock is formed.

Australia experiences relatively small earthquakes, weaknesses and movement within the Indo-Australian Plate. Most of Australia's volcanoes erupted millions of years ago. The most recent eruption on the continent was Mount Gambier in South Australia, about 4500 years ago.

8.3.5 Hotspots

Australia's volcanoes have formed over **hotspots**, regions in the crust where rocks in the upper mantle melt and magma surges upwards into the crust. Hotspots occur within plates. The active volcanoes in the Hawaiian Islands are caused by hotspots. The old volcanoes of Australia are also the result of hotspots.

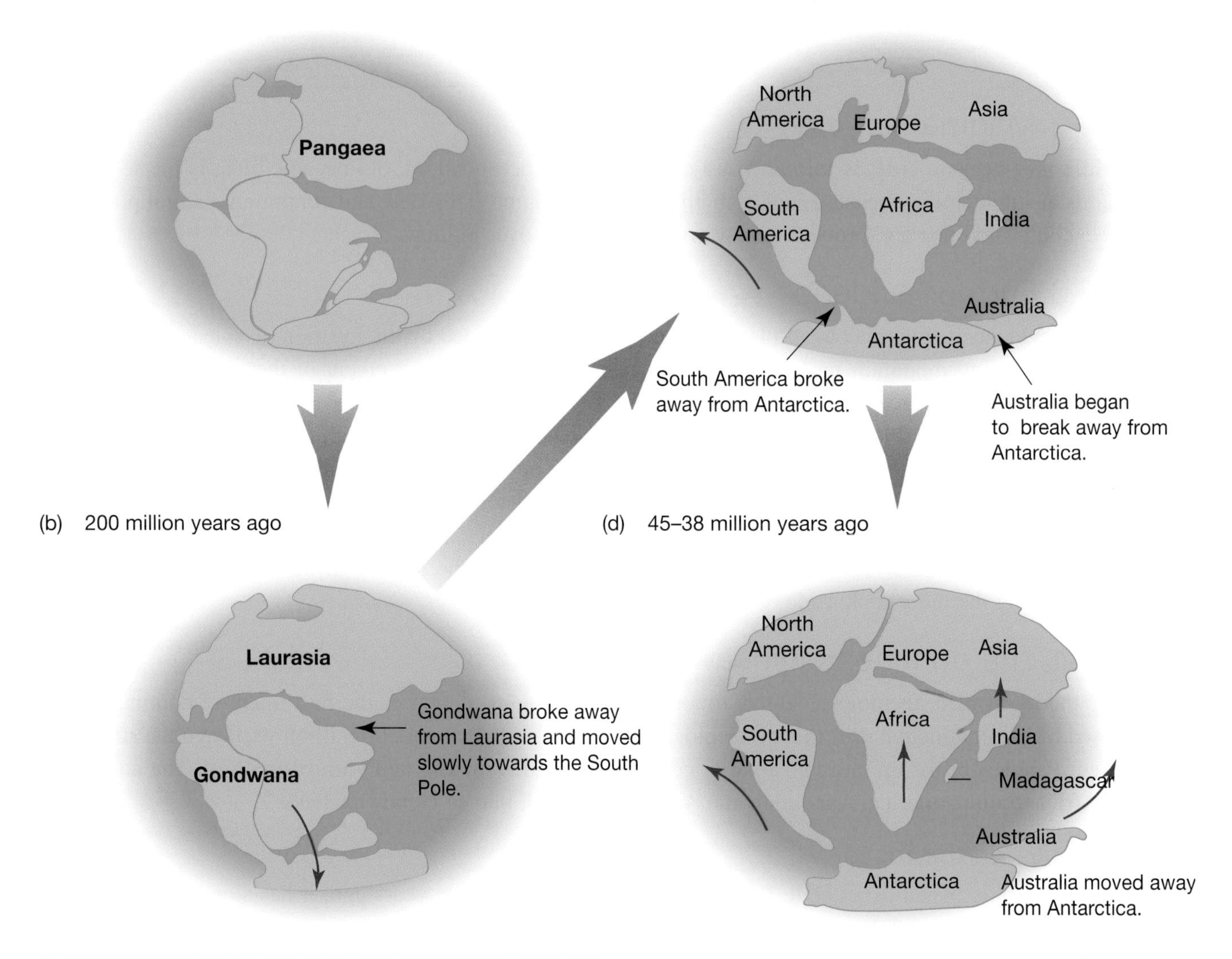

HOW ABOUT THAT!

Fossils of ancient sea creatures can be found at the top of the Himalayas, thousands of metres above sea level. How did they get there?

If you look at a map of the world, you will notice that India is joined to Asia. But that was not always the case. India has been moving towards Asia since it broke away from Gondwanaland millions of years ago. At first, seas separated the two lands. But now, the Indo-Australian Plate and the Eurasian Plate have collided. The current edges of these plates are both made of continental crust, so one plate will not easily slide under the other. Instead, the two are crumpling against each other, forming the Himalaya mountains. Sediments that once lay at the bottom of the sea between the two landmasses have been forced upward and can be found at the peaks of the mountain range.

8.3 Exercises: Understanding and inquiring

To answer questions online and to receive **immediate feedback** and **sample responses** for every question, go to your learnON title at www.jacplus.com.au. *Note:* Question numbers may vary slightly.

Remember

1. What is the theory of plate tectonics?
2. If the Earth's surface consists of moving plates, what are the plates moving on?
3. How is oceanic crust different from continental crust?

4. Describe what happens at the boundaries between plates when the plates:
   (a) slide past each other
   (b) push against each other
   (c) move away from each other.
5. State one location on Earth where:
   (a) two plates are sliding past each other
   (b) subduction is occurring
   (c) volcanoes have formed away from the edges of plates.
6. What is Gondwanaland?
7. What is a hotspot?

Think

8. Explain why earthquakes are common in the regions surrounding the Himalayas.
9. What is the Ring of Fire and why, according to the theory of plate tectonics, does it exist?
10. List, in point form, the evidence that supports the theory of plate tectonics.
11. The theory of continental drift was first proposed in 1912, over 50 years before the theory of plate tectonics was put forward. The evidence for the theory of continental drift also supports the theory of plate tectonics. Explain the difference between the two theories.
12. The illustration on the right represents part of a plate boundary.
    (a) Identify the type of boundary shown.
    (b) Describe the movement of the plates on either side of the plate boundary.
    (c) Should this boundary be described as a constructive or a destructive boundary? Explain your answer.

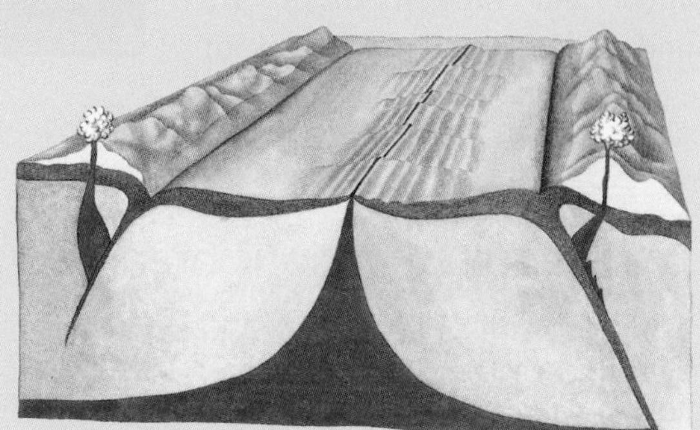

13. Explain why the climate of most of the Australian continent has changed from cold to hot and dry during the past 65 million years.
14. Explain why there are volcanoes in Australia, even though it is not on a plate boundary.

Investigate

15. The plates that make up the Earth's crust move only a few centimetres each year. Do some calculations to see whether it is really possible that Australia could have moved as far from South America as it is today.
16. Research and report on the use of sonar in mapping the ocean floor.
17. Find out how climate change, whether due to the northward movement of the Indo-Australian Plate or global warming, is likely to affect amphibians such as frogs and toads.

Create

18. Construct a model of converging continental crust. Use two piles of paper to represent the two sections of crust. Push the two piles of paper together.
    (a) What happens at the point where the paper piles meet?
    (b) Describe how this relates to the way the Himalayas have formed.
    (c) What would you expect the rocks to look like in a mountain range formed by converging plates?
19. Create your own model of sea-floor spreading using paper or other readily available materials.

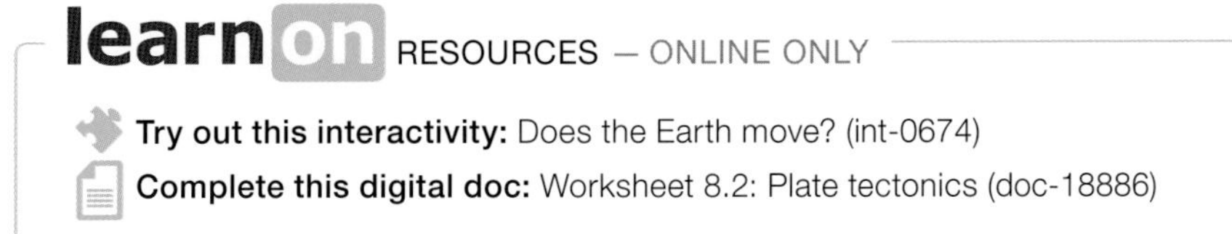

learnon RESOURCES — ONLINE ONLY

Try out this interactivity: Does the Earth move? (int-0674)

Complete this digital doc: Worksheet 8.2: Plate tectonics (doc-18886)

8.4 Rocks under pressure

8.4.1 Rocks under pressure

As the plates that make up the Earth's crust slowly move, solid rock is pushed, pulled, bent and twisted. The forces on the rocks in the crust are huge — large enough to fold them into rolling hills and valleys and large enough to crack them and move them up, down or sideways. Folded and broken rocks are found well beyond the edges of the plate boundaries.

Tightly folded rock strata in the walls of Hamersley Gorge National Park, Western Australia

8.4.2 Bending without breaking

If you hold a sheet of paper with one hand on each end and move the ends towards each other, the paper bends upwards or downwards.

The forces beneath the Earth are so large that layers of rock bend and crumple without breaking, just as the paper does. This process is known as **folding**.

Most of the major mountain ranges around the Earth have been shaped in this way. The shape of the Himalayas is the result of the folding of rock as two of the plates that make up the Earth's crust slowly collide. They are still rising as the plates continue to grind into each other.

Folds that bend upwards are called **anticlines**. Those that bend downwards are called **synclines**. Generally anticlines and synclines are formed well below the surface of the Earth and are not visible unless they are exposed by erosion of softer rock. They can often be seen in road cuttings or in cliffs formed by fast-flowing streams.

The folding of the Himalayas continues as two parts of the Earth's crust collide with each other.

Forces on solid layers of rock fold them into anticlines and synclines.

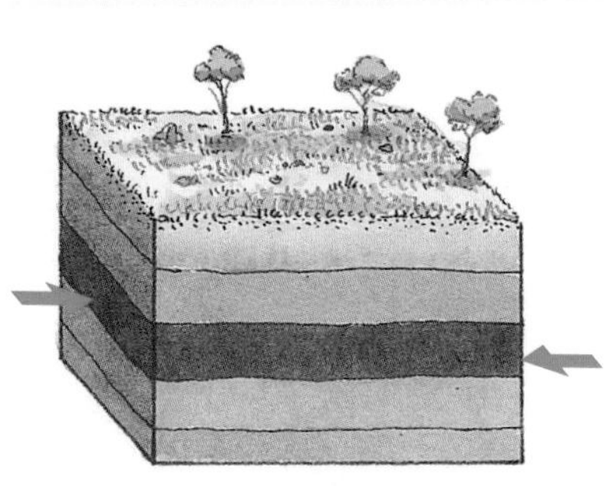

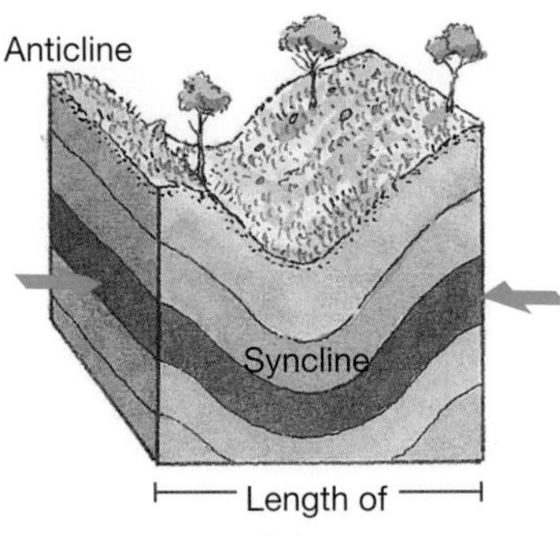

INVESTIGATION 8.2

Modelling faults

AIM: To model normal and reverse faults

Materials:

3 or 4 pieces of differently coloured plasticine
a thin sheet of polystyrene
knife or blade

(a) Normal fault (b) Reverse fault

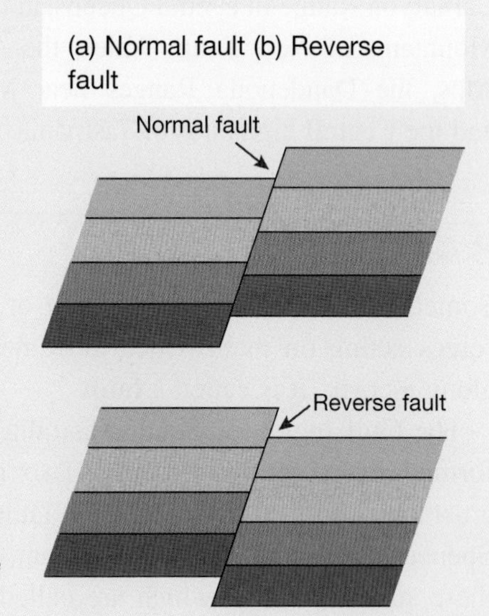

Method and results

- Place the first piece of plasticine on the bench and flatten it into a rectangular shape. Do not make it too thin. Cut a piece of polystyrene the same size, and fit it over the plasticine rectangle.
- Add two or three more layers of plasticine with a layer of polystyrene between each layer.
- Cut through the layers on an angle as shown in the diagram at right. Use the two parts to model each of the two types of fault shown at right.

1. Draw a diagram of each fault. Label it with arrows to show the direction in which each block moved to create the fault.

Discuss and explain

2. Describe the plate movement that could be responsible for each type of fault.

INVESTIGATION 8.3

Modelling folds

Rocks are usually folded well below the Earth's surface. The anticlines and synclines can be seen only along road cuttings or where erosion has exposed the layers of rock. A model is a useful way to describe how folded rocks would appear under the surface.

AIM: To model the folding of rocks

Materials:

3 or 4 pieces of differently coloured plasticine
knife or blade
board

Method and results

- Roll the pieces of plasticine into 1 cm thick layers.
- Place the layers of coloured plasticine on top of each other. Press down lightly on the layers, so that they stick together.
- With the palms of your hands, very gently compress the layers from the side.
- Model the processes of weathering and erosion on your plasticine layers.

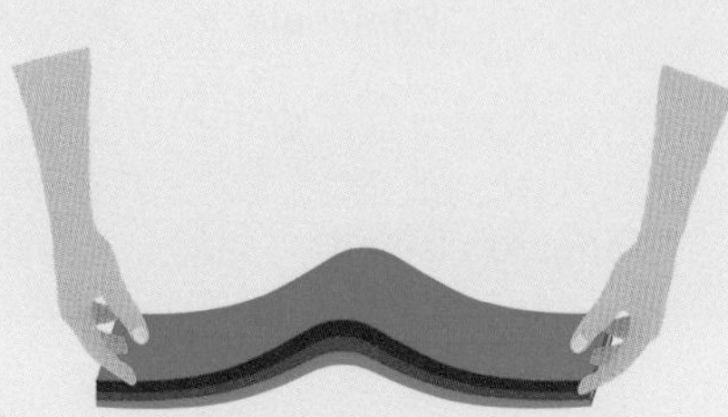

1. Describe the appearance of the plasticine when the layers are compressed.
2. Draw a diagram of the plasticine after compression, labelling anticlines and synclines.

Discuss and explain

3. Imagine that the rock layers are eroded at the Earth's surface. Draw diagrams of the eroded layers when viewed from above and when viewed from the side. Label the oldest and youngest layers. (Remember that the oldest layers are deposited before the younger ones.)

8.4.3 The Great Dividing Range

Australia's Great Dividing Range, which stretches all the way from northern Queensland to Tasmania, was formed by folding. It is actually a chain of separate mountain ranges, including the Carnarvon Range in central Queensland, the Blue Mountains of New South Wales, the Australian Alps, the Dandenong Ranges near Melbourne and the Central highlands of Tasmania.

Australia's Great Dividing Range was formed by folding as a result of the movement of the Earth's tectonic plates.

8.4.4 Faults

Sometimes rocks crack as a result of the huge forces acting on them. Once movement occurs along a crack, it is called a **fault**.

The Gulf region of South Australia has been formed by a series of faults. Two blocks of crust have dropped down between faults to form Spencer Gulf and Gulf St Vincent. Between these sunken blocks, which are called **rift valleys**, is a block that has been pushed upwards by the forces below. This block, called a **horst**, has formed Yorke Peninsula. The movement along these faults is responsible for the occasional earthquakes in the Adelaide area.

If movement along a fault is sideways, that is, where the blocks of crust slip horizontally past each other, it is termed a **slip fault**. The San Andreas Fault in California is a slip fault. It stretches about 1200 kilometres along the coast, passing through San Francisco and to the north of Los Angeles. A large movement

The Gulf region of South Australia has been formed by a series of faults.

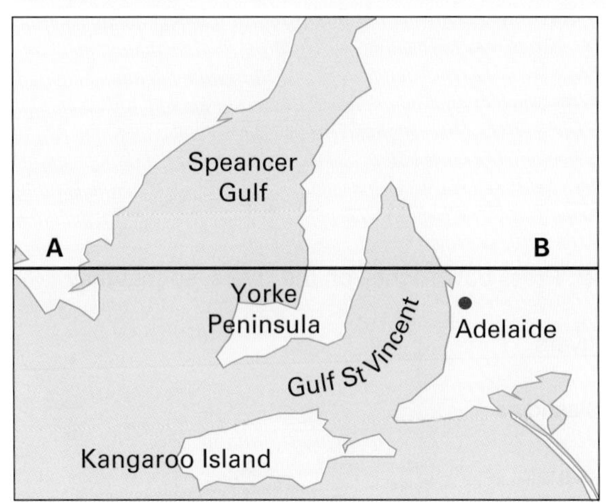

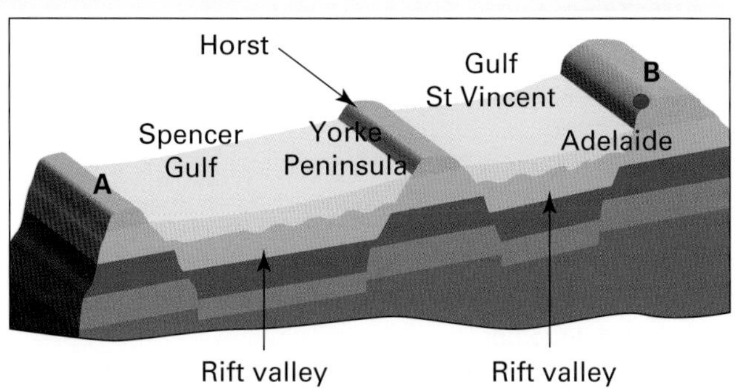

of the blocks on either side of this fault line in 1989 created a major earthquake in San Francisco, killing at least 62 people. The earthquakes experienced in this area in recent years appear to be caused by a buildup of pressure along the fault. Scientists believe that it will not be long before the pressure is relieved through a catastrophic earthquake.

8.4 Exercises: Understanding and inquiring

To answer questions online and to receive **immediate feedback** and **sample responses** for every question, go to your learnON title at www.jacplus.com.au. *Note:* Question numbers may vary slightly

Remember

1. What is *folding* and how is it caused?
2. Explain why the Himalayas are still growing in height.
3. Explain the difference between a syncline and an anticline.
4. What causes earthquakes along the San Andreas Fault?
5. Explain the difference between a reverse fault and a normal fault.

Think

6. Explain with the aid of some labelled diagrams how mountains could be formed by faulting.
7. There is a lot of faulting as well as folding in the Himalayas. Explain how it is possible for folding mountains to develop faults later in their geological history.

Create

8. Construct a model to demonstrate the formation of the Gulf region of South Australia.

Investigate

9. Imagine that you were offered the chance to spend a year in a high school in a leafy northern suburb of Los Angeles, just two kilometres from the San Andreas Fault. Would you accept the offer? Explain your response.

RESOURCES — ONLINE ONLY

Complete this digital doc: Worksheet 8.3: Folding and faulting (doc-18887)

8.5 Shake, rattle and roll

Science as a human endeavour

8.5.1 Earthquakes and tremors

Earthquakes result from movements in the Earth's crust up to 700 kilometres below the surface. These movements cause vibrations or **tremors** on the Earth's surface. Fortunately most of the tremors are too weak to be felt. When the tremors are sudden and strong, they are called earthquakes.

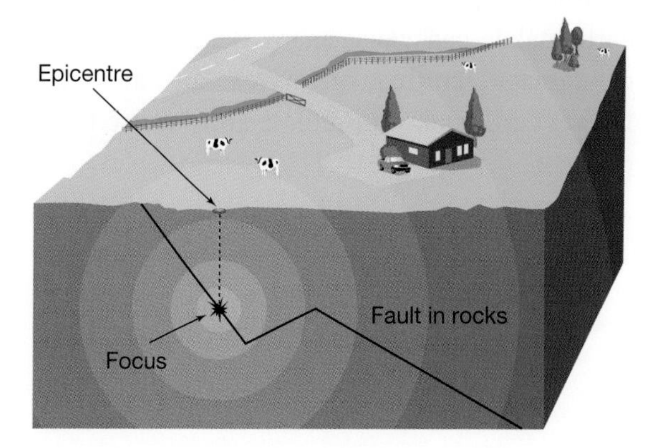

Major earthquakes occur at or near the plate boundaries where plates are:

- pushing against each other in subduction zones
- spreading apart to form ocean ridges and new underwater volcanoes or volcanic islands
- slipping and sliding against each other in sudden jolts.

Tremors and minor earthquakes can take place wherever there is a weakness in the Earth's crust, especially along faults.

The **epicentre** of an earthquake is directly above the point below the surface where the movement in the crust began. The point at which the earthquake begins is called the **focus**.

By using readings from at least 3 different seismic stations, the position of the epicentre can be determined. This process is known as **triangulation**.

Locating the epicentre using triangulation

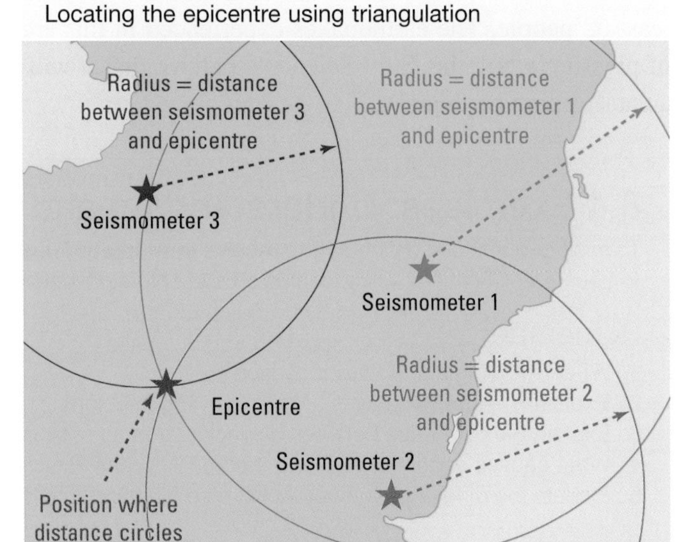

8.5.2 Measuring earthquakes

Movements in the Earth are recorded with a **seismograph**. The graph at bottom right shows what a seismograph records during an earthquake. The strength of an earthquake can be measured in a number of ways. The most well-known way of measuring the strength of earthquakes is the **Richter scale**.

The Richter scale

The Richter scale is a measure of the amount of energy released by an earthquake. Earthquakes measuring less than 2.0 on the Richter scale are called microquakes and are rarely detected by people. Earthquakes of magnitude 5.0 on the Richter scale are detectable and may even cause objects on shelves or in cupboards to rattle. Each increase of 1.0 on the scale represents a 30-fold increase in the amount of energy released. So an earthquake of magnitude 6.0 releases 30 times as much energy as one of magnitude 5.0. That means that an earthquake of magnitude 7.0 releases 900 times as much energy as one of magnitude 5.0.

The Richter scale is not always a good indication of the destructive power of an earthquake. In a crowded city, small earthquakes can cause many deaths, injuries and a great deal of damage, cutting off water, gas and electricity supplies. Larger earthquakes in remote areas cause few injuries and little damage.

An earthquake recorded on a seismograph. A strip of paper moves past a stationary pen. When the seismograph is jolted by the earthquake it jumps up and down. However, the pen is attached to a heavy mass which does not move. As the paper turns on the drum, the pen leaves a record of the vibrations.

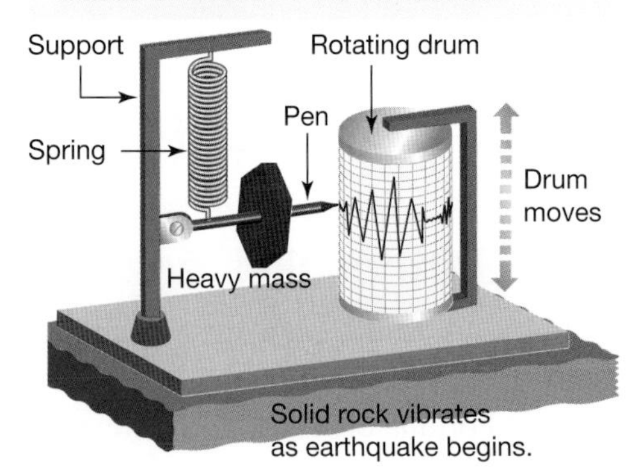

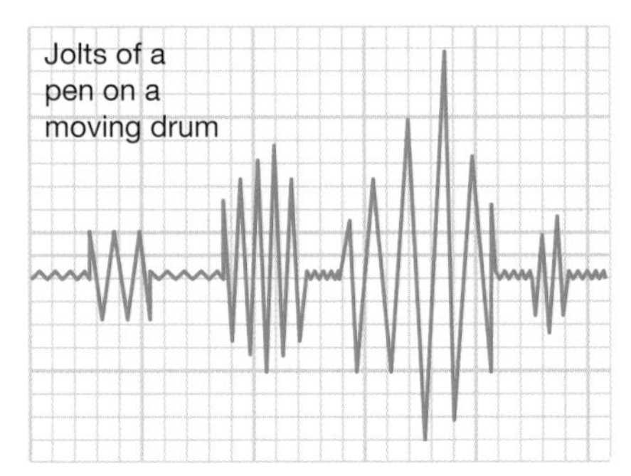

8.5.3 Destructive power

The destructive power of an earthquake in any location also depends on its distance from the epicentre. For example, the Tennant Creek earthquake of 1988 in the Northern Territory had a Richter magnitude of 6.7; however, only two buildings and the natural gas pipeline were damaged. The epicentre of the earthquake was 40 kilometres north of the town. Yet the smaller earthquake that devastated Newcastle in New South Wales in 1989 registered 5.6 on the Richter scale, killed 13 people, hospitalised 160 others and demolished 300 buildings. The epicentre of that earthquake was only five kilometres west of the city.

Some of the destruction caused in Christchurch, New Zealand, by a magnitude 6.3 earthquake in February 2011. The earthquake destroyed many buildings and homes, injured thousands, and killed 185 people.

8.5.4 Seismic waves

Energy released during an earthquake travels in the form of waves. There are three main types of wave that are generated by earthquakes: P-waves, S-waves and L-waves. These waves differ in their speed and the regions of the Earth through which they travel.

P-waves (or **primary waves**) are compression waves, moving through the Earth in the same way that sound waves move through air. They are the fastest of the seismic waves.

After the P-waves, the second sets of waves to be detected are the **secondary** waves or **S-waves**. These travel in the form of transverse waves.

P-waves travel through the earth as compression waves while S-waves are transverse waves.

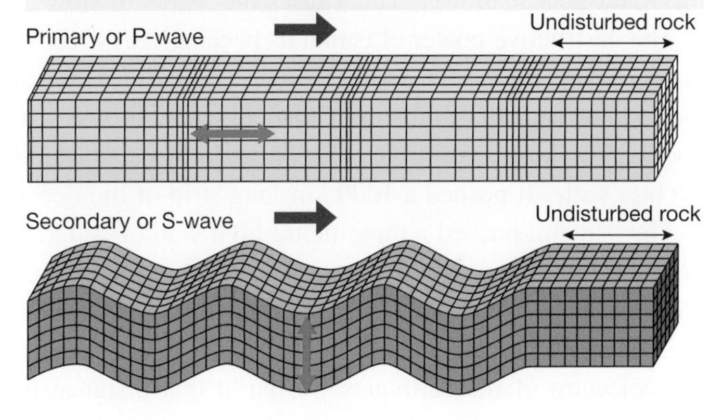

Both P-waves and S-waves are **body waves**, because they travel through the 'body' of the Earth, rather than the surface.

L-waves are **surface waves** and travel around the Earth. While they travel much more slowly than either the P-waves or S-waves, it is these surface waves that are responsible for the majority of an earthquake's destructive power. This is because all of the L-wave energy is distributed across the surface of the Earth rather than being spread out through the Earth's interior like P- and S-waves.

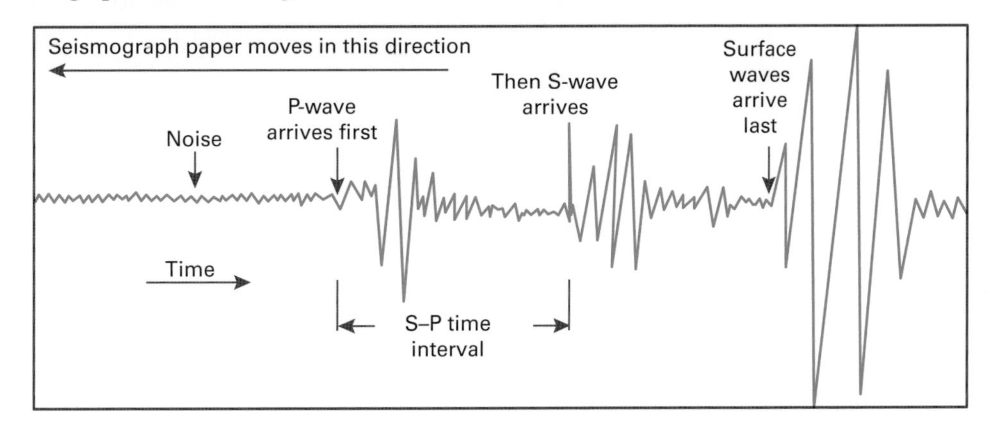

8.5.5 Living on the edge

For the people living near the plate boundaries, particularly on the edges of the Pacific Ocean, the ability of scientists to predict earthquakes and tsunamis is critical. The scientists who study earthquakes are called **seismologists**.

Although it is difficult to predict the time, location and size of earthquakes, seismologists can use sensors to monitor movement along plate boundaries and fault lines. When pressure build-ups occur they can at least warn authorities that an earthquake is likely. As yet there is no reliable early warning system in place. However, seismologists are experimenting with a variety of methods using satellites, Earth-based sensors and even animal behaviour.

P-waves are able to travel through the inner and outer core as well as the mantle; S-waves travel in the mantle; L-waves travel on the Earth's surface.

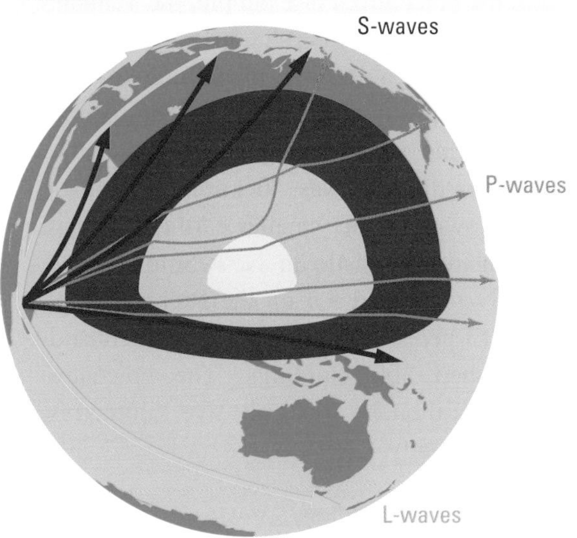

8.5.6 Waves of destruction

Earthquakes occurring under the water or near the coast can cause giant waves called **tsunamis**. These huge waves travel through the ocean at speeds of up to 900 kilometres per hour. When the waves approach land the water gets shallower. This causes the waves to slow down and build up to heights of up to 30 metres.

The destructive power of tsunamis became very clear on 26 December 2004 when about 300 000 people across South-East Asia, southern Asia and eastern Africa died. Millions more lost their homes. The tsunami, known as the Sumatra–Andaman tsunami, was caused by a huge earthquake under the ocean floor about 250 kilometres off the coast of the Indonesian island of Sumatra. The earthquake measured 9.0 on the Richter scale. It pushed a 1000 km-long strip of the ocean floor about 30 metres upwards.

The tsunami pushed a three-metre-high wall of water, mud and debris a distance of 10 kilometres inland near the Sumatran city of Banda Aceh. Thousands were killed in Sri Lanka, India and Thailand as well. Death and destruction also occurred in Malaysia, Myanmar, Bangladesh and the Maldives. More than eight hours after the earthquake, the tsunami arrived at the east coast of Africa, more than 5000 kilometres from the epicentre of the earthquake. Even at that distance from the earthquake, the tsunami caused flooding which killed more than 160 people on the coasts of Somalia, Kenya and Tanzania.

This map shows the huge area affected by the tsunami on 26 December 2004.

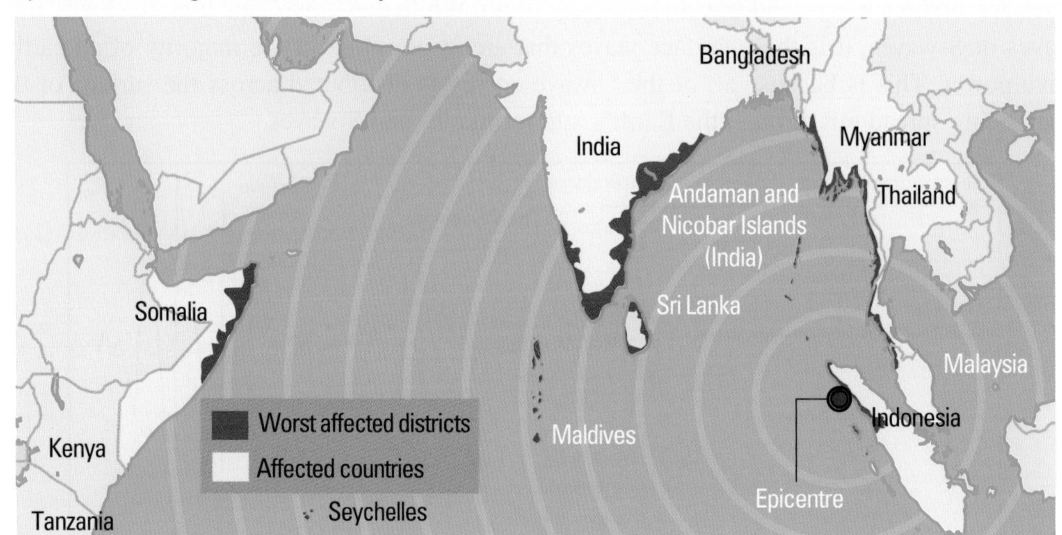

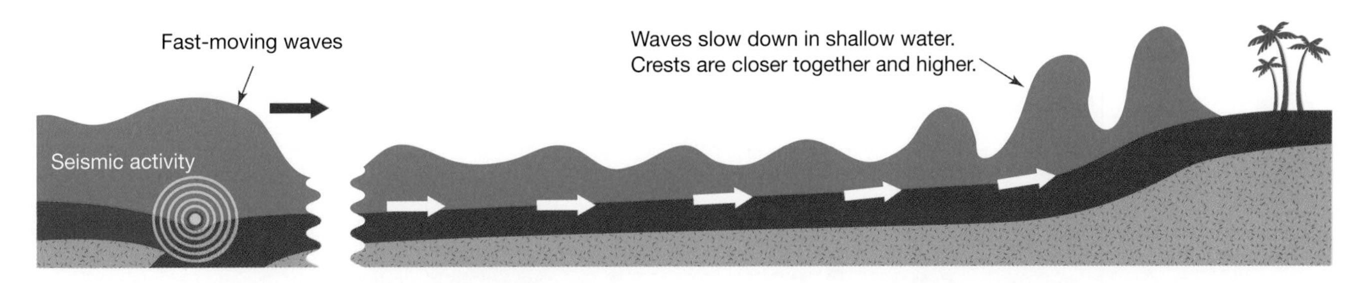

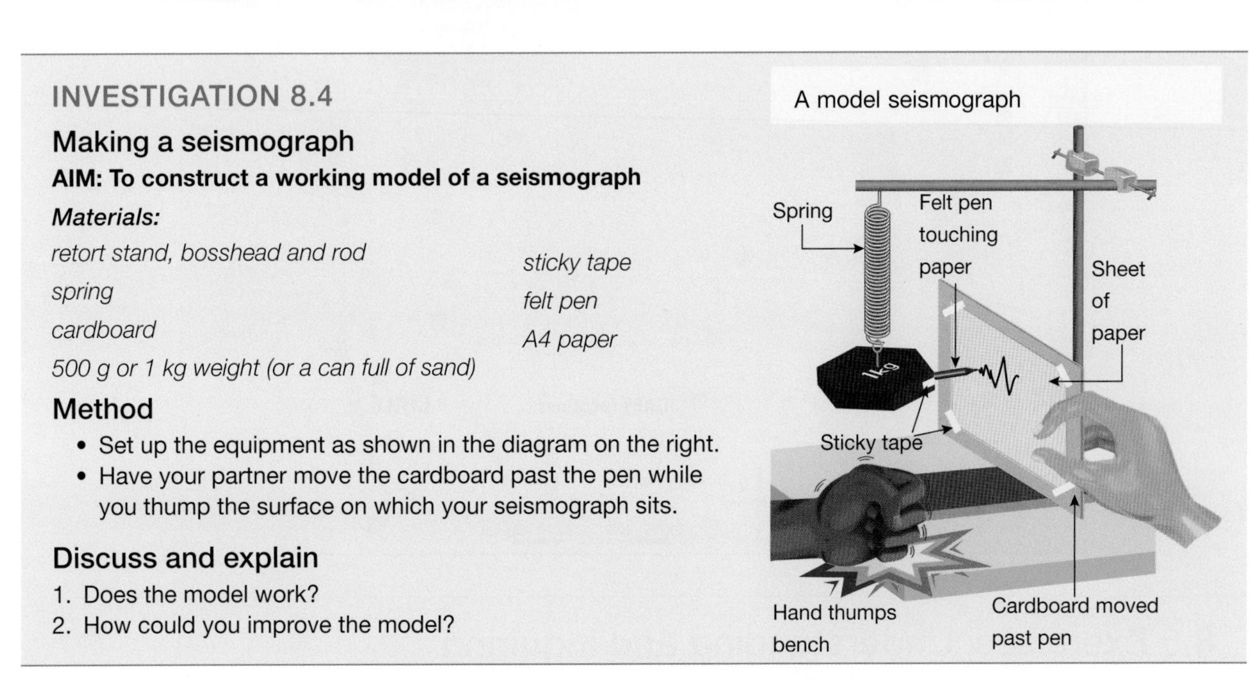

INVESTIGATION 8.4

Making a seismograph

AIM: To construct a working model of a seismograph

Materials:

retort stand, bosshead and rod
spring
cardboard
500 g or 1 kg weight (or a can full of sand)
sticky tape
felt pen
A4 paper

Method

- Set up the equipment as shown in the diagram on the right.
- Have your partner move the cardboard past the pen while you thump the surface on which your seismograph sits.

Discuss and explain

1. Does the model work?
2. How could you improve the model?

The world was reminded of the destructive power of tsunamis in March 2011, when an earthquake struck that was of the same magnitude as the 2004 Sumatra–Andaman earthquake. The epicentre of this earthquake was only 70 kilometres off the coast of the Japanese island of Honshu. The nearest major city was Sendai, where the port and airport were almost totally destroyed. In that city at least 670 people were killed and about 2200 were injured. Around 6900 houses were totally destroyed, with many more partially destroyed. Waves of up to 40 metres in height were recorded on the coast and some caused damage as far as 10 kilometres inland.

Several nuclear reactors were shut down immediately following the earthquake that caused the tsunami. However, that wasn't enough to prevent **meltdowns** in three reactors at the Fukushima Daiichi Power Plant, resulting in explosions and the leakage of radiation into the atmosphere, water and soil.

Early warning

Tsunami early warning systems rely on the early detection of earthquakes and a system of buoys placed around the Pacific and Atlantic Oceans. This system is called DART (Deep-ocean Assessment and Reporting of Tsunamis). Sudden rises in sea level are detected by the buoys and alerts are sent to tsunami warning centres.

The locations of DART buoys

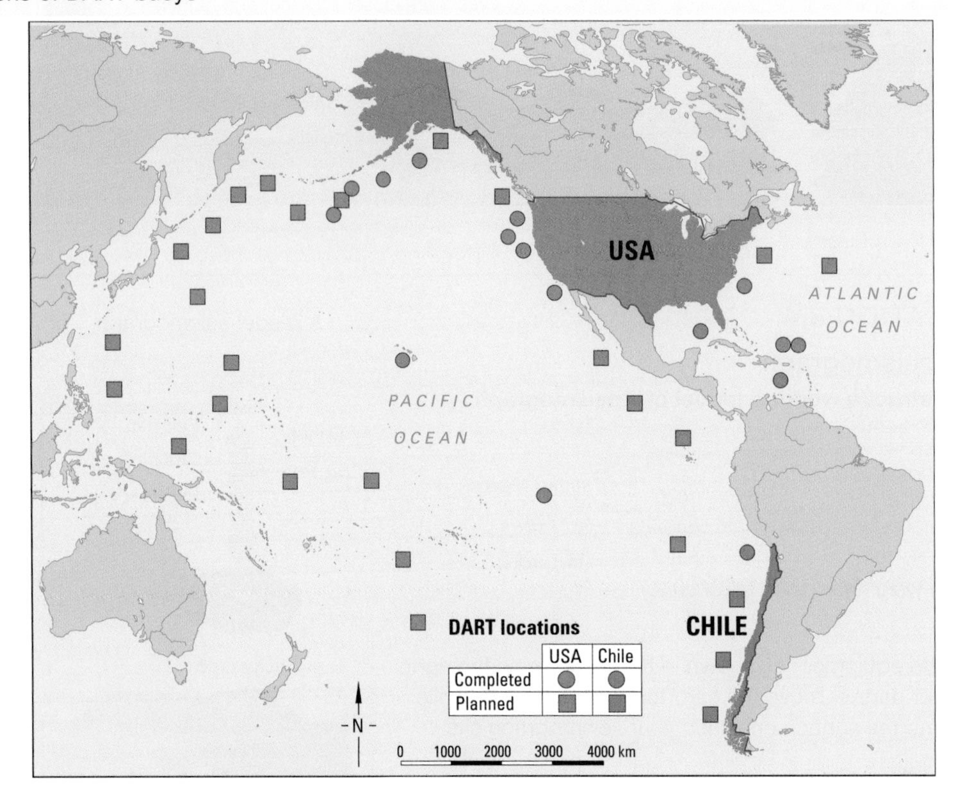

8.5 Exercises: Understanding and inquiring

To answer questions online and to receive **immediate feedback** and **sample responses** for every question, go to your learnON title at www.jacplus.com.au. *Note:* Question numbers may vary slightly

Remember

1. How are earthquakes caused?
2. Distinguish between an Earth tremor and an earthquake.
3. What name is given to the point at which an earthquake begins?
4. Where is the epicentre of an earthquake?
5. What quantity is the Richter scale a measure of?
6. Explain why a tsunami only a few metres high in open ocean can reach heights of up to 30 metres by the time it reaches land.
7. Explain how seismologists are able to make predictions about the likelihood of an earthquake.

The tsunami of March 2011 in Japan destroyed entire cities.

Analysing data

8. The table on the next page shows the number of people killed in some of the major earthquakes in recent years.
   (a) List two pairs of earthquakes that provide evidence that the Richter scale does not indicate the loss of life in earthquakes.
   (b) What factors, apart from the Richter scale measurement, affect the number of deaths in an earthquake?
   (c) How much more energy was released by the 2004 Sumatra earthquake than the 2010 Haiti earthquake?
   (d) Suggest why there were more fatalities as a result of the Haiti earthquake.

Year	Location	Number of deaths (approx.)	Richter scale magnitude
1994	Los Angeles, USA	57	6.6
1995	Kobe, Japan	6 400	7.2
1999	Iznit, Turkey	17 000	7.4
2001	Gujarat, India	20 000	7.9
2003	Bam, Iran	26 000	6.6
2004	Sumatra, Indonesia	230 000	9.0
2008	East Sichuan, China	90 000	7.9
2010	Haiti (Caribbean Sea)	316 000	7.0
2011	Sendai, Japan	21 000	9.0

Think

9. Explain why Indonesia is more likely to experience major earthquakes than Australia.
10. Outline some of the long-term consequences of the damage done to nuclear power stations by the Sendai tsunami.

Investigate

11. Use the internet or other resources to research and compare the 2004 Sumatra earthquake and the 2011 Japan earthquake. Write a report about the differences between the earthquakes and their consequences.
12. Learn more about the 2004 Sumatra–Andaman tsunami.

learnon RESOURCES — ONLINE ONLY

Complete this digital doc: Worksheet 8.4: Earthquakes (doc-18894)

Complete this digital doc: Worksheet 8.5: Plotting earthquake activity (doc-18895)

8.6 Mountains of fire

8.6.1 Volcanoes

Although most changes in the Earth's crust are slow and not readily observable, the eruption of volcanoes provides evidence that the changes can also be explosive, fiery and spectacular.

Volcanoes are formed when molten rock, or **magma**, from the Earth's mantle bursts through a weakness in the Earth's crust. The eruption of a volcano is usually spectacular. The red-hot molten rock released is called **lava**. It is a mixture of magma and gases, including steam, carbon monoxide and hydrogen sulfide ('rotten egg' gas). A scientist who studies volcanoes is called a vulcanologist.

8.6.2 What comes out?

As the pressure builds up in the magma chamber, ash and steam emerge from the vents of a volcano. When the volcano erupts, lava flows from the vents and red-hot fragments of rock, dust and ash, steam and other gases shoot out of the crater. Exploding gases often destroy part of the volcano. The larger fragments of rock blown out of the crater are called **volcanic bombs** or **lava bombs**.

The lava flowing from a volcano can be runny or pasty like toothpaste. If it is runny, it can flood large areas, cooling to form large basalt plains like those in Victoria's western district and in Melbourne, and to the city's north and south.

Thick, pasty lava builds up on the sides of volcanoes and can also block the vents as it cools. When this happens, gases build up in the magma below. As the pressure increases, the volcano can bulge and 'blow its top', thrusting rocks, gases and hot lava high into the air.

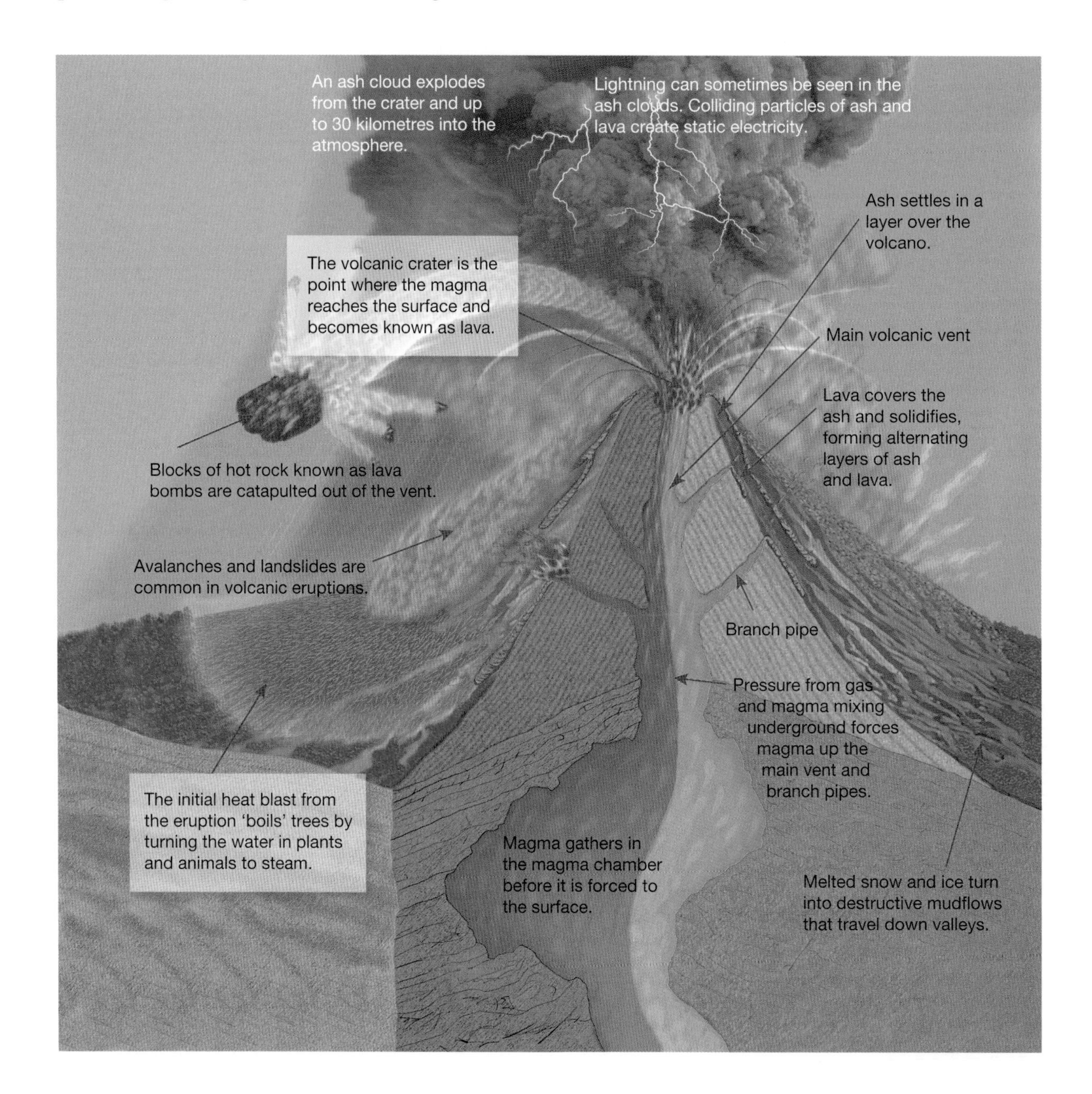

8.6.3 Birth of a volcano

On a cool winter's day in 1943, a small crack opened up in a field of corn on a quiet, peaceful Mexican farm. When red-hot cinders shot out of the crack, the shocked farmer tried to fill it with dirt. The next day, the crack had opened up into a hole over two metres in diameter. A week later, the dust, ash and rocks erupting from the hole had formed a cone 150 metres high! Explosions roared through the peaceful country-side and molten lava began spewing from the crater, destroying the village of Paricutin. The eruptions continued and, within a year, the new mountain, named Paricutin, was 300 metres high. When the eruptions stopped in 1952, Paricutin was 410 metres high.

8.6.4 Hot spots

Although most of the world's volcanoes are found at the edges of the plates of the Earth's crust, some lie over hot spots. These hot spots are regions of the crust where the mantle below is extremely hot. The volcanoes of western Victoria and Queensland have formed over hot spots.

8.6.5 Dead or alive?

Volcanoes that are erupting or have recently erupted are called **active** volcanoes. Mount Pinatubo in the Philippines, which erupted in June 1991 killing 300 people, is an active volcano. There was so much smoke and ash coming from Mount Pinatubo that scientists believe that the Earth's weather was cooler for over a year. The cloud of smoke and ash was believed to have blocked out about four per cent of the heat from the sun.

Extinct volcanoes are those that have not erupted for thousands of years. They are effectively dead and are most unlikely to erupt again.

8.6.6 Land ahoy!

Active volcanoes also erupt under the sea. An active volcano below the sea is generally not visible. If layers of lava build up, however, they may eventually emerge from the sea as a volcanic island. Lord Howe Island, off the coast of New South Wales, was formed in this way about 6.5 million years ago. A more recent example is the island of Surtsey, which emerged from the sea off the coast of Iceland in 1963.

There are many extinct volcanoes in Australia. The Glasshouse Mountains of Queensland are the remains of cooled lava trapped in the central vents of volcanoes. Mount Gambier is an extinct volcano whose collapsed craters have filled with water to form beautiful clear lakes. There are many extinct volcanoes in Victoria. Tower Hill, near Warrnambool, is just one example, and there are many others just to the north of Melbourne.

Volcanoes that have not erupted for over 20 years and are not considered to be extinct are called **dormant** volcanoes. Dormant means 'asleep' and these volcanoes could 'wake up' at any time and erupt. Mount Pinatubo was a dormant volcano before its eruption in 1991.

The lack of recent volcanic eruptions and the small number of earthquakes in Australia is due to the fact that most of it is close to the centre of the Indo-Australian Plate (see the map in section 8.3.1). It is far away from the plate boundaries where new mountains form, major earthquakes occur and new crust is formed.

8.6 Exercises: Understanding and inquiring

To answer questions online and to receive **immediate feedback** and **sample responses** for every question, go to your learnON title at www.jacplus.com.au. *Note:* Question numbers may vary slightly.

Remember

1. What can cause a volcano to erupt?
2. List the substances that emerge from a volcanic crater during an eruption.
3. Explain the difference between a dormant volcano and an extinct volcano.
4. What is a hot spot?
5. Explain in terms of the plates that form the Earth's crust why Australia experiences little volcanic or earthquake activity.

Think

6. Use a Venn diagram to show the differences and similarities between magma and lava.
7. How do you know that many of the volcanoes in the western district of Victoria had runny lava?

8. Explain how a volcano can affect the world's weather.
9. Should Mount Gambier be described as an extinct or dormant volcano? Explain your answer.

Imagine

10. Imagine that you are the Mexican farmer who found the crack that gave birth to the volcano Paricutin. Write an account of what you saw and how you felt during the week after you first found the crack.

Create

11. Write a short story about an underwater volcano entitled *The birth of an island*.
12. Compile a collection of newspaper articles about recent volcanic eruptions. Display the collection in your classroom.
13. Create a papier-mache model volcano. Shape some chicken wire into a cone with a small crater at the top. Soak small pieces of newspaper in a pasty mixture of flour and water and attach the sticky newspaper to your wire cone. You will need to apply several layers of newspaper. Use colour to brighten up your model.

Investigate

14. Find out the name and location of a dormant or extinct volcano that is close to your school.
15. Two of the most famous volcanoes in the world are Mount Vesuvius and Mount Krakatoa. Find out where they are, when they erupted and what damage they caused.
16. Write an account of a recent major volcanic eruption. Some that you might choose from are:
    - Mount Ruapehu, New Zealand, 1994 and 1996
    - Mount Tavurvur, Papua New Guinea, 1994
    - Eyjafjallajökull, Iceland, 2010.

learnon RESOURCES — ONLINE ONLY

Complete this digital doc: Worksheet 8.6: Volcanic activity (doc-18896)

Complete this digital doc: Worksheet 8.7: Geological activity (doc-18897)

8.7 Affinity diagrams and double bubble maps

8.7.1 Affinity diagrams and double bubble maps

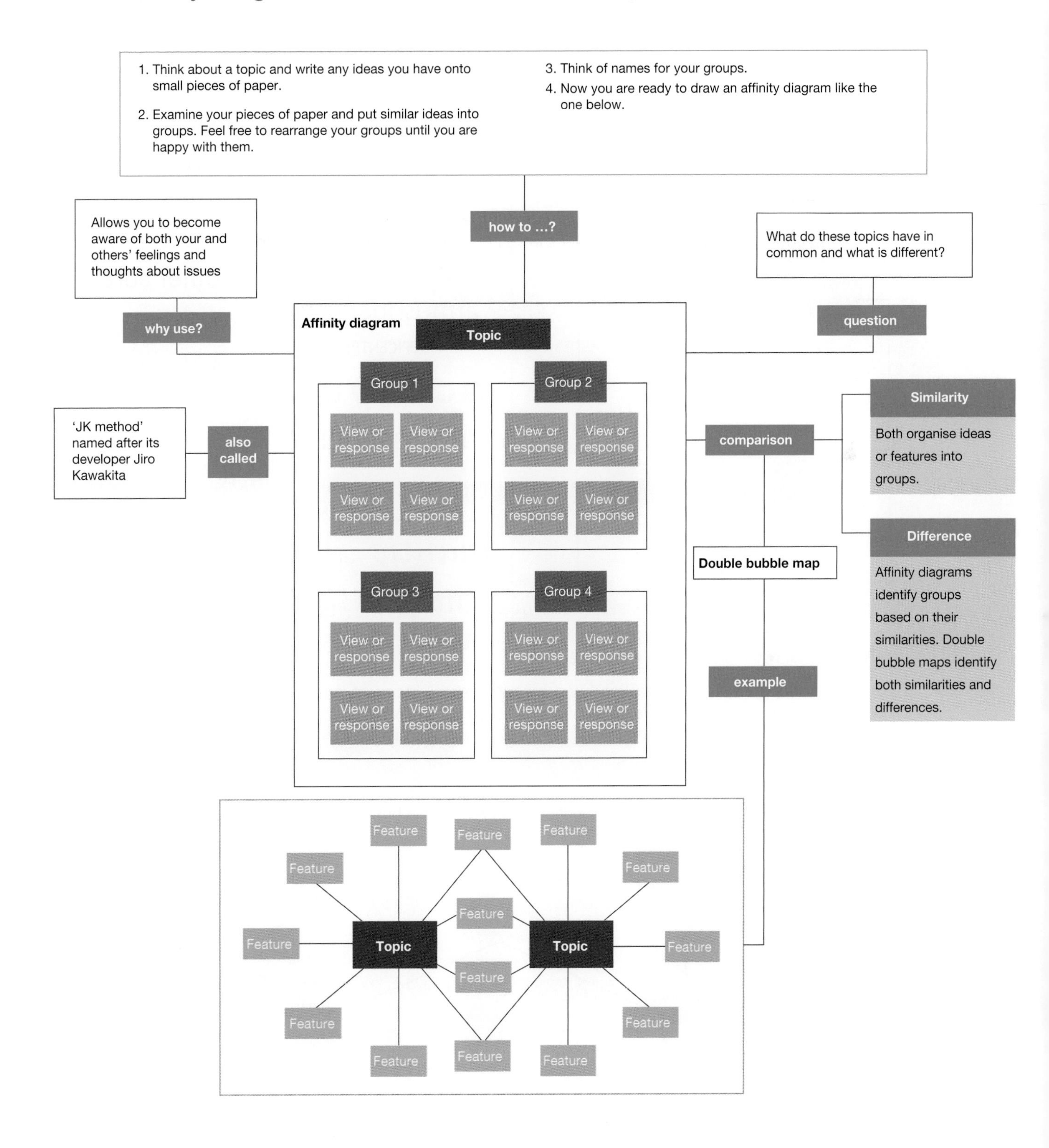

8.7 Exercises: Understanding and inquiring

To answer questions online and to receive **immediate feedback** and **sample responses** for every question, go to your learnON title at www.jacplus.com.au. *Note:* Question numbers may vary slightly.

Think and create

1. (a) Write each of the ideas listed on the right on a small card or sticky note.
   (b) Arrange the ideas on the cards into four categories in an affinity diagram like the one shown below. Write a title for each category. If any of the ideas fit into more than one category, choose the single category that best suits it.

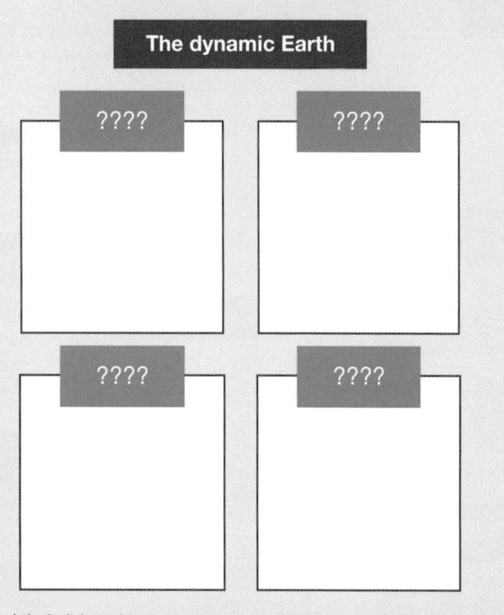

   (c) Add at least two more ideas to each category.

2. Use your affinity diagram from question 1 and any other relevant terms to create double bubble maps that illustrate the similar and different features of the following pairs of topics:
   (a) folding and faulting
   (b) earthquakes and volcanoes
   (c) continental drift and plate tectonics.

   The figure below can be used to help you get started on a double bubble map for folding and faulting.

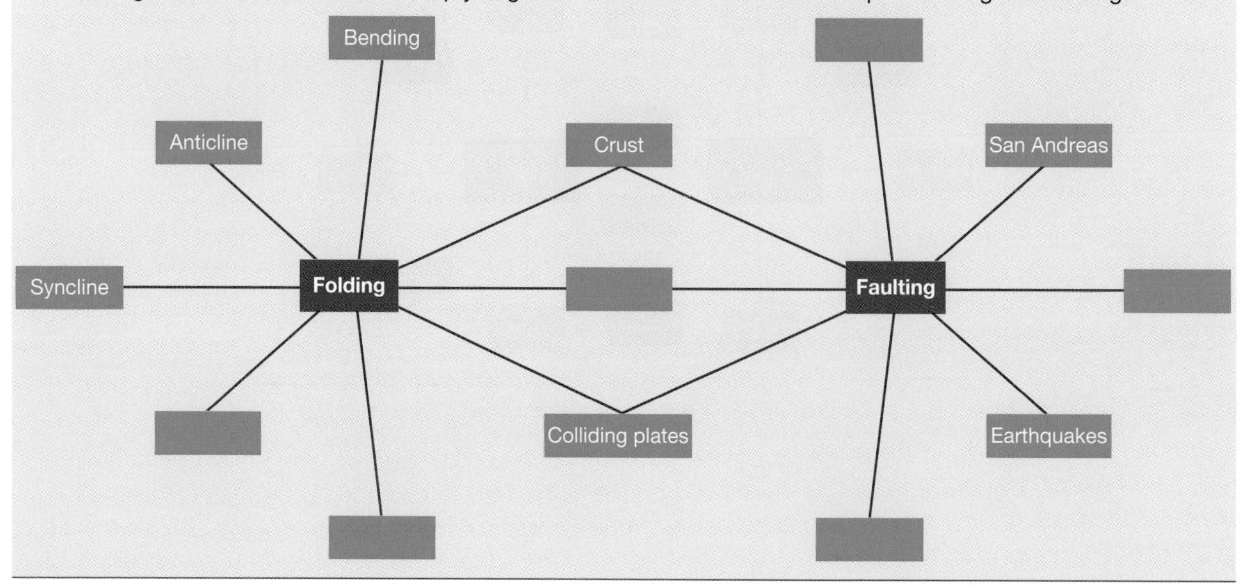

learnon

8.8 Project: Disaster-proof

Scenario

Earthquakes occur when pressure built up between adjacent sections of rock in the Earth's crust is suddenly released. The bigger the earthquake's magnitude, the greater the amount of energy that shakes the Earth. However, the magnitude of the earthquake is not necessarily a good indication of how deadly it will be. The May 2006 earthquake in Java had a magnitude of 6.2 and caused the deaths of nearly 6000 people, yet the 2004 Guadalupe earthquake was the same size but killed only 1 person. In some cases, magnitude 5.3 earthquakes have killed more people than those with magnitude 8.1. In fact, the key predictors other than magnitude of how deadly an earthquake will be are how heavily populated the area is and what type of buildings are there. Sadly, the majority of people who die in earthquakes do so because the buildings around them fail.

Unlike the more earthquake-prone regions of the world, Australia is not near a plate boundary, but we are not out of danger. The 1984 Newcastle earthquake had a magnitude of 5.6 and resulted in 13 deaths, 160 injuries and damage to over 60 000 buildings. With this in mind, your company — Shakeless Seismic Solutions — has been approached by a wealthy client who wishes to build an earthquake-proof five-storey office block in Perth. However, yours is not the only company that she has approached. In order to determine which business she will award the contract to, she is asking each company not only to come up with a design, but also to have a scale model of their design tested on a shake-table earthquake simulator.

Your task

Your group will use research, ingenuity and online simulators to design a five-storey office block that will survive an earthquake. You will build a scale model of your design and compete with other groups to

determine which model/design is able to withstand the most energetic shaking on the simulator. Your model will need to fulfil the following criteria:

- It should have a total mass of no more than 1.5 kg.
- It should have a base area no bigger than 20 cm × 20 cm and should have a height of at least 50 cm.
- No glue, staples, nails or pins are allowed; however, you may use interlocking pieces.
- It must be freestanding (it may not be stuck to the table in any way).

Before testing, you will be required to explain the main aspects of your design to the client (your teacher) and describe what makes the model and the real building earthquake-proof.

8.9 Review

8.9.1 Study checklist

The theory of plate tectonics

- describe the Earth's crust and compare it with other layers below and above the Earth's surface
- describe evidence supporting the theory of plate tectonics, including the location of volcanic activity and earthquakes
- recognise the major plates on a map of the Earth
- explain the movement of plates in terms of heat and convection currents in the Earth's mantle
- describe, compare and model the processes of subduction and the formation of ocean ridges
- distinguish between constructive and destructive plate boundaries

Folding and faulting

- describe and model the processes of folding and faulting
- relate folding to the movement of tectonic plates and the formation of mountain ranges
- explain the formation of faults in terms of the forces acting within the Earth's crust and the movement of plates relative to each other

Earthquakes and volcanoes

- relate the occurrence of major earthquakes and volcanoes to the movements along plate boundaries
- compare the energy released by earthquakes with different values on the Richter scale
- associate tsunamis with earthquakes and volcanic activity
- identify the main features of a volcano
- distinguish between lava and magma
- describe and compare the characteristics of active, dormant and extinct volcanoes
- relate the age and stability of the Australian continent and its lack of volcanic and major earthquake activity to its location away from plate boundaries

Science as a human endeavour

- explain how the theory of plate tectonics developed from the earlier theory of continental drift and further evidence
- describe the use of scientific ideas and technology in the development of the theory of plate tectonics
- describe the role of seismologists and vulcanologists in the investigation of the Earth's crust
- explain the importance of early warning systems to people living near plate boundaries, particularly on the edges of the Pacific Ocean.

Individual pathways

ACTIVITY 8.1	ACTIVITY 8.2	ACTIVITY 8.3
Revising the dynamic Earth doc-8452	Investigating the dynamic Earth doc-8453	Analysing the dynamic Earth doc-8454

learnON ONLINE ONLY

8.9 Review 1: Looking back

To answer questions online and to receive **immediate feedback** and **sample responses** for every question, go to your learnON title at www.jacplus.com.au. *Note:* Question numbers may vary slightly.

1. Which layers of the Earth have the following characteristics?
   (a) Completely molten
   (b) Partially molten
   (c) Includes solid rock, soil and landforms
   (d) Solid and mostly made up of iron
   (e) Lies above the surface
2. Describe two pieces of evidence that supported Wegener's theory of continental drift.
3. Explain how scientists know about what lies deep below the surface of the Earth without going there.
4. According to the theory of plate tectonics, the Earth's crust is divided into a number of slowly moving plates.
   (a) What makes the plates move?
   (b) What can happen when two plates slide past each other?
   (c) How does the plate tectonics theory explain the increasing height of the Himalayas?
5. What is the major difference between the continental drift theory and the theory of plate tectonics in terms of what makes up the Earth's crust?
6. Where on Earth is the Ring of Fire and why does it exist?
7. How is an ocean ridge different from a subduction zone?
8. When oceanic crust pushes against continental crust, why does the oceanic crust slide underneath the continental crust?
9. Describe the movements in the Earth's crust that cause the folding of rock that has shaped most of the Earth's mountains.
10. Explain how faults are created.
11. Copy the diagrams below and label them using the following words: anticline, continental crust, magma, normal fault, oceanic crust, reverse fault, solid upper mantle, syncline.

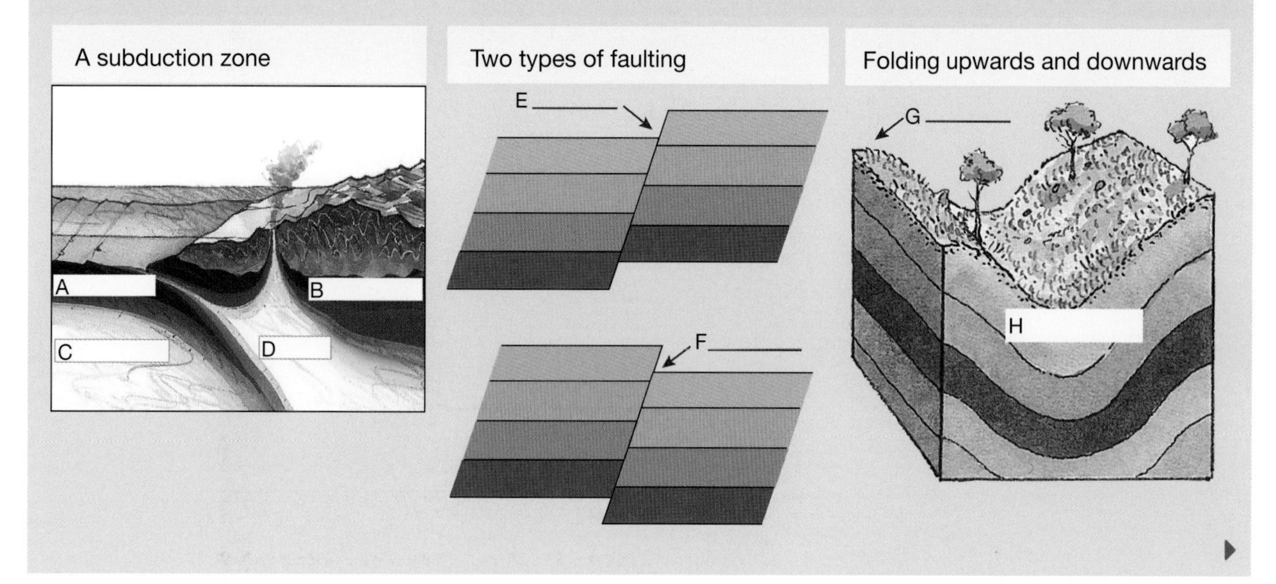

12. The San Andreas Fault makes much of coastal California, including the cities of Los Angeles and San Francisco, susceptible to earthquakes.
   (a) Explain why the San Andreas Fault is called a slip fault.
   (b) What causes major earthquakes along this fault?
13. Distinguish between the epicentre of an earthquake and its focus.
14. What is a seismograph used to measure?
15. How much more energy is released by an earthquake that registers 6.0 on the Richter scale than by one that registers 7.0 on the Richter scale?

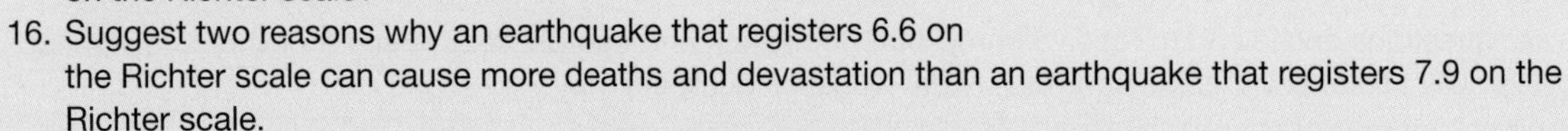

16. Suggest two reasons why an earthquake that registers 6.6 on the Richter scale can cause more deaths and devastation than an earthquake that registers 7.9 on the Richter scale.
17. Explain why Australia is less likely to experience volcanic activity and major earthquakes than New Zealand.
18. Before a volcano erupts, its vents are blocked with thick, pasty lava.
   (a) What change takes place to cause the volcano to erupt?
   (b) How is the lava emerging from a volcano different from magma?
19. Name three gases present in the lava that are released from a volcano.
20. Distinguish between a dormant volcano and an extinct volcano.
21. Identify two causes of tsunamis.

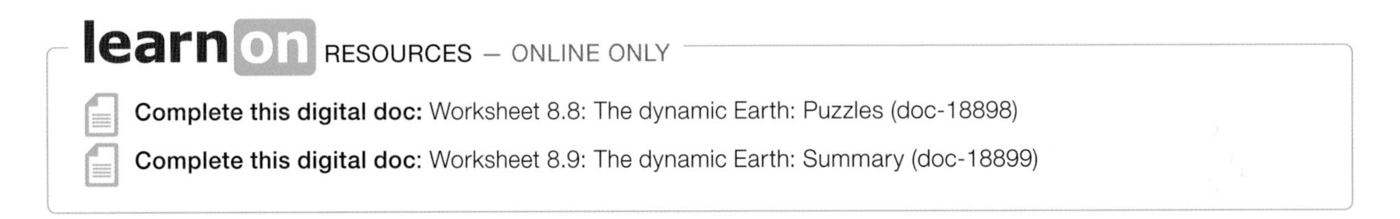

learnON RESOURCES — ONLINE ONLY

Complete this digital doc: Worksheet 8.8: The dynamic Earth: Puzzles (doc-18898)

Complete this digital doc: Worksheet 8.9: The dynamic Earth: Summary (doc-18899)

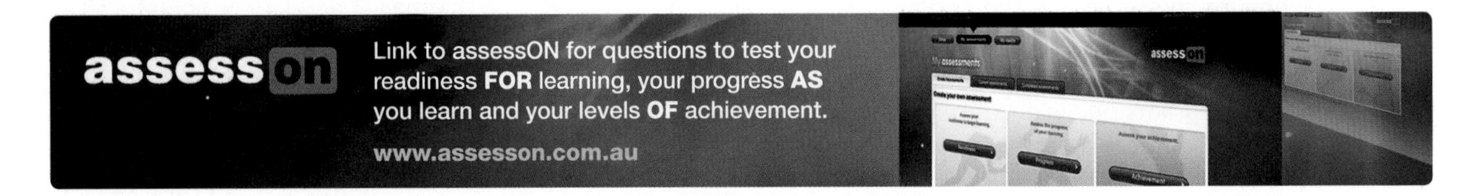

TOPIC 9
Energy transmission

9.1 Overview

Sound energy is transmitted through the air near sea level at about 1200 km/h. When a jet aircraft reaches the speed of sound, it is said to break the sound barrier. A shockwave known as a sonic boom is heard as the jet crashes through the wall of compressed air that it has created. A lesser-known phenomenon that occurs when planes fly at or near the speed of sound is this cone-shaped condensation cloud.

9.1.1 Think about energy transmission

assesson

- What travels through the air at one million times the speed of sound?
- Why do your legs look shorter when you stand in water?
- Why is the lens in your eye more like jelly than glass?
- Analogue and digital — what's the difference?
- Why are mobile phones also called cellular phones?
- What are microwaves used for apart from heating food?
- How can you stop ice-cream from melting in a 230° oven?
- Why do dogs pant?
- How does ceiling insulation keep your house warmer in winter?

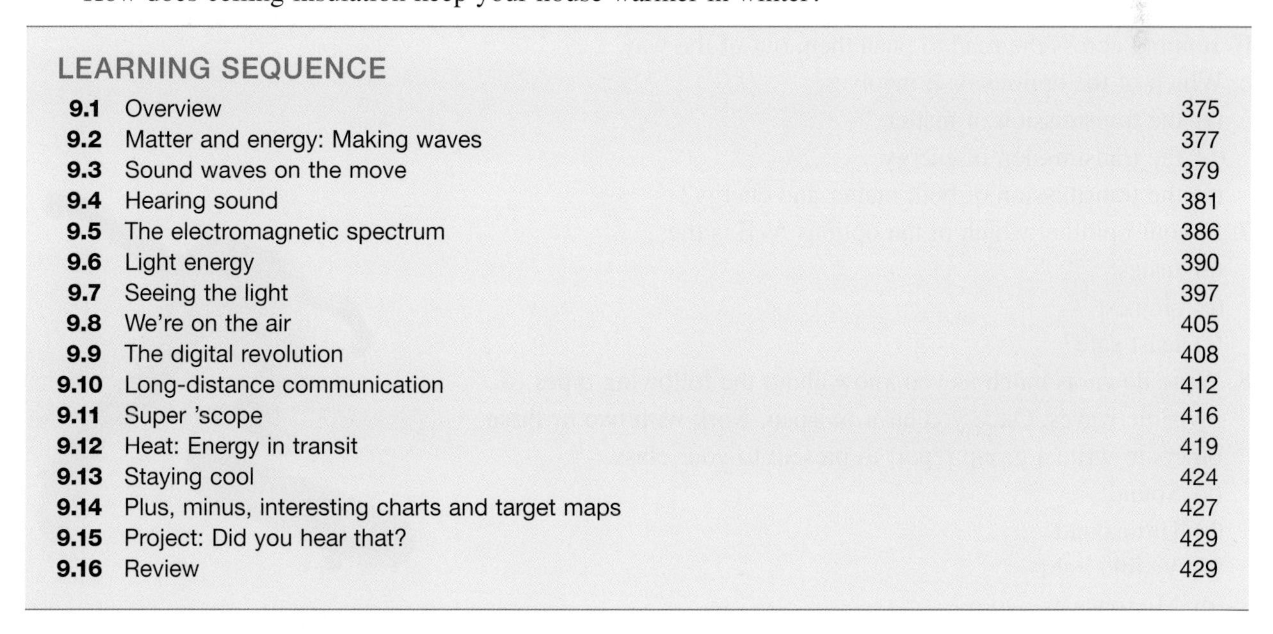

LEARNING SEQUENCE

Numerous **videos** and **interactivities** are embedded just where you need them, at the point of learning, in your learnON title at www.jacplus.com.au. They will help you to learn the content and concepts covered in this topic.

9.1.2 A matter of communication

In March 1791, Captain Arthur Phillip, Governor of New South Wales, wrote a letter to his employer, King George III in London, asking for some time off work. The only way to get the letter to London was by sailing ship. The letter took eight months to get to King George III, and his reply took a further eight months to reach Sydney.

Today, a reply from London to a request from Sydney for some time off work could be received within seconds by telephone or email. There is no longer any need for matter, such as letters, to be transported. The request and reply travel between Sydney and London via the transmission of energy at the speed of light — 300 000 kilometres per second. Over long distances, there are many advantages of energy transmission without the transmission of matter.

Although more than 200 years later the transmission of matter is a lot faster, it is still relatively slow. A hard copy of the same letter requesting leave sent from Sydney to London by air mail today would take about five days to arrive — and there would be at least another five days to wait for a reply to get back to Sydney.

Think

1. What has changed since 1791 to reduce the communication time from Sydney to London and back from 16 months to 10 days?
2. What are all of the options now available for sending a request for time off work from Sydney to London?
3. Which options do not require the movement of matter from Sydney to London and back?
4. How long would the sender have to wait for a reply if the fastest method of communication was used?
5. If matter does not move from one place to another when a message is sent over a long distance, what does move?

Even over a short distance the transmission of energy is faster than the transmission of matter. Imagine that you want to warn a couple of friends that they are about to be hit by an out-of-control skateboarder. Your options for saving them include:

A. yelling at them to get out of the way
B. yelling and pointing at the skateboarder
C. waving your arms in the air and pointing
D. holding up a sign that says 'Watch out!'
E. running across the road to push them out of the way.

6. Which of the options A–E involves:
   (a) the transmission of matter
   (b) the transmission of energy
   (c) the transmission of both matter and energy?
7. In your opinion, which of the options A–E is the:
   (a) fastest
   (b) slowest
   (c) least safe?
8. Write down as much as you know about the following types of invisible waves. Once you have finished, work with two or three others to write a group report to present to your class.
   (a) Sound
   (b) Ultrasound
   (c) Visible light
   (d) Microwaves
   (e) Infra-red
   (f) Radio waves

9.2 Matter and energy: Making waves

9.2.1 Transmitting energy with waves

When a wave is made in a still lake by dropping a rock into it, the wave spreads out. However, the particles of water do not spread out — they just move up and down. A **wave** is able to transmit energy from one place to another without moving any matter over the same distance.

9.2.2 Two types of waves

Waves on water are called **transverse waves**. Transverse waves can also be made on a slinky. As shown in the diagram below, the moving particles in a transverse wave travel at right angles to the direction of energy transfer. The diagram also shows that in a **compression wave**, the moving particles move backwards and forwards in the same direction as the energy transfer. Compression waves are also known as **longitudinal waves**.

The material through which the waves travel is called the **medium** (the plural is 'media').

Two types of energy transfer: a transverse wave (top) and a compression wave (bottom). The transfer of sound energy can be modelled using compression waves in a slinky.

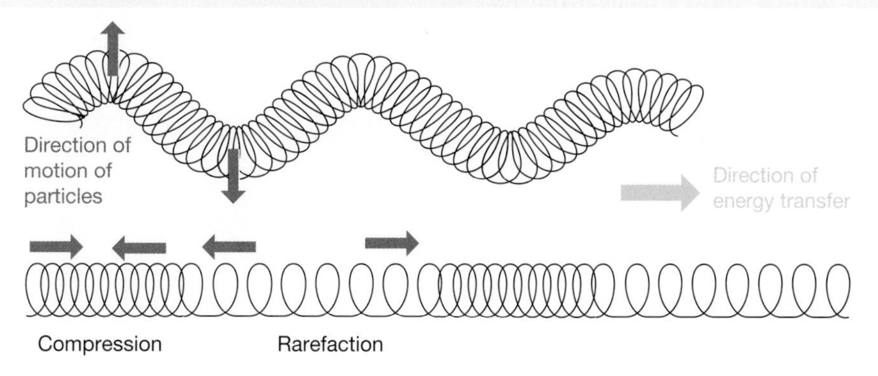

9.2.3 Sound waves

Sound is a compression wave. All sounds are caused by **vibrations**. Vibrations cause air to compress like the lower wave pattern shown in the diagram above. The diagram on the right shows how a vibrating ruler makes compression waves in air. As the ruler moves up, a **compression** is created as air particles above the ruler are pushed together. Air particles below the ruler are spread out, creating a **rarefaction**. When the ruler moves down, a rarefaction is created above the ruler, while a compression is created below it. Each vibration of the ruler creates new compressions and rarefactions to replace those that are moving through the air.

Sound is a compression wave caused by vibrations.

Compression
Rarefaction
Compression
Rarefaction
Compression
Rarefaction
Ruler
Compression
Rarefaction
Compression

9.2.4 Some jargon to learn

The **frequency** of a vibration or wave is the number of complete vibrations or waves made in one second. The frequency of a sound wave is given by the number of compressions made in a second. The note middle C, for example, creates 256 vibrations, or compressions, every second. Frequency is measured in **hertz** (Hz), a unit named after Heinrich Hertz, the German physicist who, in 1887, was the first to detect radio waves. One hertz is equal to one vibration per second. Therefore, middle C has a frequency of 256 hertz. The frequency of a sound determines its **pitch**. High-frequency vibrations produce high pitch, and low-frequency vibrations produce low pitch.

In the case of transverse waves, for example waves on water, the **wavelength** is the distance between two crests, or two troughs, or the distance between any two corresponding points on neighbouring waves. In the case of a compression wave, the wavelength is the distance between the centre of two neighbouring compressions, or two neighbouring rarefactions, as shown in the diagram below.

The distance between compressions — the wavelength — of the sound of the note middle C is about 1.3 metres. The wavelength of sound made during normal speech varies between about 5 centimetres and about 2.5 metres.

As the frequency of a sound gets higher — that is, as more compressions are produced per second — the compressions become closer together.

The pitch you hear is higher. Lower frequencies produce longer wavelengths and thus lower-pitched sounds.

The **amplitude** of a wave is the maximum distance that each particle moves away from its usual resting position. The amplitude of two different types of waves is also shown in the diagram on the right. Higher amplitudes correspond with louder sounds.

Even though sound is a compression wave, it can be modelled with water waves.

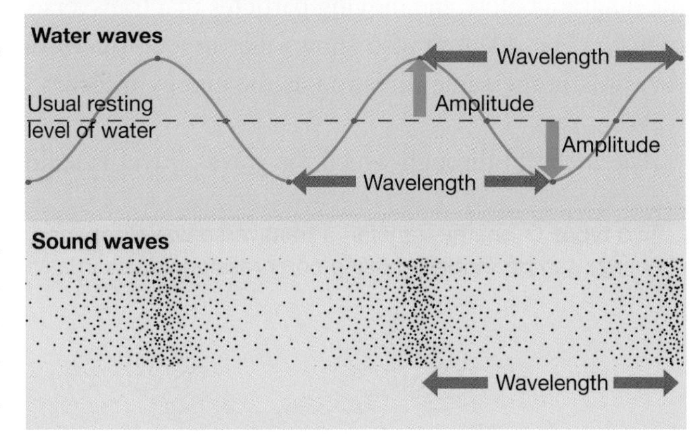

INVESTIGATION 9.1

Moving energy without matter

AIM: To model sound using waves on water and a slinky

Materials:

deep tray	*ribbon*
small cork	*slinky*
eye dropper	*water*

Method and results

- Half fill the tray with water and place a small cork on the water's surface. Use the eye dropper to release drops of water near the cork. Observe the motion of the cork and the motion of the small waves made by the drops.
- Tie a ribbon around a coil near the centre of the slinky. Firmly hold one end of the slinky while your partner, holding the other end, stretches it to about the length of the room. Make a wave by rapidly flicking one end of the slinky to one side. Observe the ribbon as the wave passes.
- Make a different type of wave by pulling about ten coils of the slinky together at one end and then releasing this compressed section. Observe the ribbon as the wave passes.

Discuss and explain

1. Describe the motion of the cork on the small waves.
2. Is there any evidence to suggest that any water moves in the same direction as the waves?
3. Describe the motion of the ribbon as the waves made by flicking move along the slinky.
4. Describe the motion of the ribbon as the compression wave moves along the slinky.
5. In each of the slinky waves produced in this experiment, energy is transferred from one end of the slinky to the other.
   (a) Where is the ribbon after the wave has passed in each case?
   (b) Has any particle on the slinky moved from one end to the other?
6. Which properties of sound waves can be modelled by waves on water?

9.2 Exercises: Understanding and inquiring

To answer questions online and to receive **immediate feedback** and **sample responses** for every question, go to your learnON title at www.jacplus.com.au. *Note:* Question numbers may vary slightly.

Remember

1. What causes all sound waves?
2. Describe the difference between a compression and a rarefaction.
3. What is the unit of frequency and what does it measure?
4. What quality of sound does the frequency determine?

Think

5. Is a Mexican wave, as seen among the crowds at some sporting events, a transverse wave or a compression wave?
6. What is the wavelength and amplitude of the transverse wave shown in the diagram at right?

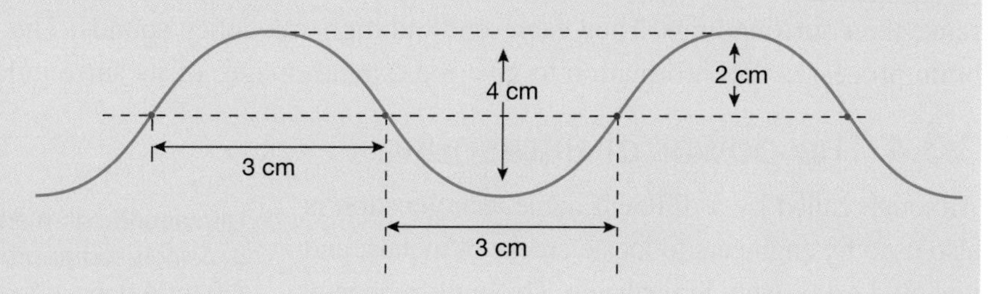

9.3 Sound waves on the move

9.3.1 The need for particles

Imagine that you are on a spacecraft on the way to Mars and a passing asteroid explodes. Would you hear the explosion before or after you saw it — or would you even hear it at all?

Because sound is transmitted as a compression wave, it can travel only through a medium that contains particles that can be forced closer together or further apart. Sound cannot be transmitted in a vacuum because there are no particles to push closer together or spread out.

As sound travels through a medium, some of its energy is absorbed by the particles in the medium and is not transmitted to neighbouring particles. Sound travels more efficiently through materials that are elastic; that is, materials in which the particles tend to come back to their original positions without losing much energy.

9.3.2 Speed of sound

The speed of sound in a particular medium depends on how close the particles are to each other and how easy they are to push closer together. In liquids and solids, the speed is much greater because the particles are more closely bound together. The table below shows the speed of sound in some common substances at 0 °C.

Speed of sound in some common substances

Substance	Speed of sound (metres per second)
Carbon dioxide (at 0 °C)	260
Dry air (at 0 °C)	330
Hydrogen (at 0 °C)	1300
Water	1400
Sea water	1500
Wood	4000–5000
Glass	4500–5500
Steel	5000
Aluminium	5000
Granite	About 6000

The speed of sound in air is greater at higher temperatures. At sea level in dry air at 0 °C, it is about 330 metres per second. At a temperature of 25 °C, it is about 350 metres per second. The speed of sound in air is lower at higher altitudes. At an altitude of 10 kilometres above sea level, it is about 310 metres per second.

9.3.3 Hear an echo?

A knowledge of the speed of sound is used in **sonar**. Sonar (SOund Navigation and Ranging) is used on ships to map the ocean floor, detect schools of fish, and locate other underwater objects such as shipwrecks and submarines. High-frequency sound is transmitted from the ship. By measuring the time taken for the echo to return to the ship, and using the speed of sound in water, the distance to the floor of the ocean or to the underwater object can be calculated.

Sonar is just one example of **echolocation**. Whales, dolphins, porpoises and bats use echolocation to sense their surroundings. They each send out high-frequency sounds. The echo is detected and the animal's brain processes the information to give it a 'sound image' of its surroundings.

9.3.4 The power of ultrasound

Although called by a different name, echolocation is also used by engineers to locate cracks in metals; and it is used extensively in medicine. The high-frequency sound used in industry and in medicine is called **ultrasound**. Ultrasound has frequencies higher than humans can hear. Echolocation with ultrasound is used in medicine to produce images of unborn babies in the womb during pregnancy, and to search for gallstones, circulation problems and cancers. Ultrasound also has uses other than echolocation. It is used to remove some cancers, treat an eye condition called glaucoma, shatter kidney stones and gallstones in a process called shockwave therapy, speed up the healing of muscle damage, clean surfaces, mix paint, homogenise milk and cut into glass and steel.

Ultrasound has many applications. Here, ultrasound is being used to produce an image of the face of a full-term baby in the womb.

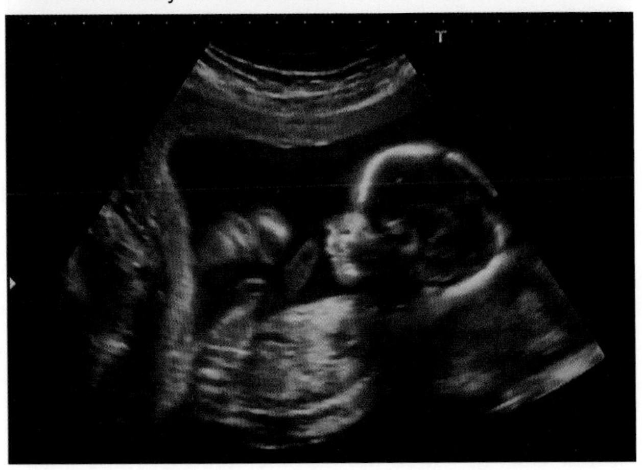

INVESTIGATION 9.2

Sound in different media

AIM: To investigate the transmission of sound in different media

Materials:

ticking watch
metre ruler
teaspoon (or spatula)
cotton thread (or light string)

Transmitting sound

Method and results

- Place a ticking watch against your ear and listen to the tick. Have your partner slowly move the watch away from your ear until you can no longer hear the ticking.

1. Measure and record the distance from your ear to this point.

- Place a metre ruler gently against the same ear and rest the watch on it against the ear. Have your partner slowly slide the watch along the ruler to a point where you can no longer hear the ticking.

2. Measure and record the distance from your ear to this point.

- Tie about 80 cm of cotton thread to a teaspoon. Swing the teaspoon slowly so that it gently strikes the side of a bench, wall or cupboard. Listen to the sound made.

- Place the free end of the cotton thread carefully against your ear and again gently strike the teaspoon against the same surface. Listen to the sound made.

Discuss and explain

3. What effect did the ruler have on the distance over which you could hear the sound of the ticking watch?
4. What difference does the cotton thread make to the sound heard when the spoon strikes a surface?
5. Is sound conducted better through air or through solids?
6. What property of the solids do you think makes the difference?

9.3 Exercises: Understanding and inquiring

To answer questions online and to receive **immediate feedback** and **sample responses** for every question, go to your learnON title at www.jacplus.com.au.

Remember

1. Why are sound waves unable to travel through a vacuum?
2. What is ultrasound and how is it useful?

Using data

The following questions refer to the data given in the table on page 379.

3. In general, how does the speed of sound in solids compare with that in liquids and gases?
4. Why do you think such a large range of speeds is given for wood and glass?
5. Suggest a reason why the speed of sound in most woods is generally lower than the speed of sound in steel and aluminium.
6. What does the data suggest about the effect of the density of a gas on the speed of sound?

Think

7. Answer the question posed at the beginning of this subtopic.
8. Suggest why the speed of sound depends on altitude and temperature.

Imagine

9. Imagine that you are one of two astronauts walking on the moon.
   (a) Would you be able to conduct a conversation with your partner without radios? Explain why.
   (b) Imagine that both of your radios stopped working because you forgot to recharge the batteries. Explain how you would still be able to communicate with your partner.
10. Imagine that sounds of different frequencies had different speeds in air. How would that affect what you heard if you were in the back row of a stadium listening to a band?
11. Imagine that you are standing near a steep, rocky cliff. You shout 'Hello' and one second later you hear the echo. The air temperature is about 25 °C, so you estimate that the speed of sound is about 350 m/s. How far are you from the cliff?

9.4 Hearing sound

9.4.1 The ear

The energy of sound waves is transformed by your ear into electrical signals that are sent to your brain. Each of your ears has three distinct parts — the outer ear, middle ear and inner ear. Each part has its own special job to do.

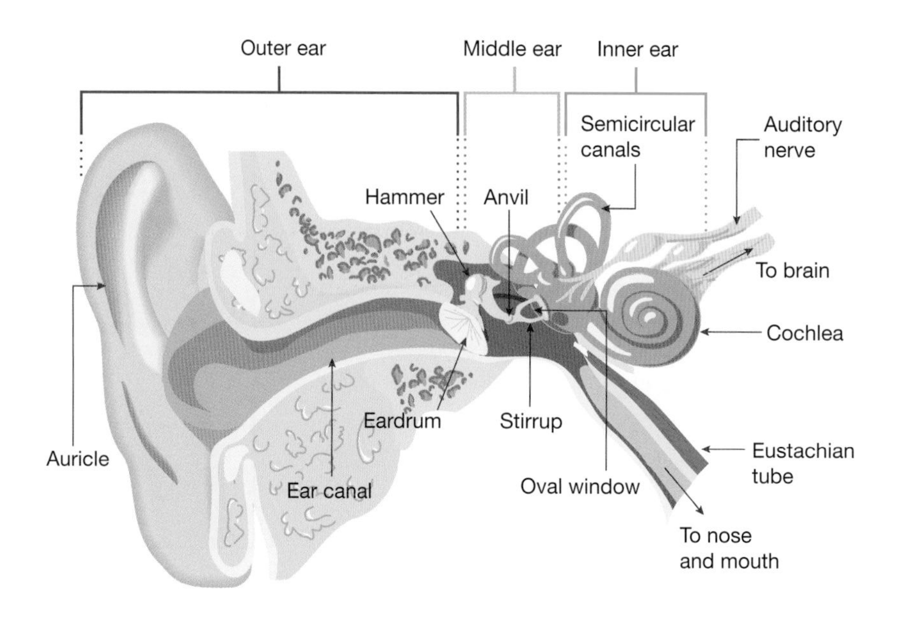

9.4.2 Outer ear

The outer ear funnels the energy of the vibrating air through the **ear canal** to the **eardrum**. The eardrum is a thin flap of skin, or **membrane**, which vibrates in response to the vibrating air particles. The fleshy, outer part of the ear is called the **auricle**.

9.4.3 Inner ear

The inner ear contains the **cochlea** and the **semicircular canals**. The cochlea is a spiral-shaped system of tubes full of fluid. When vibrations are passed through the oval window by the stirrup, the fluid moves tiny hairs inside the cochlea. These hairs are attached to the receptor nerve cells that send messages on their way to the brain through the **auditory nerve**. The semicircular canals also contain a fluid. However, they are not involved in hearing sound. When you move your head, the fluid in the semicircular canals moves hairs that send signals to your brain. The signals provide your brain with information to help you keep your balance.

The decibel scale

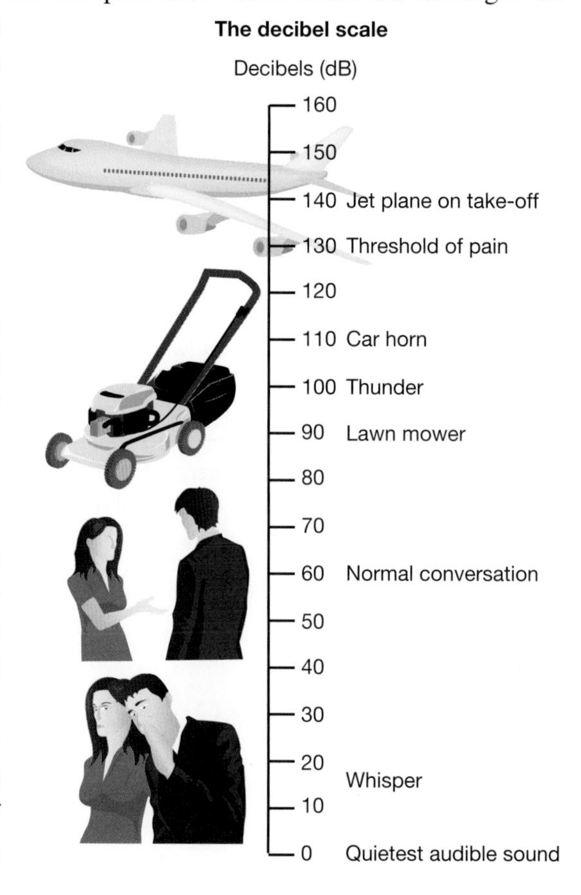

9.4.4 Middle ear

The middle ear contains three small bones called the hammer, the anvil and the stirrup. These three tiny bones (known as the **ossicles**) pass on the vibrations to the inner ear through the **oval window**.

9.4.5 Too loud

Sound makes your eardrums vibrate. But if the sound is too loud, the vibrations can cause pain and even permanently damage your ear. That's because loud sounds disturb the air — and your eardrums — more than soft sounds.

Although loudness can be a matter of opinion, the disturbance to the air can be measured. The measurement is called the **relative intensity**, or **sound level**. The unit of measurement is the **decibel (dB)**. The number of decibels gives a good indication of the loudness of a sound. It's not a perfectly accurate measure of loudness, because your ear is more sensitive to some pitches than others.

The **threshold of hearing** is the smallest sound level that can be heard when the air is vibrating at 1000 Hz. For most people it's about 0 dB. Sound levels of more than about 130 dB can cause pain and permanent ear damage. The smallest sound level that causes pain is called the **threshold of pain**. Sound levels of even 80 dB can cause damage to your ears if you are exposed to the sound for long enough.

Ear protection is needed when working with noisy machinery, including racing cars.

HOW ABOUT THAT!

When you are landing or taking off in a plane, your ears 'pop'.

If you climb steeply, the air pressure inside your middle ear remains the same while the air pressure outside drops. The air inside pushes on the eardrum causing an uncomfortable feeling.

The 'popping' is caused as the Eustachian tube, which is normally closed, opens. This allows air to rush out of your middle ear to your nose and mouth. The pressure is then the same on both sides of the eardrum.

When you descend quickly, the 'popping' occurs as the air rushes into your middle ear to balance the increasing pressure outside. If you swallow hard, you can make the 'popping' happen sooner.

Noise, noise, noise

One of the 'side effects' of living in an industrialised world is noise. Some of this noise is loud enough to damage your ears. But some of it is just annoying. The offending noises come from sources that include:

- transport, such as aeroplanes, trains, trams, trucks, cars, buses and cars
- heavy machinery, such as tractors, bulldozers, harvesters and jackhammers
- entertainment venues, such as rock concerts, nightclubs and sporting events
- domestic sources, such as mowers, power tools and much, much more.

The effect of freeway noise on nearby residents is reduced by barriers that absorb and reflect sound energy.

With good planning and zoning, the noise of traffic, factories and entertainment venues can be kept away from residential areas, hospitals and schools. Sound barriers are built and trees planted beside freeways to reduce the noise heard by nearby residents. Sound barriers are designed to reflect and absorb sound energy. Trees are great natural absorbers of sound energy.

State and local government laws restrict times when you can use mowers, power tools and other noisy items like air-conditioners and swimming-pool pumps. These laws vary from state to state and between local councils.

9.4.6 Too soft

Hearing aids have been used for many years to make sounds louder for those with impaired hearing. The battery-operated hearing aid that some people wear amplifies the vibrations so that they can reach a properly working cochlea. Another type of hearing aid 'bends' the vibrations so that they go through a bone behind the ear to the cochlea.

HOW ABOUT THAT!

If you've ever been to a really loud concert, you may have experienced ringing in your ears afterwards. You would also have had trouble hearing. Even after you had gone home to bed and the house was silent, the ringing would still have been there.

This ringing in your ears is called **tinnitus** (sometimes pronounced tin-eye-tus). Some of the cells in your inner ear — the ones that detect vibrations — have been damaged. Fortunately, your ears are likely to recover. The ringing will stop and your hearing will return to normal — hopefully in a few hours, but maybe in a day or two. If you listen to loud music for too long or too often, the cells don't recover. Your hearing can be permanently damaged. It's a good idea to avoid this by wearing earplugs at loud concerts.

The cochlear implant

Many people who have severely or profoundly impaired hearing are unable to benefit from hearing aids. Profoundly hearing-impaired people hear no sounds at all.

Australian scientists have, in recent years, developed a device that has allowed some people who are profoundly hearing impaired to detect sound for the first time in their lives. The **cochlear implant**, or **bionic ear**, is surgically placed inside the ear. A microphone worn behind the ear detects sound and sends a signal to the speech processor (a small computer worn in a pocket or on a belt). It converts the sound into an electrical signal that

Bionic ear headset and speech processor

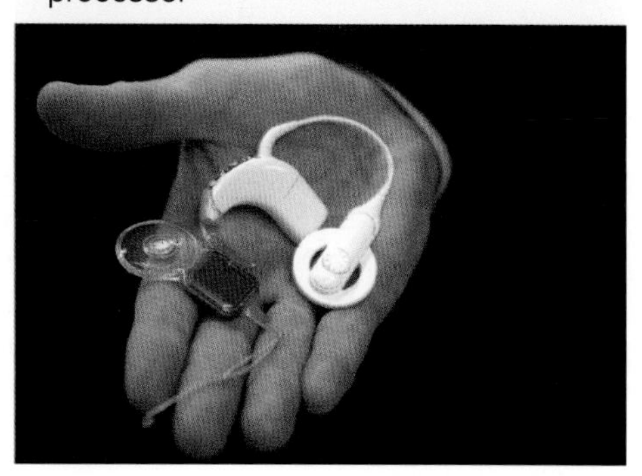

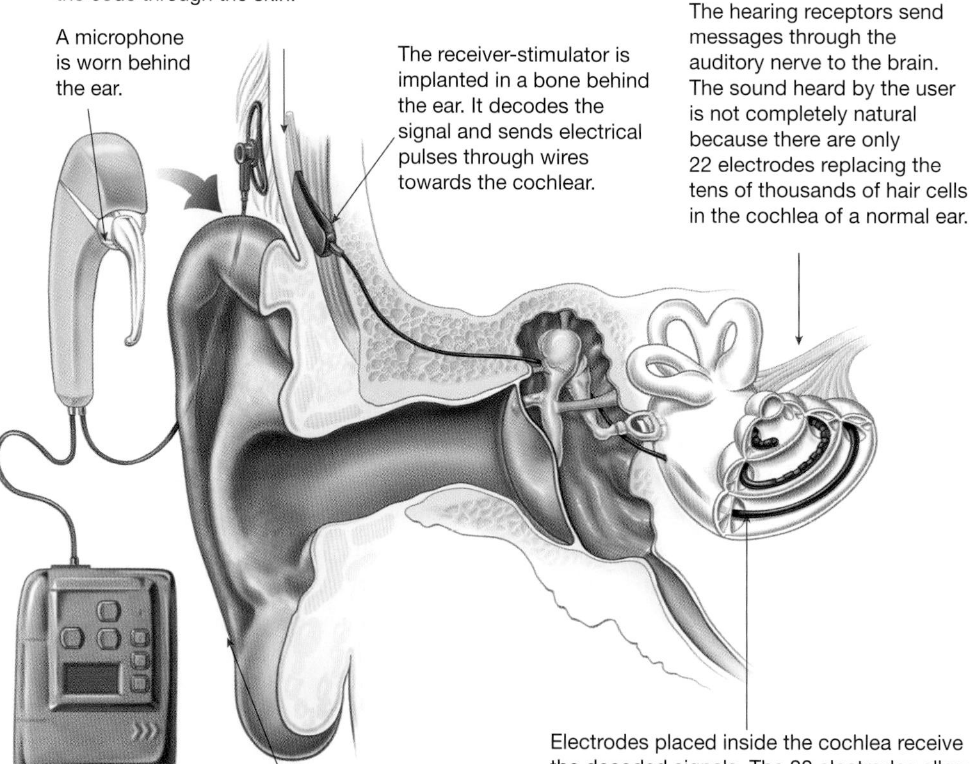

is sent to a receiver behind the ear and on to the implant in the cochlea. The signal then travels along the auditory nerve to the brain.

INVESTIGATION 9.3

Making it seem louder

AIM: To investigate a method of making a sound seem louder

Materials:

ticking watch
sheet of paper, about A4 size
metre ruler
blindfold

CAUTION

Take care not to put the funnel into the car canal.

Method and results

- Have your blindfolded partner sit on a chair. Hold a ticking watch close to your partner's right ear. The left ear should be covered with an open palm.
- Move slowly away until your partner indicates that the sound of the ticking watch can no longer be heard.

1. Measure and record the approximate distance from the watch to your partner's right ear.

- Make a funnel with a sheet of paper. Place the narrow end of the funnel close to, but not touching, your partner's right ear. Your partner should be able to hold it in place.
- Again, move the ticking watch slowly away from your partner, starting near the wide end of the funnel, until it can no longer be heard.

2. Measure and record the approximate distance between the watch and your partner's right ear.

Discuss and explain

3. What difference does the funnel make?
4. How does the funnel work?
5. Look at your own ears. Why do you think they are that shape?

INVESTIGATION 9.4

Sound proofing

AIM: To investigate the effect of different materials on the transmission of sound

Materials:

variety of materials to test (such as wood, fabric, glass and cardboard)
source of sound (such as an mp3 player)
sound level meter or data logger and sound probe

Method and results

- Design an experiment to investigate the most effective material to insulate against noise.

1. Record your results in a suitable table and graph.

Discuss and explain

2. Analyse your results to draw an appropriate conclusion.

9.4 Exercises: Understanding and inquiring

To answer questions online and to receive **immediate feedback** and **sample responses** for every question, go to your learnON title at www.jacplus.com.au. *Note:* Question numbers may vary slightly.

Remember

1. Complete the following table to describe some of the important structures of the human ear.

Structure	Description	Purpose
Eardrum		
	Three small bones in the middle ear	
	An opening into the inner ear	Allows vibrations to pass into the cochlea
Cochlea		Contains receptor cells for hearing
		Allows air to move between the middle ear and the mouth and nose

2. Explain why the sound level measured in decibels is not regarded as a completely accurate measure of loudness.
3. Define the threshold of hearing.
4. Describe tinnitus and explain how it can be avoided.
5. Explain how the bionic ear is different from a normal hearing aid.

Think

6. When you clap your hands, a sound is heard. Explain how the energy of the sound gets from your hands, through the air, through your ear and finally to your brain.
7. Two astronauts working outside a space station in orbit are unable to hear each other, no matter how loudly they speak or shout. However, when they are inside the space station, even with their helmets still on, they can hear each other easily. Why is there a difference?
8. Explain why luggage handlers who work on the tarmac at airports are required to wear ear protection.
9. Why should you wear ear protection when mowing or trimming the lawn?
10. Explain why trees are good absorbers of sound energy.

Create

11. Create a poster to warn people about the dangers of loud noise.

Brainstorm

12. Form a group of three or four and brainstorm a list of domestic noises that could damage neighbours' ears, disturb their sleep or annoy them.

Investigate

13. If a sound level meter is available, measure the sound level at a number of different locations around your school. Create a graph to display your measurements.
14. Find out when the cochlea implant was invented and by whom.
15. Find out more about the following careers that involve using and understanding sound energy.
    (a) Audiologist
    (b) Acoustic engineer
    (c) Audio engineer

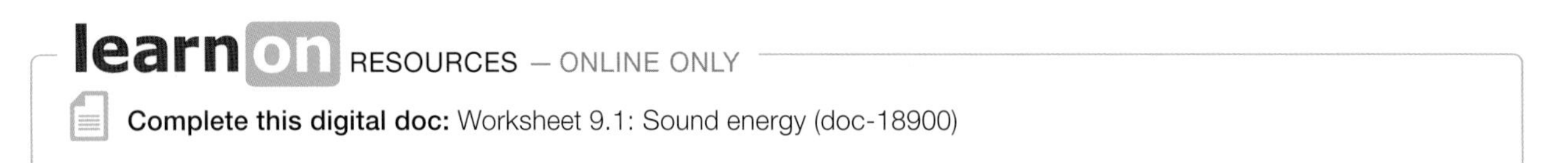

9.5 The electromagnetic spectrum

9.5.1 The electromagnetic spectrum

Isn't it great to be able to sit back in a comfortable chair and use a remote control to change the television channel or skip to your favourite scene on a DVD?

When you push that remote control button, a beam of invisible infra-red radiation travels at 300 million metres per second towards a detector inside your television set or DVD player. The detector converts the energy of the infra-red radiation into electrical energy which, in turn, operates the controls.

9.5.2 A family of waves

Infra-red radiation is just one part of the family of waves that make up the **electromagnetic spectrum**. Let's consider the whole spectrum, starting with the least 'energetic' members and finishing with those that are the most 'energetic'.

Radio waves

Radio waves include the low-energy waves that are used to communicate over long distances through radio and television. They also include radar and the microwaves used in microwave ovens for cooking.

Infra-red radiation

Infra-red radiation, invisible to the human eye, is emitted by all objects and is sensed as heat. The amount of infra-red radiation emitted by an object increases as its temperature increases.

A gentle push of the button sends infra-red radiation to the television set at 300 million metres per second.

Visible light

Visible light represents only a very small part of the electromagnetic spectrum. It is necessary for the sense of sight. The process of photosynthesis in green plants cannot take place without visible light.

Ultraviolet radiation

Like infra-red radiation, **ultraviolet radiation** is invisible to the human eye. It is needed by humans to help the body produce vitamin D; however, too much exposure to ultraviolet radiation causes sunburn.

The electromagnetic spectrum

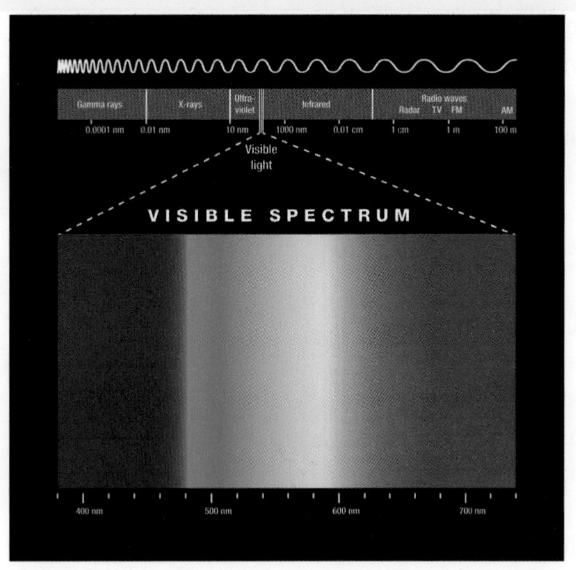

WHAT DOES IT MEAN?

The words *ultraviolet* and *ultrasound* are derived from the Latin term *ultra,* meaning 'beyond'. Ultraviolet radiation has frequencies beyond those of the colour violet, and ultrasound has frequencies beyond those of sounds we can hear.

X-rays

X-rays have enough energy to pass through human flesh. They can be used to kill cancer cells, find weaknesses in metals and analyse the structure of complex chemicals. X-rays are produced when fast-moving electrons give up their energy quickly. In X-ray machines, this happens when the electrons strike a target made of tungsten.

Some parts of the human body absorb more of the energy of X-rays than others. For example, bones absorb more X-ray energy than the softer tissue around them. This makes X-rays useful for obtaining images of bones and teeth. To obtain an image, X-rays are passed through the part of the body being examined. The X-rays that pass through are detected by photographic film on the other side of the body. Because bones, teeth and hard tissue such as tumours absorb more energy than soft tissue, they leave shadows on the photographic film, providing a clear image.

CT scanners (or CAT scanners) consist of X-ray machines that are rotated around the patient being examined.

X-ray showing a fracture of a radius and ulna in a forearm

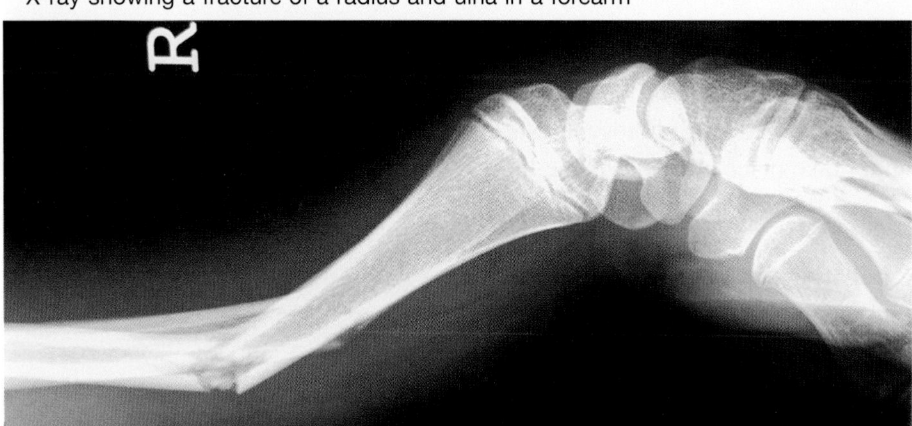

Gamma rays

Gamma rays have even more energy than X-rays and can cause serious damage to living cells. They can also be used to kill cancer cells and find weaknesses in metals. Gamma rays are produced when energy is lost from the nucleus of an atom. This can happen during the radioactive decay of nuclei or as a result of nuclear reactions.

Gamma cameras are used in PET scans to obtain images of some organs. To obtain a PET scan, a radioactive substance that produces gamma rays is injected into the body (or in some cases, inhaled). As it passes through the organ being examined, it produces gamma rays, which are detected by the camera.

A patient undergoing a PET scan of her brain

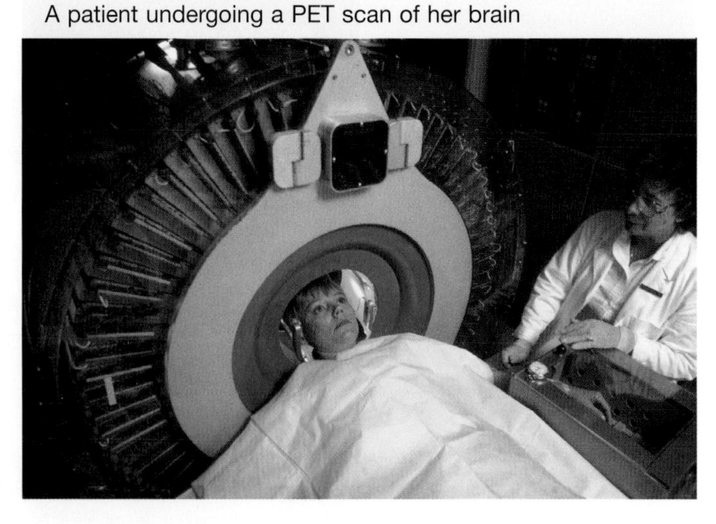

9.5.3 Electromagnetic waves

Like all waves, **electromagnetic waves** transmit energy from one place to another. All electromagnetic waves travel through air at 300 000 000 metres per second. Unlike sound waves and water waves, electromagnetic waves can travel through a vacuum. The waves consist of a repeating pattern of electric and magnetic forces. These forces are generated by changes in the speed or direction of moving electric charge. The frequency of electromagnetic waves is a measure of the number of times per second that a new electric or magnetic force is generated. The wavelength is the distance between adjacent electric or magnetic forces. The diagram on the right shows that electromagnetic waves are transverse waves. The electric and magnetic forces are at right angles to (that is, across) the direction of motion of the wave. The direction of the electric force is defined as the direction in which it would push on a positive electric charge. The direction of the magnetic force is defined as the direction in which a compass needle would point.

A representation of part of an electromagnetic wave

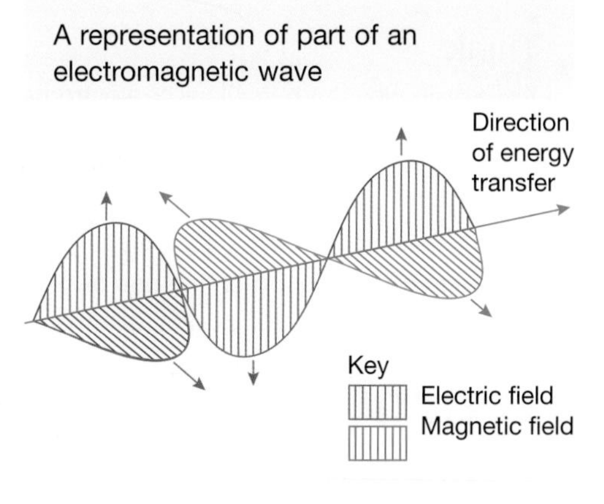

9.5.4 It's natural

Some electromagnetic radiation is emitted by all objects, including the sun. The higher energy waves, like ultraviolet radiation and X-rays are emitted naturally by stars. Our own sun emits ultraviolet radiation and X-rays. Gamma rays are emitted by radioactive substances and larger stars. All types of electromagnetic radiation can be produced artificially.

9.5.5 What's the difference?

Some differences between sound waves and electromagnetic waves are summarised in the following table.

Sound waves	Electromagnetic waves
Compression (longitudinal) waves	Transverse waves
Travel through all solids, liquids and gases, but are unable to travel through a vacuum	Unable to travel through some substances but travel through a vacuum
Speed in air between about 330 m/s and 350 m/s, depending on temperature	Speed in air about 300 000 000 m/s

9.5 Exercises: Understanding and inquiring

To answer questions online and to receive **immediate feedback** and **sample responses** for every question, go to your learnON title at www.jacplus.com.au. *Note:* Question numbers may vary slightly.

Remember

1. Are electromagnetic waves transverse waves or longitudinal waves?
2. List three properties that all electromagnetic waves have in common.
3. List three differences between sound waves and electromagnetic waves.
4. List the types of electromagnetic radiation that have more energy than visible light.

Using data

5. The sound of a starting pistol can be heard easily by an observer one kilometre away. From that distance, the sound of the pistol is heard some time after the smoke is seen. Calculate to the nearest second:
   (a) how long it takes sound to reach the observer
   (b) how long it takes light scattered from the smoke to reach the observer.
6. The speed of sound in air at a temperature of 25 °C is about 350 m/s.
   (a) How long would it take for sound waves to travel from Melbourne to Sydney, a straight distance of about 700 km, when the air temperature is 25 °C?
   (b) Why doesn't it take this long for the sound to reach you by telephone, television or radio?

Think

7. Explain why the behaviour of electromagnetic waves cannot be modelled using compression waves in a gas or a slinky.
8. Explain why you always hear thunder after you see the lightning that caused it.

Investigate

9. Research and report on the use of radiation therapy in the treatment of disease.

learnon RESOURCES — ONLINE ONLY

- **Complete this digital doc:** Worksheet 9.2: Reflection and scattering of light (doc-25104)
- **Complete this digital doc:** Worksheet 9.3: Curved mirrors (doc-25105)
- **Complete this digital doc:** Worksheet 9.4: Refraction (doc-25106)

9.6 Light energy

9.6.1 Tracing the path of light

You can get it at the flick of a switch or by striking a match. It travels through space at a speed of about 300 000 kilometres per second. It makes life on Earth possible and provides us with some beautiful images such as rainbows and spectacular sunsets. That's light!

Light travels in straight lines as it travels through empty space or through a uniform substance like air or water. The lines that are used to show the path of light are called **rays**. You cannot see a single light ray. A stream of light rays is called a **beam**. You can see beams of light only when particles in substances like air scatter the light as shown in the photograph on the right. Some of the scattered light enters your eye, allowing you to see the particles within the beam.

The ray box shown in the photograph below right provides a way of tracing the path of light. It contains a light source and a lens that can be moved to produce a wide beam of light that spreads out, converges or has parallel edges. The light box is placed on a sheet of white paper, making the beam visible as some of the light is reflected from the paper into your eyes.

Black plastic slides can be placed in front of the source to produce a single thin beam or several thin beams. The beams are narrow enough to trace with a fine pencil onto the white paper. The fine pencil line can be used to represent a single ray.

A beam of light can be seen if there is smoke or fog in the air. Light is scattered by the tiny particles. Some of the scattered light enters your eye, allowing you to see the particles within the beam.

A ray box. It provides a way of tracing the path of light.

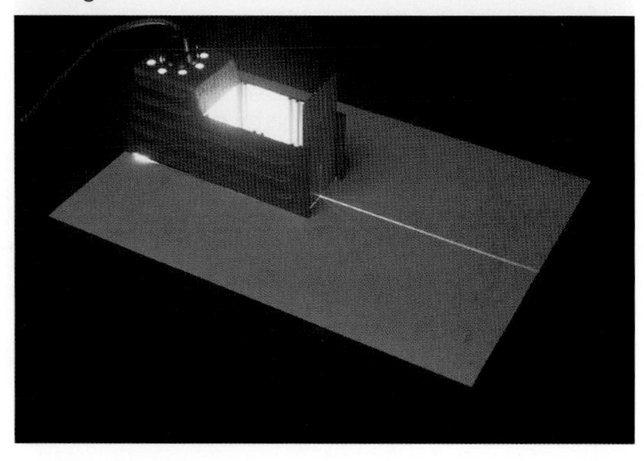

9.6.2 Crossing boundaries

When light meets a boundary between two different substances, a number of things can happen.

On the rebound

The light may bounce off the surface of the substance. This is called **reflection**, and is what allows you to see non-luminous objects. Most of the objects that you see are non-luminous. **Luminous** objects are those that emit light. The sun and the flame of a burning match are examples of luminous objects.

Light can also be reflected from particles within substances. This is called **scattering** because the light bounces off in so many different directions. Light is scattered by the particles of fog, dust and smoke in the atmosphere. Scattering is also evident in cloudy water. A luminous object in very deep or dirty water is not visible from the surface because all of the light is scattered before it can emerge. The same object is more likely to be visible on the surface of shallower or cleaner water because less light would be scattered.

Just passing through

The light may travel through the substance. Some light is always reflected when light crosses a boundary between two substances. If most of the light travels through the substance, the surface is called **transparent** because enough light gets through for you to be able to see objects clearly on the other side. Some materials let just enough light through to enable you to detect objects on the other side, but scatter so much light that you can't see them clearly. The frosted glass used in some shower screens is an example. Such materials are said to be **translucent**.

Lost inside

The light may be absorbed by the substance, transferring its energy to the particles in the substance. Substances that absorb or reflect all the light striking them are said to be opaque. Most objects in your classroom are **opaque**.

(a) Transparent

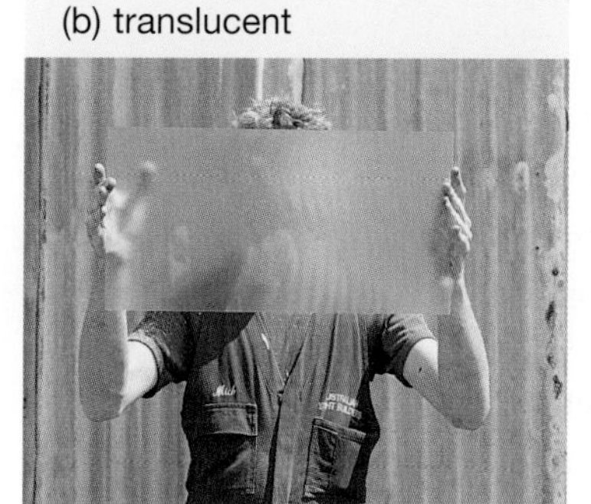
(b) translucent

(c) opaque materials

9.6.3 Reflections

When you look in a mirror you see an image of yourself. If the mirror is a plane or flat mirror, the image will be very much like the real you. If the mirror is curved, the image might be quite strange, like the one in the photograph on the right.

The images in mirrors are formed when light is reflected from a very smooth, shiny metal surface behind a sheet of glass. Images can also be formed when light is reflected from other smooth surfaces, such as a lake.

What does this person really look like?

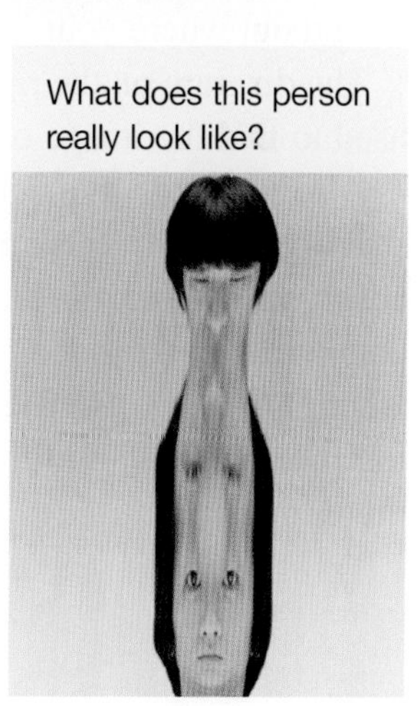

INVESTIGATION 9.5

Seeing the light

AIM: To investigate the reflection of light and its transmission through a prism and lens

Materials:
ray box kit
power supply
several sheets of white paper
ruler and fine pencil

Method and results

- Connect the ray box to the power supply. Place a sheet of white paper in front of the ray box. Move the lens backwards and forwards until a beam of light with parallel edges is projected.
- Use one of the black plastic slides to produce a single thin beam of light that is clearly visible on the white paper.

1. Trace the path of this single beam of light as it meets the lens, prism or one of the mirrors shown in the diagram to the right.
   The path can be traced by using pairs of very small crosses along the centre of the beam before and after meeting each 'obstacle'. Trace and label the shape of each 'obstacle' before you trace the light paths.

- Change the slide in the ray box so that you can project several parallel beams towards each of the 'obstacles'.

2. Use a ruler to draw a small diagram showing the path followed by the parallel beams when they meet each of the 'obstacles'.

Tracing the path of a beam of light

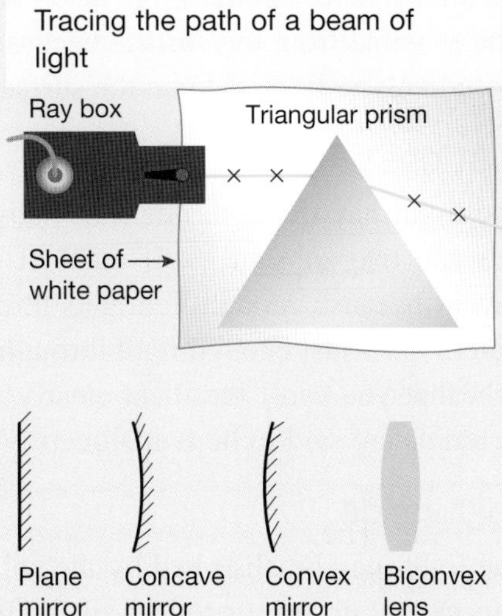

Discuss and explain

3. What happens to a beam of light when it meets a perspex surface:
   (a) 'head on'
   (b) at an angle?
4. What happens to a beam of light when it meets a plane mirror surface:
   (a) 'head on'
   (b) at an angle?
5. Describe your observations of the path followed by the three parallel beams when they meet each of the mirrors and the lens.

Where is the image?

Whenever light is reflected from a smooth, flat surface, it bounces away from the surface at the same angle from which it came. This observation is known as the Law of Reflection. This law can be used to find out where your image is when you look into a mirror.

The diagram on the right shows how the Law of Reflection can be used to find the image of the tip of your nose.

Almost all of the light coming from the tip of your nose and striking the mirror is reflected. (A very small amount of light is absorbed by the mirror.) All of the reflected light appears to be coming from the same point behind the mirror; and that is exactly where the image is. The image of the tip of your nose is the same distance behind the mirror as the real tip of your nose is in front of the mirror.

The reflected light appears to be coming from just one place. That's where the image is.

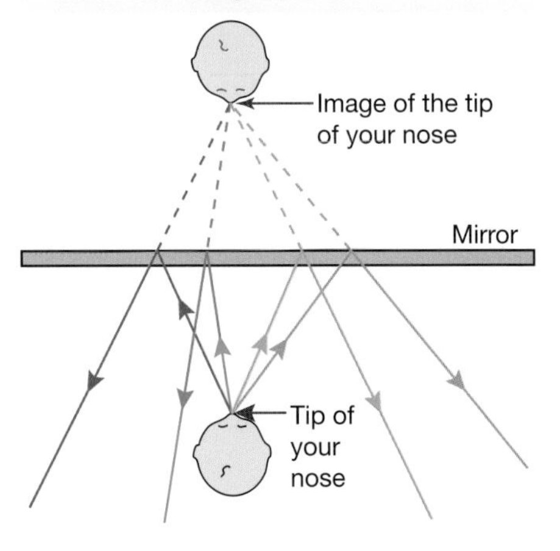

Reflection from curved mirrors

Flat mirrors are commonly found in the home. Curved mirrors have many applications too, including make-up mirrors, security mirrors in shops and safety mirrors at dangerous street intersections. Curved mirrors may be **concave** (curved inwards) or **convex** (curved outwards). Light reflecting from concave and convex mirrors also follows the law of reflection, such that the parallel rays of light are reflected to a **focal point** as shown on the right.

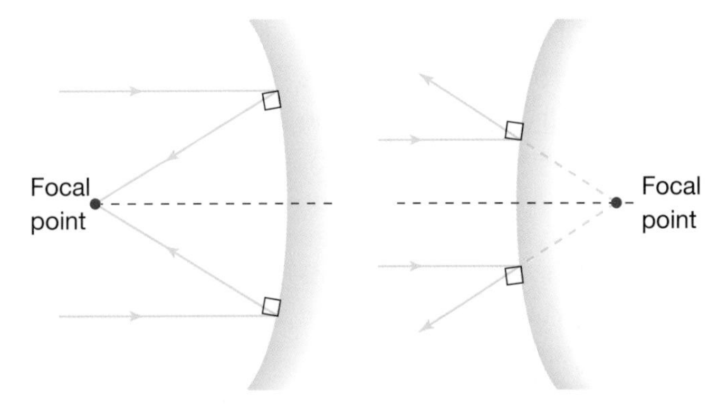

Reflected light rays converge from a concave mirror.

Reflected light rays diverge from a convex mirror.

Why can't you see your image in a wall?

When you look very closely at surfaces like walls, you can see that they are not as smooth as the surface of a mirror. The laws of reflection are still obeyed, but light is reflected from those surfaces in all directions. It doesn't all appear to be coming from a single point. There is no image.

Lateral inversion

The sideways reversal of images that you see when you look at yourself in a mirror is called **lateral inversion**. The sign on the ambulance in the photograph on the right is printed so that drivers in front of it can easily read the word 'AMBULANCE' in their rear-view mirrors.

Why is the word 'AMBULANCE' printed in reverse?

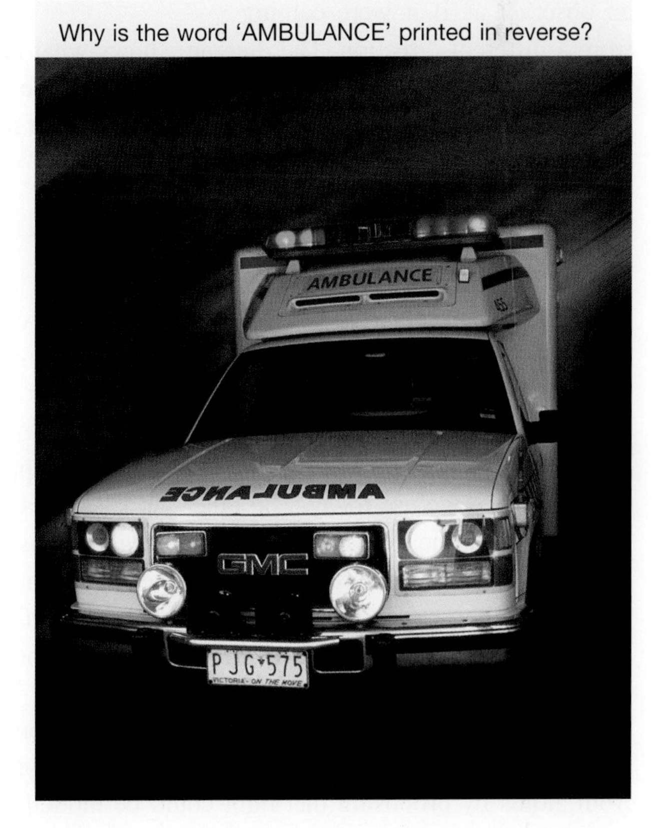

9.6.4 A change of direction

When light travels from one substance into another substance that is transparent or translucent, it can slow down or speed up. This change in speed as light travels from one substance into another is called **refraction**. Refraction causes light to bend, unless it crosses at right angles to the boundary between the substances.

The best way to describe which way the light bends is to draw a line at right angles to the boundary. This line is called the **normal**. When light speeds up, as it does when it passes from water into air, it bends away from the normal. When light slows down, as it does when it passes from air into water, it bends towards the normal.

What happened to my legs?

Looks can be deceiving! The people in the photograph do not have unusually short legs. Everything you see is an **image**. An image of the scene you are looking at forms at the back of your eye. When light travels in straight lines, the image you see provides an accurate picture of what you are looking at. However, when light bends on its way to your eye, the image you see can be quite different.

The light coming from the swimmers' legs in the photograph bends away from the normal as it emerges from the water into the air. The light arrives at the eyes of an observer as if it were coming from a different direction. The diagram shows what happens to two rays of light coming from the swimmer's right toe. To the observer, the rays appear to be coming from a point higher than the real position of the toe. It can be seen by looking at the diagrams that the amount of bending depends on the angle at which the light crosses the boundary.

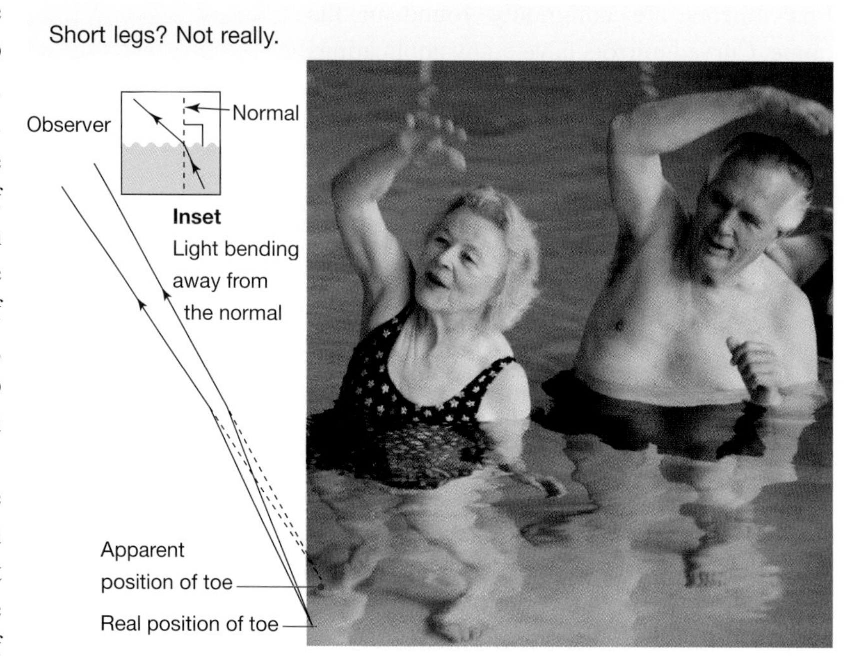

Short legs? Not really.

9.6.5 Models of light

The representation of visible light and the rest of the electromagnetic spectrum as waves is an example of a *model*. A model provides a useful way of investigating the properties and behaviour of something that you can't see. For example, the particle model of matter allows us to explain the properties and behaviour of solids, liquids and gases.

During the seventeenth century, there were two 'opposing' models of light. One was a wave model similar to the one we use today. The other model, a particle model proposed by Sir Isaac Newton, was more popular at the time. Newton proposed that light consisted of a stream of tiny particles that he called corpuscles. This model successfully explained the reflection of light. However, the only way that Newton's model could explain light bending towards the normal when it moves from air to water was if it travelled faster in water. Of course, at that time there was no way to measure the speed of light in water. We now know that light travels more slowly in water.

The wave model is successful at explaining most properties of light. However, in the early twentieth century, Albert Einstein, more famous for his theories of relativity, explained how light could eject electrons from atoms by proposing that light could be modelled as 'quantums', which were later named **photons**. Photons are not particles and have no mass. They are 'packets of energy' that have properties of both particles and waves. The accepted models of light and other electromagnetic radiation have been changing for more than 400 years and there is no reason to doubt that they will continue to change.

INVESTIGATION 9.6

Looking at images

AIM: To observe and compare the reflection of light from plane mirrors and curved mirrors

Materials:

plane mirror
shiny tablespoon or soup spoon

Method and results

- Look at your image in the back of a spoon. This surface is **convex**. Convex means curved outward. Move the spoon as close to your eyes as you can and then further away.

1. Record your observations in a table like the one below. Is the image small or large? Right-side up or upside down? Is there anything strange about the image?

- Look at your image in the front of the spoon. This surface is **concave**. Concave means curved inward. Move the spoon closer to you and then further away.

2. Record your observations in the table.

- Look at the image of your face in a plane mirror. Wink your right eye and take notice of which eye appears to wink in the image.

3. Write the word IMAGE on a piece of paper and place it in front of the mirror so that it faces the mirror. Write down the word as you see it in the image.
4. Write down how you think an image of the word REFLECTION would look in the mirror.

Discuss and explain

5. Which eye in the plane mirror image appears to wink?
6. Which letters in the image of the word IMAGE look different? Which look the same?
7. Test your hypothesis about the image of the word REFLECTION. Was your hypothesis correct?
8. List some places where you have seen curved mirrors. State whether the mirrors were convex or concave and explain why they are used.

	Observations of image		
	First observation	When you move closer	When you move further away
Convex side			
Concave side			

INVESTIGATION 9.7

Floating coins

AIM: To observe the refraction of light

Materials:

2 beakers
evaporating dish
coin

Method and results

- Place a coin in the centre of an evaporating dish and move back just far enough so you can no longer see the coin. Remain in this position while your partner slowly adds water to the dish.

1. Make a copy of the diagram on the right. Use dotted lines to trace back the rays shown entering the observer's eye to see where they seem to be coming from. This enables you to locate the centre of the image of the coin.
2. Is the image of the coin above or below the actual coin?
3. What appears to happen to the coin while water is added to the evaporating dish?

The image of the coin is not in the same place as the actual coin.

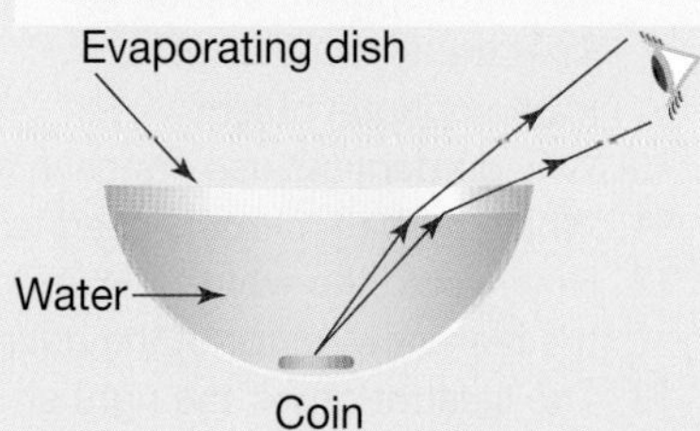

INVESTIGATION 9.8

How much does it bend?

AIM: To investigate the refraction of light as it travels into and out of a rectangular prism

Materials:

ray box kit
power supply
sheet of white paper

Ray box
Rectangular perspex prism
Normal
Thin beam
Sheet of white paper

Method and results

- Connect the ray box to the power supply. Place a sheet of white paper in front of the ray box. Project a single thin beam of white light towards a rectangular perspex prism as shown in the diagram above right.

1. Does the light bend towards or away from the normal as it enters the perspex? (Remember that the normal is a line that can be drawn at right angles to the boundary. It is shown as a dotted line in the diagram. You don't need to draw it — just imagine that it's there.)
2. Imagine a normal at the boundary where the light leaves the perspex to go back into the air. Which way does the light bend as it re-enters the air — towards or away from the normal?
3. Does all of the light travelling through the perspex re-enter the air? If not, what happens to it?
4. Look at the light beam as it enters and leaves the perspex. What do you notice about the direction of the incoming and emerging beam?
5. Turn the prism without moving the ray box so that the light enters the perspex at different angles.
   (a) How can you make the incoming light bend less when it enters the perspex?
   (b) How can you make the incoming light bend more when it enters the perspex?

9.6 Exercises: Understanding and inquiring

To answer questions online and to receive **immediate feedback** and **sample responses** for every question, go to your learnON title at www.jacplus.com.au. *Note:* Question numbers may vary slightly.

Remember

1. You cannot usually see light as it travels through the air. What makes it possible to see a beam of light?
2. What happens to light when it travels through air and meets:
   (a) a transparent surface
   (b) a translucent surface
   (c) an opaque surface?
3. What does a mirror do to light in order to form an image?
4. In which type of mirror can your image be upside down?
5. How is your image in a plane mirror different from the real you?
6. What is refraction?
7. Which way does light bend when it slows down while passing from air into water: towards or away from the normal?
8. Describe one weakness of Sir Isaac Newton's particle model of light.

Think

9. List one example of each of the following.
   (a) A transparent object
   (b) A translucent object
   (c) An opaque object
10. Why do dentists use concave mirrors to examine your teeth?
11. Which type of mirror is used to help you see around corners?
12. How would the word 'TOYOTA' on the front of a van look in the rear-view mirror of the driver in front of it?
13. The illustration on the right shows a ray of light emerging from still water after it has been reflected from a fish.

Should the spear be aimed in front of or behind the image of the fish? Use a diagram to explain why.

14. When you look down on a coin at the bottom of a glass of water it looks closer to you than it really is.
    (a) Draw a diagram to show why it looks closer.
    (b) In what other way is the image of the coin different from the real coin?
15. Explain how sound waves can be modelled with water waves, which are transverse waves. Use a diagram to illustrate your answer.
16. Are photons particles or waves? Explain your answer.

Imagine

17. Imagine that the world is plunged into darkness by a mysterious cloud of dust. What problems would be caused by the lack of visible light if the cloud lingered for:
    (a) one hour
    (b) three days
    (c) six weeks?
18. Imagine that you are the fish from question 13 in the illustration below.
    (a) Will the image of the girl's head be higher or lower than her real head?
    (b) Draw a sketch of how the girl might appear to you.

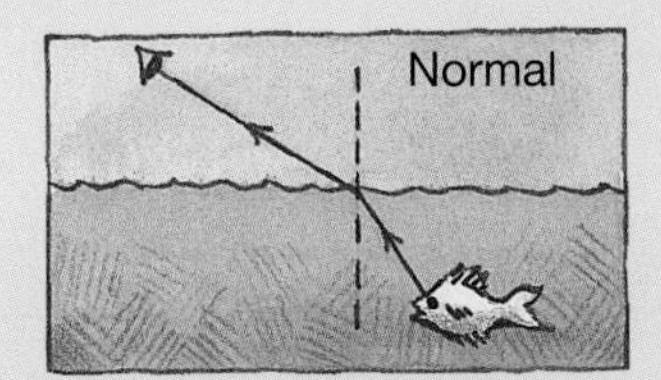

Design and create

19. Design and build a simple periscope like the one shown in the diagram on the right. You will need stiff card, scissors, two small mirrors, sticky tape or glue, a pencil and a ruler. Explain, with the aid of a diagram, how it works.

A periscope uses mirrors to enable you to see around corners or over objects.

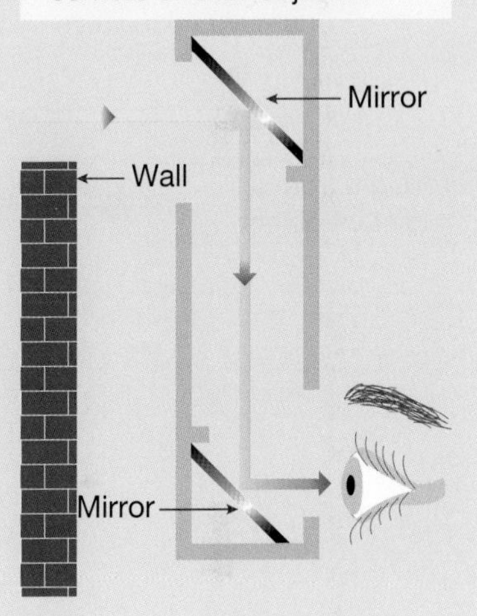

learn on RESOURCES — ONLINE ONLY

- **Watch this eLesson:** Twinkle, twinkle (eles-0071)
- **Watch this eLesson:** Galileo and the telescope (eles-1765)
- **Try out this interactivity:** Bend it (int-0673)
- **Complete this digital doc:** Worksheet 9.5: The eye (doc-25107)

9.7 Seeing the light

Science as a human endeavour

9.7.1 The human eye

Everything that you see is an image, created when the energy of light waves entering your eyes is transmitted to a 'screen' at the back of each eye.

This screen, called the **retina**, is lined with millions of cells that are sensitive to light. These cells respond to light by sending electrical signals to your brain through the **optic nerve**.

The human eye

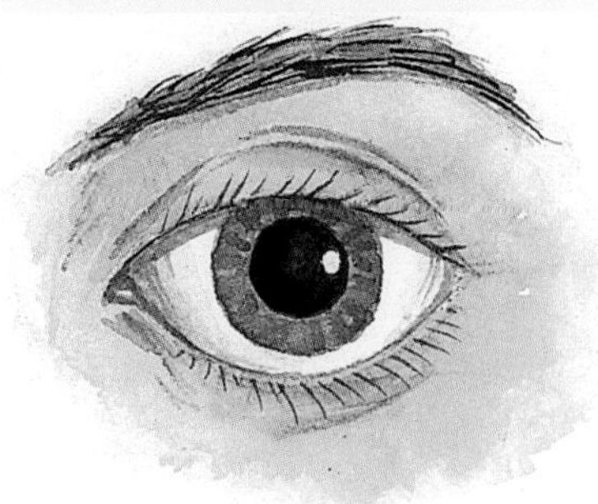

Some of the light reflected from your surroundings, along with light emitted from luminous objects such as the sun, enters your eye. It is refracted as it passes through the outer surface of your eye. This transparent outer surface, called the **cornea**, is curved so that the light converges towards the **lens**. Most of the bending of light done by the eye occurs at the cornea.

On its way to the lens, the light travels through a hole in the coloured **iris** called the **pupil**. The iris is a ring of muscle that controls the amount of light entering the lens. In a dark room the iris contracts to allow as much of the available light as possible through the pupil. In bright sunlight the iris relaxes, making the pupil small to prevent too much light from entering. The clear, jelly-like lens bends the light further, ensuring that the image formed on the retina is sharp. The image formed on the retina is inverted. However, the brain is able to process the signals coming from the retina so that you see things the right way up.

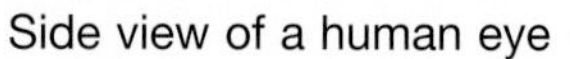

Side view of a human eye

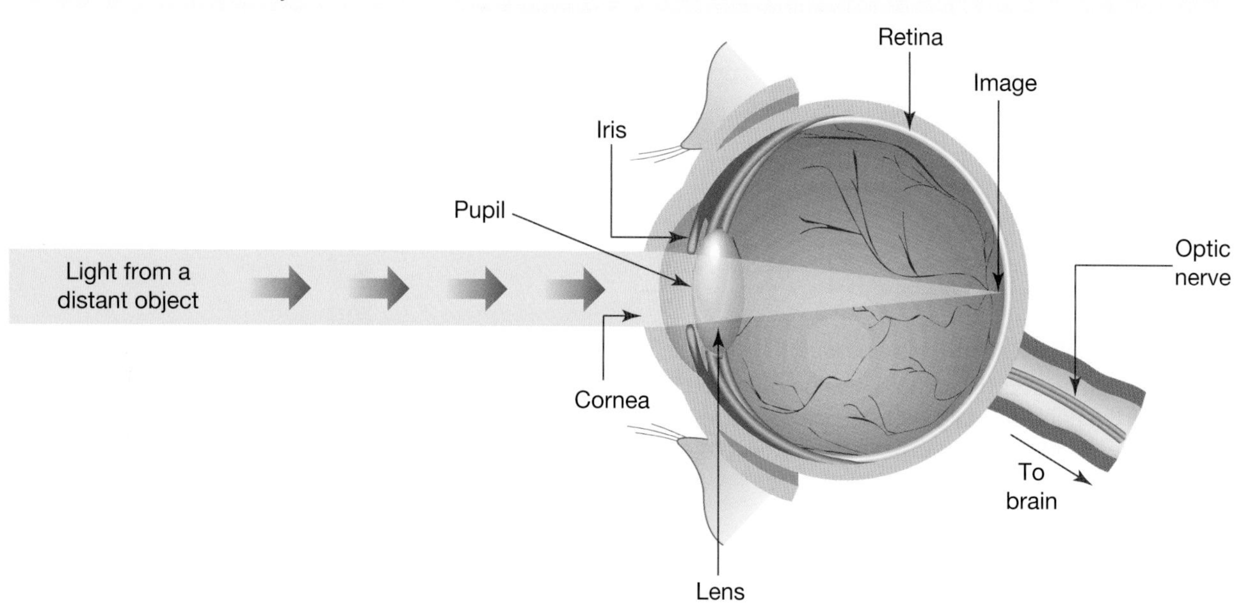

The image formed on the retina is upside down.

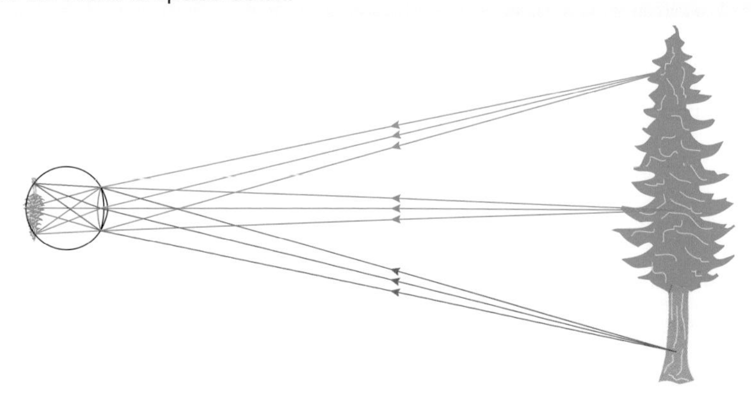

9.7.2 Getting things in focus

Although most of the bending of light energy done by the eye occurs at the cornea, it is the lens that ensures the image is sharp.

Two types of lenses

The lens in each of your eyes is a **converging lens**. Its shape is **biconvex** — that means it is curved outwards on both sides. A beam of parallel rays of light travelling through a biconvex lens 'closes in' (converges) towards a point called the **focal point**, or focus.

Another type of lens is a **diverging lens**, which spreads light outwards because of its biconcave shape. A biconcave lens does not have a real focal point. When the parallel light rays emerge from a biconcave lens, they do not converge to a focal point. However, if you trace the rays back to where they are coming from, you find that they do appear to be coming from a single point. That point is called the **virtual focal point**, or virtual focus.

The light coming from a nearby object needs to be bent more than the light coming from a distant object. The lens in your eye becomes thicker when you look at nearby objects.

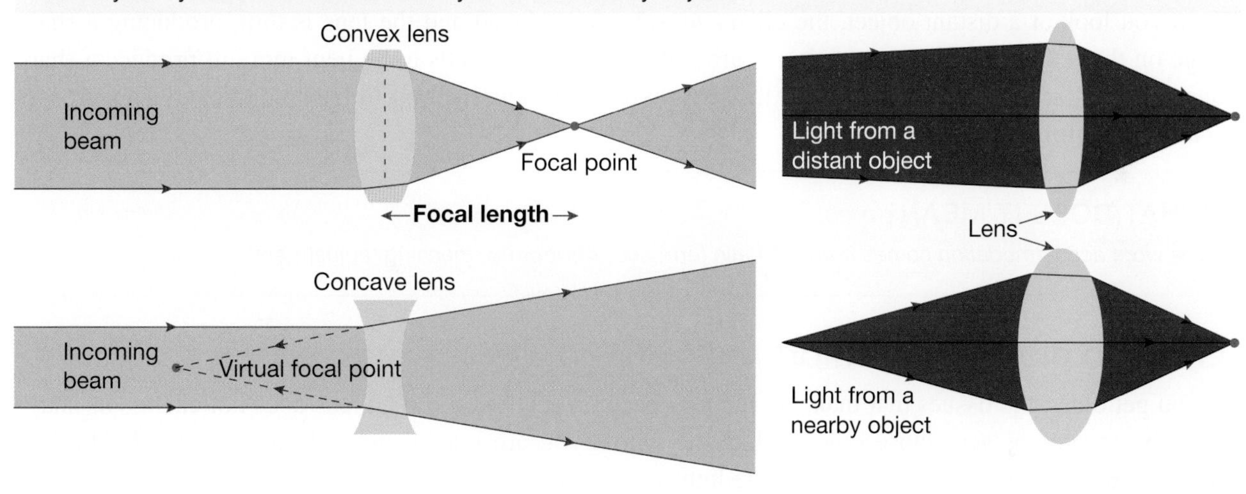

INVESTIGATION 9.9

Focusing on light

AIM: To investigate the transmission of light through different lenses

Materials:

ray box kit
sheet of white paper
12 V DC power supply
ruler and fine pencil

Method and results

- Connect the ray box to the power supply and place it on a page of your notebook.

Part A: Biconvex lenses

- Place the thinner of the two biconvex lenses in the kit on the page and trace out its shape. Project three thin parallel beams of white light towards the lens.

1. Trace the paths of the light rays as they enter and emerge from the lens. Remove the lens from the paper so that you can draw the paths of the light rays through the lens.

- Replace the thin biconvex lens with a thicker one and repeat the previous steps.

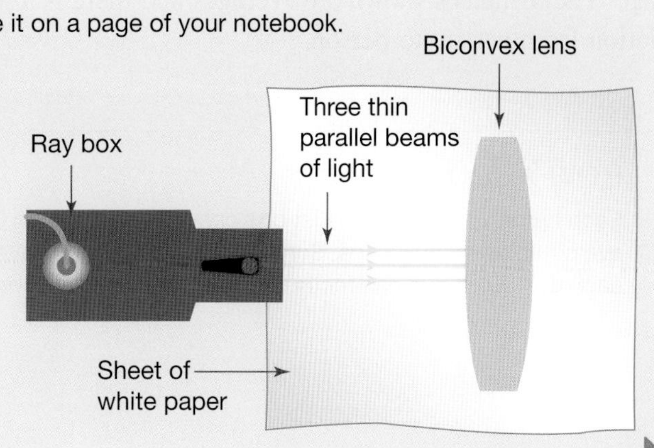

Part B: Biconcave lenses

- Place the thinner of the two biconcave lenses on your notebook page and trace out its shape.

2. Trace the path of each of the three thin light beams as they enter and emerge from the lens. Remove the lens from the page so that you can draw the paths of the light beams through the lens.

Discuss and explain

3. State the focal length (distance from the focal point to the centre of the lens) for each lens.
4. Which of the biconvex lenses bends light more, the thin one or the thicker one?
5. Explain why the middle light ray does not bend.
6. How many times do each of the other rays bend before arriving at the focal point?
7. Do the diverging rays come to a focus?
8. Do the diverging rays appear to be coming from the same direction? Use dotted lines on your diagram to check.
9. Predict where the diverging rays will appear to come from if you use a thicker biconcave lens. Check your prediction with the thicker biconcave lens in the ray box kit.

Accommodation

The exact shape of the clear jelly-like lens in your eye is controlled by muscles called the **ciliary muscles**. When you look at a distant object, the ciliary muscles are relaxed and the lens is thin, producing a sharp image on the retina. When you look at a nearby object, the light needs to be bent more to produce a sharp image. The ciliary muscles contract and the jelly-like lens is squashed up to become thicker. This process is called **accommodation**.

WHAT DOES IT MEAN?

The word *accommodation* comes from the Latin term ***accommodatio***, meaning 'adjustment'.

9.7.3 Too close for comfort

As you get older, the tissues that make up the lens become less flexible. The lens does not change its shape as easily. Images of very close objects (like the words you are reading now) become blurred. The lens does not bulge as much as it should and the light from nearby objects converges to a point behind the retina instead of on the retina. You may have to hold what you are reading further away in order to obtain a clear image.

This change in accommodation with age is a natural process. Some people are not inconvenienced at all while others need to wear reading glasses so that they can read more easily and comfortably. The table on the right shows how the smallest distance at which a clear image can be obtained changes with age. The distances shown are averages and there is a lot of variation from person to person.

How the average smallest distance at which a clear image can be obtained changes with age

Age (years)	Distance (cm)
10	7.5
20	9
30	12
40	18
50	50
60	125

HOW ABOUT THAT!

Each human eye contains just one convex lens. Insects have compound eyes. Each eye contains many lenses. Some types of dragonfly have more than 10 000 lenses in each eye. Each eye can focus light coming from only one direction.

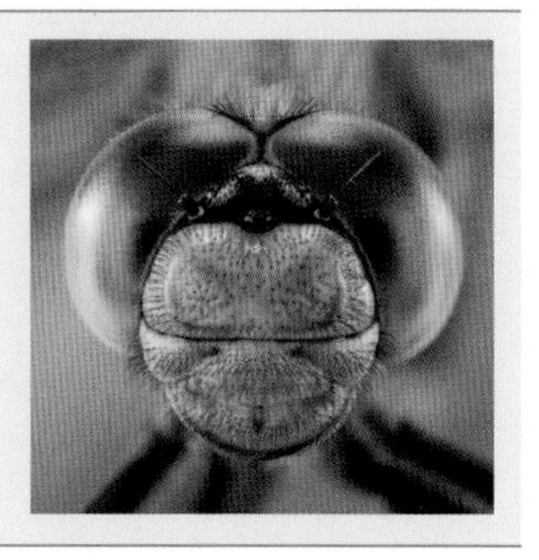

INVESTIGATION 9.10

Getting a clear image

AIM: To investigate accommodation

Materials:
ruler

Method and results

- Look closely at the **X** printed here from the smallest distance at which you can see it clearly and sharply with comfort. Quickly look away and focus on a distant object for a second or two and then focus on the 'X' again from the smaller distance.
- Try to feel the action of the muscles that allow you to see a sharp image of the 'X'.
- Use the following procedure to estimate the smallest distance at which you can obtain a clear image of a nearby object. (If you are wearing glasses, remove them during this part of the experiment.)
- Hold this book vertically at arm's length from your eyes and focus on it. Move the book to a position about three or four centimetres from your eyes and then gradually move the book further away until you can see the print clearly and sharply.
- Have a partner use the ruler to estimate the distance between the page and your eyes. The ruler should be placed carefully beside your head for this measurement.

1. Record the distance measured.
2. Collate the results for the whole class and determine the average smallest distance at which a clear image could be obtained.

Discuss and explain

3. How does your result compare with the average smallest distance for your class?
4. Write down the highest single result and lowest single result for your class. Comment on the range of results.

9.7.4 A look inside

The photograph on the right shows the inside of a human stomach. It has been photographed through a long, flexible tube called an **endoscope**. Inside the endoscope are two bundles of narrow glass strands called **optical fibres**. The glass in optical fibres is made so that light is unable to emerge from the glass.

A beam of bright light is directed through one bundle of fibres, illuminating the inside of the stomach. Some of the light is reflected through the other bundle of fibres. A lens at the end of this bundle allows an image to be viewed, photographed or videotaped.

Endoscopes can be used to look at many different parts of the body. Different types of endoscopes include:

- gastroscopes, which are used to examine the stomach and other parts of the digestive system
- arthroscopes, which are used to search for problems in joints like shoulders and knees
- bronchoscopes, which are used to see inside the lungs.

Optical fibres allow us to see inside the human body via an endoscope.

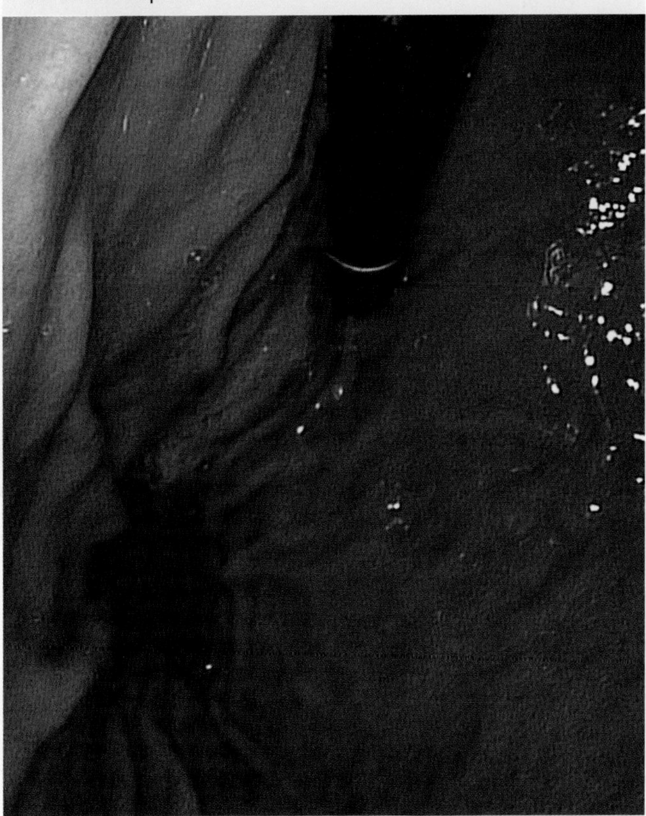

Endoscopes can also be used in laser surgery. Intense laser beams can be directed into the optical fibres. The heat of the laser beams can be used to seal broken blood vessels or destroy abnormal tissue inside the body.

The glass in optical fibres is made so that light is unable to emerge from the glass. As light travels from a substance such as glass into air, it bends away from the normal (see subtopic 9.6). If the light strikes the boundary at a small enough angle, it bends so much that instead of leaving the glass, it is reflected back into it. This process is called **total internal reflection**. The diagram below right shows how total internal reflection occurs in an optical fibre.

A gastroscope being used to look inside a patient's stomach

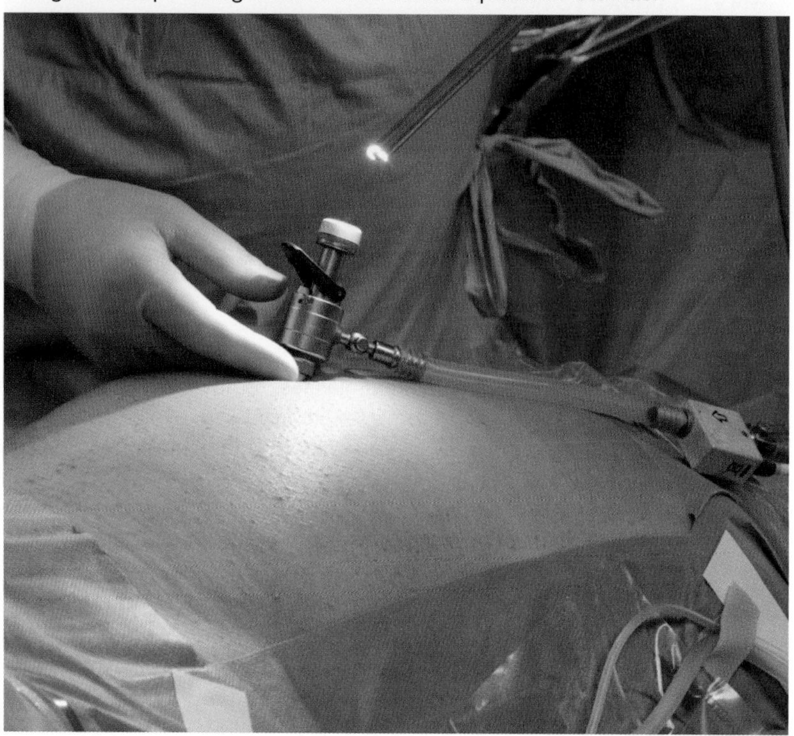

A bundle of optical fibres. The light can be seen through the ends.

Total internal reflection in an optical fibre

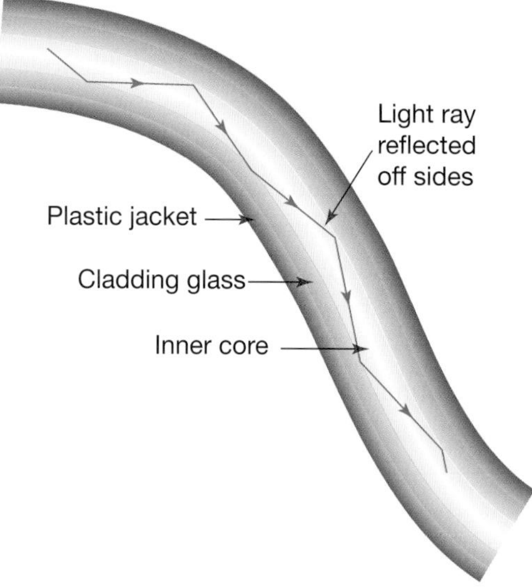

INVESTIGATION 9.11

Total internal reflection

AIM: To investigate total internal reflection in a triangular prism

Materials:

ray box kit
12 V DC power supply
perspex triangular prism

Method and results

- Connect the ray box to the power supply. Place the ray box over a page of your notebook. Use one of the black plastic slides in the ray box kit to produce a single thin beam of light which is clearly visible on the white paper.
- Place a perspex triangular prism on your notebook and direct the thin beam of light towards it as shown in the diagram on the right. Observe the beam as it passes through the prism.
- Turn the prism slightly anticlockwise, closely observing the thin light beam as it travels from the perspex prism back into the air. Continue to turn the prism until the beam no longer emerges from the prism.

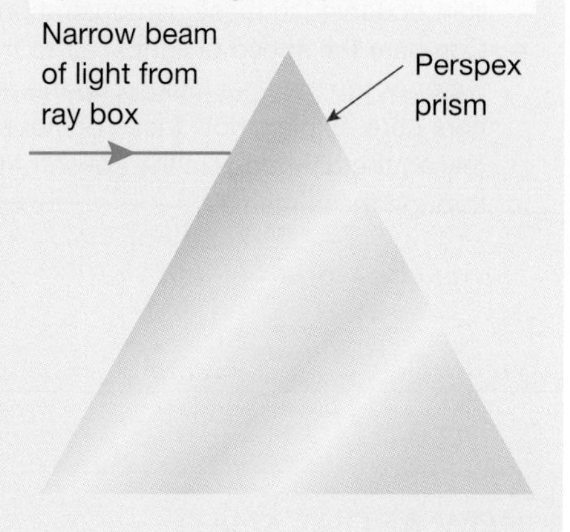

1. Describe what happens to the thin light beam as it passes from air into the perspex prism and back into the air.
2. What happens to the beam of light when it no longer emerges from the prism?

Discuss and explain

3. Draw a series of two or three diagrams showing how the path taken by the beam of light changed as you turned the prism.
4. Explain how the amount of light reflected changes as the prism is turned.

INVESTIGATION 9.12

Optical fibres

AIM: To model the transmission of light through an optic fibre

Materials:

transparent 2–3 L fruit juice container
large nail
laser (class 1)
demonstration optical fibre cable or light pipe

Method and results

- Use the nail to poke a narrow 5 mm hole into the front of a fruit juice container approximately 5 cm from the bottom.
- Darken the room.
- Fill the container to the top with water and position it on the edge of a sink so that a thin stream of water flows from the container into the sink.
- Direct a laser beam into the container and out through the centre of the stream of water.

CAUTION

Class 1 and class 2 lasers have a relatively low power output and so are safe for classroom use under direct supervision of the teacher. Laser beams should not be pointed towards others in the room because of the sensitivity of the retina of the eye. Ensure that those viewing this demonstration are positioned on either side of the stream of water to eliminate the possibility of the laser beam being directed towards them.

1. Describe the path of the laser beam.
   - Shine a laser beam down a length of 'light pipe' or loop of optical fibre.
2. Describe your observations.

Discuss and explain

3. Explain why the laser beam took the path of light observed in these demonstrations.
4. Compare the speed of light in air to that in water or the material making up the optical fibre core. Explain how these demonstrations rely on the difference in the speed of light through these media.

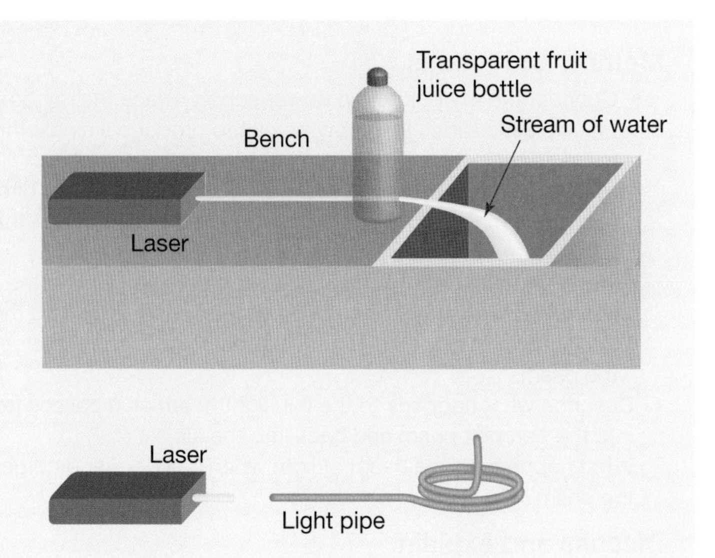

HOW ABOUT THAT!

Diamonds can sparkle with coloured light, each of its surfaces producing a dazzling display. Diamond is the most optically dense, naturally occurring material on Earth. This means that light entering a diamond through each of its facets (or geometrically cut sides) is refracted by a huge angle, causing light inside the gemstone to bounce back and forth several times before it strikes a facet with an angle straight enough to escape. Because the light has travelled so far, the spectrum of colours that make up light have dispersed (or separated) so significantly that a stunning display of colours is produced.

9.7 Exercises: Understanding and inquiring

To answer questions online and to receive **immediate feedback** and **sample responses** for every question, go to your learnON title at www.jacplus.com.au. *Note:* Question numbers may vary slightly.

Analyse and evaluate

1. Use the data in the table on page 400 to draw a line graph to show how the ability to focus on nearby objects changes with age.
   (a) Use your graph to predict the smallest distance at which a clear image can be obtained by an average person of your age.
   (b) At what age does the decrease in focusing ability appear to be most rapid?

Remember

2. At which part of the human eye does most of the bending of light occur?
3. Describe the function that the iris and pupil work together to perform.
4. Name and sketch the shape of a lens that:
   (a) converges a beam of light to a single point
   (b) makes the rays in a beam of light diverge.
5. What is the focal length of a converging lens a measure of?
6. What is accommodation?
7. What is the name given to the shape of the lens in the human eye?
8. Sketch the shape of the lens in the eye when you are viewing:
   (a) a nearby object
   (b) a distant object.
9. How does the lens change its shape?
10. Why is it common to see older people holding a newspaper at arm's length while they are reading it?
11. How are messages sent from the eye to your brain?
12. Explain how an endoscope works.
13. List three medical uses of endoscopes.
14. Explain how optical fibres allow light to travel along a bent tube.

Think

15. Does light slow down or speed up when it passes from the air into the cornea? How do you know this? (*Hint:* Refer to subtopic 9.6.)
16. Explain why the focal point of a diverging lens is called a virtual focal point.
17. Why does the lens need to be thicker for viewing nearby objects?
18. Can total internal reflection occur when light travels from air into glass? Explain your answer.

Create and investigate

19. Use two or more lenses and lens holders to make a model microscope or telescope on a laboratory workbench. Investigate the effect of changing the distance between the lenses on the magnification and write a report on your findings.

Investigate

20. Use the internet to research and report on the development of the bionic eye by Australian scientists. Include in your report information about:
    - macular degeneration
    - which patients it is designed to benefit
    - how it works
    - a comparison with the bionic ear.

9.8 We're on the air

Science as a human endeavour

9.8.1 Riding on a radio wave

The human desire to continually improve communications seems infinite.

Since the Stone Age, humans have been finding new ways to communicate with more people over greater distances at a greater speed. After the discovery of radio waves by Heinrich Hertz in 1887, and a few years of clever work by the Italian engineer Guglielmo Marconi, it became possible to send messages over long distances at the speed of light.

Radio waves are emitted naturally by stars. They can also be produced artificially when electrons in a metal rod are made to vibrate rapidly. This metal rod is called a **transmitting antenna** or transmitter. These vibrations cause radio waves to travel through the air (at about 300 000 kilometres per second). The radio waves can be detected by a **receiving antenna**, which is a metal rod just like the transmitter. The radio waves cause electrons in the receiving antenna to vibrate rapidly, producing an electrical signal.

Compare the sizes of these transmitting and receiving antennae. Why are they different?

AM radio

Each AM radio station has a particular frequency of radio waves on which it transmits sound signals. The sound signal must firstly be changed to an electrical signal. This electrical signal is called an **audio** signal. The waves on which messages are sent are called **carrier waves**. The audio signal is added to the carrier wave, producing a modulated wave, as shown in the diagram on the following page. The receiving antenna of your radio detects the modulated wave. Your radio then 'subtracts' the carrier wave from the signal, leaving just the audio signal. The audio signal is amplified by an audio amplifier inside the radio and sent to speakers. In the speakers, the changing electric current is used to make the surrounding air vibrate to produce sound.

The carrier signals for AM radio stations have frequencies ranging from about 540 kilohertz up to about 1600 kilohertz. When you tune in your radio, you are selecting the frequency of the carrier wave that you wish to receive. For example, if you tune to ABC Local Radio Melbourne, you are selecting the carrier wave with a frequency of 774 kilohertz.

AM stands for **a**mplitude **m**odulation: the audio signal changes the amplitude of the carrier wave.

FM radio

Like AM radio stations, each FM radio station has its own carrier wave frequency. However, the carrier frequencies are much greater — between 88 megahertz and 108 megahertz (1 megahertz equals 1 million hertz). The other major difference between AM and FM radio is the way that the audio signal is carried on the carrier wave. Instead of adding the audio signal to the carrier wave (which changes the amplitude of the wave), the audio signal changes the frequency of the carrier wave as shown in the diagram below right. FM stands for **f**requency **m**odulation.

As with AM radio, when you tune in your radio to FM you are selecting the frequency of the carrier wave that you wish to receive. For example, if you tune to Triple J in Melbourne, you are selecting the carrier wave with a frequency of 107.5 megahertz.

FM radio waves are affected less by electrical interference than AM radio waves and therefore provide a higher quality transmission of sound. However, they have a shorter range than AM waves and are less able to travel around obstacles such as hills and large buildings.

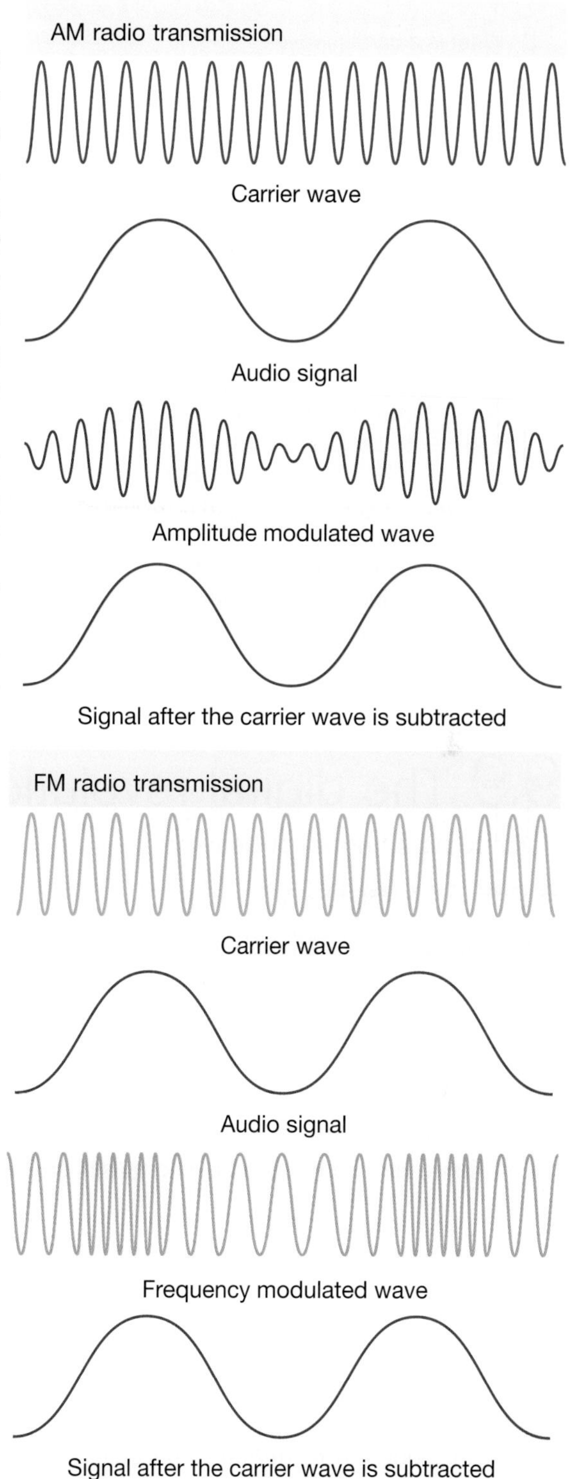

Digital radio

Digital radio began in 2009. Digital signals are different from AM or FM radio signals, and are discussed in subtopic 9.9.

9.8.2 Television

Before 2014, television signals were transmitted on two separate carrier waves. The visual signal was added onto one carrier wave using amplitude modulation. The audio signal was carried on a separate carrier wave using frequency modulation. When you tuned your television set to a particular channel, you were selecting the visual and audio carrier waves that you wished to receive. Your television set then completed the task of removing the carrier waves and translating the signals sent into a picture and sound. This was quite a complex task, as you might imagine!

9.8 Exercises: Understanding and inquiring

To answer questions online and to receive **immediate feedback** and **sample responses** for every question, go to your learnON title at www.jacplus.com.au. *Note:* Question numbers may vary slightly.

Remember

1. How are radio waves produced artificially?
2. Sound waves cannot be directly transmitted through the air over long distances. What has to happen to them before they can be transmitted on radio waves?

3. What is a carrier wave?
4. Explain the difference between the way that AM and FM radio waves carry audio signals.
5. List one advantage and one disadvantage of FM radio over AM radio.
6. How are television signals carried by radio waves?

Using data

7. Express the frequency of the following radio stations in Hz. (A frequency of 1 Hz corresponds to one complete wave being produced each second.)
   (a) Triple M (FM), Melbourne: 105.1 MHz
   (b) 3CV (AM), Bendigo: 1071 kHz
   (c) Triple J (FM), Shepparton: 94.5 MHz
   (d) Triple J (FM), Latrobe Valley: 96.7 MHz
   (e) ABC (AM), Sale: 828 kHz
   1 kHz = 1000 Hz (or 10^3 Hz)
   1 MHz = 1 000 000 Hz (or 10^6 Hz)
8. The wavelength (λ) of a wave is related to the frequency (f) of the wave by the equation:
$$v = f\lambda$$
   where v is the speed of the wave.
   The speed of radio waves in air is 300 000 000 m/s.
   Use this equation to calculate the wavelength of the carrier waves used by radio stations Triple M, Melbourne, and 3CV, Bendigo.

9.9 The digital revolution

Science as a human endeavour

9.9.1 Analogue vs digital

Analogue quantities are those that can have any value and can change continually over time. **Digital** quantities are those that can have only particular values and are represented by numbers.

Whereas analogue radio and television signals are carried as continuously changing amplitudes or frequencies, digital signals are carried as a series of 'on' and 'off' pulses. The signals can have only two values — 'on' or 'off'. The original audio and video (sound and vision) are sampled and converted into pulses. Audio signals are sampled about 40 000 times every second. Video signals are sampled more than 13 million times every second. The pulses are added to the carrier waves for transmission.

9.9.2 Out with the old, in with the new

Digital television commenced transmission in Australian capital cities in 2001. The phasing out of analogue signals began in 2010 and was completed in December 2014. A digital TV set-top box can be used to convert the digital signals back to an analogue form. This means that nobody had to replace their old analogue TV sets with digital TV sets unless they chose to.

The introduction of digital radio began in 2009, but there are no plans to phase out existing analogue radio services.

Analogue or digital — smooth or in bits

You can read the time from an analogue clock with hands that continuously rotate, or from a digital clock with LEDs (light-emitting diodes) or liquid crystals that simply turn off and on.

All physical quantities such as time, speed, weight and pH can be represented in analogue or digital form. Likewise, invisible waves like sound and radio waves can be transmitted in analogue or digital form.

- Analogue forms change smoothly if the quantity being measured changes smoothly.
- Digital forms display or transmit quantities as a limited series of numbers or pulses. Digital devices are usually electronic. Their displays are made from devices that can only ever be 'on' or 'off'. For example, each number display of a digital measuring device is made up of seven devices. Each of the seven devices can be either 'on' or 'off'. The arrangement of the seven devices allows all of the numbers from 0 to 9 to be displayed.

Each number in this digital display is made up of seven LEDs, each of which can be either 'on' or 'off'.

An analogue clock represents time as a quantity that changes smoothly.

9.9.3 What's the advantage?

Both analogue and digital television signals fade away as they travel through the air. Like all other waves, the energy they carry spreads out. So, as they travel over distance their intensity, or strength, decreases. As the continuous analogue becomes weaker, the background radiation and signals from other sources have a greater effect on the amplitude of the wave. It becomes distorted. The result is a fuzzy picture and poor quality sound. Because digital signals can be only 'on' or 'off' pulses, background radiation and signals from other sources cannot interfere with them — even as they become weaker. The rapidly pulsating signals are still either 'on' or 'off' until the 'on' signals have faded away to nothing.

As a result, digital television has several advantages over analogue television. It provides:

- sharper images and 'ghost free' reception
- widescreen pictures
- better quality sound
- capability of 'surround' sound
- access to the internet and email
- capability of interactive television, allowing the viewer to see different camera views or even different programs on the same channel
- Electronic Program Guides (EPGs) which can provide 'now' and 'next' information about programs.

9.9.4 Going mobile

Since the first major mobile phone service was introduced in Australia in 1987, millions of Australians have purchased mobile phones.

How mobile phones work

Domestic and business telephones are connected by cable to the network of microwave and radio links, coaxial cables and optical fibres. Mobile phones transmit signals on radio waves to a **base station**, which consists of several antennas at the top of a large tower or on top of a tall building. The base station is connected to a **switching centre**. Each switching centre is, in turn, connected to many base stations. The switching centre switches the call to other mobile phones through the **cellular system** or the fixed telephone system.

A network of cells

Mobile phones are also called **cellular phones**. That is because the base stations are set up in a network of hexagonal cells as shown in the diagram below left. The cells range in size from 100 metres across to over 30 kilometres across. The base stations receive and transmit mobile phone signals from the cells that adjoin them. A mobile phone signal moves from cell to cell until it reaches its destination base station.

The base stations are placed at the intersection of three cells.

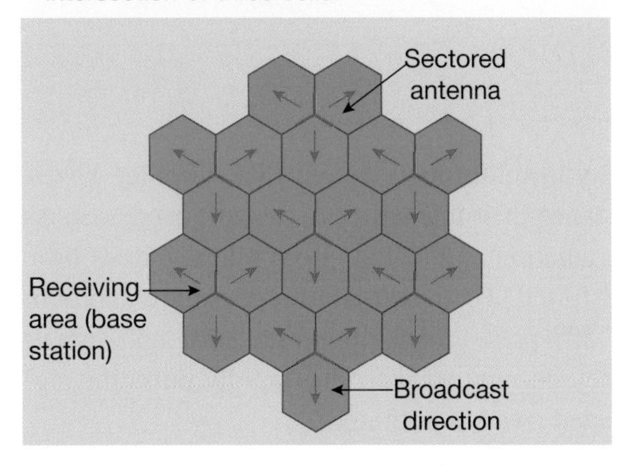

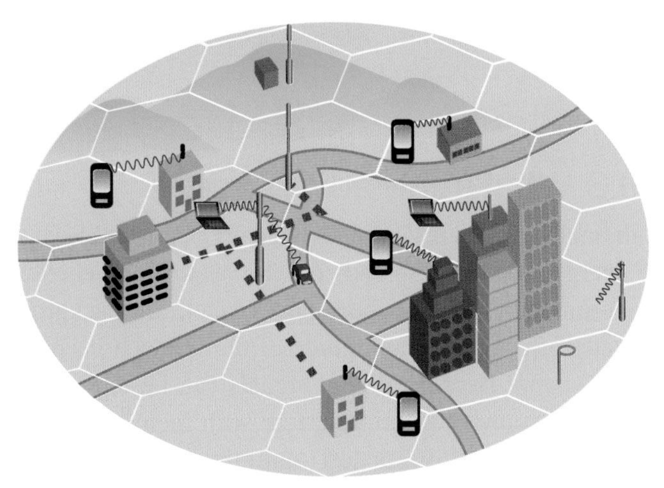

From analogue to digital

The first mobile phone service operated on an analogue system. The analogue system ceased operating at the end of 1999 and was totally replaced with a digital system. The digital system carries voice signals as a series of 'on' and 'off' pulses that are added to the carrier wave 8000 times every second.

An early mobile phone, used in the late 1980s

A landline is part of the communications network.

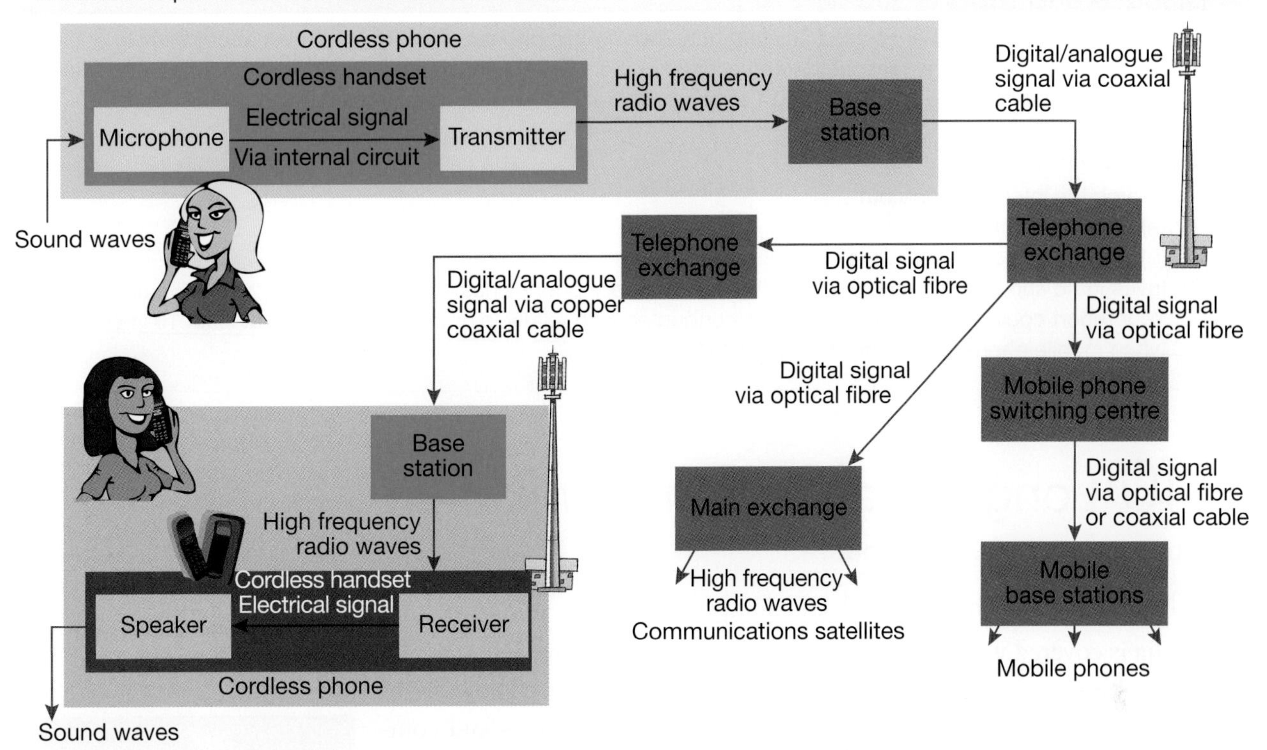

9.9 Exercises: Understanding and inquiring

To answer questions online and to receive **immediate feedback** and **sample responses** for every question, go to your learnON title at www.jacplus.com.au. *Note:* Question numbers may vary slightly.

Remember

1. Summarise the differences between the digital and analogue signals that are added to carrier waves for television transmission.
2. Do you have to have a digital TV set to watch programs now that analogue transmission has ceased? Explain your answer.
3. Digital television has several advantages over analogue television.
   (a) List five of the advantages.
   (b) Explain why digital signals have these advantages.
4. How are all mobile phones different from fixed telephones in the way that they transmit and receive voice messages?
5. Why are mobile phones also known as cellular phones?

Think

6. Why do digital TV signals provide the capability of interactive TV whereas analogue signals do not?
7. If you could choose an analogue device or a digital device to measure your weight, which would you choose? State the reasons for your choice.
8. Write the name of an analogue device (or substance) that measures:
   (a) time
   (b) speed
   (c) weight
   (d) pH.
9. Use a two-column table to list the advantages and disadvantages of mobile phones over fixed lines.

Discuss and report

10. About one in three primary school children now own mobile phones. What do you think about this? In a small group, discuss the reasons for and against primary school children owning mobile phones. After the group discussion, write your own report of approximately 500 words about your opinion on this issue.

Investigate

11. Search the internet to find out:
    (a) what multichannelling is
    (b) the benefits of interactive television.
12. Investigate and report on the way that the use of mobile phones has changed the way people live and work. Your report could be based on interviews conducted with relatives or friends who were at least 10 years old when mobile phones were introduced in 1987.

9.10 Long-distance communication

Science as a human endeavour

9.10.1 Our communication network

Australia is covered with a network of microwave and radio repeater towers, coaxial cables, optical fibres and satellite dishes. This network allows us to transmit television and radio signals, telephone calls, facsimiles and computer data across our massive continent.

The repeater stations are towers with dish-shaped antennas.

9.10.2 Wireless technology

Television and radio signals, computer data and telephone messages can be transmitted over long distances using **microwaves**. Microwaves can carry many signals at the same time. However, **repeater stations** need to be used so that the signal does not fade away before it reaches its destination. Antennas on the repeater stations receive the microwave signals and send them on to the next station.

Each repeater tower needs to be within sight of the next one because microwaves, like visible light, travel in straight lines. So, the repeater towers are built on the top of hills wherever possible.

9.10.3 The electric way

Coaxial cables allow sound, pictures and data to be transmitted as pulses of electric current rather than as electromagnetic waves. The signals are carried along conducting wires inside tubes. The thin wire is held in the centre of the tube by a plastic insulating disc. Most Australian coaxial cables contain four, six or twelve tubes. Smaller conductor wires in the cable are used to provide links to small towns along the length of the cable. They are also used to control the system. Coaxial cables are buried under the ground or laid on the ocean floor.

The first major coaxial cable in Australia was laid between Sydney, Melbourne and Canberra in 1962. Coaxial cables can simultaneously transmit many more telephone calls and television signals than earlier cables were able to. As with the microwave system, repeater stations need to be used along the length of the coaxial cable so that the signal does not fade away. Coaxial cable repeater stations need to be even closer together than microwave repeater stations.

9.10.4 The light fantastic

The table below right shows that optical fibres can transmit more messages at one time than coaxial cable or microwaves. Electrical signals from a microphone, television camera, computer or facsimile machine are converted into pulses of light. These pulses are produced when an electrical signal is used to turn the light on and off millions of times per second.

The light pulses received at the other end are converted back into electrical signals that can be fed into speakers, a television set, computer or facsimile machine. The messages can also be retransmitted as microwaves or radio waves if necessary.

A coaxial cable contains many conducting wires in up to 22 tubes. Coaxial cables can carry television signals as well as telephone calls and facsimile messages. They are designed to minimise interference from outside the cable and to prevent the many signals being carried from interfering with each other.

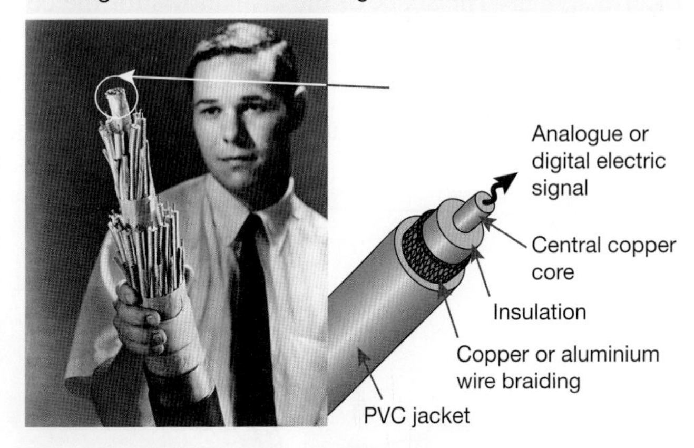

Options for long-distance communication over land

Type of link	Number of two-way conversations at once
Copper cable	600
Coaxial cable	2700
Microwave	2000
Optical fibres	30 000

The idea of using visible light energy to transmit messages over long distances was not feasible until the invention of the laser in 1958. The word 'laser' is an acronym, standing for light amplification by stimulated emission of radiation. A laser produces an intense light beam of one pure colour. As the beam travels through the optical fibre, the glass absorbs some energy. Repeaters are needed every 35 to 55 kilometres along optical fibre cables to boost the signal. The light loses energy less quickly than normal light would, because a laser beam spreads out very little.

The first successful glass optical fibres were made in 1973. The advantages of optical fibres are so great that Australia already has a network of fibre cables between all capital cities. Optical fibres can be laid under the ground or under water. They are smaller, lighter, more flexible and cheaper than the electrical cables previously used for long-distance telephone, radio and television communication. Light pulses cannot be interfered with by radio waves, so there is no 'static'.

HOW ABOUT THAT!

The first microwave communications system in Australia was opened in 1959. It provided a link between Melbourne and Bendigo. This link signalled the beginning of the end for the overhead telegraph wires that hung between wooden poles.

9.10.5 Satellites

Communications **satellites** allow radio waves and microwaves to be transmitted at the speed of light from continent to continent. In Australia, satellites are used to transmit radio, television and telephone signals from cities to remote areas.

Signals are transmitted to a satellite in **geostationary** orbit. The signals are then sent back to other parts of Australia, or to other satellites which, in turn, transmit the signals to other continents. The energy needed to amplify and retransmit the signals is mostly provided by the sun. Solar panels on the satellite collect solar energy, which is either used straight away or stored in batteries for later use.

A geostationary satellite is one that orbits the Earth once every 24 hours, thus remaining over the same point on Earth at all times. In order for the satellite to orbit at that rate, it must be located about 36 000 kilometres above the equator. Tracking stations on Earth use radio signals to activate small rockets on the satellites to keep them in the correct orbit.

Dish antennas, such as the ones in the photograph below left, are aimed at a particular satellite ready to receive signals. The shape of the dish allows for the collection of a large amount of electromagnetic energy, which is focused towards the antenna.

These antennas receive signals that have been retransmitted by a geostationary satellite.

A geostationary satellite relays radio signals to other locations in Australia, or to other continents.

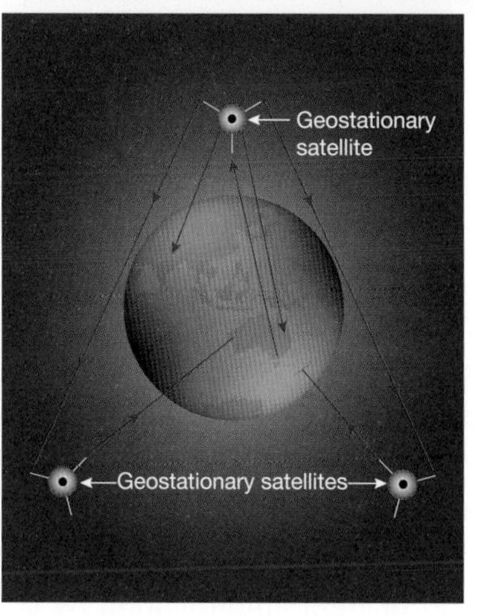

Navigating by satellite

The Global Positioning System (GPS) makes use of up to 32 satellites orbiting the Earth twice each day. GPS satellites orbit the Earth on different paths, all located about 20 000 kilometres above the Earth's surface. A GPS receiver uses radio signals from at least four of these satellites to accurately calculate and map your position. It can also calculate your speed, direction of movement and the distance to your destination.

Since 1993, the GPS system has been available for use by the public free of charge and has many applications. With a GPS receiver you can now find your way around the road system, locate places of interest and even find lost dogs.

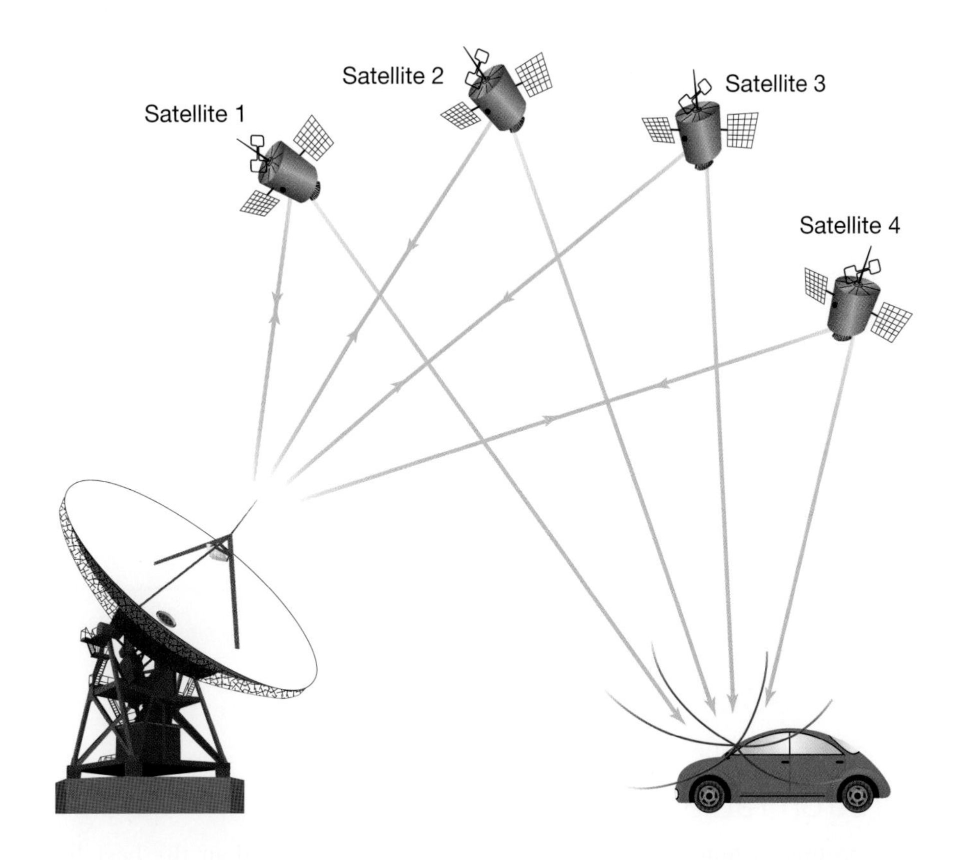

HOW ABOUT THAT!

The GPS system was originally developed for military purposes by the US Department of Defense. The first GPS satellite was launched in 1978 and the general public became aware of its existence when it was used extensively during the Gulf War (1990–1992). It was successfully used by soldiers to locate their own position and that of other soldiers plus tanks and other vehicles.

9.10 Exercises: Understanding and inquiring

To answer questions online and to receive **immediate feedback** and **sample responses** for every question, go to your learnON title at www.jacplus.com.au. *Note:* Question numbers may vary slightly.

Remember

1. List the different types of electromagnetic waves that are used for long-distance communication across Australia.
2. Why are repeater stations necessary for the transmission of microwaves and other radio waves?
3. What is total internal reflection?
4. How can light be used to transmit the electrical signals from microphones, television cameras, computers or facsimile machines?
5. From where do communication satellites get the energy required to amplify and retransmit signals from Earth?

Think

6. Why are repeater stations necessary along coaxial cables?
7. Why are microwaves and other radio waves preferred for communication in the outback rather than optical fibres or coaxial cables?
8. Why are communication satellites placed in geostationary orbit?

9. The term 'global village' has been used to describe the Earth in recent times.
   (a) Why do you think this term has been used?
   (b) How has the development of long-distance communication changed the lifestyles of Australians during the past 40 years?

Investigate

10. Research and report on how outback Australians communicate with their neighbours and the rest of the world.

learnon RESOURCES — ONLINE ONLY

Complete this digital doc: Worksheet 9.7: Communications (doc-18902)

Complete this digital doc: Worksheet 9.8: A step back in time (doc-18903)

9.11 Super 'scope

Science as a human endeavour

9.11.1 Synchrotron radiation

Imagine a microscope that is tens of millions of times more powerful than the best light microscopes. It already exists. It is called a **synchrotron** and it is much, much larger than any light microscope.

There are now more than 50 of them throughout the world, including one in Australia. The Australian Synchrotron in the Melbourne suburb of Clayton is about 70 metres in diameter. The building that houses it is not much smaller than the Melbourne Cricket Ground.The Australian Synchrotron is in the Melbourne suburb of Clayton.

The energy directed at the target in a synchrotron is, like visible light, electromagnetic radiation. However, it is very different from the light used in a conventional microscope. Synchrotron radiation:

- can range from the low energy, long wavelength infra-red part of the electromagnetic spectrum up to high energy, short wavelength X-rays. The radiation can be 'tuned' to the energy and wavelength most suited for the purpose for which it is being used.
- is hundreds of thousands of times as intense as the radiation produced by conventional X-ray tubes. Intensity is a measure of the amount of power delivered to the target.
- is usually emitted in short pulses that last less than a billionth of a second
- is highly polarised.

Together, these characteristics allow a synchrotron to produce data describing objects as small as a single molecule.

Unlike a conventional microscope, you can't actually see an image. The image has to be created from the data obtained when the radiation strikes the target with the aid of computers.

How it works

1. Electrons are fired from a heated tungsten filament with the aid of a voltage of 90 000 volts. They reach a speed of about 159 million metres per second, 53 per cent of the speed of light.
2. The **linear accelerator** (linac) uses an even higher voltage (100 million volts) to accelerate the electrons to a speed of 99.9987 per cent of the speed of light.
3. The booster ring is used to increase the energy of the electrons before they are transferred into the storage ring.

4. In the storage ring, large magnets are used to steer the electrons. As the electrons change direction, they emit electromagnetic radiation for many hours. Magnetic fields are used to replace the energy lost by the electrons during each ‘lap’ of the ring.
5. The synchrotron radiation is directed into a **beamline**. The beam passes through a silicon **monochromator**, which allows only the required wavelengths to pass through.
6. The experimental station contains the target object, which is rotated so that a complete, clear image is obtained.

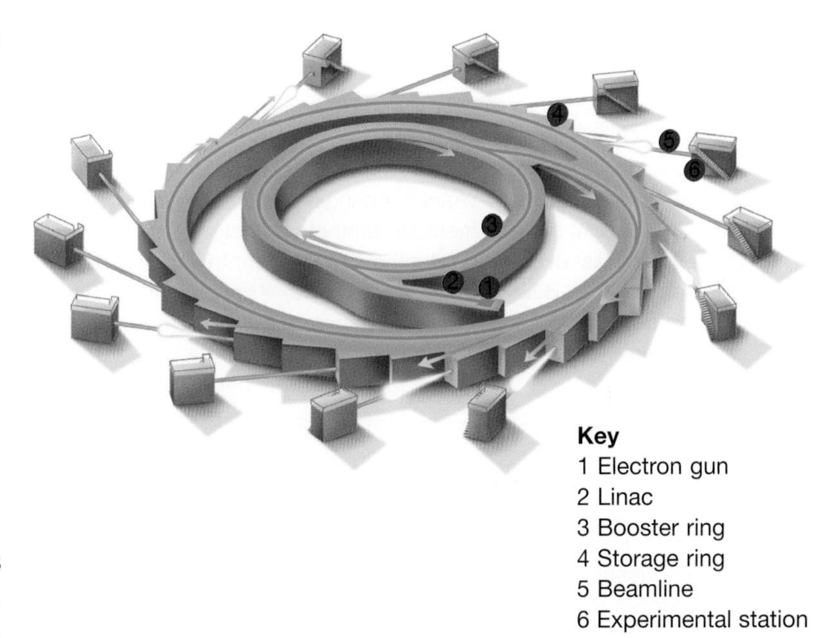

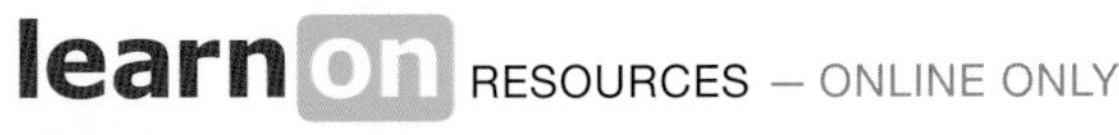

RESOURCES — ONLINE ONLY

Watch this eLesson: Australian Synchrotron (eles-1088)

9.11.2 The benefits of a super ’scope

Synchrotron radiation has a wide range of applications in many areas of science and technology, including medicine, nutrition, environmental science, mining, materials, transportation, forensic science and archaeology. Some examples of the use of synchrotrons are:

- imaging proteins in the influenza virus to help develop a drug to stop it from multiplying
- the development of an artificial substance to coat the lungs of premature babies so that they can breathe more easily
- producing X-ray images of human tissue that have much more detail than conventional X-ray images. These images are being used in the fight against heart disease, breast cancer, brain tumours and many other diseases.
- the detection of weaknesses and cracks in materials used in aircraft and spacecraft
- the analysis of drill core samples in mineral exploration
- assisting in criminal investigations by more accurately identifying substances such as biological fluids, poisons, fibres and paint pigments
- identifying substances in archaeological finds such as ancient armour
- imaging molecules in chocolate to help in the production of smoother, creamier chocolate with a longer shelf life.

Images from the Sychrotron were used to develop the drug Relenza™, which is effective in the fight against all known strains of influenza, including bird flu.

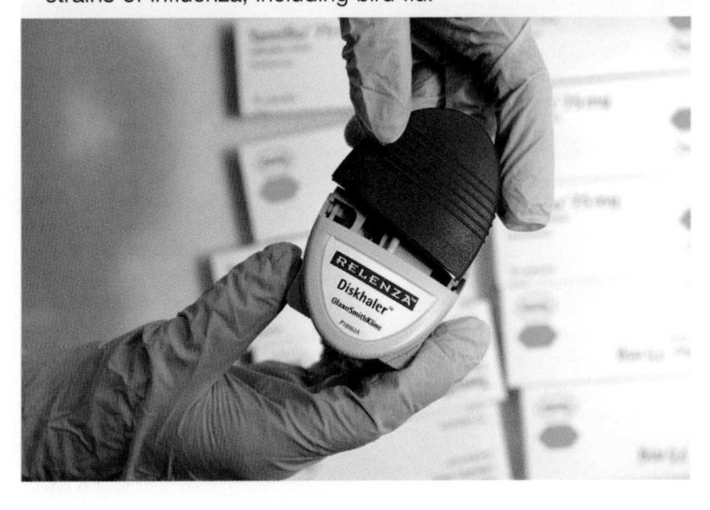

HOW ABOUT THAT!

A synchrotron was used to help solve the mystery of the death of the German composer Ludwig van Beethoven (1770–1827). During his lifetime, Beethoven suffered from loss of hearing, cramps, fevers, chronic abdominal pain, irritability and depression. Following his death, a 15-year-old fellow musician cut off a small lock of Beethoven's hair. After many years of changing hands, the lock of hair was found in a London auction house, where it was bought by the American Beethoven Society, who allowed it to be examined at a synchrotron in Chicago. It was discovered that Beethoven's hair contained about 100 times more of the heavy metal lead than today's normal level. Beethoven's symptoms were consistent with severe lead poisoning. No-one knows where the lead came from, but it could have been from the lead that was used in serving dishes, flasks and crystal glasses during Beethoven's lifetime.

9.11 Exercises: Understanding and inquiring

To answer questions online and to receive **immediate feedback** and **sample responses** for every question, go to your learnON title at www.jacplus.com.au. *Note:* Question numbers may vary slightly.

Remember

1. Which parts of the electromagnetic spectrum can synchrotron radiation consist of?
2. Apart from having a wider range of wavelengths, how is synchrotron radiation different from visible light?
3. From where do the electrons used to produce synchrotron radiation come?
4. In the Australian Synchrotron, what is the main purpose of the:
   (a) linear accelerator (linac)
   (b) booster ring
   (c) silicon monochromator?
5. What happens to electrons in the storage ring of the Australian Synchrotron?

Think

6. What makes the synchrotron such a useful tool in the fight against diseases such as influenza?

Calculate

7. According to the US National Institute of Standards and Technology, the speed of light in a vacuum is 299 792 458 metres per second. In a synchrotron, electrons reach a speed of 99.9987 per cent of the speed of light in the booster ring.
   (a) Calculate this speed in metres per second.
   (b) How much slower, in metres per second, than the speed of light are the electrons travelling?

Investigate

8. Use the internet to search for case studies on the use of synchrotrons. Make a list of examples of the use of the synchrotron other than those mentioned on this page.

9.12 Heat: Energy in transit

9.12.1 Heat

Heat is energy in transit from one region to another region with a lower temperature. There are three different ways in which heat can move from one place to another — by **conduction**, **convection** or **radiation**.

Conduction and convection are best explained using the particle model of matter. Radiation, however, is the transfer of electromagnetic energy and is well explained using a wave model.

Metals are excellent conductors of heat. The plastic or wood used to make the handles of utensils are insulators and prevent conduction of heat to the user's hands. Metal handles can be very hot!

9.12.2 Conduction

Conduction is the transfer of heat through a substance as a result of neighbouring vibrating particles. The particles in the region of high temperature are vibrating more quickly than those in the region of lower temperature and therefore have more **kinetic energy**. They collide with less energetic particles, giving up some of their kinetic energy. This transfer of kinetic energy from particle to particle continues until both regions have the same temperature.

Most solids are better conductors than liquids and gases because their particles are more tightly bound and closer together than those of liquids and gases. Metals are the best conductors of heat. The electrons of metals are more free to move than those of other solids and are therefore able to transfer their kinetic energy more readily to neighbouring electrons and atoms.

Materials that are poor conductors are called **insulators**. Materials such as polystyrene, foam, wool and fibreglass batts are effective insulators because they contain pockets of still air. Air is a very poor conductor of heat.

9.12.3 Convection

Unlike the particles that make up solids, those of liquids and gases are able to move around. In liquids and gases, heat can be transferred from one region to another by the actual movement of particles. This type of heat transfer is called convection. In solids, the particles vibrate about a fixed position and convection does not occur.

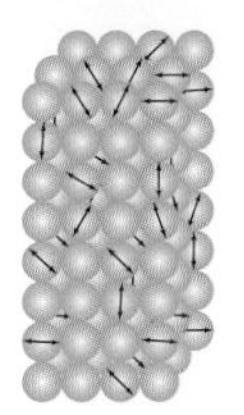

The particles in a solid are packed closely together. If some particles receive heat energy and begin to move faster, they collide easily with other particles nearby and pass the heat energy along.

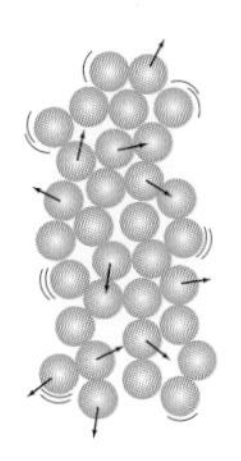

The particles in liquids are further apart than the particles in solids. When some particles receive heat energy and start to move faster, they collide with other particles. But the distance between the particles means that there are fewer collisions. So, heat is transferred by conduction more slowly in a liquid than in a solid.

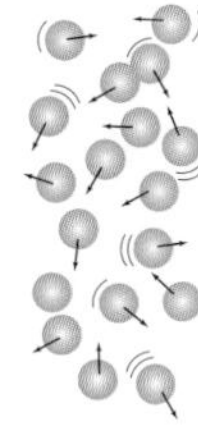

The particles in a gas are far apart. Heat does not travel easily by conduction through gases.

Modelling heat transfer in air

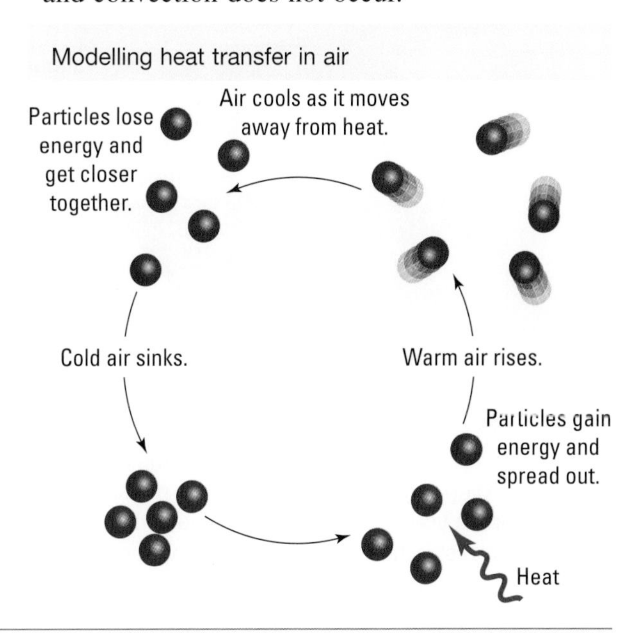

The diagram on the right shows how convection takes place in air. When air is heated, the particles near the source of heat gain energy, move faster and spread out. The particles of the cooler air nearby are moving more slowly and are less spread out. This cooler air is denser and sinks, forcing the warmer, less dense air upwards. As the warmer air rises, it cools and becomes denser and eventually sinks, replacing the newly heated air. The movement of particles during convection forms a **convection current**.

Home heating systems create convection currents that move warm air around. When ducted heating vents are in the floor, warm air rises and circulates around the room until it cools and sinks, being replaced with more warm air. Powerful fans are not necessary. Gas wall heaters have fans to push warm air across the room near floor level so that it heats the entire room. Ducted heating vents in the ceiling require powerful fans to push the warm air downwards so that it can circulate more efficiently.

Convection currents circulate warm air pushed out by heaters around the room.

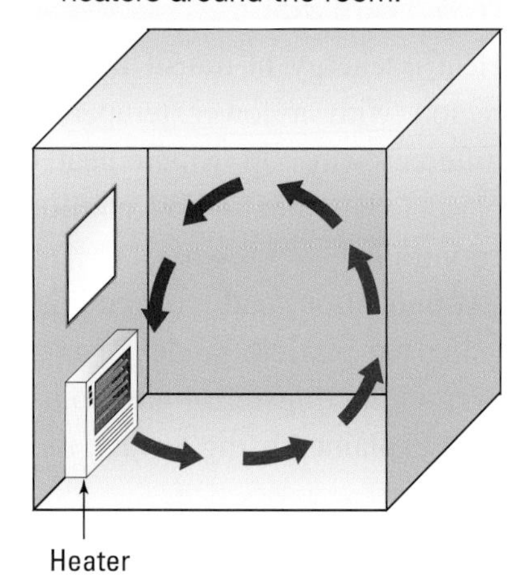

INVESTIGATION 9.13

Heat conduction in solids

AIM: To compare the conduction of heat through different metals

Materials:

set of three or four metal rods, different metals but identical in size; or heat conduction apparatus

wax candle	*Bunsen burner and heatproof mat*
matches	*tripod*
ruler	*stopwatch*

Method and results

- Set up the tripod and rods or heat conduction apparatus as shown below.
- Light the candle and melt a blob of wax onto one end of each rod. Ensure that each wax blob is the same distance from the end that will be heated by the Bunsen burner flame.
- Use the blue flame of the Bunsen burner to heat the end of each rod. Start the stopwatch at the instant that heating begins.

1. Record the time taken for each blob to produce its first droplet of wax.
2. Present your data in a table and draw a bar or column graph in which to display them.

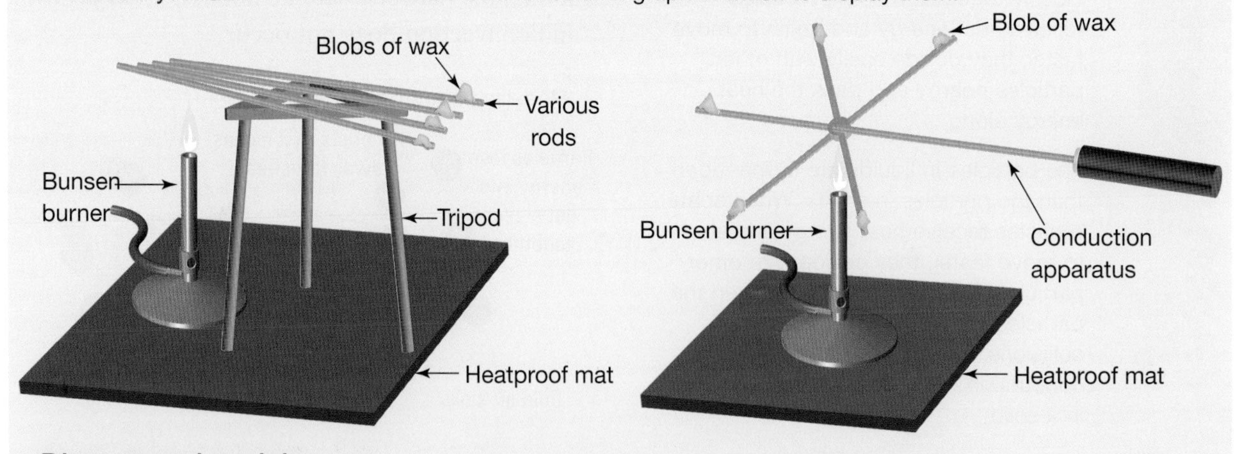

Discuss and explain

3. According to your data, which of the metals is the best conductor of heat?
4. According to your data, which of the metals is the poorest conductor of heat?
5. Compare your data with that of others in your class. Comment of the consistency of the conclusions within your class. If there was inconsistency, suggest one or more reasons for it.

9.12.4 Radiation

Heat can be transferred without the presence of any particles at all, as electromagnetic radiation. Heat transferred in this way is called **radiant heat**. Heat from the sun reaches the Earth by radiation, most of it in the form of infra-red radiation. There are not enough particles between the sun and Earth for heat transfer by either conduction or convection.

Like all electromagnetic radiation, radiant heat can be reflected, transmitted or absorbed. How much energy is reflected, transmitted or absorbed depends on the properties, including colour, of the surface.

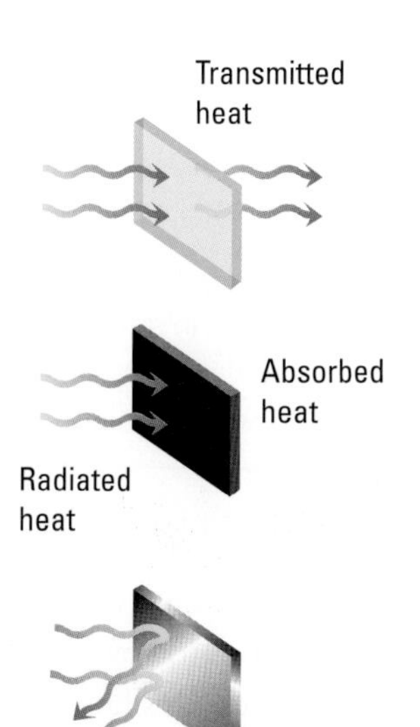

Transmitted radiant heat
Clear objects, like glass, allow light and radiant heat to pass through them. The temperature of these objects does not increase quickly when heat reaches them by radiation.
Absorbed radiant heat
Dark-coloured objects tend to absorb light and radiant heat. Their temperatures increase quickly when heat reaches them by radiation.
Reflected radiant heat
Shiny or light-coloured surfaces tend to reflect light and radiant heat away. The temperature of these objects does not change quickly when heat reaches them by radiation.

INVESTIGATION 9.14

Radiating and absorbing radiant heat

AIM: To compare the radiation and absorption of heat through black surfaces and shiny surfaces

Materials:
heater or microscope lamp
2 identical shiny, empty, soft-drink cans
matt black paint and paintbrush
2 thermometers or data logger and 2 temperature sensors

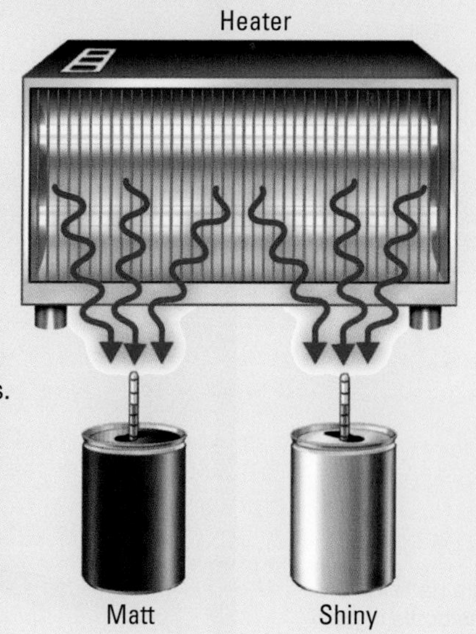

Method and results

- Paint one of the cans matt black and leave the other as it is.

1. Construct a table or spreadsheet headed 'Radiating heat' in which to record the temperature every two minutes for up to 14 minutes.
2. Make a prediction about your results and write down a hypothesis.
   - Pour equal amounts of hot water into each can.
3. Place a thermometer or temperature probe into each can. Measure the initial temperature of the hot water and again every two minutes.
4. Enter your data into the table or spreadsheet.
   - Empty the cans and pour equal amounts of cold tap water into each can.
5. Construct a table or spreadsheet headed 'Absorbing radiant heat' in which to record the temperature every two minutes for up to 14 minutes.
6. Make a prediction about your results and write down a hypothesis.
7. Place a thermometer or temperature probe into each can. Measure the initial temperature of the water.
   - Place the two cans at equal distances from a heater or microscope lamp.

8. Measure and record your data into the table or spreadsheet every two minutes.
9. Plot line graphs that show how the temperature changed over 14 minutes during the cooling and heating of the cans. You may wish to plot both graphs on the same set of axes.

Discuss and explain

10. Which can radiated heat more quickly?
11. Which can absorbed heat more quickly?
12. Were each of your hypotheses supported by the data?
13. Why was it important to use cans that were identical in size and shape?
14. What other variables had to be controlled during this experiment?

9.12.5 Reducing heat transfer: The vacuum flask

The vacuum flask, also known as a thermos flask, invented by British chemist Sir James Dewar, was originally designed for the cold storage of liquefied gas. However, Dewar quickly realised that it could also be used for keeping liquids warm. The vacuum flask reduces heat transfer in or out by conduction, convection and radiation.

A vacuum flask reduces heat transfer by conduction, convection and radiation.

Silvered facing surfaces of glass bottle reduce heat transfer by radiation.

Stopper made of plastic or cork reduces heat transfer by conduction and convection.

Protective case

Partial vacuum between facing surfaces of glass bottle reduces heat transfer by conduction and convection.

Double-walled bottle made of glass reduces heat transfer by conduction.

Foam pads keep glass bottle in place and absorb impacts.

Shock absorber

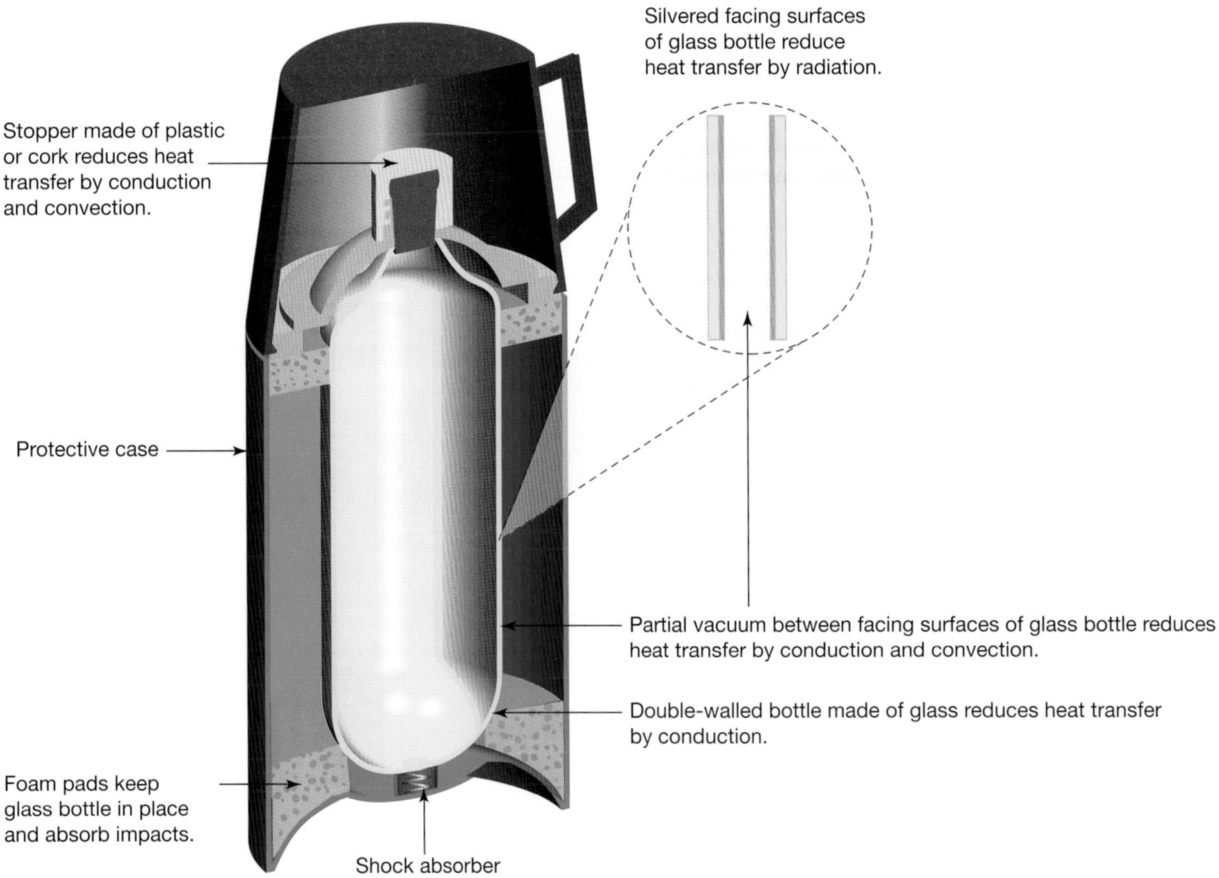

9.12.6 Hot summer days by the sea

During hot summer days, radiant energy from the sun heats the land and the sea. However, as a result of the different properties of the land and water, after a few hours the land has a higher temperature than the sea. The air near the land becomes hot as a result of conduction. As this air gets hot it expands, becoming less dense than the cooler, denser air over the sea. The air over the sea rushes in towards the land, replacing the rising warm air, causing what is known as a **sea breeze**. Coastal areas usually experience less extreme maximum temperatures on hot summer days as a result of sea breezes. At night, if the sea temperature is higher than the land temperature, the convection currents move in the opposite direction, creating a flow of air towards the sea.

A sea breeze is caused by convection currents in the air during warm summer days. At night the convection currents are reversed.

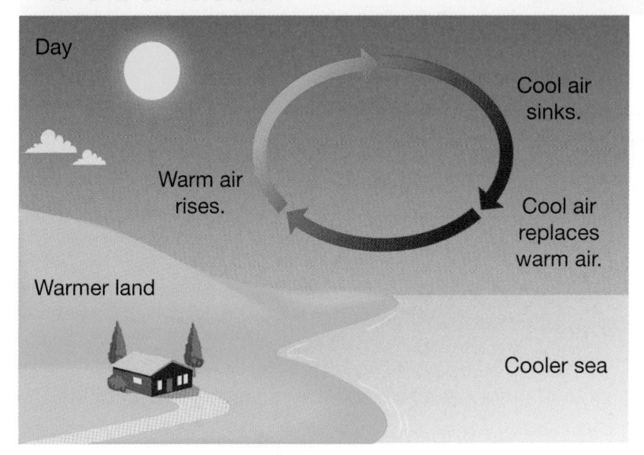

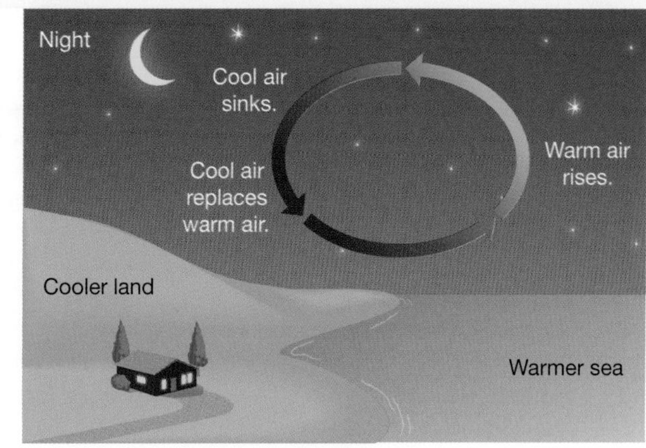

HOW ABOUT THAT!

Imagine putting a scoop of ice-cream into a 230 °C oven for three minutes. That's exactly what you do when you make Bombe Alaska, a dessert with a solid ice-cream centre on a sponge cake, covered with meringue. Bombe Alaska — ice-cream and all — is baked in a preheated 230 °C oven for three minutes. Yet the ice-cream doesn't melt! The secret to this is the insulating properties of the sponge and meringue. The bombe pictured here has strawberry (pink) and orange ice-cream on top of the sponge cake. The white meringue has been cooked and has been changed to a brown colour by the heat of the oven.

Bombe Alaska, also known as Baked Alaska

9.12 Exercises: Understanding and inquiring

To answer questions online and to receive **immediate feedback** and **sample responses** for every question, go to your learnON title at www.jacplus.com.au. *Note:* Question numbers may vary slightly.

Remember

1. Which form of energy do particles transfer to each other as heat flows through a conductor?
2. Explain why solids such as polystyrene foam and wool are poor conductors of heat.
3. Explain why air near a wall furnace rises when it gets warmer.
4. What is a convection current?
5. Which form of electromagnetic radiation from the sun is responsible for most of the radiant heat reaching the Earth?
6. List three things that can happen to radiant heat when it arrives at any surface.

7. Identify the features of a vacuum flask that reduce heat transfer by:
   (a) conduction
   (b) convection
   (c) radiation.
8. Explain, with the aid of a diagram, how a coastal sea breeze results from convection currents.

Think

9. Suggest why metal saucepans usually have plastic or wooden handles.
10. Use the particle model of matter to explain why heat can't travel through metals and other solids by convection.
11. Why is air a poor conductor of heat?
12. When you heat water in a saucepan on a gas or electric hotplate, all of the water gets hot and not just the lower part near the heat source. How does this happen, given that water is a poor conductor of heat?
13. Why do many sportspeople wear white clothing when competing on hot summer days?
14. To which form of heat transfer do the following statements apply?
    (a) Energy is transferred at the speed of light.
    (b) Particles move from one place to another.
    (c) No particles are required for energy transfer.
    (d) This form of heat transfer is best explained by a wave model.
15. Advertisements in the media often describe all products designed to reduce heat transfer to keep homes warm in winter or cool in summer as insulators. Explain why it is incorrect to describe reflective metal foil as insulation.

Create

16. Create a design brief for a spacesuit to be used for walking on Mars. The spacesuit must protect the wearer from the blistering heat of day and the brutally cold nights. In your design brief, list the features of the spacesuit that will reduce heat transfer in and out by conduction, convection and radiation.

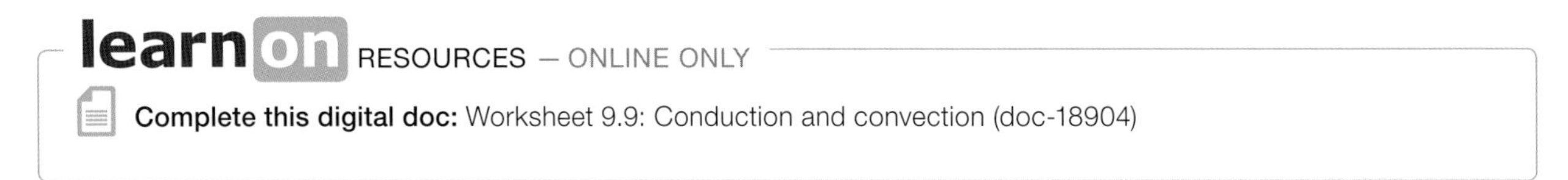

9.13 Staying cool

Science as a human endeavour

9.13.1 Body temperature

A healthy human body has a **core body temperature** of about 37 °C. With the right clothing and shelter, the inner parts of your body remain at this temperature almost anywhere on Earth.

The energy needed to keep your core body temperature at 37 °C is converted from the chemical energy in food. However, when you exercise, your body needs more oxygen and more energy. This extra energy is converted from food more quickly than when you are resting. Much of the converted energy causes an increase in body temperature.

In cool weather, your body is able to cool itself by transferring the extra heat into the surrounding air. However, in warm weather, it is much more difficult for your body to cool down.

9.13.2 Natural cooling

Your body can protect itself from getting too hot by transferring heat from your skin to the cooler air surrounding it. This happens in three different ways.

- As long as the surrounding air is cooler than your skin, most of the heat is transferred by radiation. The bigger the difference between your skin temperature and the air temperature, the more heat is radiated.

When your core body temperature rises, the blood vessels beneath your skin get larger. This brings hot blood closer to your skin so that more heat is radiated away.

- Some heat is transferred from your body by convection. Heat rises from the warm layer of air just above your skin into cooler air. The photo below shows how heat is transferred by convection from an unclothed person in a standing position. By wearing light, loose-fitting clothing when you exercise, your body can still lose heat by convection. This is especially important in warm weather.
- A small amount of heat is lost by conduction from your skin to the air. The amount is small because there is a thin layer of still air above your skin. There is also an insulating layer of fat beneath your skin.

If the surrounding air is not cooler than your skin, no heat can be transferred by radiation, convection or conduction. These methods of heat transfer from your body can work only if the air temperature is lower than your skin temperature.

Insulating your body

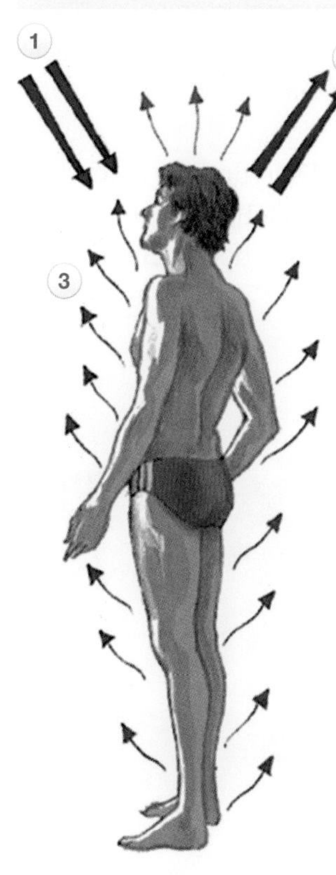

1. **Absorbing radiant heat**
   When sunlight strikes any object, including the human body, the object absorbs some radiant heat.
2. **Losing radiant heat**
   All objects give off some radiant heat. The amount depends on the temperature around the object. On a hot day, an object does not transfer as much radiant heat away as in the cold weather.
3. **Convection**
   Convection currents form when the heat from your body warms the air next to it. The air rises, taking some of your body heat away with it. Convection currents can form only in air that is free to move.

HOW ABOUT THAT!

Body temperature

In cold weather, only the core of your body remains at 37 °C. The core is where all of your most important organs are. The other parts of your body can be quite a lot cooler. In warm weather, the temperature of your body is more even. Most of it remains within one degree of 37 °C.

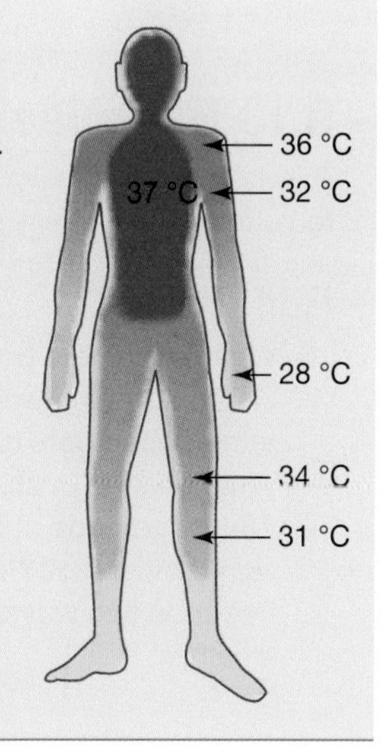

9.13.3 Heat transfer by evaporation

When your core body temperature increases, sweat glands under your skin produce **perspiration**. The water in the perspiration evaporates. The energy needed to change the liquid water into water vapour comes from your skin. In other words, heat is transferred from your skin to water on your skin.

When the air temperature is higher than your skin temperature, evaporation of water is the only way your body can reduce its temperature naturally. If you wear tight-fitting or too much clothing in hot weather or while exercising, you limit the evaporation of water from your skin. Your body is in danger of overheating.

9.13.4 Just too hot!

In hot weather and during strenuous exercise, you need to take care. The human body does not cope well with high body temperatures. If the core body temperature reaches 39 °C, **heat exhaustion** occurs. The body feels cold and sweaty. There is a danger of fainting. When the core body temperature reaches about 41 °C, the body's natural cooling methods fail. **Heatstroke** sets in and the body's vital organs begin to suffer damage. The early signs of heatstroke include irritability, confusion and dizziness. The skin usually becomes hot and dry because perspiration stops.

One danger during hot weather, especially while exercising, is **dehydration**. Dehydration is the loss of water from the body. When evaporation is the only way for the body to cool itself, a large amount of water is lost. If the lost water is not replaced, perspiration stops and the core body temperature increases.

HOW ABOUT THAT!

Dogs have sweat glands only in their feet. These glands don't do much to help dogs reduce their body temperature. A dog controls its body temperature by sticking out its tongue and panting. Air passing over its tongue and mouth evaporates its saliva. When a dog pants, some water evaporates from its throat and lungs. The energy transferred from the dog to cause all this evaporation keeps it cool.

9.13 Exercises: Understanding and inquiring

To answer questions online and to receive **immediate feedback** and **sample responses** for every question, go to your learnON title at www.jacplus.com.au. *Note:* Question numbers may vary slightly.

Remember

1. What is the core body temperature of a healthy human body?
2. List four natural ways in which the body can transfer heat away from the skin when the core body temperature gets too high.
3. What change of state takes place on your skin as you perspire?
4. Which methods of heat transfer from your body cannot take place if the air temperature is greater than your skin temperature? Give a reason for your answer.
5. Describe two ways in which tight-fitting clothing prevents heat loss from your body while exercising in hot weather.

6. What causes heat exhaustion?
7. What are the early signs of heatstroke?
8. Hot weather is one of many causes of dehydration.
   (a) What is dehydration?
   (b) How is water lost during hot weather and while exercising vigorously?

Think

9. Why does the skin become red during vigorous exercise?
10. List at least two ways of reducing the chances of becoming dehydrated during hot weather.
11. Antiperspirants are used to reduce perspiration and body odour. Construct a PMI chart about antiperspirants.

Think and create

12. Imagine that you are the coach of a long-distance runner competing in a 21-kilometre race on a warm day. Make a list of 'dos and don'ts' for the runner so that dehydration, heat exhaustion and heatstroke are avoided. Present your list as a poster or PowerPoint presentation.

9.14 Plus, minus, interesting charts and target maps

9.14.1 Plus, minus, interesting charts and target maps

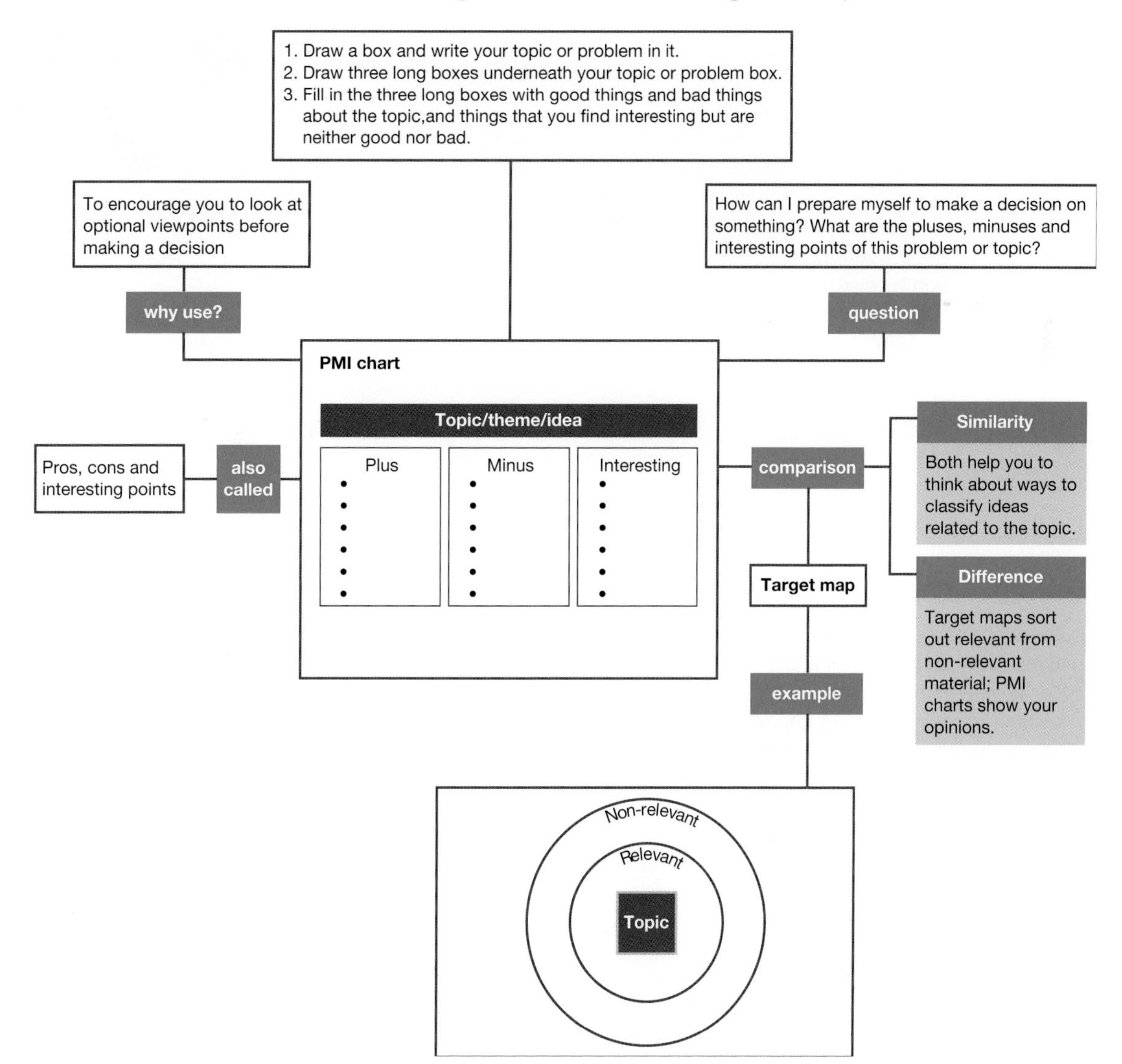

9.14 Exercises: Understanding and inquiring

To answer questions online and to receive **immediate feedback** and **sample responses** for every question, go to your learnON title at www.jacplus.com.au.
Note: Question numbers may vary slightly.

Think and create

1. With a partner or in a small group, discuss the positive, negative and interesting aspects of each of the following forms of energy transmission. Work together to create a PMI chart for each type of energy transmission to summarise your discussions.
   (a) Sound
   (b) Light
   (c) Microwaves
   (d) X-rays
2. Create a PMI chart to illustrate the positive, negative and interesting aspects of each of the following methods of communication.
   (a) 'Snail' mail
   (b) Mobile phones
   (c) Landlines
   (d) Internet
3. Create a target map on each of the following topics using the words in the box below right and at least three additional words that are relevant.
   (a) Sound
   (b) Radio waves
   (c) Long-distance communication
   (d) Medical diagnosis or treatment
4. Work with a partner or in a small group to create a large target map on the topic of digital communication using both words and images. Start by brainstorming a collection of words and pictures related to both digital and analogue communication.
5. Create a timeline to illustrate the development of communication by electromagnetic radiation, beginning with the discovery of radio waves by Heinrich Hertz in 1887.

'Snail mail' may be slow, but it has some advantages over the faster methods of sending and receiving information such as phone, email and chatting over the internet. Creating PMI charts can help you reflect on the different ways of communicating.

light longitudinal waves reflection
pitch cochlea ultraviolet infra-red
retina sound waves Lens compression
300 000 km/s rarefaction X-rays
transverse waves particles
ultrasound frequency wavelength
microwaves
gamma rays AM radio FM radio

HOW ABOUT THAT!

- The African elephant's ears enable it to hear low-pitched sounds from other elephants over four kilometres away. They also use their giant ears to release heat, sometimes flapping them to cool down more quickly.
- Some insects have ears but they are not on their heads. The ears are membranes like eardrums on the surface of their bodies. A cricket has an ear just below the knee of each of its front legs. A grasshopper has an ear on each side of its body just below the wing. Most insects, however, do not have ears but detect vibrations with sensitive hairs on their antennae or other parts of their bodies.

HOW ABOUT THAT!

The human ear is capable of detecting frequencies between about 20 Hz and about 20 000 Hz. Dogs have a much greater range of hearing and can detect frequencies between 15 Hz and 50 000 Hz. Cats can hear even higher frequencies — up to 60 000 Hz.

learn on

9.15 Project: Did you hear that?

Scenario

Since the invention of the Walkman — a portable cassette tape player — in 1979, through to the modern iPod, we have loved to carry our favourite music around with us everywhere we go. Wherever you look, you'll see people walking the dog, riding the bus, going for a run, hitting the books or just sitting around hooked in to an audio device of some form. With more than 220 million iPods alone sold since their release in 2001 and the increasing affordability of personal music players in general, more and more people are spending time plugged in. But for every person who loves their mp3 player, there's another who'll be warning them that channelling all that sound directly into their ears will have long-term effects on their hearing.

Your fifty-year-old principal wonders whether there aren't short-term effects as well, because she finds it difficult to hear her mobile ringing for about ten minutes after she has stopped listening to music on her iPod. She comes to your science class (known for their cleverness) for some possible answers. One clever classmate suggests that maybe the type of music she was listening to had lots of high frequency sounds in it and that this had somehow affected her ear's ability to pick up the high frequencies of her mobile ring tone. Another clever classmate thinks that maybe she had the volume up too high on her iPod and that this might have caused some temporary deafness. A cheeky classmate suggests that maybe she can't hear it because she's old! Somewhat grumpy with that last comment, your principal decides that maybe she should just ban all personal music players in the school unless you can provide her with some thorough scientific evidence that something other than age can have short-term effects on hearing range after iPod use.

Your task

Using personal music players and online hearing tests, your group will perform a series of scientific investigations to explore the short-term effects of personal music players (such as iPods) on hearing range. You will then present your findings in the form of a scientific report suitable for sending to the principal.

Suggested factors to consider include:

- volume used
- hearing range differences between males and females
- type of music (for example, classical, jazz, R&B or pop).

Note that you will need to minimise any risk of permanently causing damage to the hearing of your human subjects by ensuring that the volume does not exceed 90 dB and limiting trial durations to a few minutes.

9.16 Review

9.16.1 Study checklist

Energy transmission in waves

- describe examples of the transmission of energy
- compare the transmission of waves through different media

Sound energy

- distinguish between transverse waves and compression waves
- describe the transmission of sound energy
- relate the pitch and loudness of sound to the properties of sound waves
- explain how sound is transmitted through the human ear and detected by the brain

Electromagnetic energy

- describe the properties and uses of the waves that make up the electromagnetic spectrum
- describe the transmission, reflection and refraction of light
- describe photons and compare the wave and particle models of light

Energy for communication

- describe the transmission of radio waves in terms of carrier waves and audio and visual signals
- describe the advantages of digital over analogue signals
- explain the basic operation of the digital mobile phone network
- compare methods of long-distance communication

Heat transfer

- describe, compare and explain the transfer of heat by conduction, convection and radiation
- contrast the properties of conductors and insulators
- investigate the radiation and absorption of radiant heat
- explain sea breezes in terms of heat transfer

Science as a human endeavour

- describe examples of the use of ultrasound by engineers and in medicine and industry
- explain the operation and impact of the cochlear implant
- describe the use of endoscopes and X-rays in medical diagnosis
- outline the impact on daily life of communication technology, including mobile phones and the GPS
- examine the application of radiation in the Australian Synchrotron to science and technology
- explain how the unwanted transfer of heat can be decreased to reduce energy consumption by householders
- explain climate patterns in terms of heat transfer

Individual pathways

ACTIVITY 9.1	ACTIVITY 9.2	ACTIVITY 9.3
Revising energy transmission doc-19657	Investigating energy transmission doc-19658	Investigating energy transmission further doc-19659

learnon ONLINE ONLY

9.16 Review 1: Looking back

To answer questions online and to receive **immediate feedback** and **sample responses** for every question, go to your learnON title at www.jacplus.com.au. *Note:* Question numbers may vary slightly.

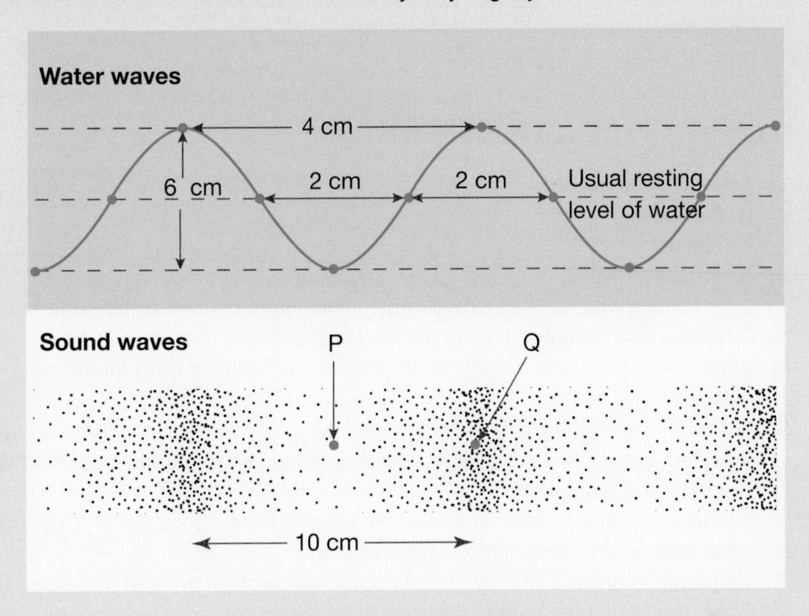

1. Explain the difference between a transverse wave and a compression wave. List two examples of each type.
2. Refer to the water wave and sound wave shown in the figure on the right to answer the following questions.
   (a) What is the amplitude of the water wave?
   (b) What is the wavelength of the water wave?
   (c) What is the wavelength of the sound wave?
   (d) Which of the points P and Q is in the centre of a rarefaction?
3. How are ultrasound waves different from the sound waves that you can hear?
4. List some of the uses of ultrasound.
5. Replace each of the following descriptions with a single word.
   (a) Regions of air in which the particles that sounds bring closer together than usual
   (b) Regions of air in which the particles that sounds are moved further apart than usual
   (c) The effect of sound reflected from a hard surface over and over again
   (d) What you see when you look in a mirror — even when you are not directly in front of it
6. When an object vibrates faster, what happens to the pitch of the sound it produces?
7. Explain why a hearing aid is of no use to some hearing-impaired people.
8. (a) Copy and complete the table below.

Electromagnetic wave type	Wavelengths (m)	Properties
Infra-red radiation		
Gamma rays		
Ultraviolet radiation		
Light		
X-rays		
Radio		

   (b) State three differences between sound waves and all of the waves listed in the table.
   (c) What two properties do all of the waves listed in the table have in common?
   (d) To which type of electromagnetic waves listed in the table do microwaves belong?
   (e) Which of the electromagnetic waves listed in the table:
      (i) can be produced artificially
      (ii) transmits the most energy?
9. Which aspect of sound and light can easily be modelled with both particles and waves?
10. Which type of electromagnetic radiation is used in remote control devices?
11. What is the major use to society of:
    (a) X-rays
    (b) ultraviolet radiation
    (c) gamma rays?
12. If there were no visible light coming from the sun, it is obvious that we wouldn't be able to see. But the lack of visible light would cause a much greater problem. What is that problem?

13. Explain the difference in the meaning of each of the following pairs of words.
    (a) Ray and beam
    (b) Reflection and scattering
14. When a light ray passes from air to glass and back into air again, how does its speed change when it:
    (a) enters the glass
    (b) emerges back into the air?
15. Use a diagram to explain why your legs appear to be shorter when you stand in clear, shallow water.
16. Describe the role of each of the following parts of the eye.
    (a) Cornea
    (b) Iris
    (c) Lens
    (d) Retina
    (e) Ciliary muscles
17. Use labelled diagrams to explain how visible light is used to transmit sound along optical fibres.
18. Describe how digital radio signals are different from analogue radio signals.
19. What does the digital transmission of television signals have in common with the analogue transmission of television signals?
20. Although electromagnetic radiation has many uses in society, there are also dangers associated with it.
    (a) What danger does ultraviolet radiation pose to the human body and what measures should be taken to protect against it?
    (b) Find out what precautions must be taken by the operators of X-ray equipment in hospitals.
21. What is synchrotron radiation and how is it 'created'?
22. List some examples of how images obtained through the use of a synchrotron can benefit medical science.
23. Briefy describe how heat is transferred from one region to another by:
    (a) conduction
    (b) convection
    (c) radiation
24. Identify the main method or methods by which heat is transferred to the human body by:
    (a) a gas wall furnace
    (b) the sun
    (c) holding a hot plate
    (d) an open fireplace
    (e) walking on hot coals.
25. Explain why cooks often cover meat with aluminium foil instead of plastic.
26. Explain why solids such as polystyrene, foam, wool and breglass batts do not conduct heat as well as most other solids.
27. Heat is always transferred from a region of high temperature to a region with a lower temperature. Explain how your body is able to keep its core temperature at 37 °C even when the air temperature is greater than 37 °C.
28. Explain how the wearing of light, loose-fitting clothes protects your body from overheating in hot weather.
29. Why do your blood vessels get larger in hot weather?
30. Explain how fibreglass batts are able to reduce the loss of heat through the ceiling by both conduction and convection.
31. Why do coastal areas experience less extreme high and low temperatures than inland regions?
32. What are the two main causes of ocean currents?
33. Explain how mountain ranges are often almost permanently surrounded by clouds.

TOPIC 10
Electricity at work

10.1 Overview

We use electric circuits every day. They can be as simple those in a plastic torch or as complex as those on the circuit boards like the one at right in your computers, mobile phones, TVs and numerous other electronic devices. All of these devices are powered by batteries or the electricity from a power point.

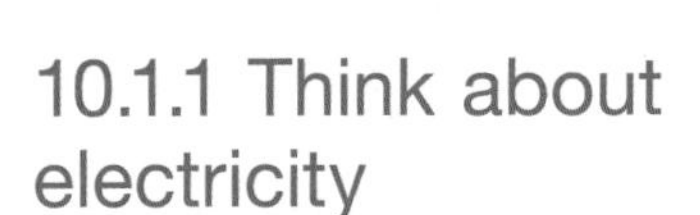

10.1.1 Think about electricity

- What actually moves when an electric current flows?
- What is the difference between current and voltage?
- Why don't all the lights in your house go out when one light globe or fluorescent tube breaks?
- If a car battery is rechargeable, why doesn't it last forever?
- How many electronic components can be fitted onto a silicon chip?
- What is an LED and how is it different from a torch or light globe with a filament?
- What's the connection between electricity and magnetism?
- How is AC different from DC?
- Why do you need a safety switch as well as fuses or circuit breakers in your meter box?
- Why is it dangerous to use electrical appliances in the wet or near water?

LEARNING SEQUENCE

Numerous **videos** and **interactivities** are embedded just where you need them, at the point of learning, in your learnON title at www.jacplus.com.au. They will help you to learn the content and concepts covered in this topic.

10.1.2 Electricity — it's everywhere!

Think

Work in a small group to compile answers to each of the following questions.

1. Make a list of all of the electrical devices used in and around the home.
2. Which devices need to be connected to a power point while they are in normal operation?
3. Which of the devices that need to be connected to a power point have
   (i) plugs with 2 pins?
   (ii) plugs with 3 pins?
4. Suggest why some plug-in devices have three pins while others have two.
5. Which devices are powered by rechargeable batteries?
6. Which devices are powered by non-rechargeable batteries?

10.1.3 What's the connection?

You probably already know quite a lot about electricity and magnetism. Use what you already know to answer the following questions.

7. What is a magnetic field and how do you know when it is present?
8. What is an electric field and how can you demonstrate its existence?
9. Describe at least one example of evidence that there is a connection between electricity and magnetism.

INVESTIGATION 10.1

Making the right connections

AIM: To connect a battery to a light globe so that it lights up

Materials:

a 2.5-volt torch light globe
a 1.5-volt battery
two connecting leads

Method and results

- Connect one or two connecting leads, a 2.5-volt light globe and a 1.5-volt battery to make the globe light up.
- Try different arrangements to see whether there is more than one way to make the globe light up.

1. In which of the following electric circuits are the components correctly arranged so that the light globe will work?
2. Describe, with the aid of a diagram, any other arrangements that cause the globe to light up.

A B C D

Discuss and explain

3. Draw a flowchart to show the energy transformations that take place when the globe lights up.
4. Are all the energy transformations that take place useful? Explain your answer.

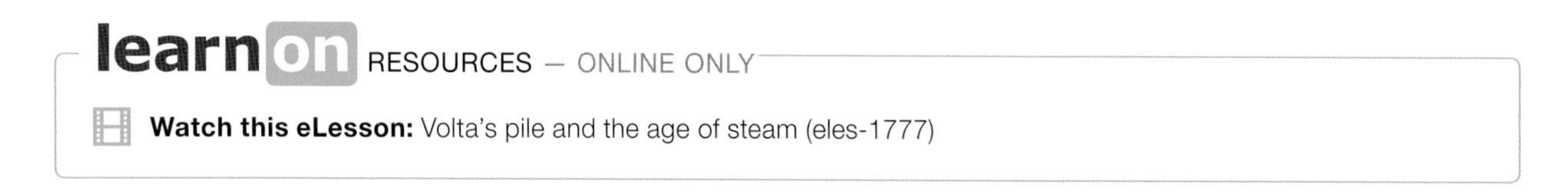

10.2 Electricity in transit

10.2.1 Electric charge in motion

A bolt of lightning is clear evidence that electrical energy can move from one place to another. But to make energy useful it needs to be controlled. When you switch on a television, a computer or an mp3 player you are completing a pathway along which electrical energy flows. The pathway is called an **electric circuit**, in which electrical energy is transformed and used for lighting, heating, cooling and much, much more.

The movement of electrical energy from one place to another, whether rapidly in the form of lightning or more slowly in an electric circuit, depends on the movement of **electric charge**.

The cords to earbud headphones carry electricity, not sound. Each cord actually contains two wires. The jack that plugs into the player is composed of four wires, two per earbud.

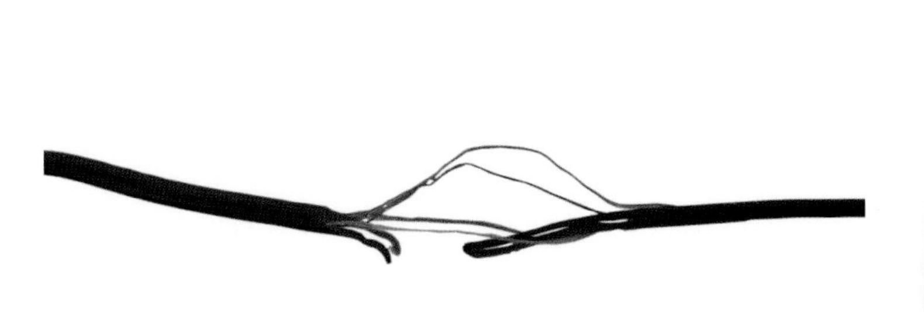

WHAT IT MEANS TO BE CHARGED

All matter is made up of atoms. At the centre of each atom is a heavy **nucleus**. Surrounding the nucleus is a lot of empty space and tiny particles called **electrons**. Electrons are constantly moving around the nucleus. Each electron carries a **negative electric charge**.

Inside the nucleus are two different types of particles. The **protons** inside the nucleus are much heavier than electrons. Each proton carries a **positive electric charge**. The **neutrons** inside the nucleus are similar to protons but carry no electric charge. The positive electric charge of a proton exactly balances the negative charge of an electron. Atoms usually contain an equal number of electrons and protons.

Any particle or substance that has more protons than electrons is said to be **positively charged**. Any particle or substance that has more electrons than protons is said to be **negatively charged**. Any particle or substance that has equal amounts of positive and negative charge is said to be **neutral**. The term 'uncharged' is also used to describe neutral particles or substances.

Substances usually become charged by the addition or removal of electrons.

A neutral atom contains an equal number of protons and electrons. (Two of the protons are hidden in this diagram.) This diagram represents a carbon atom. The number of neutrons is not always the same as the number of protons.

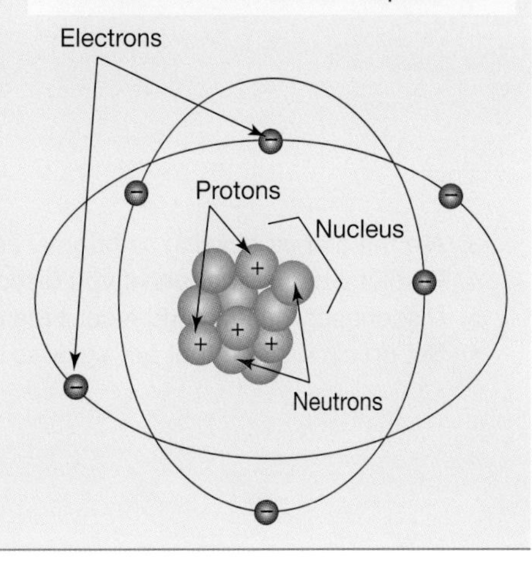

10.2.2 Static electricity

The electric charge that builds up in clouds and objects such as plastic rulers and balloons is called **static electricity**; that is, it is not moving. When enough static electrical energy builds up it can move very quickly, as it does when lightning strikes or when you get a small electric shock when you step out of a car or touch a metal door handle. Static charge can leak away from substances such as air and rubber. These substances are called electrical insulators because they don't allow electrical charge to pass through them quickly.

INVESTIGATION 10.2

Making changes

AIM: To observe and compare the flow of electrical energy through some simple circuits

Materials:

power supply (set to 6 volts)
three 6-volt light globes and holders
6 connecting leads with alligator clips or banana plugs
piece of plastic
iron nail

Method and results

- In this experiment you will be setting up each of the circuits shown in the diagrams below. Draw each circuit diagram in your workbook and record your observations of the globes next to them.

1. Set up circuit 1. Record the change that takes place in the globe.

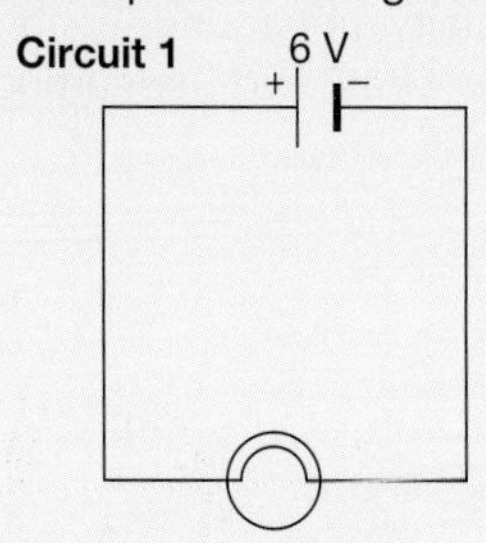

2. Set up circuit 2. Record the change that takes place in the globe when the plastic is replaced with an iron nail.

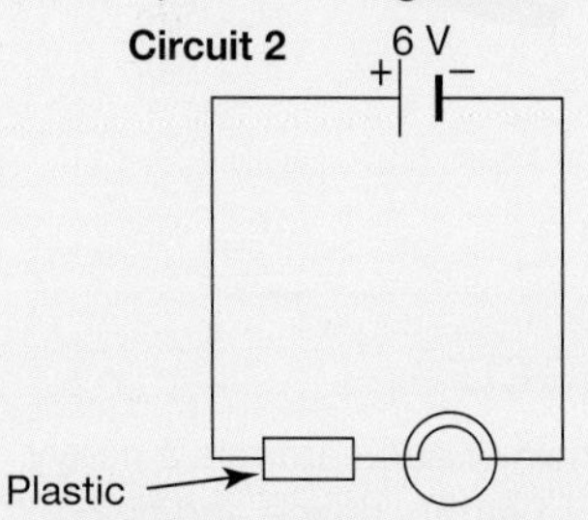

- Set up circuit 3 so that both globes light up. The two globes are connected in series.

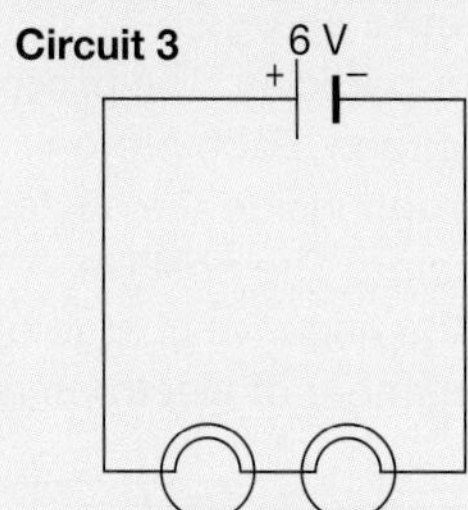

3. Are the globes glowing as brightly as the globe in circuit 1?
4. Predict what will happen if you disconnect the lead between the two globes in circuit 3.
5. Disconnect the lead and record the changes that take place in each of the globes.

- Set up circuit 4 so that both globes light up. The two globes are connected in parallel.

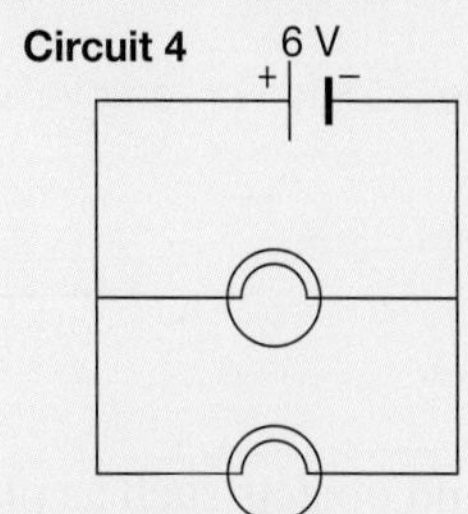

6. Are the globes glowing as brightly as the globes in circuit 1?
7. Predict what will happen if you disconnect one of the leads between the two globes in circuit 4.
8. Disconnect the lead and record the changes that take place in each of the globes.

Discuss and explain

Circuit 1

9. Explain why the change in the globe took place.

Circuit 2

10. Explain the change that took place when the piece of plastic was replaced with the iron nail.

Circuit 3

11. Explain why the globes behaved differently from the way the globe behaved in circuit 1.
12. Was your prediction about what would happen when you disconnected the lead correct?
13. Explain the change that took place when you disconnected the lead.

Circuit 4

14. Was your prediction about what would happen when you disconnected the lead correct?
15. Explain the change that took place when you disconnected the lead.

10.2.3 Electric circuits

All electric circuits consist of three essential items:

- a **power supply** to provide the electrical energy
- a **load** (or loads) in which electrical energy is converted into other useful forms of energy
- a **conducting path** that allows electric charge to flow around the circuit.

This electric circuit is a model of an electric kettle. The wire coil, the load, will heat up when electrical energy is passed through it.

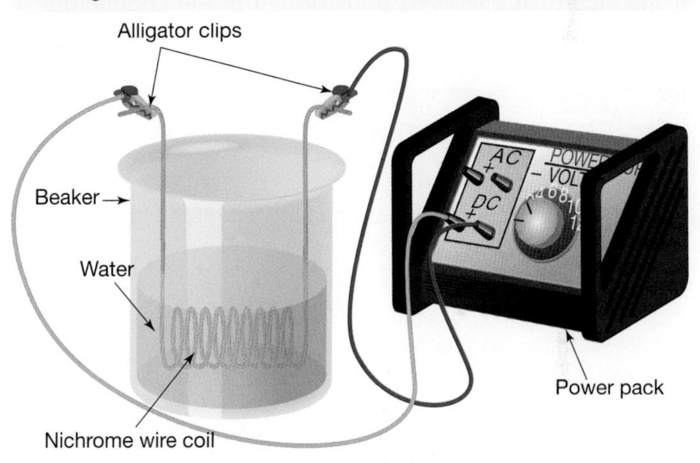

Where it all begins

The batteries (cells) used in torches and many other devices store chemical energy in the substances inside them. The chemical energy is transformed into electrical energy when a chemical reaction takes place inside the cell.

The electrical energy that is used when you turn on a light switch or an appliance connected to a power point comes from a power station.

10.2.4 Current and voltage

The flow of electric charge is called **electric current**. It is a measure of the amount of electric charge passing a particular point in an electric circuit every second. Even though the electric current in wires and most electrical devices is caused by the flow of negatively charged electrons, electric current is defined as the direction of the movement of positive charge. The electric current is said to flow through the circuit from the positive terminal of the power supply to the negative terminal.

Voltage is a measure of the amount of electrical energy gained or lost by electric charge as it moves through the circuit. Voltage is also known as potential difference because it

The filament in this light globe is an example of a load in an electric circuit. The load is where energy is converted — in this case, from electrical energy into light (and thermal) energy.

measures a change in the potential (stored) energy of charge as it moves between one place and another. When electric charge moves through a circuit, it gains electrical energy as it passes through the power supply. It loses the same amount of energy as it passes through the rest of the circuit. That is, the voltage gain across the terminals of the power supply is equal to the total voltage drop across the rest of the circuit.

10.2.5 Where the action is

The load in an electric circuit is an energy converter. It is here that most of the electrical energy carried by electric charge is transformed into useful forms of energy such as light, heat, sound and movement. In a simple torch the load is the **filament**, a coiled tungsten wire inside the globe. The filament glows brightly when it gets hot. In a hairdryer there are two loads: a heater and a fan.

10.2.6 Switched on

The electrical energy provided by batteries and power outlets is transformed into other forms of energy only when the conducting path is complete. A **switch** in an electric circuit allows you to have control over whether or not the conducting path is complete.

Completing the path

A complete conducting path allows electric charge to flow through the circuit. In an efficient electric circuit, most of the electrical energy provided by the power supply is transformed in the load. However, some of the electrical energy is transformed in the conducting path, heating the path and its surroundings. The conducting paths in the electric circuits in appliances are usually made of metals such as copper so that

INVESTIGATION 10.3

Switched-on circuits

AIM: To compare the effect of a single switch with two switches connected in parallel

Materials:

2.5-volt globe and holder
1.5-volt battery and holder
5 connecting leads with alligator clips or banana plugs
2 tapping switches

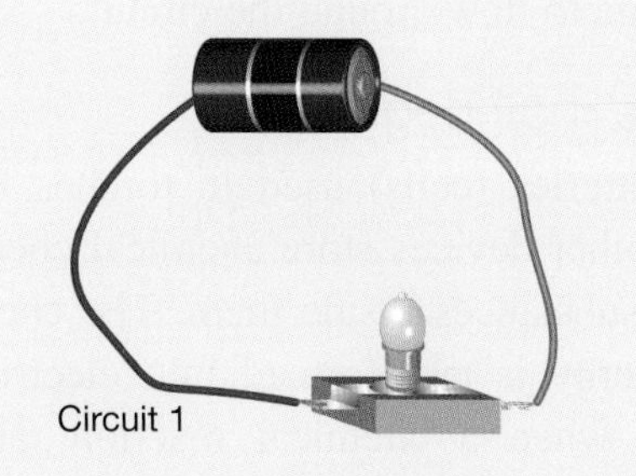

Method and results

- Connect circuit 1 as shown.
1. How can you stop the globe in circuit 1 from glowing?
- Connect circuit 2 as shown.
- Close the switch.
- If nothing happens, open the switch, check that your circuit is connected properly and try again. If nothing happens this time, replace the globe.
2. Describe what happens to the globe in circuit 2 when the switch is closed.
- Add a second switch as shown in circuit 3.
3. Describe what happens to the light globe in circuit 3 when:
   (a) neither of the switches is closed
   (b) either one of the switches is closed
   (c) both of the switches are closed.

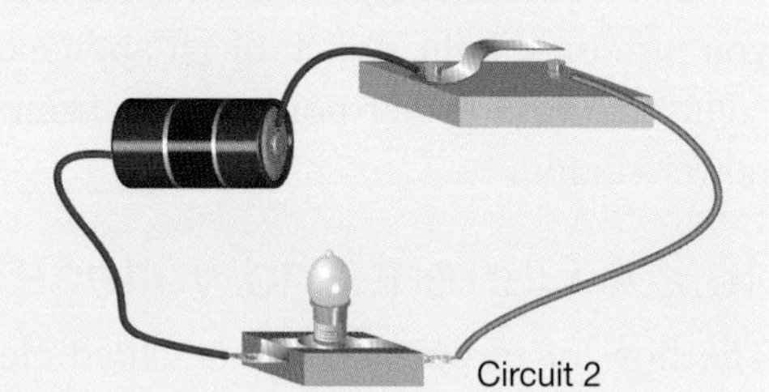

Discuss and explain

4. Explain what happens to the light globe in circuit 3 when:
   (a) neither of the switches is closed
   (b) either one of the switches is closed
   (c) both of the switches are closed.

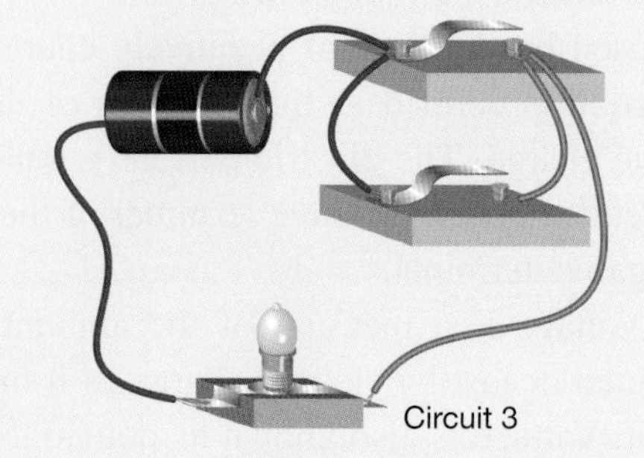

they have little **resistance** to the flow of electric charge. The conducting path in a torch consists of copper wires covered with an insulating layer of plastic. If the connecting wires had too much resistance to the flow of charge, a lot of electrical energy would be wasted in heating the wires, the plastic covering and the air surrounding them.

10.2.7 Electric maps

Maps of electric circuits need to be drawn so that people all over the world can read them. These maps are called **circuit diagrams**. Some of the standard symbols used in circuit diagrams are shown in subtopic 10.3.

10.2 Exercises: Understanding and inquiring

To answer questions online and to receive **immediate feedback** and **sample responses** for every question, go to your learnON title at www.jacplus.com.au. *Note:* Question numbers may vary slightly.

Remember

1. Explain the difference between the transfer of electrical energy in a bolt of lightning and the transfer of electrical energy in an electric circuit.
2. What are the three essential features of all electric circuits?
3. Contrast what electric current is a measure of with what voltage is a measure of.
4. What is the purpose of a switch in an electric circuit?
5. Why are connecting wires usually made of copper?
6. Draw a circuit diagram showing:
   (a) a cell connected to two light globes connected in series
   (b) a cell connected to two light globes connected in parallel.
7. Why is voltage also known as potential difference?

Think

8. From which form of energy is electrical energy transformed in a television remote control?
9. Name a load that is designed to convert electrical energy into:
   (a) light
   (b) sound
   (c) heat
   (d) movement.
10. Redraw the circuit diagram in question 6(b), adding a switch that turns both light globes on or off at the same time.
11. Make a steady-hand tester. You will need: an old coathanger; a loop of thin wire; wirecutters; battery; electric bell or light globe; connecting wires; and a shirt box, shoe box or cereal packet for the base.
    The 'alarm' can be a bell hidden in the base or a globe attached to the base. Hide as much of the connecting wires as you can.

A steady-hand tester

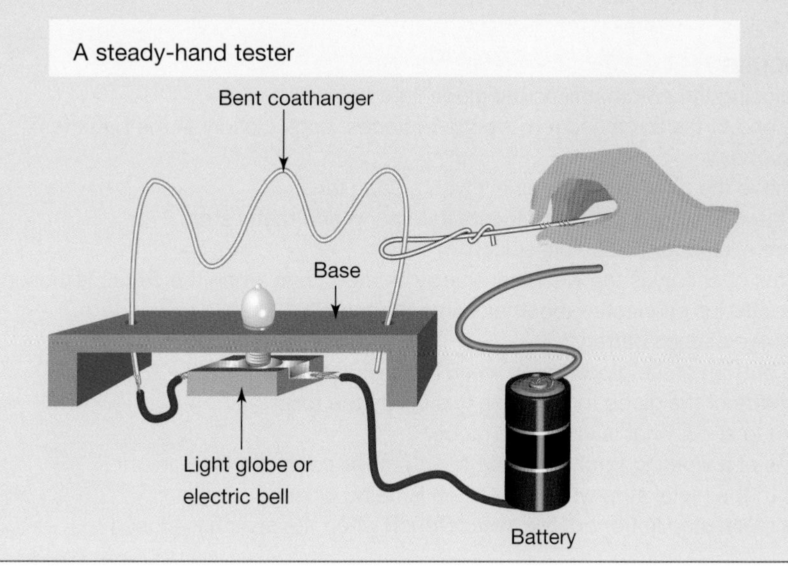

learn on RESOURCES — ONLINE ONLY

Complete this digital doc: Worksheet 10.1: Conductors and insulators (doc-18906)

10.3 A light in the dark

10.3.1 Circuit diagrams: a common language

Many battery-operated devices use more than one battery. In torches, mp3 players and cordless power tools, two or more batteries are connected in series.

They are connected end-to-end as shown in the diagram in this section. It is important to ensure that the positive end of one battery is connected to the negative end of the other.

Diagrams of electric circuits need to be drawn so that people all over the world can read them. Circuit diagrams use straight lines for connecting leads and symbols for other parts of circuits.

Some circuit diagram symbols

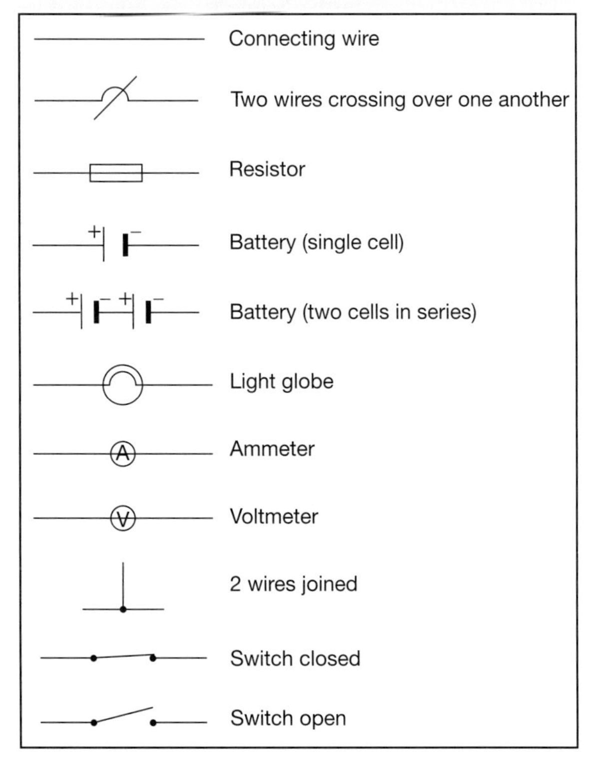

HOW ABOUT THAT!

Even though batteries were invented before 1800, it took until 1898 for the modern torch to be invented. This is mainly because early batteries contained a lot of liquid and were too heavy to carry around.

INVESTIGATION 10.4

What's inside a torch?

AIM: To investigate the electric circuit in a torch

Materials:

torch fitted with two 1.5-volt batteries
hand lens

Method and results

- Check that closing the switch makes the globe light up.
- Unscrew the end of the torch and remove the batteries. Look closely at the batteries.
- Look at the globe.
- Carefully remove the globe and examine it with a hand lens.
- Look inside the case of the torch and locate the spring and metal strip.
- Locate the metal strip and close the switch.

1. What other forms of energy is the electrical energy changed into when the circuit is closed?
2. How were the batteries connected together inside the torch?
3. What is the voltage of each battery?
4. What does the bottom of the globe touch when it is inside the torch?
5. What does the side of the globe touch when it is inside the torch?
6. Draw a diagram to show what is inside the globe.
7. Which two parts of a working torch does the spring make contact with?
8. What happens to the metal strip while the switch is being closed?
9. What does the metal strip in front of the switch touch when the switch is closed?

Inside a torch. When the switch is closed, electric current flows.

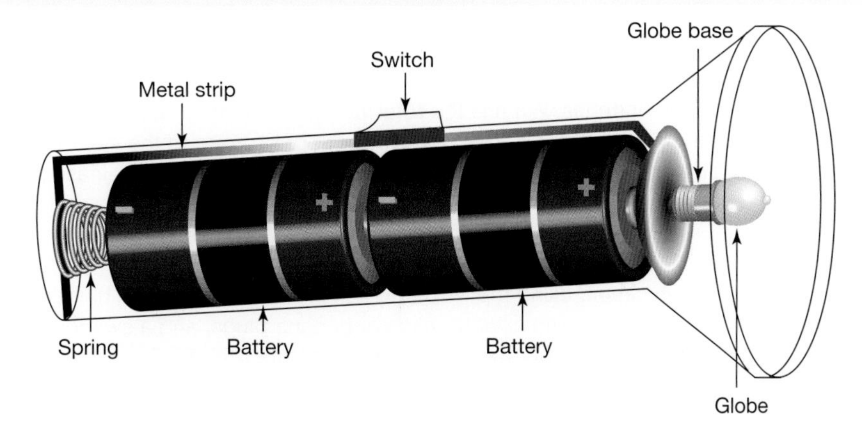

10.3.2 The torch circuit

The power supply of a torch usually consists of two or more 1.5-volt batteries connected in series. When two 1.5-volt batteries are connected in series, the total voltage is 3.0 volts. Twice as much electrical energy is available to move the electric charge around the circuit.

The load in a torch circuit is the globe. When the switch is closed, electric current flows around the circuit. As electric charge passes through the globe, its electrical energy is released as heat in the filament. The filament is the coiled wire inside the globe. It is made of the metal tungsten and glows brightly when it gets hot.

WHAT DOES IT MEAN?

The word *filament* comes from the Latin term *filamentum*, meaning 'spin'.

Circuit diagram for a torch

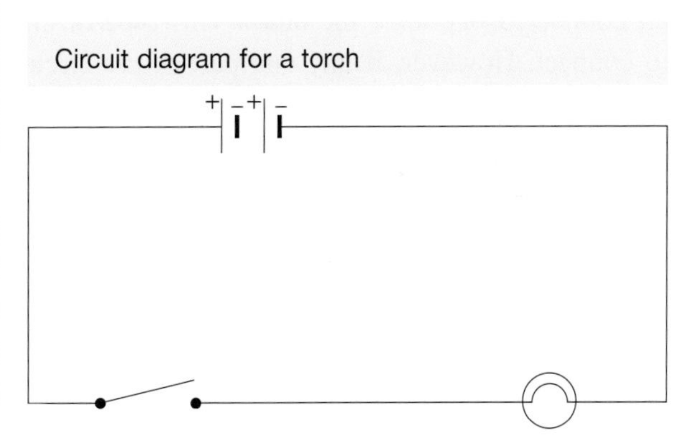

The conducting path in a torch consists of the spring that pushes the battery against the base of the globe (or a metal globe holder) and the metal strip that includes the switch. When the switch is open, the metal strip does not make contact with the globe and the circuit is not complete.

10.3 Exercises: Understanding and inquiring

To answer questions online and to receive **immediate feedback** and **sample responses** for every question, go to your learnON title at www.jacplus.com.au. *Note:* Question numbers may vary slightly.

Remember

1. Draw a diagram to show how two batteries can be connected in series. Label the positive and negative terminals of each battery.
2. Which features of a torch provide the following parts of an electric circuit?
   (a) The power supply
   (b) The load
   (c) The conducting path

3. Describe how a torch works. Ensure that the words 'current', 'energy' and 'circuit' appear in your description.
4. Construct a circuit diagram of a two-battery torch with a closed switch.

Think

5. Some torches use three 1.5-volt batteries. What is the total voltage of such torches?
6. A CD player contains more than one electric circuit. The power supply provides the electrical energy for all of these circuits. Name two different parts of a CD player in which electrical energy is released as other forms of energy. These parts represent separate loads.

Create

7. Construct your own model torch circuit, using the following items; a torch globe and holder; two 1.5-volt batteries and holders; connecting leads with alligator clips or banana plugs; and a switch.
   Use other available materials to make your model torch circuit more realistic.

Investigate

8. Thomas Edison is credited with the invention of the electric light globe. Research and report on one of the following topics.
   (a) The life and inventions of Thomas Edison
   (b) The first electric light globe and how light globes have changed since their invention
   (c) How the invention of the light globe changed the way people lived

10.4 Series and parallel circuits

10.4.1 In series: one after the other

The parts of the torch circuit shown in subtopic 10.3 — the batteries, the switch and the globe — were all connected one after the other. This type of circuit is a **series circuit**. Series circuits are usually easy to connect. However, if any one part of the circuit is faulty, the connecting path is broken and nothing in the circuit will work. For example, if the Christmas tree lights in the photo at right were connected in series, a single faulty globe would cause all of the globes to stop glowing.

Christmas lights — if these lights are connected in parallel, when one light blows out, the others still work.

10.4.2 In parallel: side by side

In a **parallel circuit**, each component is connected in a separate conducting path. This means that if one part of the circuit is faulty, the other parts will still work. If the Christmas tree lights shown in the photograph were connected in parallel, if one globe was faulty the other globes will still glow. Their conducting paths would not be affected.

A parallel circuit has more than one conducting path. If one of the components is faulty, its conducting path is broken. The other conducting paths are still intact and all other components will still work.

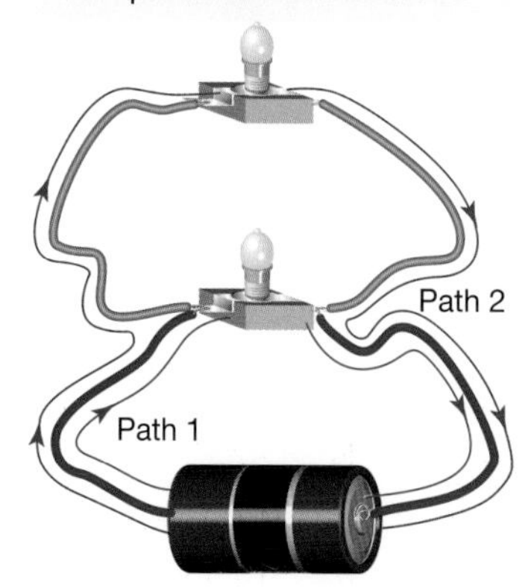

10.4.3 Bright lights

The brightness of light globes in electric circuits depends on how quickly energy flows through the globe. There are two factors that affect how quickly the energy flows.

- The amount of energy each electron flowing through the globe has. The voltage across the globe is a measure of this.
- The number of electrons passing through the globe each second. The electric current passing through the globe is a measure of this.

Identical globes in a series circuit have the same brightness because they share the voltage equally and all globes have the same electric current passing through them. Identical globes in a parallel circuit have the same brightness because they all have the same voltage across them and they equally share the electric current passing through the power supply.

These lights are connected in parallel. Why?

The lights in your home are connected in parallel. Each light has the same voltage across it. Each electron gets the same amount of energy from the power supply. But different globes, bulbs and fluorescent tubes allow different amounts of electric current through. Because the brightness depends on both voltage and electric current, the brightness of the lights can differ.

INVESTIGATION 10.5

In a line or side by side

AIM: To compare series and parallel circuits

Materials:
three 2.5-volt or 3.0-volt torch globes
6 connecting leads

Part A: Series circuits

Method and results

- Connect one globe and the battery together with wire leads so that the globe lights up.
- Add a second globe in series with the first globe as shown in the diagram at right.
- Remove one globe from its holder.
- Replace the globe that was removed, and then remove the other one.

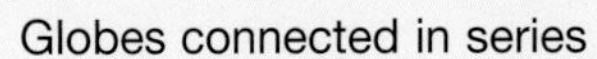

Globes connected in series

1. Draw a circuit diagram to represent the circuit that you have connected.
2. How does the brightness of the two globes compare with the brightness of a single globe connected to the same battery?
3. What effect does the removal of one globe have on the other globe when the battery is connected?
4. Does it matter which globe is removed?

Discuss and explain

5. What would be the effect on the other globes if a third globe were added in series? Test your prediction.
6. Can electric current flow in this series circuit when either globe is removed?
7. Would it be sensible to have all of the ceiling lights in your home connected in series? Give a reason for your answer.

Part B: Parallel circuits

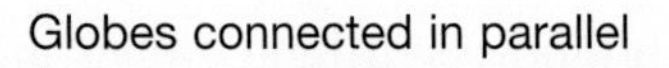

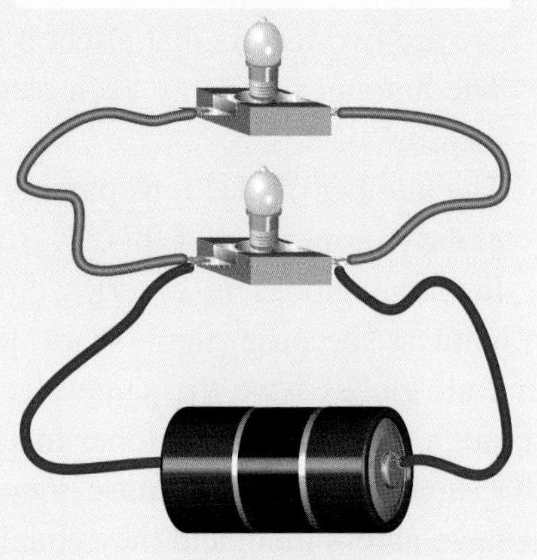

Method and results

- Connect the two globes, battery and wire leads as shown in the diagram at right.
- Remove one globe from its holder.
- Replace the globe that was removed, and then remove the other one.

8. Draw a circuit diagram to represent the circuit that you have connected.
9. How does the brightness of the two globes compare with the brightness of a single globe connected to the same battery?
10. What effect does the removal of one globe have on the other globe?
11. Does it matter which globe is removed?
12. Outline whether the removal of one globe has any effect on the other globe.
13. What would be the effect on the other globes if a third globe were added in parallel? Design a circuit to test your prediction.

Discuss and explain

14. Can electric current flow in this parallel circuit when either globe is removed?
15. Would it be sensible to have all of the ceiling lights in your home connected in parallel? Give a reason for your answer.

10.4 Exercises: Understanding and inquiring

To answer questions online and to receive **immediate feedback** and **sample responses** for every question, go to your learnON title at www.jacplus.com.au. *Note:* Question numbers may vary slightly.

Remember

1. Explain, in words and without the use of a diagram, the difference between a circuit with two light globes in series and a circuit with two light globes in parallel.
2. Copy and complete the following sentences, by choosing the correct word from the pair of words in italics.
   (a) When light globes are connected in *series/parallel*, the same electric current flows through each globe. The globes share the voltage of the power supply.
   (b) When light globes are connected in *series/parallel*, the electric current splits to be shared by the globes. Each globe uses the same voltage.

Think

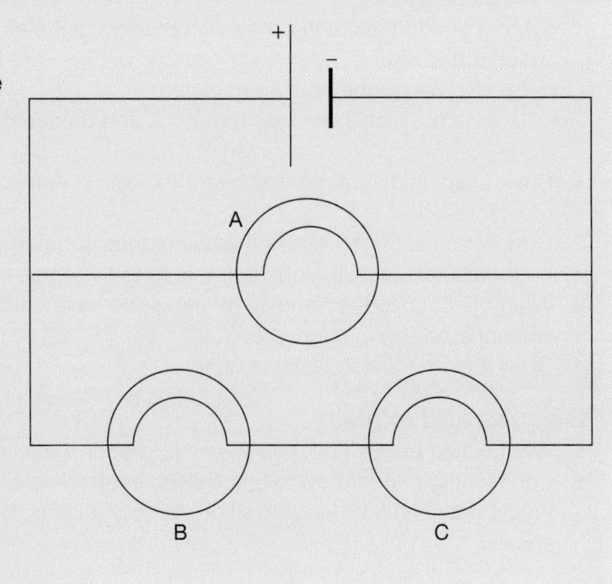

3. Draw a circuit diagram of each of the circuits shown in Investigation 10.5. Make sure that you include the switch.
4. In a small shop, the six light globes are in series. One switch is used to switch all of the lights on or off at once. Draw a circuit diagram of this circuit.
5. Examine the circuit diagram shown at right.
   (a) If the filament of globe A breaks, do globes B and C remain lit or do they stop working also?
   (b) If the filament of globe B breaks, which globe or globes (if any) remain lit?
   (c) If the filament of globe C breaks, which globe or globes (if any) remain lit?
6. In a house, six light globes are in parallel. However, the lights are in separate rooms. This means that a separate switch is needed for each globe. Draw a circuit diagram of this circuit.

7. Examine the circuit shown at right.
   (a) If the two globes are identical, how much of the current that flows through the battery flows through each globe?
   (b) In what way is this circuit similar to the one in part B of Investigation 10.5?
   (c) In what way is this circuit different from the one in part B of Investigation 10.5?

Is this really the same circuit as the one in part B of Investigation 10.5?

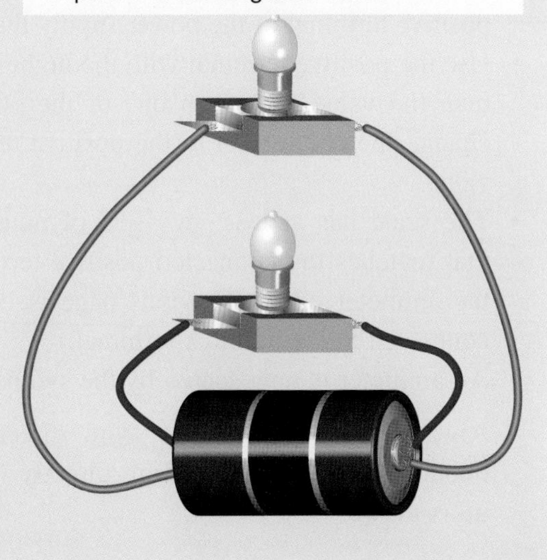

Create

8. Design a circuit with two switches and an electric bell, so that the bell rings when either one (or both) of the two switches is closed. Draw a picture and circuit diagram of your circuit. Invent your own symbol for the bell. If a bell is not available, use a light globe instead.
9. Design a circuit with two switches and an electric bell, so that the bell rings only when both switches are closed. Draw a picture and circuit diagram of your circuit. Invent your own symbol for the bell. If a bell is not available, use a light globe instead.

learnon RESOURCES — ONLINE ONLY

Watch this eLesson: The hydraulic model of current (eles-0029)

Complete this digital doc: Worksheet 10.2: Simple circuits (doc-18907)

Complete this digital doc: Worksheet 10.3: Series and parallel circuits (doc-18908)

10.5 Scale and measurement: Made to measure

10.5.1 Electric currents

Like the currents of water in rivers and the sea, electric current can be measured.

Water currents in a river or the sea can be measured by determining the amount of water that passes a particular point every second. Likewise, the size of the electric current in an electric circuit can be measured by determining the amount of electric charge passing a particular point in an electric circuit every second.

10.5.2 Using an ammeter

An **ammeter** is used to measure the size of electric current flowing in an electric circuit. An ammeter measures electric current in amperes (A) or in one-thousandths of an ampere, which are called milliamperes (mA).

An ammeter is used to measure electric current.

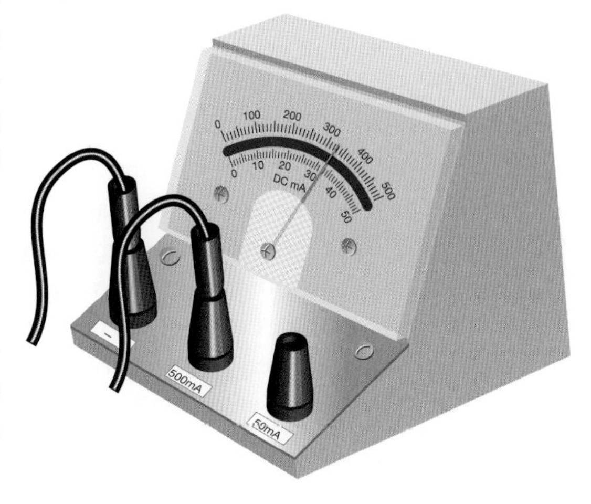

Most ammeters used in school laboratories have one (black) negative terminal and two or more (red) positive terminals. Remember the following points when using ammeters.

- The positive terminal of the ammeter should always be connected in series so that it is closer to the positive terminal of the power supply than the negative terminal of the power supply.
- Use the positive terminal with the highest value first. If the measured current in your circuit is smaller than the value shown on one of the other terminals, you may change the connection to the positive terminal with the smaller value.
- The scale has at least two sets of numbers on it. Use the set that matches the connected positive terminal. (The top scale of the ammeter on the previous page is used because the lead is connected to the 500 mA terminal.)
- An ammeter is represented by the symbol —(A)—.
- Always read an ammeter from directly in front. The error obtained by not reading from directly in front is called a **parallax error**.

Circuit diagram showing how an ammeter is used to measure the electric current through a light globe

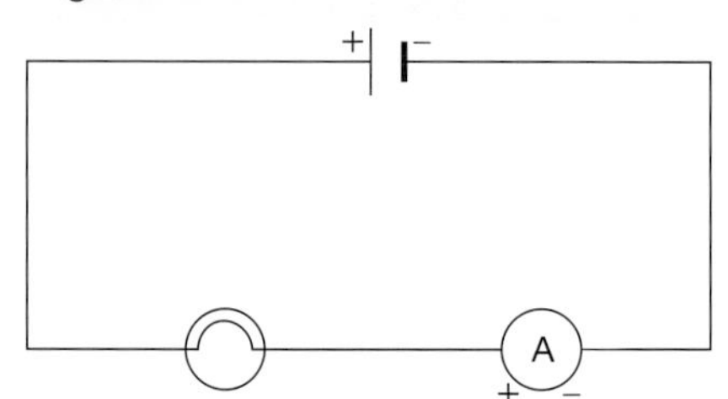

10.5.3 Using a voltmeter

A **voltmeter** is used to measure the voltage gain across the terminals of a power supply or voltage drop across parts of an electric circuit. Voltage is (not surprisingly) measured in volts (V).

Like ammeters, most voltmeters used in school laboratories have one (black) negative terminal and two or more (red) positive terminals. Remember the following points when using voltmeters.

A voltmeter is used to measure the voltage gain or drop across two parts of an electric circuit. This voltmeter is being used to measure the voltage across the light globe.

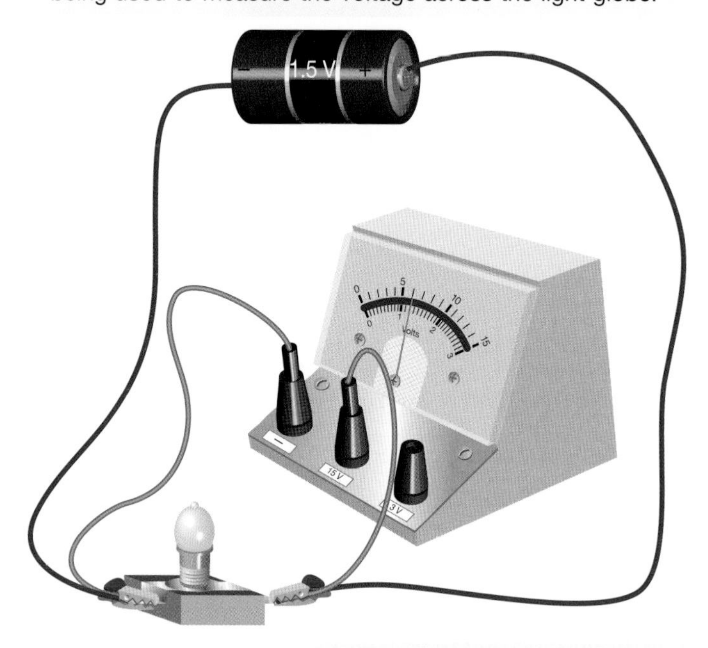

- A voltmeter should be connected in parallel with the part of the circuit across which the voltage is being measured. The positive terminal should always be connected so that it is closer to the positive terminal of the power supply than the negative terminal of the power supply.
- Use the positive terminal with the highest value first. If the measured voltage in the circuit is smaller than the value shown on one of the other terminals, you may change the connection to the positive terminal with the smaller value.
- The scale has at least two sets of numbers on it. Use the set that matches the connected positive terminal.
- A voltmeter is represented by the symbol —(V)—.
- Always read a voltmeter from directly in front to avoid parallax error.

Circuit diagram showing how a voltmeter is used to measure the voltage drop across a light globe

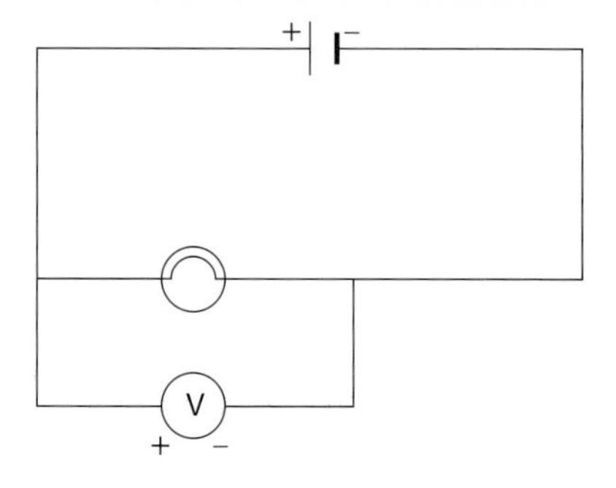

10.5.4 Errors of measurement

There is always a degree of error when reading a scale like this.

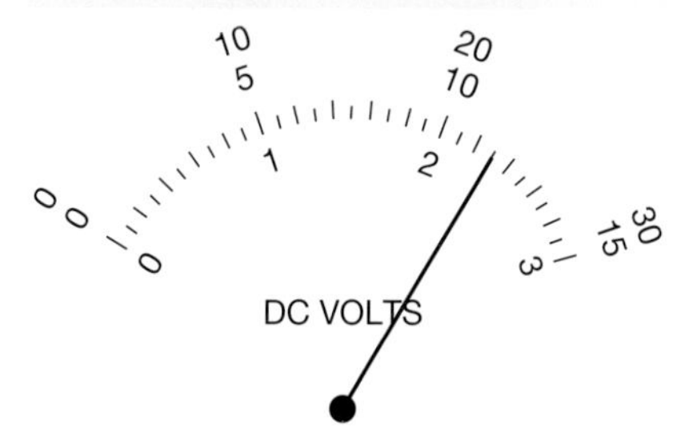

No matter how much care you take with your measurements, there will always be errors because of limitations of your equipment. In addition, when you are reading scales such as those on a ruler, a mercury or alcohol thermometer, an ammeter or a voltmeter, you always have to make an estimate. You should be able to read a scale to about one-tenth of the smallest division marked on it.

For example, on the 3-volt scale of the voltmeter shown at right, the smallest division is 0.1 volt. With care, the scale can be read with an uncertainty of about 0.01 volt. The reading appears to be 2.30 volts, but because of the thickness of the needle and the difficulty in making an estimate it could be read as 2.31 volts or 2.29 volts.

Random errors

Errors that occur due to estimation when reading scales are called **random errors**. Random errors also occur when the quantity being measured changes randomly. For example, when the temperature of the water in a saucepan is being measured, it may increase or decrease slightly due to the convection currents in the water.

Systematic errors

Errors that are consistently high or low due to the incorrect use or limitations of equipment are called **systematic errors**. Parallax errors caused by consistently reading the scale of an ammeter or voltmeter from one side instead of directly in front are systematic errors. An incorrect zero reading when there is no current or voltage, or uneven scales, are also systematic errors.

Reducing errors

Random errors can be reduced by repeating measurements numerous times and calculating an average. But this is not always possible or practical. Such errors can never be totally eliminated. Some systematic errors can be eliminated by knowing how to use the equipment correctly. If a measuring instrument does not read zero when it should, the error can be eliminated by adding or subtracting the 'zero error'. But there will always be systematic errors, because no scale or measuring instrument is perfect.

INVESTIGATION 10.6

Probing a simple circuit

AIM: To investigate the current and voltage within an open and closed circuit

Materials:

power supply (set to 6 volts)
6-volt light globe and holder
6 connecting leads with alligator clips or banana plugs
very long connecting lead (at least 2 m long)
switch
ammeter
voltmeter

Method and results

1. Set up the circuit shown in the diagram at right. You should be able to set it up using only three connecting leads. Make a copy of the table on next page in which to record your measurements.

Probing a simple circuit

Currents and voltages around a simple circuit

	Using the ammeter		Using the voltmeter	
	Location in circuit	Electric current (mA)	Item	Voltage (V)
Switch closed	A		Power supply	
	B		Light globe	
	C		Switch	
	D		Connecting lead	
Switch open	A		Power supply	
	B		Light globe	
	C		Switch	
	D		Connecting lead	

Part A

- Use the ammeter to measure the electric current at each of the points A, B, C and D.

2. Record your measurements in the table.
   - Remove the ammeter from the circuit.
3. With the switch closed, use the voltmeter to measure the voltage across:
   - the power supply (across points A and D)
   - the light globe (across points B and C)
   - the switch (across points C and D)
   - one of the connecting leads (across points A and B).

CAUTION

Check that the ammeter is connected properly before closing the switch. Ask your teacher if you are not sure.

Discuss and explain

4. Is there any difference between the amount of current travelling through the points A, B, C and D?
5. How does the voltage across the terminals of the power supply compare with the voltage across the light globe when the switch is closed?
6. Where is most of the electrical energy generated by the power supply lost?

Part B

7. With the switch open, use the ammeter to measure the electric current at each of the points A, B, C and D.
   - Before you connect the ammeter, make a prediction of the electric current at each of the four points.
8. With the switch open, use the voltmeter to measure the voltage across:
   - the power supply (across points A and D)
   - the light globe (across points B and C)
   - the switch (across points C and D)
   - one of the connecting leads (across points A and B).

- Before you connect the voltmeter, make a prediction of the voltage across each of the four items.

Discuss and explain

9. Were your predictions correct?
10. Why has the voltage across the switch changed so much?
11. Explain how a voltage drop can occur even though the circuit is not closed. (*Hint*: Think about what voltage measures.)

10.5 Exercises: Understanding and inquiring

To answer questions online and to receive **immediate feedback** and **sample responses** for every question, go to your learnON title at www.jacplus.com.au. *Note:* Question numbers may vary slightly.

Remember

1. Which physical quantity is measured by:
   (a) an ammeter
   (b) a voltmeter?

2. Explain how an ammeter must be connected in an electric circuit and state which terminal of the ammeter must be connected closest to the positive terminal of a battery.
3. Describe how a voltmeter must be connected to the part of a circuit across which the voltage is to be measured.
4. Describe two causes of random errors that occur when reading scales to measure any physical quantity.
5. Describe one way in which random errors can be reduced.
6. Describe two causes of systematic errors when using a voltmeter or ammeter.

Think

7. Explain why voltmeters and ammeters have two or three scales.
8. Express an electric current of 0.350 A in mA.
9. Identify two errors in the circuit shown above right.
10. Explain why the voltmeter in Investigation 10.6 must be connected in parallel with the light globe.
11. When using an ammeter, you are advised to use the positive terminal with the highest value scale first. If you can choose between a 50 mA and a 500 mA scale, which one should you connect first? Explain your answer.
12. (a) Why can't you accurately read the electric current measure on the ammeter illustrated at the start of this subtopic?
    (b) Estimate the reading.
13. Is a parallax error a random or systematic error? Explain your answer.
14. Can random errors be eliminated by using digital measuring instruments?
15. Explain why it is never possible to totally eliminate errors when measuring physical quantities.
16. Read the measurement on the ammeter at right accurately.

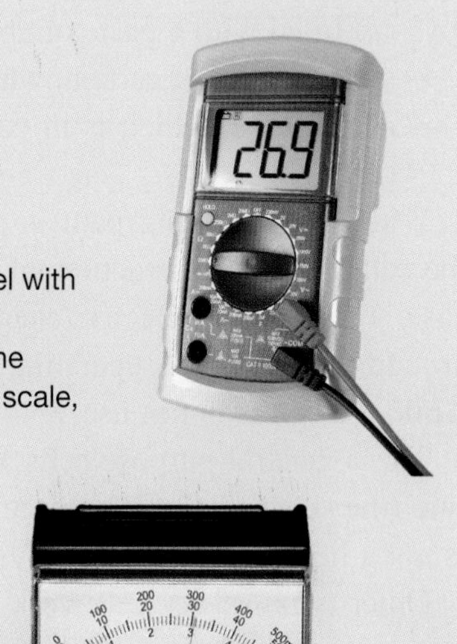

Investigate

17. Research and report on the life and contributions to the investigation of electricity of either André-Marie Ampère or Alessandro Volta.

RESOURCES — ONLINE ONLY

Complete this digital doc: Worksheet 10.4: Ammeters and voltmeters (doc-18909)

10.6 Electricity in a packet

10.6.1 Batteries

The great thing about batteries is that they are light and portable. They are used mostly in devices that need to be moved about. Imagine the disadvantages if you could only get electricity for a torch, a mobile phone or an mp3 player by plugging it into a power point.

Batteries are also used in devices such as clock radios and DVD players as a backup in case of power failure.

A battery is made up of two or more **cells** connected in series. However, in everyday language the word battery is used for a single cell. The batteries used in a torch are actually single cells. An electric cell consists of two **electrodes** and a substance through which electric charge can flow. When the two electrodes are joined together by a conducting path, a **chemical reaction** takes place inside the cell, releasing electric charge and allowing current to flow.

10.6.2 Dry cells

The general-purpose cells used in torches, clocks, smoke detectors and toys are filled with a paste of chemicals. The two electrodes are:

- a central rod of carbon, which is attached to the positive terminal
- a zinc case, which is in contact with the negative terminal of the cell.

When a conducting path is provided between the two terminals of the cell, a chemical reaction takes place between the paste and the zinc case. This releases electric charge, allowing an electric current to flow around the circuit. A separating layer stops the chemicals from reacting while the cell is not in use.

These general-purpose cells are called **dry cells** because the **electrolyte** (the substance inside the cell through which electric charge moves) is not a liquid.

Other types of dry cell work in the same way but use different electrodes or electrolytes.

A general-purpose dry cell

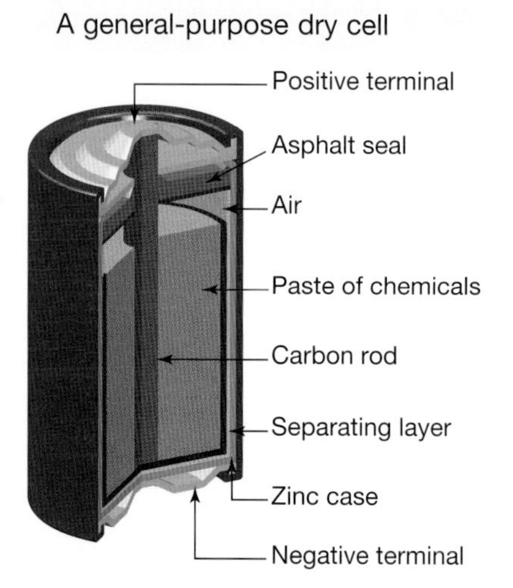

HOW ABOUT THAT!

Have you ever experienced a sharp pain in your mouth when a piece of aluminium foil, probably from a sweet wrapper, contacts a filling? The cause of the pain is an electric cell that you have accidently created in your mouth. The aluminium foil acts as the negative electrode giving up electrons to the filling, with your saliva acting as the electrolyte. The metal in your filling acts as the positive electrode. Contact between the aluminium foil and the filling short-circuits the cell, causing a weak electric current to flow between the electrodes. You feel this as a sharp pain.

Alkaline cells contain an electrolyte that allows a greater electric current to flow. They are ideal for heavy-duty torches, battery-operated shavers, portable CD players and digital cameras.

Mercury cells produce a voltage that is much steadier than other dry cells. Their steady output makes them ideal for pagers, hearing aids, watches, calculators and measuring instruments.

10.6.3 Fruity cells

Citrus fruits such as lemons, oranges and grapefruit can be used to make a battery. When a conducting path is provided between different metals inserted into the fruit, a chemical reaction takes place with the acids and a small electric current flows.

Different types of batteries

A car battery

A button battery

Akaline zinc/manganese dioxide batteries

A rechargeable battery

10.6.4 Car batteries

Car batteries consist of six cells connected in series. Each cell has two lead electrodes, one of which is coated with a paste of lead dioxide. The electrodes are surrounded by a sulfuric acid solution. When the battery is in use, a chemical reaction occurs between the electrodes and the sulfuric acid. One of the products of the reaction is lead sulfate. Once the engine is running, the chemical reaction is reversed and the battery recharges. The lead sulfate is converted back to lead and lead dioxide. After a few years, the lead sulfate builds up on the electrodes and becomes so hard that the reverse reaction cannot take place. The battery cannot be recharged and needs replacing.

Nickel–cadmium cells, such as those used in mobile phones, can also be recharged. A battery charger can be used to reverse the chemical reaction that causes electric current to flow.

HOW ABOUT THAT!

The very first working battery, made by Alessandro Volta more than 200 years ago, was a tall pile of silver and zinc discs with pieces of cloth soaked in salty water between the discs. This structure became known as a voltaic pile.

INVESTIGATION 10.7

A lemon battery

AIM: To use lemons to create a battery

Materials:

3 lemons
3 galvanised nails
three 5 cm lengths of uninsulated copper wire
microammeter
4 connecting leads

Method and results

- Squeeze all three lemons to break up some of the pulp inside.
- Insert a galvanised nail and a piece of copper wire into one of the lemons. The nail and wire should be about 3 cm apart.
- Use connecting leads to connect the negative terminal of the microammeter to the nail and the positive terminal to the copper wire.

1. Record the electric current.
2. Add a second lemon in series and record the electric current again.
3. Add a third lemon in series and record the electric current.

- Investigate the effect on the electric current of:
  - pushing the electrodes further into the lemons
  - changing the distance between the nail and the copper wire in each lemon.

Discuss and explain

4. What is the electrolyte in this lemon battery?
5. How did the adding of a second and third lemon in series affect the electric current?
6. How did changing the depth of the electrodes and the distance between them affect the electric current?

10.6 Exercises: Understanding and inquiring

To answer questions online and to receive **immediate feedback** and **sample responses** for every question, go to your learnON title at www.jacplus.com.au. *Note:* Question numbers may vary slightly.

Remember

1. What is the difference between a cell and a battery?
2. What takes place inside a cell to cause an electric current to flow?
3. What substances are the electrodes of a general-purpose dry cell made of?
4. How do alkaline cells differ from general-purpose dry cells?
5. If a car battery can be recharged, why can't it last for ever?

Think

6. Make a list of all the devices you can think of that use batteries.
7. What substance makes up the electrolyte in a car battery?
8. Explain why mercury cells are ideal for watches, hearing aids and measuring instruments.
9. Why does a car battery need replacing sooner if it is not used very often?

Imagine

10. Imagine that batteries were no longer available. Consider each of the devices you listed in question 6, and state:
    (a) what could replace the battery
    (b) how you would be affected if the battery could not be replaced.

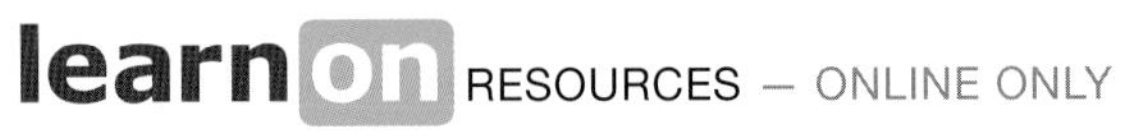

Complete this digital doc: Worksheet 10.5: Testing batteries (doc-18910)

10.7 Driving on batteries

10.7.1 Changing times

What sort of car do you expect to be driving thirty years from now? Will it be just a newer, sleeker, lighter version of the cars you see on the road today? How much will petrol cost: $2 per litre or $20 per litre? Most medium-sized cars have petrol tanks that hold between 50 and 80 litres. How much will it cost to fill the tank? Will you have trouble breathing the polluted air in traffic-clogged cities?

It is unlikely that you will be driving a car with an engine powered by petrol. There are several reasons for this:

- Petrol is made from oil and the world's oil supply is rapidly decreasing. At the same time, the amount of oil being used is increasing. It has been predicted that the world's oil reserves will run out in less than fifty years.

The Tesla Roadster is a high performance electric car now available in Australia. It can accelerate from 0 to 100 km/h in 4 seconds, has a top speed of 200 km/h and can travel 350 km before the batteries need recharging.

- Petrol is becoming more expensive. This is partly due to attempts by politicians to conserve the world's oil reserves. As the cost of petrol increases, alternative fuels become more attractive. LPG (liquefied petroleum gas) is already increasing in popularity as a fuel for cars.
- Petrol-driven car engines cause air pollution. Gases released from car exhausts include carbon monoxide (a poisonous gas), carbon dioxide (a major cause of the 'greenhouse effect' and a probable cause of climate change) and nitrogen oxides (which lead to smog and acid rain).

learnon RESOURCES — ONLINE ONLY

Watch this eLesson: The Australian International Model Solar Challenge (eles-0068)

10.7.2 Electric cars

One of the most attractive alternatives to the petrol-driven car is a car powered by rechargeable batteries. Electrical energy from the batteries is used to drive a motor which turns the wheels. The batteries can be recharged while the driver is at home or at work.

Electric cars have three main benefits:

1. Their use will reduce the demand for oil. The world's oil reserves will last longer.
2. They do not release exhaust gases. This would reduce air pollution in large cities.
3. They are very quiet.

An additional benefit is that electric cars can be designed so that their batteries can be fully or partially charged by solar energy.

There are also some drawbacks to electric cars:

1. Electric cars can travel only about 160 kilometres before the batteries need recharging. A tank of petrol would allow most cars to travel 500–800 km before refuelling.
2. The batteries, which are very expensive, need replacing after a few years.
3. Electric cars do not accelerate quickly and can usually reach speeds of only 100 km per hour.
4. Electric cars are more expensive to buy than petrol-driven cars.
5. If everyone owned electric cars, power stations would need to supply more energy for recharging batteries from power points. Although air pollution in cities would be reduced, the air pollution around the power stations would be increased.

10.7.3 Solving the problems

Some of the disadvantages of electric cars will be overcome as the need to replace petrol-driven cars becomes more urgent.

- Automotive engineers are using scientific knowledge, together with computer techniques and models, to design lighter cars. They are also designing car bodies and tyres to reduce the friction caused by the air and road surfaces. These changes will reduce the amount of energy needed to keep cars running. Batteries will last longer.
- Electric cars that have been converted from petrol-driven cars need at least 12 standard lead–acid car batteries (connected in series) to run at normal speeds. These batteries are very heavy. Lighter, smaller and more powerful batteries are currently being developed.
- As more electric cars are made, the cost of each car will decrease. Also, as petrol becomes more expensive, the higher cost of electric cars will be less of a problem.

New car designs, better batteries and decreasing costs make electric cars a very likely alternative to petrol-driven cars.

10.7.4 The hybrid car

The hybrid car combines a bank of rechargeable batteries with a small petrol engine. It provides most of the benefits of a totally electric car. In the car illustrated on the next page, the petrol engine generates energy to recharge the batteries while the car is being driven. In some hybrid cars, the petrol engine is connected directly to the motor that turns the wheels. This means that less energy is required from the batteries. The exhaust fumes of hybrid cars still contribute to air pollution, but to a much lesser extent than current petrol-driven cars.

Hybrid cars, such as Toyota's Prius and Camry, are gaining popularity in Australia as the increased price of petrol puts pressure on car owners to use less of it. Hybrid car manufacturers are struggling to keep up with demand.

In Australia, the CSIRO is working with car manufacturers to produce lighter and more compact lead–acid batteries that will require less frequent recharging.

A hybrid car combines a bank of rechargeable batteries with a small petrol-driven engine.

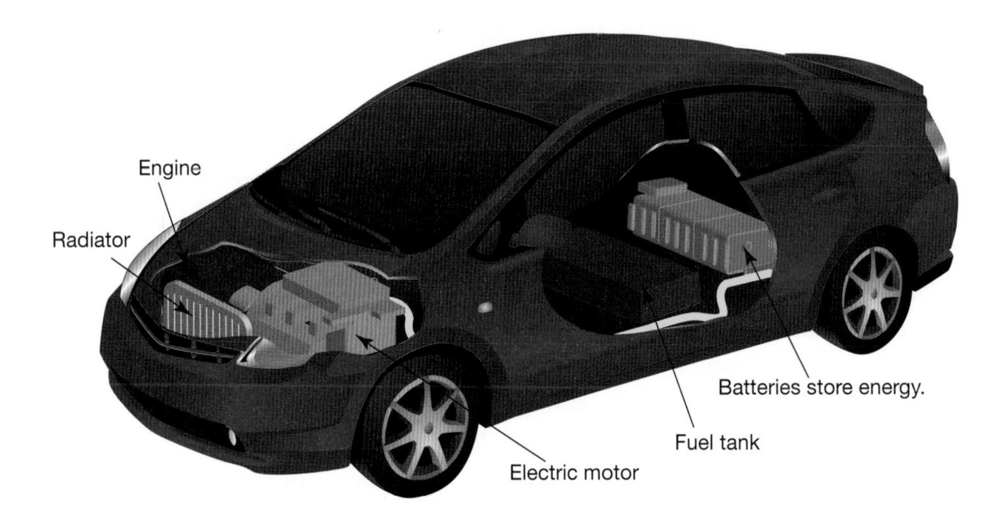

HOW ABOUT THAT!

Battery-powered electric cars are not new. In fact, before 1900 there were more electric cars than petrol-driven cars. However, petrol-driven cars were more powerful and could travel for longer distances without having to stop. Using petrol was also cheaper than replacing and recharging batteries. By 1930, electric cars had been replaced almost entirely by petrol-driven cars.

An early electric car

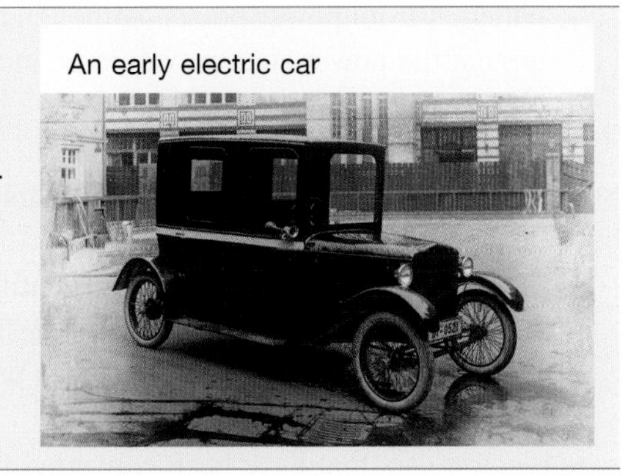

10.7 Exercises: Understanding and inquiring

To answer questions online and to receive **immediate feedback** and **sample responses** for every question, go to your learnON title at www.jacplus.com.au. *Note:* Question numbers may vary slightly.

Remember

1. List three benefits of electric cars.
2. State five disadvantages of electric cars at the current time.
3. What is a hybrid car?

Think

4. Why are electric cars likely to become popular after being ignored for over 60 years?
5. Suggest why most electric cars are currently able to reach speeds of only 100 kilometres per hour.
6. Which disadvantages of electric cars do hybrid cars partially or entirely overcome?
7. Advertisements for hybrid cars highlight their advantages and sometimes make claims without evidence.
   (a) List some disadvantages of hybrid cars.
   (b) What type of evidence would you expect from automotive engineers before purchasing a hybrid vehicle?
8. Do you think the government should force car manufacturers to stop making petrol-driven cars and replace them with electric or hybrid cars? State reasons for your opinion.

Investigate

9. Research and report on how solar cars work. Include in your report a discussion about the possibility of solar cars replacing petrol-driven cars on Australia's roads.

Imagine

10. If you could buy an electric car for the same price as a petrol-driven car of the same make and model, which would you choose? Why?

Create

11. Draw a diagram of a car of the future. Label the features of the car that will reduce the amount of energy needed to make it go fast.

10.8 A question of resistance

10.8.1 Pushing through the atoms

Have you ever had to push your way through a crowd of people to get somewhere quickly? Frustrating, isn't it? It slows you down and saps your energy. Well, if the electrons moving in an electric circuit had feelings, they would understand how you felt.

The negatively charged electrons moving in an electric circuit have to make their way past the atoms in the connecting leads and devices that make up the circuit. Electrical **resistance** is a measure of how difficult it is for electrons to flow through part of a circuit. The resistance to the flow of electric charge limits the electric current, just as the resistance of a narrow and crowded corridor limits the number of students that can pass through in a given time interval. Electrical resistance also determines how much energy is lost by electric charge as it moves through a circuit.

10.8.2 The value of resistance

Conductors have very little resistance. They allow large electric currents to flow with little loss of energy. **Insulators** have a very large electrical resistance. They allow very little electric current to flow.

The letter R is used to represent resistance and its unit is the ohm (Ω). The value of the resistance of part of an electric circuit is defined by the following formula, where V is the voltage drop in volts and I is the electric current in amperes.

$$R = \frac{V}{I}$$

A torch globe carrying an electric current of 200 mA with a voltage drop of 3 volts therefore has a resistance of:

$$R = \frac{V}{I} = \frac{3}{0.2} = 15\Omega$$

10.8.3 When resistance is constant

In 1827, a German physicist, Georg Simon Ohm, discovered that the electric current in metallic conductors was proportional to the voltage drop across the conductor. That is, if the voltage was doubled, the current doubled. If the voltage was tripled, the current tripled. This discovery has become known as **Ohm's Law**.

Materials that obey Ohm's Law are said to be **ohmic**. Metals and carbon are ohmic materials as long as the temperature remains fairly constant. The filament in a light globe is not ohmic.

One way of working out whether a material is ohmic is to draw a graph of voltage drop versus electric current. Recall that resistance is defined as:

$$R = \frac{V}{I}$$

$$\therefore\ V = RI$$

Graph of voltage drop (V) versus electric current (I) for an ohmic conductor

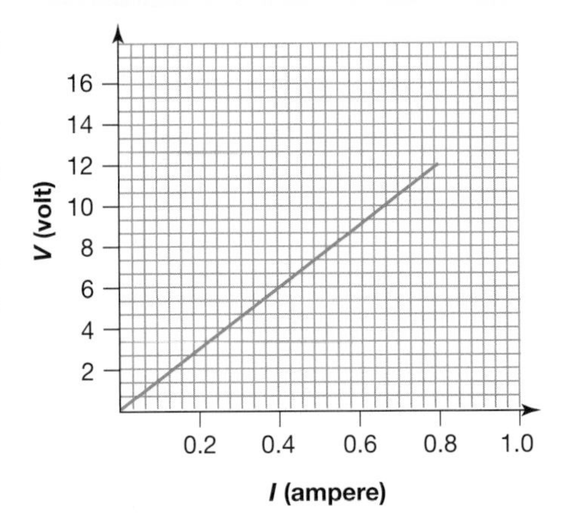

If the material is ohmic, R is constant. A graph of V versus I yields a straight line.

10.8.4 Controlling the flow

When you turn down the volume of a radio or television, you are changing the voltage across and current flowing through parts of the electric circuits inside. The volume dial or sliding knob is part of a **variable resistor**.

Resistors are used in electric circuits to control the voltage and current. They can have a fixed resistance or can be variable like those in volume controls.

Three different types of carbon resistors are illustrated below left. The two can-shaped resistors below are a type of variable resistor used in volume dials.

There are several symbols that can be used to represent variable resistors.

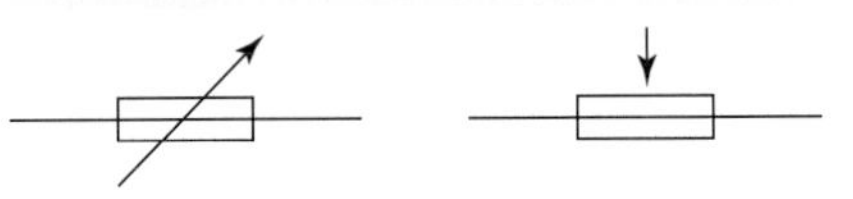

A range of carbon resistors

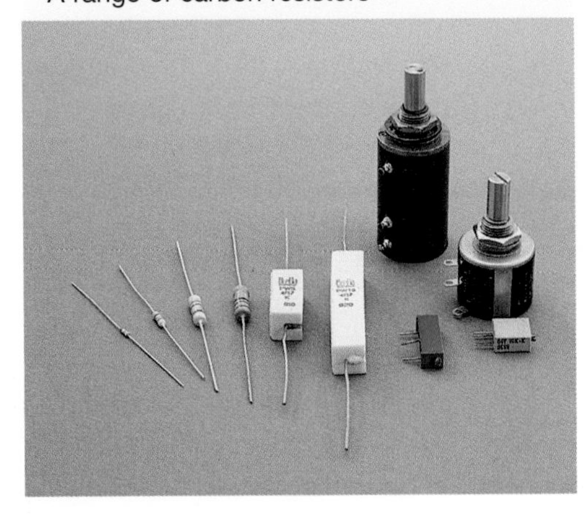

This variable resistor consists of coils of wire touched by a slider. The slider is connected to the coil, and controls the number of coils through which the current flows and, therefore, the resistance.

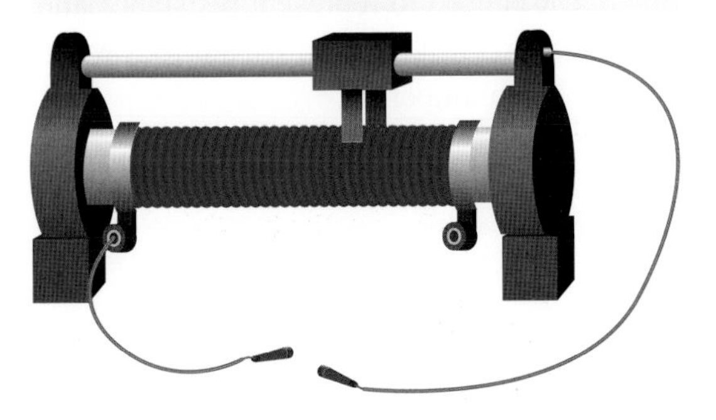

Taking a dimmer view of things

A variable resistor can be used in a circuit like the one in Investigation 10.9 to control the voltage across the light globe and the electric current flowing through the light globe. As the resistance of the variable resistor increases, the total resistance in the entire circuit increases. This increase in resistance causes a decrease in the amount of current flowing through the circuit.

In addition, the amount of available electrical energy transformed in the globe also decreases. As the resistance of the variable resistor increases, more electrical energy is lost in heating the resistor and the surrounding air. Consequently, the globe glows less brightly because not only does less electric charge pass through it every second, but each electric charge has less energy to heat the globe's filament.

Resistors in sensors

In sensors, such as those used in lights that automatically turn on when it gets dark, light dependent resistors (LDRs) are used to switch lights off and on. The resistance of LDRs changes with the amount of light falling on them, changing the voltage to turn lights off or on. Most LDRs are made of substances that have less resistance when the light intensity increases.

Thermistors, such as those used in the heat sensors in air conditioners, refrigerators, car engine cooling systems and fire alarms have a resistance that changes as the temperature increases or decreases.

Infra-red sensors are used in a number of devices and recent-model cars to measure distances to nearby objects. An infra-red beam is directed away from the device and the sensor detects the signal reflected from the object. A computer in the device calculates the distance and send a warning signal or even take action to avoid the object.

INVESTIGATION 10.8

Changing resistance

AIM: To investigate the relationship between voltage and resistance of a light globe

Materials:

power supply (variable)
9-volt light globe and holder
6 connecting leads with alligator clips or banana plugs
switch
ammeter
voltmeter

Method and results

- Set up the circuit shown in the diagram at right and leave the switch open.

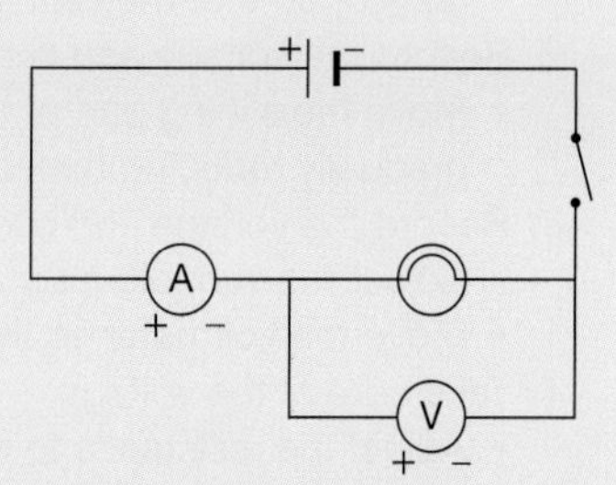

1. Construct a table like the one on the next page in which to record your measurements of the electric current flowing through the light globe, the voltage drop across the globe and the calculated resistance.
- Set the power supply to 2 volts.
2. Close the switch and quickly read the meters, recording the electric current and voltage drop in your table. Ensure that the electric current is recorded in amperes (not milliamperes).
- With the switch closed, set the power supply to 4 volts.
3. Quickly measure and record the electric current and voltage displayed on the meters.
4. Set the power supply to 6 volts (again not opening the switch) and quickly measure and record the electric current and voltage displayed on the meters.
5. Calculate the resistance of the globe for each of the three power supply settings and record them in your table.
6. Plot a graph of voltage drop (V) versus electric current (I) for the light globe.
7. Repeat the experiment, constructing a new table in which to record your data. This time, however, start with the power supply set to 6 volts. Then decrease it to 4 volts and then 2 volts.

8. Plot a second graph of the voltage drop versus electric current on the same set of axes as the first graph in a different colour.

Discuss and explain

9. Does the resistance increase or decrease during the first part of the experiment, when the power supply setting is being increased?
10. How is the change in resistance different during the second part of the experiment, when the power supply setting is being decreased?
11. What changing property of the filament do you think caused the resistance of the light globe to change?
12. Explain any difference between the shape of the first graph and that of the second graph.

The characteristics of a light globe

Power supply setting (volts)	Electric current (amperes)	Voltage drop (volts)	Resistance (ohms)
2			
4			
6			

INVESTIGATION 10.9

Making the change

AIM: To investigate the effect of a variable resistor on a light globe connected with it in series

Materials:

power supply (set to 6 volts)
6-volt light globe and holder
variable resistor
6 connecting leads with alligator clips or banana plugs
ammeter and voltmeter

Method and results

1. Construct a table in which you can record five sets of measurements of voltage across the light globe and electric current flowing through the light globe.
   - Set up the circuit shown in the diagram at right. The variable resistor is connected in series with the light globe. Move the sliding part of the variable resistor so that the voltage drop across the light globe is at a maximum.
2. Record the voltage and current shown on the meters.
   - Move the sliding part of the variable resistor to four different positions, gradually reducing the voltage across the light globe.
3. Record the voltage and current in your table for each position.
   - Adjust the variable resistance so that the globe is at its brightest.
   - Move the voltmeter so that it measures the voltage across the variable resistor.
4. Take note of the voltage.
   - Adjust the resistance to make the globe dimmer and dimmer.
5. Take note of how the voltage across the variable resistor changes.

Controlling current and voltage with a variable resistor

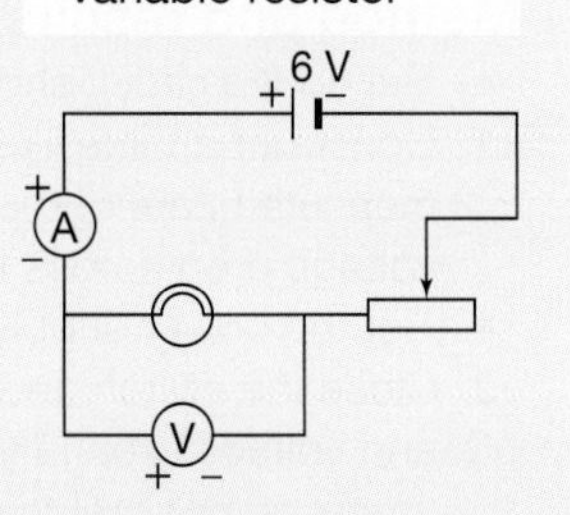

Discuss and explain

6. What would you expect the resistance of the variable resistor to be when the voltage drop across the light globe is at a maximum?
7. What happens to the electric current flowing through the light globe as the resistance of the variable resistor increases?
8. What happens to (a) the voltage across the light globe and (b) the brightness of the globe as the resistance of the variable resistor increases?
9. When the globe is at its brightest, what is the voltage across the variable resistor?
10. How does the voltage across the variable resistor change when the globe is made dimmer?
11. What would you expect the sum of the voltage across the light globe and the voltage across the variable resistor to be?

10.8 Exercises: Understanding and inquiring

To answer questions online and to receive **immediate feedback** and **sample responses** for every question, go to your learnON title at www.jacplus.com.au. *Note:* Question numbers may vary slightly.

Using data

Answer these questions about the ohmic conductor described by the graph of voltage drop versus electric current at the beginning of this subtopic.

1. What is the voltage drop across the conductor when the electric current is 1.0 ampere?
2. What electric current (in mA) flows through the conductor when the voltage drop across it is 6 volts?
3. Calculate the resistance of the conductor.
4. Explain how you know that the conductor is ohmic.

Remember

5. Why does very little current flow through insulators?
6. What does Ohm's Law say about the relationship between electric current in and voltage across metallic conductors?
7. Which two properties of a working electric circuit change when the resistance of a variable resistor changes?
8. What happens to the electric current flowing through a light globe when the resistance of a variable resistor in series with the globe increases?

Think

9. What is the voltage drop across a 100 Ω resistor when the electric current flowing through it is measured at 250 mA?
10. The electric current flowing through a light globe is measured to be 200 mA when the voltage across the globe is 1.5 volts. When the voltage is increased to 3.0 volts, the current is measured to be 360 mA.
    (a) What is the resistance of the light globe when the electric current is 200 mA?
    (b) Is the light globe ohmic?
    (c) If the light globe were ohmic, what would happen to the electric current flowing through that light globe if the voltage across it were doubled?

Investigate

11. Take a close look at a variable resistor and explain what causes its resistance to increase when a slider is moved or the dial turned.
12. Research and report on the materials that are used to make light dependent resistors.
13. Robot vacuum cleaners use a variety of sensors to clean a room. Find out what these sensors are and how they work?

10.9 Electronics—it's a small world

10.9.1 Components

Science as a human endeavour

Electronics involves the use of electric circuits to control devices that make life easier, safer and a lot more enjoyable. Computers, mobile phones, televisions and remote controls are all examples of electronic devices. The parts used in the electric circuits used in electronic devices are called components.

Larger electric appliances such as washing machines, air conditioners, clothes dryers and dishwashers — in fact any appliances that are programmable — contain electronic components. Regretfully the same electronic components are used in weapons such as bombs and missiles.

Scaling down

The very first electronic components, invented about 100 years ago, were very large glass tubes from which most of the air was removed. They were called vacuum tubes and for many years were used in devices such as movie projectors, radios, televisions, amplifiers and radar. The problem was that vacuum tubes were bulky, heavy and easily broken.

> **HOW ABOUT THAT!**
>
> The first fully electronic general purpose computer, built in 1946, weighed 27 tonnes and occupied three floors of an office building because of the weight and size of its components. The same functions can now be performed faster by computers that fit on your desk, in a small case or even inside a mobile phone.

Most electronic components are now too small to see. **Integrated circuits** that contain thousands of tiny electronic components are etched onto thin pieces of silicon called **chips**.

The first silicon chip was developed in 1958 and, by 1965, most chips could hold about 30 electronic components. In 1975, a similar sized chip could hold about 30,000 components, which allowed the development of desktop computers. Now, silicon chips no larger than the fingernail on your little finger may now contain millions of electronic components.

The silicon chips on the memory cars in personal computers contain millions of different electronic components. These components can be smaller than a millionth of a metre across and can be seen only with powerful microscopes.

10.9.2 Electronic building blocks

Apart from resistors, the most common electronic components are capacitors, diodes and transistors.

Capacitors—storing charge for a while

Capacitors store electric charge for a short time before allowing it to flow to other parts of a circuit. The amount of charge that can be stored for each volt across a capacitor is called its capacitance. Capacitance is measured in units of farad (F) or microfarad (μF). A microfarad is one-millionth of a farad.

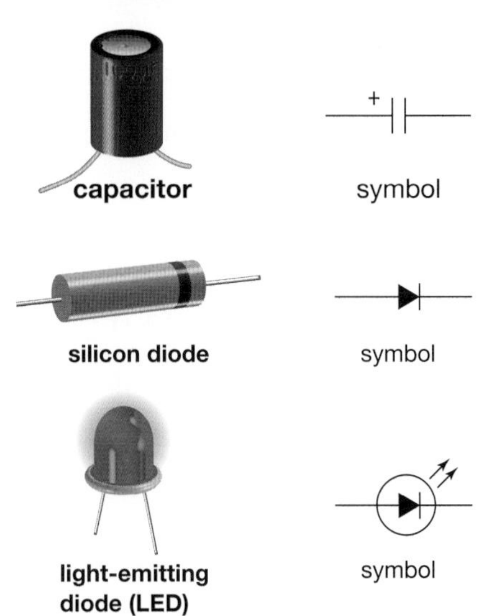

Diodes—one-way streets

Diodes allow electric current to travel through them in only one direction. They look like small resistors but have a single band at one end. This end of the diode is the negative end and should be connected closer to the negative terminal of the power supply than the positive terminal.

Light-emitting diodes (LEDs) also allow current to flow in only one direction but transform electrical energy into light energy. LEDs are often used as indicator lights in electrical appliances. An arrangement of seven LEDs can be used in devices like watches, clocks, calculators and digital meters to display any number between 0 and 9. The display circuit is designed so that the LEDs light up in different combinations. Liquid crystals are often used instead of LEDs for the same purpose. Small voltages cause the molecules in liquid crystals to rearrange themselves, changing the colour of the crystals.

LEDs are now used extensively in household and commercial building lighting because they use less energy than older lighting devices.

An arrangement of light-emitting diodes can be used to display numbers.

Transistors—switches on three legs

Transistors act like switches, changing the size or direction of electric current as a result of very small changes in the voltage across them. This makes them ideal for use in devices that amplify sound. However, they have many other uses and most electronic devices contain chips that hold many microscopic transistors.

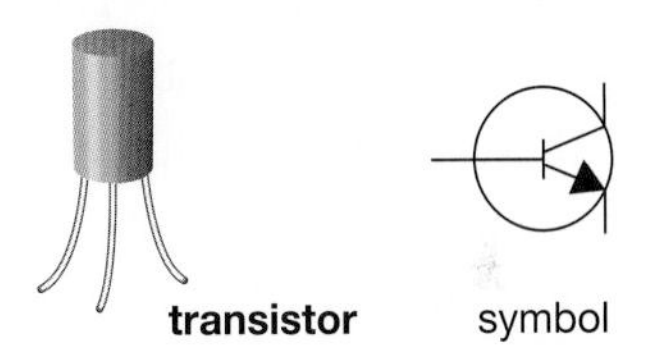

10.9.3 Microprocessors

A silicon chip that is able to store information, process it and control other electric circuits is called a **microprocessor**. Since their development in 1971, microprocessors have been used in calculators and computers. As they became less expensive they began to be used in household appliances like microwave ovens, televisions and washing machines. The microprocessors made them 'programmable' and able to perform tasks with little human effort. They are now used in robots, cars, phones and many other devices that store and process information.

10.9 Exercises: Understanding and inquiring

To answer questions online and to receive **immediate feedback** and **sample responses** for every question, go to your learnON title at www.jacplus.com.au. *Note:* Question numbers may vary slightly.

Remember

1. What is on the pieces of silicon that make silicon chips so useful?
2. How is a microprocessor different from other silicon chips?
3. Copy and complete the table below.

Component	Circuit symbol	Function
capacitor		
diode		
LED		
transistor		

Think

4. Describe three advantages that today's electronic components have over the vacuum tubes that were first used in devices such as televisions and movie projectors.

Imagine

5. Make a list of devices and human achievements that could not have been possible without the 'miniaturisation' of electronic circuits that began with the development of integrated circuits and chips.

Investigate

6. Find out what nanoprocessors are and how they are different from microprocessors.
7. Find out what a pill-cam is, what it is used for and how it works.

10.10 Magnetic effects of electricity

10.10.1 Power plants

When you are at home, most of the electricity you use is obtained by simply plugging a lead into a power point and flicking a switch or two. The electric current that flows from power points in most homes is generated by power plants that are many kilometres away. The generation of electricity by power plants, whether they are coal-fired, gas-fuelled or hydroelectric depends on the close relationship between electricity and magnetism.

The generation of household electricity in power plants and by wind turbines depends on the close relationship between electricity and magnetism.

10.10.2 A magnetic attraction

Magnets attract iron and alloys containing iron. They also attract alloys containing nickel and cobalt, which, like iron, have magnetic properties. All magnets, no matter what their shape, have a **north pole** and a **south pole**. Like poles of magnets repel while unlike poles attract. **Permanent magnets** retain their magnetism at all times, while **temporary magnets** (like the nail suspended from the permanent magnet in the diagram at right) lose their magnetism when removed from other magnets.

The region in which a magnetic force exists is called a **magnetic field**. A 'map' of a magnetic field around a bar magnet is shown on the far right. It shows the direction of the magnetic force around the magnet. The direction of the magnetic force is defined as flowing away from the north pole and towards the south pole. A compass needle is a small magnet that lines itself up with the magnetic field. It always points in the direction of the magnetic force.

The nail is a temporary magnet while it is in contact with the permanent magnet.

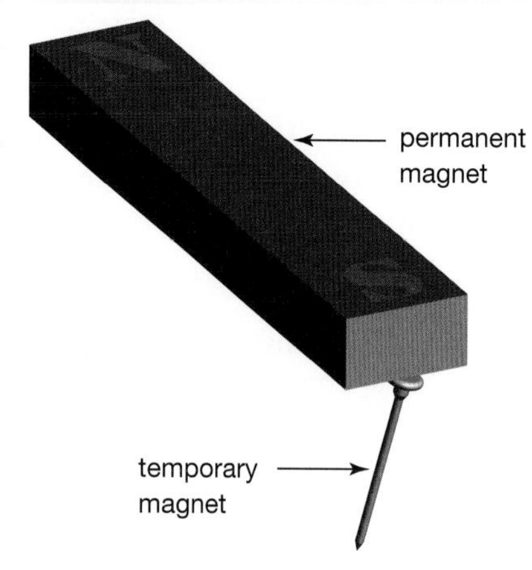

The magnetic field around a bar magnet. The closeness of the lines gives an indication of the relative strength of the magnetic force, which is strongest near the ends of the magnet.

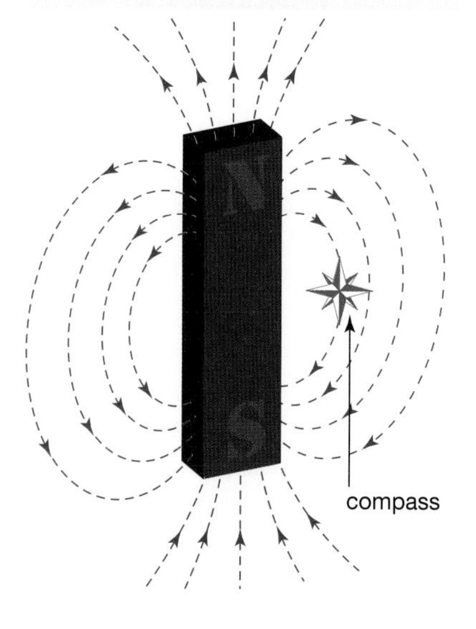

10.10.3 Switched on magnets

An **electromagnet** is a temporary magnet consisting of a coil of wire and an iron core. The coil of wire is called a **solenoid**. When electric current flows in the coil, a magnetic field is produced in and around the solenoid. Without the iron core, a magnetic field would still be created. However, it would not be as strong. The iron core increases the strength of the magnetic field while the current is flowing. The diagram at right shows that, just like a bar magnet, one end of a solenoid is a north pole and the other end is a south pole.

The magnetic field of a solenoid. An iron core inside the solenoid increases the strength of the magnetic field.

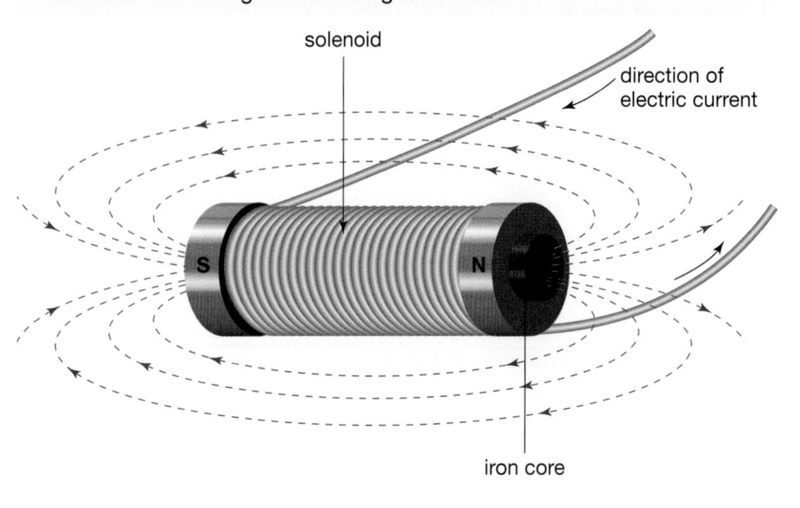

A big advantage of electromagnets over other magnets is that the magnetic field can be turned on or off at the flick of a switch or the push of a button.

INVESTIGATION 10.10

Mapping the magnetic field

AIM: To map the magnetic field around magnets

Materials:

horseshoe magnet in a plastic bag
overhead transparency
2 bar magnets in plastic bags
iron filings
sheet of A4 paper
small compass

Method and results

- Place a bar magnet in the centre of a sheet of white paper. Cover the paper and magnet with an overhead transparency.
- Carefully sprinkle iron filings over the transparency, gently tapping it to spread the filings out. Take care not to let iron filings get under the transparency.

1. Draw a diagram of the pattern made by the iron filings. Label the north pole and south pole of your magnet on the diagram. The pattern in your diagram is a map of the magnetic field around the bar magnet.
   - Use the iron filings to investigate the magnetic fields around a horseshoe magnet and the pairs of magnets shown in the figure on the right.
2. Place a compass at several positions around the magnet. The direction in which the compass needle points shows the direction of the magnetic field lines. Add arrows to your diagram to show the direction of the magnetic field.
3. Do the magnetic field lines run from north pole to south pole or from south pole to north pole?
4. Draw diagrams of the magnetic fields around the magnets in the figure at right. Use your compass to help you decide which way the arrows should go on your diagram.

Discuss and explain

5. Where does the magnetic field appear to be strongest? How do you know this?
6. What happens to the strength of the magnetic field as you get further from the magnet?

Use the iron filings to investigate the magnetic fields around these magnets.

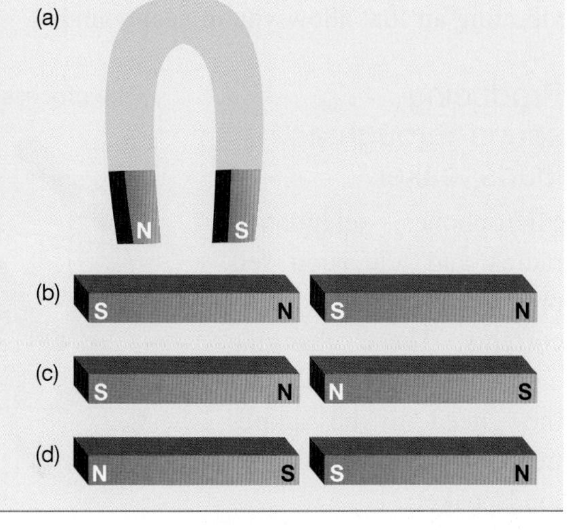

INVESTIGATION 10.11

A look at the field

AIM: To investigate the magnetic field in and around an electromagnet

Materials:

insulated copper wire
power supply
switch
compass
large iron nail

Method and results

Circuit for an electromagnet

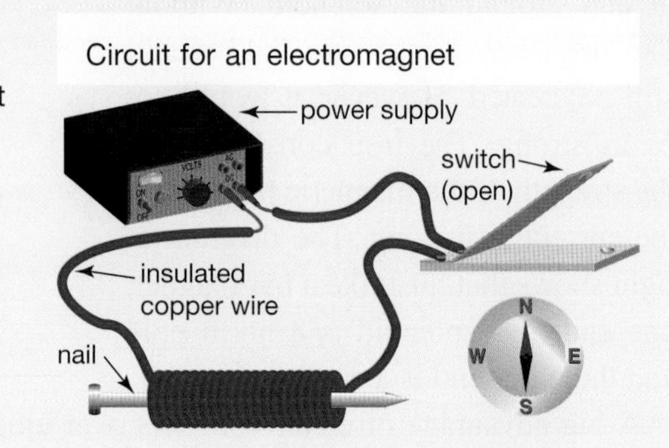

- Set up the circuit shown at right. Wind 20 turns of wire around the nail. There will be a lot of wire left over but don 't cut it.
- Set the power supply to two volts.
- Use the compass to find north and line the nail up so that the sharp end points to the east.
- Place the compass at the sharp end of the nail as shown in the diagram and close the switch just long enough to observe any change in direction of the compass needle.

1. State whether the sharp end of the nail is the north or south pole of the electromagnet. (Remember that the end of the compass that originally pointed north now points in the direction of the magnetic field.)
   - Place the compass at the blunt end of the nail, close the switch briefly and observe the change in direction of the compass needle.
   - Place the compass beside the nail, close the switch briefly and observe the change in direction of the compass needle.
2. Draw a diagram of the magnetic field in and around your home-made electromagnet. Label the north and south poles.
   - Reverse the connections to the terminals of the power supply so that the electric current travels in the opposite direction.
   - Repeat your tests with the compass to find which end of the electromagnet is the north pole and which end is the south pole.
3. What happens to the electromagnet when the direction of the electric current is reversed?

10.10.4 Moving coils

Because a coil of wire carrying electric current produces a magnetic field, it is not surprising that the same coil of wire — as long as it carries an electric current — moves when it is in the region of another magnet. Just as two magnets attract or repel each other, the coil can be attracted to or repelled by a magnet. Loudspeakers rely on this interaction to convert the changing electric currents into the kinetic energy of vibrating air that allow you to hear sound.

Producing sound through a loudspeaker

Microphones, telephones, radios and television sets produce changing electric currents. These changing currents are so small that they need to be amplified before being sent to a loudspeaker.

The process of producing or reproducing sound through a loudspeaker

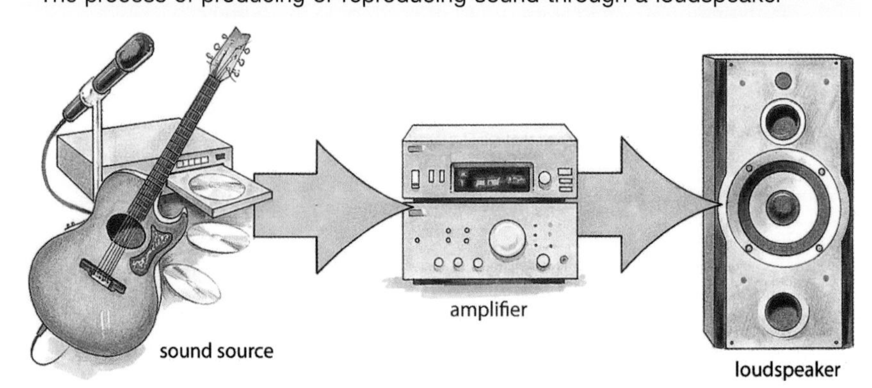

Inside a speaker

Inside a speaker, the electric current flows through the copper wire coil that is tightly wound around the base of the **cone**. The coil, known as the **voice coil**, and the base of the cone sit inside a cylindrical permanent magnet. Like any solenoid with electric current flowing through it, the coil produces its own magnetic field. However, in a speaker the electric current is rapidly changing direction. The coil moves backwards and forwards as it is alternately repelled and attracted to the base of the cylindrical magnet. The cone of the speaker, also known as the **diaphragm**, vibrates as the coil moves, causing the air nearby to vibrate, producing sound.

10.10.5 Phone talk

Your telephone handset or mobile phone has two parts: an earpiece and a mouthpiece. When you have a conversation on a telephone, a microphone in the mouthpiece converts sound into electrical energy so that a signal can be sent to the other telephone. A speaker in the earpiece converts the electrical energy received from the other telephone into sound energy.

The mouthpiece of a telephone handset contains a small metal plate (the diaphragm), which vibrates when you speak. A crystal beneath the diaphragm gets pushed out of shape when the diaphragm vibrates. The electric current flowing from the microphone changes with the shape of the crystal.

The earpiece of a telephone handset contains a speaker. It consists of a permanent magnet, a voice coil wound around a soft iron core and a plastic disc (the diaphragm). The plastic disc is pulled backwards and forwards by the iron core in the voice coil as the electric current in the coil changes.

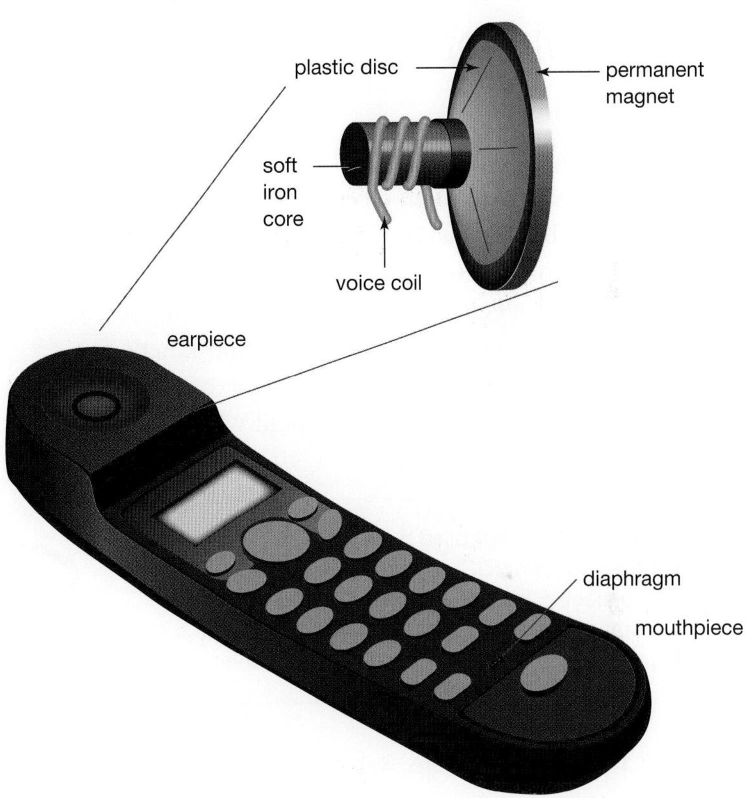

10.10 Exercises: Understanding and inquiring

To answer questions online and to receive **immediate feedback** and **sample responses** for every question, go to your learnON title at www.jacplus.com.au. *Note:* Question numbers may vary slightly.

Remember

1. How is a temporary magnet different from a permanent magnet?
2. What is a magnetic field?
3. Explain the difference between an electromagnet and a solenoid.
4. What advantage do electromagnets have over other types of magnets?
5. What is the role of a permanent magnet in a loud speaker?
6. Explain why the voice coil in a speaker is attached to the cone.

Think

7. List some common devices that make use of electromagnets.
8. Why does a compass needle line up with a magnetic field?

9. The Magnetic North Pole of the Earth can be considered as one pole of a bar magnet. Is it a south pole or a north pole? Explain your answer.
10. What two important energy transformations take place inside the microphone of a mobile phone before a signal is transmitted through the air?

Create

11. Construct a device that uses an electromagnet.

Try out this interactivity: Magnetic flux and Lenz's Law (int-0050)

10.11 Motoring along

10.11.1 Motors

What do a hair dryer, a DVD player, a food processor, a clothes dryer and an electric drill all have in common? The obvious answer is that they all use electrical energy. Another thing that these appliances have in common is an **electric motor**. An electric motor is a device that converts electrical energy into kinetic energy. An electric motor turns because it contains coils of wire that produce a magnetic field when electric current flows through them.

An electric motor converts electrical energy into kinetic energy. This conversion can only take place because of the magnetic effects of electric current.

The electric current supplied by a cell or battery is called *direct current* (DC). It flows in one direction only. The electric current provided to your home by power stations is called *alternating current* (AC). It changes direction about 100 times every second.

10.11.2 How a DC electric motor works

The armature

This is the turning part of the motor on which coils of wire are wound. The coils are called rotor coils because they cause the **armature** to rotate.

The rotor coils

When electric current flows through the **rotor coils**, a magnetic field is produced. The magnetic field produced by these coils interacts with the magnetic field produced by the **field magnets**. The repulsive and attractive forces acting on the rotor coils cause them to turn.

The field magnets

The field magnets are permanent magnets that do not move. In larger commercial motors they are replaced with a separate coil (called a field coil) which provides a stationary electromagnet.

The brushes

These **brushes** are connected to the power supply and lightly touch the commutator as the armature turns. This allows current to travel through the rotor coils.

The shaft

This part of the motor is attached to the device the motor is turning, like a fan or gear wheel. As the armature turns, the **shaft** turns.

The commutator

As each rotor coil turns through 180 degrees to face the opposite field magnet, the force on it would change direction, turning it back the other way. The **commutator** consists of a split metal ring. As the armature turns, the commutator turns with it while the brushes remain still. When the armature has turned through 180 degrees, the opposite side of the commutator makes contact with the brush connected to the positive terminal of the power supply. The direction of the current in each rotor coil reverses. This allows the armature to keep rotating in the same direction, rather than spinning first one way, then the other.

A commercial DC motor

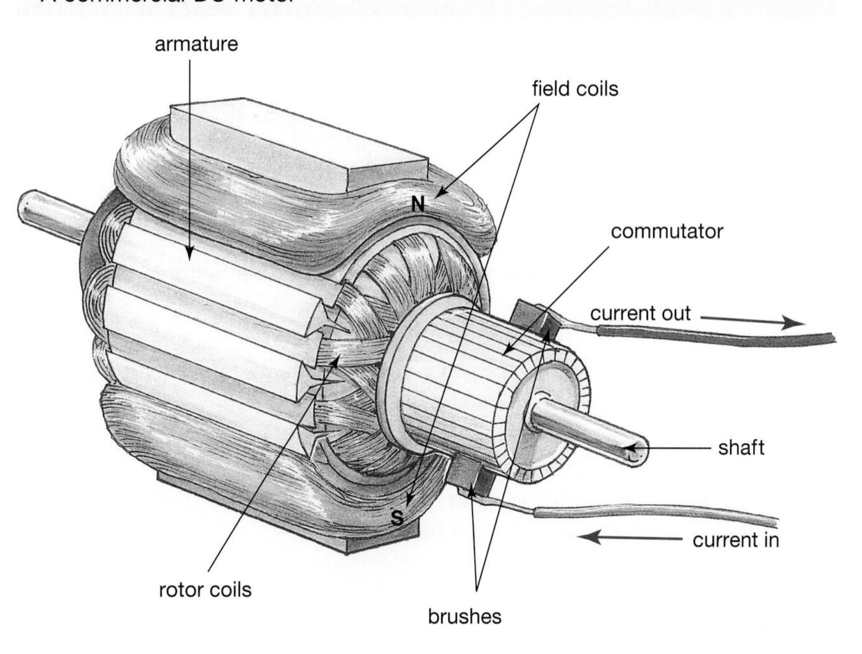

A simplified diagram of a DC electric motor

TRY THIS

Tinkering with an electric motor

You will need:

electric motor kit and/or DC electric motor

- Use an electric motor kit to build an electric motor. Follow the instructions that come with the kit very carefully.

1. Draw a diagram of the electric motor that you have built and label as many of the parts shown in the diagram at right as you can.

- If an electric motor is available, examine it closely.

2. Draw a diagram of the DC electric motor and label as many of the parts shown in the diagram at right as you can.

10.11 Exercises: Understanding and inquiring

To answer questions online and to receive **immediate feedback** and **sample responses** for every question, go to your learnON title at www.jacplus.com.au. *Note:* Question numbers may vary slightly.

Remember

1. Into what form of energy do motors convert electrical energy? How does this differ from the energy transformations in microphones and speakers?
2. Explain the difference between AC and DC.
3. Construct a table to group all the parts of an electric motor into 'moving' and 'non-moving' parts.

Think

4. List as many appliances or toys as you can that contain electric motors.
5. Why doesn't an AC motor need a commutator?

Imagine

6. Imagine you were advertising in the employment pages of a newspaper for the parts of a DC electric motor. Write a job description for each part.

Investigate

7. The needles of moving coil ammeters and galvanometers are connected to a coil of copper wire that lies between the poles of a magnet. Take a close look at one of these instruments and write an account of what makes the needle move when a current passes through the coil.

10.12 Generating electricity

10.12.1 It's all relative

Several years after Danish scientist Hans Ørsted discovered that electric currents produced magnetic fields, British scientist Michael Faraday suggested that perhaps a magnetic field might produce an electric current. In 1831, Faraday succeeded in generating an electric current by moving a coil of wire through a magnetic field. He had made the very first **electric generator**, also known as a **dynamo**.

Current flows in a generator as long as there is relative movement between the magnetic field and the coil. It does not matter if it is the coil that moves while the magnet remains still, or the magnet that moves while the coil remains still. In a bicycle generator, a magnet spins while the coil of wire remains still. In the generator of a motor car, many coils spin around inside a stationary electromagnet. The coils are turned by the car's engine. The electromagnet is connected to the car battery. The car's generator is used instead of the battery to supply electric current to other parts of the car while the engine is running.

A bicycle dynamo generates electricity to keep a headlamp glowing without the need for a battery. As the wheel turns, a magnet spins inside a stationary coil of wire. The relative movement between the magnetic field and the coil results in an electric current in the coil which powers the headlamp.

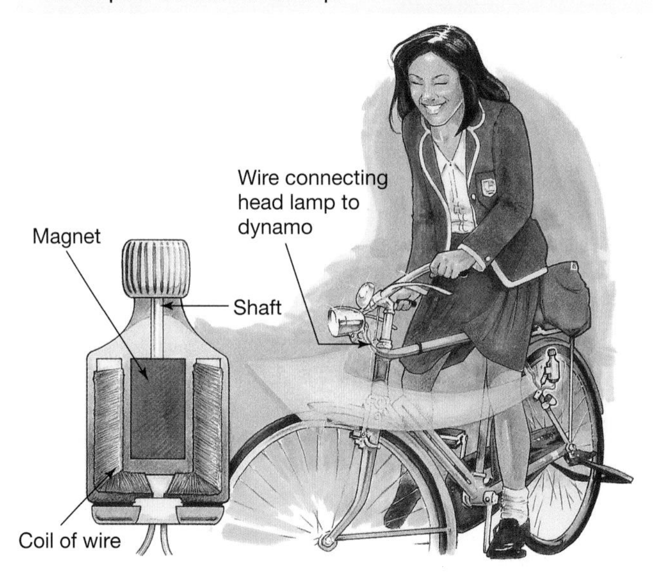

Increasing the current

The size of the electric current produced by a generator can be made larger by increasing:

- the number of turns of wire in the coil
- the strength of the magnet
- the speed of the relative movement between the coil and the magnetic field.

This car generator looks very much like an electric motor. However, instead of converting electrical energy to kinetic energy, it converts kinetic energy into electrical energy. Each slip ring is connected to one end of each coil of wire. Like the bicycle dynamo shown on the previous page, it generates an alternating current (AC).

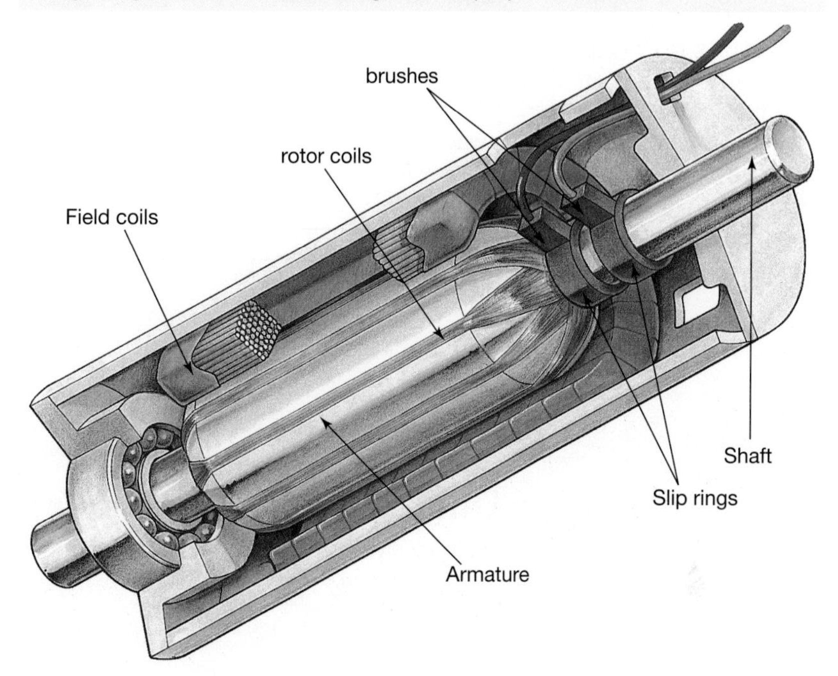

10.12.2 AC or DC?

The current flowing from an AC generator changes direction after every half turn. In a DC generator, the slip rings are replaced with a commutator. The electric current generated then flows in the same direction while the armature is turning.

INVESTIGATION 10.12

Electrical energy from kinetic energy

AIM: To investigate the generation of electric current by the movement of a magnet inside a coil of wire

Materials:

bar magnet
length of insulated copper wire
cardboard tube
masking tape
galvanometer
large iron nail

Method and results

- Make a solenoid by winding the insulated copper wire evenly around the cardboard tube. Tape the wire down so that it cannot unwind itself. Connect the free ends of the wire to the galvanometer.
- Place the magnet inside the solenoid so that the end you are holding is just inside the cardboard tube.

1. A galvanometer is used to detect and measure small electric currents. What is the reading on the galvanometer while the magnet is inside the solenoid?
   - Watch the needle on the galvanometer while you rapidly pull the magnet out of the solenoid.
2. Describe what happens to the needle of the galvanometer while the magnet is being pulled out.
   - Watch the galvanometer needle while you rapidly push the magnet back into the solenoid.
3. Describe what happens to the needle of the galvanometer while the magnet is being pushed in.
4. Predict which way the needle of the galvanometer will move if the magnet is reversed and pulled out of the solenoid and then pushed into the solenoid.
   - Test your predictions about the movement of the galvanometer needle with the magnet reversed.
5. Were your predictions correct?
6. Does a stationary magnetic field inside a solenoid produce an electric current in the solenoid?

7. Does a moving magnetic field inside a solenoid produce an electric current in the solenoid?
   - Find out how the size of the electric current changes if the magnet is moved faster or slower, if the solenoid is pulled away from the magnet or if a larger magnet is used.

Discuss and explain

8. How is the current affected if the magnet is moved into or out of the solenoid faster?
9. Does pulling the solenoid away from the magnet have the same effect as pulling the magnet away from the solenoid?

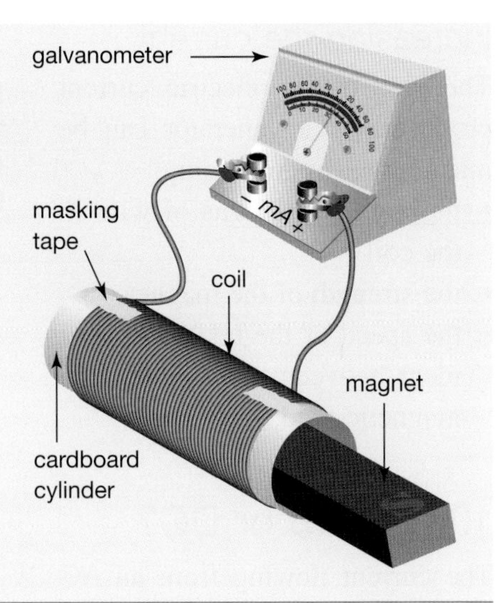

10.12 Exercises: Understanding and inquiring

To answer questions online and to receive **immediate feedback** and **sample responses** for every question, go to your learnON title at www.jacplus.com.au. *Note:* Question numbers may vary slightly.

Remember

1. How did Michael Faraday generate an electric current?
2. How is the electric current produced by an AC generator different from the electric current produced by a DC generator?
3. What changes can be made to a dynamo to increase the size of the electric current generated?

Think

4. Examine the diagram of the car generator on the previous page.
   (a) How is the AC generator similar to an electric motor?
   (b) Describe the differences between an AC generator and a DC motor
   (c) The electric generator is sometimes described as the reverse of an electric motor. Explain why this is so.
5. The speakers used in some fast food 'drive thru's' also act as microphones. They contain a permanent magnet, a voice coil and a diaphragm just as normal speakers do. (See pages 464 and 465.) Explain how it might be possible to use the speaker to change sound energy into electrical energy.

Create

6. Use readily available materials to make your own model generator. Use a sensitive galvanometer to check that it works. Remember that your main objective is simply to create a model that allows relative movement between a coil of wire and a magnet.

Investigate

7. Investigate the operation of a hand-operated generator.
   (a) Use a stopwatch, a voltmeter and two connecting leads to investigate how the number of turns of the handle per minute affects the voltage output.
   (b) Draw a diagram of a hand-operated generator and label the magnet, the coil, the slip rings (or commutator), the shaft and the brushes.
   (c) Explain in your own words how a hand-operated generator produces an electric current.

10.13 Electricity at home

Science as a human endeavour

10.13.1 The generator

The method used to generate the electricity used in your home, school or workplace is not very different from that used by a single hand-operated generator or bicycle dynamo. Of course, the scale is very much larger.

In coal-fired power stations electrical energy is transformed from the chemical energy stored in coal. When it is burned, chemical energy stored in the coal is used to boil water to create high pressure steam. The steam pushes on the blades of fan-like turbines, which then rotate coils of wire rapidly inside huge electromagnets. The motion of the coils in the magnetic field produces a large electric current.

In nuclear power stations the energy required to boil water to produce the steam that turns the turbine blades is released in nuclear reactions. In hydro-electric power stations, the energy used to turn the turbines is transformed from gravitational potential energy. Water falling from a great height turns the turbines directly with no need for high pressure steam. About 90 per cent of Australia's electrical energy is generated by coal-fired power stations. Most of the remaining 10 per cent is provided by hydro-electric power stations. A very small number of Australian power stations use gas as a fuel.

A superheated steam turbine generator during construction

WHAT DOES IT MEAN?

The term *hydro* in hydro-electric comes from the Greek word *hydor*, meaning 'water'.

The electrical energy supplied by coal-fired power stations is transformed from chemical energy.

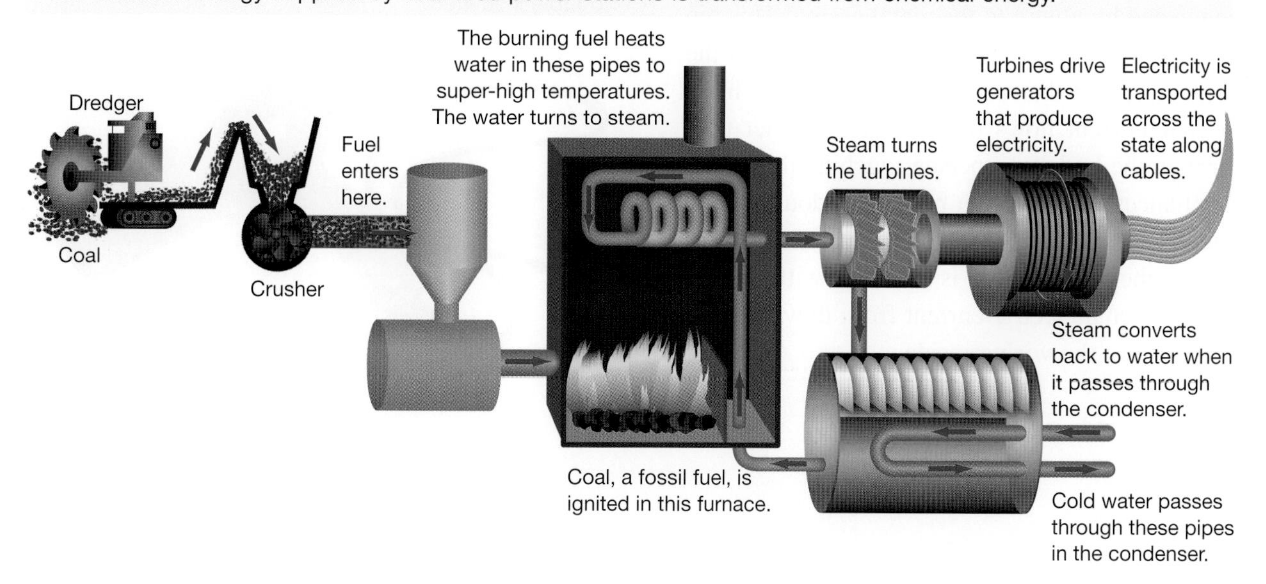

10.13.2 AC or DC — what's the difference?

The electric current supplied by a cell or battery is called **direct current** (DC). It flows in one direction only. The electric current provided to your home by power stations is called **alternating current** (AC). It changes direction about one hundred times every second. In household lights and appliances, electrical energy is transformed into other forms of energy as electrons move backwards and forwards.

Alternating current, rather than direct current, is supplied by power stations because it is easier and cheaper to generate. It is also easier and cheaper to distribute widely over large distances. In Australia, electricity is supplied to homes at a voltage of 240 volts.

10.13.3 It's in the box

The electric cable that carries alternating current to your home holds two wires. When the **main switch** in your home's meter box is open, current doesn't flow through the wires in the cable. The circuit is not complete. When the main switch in your meter box is closed, electric current is able to flow through these wires. However, it flows only when other switches inside your house (such as light and appliance switches) are closed. When the main switch is open, the current also flows through the meter in your meter box.

10.13.4 Why there are three

Power points have three sockets. When you plug in an electrical device and switch on the power, alternating current flows between the top two sockets through the appliance. The third socket, called the **earth socket**, is connected to a metal pipe in the ground.

If an electrical device has an uninsulated metal casing, its plug has three pins. The bottom pin is connected by a wire to the metal casing. This pin fits into the earth socket. If there is a fault in the appliance, and the metal casing becomes 'live', electric current flows to the ground, rather than through the body of a person touching the metal. Appliances with two-pin plugs are 'double insulated' to make them safe. Any metal on the outside of these appliances is insulated with plastic. This prevents electric current from flowing from the metal to the wiring inside.

Hand-held electric appliances like this hairdryer are double insulated.

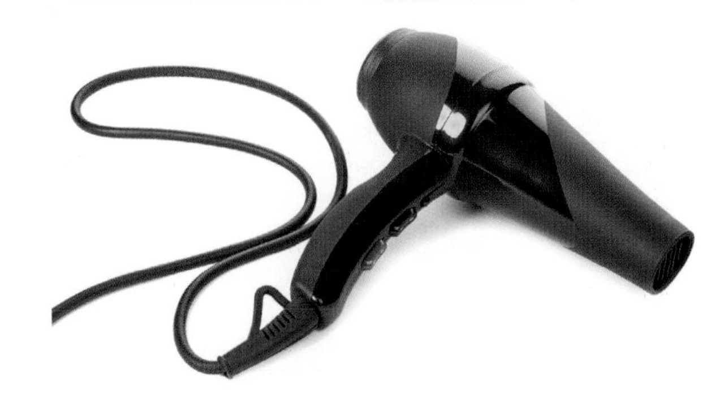

10.13.5 Stepping down

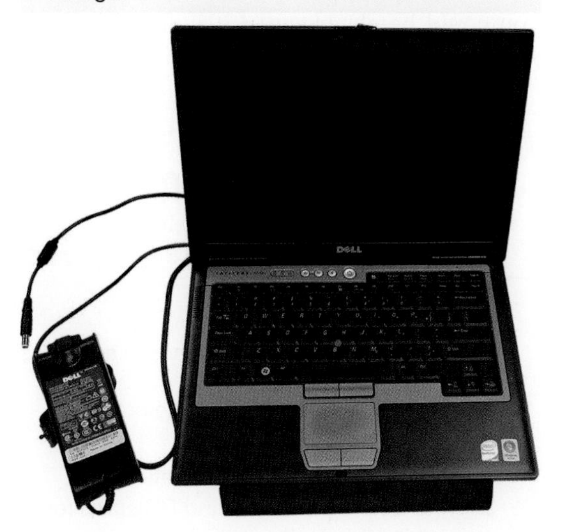

Some electrical devices require less than the 240 volts supplied by power points. A transformer is used to 'step down' the voltage.

Many of the electrical devices that you plug into power points are not designed to use a 240-volt power supply. For example, devices like computer printers and electronic games require voltages of 9 volts or 12 volts. A **transformer**, usually attached to the lead, is plugged into the power point. The transformer reduces the voltage from 240 volts to the 9 volts or 12 volts required.

Some devices that use DC current, such as mobile phone battery chargers, are plugged into power points. The black box that is plugged in contains a transformer, which 'steps down' the voltage, and another circuit, called a **rectifier**, that changes the alternating current into direct current.

10.13.6 Keeping circuits safe

The appliances and lights in your home are all connected in parallel. There are normally two separate parallel circuits in a house, one for the lights and one for the power points. If too many appliances or lights are turned on at once, the total current is too large and a **fuse** or **circuit breaker** can open the main circuit and stop the flow of current. A set of fuses or circuit breakers can be found in your meter box at home. A fuse is a short piece of wire that melts if the current gets too high. A circuit breaker is a special switch that opens automatically if the electric current gets too high.

Safety switches

The switch on the left is a safety switch. The other switches are circuit breakers.

Fuses and circuit breakers open circuits before they overheat. But they work too slowly to stop people from getting an electric shock if a short circuit causes a dangerously high current. Safety switches (also known as residual current devices) can turn off the power much more quickly — in less than one-thirteenth of a heartbeat — thus reducing the risk of death due to electric shock in the home.

10.13.7 Take care!

The 240-volt AC household power supply can kill. If you tamper with working appliances or electrical wiring, it is possible that electric current will flow through your body. Electrocution — death from electric shock — can be caused by electric currents as low as 0.05 amperes flowing through your body.

One of the biggest causes of electrocution in the home is the use of damaged cords and plugs. If appliance cords and plugs are frayed or damaged, exposing the smaller plastic covered wires inside, the appliance should be replaced or taken to a qualified repairer.

The bathroom can be a very dangerous place in which to use electrical appliances. Tap water contains charged particles (ions) due to substances dissolved in it. It is therefore a good conductor of electric current. Appliances like hair dryers, radiators and electric shavers should never be used when there is water in the basin or bath, or when there is water on the floor. If a working appliance was to fall into some water, you could be electrocuted if you came in contact with the appliance, either by accidentally touching it or by trying to pick it up.

10.13 Exercises: Understanding and inquiring

To answer questions online and to receive **immediate feedback** and **sample responses** for every question, go to your learnON title at www.jacplus.com.au. *Note:* Question numbers may vary slightly.

Remember

1. What is the role of the turbines in a power station?
2. What drives the turbines in:
   (a) a coal-fired power station?
   (b) a nuclear power station?
   (c) a hydro-electric power station?
3. What do the letters AC and DC stand for?
4. Explain why devices like electronic games and computer printers have heavy transformers attached to their leads or plugs.
5. What should be done when appliance cords or plugs become frayed or damaged?
6. Why does extra care need to be taken when using electrical appliances in the bathroom?
7. What important roles do both fuses and circuit breakers play?
8. Explain how circuit breakers are different from fuses.

Think

9. From which forms of energy (not just the initial form) is electrical energy converted in:
   (a) a coal-fired power station?
   (b) a hydro-electric power station?
10. How is the generation of electricity in a power station similar to the generation of electricity by a hand-operated generator or bicycle dynamo?
11. What form of energy is the electrical energy for a streetlight most likely to have been transformed from in:
    (a) Melbourne
    (b) Hobart?
12. Why do power points have three sockets even though many appliances have plugs with only two pins?
13. Not all of the chemical energy stored in coal is transformed into useful electrical energy in the home. List the ways in which energy is wasted from the time it is burned up until the time that it is used to light a room at night.

Investigate

14. Research and report on the generation of electricity by hydro-electric power stations in Australia. Find out about their location and history.
15. On the back of a mobile phone battery charger is printed 'Input: AC240V 50Hz'. Find out what '50Hz' means.
16. Find out how a step-down transformer works. Use a diagram to illustrate your answer.

10.14 Concept maps and flowcharts

10.14.1 Concept maps and flowcharts

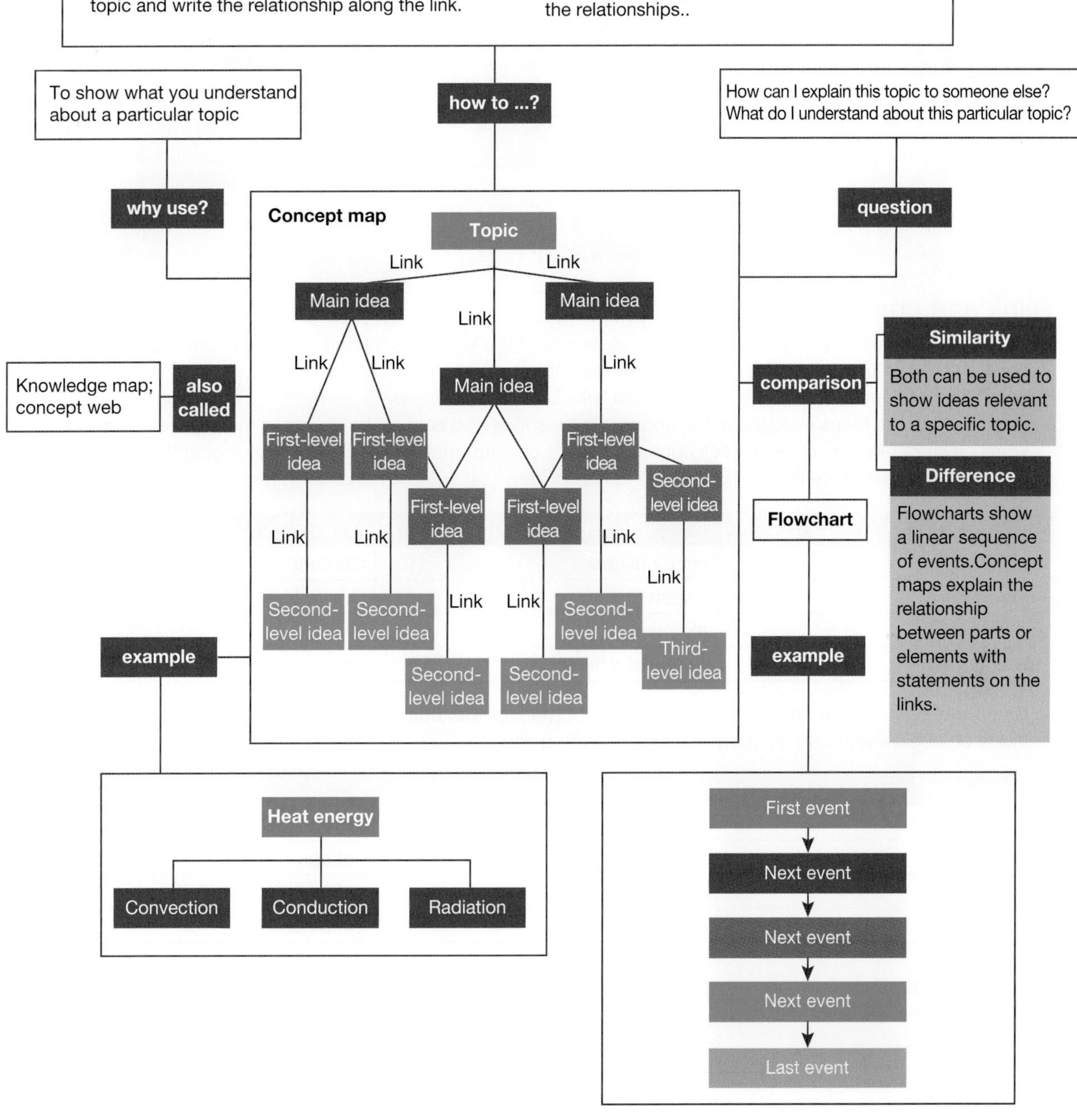

10.14 Exercises: Understanding and inquiring

To answer questions online and to receive **immediate feedback** and **sample responses** for every question, go to your learnON title at www.jacplus.com.au. *Note:* Question numbers may vary slightly.

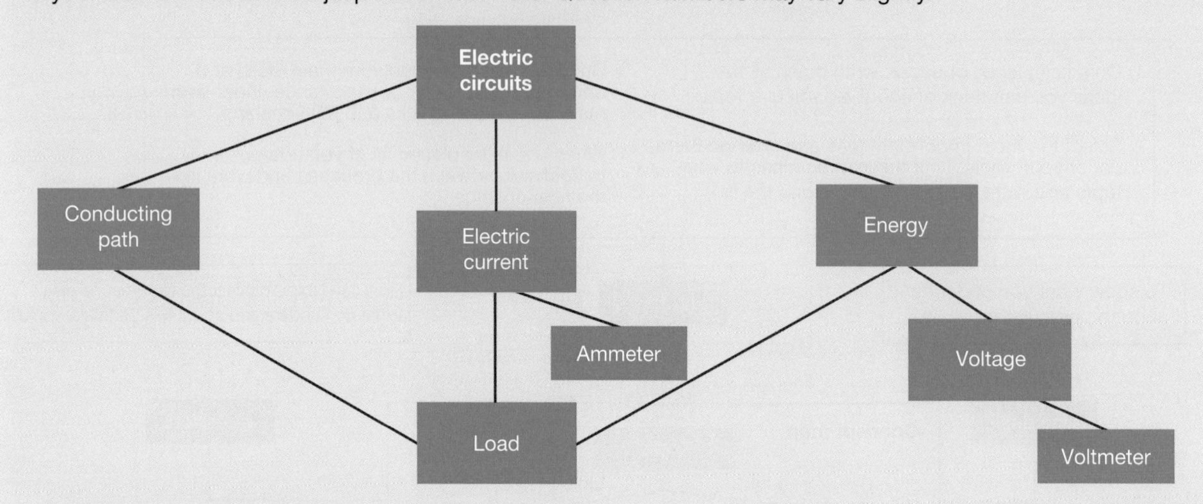

Think and create

1. The incomplete concept map above represents some of the key ideas related to electric circuits. This concept map is just one way of representing ideas about matter and how they are linked. Copy and complete the concept map by writing suitable links between the ideas.
2. Write each of the ideas included in the concept map above and each of the terms in the box below on a small piece of paper such as a 'sticky note'. Create a concept map of your own by arranging the ideas on an A3 sheet of paper. Write links in light pencil at first in case you want to make changes to your arrangement.

Electric circuit ideas		
resistance	circuit diagram	filament
in parallel	switch	Ohm's Law
in series	electrons	load
power supply	potential difference	power supply

3. Use sticky notes and A3 paper to create a concept map for the topic of household electricity. Use the ideas in the box below and add as many other ideas as you can.

Household electricity ideas		
alternating current	main switch	appliances
electric power	transformer	voltage
in parallel	rectifier	safety
direct current	electrical energy	circuit breaker

4. Create two separate flowcharts that show, step by step, how to connect each of the circuits below so that the light globes glow.

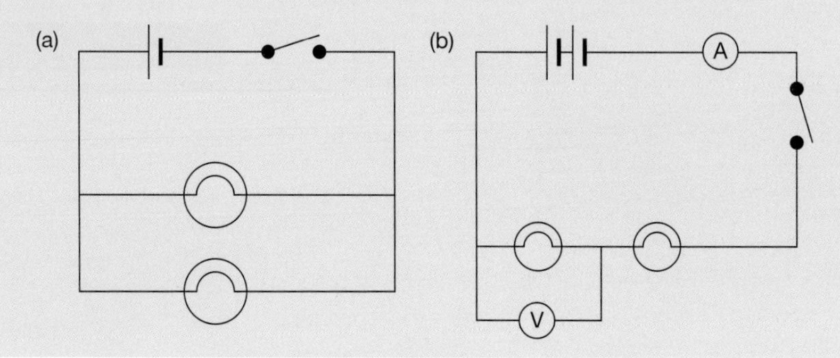

5. Create a flowchart to show how electricity is produced from fossil fuels.

learn on

10.15 Project: Go-Go Gadget online shop

Scenario

We use the term *technology* to describe the application of science to develop devices, machines and techniques to make some aspect of our lives easier. Televisions, satellites and the internet are all pretty obvious examples of technology, but small devices such as the automatic cat-flap and the humble vegetable peeler are also forms of technology. Small or specialised pieces of technology such as these are often referred to as *gadgets*. Every year, patents for thousands of such gadgets are issued to inventors. Some of them, like the NavMan, are immediate successes, while others — for example, a combination shoe-polisher and toothpick — don't make it into mass production. So what happens if you need a device to do a particular job but no-one has ever made one?

This is just what you and your partners were thinking when you decided to open the Go-Go Gadget online shop. Once established, clients would browse designs for gadgets that you have already developed or ask you to design something new for them that will do the job they need done. Maybe the client wants a hamster wheel that can drive a coffee-grinder or a signalling device that will tell a cat-owner whether their cat has come inside through the cat-flap or is still outside. They just tell you what they need and you design it for them! You then ship them the design, the parts they need to assemble it and an instruction brochure.

To get the business started, you decide to take out a business loan with the bank. The bank manager is intrigued with the idea but wants some assurance that you know what you are doing before they hand over the money.

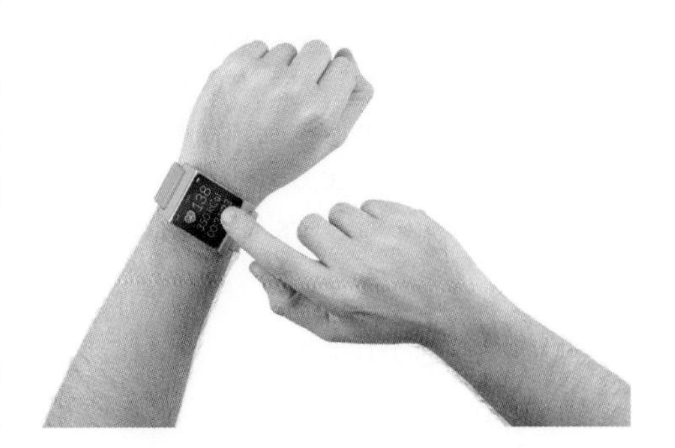

Your task

As part of your presentation to the bank, you and your business partners are to develop a design for one of the following clients.

- Taylor wants a snooping-parent device that will warn her when one of her parents is coming up the hallway that leads to her bedroom. This device will give her a silent signal so she has time to turn off her computer and open her homework books before they open the door and catch her playing computer games or surfing the net instead of working.
- Heisenberg has an office on the top floor of his house. His cat, Schrödinger, can enter the house through a cat-flap in the door downstairs. When Heisenberg is locking up the house when going out or to bed, it would save a lot of time if he could know whether the cat is already inside the house. He needs a device that is connected to the cat-flap that sends a signal to Heisenberg upstairs indicating whether the cat has come in or gone out the cat-flap.
- Felicity often works until late at night and doesn't get time to exercise her dog by taking her out for a walk. She can use her computer at work to turn on switches in her apartment, and wants a device that will allow her to exercise her dog by remote control without the dog leaving the apartment.

You will then create the following to submit to the bank in support of your loan application.

1. A brief overview (approximately 300 words) of why there is a market for the services of your online shop. To support your argument, you should include references to gadgets that have been successfully developed.
2. A brochure for the gadget you have designed that includes:
   - a diagram of your design
   - a list of parts that are included in the package sent with the brochure
   - instructions on assembly/installation of the gadget
   - a troubleshooting guide to solve problems.

10.16 Review

10.16.1 Study checklist

Electric circuits

- describe the properties and uses of electronic components such as LEDs, and temperature and light sensors
- investigate the use of sensors and microprocessors in robotics and control devices
- distinguish between static electricity and the flow of electric current in a closed circuit
- explain energy transfer in an electric circuit
- relate electric current to the flow of electric charge in a closed circuit
- relate voltage to the energy gained or lost by electric charge as it moves through an electric circuit
- draw and interpret electric circuit diagrams
- distinguish between series and parallel circuits
- measuring voltage drops across and currents through various components of electric circuits
- appreciate the errors associated with reading scales and distinguish between random and systematic errors

Cells and batteries

- compare the advantages of alkaline and mercury cells over general-purpose dry cells
- describe the operation of rechargeable batteries
- define electrical resistance and explain how it affects the electric current flowing through a circuit

- distinguish between conductors and insulators
- distinguish between ohmic and non-ohmic conductors

Electromagnetic effects

- investigate and describe the magnetic field around magnets and current-carrying wires
- describe and explain the operation of a loud speaker and an electric motor
- investigate and describe the generation of an electric current by relative movement between a wire and a magnetic field

Electricity at home

- distinguish between alternating and direct currents
- explain why household power points have three sockets
- describe the role of fuses and circuit breakers
- describe the role of transformers and rectifiers as they are used in household appliance leads

Science as a human endeavour

- use a scientific understanding of heat transfer to explain how humans maintain a safe body temperature
- investigate the contributions of Alessandro Volta and André-Marie Ampère to the study of electricity
- evaluate claims made in the media about the benefits of hybrid cars
- describe the dangers of tampering with electrical wiring, damaged cords and plugs and using electrical appliances near waternce of

Individual pathways

ACTIVITY 10.1	ACTIVITY 10.2	ACTIVITY 10.3
Revising electricity doc-23903	Investigating electricity doc-23904	Investigating electricity further doc-23905

learnON ONLINE ONLY

10.16 Review 1: Looking back

To answer questions online and to receive **immediate feedback** and **sample responses** for every question, go to your learnON title at www.jacplus.com.au. *Note:* Question numbers may vary slightly.

1. Match each term with its correct description.

Word	Description
Static electricity	A material that allows current or heat to flow through it
Electron	Positively charged particle in the nucleus of an atom
Proton	The build-up of charge on an object
Current	A material that does not allow current or heat to flow through it easily
Voltage	Particle in an atom with a negative charge
Conductor	A path that has no breaks in it
Closed circuit	The energy supplied to move electrons around a closed circuit
Insulator	The flow of electrons around a closed circuit

2. Which of the following circuits are parallel circuits and which are series circuits?

(a)

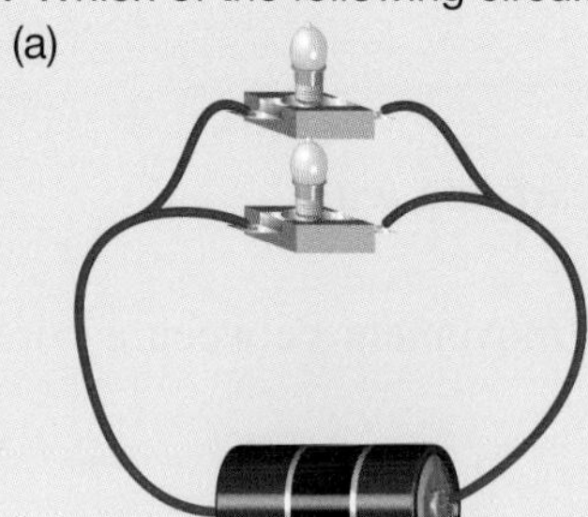

(b)

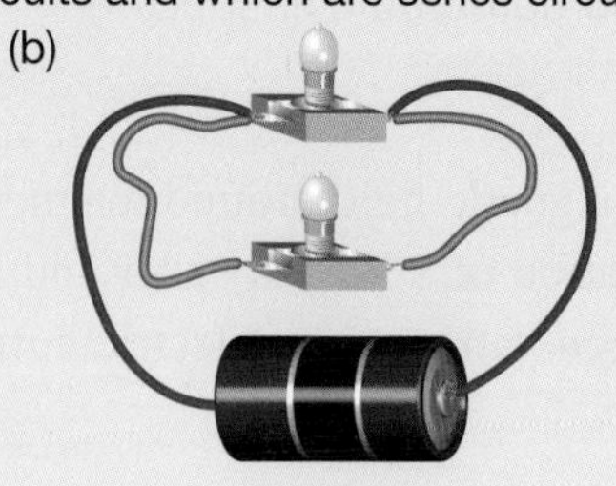

(c)

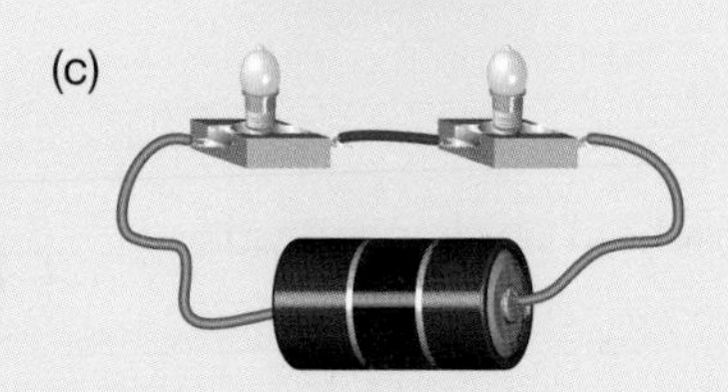

(d)

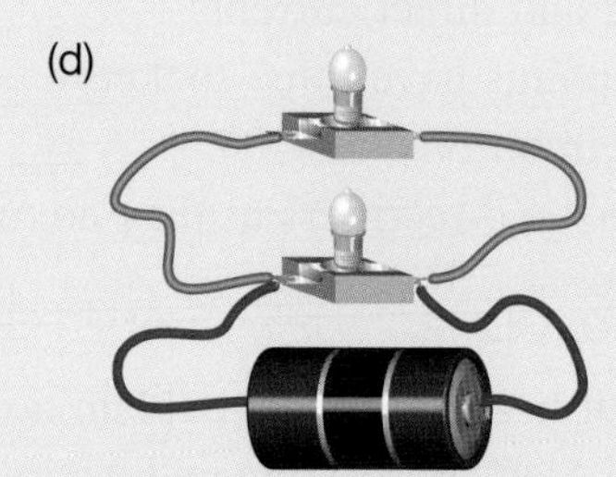

3. Use symbols to draw a circuit containing a light globe in series with an ammeter, a battery and a switch.
4. Using either a particle or a wave model, briefly describe how heat is transferred from one region to another by:
   (a) conduction
   (b) convection
   (c) radiation.
5. Identify the main method or methods by which heat is transferred to the human body by:
   (a) a gas wall furnace
   (b) the sun
   (c) holding a hot plate
   (d) an open fireplace
   (e) walking on hot coals.
6. Explain why cooks often cover meat with aluminium foil instead of plastic.
7. Explain why solids such as polystyrene, foam, wool and fibreglass batts do not conduct heat as well as most other solids.
8. (a) Heat is always transferred from a region of high temperature to a region with a lower temperature. Explain how your body is able to keep its core temperature at 37 °C even when the air temperature is greater than 37 °C.
   (b) Explain how the wearing of light, loose-fitting clothes protects your body from overheating in hot weather.
   (c) Why do your blood vessels get larger in hot weather?
9. In which one or more of the following arrangements will the globe light up?

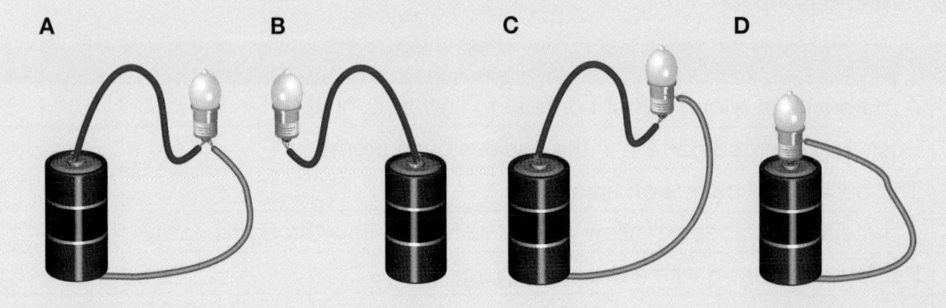

10. Write down the meanings of each of the following terms based on what you have learned from this chapter.
    (a) Conduction
    (b) Convection
    (c) Radiation
    (d) Current

(e) Insulation
(f) Energy
(g) Metal
(h) Charge
(i) Vibration
(j) Voltage
(k) Reflection
(l) Circuit

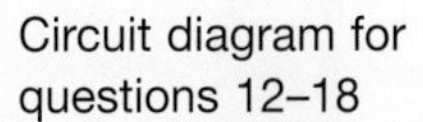
Circuit diagram for questions 12–18

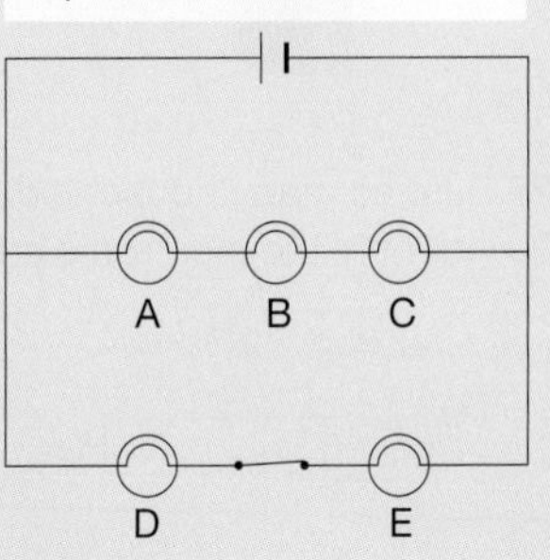

11. Which of the physical quantities electric current and voltage is a measure of a change in energy?

Questions 12–18 refer to the circuit diagram at right. The light globes, labelled A–E, are identical to each other.

12. Which two or more of the light globes are connected:
    (a) in series with globe A
    (b) in parallel with globe A?
13. If the voltage across globe C was measured to be 4 volts, what is the voltage across:
    (a) globe A
    (b) the terminals of the power supply
    (c) globe E?
14. If the electric current flowing through globe B was measured to be 200 mA and the electric current flowing through globe D was measured to be 300 mA, what is the electric current flowing through:
    (a) globe A
    (b) globe E
    (c) the power supply?
15. If the filament in globe B was to break, which of the light globes would remain glowing?
16. If the switch in the circuit was opened, which light globes would stop glowing?
17. How could you make all of the light globes stop glowing without opening the switch or turning off the power supply?
18. The voltage across globe C is measured to be 4 volts and the current flowing through it is 200 mA.
    (a) What is the electric current flowing through globe C, in amperes?
    (b) What is the resistance (in ohms) of globe C while this current is flowing?
19. Complete the table below by writing down the missing quantity, unit or abbreviation.

Electrical quantities and their units

Quantity	Unit	Abbreviation
Voltage	Volt	
Electric current		A
	Ohm	

20. Draw a circuit diagram that shows how a voltmeter and ammeter are used to measure the voltage across and current flowing through a single light globe connected to a 6-volt power supply. Label the positive and negative terminals of the power supply and each meter with + and – symbols.
21. What is the electric current being shown on the ammeter at right if the positive lead is placed in the:
    (a) 500 mA terminal
    (b) 5 A terminal?

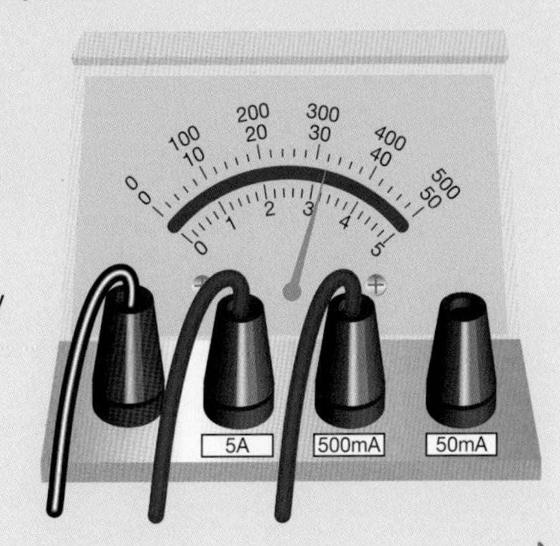

22. Describe an example of (a) a random error and (b) a systematic error when the scale on the ammeter in question 21 is read.
23. Differentiate between an electric cell and a battery.
24. In what way does a conductor that obeys Ohm's Law behave differently from one that doesn't? Sketch a graph to support your answer.
25. How does increasing the resistance of a variable resistor in series with a power supply and a lamp affect the:
    (a) electric current in the lamp
    (b) voltage across the variable resistor
    (c) voltage across the lamp?

26. Identify each of the electronic devices illustrated below.

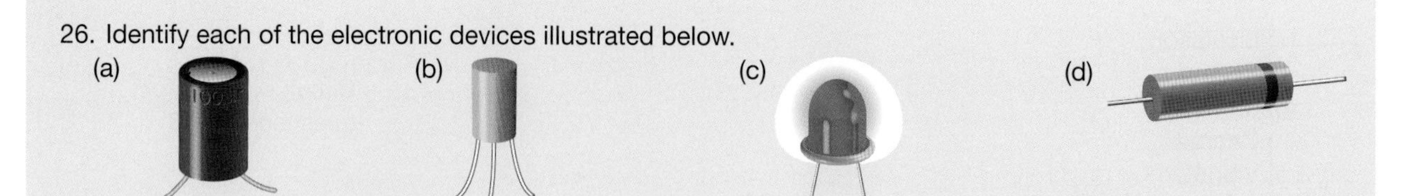

27. List as many household devices as you can that are likely to have been designed by electronic engineers.
28. Identify the electronic components in the diagrams below.

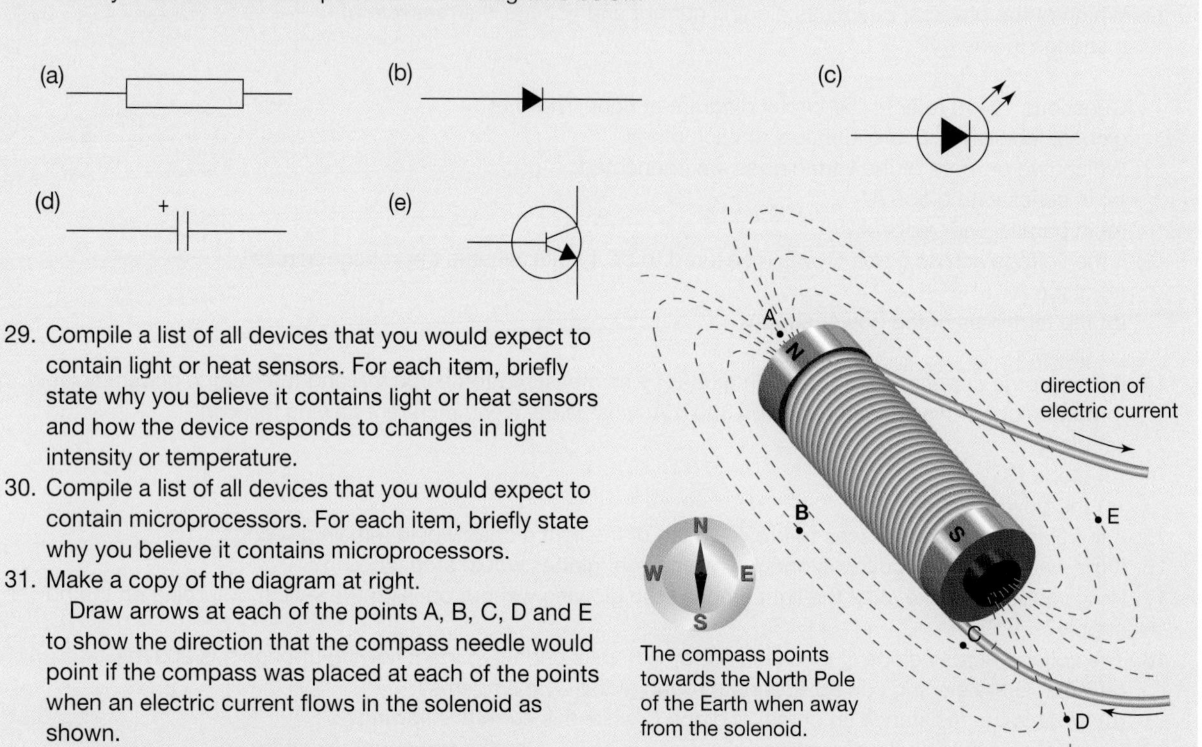

The compass points towards the North Pole of the Earth when away from the solenoid.

29. Compile a list of all devices that you would expect to contain light or heat sensors. For each item, briefly state why you believe it contains light or heat sensors and how the device responds to changes in light intensity or temperature.
30. Compile a list of all devices that you would expect to contain microprocessors. For each item, briefly state why you believe it contains microprocessors.
31. Make a copy of the diagram at right.
    Draw arrows at each of the points A, B, C, D and E to show the direction that the compass needle would point if the compass was placed at each of the points when an electric current flows in the solenoid as shown.
    At which of the points A, B, C, D or E is the magnetic field the strongest?
32. Use a flowchart to show how a moving coil speaker transforms electrical energy into the sound energy that you hear.

Cross-section of a moving coil speaker

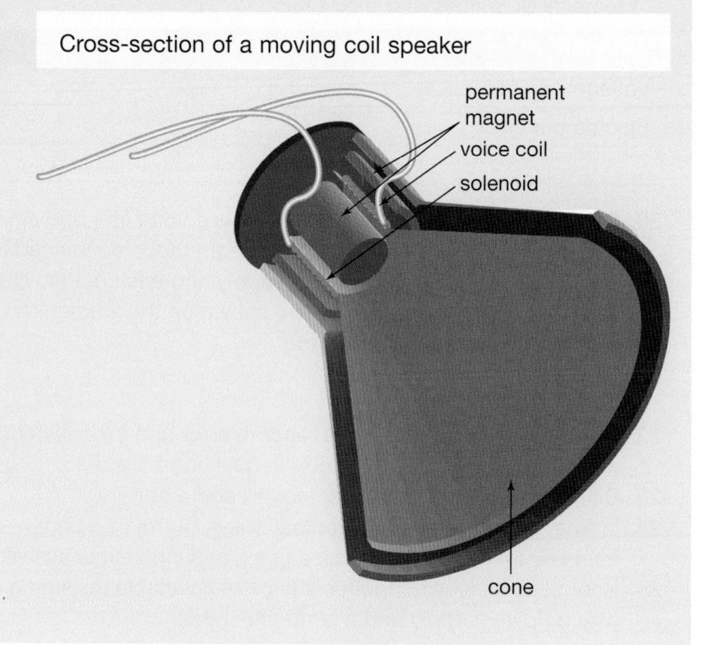

33. Make a copy of the diagram at right.
    (a) Complete the labelling of the diagram.
    (b) Copy and complete the table below to indicate the role of each of the parts of the motor listed.

Part	Purpose
field magnets	
armature	
rotor coil	
shaft	
brushes	
commutator	

A simplified diagram of a DC electric motor

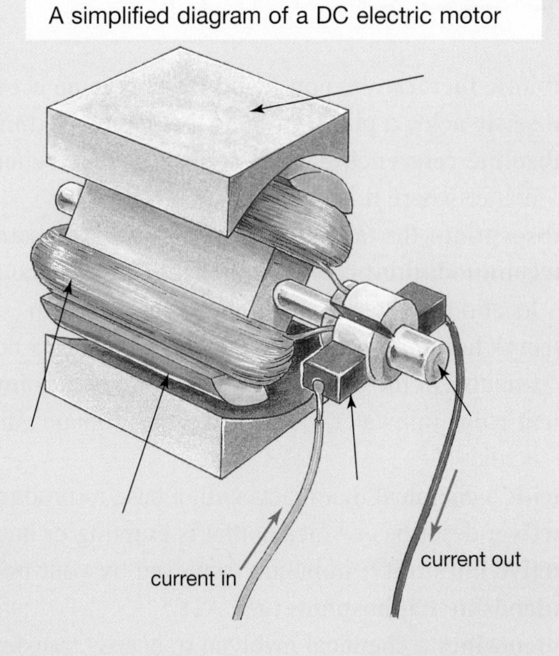

34. Many electrical devices contain electromagnets.
    (a) Is an electromagnet a permanent or temporary magnet? Explain your answer.
    (b) Which part of an electromagnet is the solenoid?
    (c) What is the role of the iron core in an electromagnet?
35. Imagine that you were given a three-metre length of wire, an A4 sheet of thin card and a bar magnet.
    (a) Explain how you could produce an electric current.
    (b) What piece of equipment would you need to demonstrate that an electric current was produced?
36. List three changes that can be made to an electric generator to increase the size of the electric current it produces.
37. What is the role of the turbines in a coal-fired power station?

GLOSSARY

abiotic factors: the non-living features in an ecosystem
abscisic acid: a plant hormone that is involved in the process of development
absolute referencing: used in a spreadsheet when a cell address in the formula remains constant, no matter where it is copied to
absorption: the taking in of a substance, for example, from the intestine to the surrounding capillaries
accommodation: changing the lens shape to focus a sharp image on the retina according to the relative location of the cell that it has been copied to
acetylcholine: a neurotransmitter that carries a nerve impulse across nerve synapses to stimulate an impulse in the next neuron. It is particularly involved in long-term memory formation.
acid rain: rainwater, snow or fog that contains dissolved chemicals, such as sulfur dioxide, that make it acidic
acid: a chemical that reacts with a base to produce a salt and water. Edible acids taste sour.
active: describes a volcano that is erupting or has recently erupted
active immunity: immunity achieved by your body making antibodies to a specific antigen
adenosine triphosphate: *see* ATP
adenosine: a chemical involved in energy transfer; may cause drowsiness and be involved in falling asleep
adrenal glands: a pair of glands situated near the kidneys that release adrenaline and other stress hormones
adrenaline: a hormone secreted in response to stressful stimuli. It readies the body for 'flight or fight' by increasing the heart and breathing rates and the blood supply to the muscles.
adult stem cells: cells that are not fully differentiated. Under the right conditions, they can develop into various types of specialised cells.
aerobic respiration: the chemical breakdown of food using oxygen. The reaction needs enzymes, occurs in all body cells and releases energy.
alcohol: a colourless volatile flammable liquid (such as ethanol, C_2H_5OH) that is made by fermentation of sugars and starches
alkalis: bases that dissolve in water
alpha (α) particles: positively charged nuclei of helium atoms, consisting of two protons and two neutrons
alpha (α) waves: waves of electrical impulses emitted by your brain at a frequency of 8–12 Hz; this type of brainwave is associated with being calm, relaxed but aware of your environment
alternating current: current that changes direction along a wire a number of times per second. The alternating current (AC) supplied to homes changes direction 50 times every second.
alveoli: tiny air sacs in the lungs at the ends of the narrowest tubes. Oxygen moves from alveoli into the surrounding blood vessels, in exchange for carbon dioxide.
ammeter: device used to measure the amount of current in a circuit. Ammeters are placed in series with other components in a circuit.
amphetamines: nervous system stimulants, such as 'speed'
amplitude: maximum distance that a particle moves away from its undisturbed position
amygdala: emotional centre of the brain. It processes primal feelings, such as fear and rage. It may also be involved in emotional memories.
amylase: an enzyme in saliva that breaks starch down into sugar
anaerobic respiration: the chemical breakdown of food without oxygen. The reaction needs enzymes, occurs in cells and releases less energy than aerobic respiration.
analogies: a similarity between two or more things on which a comparison may be based

analogue: describes quantities that can have any value and change continuously over time
analyse: examine closely to answer a question or solve a problem
anther: the part of a flower that produces pollen (the male gametes)
antibiotic: a substance derived from a micro-organism and used to kill bacteria in the body
antibodies: any of various proteins that are produced as a result of the presence of a foreign substance in the body and that act to neutralise or remove that substance
anticlines: folds that bend upwards
antigen: substance that stimulates the production of antibodies
anus: the final part of the digestive system, through which faeces are passed as waste
aorta: a large artery through which oxygenated blood is pumped at high pressure from the left ventricle of your heart to your body
argument: value that a function in a spreadsheet will operate on
armature: the turning part of an electric motor on which coils of wire are wound
arteries: hollow tubes (vessels) with thick walls carrying blood pumped from the heart to other body parts
arterioles: vessels that transport oxygenated blood from the arteries to the capillaries
Asian influenza: a strain of influenza caused by the H2N2 subtype of influenza virus; a pandemic spread across parts of the world in 1956–58
association: linking new knowledge to previous knowledge
ATP: adenosine triphosphate: a form of energy released during cellular respiration that provides cells with the energy needed to perform their functions
audio: waves with a frequency range of sounds audible to people
auditory nerve: a large nerve that sends signals to the brain from the hearing receptors in the cochlea
auricle: the fleshy outside part of the ear
autotroph: *see* producer
auxin: a plant hormone that regulates or modifies the growth of plants in the presence of light
avian influenza: a strain of influenza caused by the H5N1 subtype of influenza virus; a form of this virus is highly contagious in birds and has caused over 300 fatalities in humans since 2003
axon: an appendage of the neuron that nervous impulses travel along to the next neuron or to an effector organ (muscle or gland)
B lymphocyte: a type of lymphocyte that produces antibodies that assist in the destruction of invading pathogens; also known as plasma cells
bacillus: a rod-shaped bacterium
barbiturates: chemicals that inhibit or decrease synaptic transmission and are hence depressants. They are often taken to calm people down and are used as sedatives.
base station: consists of antennas on top of a large tower that transmit signals from mobile phones to a switching centre
base: a chemical substance that will react with an acid to produce a salt and water. Edible bases taste bitter.
beam: wide stream of light rays, all moving in the same direction
beamline: part of a synchrotron that directs radiation through a monochromator and into an experimental station
beta (β) particles: charged particles (positive or negative) with the same size and mass as electrons
beta (β) waves: waves of electrical impulses emitted by your brain at a frequency of 13–30 Hz; this type of brainwave is associated with being strongly engaged and using many of your senses, and perhaps with anxiety
biconvex: describes a convex lens with both sides curved outwards
bile: a substance produced by the liver that helps digest fats and oils

bioaccumulation: magnification of concentrations of a substance such as a nonbiodegradable pesticide along the food chain; also known as biological magnification

biodiversity: the variety of species of biological organisms, often in relation to a particular area

biological magnification: *see* bioaccumulation

biologists: scientists who study the science of life. Biology is concerned with how species came into existence and the interaction between, and behaviour of, different organisms

bionic ear: *see* cochlear implant

biotic factors: the living things (organisms) in an ecosystem

birth rate: the number of organisms within a population that are born within a particular period of time

Black Death: *see* bubonic plague

bladder: sac that stores urine

blog: a personal website or web page where an individual can upload documents, diagrams, photos and short videos, add links to other sites and invite other people to post comments

body waves: waves that travel through the interior of the Earth; P-waves and S-waves are said to be body waves

bolus: round, chewed-up ball of food made in the mouth that makes swallowing easier

botanists: scientists who study the life of plants

boundaries: the edges of tectonic plates

brachytherapy: cancer treatment also known as internal radiotherapy. Radioisotopes are placed inside the body at, or near, the site of a cancer.

brain stem: the part of the brain connected to the spinal cord, responsible for breathing, heartbeat and digestion. Sometimes called the medulla.

brain-control interface technology: a direct communication pathway between the brain and an external device, used to decode brain wave patterns and facial movements to bring about particular responses in the external environment

bronchi: the narrow tubes through which air passes from the trachea to the smaller bronchioles and alveoli in the respiratory system. Singular = bronchus.

bronchioles: small branching tubes in the lungs leading from the two larger bronchi to the alveoli

brushes: part of an electric motor that allows current to travel through the rotor coils by being connected to the power supply and lightly touching the commutator

bubble map: a visual thinking tool that organises, analyses and compares

bubonic plague: an infectious, epidemic disease, caused by the *Yersinia pestis* bacteria and carried by fleas from rats; also known as the Black Death

caffeine: an example of an excitatory psychoactive drug that stimulates or increases synaptic transmission. It is found in coffee, tea, cocoa, chocolate and some soft drinks.

capillaries: minute tubes carrying blood to body cells. Every cell of the body is supplied with blood through capillaries.

carbohydrates: organic substances, such as sugars and starch, that are made up of carbon, hydrogen and oxygen and contain useful chemical energy

carbon dioxide: a gas in the air produced by respiration and used by plants as part of photosynthesis. The burning of fossil fuels releases carbon dioxide.

carnivore: an animal that eats other animals

carrier waves: radio waves that are altered in a precise way so that they contain an audio signal

carrying capacity: the maximum population size that a particular environment can sustain

cell body: contains the nucleus of a neuron, also called grey matter

cell membrane: structure that encloses the contents of a cell and allows the movement of some materials in and out

cell: the smallest unit of life and the building blocks of living things; electrical device containing chemicals that react to supply an electric charge

cellular immune response: immunity involving the activation of cells (phagocytes) rather than antibodies
cellular pathogen: a pathogen that is made up of cells, such as a tapeworm, fungus or bacterium
cellular phones: mobile phones; so called because base stations that receive mobile phone transmissions are arranged in a network of hexagonal cells
cellular respiration: the chemical reaction involving oxygen that moves the energy in glucose into the compound ATP. The body is able to use the energy contained in ATP.
cellular system: mobile phone system
cellulose: a natural substance that keeps the cell walls of plants rigid
central nervous system: the part of the nervous system composed of the brain and spinal cord
cerebellum: the part of the brain that controls balance and muscle action
cerebral cortex: the outer, deeply folded surface of the cerebrum
cerebral hemispheres: the left and right halves of the brain
cerebrum: the largest part of the brain (about 90 per cent of total brain volume), responsible for higher order thinking, controlling speech, conscious thought and voluntary actions. It is made up of the frontal, temporal, parietal and occipital lobes.
chemical digestion: the chemical reactions changing food into simpler substances that are absorbed into the bloodstream for use in other parts of the body
chemical energy: energy stored in chemical bonds that is released during chemical reactions
chemical process: a reaction that changes the arrangement of the atoms or molecules of the substances involved
chemical reaction: a chemical change in which one or more new chemical substances are produced
chemoreceptors: special cells within a sense organ (especially the nose and tastebuds) that are sensitive to particular chemicals, giving you the sensations of smell and taste
chemosynthetic: describes organisms that produce organic material using energy released from chemical reactions rather than light
chips: tiny pieces of silicon onto which tiny electric circuits can be etched
chlorophyll: the green-coloured chemical in plants that absorbs the light energy used in photosynthesis, which makes food from carbon dioxide and water
chloroplast: oval-shaped organelle found only in plant cells. Chloroplasts contain the pigment chlorophyll. They are the 'factories' in which carbon dioxide and water are changed by sunlight and water into food by the process of photosynthesis.
chunks: small pieces of information that are easier to remember than lots of information at once
ciliary muscles: muscles that control the shape of the lens behind the iris
circadian rhythm: the 24-hour pattern of behaviour exhibited in animals and plants even if deprived of environmental changes
circuit breaker: safety device that breaks a circuit if the current suddenly exceeds a specified size. Circuit breakers can be reset.
circuit diagram: diagram using symbols to show the parts of an electric circuit
circulatory system: the body system that circulates oxygen in blood to all the cells of the body. The circulatory system consists of the heart, blood vessels and blood.
clonal selection theory: a model for how the immune system responds to infection and how certain types of B and T lymphocytes are selected for destruction of specific antigens invading the body
coaxial cables: wires that can transmit a number of different signals as electrical pulses
cocaine: an example of an excitatory psychoactive drug that stimulates or increases synaptic transmission
coccus: a spherical bacterium
cochlea: the snail-shaped part of the inner ear. It is lined with tiny hairs that are vibrated by sound and stimulate the hearing receptors.

cochlear implant: a device implanted behind the ear that detects and processes sound, then sends signals to the auditory nerve so that severely deaf people can hear

cognition: another name for thinking or mental activity

cold thermoreceptors: a type of receptor in your skin that can detect a decrease in skin temperature below 35.8 °C

colon: the part of the large intestine where a food mass passes from the small intestine, and where water and other remaining essential nutrients are absorbed into your body

colour blindness: an inherited condition, more common in males, in which a deficiency of one or more of the different types of cones may mean that you find it difficult to see a particular colour or combinations of colours

combustion: chemical reaction in that a substance reacts with oxygen and heat is released

commensalism: relationship between organisms where one benefits and the other is unaffected

community: more than one population living in the same area at a particular time

commutator: device attached to the rotor shaft that receives the current. In most motors it is a metal ring split into sections.

competition: the struggle among organisms for food, territory and other factors

compounds: a chemical compound is the combination of two or more elements

compression wave: wave involving the vibration of particles in the same direction as energy transfer

compression: region in which the particles are closer than when not disturbed by a wave

computerised axial tomography (CAT): a medical imaging technology employing X-rays to produce a 3D image of a body using computer processing

concave: curved inwards

concept map: a visual thinking tool that shows the connection between ideas

conducting path: connected series of materials along which an electric current can flow

conduction: transfer of heat through collisions between particles

cones: sensory receptors in the retina that respond to red, green or blue light. Cones are mainly in the central part of the retina.

cones (speakers): hollow shape with a circular opening at one end and a closed point at the other

conservative plate boundaries: boundaries between sliding plates, like the San Andreas Fault in the United States

constipation: a condition of the bowels, caused by lack of dietary fibre, in which solid wastes cannot easily leave

constructive plate: plate that creates new land from cooling magma

consumer: organism that relies on other organisms for its food; also known as a heterotroph

continental crust: the plates of the Earth's crust that make up the land

continental drift: movement of the plates of the Earth's crust in relation to each other

continuum: a visual thinking tool that shows extremes of an idea or where people stand on a particular idea or issue

control: an experimental set-up in which the independent variable is not applied. It is used to ensure that the result is due to the variable and nothing else

controlled variables: the conditions that must be kept the same throughout an experiment

convection current: circular movement that occurs when warmer, less dense fluid particles rise and cooler, denser fluid particles sink

convection: transfer of heat through the flow of particles

converging lens: lens that bends rays so that they move towards each other. Converging lenses are thicker in the middle than at the edges.

convex: curved outwards

core body temperature: the operating temperature of an organism, especially near the centre of the body

cornea: the curved, clear outer covering of your eye
corpus callosum: a bridge of nerve fibres through which the two cerebral hemispheres communicate
corpus luteum: an endocrine structure that is involved in the production of progesterone
corrosive: describes a chemical that wears away the surface of substances, especially metals
cosmic radiation: naturally occurring background radiation from outer space
cross-pollination: transfer of pollen from stamens of one flower to the stigma of a flower of another plant of the same type
crude oil: liquid formed from the remains of marine plants and animals that died million of years ago—a fossil fuel. Many other fuel products are obtained from crude oil.
cycle: a visual thinking tool that shows order and sequence
cytokines: proteins secreted by cells to regulate the function of other cells, especially in the immune system
cytokinins: hormones that promote cell division in plants
cytosol: the fluid found inside cells
death rate: the number of organisms within a population that die within a particular period of time
decibel (dB): a unit of measurement of relative sound intensity
decomposer: organisms that break down organic matter into inorganic materials
dehydrated: state in which too much water has been lost from the body
dehydration: loss of water from the body
delta (δ) waves: waves of electrical impulses emitted by your brain at a frequency of 1–3 Hz; this type of brainwave is associated with being in a deep, dreamless sleep
dendrimer: a molecule that forms the basic structure of a nanoparticle
dendrite: structure that relays information towards the cell body of a neuron
density: the number of a species living within an area
deoxygenated blood: blood from which some oxygen has been removed
dependent variable: a variable that is expected to change when the independent variable is changed. The dependent variable is observed or measured during the experiment.
depressants: inhibitory psychoactive drugs that reduce or decrease synaptic transmission
destructive plate boundary: a convergent boundary where two plates collide
detritivore: animal that feeds on and breaks down dead plant or animal matter
diabetes mellitus: a medical condition in which the liver cannot effectively convert glucose to glycogen
diaphragm: cone of a loudspeaker that vibrates to produce a sound wave
digestion: breakdown of food into a form that can be used by an animal. It includes both mechanical digestion and chemical digestion.
digestive system: a complex series of organs and glands that processes food in order to supply your body with the nutrients it needs to function effectively
digital: describes quantities that can have only particular values and are represented by numbers
direct current: electric current that flows in one direction only
disease: any change that impairs the function of an individual in some way and causes harm to the individual
distribution: the area inhabited by a plant or animal species
diverging lens: lens that bends rays so that they spread out. Diverging lenses are thinner in the middle than at the edges.
DNA (deoxyribonucleic acid): a substance found in all living things that contains information about hair colour, eye colour etc. Only identical twins have the same DNA.
dopamine: a neurotransmitter involved in producing positive moods and feelings. It is also involved in the body's reward system.
dormant: describes a volcano that has not erupted for more than 20 years but is not considered extinct
drought resistant: able to store water and hence tolerate long periods of time without water

drought tolerant: able to tolerate a period of time without water
dry cells: devices containing chemicals as solids and pastes that react to supply an electric charge
dynamite: a relatively stable explosive invented by Alfred Nobel in 1866. It is created by mixing nitroglycerine with an absorbent substance such as silica, forming a paste that can be shaped into rods.
dynamo: electric generator
ear canal: the tube that leads from the outside of the ear to the eardrum
eardrum: a thin piece of stretched skin inside the ear that vibrates when sound waves reach it
earth socket: connection that provides a route for current to flow to the ground when an electrical appliance malfunctions. It is a safety device that helps to avoid the flow of current through a person using the appliance.
echolocation: use of sound to locate objects by detecting echoes
ecological niche: the position of a species or population in its ecosystem in relation to each other
ecology: the study of ecosystems
ecosystem: communities of organisms that interact with each other and their environment
ecstasy: an example of an excitatory psychoactive drug; a synthetic hallucinogenic drug (methylenedioxymethamphetamine, MDMA)
ectoparasite: parasite that lives outside the body of its host organism
effector: an organ that responds to a stimulus
electric charge: physical property of matter that causes it to experience a force when near other electrically charged matter. Electric charge can be positive or negative.
electric circuit: a path for electrons to follow, consisting of a power supply, one or more loads, and conductors joining the power supply and loads
electric current: a measure of the number of electrons flowing through a circuit every second. An increase in current means an increase in the rate of flow of electrons in the circuit.
electric generator: device that transforms kinetic energy of rotation into electrical energy
electric motor: device that converts electrical energy into kinetic energy of a rotating shaft
electrical impulses: nerve signals that pass rapidly from the receptors and along the peripheral nervous system to the central nervous system
electrode: conductor through which an electric current enters or leaves a cell
electroencephalogram (EEG): a device that detects and records the electrical activity of the brain
electroencephalography: a medical imaging technique for recording electrical activity in the brain
electrolyte: acid, base or salt that conducts electricity when dissolved in water or melted
electromagnet: magnet formed by wrapping a coil of wire around an iron core. When electricity passes through the coil, the iron core becomes an electromagnet.
electromagnetic pulse: a burst of electromagnetic activity caused by the detonation of nuclear devices
electromagnetic spectrum: complete range of wavelengths of energy radiated as electric and magnetic fields
electromagnetic waves: electromagnetic energy that is transmitted as moving electric and magnetic fields. There are many different types of electromagnetic energy, e.g. light, microwaves, radio waves.
electron transport chain reaction: part of the aerobic respiration process
electrons: negatively charged, very light particles in an atom. Electrons move around the nucleus of the atom.
embryonic stem cells: embryonic stem cells can remain unspecialised and keep dividing. If they are allowed to clump, they can spontaneously develop into groups of specialised cells, such as muscle cells and nerve cells. It is possible to control the type of cells they will develop into by providing the stem cells with exactly the right growing conditions.
emigration: the number of individuals leaving an area
emotions: feelings, such as happiness, sadness and anger, that are interpreted by the brain
empathy: the capacity to recognise and to some extent share feelings that are being experienced by other people

endocrine glands: organs that produce hormones. Endocrine glands release their hormones into the bloodstream for transport to target organs.
endocrine system: the body system of glands that produce and secrete hormones into the bloodstream in order to regulate processes in various organs
endoparasite: parasite that lives inside the body of its host organism
endorphins: hormones resembling opiates that are released by the brain when you are in pain, in danger or under other forms of stress
endoscope: long, flexible tube with one optical fibre to carry light to an area inside the body and another optical fibre to carry light from the body to a lens. The image formed by the lens is examined or recorded.
endothermic: describes chemical reactions that absorb heat energy from the surroundings
energy pyramid: a representation of the level of food energy at each level within a food chain
enzymes: special chemicals that speed up reactions but are themselves not used up in the reaction
ephemeral: lasting for only a very short time
epicentre: the point on the Earth's centre directly above the site where an earthquake originates
epidemic: a disease affecting a large number of people in a particular area in a relatively short period of time
equilibrium: a state in which conditions are balanced and there is neither growth nor a decrease in number; also known as steady state or plateau phase
erythrocytes: red blood cells
ethanol: an end product of anaerobic respiration in plants; a form of alcohol
ethics: the system of moral principles on the basis of which people, communities and nations make decisions about what is right or wrong
ethylene: a monomer in the manufacture of polymers such as polyethylene or polythene
eutrophication: a form of water pollution involving an excess of nutrients such as nitrates and phosphorus, resulting in algal blooms and possible death of fish and other organisms
excitatory psychoactive drugs: chemicals such as caffeine that increase or stimulate synaptic transmission
excretion: removal of wastes from the body
excretory system: the body system that removes waste substances from the body
exothermic: describes chemical reactions that give out heat energy to the surroundings
exponential growth: a rapid increase in number or size, represented by a J-shaped graph
external radiotherapy: cancer treatment where radiation is directed from an external machine to the site of the cancer
extinct: describes a volcano that has not erupted for thousands of years and is effectively dead
fair testing: a method for determining an answer to a problem without favouring any particular outcome; another name for a controlled experiment
fallopian tube: tubes through which ova must travel to reach the uterus. Fertilisation occurs in the fallopian tubes.
fat: an organic substance that is solid at room temperature and is made up of carbon, hydrogen and oxygen atoms
fault: a break in a rock structure causing a sliding movement of the rocks along the break
fertilisation: penetration of the ovum by a sperm
field magnets: magnets producing a magnetic field that acts on the rotor coils
filament: coil of wire made from a metal that glows brightly when it gets hot. The filaments in light globes heat up when electricity flows through them.
first-order consumer: organism that is within the second trophic level of a food chain (herbivores); also known as a primary consumer
fishbone diagram: a visual thinking tool that analyses and compares
fission: splitting of the nuclei of large atoms into two smaller atoms and several neutrons, releasing radiation and heat energy

flaccid: limp, not firm
flowchart: a visual thinking tool that shows order and sequence
flower: the sexually reproductive structure of some plants
focal point: the focus for a beam of light rays
focus: the point at which an earthquake begins
folding: buckling of rocks caused when rocks are under pressure from both sides
follicles: found in the ovary and contain a single immature ovum (egg)
follicle-stimulating hormone: regulates the development, growth and reproductive processes of the body
food chain: diagram that shows how the energy stored in one organism is passed to another
food web: diagram showing several food chains joined together to demonstrate that animals eat more than one type of food
forebrain: consists of the cerebrum, cerebral cortex, thalamus and hypothalamus
fossil fuel: substance, such as coal, oil and natural gas, that has formed from the remains of ancient organisms. Coal, oil and natural gas are often used as fuels; that is, they are burned in order to produce heat.
frequency: number of vibrations in one second, or the number of wavelengths passing in one second
FSH: *see* follicle-stimulating hormone
fuel rod: one of the rods that form the fuel source of a nuclear reactor; contains the fissile nuclides needed to produce a nuclear chain reaction
fuel: a substance that is burned in order to release energy, usually in the form of heat
fumigant: a chemical used in the form of smoke or fumes, to kill pests
function: common type of calculation built into spreadsheets
functional magnetic resonance imaging (fMRI): a type of specialised MRI scan used to measure the change in blood flow related to neural activity in the brain or spinal cord
functional magnetic resonance spectroscopy (fMRS): a medical imaging technique used to measure levels of different metabolites in body tissues; this can be used to diagnose certain metabolic disorders, especially those affecting the brain
fungi: organisms, such as mushrooms and moulds, that help to decompose dead or decaying matter. Many are parasites, and these can lead to disease.
fungicide: a chemical used to kill fungal growth
fuse: safety wire that melts when too much current flows through it. Fuse wires are designed to melt at different currents.
gall bladder: a small organ that stores and concentrates bile within the body
galvanometer: an instrument used to measure small electric currents; named after Luigi Galvani
gametes: sex cells
gamma rays: high-energy electromagnetic radiation produced during nuclear reactions. They have no mass and travel at the speed of light.
genetic modification (GM): the technique for moving genes from one plant to another, making it possible to design plants that have certain characteristics
geostationary: describes a satellite that remains above the same location of the Earth's surface
GHB: gamma hydroxybuturate, also known as liquid E or fantasy, which depresses the nervous system
gibberellins: plant hormones that regulate growth, including germination and dormancy
gland: a hormone-releasing structure
glucagon: a hormone, produced by the pancreas, that increases blood glucose levels
glucose: a simple carbohydrate and the simplest form of sugar
glycaemic index (GI): a measure of how quickly a particular food raises the level of blood sugar over a two-hour period
glycogen: the main storage carbohydrate in animals, converted from glucose by the liver and stored in the liver and muscle tissue
glycolysis: process by which glucose is converted into a simple form, during which energy is released

Gondwanaland: one of the two smaller continents created when the supercontinent Pangaea broke apart about 200 million years ago
ground zero: the centre of a nuclear weapon blast
guard cells: cells surrounding each stoma in a leaf enabling it to open or close depending on the availability of water
haemoglobin: the red pigment in red blood cells that carries oxygen
half-life: time taken for half the radioactive atoms in a sample to decay—that is, change into atoms of a different element
heart: a muscular organ that pumps blood through the circulatory system so that oxygen and nutrients can be transported to the body's cells and wastes can be transported away
heat exhaustion: a heat-related illness that occurs when the core body temperature reaches 39°C, due to loss of fluid and salt from the body
heatstroke: a serious medical condition that occurs when the core body temperature reaches 41°C and the body's internal organs begin to shut down
herbicide: a chemical used to kill unwanted plants (weeds)
herbivore: animal that eats only plants
heroin: an inhibitory psychoactive drug that decreases synaptic transmission
hertz: unit of frequency; its abbreviation is Hz. One hertz is equal to one vibration every second.
heterotroph: *see* consumer
higher order thinking: involves problem solving and decision making
hindbrain: a continuation of the spinal cord
hippocampus: part of the brain with a key role in consolidating learning, comparing new information with previous experience, and converting information from working memory to long-term storage
Hippocratic Oath: an oath historically taken by doctors that requires them to follow ethical rules and principles
homeostasis: the maintenance of a relatively constant internal physiological environment of the body or part of the body (e.g. blood glucose level, pH, body temperature) in varying external conditions
hormones: chemical substances produced by glands and circulated in the blood. Hormones have specific effects in the body.
horst: a block of the Earth's crust, with faults on either side, that has been pushed upwards by the forces below
host: organism living in a relationship with another organism. The host supplies something needed by the other organism (called the parasite).
hot thermoreceptors: a type of receptor in your skin that can detect an increase in skin temperature above 37.5°C (normal body temperature)
hotspot: a localised place where an activity occurs
humoral immune response: immunity involving the activation of antibodies rather than cells
hydrocarbons: compounds containing only hydrogen and carbon atoms
hypothalamus: monitors internal systems and controls the release of hormones to maintain the normal body state (homeostasis)
image: picture of an object
immigration: the number of individuals moving into an area
immune system: a network of interacting body systems that protects against disease by identifying and destroying pathogens and infected, malignant or broken-down cells
immunisation: *see* vaccination
immunity: resistance to a particular disease-causing pathogen
immunology: the branch of science that deals with immunity from disease
implanted electrodes: technological devices that connect directly to a biological subject's brain, often to assist people with areas in the brain that have become dysfunctional after a stroke or other head injuries

independent variable: a variable that is deliberately changed during an experiment
indicator: a substance that changes colour when it reacts with acids or bases. The colour shows how acidic or basic a substance is.
infectious disease: a disease that is contagious (can be spread from one organism to another) and caused by a pathogen
inflammation: a reaction of the body to an infection, commonly characterised by heat, redness, swelling and pain
infra-red radiation: invisible radiation emitted by all warm objects. You feel infra-red radiation as heat.
inhibitory psychoactive drugs: chemicals, such as barbiturates, that decrease synaptic transmission
insect pollination: transfer of pollen from one flower to another by insects
insecticide: a chemical used to kill insects
insulator: material that does not allow heat to move through it
insulin: hormone that removes glucose from the blood and stores it as glycogen in the liver and muscles
integrated circuits: electric circuits made up of miniature components that can be etched onto silicon chips
intermediate host: the organism that a parasite lives in or on in its larval stage; also known as secondary host
internal radiotherapy: cancer treatment also known as brachytherapy. Radioisotopes are placed inside the body at, or near, the site of a cancer.
interneuron: a nerve cell that carries nervous impulses through the central nervous system. They provide the link between sensory neurons and motor neurons.
interspecific competition: competition between organisms of the same species
intraspecific competition: competition between organisms of different species
introduced species: an organism that has been released into an ecosystem in which it does not occur naturally
iris: coloured part of the eye that opens and closes the pupil to control the amount of light that enters the eye
isotopes: atoms of the same element that differ in the number of neutrons in the nucleus
kerosene: the fuel used in jet aircraft
kidneys: body organs that filter the blood, removing urea and other wastes
kinetic energy: energy due to the motion of an object
Krebs cycle: part of the aerobic respiration process in which carbon dioxide and ATP (energy) are produced
lactate: *see* lactic acid
lactic acid: an end product of anaerobic respiration in animals; also known as lactate
large intestine: the penultimate part of the digestive system, where water is absorbed from the waste before it is transported out of the body
lateral inversion: reversed sideways
Laurasia: one of the two smaller continents created when the supercontinent Pangaea broke apart about 200 million years ago
lava bomb: *see* volcanic bomb
lava: mixture of molten rock and gases that has reached the Earth's surface from a volcano
Law of Conservation of Mass: law that states that, in a chemical reaction, the total mass of the reactants is the same as the total mass of the products
Law of Constant Proportions: law that states that, in chemical compounds, the ratio of the elements is always the same
left atrium: upper left section of the heart where oxygenated blood from the lungs enters the heart
left ventricle: lower left section of the heart, which pumps oxygenated blood to all parts of the body

lens: a transparent curved object that bends light towards or away from a point called the focus. The eye has a jelly-like lens.
liability: an obligation, responsibility, hindrance or something that causes a disadvantage
light energy: energy from the light of the sun, absorbed by plants and used in photosynthesis
limbic system: a collection of structures within your brain involved in memory, controlling emotions, decision making, motivation and learning
linear accelerator: part of a synchrotron that uses extremely high voltages (100 million volts) to accelerate electrons to 99.9987 per cent of the speed of light
lipases: enzymes that break fats and oils down into fatty acids and glycerol
lipids: type of nutrient that includes fats and oils
liver: largest gland in the body. The liver secretes bile for digestion of fats, builds proteins from amino acids, breaks down many substances harmful to the body and has many other essential functions.
load: device that uses electrical energy and converts it into other forms of energy
logbook: a complete record of an investigation from the time a search for a topic is started
longitudinal wave: *see* compression wave
long-sightedness: the condition of not being able to see clearly things that are close
luminous: releasing its own light
lungs: the organ for breathing air. Gas exchange occurs in the lungs.
luteinising hormone: a hormone, sometimes called LH, that rises sharply just before ovulation
lymphatic system: the body system containing the lymph vessels, lymph nodes, lymph and white blood cells that is involved in draining fluid from the tissues and helping defend the body against invasion by disease-causing agents
lymphocyte: small, mononuclear white blood cells present in large numbers in lymphoid tissues and circulating in blood and lymph. They combat microbial invasion, fight cancer and neutralise toxic chemicals.
lysozyme: a chemical in human teardrops able to kill some types of bacteria without harming the body's natural defences; discovered by Scottish bacteriologist Sir Alexander Fleming
magma: a very hot mixture of molten rock and gases, just below the Earth's surface, that has come from the mantle
magnetic field: area where a magnetic force is experienced by another magnet. The direction of the magnetic force is shown by drawing field lines; the size of the force is shown by how close together the lines are.
magnetic resonance imaging (MRI): a medical imaging technique employing a powerful magnetic field and radio waves to produce a 3D image of a body
magnetoencephalography (MEG): a medical imaging technique for mapping brain activity by recording magnetic fields produced by electrical currents occurring naturally in the brain
main switch: control switch that turns all the household circuits on or off
mantle: thick layer inside the Earth, below the crust. Most of the mantle is solid rock, although the upper part is molten rock called magma.
marijuana: plant in which the active ingredient is an inhibitory psychoactive drug that reduces or decreases synaptic transmission; sometimes called cannabis
mark, release and recapture: a sampling method used to determine the abundance of mobile species
matrix: a visual thinking tool that organises, analyses and compares
mechanical digestion: digestion that uses physical factors such as chewing with the teeth
mechanoreceptors: special cells within the skin, inner ear and skeletal muscles that are sensitive to touch, pressure and motion, enabling you to balance, hear and sense pressure and movement
medium: material through which a wave moves
medulla oblongata (medulla): a part of the brain developed from the posterior portion of the hindbrain and continuing with the spinal cord; *see also* brain stem

melatonin: hormone produced by the pineal gland that is involved in sleepiness
meltdown: the melting of a nuclear-reactor core as a result of a serious nuclear accident
membrane: a thin layer of tissue
memory cells: cells that may be formed from lymphocytes after infection with a pathogen — they 'remember' each specific pathogen encountered and are able to mount a strong and rapid response if that pathogen is detected again
menstrual cycle: cyclic changes in the ovaries and lining of the uterus as a result of changing hormone levels in the blood
menstruation: for about 5 days each month, women menstruate (have their period)
metabolism: the chemical reactions occurring within an organism that enable the organism to use energy and grow and repair cells
metaphor: a figure of speech in which something is spoken of as if it were something else
micrometre: one millionth of a metre
microprocessor: electronic central processing units of computers on a microchip
microwave: an electromagnetic wave of very high frequency, with a wavelength range from 50 cm to 1 cm
middle ear: the section of the ear between your eardrum and the inner ear, containing the ossicles
mineral: any of the inorganic elements that are essential to the functioning of the human body and are obtained from foods
mirror neurons: group of neurons that activate when you perform an action and when you see or hear others performing the same action
mitochondria: small rod-shaped organelles that supply energy to other parts of the cell. They are usually too small to be seen with light microscopes. Singular = mitochondrion.
mnemonic: a strategy to help you to remember things
monochromator: material that allows only specific wavelengths of radiation to pass through
motor neuron disease: a medical condition that progressively destroys motor neurons, resulting in progressive paralysis but leaving the brain and sense organs unaffected. The condition is eventually fatal.
motor neuron: the nerve cell that causes an organ, such as a muscle or gland, to respond to a stimulus
moulds: types of microscopic fungi found growing on the surface of foods
multicellular organism: an organism that is composed of many cells. Most plants and animals are multicellular.
multipotent: multipotent stem cells can differentiate into only a few cell types, whereas pluripotent stem cells can differentiate into any of the types of cells found in an adult organism
mutualism: relationship between two different organisms in which both benefit
myelin: a fatty, white substance that encases the axons (connecting branches) of the neurons in the nervous system
myelination: the process of neurons becoming coated in a myelin sheath
nanoparticle: a microscopic particle about 0.1–100 nanometres in size
natural gas: gas formed from the remains of plants and animals that died millions of years ago; a fossil fuel
negative electric charge: the charge on an atom or object with more electrons than protons
negative feedback reaction: a reaction in which the response is in the opposite direction to the stimulus — for example, if levels of a particular chemical in the blood were too high, then the response would be to lower them
negatively charged: having more electrons than protons (more negative charges than positive charges)
nephrons: the filtration and excretory units of the kidney
nerve: a bundle of neurons

nervous system: the system of nerves and nerve centres in an animal in which messages are sent as an electrical and then a chemical impulse. It comprises the central nervous system and the peripheral nervous system.
neural prostheses: technological devices that connect directly to a biological subject's brain, often to assist people with areas in the brain that have become dysfunctional after a stroke or other head injuries
neurogenesis: the creation of neurons
neuron: nerve cell
neurotoxic: toxic to neurons
neurotransmitter: chemical released from the axon terminals into the synapse between your nerve cells (neurons) during a nerve impulse
neutral: having the same number of protons and electrons
neutralisation: a reaction between an acid and a base. A salt and water (a neutral liquid) are the products of this type of reaction.
neutrons: particles with no electrical charge that are found in the nucleus of an atom
nitroglycerine: a highly explosive liquid
nocturnal: active only at night
non-cellular pathogen: a pathogen that is not made up of cells, such as a virus, prion or viroid
non-infectious disease: a disease that cannot be spread from one organism to another
norepinephrine: also called noradrenaline; common neurotransmitter involved in arousal states
normal: a line drawn perpendicular to a surface at the point where a light ray meets it
north pole: end of the magnet that when free to rotate, points to the north pole of the Earth. The magnetic forces are strongest at the poles of a magnet.
nuclear fallout: irradiated dust blasted high into the atmosphere during detonation and the formation of the mushroom cloud. In the weeks following the nuclear explosion, these come back down to Earth as nuclear fallout.
nuclear radiation: radiation from the nucleus of an atom, consisting of alpha or beta particles, or gamma rays
nuclear reactor: power plant where the radioactive properties of uranium are used to generate electricity
nucleus: central part of an atom, made up of protons and neutrons; roundish structure inside a cell that acts as the control centre for the cell. Plural = nuclei.
obligate anaerobes: organisms that can respire only anaerobically (in the absence of oxygen)
obligate intracellular parasite: a parasite that needs to infect a host cell before it can reproduce
ocean ridge: an area where the tectonic plates move apart, allowing magma from the mantle to rise, forming underwater volcanoes and creating new oceanic crust as it is cooled and solidified by sea water
oceanic crust: one of the types of crust that makes up the Earth's outer layer. Oceanic crust is thinner than continental crust and made up of dense, heavy rocks such as basalt.
octane: the major component of petrol
oesophagus: part of the digestive system composed of a tube connecting the mouth and pharynx with the stomach
oestrogen: a female hormone
Ohm's Law: statement relating the change in voltage across a conductor to the change in current. It states that the voltage across a particular conductor divided by the current through it is constant.
ohmic: describes conductors that obey Ohm's Law
olfactory nerve: nerve that sends signals to the brain from the chemoreceptors in the nose
omnivore: animal that eats plants and other animals
opaque: describes a substance that does not allow any light to pass through it
opiates: drugs derived from the opium poppy that involve the neurotransmitter dopamine in stimulating pleasure centres in the brain; they may also induce sleep and alleviate pain
optic nerve: large nerve that sends signals to the brain from the sight receptors in the retina

optical fibres: narrow strands made of two concentric glass layers so that the light is internally reflected along the fibres
optimum range: the range, within a tolerance range for a particular abiotic factor, in which an organism functions best
organelle: small structure in a cell with a special function
organism: living thing
ossicles: a set of three tiny bones that send vibrations from the eardrum to the inner ear. They also make the vibrations larger.
ova: female gametes (eggs) or sex cells
oval window: an egg-shaped hole covered with a thin tissue. It is the entrance from the middle ear to the outer ear.
ovary: in plants, the hollow, lower end of the carpel containing the ovules (the female egg cells); in animals, the female organ that produces ova and reproductive hormones
ovulation: the release of an ovum
ovum: females produce ova (eggs) in their ovaries
oxidation: chemical reaction involving the loss of electrons by a substance
oxygen: a gas in the air (and water) that animals need to breathe; made up of particles with two oxygen atoms. Plants produce oxygen as part of photosynthesis.
oxytocin: hormone secreted from the pituitary gland that assists in the formation of bonds between mothers and their babies, and perhaps between people in close relationships
P-waves (or primary waves): compression waves that move through the Earth in the same way that sound waves move through air
pain receptors: special cells located throughout the body (except the brain) that send nerve signals to the brain and spinal cord in the presence of damaged or potentially damaged cells, resulting in the sensation of pain
pancreas: a large gland in the body that produces and secretes the hormone insulin and an important digestive fluid containing enzymes
pandemic: a disease occurring throughout an entire country or continent, or worldwide
Pangaea: a super-continent that existed about 225 million years ago. All of the landmasses that existed at this time were joined together to form this super-continent.
Panthalassa: the vast sea surrounding the supercontinent of Pangaea
Pap test: a test used in women to detect abnormal cervical cells that can lead to cancer
papilla: bumps on your tongue that are thought to contain tastebuds
parallax error: error caused by reading a scale at an angle rather than placing it directly in front of the eye
parallel circuit: a circuit that has more than one path for electricity to flow through. If one of the paths has a break in it, the others will still work.
paralysis: loss of the ability to move
parasite: organism that lives in or on another organism. The parasite benefits while usually harming the host organism.
parasitism: an interaction in which one species (the parasite) lives in or on another species (the host) from which it obtains food, shelter and other requirements
passive immunity: immunity achieved by your body recieving antibodies from an outside source, such as from your mother's milk or through vaccination
pathogen: a disease-producing organism
penicillin: a powerful antibiotic substance found in moulds of the genus *Penicillium* that kills many disease-causing bacteria without harming the body's natural defences
perennial: lasting for three or more years

peripheral nervous system (PNS): made up of sensory and motor neurons. It connects the central nervous system to the rest of the body, and detects and responds to change.
peristalsis: the process of pushing food along the oesophagus or small intestine by the action of muscles
permanent magnet: magnet that retains its magnetic effect for many years
perspiration: the salty fluid produced by sweat glands under the skin
pH scale: scale from 1 (acidic) to 14 (basic) that measures how acidic or basic a substance is
phage: a type of virus that infects and kills bacteria
phagocyte: white blood cell that ingests and destroys foreign particles, bacteria and other cells
phagocytosis: the ingestion of solid particles by a cell
phloem vessels: long, narrow, living cells that are joined together to form long tubes in a plant. The tubes move the food made in the leaves to other parts of the plant, such as the roots and storage areas.
photon: a particle such as a quantum of light or electromagnetism
photoreceptor: a receptor cell located in your eye that is stimulated by light, converting it to electrical energy that is sent to the brain, giving you the sensation of light
photosynthesis: the food-making process in plants that takes place in chloroplasts within cells. The process uses carbon dioxide, water and energy from the sun.
pickling: preserving food by storing it in vinegar (acetic acid)
pie chart: a diagram using sectors of a circle to compare the sizes of parts making up a whole quantity
pineal gland: gland that produces the hormone melatonin, which can make you feel drowsy
pitch: highness or lowness of a sound. The pitch that you hear depends on the frequency of the vibrating air.
pituitary gland: a small gland at the base of the brain that releases hormones
plagues: contagious diseases that spread rapidly through a population and results in high mortality (death rates)
plasma cell: *see* B lymphocyte
plate tectonics: the theory concerning the movement of the continental plates
plateau phase: *see* equilibrium
pluripotent: pluripotent stem cells can differentiate into any of the types of cells found in an adult organism
PMI chart: visual thinking tool that helps you look at something from different perspectives
poliomyelitis: a highly infectious disease caused by the *Picornaviridae* virus that can have consequences including complete recovery, limb and chest muscle paralysis, or death
pollen grains: the male gametes of a flower
pollination: transfer of pollen from the stamen (the male part) of a flower to the stigma (the female part) of a flower
pollinator: something that transfers pollen from one flower to another
pons: part of the brain involved in regulating sleep, arousal and breathing, and coordinating some muscle movements
population: a group of individuals of the same species living in the same area at a particular time
positive electric charge: the charge on an atom or object with fewer electrons than protons
positive feedback reaction: a reaction in which the response is in the same direction as the stimulus — for example, during childbirth the onset of contractions activates the release of the hormone oxytocin, which stimulates further contractions
positively charged: having more protons than electrons (more positive charges than negative charges)
positron emission tomography (PET): a nuclear medicine imaging technique employing gamma rays to produce a 3D image of a body or functional processes in the body
power supply: a device that can provide an electric current
predator–prey relationship: relationship between organisms in which one species (the predator) kills and eats another species (the prey)
primacy: involves remembering the first time that you do something, or the beginning of something
primary consumer: the first consumer in a food chain; also known as a first-order consumer

primary host: the organism that a parasite lives in or on in its adult stage
prion: a non-cellular pathogen
priority grid: a visual thinking tool that quantifies and ranks
producer: organism at the base of the food chain that does not need to feed on other organisms; also known as an autotroph
products: chemical substances that result from a chemical reaction
progesterone: progesterone is involved in the menstrual cycle but it is also involved in pregnancy
proteases: enzymes that break proteins down into amino acids
protein: chemical made up of amino acids needed for the growth and repair of the cells in living things
protons: positively charged particles found in the nucleus of an atom
pseudoscience: an apparently scientific approach to a theory that on close analysis is shown to have no scientific validity
psychoactive drugs: chemicals that decrease synaptic transmission (such as barbiturates) or increase synaptic transmission (such as caffeine)
pulmonary artery: the vessel through which deoxygenated blood, carrying wastes from respiration, travels from the heart to the lungs
pulmonary vein: the vessel through which oxygenated blood travels from your lungs to the heart
pupil: a hole through which light enters the eye
pyramid of biomass: a representation of the dry mass of organisms at each level within a food chain
pyramid of numbers: a representation of the population, or numbers of organisms, at each level within a food chain
quadrat: a sampling area, often one square metre, in which the number of organisms in that area is counted and recorded
quarantine: strict isolation of sick people from others for a period of time (originally 40 days) in order to prevent the spread of disease
radiant heat: heat that is transferred from one place to another by radiation
radiation: a method of heat transfer that does not require particles to transfer heat from one place to another
radio waves: low energy electromagnetic waves with a much lower frequency and longer wavelength than visible light
radiocarbon dating: a method of determining the age of a fossil using the remaining amount of unchanged radioactive carbon. All living organisms have the same proportion of radioactive and nonradioactive carbon atoms, but this changes after they die.
radioisotope: a radioactive form of an isotope
radiometric dating: determining the ages of rocks and fossils based on the rate of decay or half-life of particular isotopes
random error: an error that occurs due to estimation when reading scales, or when the quantity being measured changes randomly
rarefaction: region in which the particles are further apart than when not disturbed by a wave
rays: narrow beams of light
reactants: the original substances present in a chemical reaction
receiving antenna: metal structure in which electrons are made to vibrate by radio waves or microwaves in the atmosphere
recency: involves remembering the last time that you do something, or the end of something
receptors: special cells that detect energy and convert it to electrical energy that is sent to the brain
rectifier: device that changes alternating current to direct current
rectum: the final section of the digestive system, where waste food matter is stored as faeces before being excreted through the anus

red blood cells: living cells in the blood that transport oxygen to all other living cells in the body. Oxygen is carried by the red pigment haemoglobin.

reflection: bouncing off the surface of a substance

reflex action: a quick response to a stimulus. Reflex actions do not involve thought.

reflex arc: a nervous pathway involving a small number of neurons. A reflex occurs when nervous impulses travel from the receptor to the spinal cord and then to the effector organ.

refraction: change in the speed of light as it passes from one substance into another. It usually involves a change in direction.

regulating: the process by which the brain detects and responds to the body's internal and external environments

relative intensity: a measure of how loud a sound is using a sound meter

relative referencing: used in a spreadsheet when the cell address in the formula is changed

repeater station: equipment that retransmits communication signals with increased energy so the signal does not fade away

repetition: the act of regularly reviewing information

resistance: measure of the electrical energy required for an electric current to pass through an object. The energy is changed to heat.

resistors: circuit component that has resistance

respiration: the process by which your body gains energy by breaking down glucose, using oxygen and creating carbon dioxide and water; a slow combustion reaction

respiratory: concerning the airways or breathing

respiratory system: the body system involving the lungs and associated structures, which take in air and supply the blood with oxygen to deliver to the body's cells so they can carry out their essential functions; it also performs gas exchange to remove the waste gas carbon dioxide

reticular formation: a network of neurons that opens and closes to increase or decrease the amount of information that flows into and out of the brain

retina: curved surface at the back of the eye. It is lined with sight receptors.

Richter scale: a scale that measures the amount of energy released during an earthquake

rift valley: a sunken area where two blocks of crust have dropped down between faults

right atrium: upper right section of the heart where deoxygenated blood from the body enters

right ventricle: lower right section of the heart, which pumps deoxygenated blood to the lungs

rods: sensory receptors in the retina that respond to low levels of light and allow you to see in black and white in dim light

root hairs: tube-like outgrowths of cells on the surface of roots. They have thin walls, which allow water and dissolved substances to move into the root.

rotor coils: coils of a motor that turn when a current flows through them

S-waves: the second set of waves to be detected after P-waves. During seismic activity, secondary waves or S-waves travel in the form of transverse waves.

saliva: watery substance in the mouth that moistens food before swallowing

salivary glands: glands in the mouth that produce saliva

sampling methods: techniques used to determine the density and distribution of various populations and communities within an ecosystem

satellite: object that orbits another object. The moon is the Earth's satellite. Scientists have made and launched many artificial satellites.

scattering: describes light sent in many directions by small particles within a substance

sea breeze: the breeze that occurs when differences in air pressure cause air particles to flow from the ocean towards the land

secondary consumer: the second consumer in a food chain; also known as a second-order consumer

secondary host: *see* intermediate host

second-order consumer: organism that is within the third trophic level of a food chain (carnivores); also known as a secondary consumer
seismic waves: waves of energy that travel through the Earth's crust, caused by earthquakes
seismograph: an instrument used to detect and measure the intensity of an earthquake
seismologist: a scientist who studies earthquakes
self-pollination: transfer of pollen from the flower's own stamen to its stigma
semicircular canals: three curved tubes, filled with fluid, in the inner ear that control your sense of balance
sense organ: a specialised structure that detects stimuli (such as light, sound, touch, taste and smell) in your environment
sensory neuron: a nerve cell in the sense organs. It detects change in the environment and sends a message to the central nervous system.
sensory register: a part of the memory process that enables you to remember and retain information you consider to be important
series circuit: a circuit with the components joined one after the other in a single continuous loop
serotonin: a common neurotransmitter involved in producing states of relaxation and regulating sleep and moods
shaft: central rotating rod of the motor that transmits the kinetic energy
short-sightedness: the condition of not being able to see clearly things that are far away
sigmoid: the shape of a graph that shows a population increasing in number then reaching a plateau
signalling molecule: a chemical involved in transmitting information between cells
skin: external covering of an animal body
slip fault: a geological feature where movement along a fault is sideways — that is, where the blocks of crust slip horizontally past each other
small intestine: the part of the digestive system between the stomach and large intestine, where much of the digestion of food and absorption of nutrients takes place
solenoid: coil of wire able to pass a current
somatic stem cells: undifferentiated cells that are found in adults; they can generate all the cell types
sonar: use of reflected sound waves to locate objects under water (sound navigation and ranging)
sound level: the energy of sound that is an indication of the loudness of a sound
south pole: end of a magnet opposite the north pole and attracted to the north pole of another magnet
Spanish influenza: a strain of influenza caused by the H1N1 subtype of influenza virus; a Spanish flu pandemic spread across the world in 1918–1920
species: a group of living organisms capable of interbreeding with each other but not with members of other species
sperm: male sex cell
spirochaete: a spiral-shaped bacterium
stable: describes a nucleus that does not change spontaneously. The protons and neutrons in the nucleus are held together strongly.
static electricity: a build-up of charge in one place
steady state: *see* equilibrium
stigma: the female part of a flower, at the top of the carpel, that catches the pollen during pollination
stimulants: excitatory psychoactive drugs, such as caffeine and amphetamines, that increase or stimulate synaptic transmission
stimuli: changes in the environment that can be detected and responded to
stimulus–response model: a system in which any changes or variations (stimuli) in the internal environment are detected (by receptors); if a response is required, this is communicated to effectors to bring about some type of change or correction so the conditions can be brought back to normal

stoma: small openings through which water transfer occurs, located mainly on the lower surface of leaves. These pores are opened and closed by guard cells. Plural = stomata

stomata: pores that exchange gases found on the surface of leaves. They are bordered by guard cells that change the size of the opening of the stomata.

storyboard: a visual thinking tool that shows order and sequence

subduction: process in which two tectonic plates push against each other, and oceanic crust sinks below the less dense continental crust

suprachiasmatic nucleus (SCN): the biological clock, located in the hypothalamus near where the optic nerves cross

surface protein: a protein molecule occurring on the surface of a virus

surface waves: (or L-waves) earthquake waves which travel only through the Earth's crust; they are responsible for the majority of an earthquake's destructive power

swine flu: a strain of influenza caused by the H1N1 subtype of influenza virus, containing a combination of genes from the swine, human and avian flu viruses. A swine flu pandemic killed several thousand humans in 2009.

switch: device that opens and closes the conducting path through which a current flows

switching centre: switches mobile phone calls to other base stations or to a fixed telephone system

symbiotic relationship: very close relationship between two organisms of different species. It may benefit or harm one of the partners.

synaesthesia: a condition in which a sensation is produced in one physical sense when a stimulus is applied to another; affected people may associate letters with a flavour, numbers with a gender or sounds with colour

synapse: the gap between adjoining neurons across which electrical nervous impulses are sent

synaptic pruning: the elimination of the least used and hence weakest synapses (connections between neurons) in the brain during adolescence

synchrotron: a building-sized device that uses electrons accelerated to near the speed of light to produce intense electromagnetic radiation. This is used to produce data that can describe objects as small as a single molecule.

synclines: folds that bend downwards

systematic errors: errors that are consistently high or low due to the incorrect use or limitations of equipment

T cell: *see* T lymphocyte

T lymphocyte: a type of lymphocyte that destroys invading pathogens by attacking them

target cells: cells in the body that respond to a particular hormone

target map: a visual thinking tool that quantifies and ranks

tastebuds: nerve endings located in your tongue that contain receptors sensitive to sweet, salty, bitter, sour and savoury chemicals, allowing you to experience taste

temporary magnet: magnet that stays magnetic while it touches a permanent magnet, or one that is magnetic for a very short time

testes: organs that produce sperm cells

testosterone: male sex hormone

thalamus: part of the brain through which all sensory information from the outside (except smell) passes before going to other parts of the brain for further processing

THC: the active ingredient in marijuana; also known as delta-9 tetrahydrocannobinol

thermal flash: enormous amounts of heat and radiation that spread out from the centre of a nuclear blast; *see also* ground zero

thermoreceptors: special cells located in your skin, part of your brain and body core that are sensitive to temperature

thermoregulation: the control of body temperature

thermostat: a device that establishes and maintains a desired temperature automatically
threshold of hearing: the lowest level of sound that can be heard by the human ear
threshold of pain: the lowest level of sound that causes pain to the human ear
timeline: a visual thinking tool that shows order and sequence
tinnitus: a ringing or similar sensation of sound in the ears, caused by damage to the cells of the inner ear
tolerance range: range of an abiotic factor in the environment in which an organism can survive
tongue: the sense organ responsible for taste
total internal reflection: complete reflection of a light beam that may occur when the angle between a boundary and a beam of light is small
totipotent: a type of cell that can form an entire organism
trachea: narrow tube from the mouth to the lungs through which air moves
transect: a sampling area along a straight line in which the number of organisms in that area is counted and recorded
transformer: device that can increase or decrease voltages for alternating current
translocation: transport of materials, such as water and glucose, in plants
translucent: allowing light to come through imperfectly, as in frosted glass
transmissible spongiform encephalopathy (TSE): a degenerative neurological disease caused by prions
transmitting antenna: metal structure in which vibrating electrons cause radio waves to travel through the air
transparent: describes a substance that allows most light to pass through it. Objects can be seen clearly through transparent substances.
transpiration stream: movement of water through a plant as a result of loss of water from the leaves
transpiration: loss of water from plant leaves through their stomata
transverse wave: wave involving the vibration of particles perpendicular to the direction of energy transfer
tremor: vibration on the Earth's surface caused by an earthquake
triangulation: finding a location by using at least three different sources of detection
trophic level: a level within a food chain, food web or food pyramid
tsunami: a powerful ocean wave triggered by an undersea earth movement
turgid: firm, distended
Type 1 diabetes mellitus: a disease in which the pancreas stops producing insulin. Without insulin, the body's cells cannot turn glucose (sugar) into energy, and so daily insulin injections are required.
Type 2 diabetes mellitus: the most common form of diabetes, where the pancreas makes some insulin but does not produce enough. It can often be managed with regular exercise and healthy eating.
ultrasound: sound with frequencies too high for humans to hear
ultraviolet radiation: invisible radiation similar to light but with a slightly higher frequency and more energy
umbilical cord stem cells: stem cells that can develop into a few types of cells (mainly blood cells and cells involved in fighting disease) and are also being used to treat leukaemia and other blood disorders
universal indicator: a mixture of indicators that changes colour as the strength of an acid or base changes, indicating the pH of the substance
unstable: describes an atom in which the neutrons and protons in the nucleus are not held together strongly
ureters: tubes from each kidney that carry urine to the bladder
urine: yellowish liquid, produced in the kidneys. It is mostly water and contains waste products from the blood such as urea, ammonia and uric acid.
uterus: the organ in which a baby grows and develops
vaccination: administering a vaccine to stimulate the immune system of an individual to develop immunity to a disease; also known as immunisation

valid: sound or true. A valid conclusion can be supported by other scientific investigations.
variable resistor: device for which the resistance can be altered
variable: quantity or condition that can be changed, kept the same or measured during an experiment
variolation: deliberate infection of a person with smallpox (Variola) in a controlled manner so as to minimise the severity of the infection and also to induce immunity against further infection
vascular bundle: group of xylem and phloem vessels within a plant stem
vector: an organism that carries a pathogen between other organisms without being affected by the disease the pathogen causes; an organism that carries and disperses reproductive structures (e.g. pollen) of a different species
veins: blood vessels that carry blood back to the heart. They have valves and thinner walls than arteries.
vena cava: large vein leading into the top right chamber of the heart
Venn diagram: a visual thinking tool that analyses and compares
venules: small veins
vesicle: a small cavity, usually filled with fluid
vibrations: repeated fast back-and-forth movements
villi: tiny finger-like projections from the wall of the intestine that maximise the surface area of the structure to increase the efficiency of nutrient absorption. Singular = villus.
virtual focal point: common point from which rays appear to have come before passing through a concave lens
virus: a non-cellular pathogen
visible light: very small part of the electromagnetic spectrum to which our eyes are sensitive
vitamin D: a nutrient that regulates the concentration of calcium and phosphate in the bloodstream and promotes the healthy growth and remodelling of bone, along with other functions
vitamin deficiency disease: diseases caused by a lack of any of the 13 vitamins in the diet
vitamins: organic nutrients required in small amounts. They include vitamins A, B, C, D and K.
voice coil: coil of wire in a loudspeaker that vibrates at the frequency of the electrical signal. This frequency matches the frequency of the sound produced by the cone.
volcanic bomb: large rock fragment that is blown out of erupting volcanoes; also known as a lava bomb
volcano: natural opening in the Earth's crust connected to areas of molten rock deep inside the crust
voltmeter: device used to measure the voltage across a component in a circuit. Voltmeters are placed in parallel with the components.
wave: transmitter of energy without the movement of particles from place to place. The vibration of particles or energy fields is involved.
wavelength: distance between two neighbouring crests or troughs of a wave. This is the distance between two particles vibrating in step.
white blood cells: living cells that fight bacteria and viruses. They are part of the human body's immune system.
wind pollination: transfer of pollen from one flower to another by the wind
xerophyte: plant that is adapted to survive in deserts and other dry habitats
X-rays: high-energy electromagnetic waves that can be transmitted through solids and provide information about their structure
xylem vessels: long narrow cells that are joined together to form long tubes in a plant. The tubes, made from xylem cells, move water and dissolved minerals up from the roots to the stem and leaves. The wood in a tree trunk consists mostly of dead xylem cells.
Y chart: a visual thinking tool that helps you visualise and reflect
zero population growth: the point at which birth and death rates balance each other out

INDEX

E

F